Remote Sensing of Hydrometeorological Hazards

Remote Sensing of Hydrometeorological Hazards

Edited by
George P. Petropoulos
Tanvir Islam

CRC Press
Taylor & Francis Group
Boca Raton London New York

CRC Press is an imprint of the
Taylor & Francis Group, an **informa** business

Cover: The SEVIRI satellite shown on the cover is provided by EUMETSAT.

CRC Press
Taylor & Francis Group
6000 Broken Sound Parkway NW, Suite 300
Boca Raton, FL 33487-2742

First issued in paperback 2020

© 2018 by Taylor & Francis Group, LLC
CRC Press is an imprint of Taylor & Francis Group, an Informa business

No claim to original U.S. Government works

ISBN-13: 978-0-367-57273-0 (pbk)
ISBN-13: 978-1-4987-7758-2 (hbk)

Library of Congress Cataloging-in-Publication Data

Names: Petropoulos, George P., author. | Islam, Tanvir, author.
Title: Remote sensing of hydrometeorological hazards / George P. Petropoulos and Tanvir Islam.
Description: Boca Raton, FL : Taylor & Francis, 2018. | Includes bibliographical references and index.
Identifiers: LCCN 2017019423 | ISBN 9781498777582 (hardcopy : acid-free paper)
Subjects: LCSH: Hydrometeorology--Remote sensing. | Environmental monitoring--Remote sensing. | Droughts--Remote sensing. | Floods--Remote sensing | Forest fires--Remote sensing.
Classification: LCC GB2801.72.R42 P47 2018 | DDC 551.57028/4--dc23
LC record available at https://lccn.loc.gov/2017019423

Visit the Taylor & Francis Web site at
http://www.taylorandfrancis.com

and the CRC Press Web site at
http://www.crcpress.com

Dedication

This book is dedicated to my sister, Konstantina, for her enduring love and support in all my endeavors. I wish her to have every happiness in her life with her family.

—George P. Petropoulos

To my inspiring parents, encouraging brother, caring wife, and all the true well-wishers.

—Tanvir Islam

Contents

SECTION I Remote Sensing of Drought

SECTION II Remote Sensing of Frost and Sea Ice Hazards

SECTION III   Remote Sensing of Wildfires

SECTION IV   Remote Sensing of Floods

SECTION V Remote Sensing of Storms

SECTION VI Remote Sensing of Landslides

Preface

AIMS AND SCOPE

It is widely acknowledged today that extreme weather and climate change aggravate the frequency and magnitude of disasters. Facing atypical and more severe events, existing early warning and response systems become inadequate both in scale and scope, while planning and coordination schemes need to scale-up for a timely and adequate reaction. Too often, response to extreme weather events is hampered by the lack of information, long distances, and coordination difficulties, leaving citizens in great peril. Instead of being immediate and targeted, response lingers and misses. Simultaneously, emerging technologies open up for an improved emergency response. A special category of hazards includes hydrometeorological hazards, and in this subgroup some of the main hazards included are droughts, frost, floods, landslides, and storms/cyclones. Today, Earth observation (EO) provides one of the most promising avenues for providing information at global, regional, and even basin scales related to hydrometeorological hazards. The general circumstances that make EO technology attractive for this purpose in comparison to traditional techniques include their ability to provide inexpensive, repetitive, and synoptic views of large areas in a spatially contiguous fashion without a disturbing influence on the area to be surveyed and without site accessibility issues.

The preparation of this book is motivated by the scientific challenges emerging from the requirements to develop a capability for predicting and mapping hydrometeorological hazards to support research, practical applications, and decision-making from local to larger scales. In all cases, the need for improving relevant observations and modeling capabilities of parameters related to those hazards is mandatory in order to overcome the current drawbacks and the limitations faced by the scientific and operational communities.

This book provides readers an all-inclusive critical overview of the state of the art in different algorithms and techniques applied today in hydrometeorological hazards exploiting the EO technology datasets. In particular, it focuses on covering the following types of hydrometeorological hazards: droughts, frost, wildfires, floods, storms, and landslides. This book aims at providing readers an overview of examples of case studies in which EO data have been used in each of the aforementioned groups of hydrometeorological hazards. In each of those case studies, readers are provided with a deeper understanding of the operation and principles of widely applied recent approaches in each hazard case study. What is more, this book allows readers to value the added importance of EO in hydrometeorological hazards in comparison to conventional techniques applied today and also become aware of any operationally distributed relevant EO-based products available today, as well as from ground-installed operational networks that could be used in such studies. Such a book is needed due to the importance of hydrometeorological hazards today globally, having devastating effects in human lives as well as on global economies, with their frequency being dramatically increased over the past decade or so.

Evidently, this book integrates decades of research conducted by leading scientists in the field, and it has been designed with different potential users in mind. As such, it promotes the synergistic and multidisciplinary activities among scientists. Potential readers of the book may come from a wide spectrum of scientific backgrounds, such as environmental sciences, hydrology, meteorology, ecology, agricultural sciences, and geography. This book preparation has been possible because of the extensive and valuable contributions from interdisciplinary experts from all over the world in the fields covered within it. On account of the unique way it is structured, consisting of a series of independent parts, its use can be adapted to meet the specific needs of different readers, leading to its adoption for teaching and research purposes alike. The different chapters can be perceived as even smaller units that can be combined with other materials if required each time.

SYNOPSIS OF THE BOOK

This book is divided into the following six sections:

Section I focuses on droughts. Chapter 1 by Dalezios reviews the use of EO in droughts. Chapter 2 by Martinez-Fernandez et al. presents an approach for modeling droughts for agricultural crops using soil moisture estimates from space and demonstrates the use of this approach in the Iberian Peninsula. Chapter 3 by Texeira et al. provides a method for drought assessment via the coupling of MODIS satellite and weather data and testing their approach for a region in Brazil. Chapter 4 by Enekel et al. furnishes a reflective discussion on the added value of satellite soil moisture for agricultural insurance assessment. In Chapter 5, Yagci et al. demonstrate the use of drought indicators derived from Moderate Resolution Imaging Spectroradiometer (MODIS) data for detecting drought conditions in the Southern United States.

Section II exemplifies recent advances in modeling approaches utilizing EO data for frost conditions. In Chapter 6, Dalezios and Petropoulos summarize the main frost types and properties defining frost conditions and provide a critical overview of the use of EO in this domain. Chapter 7 by Gupta presents a review of the use of remote sensing in ice hazard, whereas Chapter 8 by Youngwook et al. discusses and demonstrates the use of satellite microwave remote sensing of landscape freeze/thaw status related to frost hazard monitoring. In Chapter 9, Louka et al. provide the usefulness of the coupling of remote sensing data, thermal mapping, and geographic information system (GIS) techniques for mapping temperature fluctuation and frost risk on a road network.

Section III mainly focuses on the use of EO in wildfires. In Chapter 10, Dalezios et al. furnish a systematic overview of the use of this technology in this domain, providing a few examples as case studies. Subsequently, Chapter 11 by Piles et al. critically offers a detailed overview of the state of the art specifically in European remote sensing activities in wildfire prevention. Chapter 12 by Zhao et al. furnishes a discussion on recent advances in burnt areas and burn severity mapping from remote sensing, providing relevant examples of recently published studies. Chapter 13 by Hassan et al. uses as an example wildfires in Alberta, Canada, and discusses the relationships between topographical elements and the occurrence of forest fires. Chapter 14 by Mills and Colton discusses the use of operational products in burnt area mapping and presents results from the use of such products in quantifying the interannual variability of wildfires across Portugal.

Section IV focuses on remote sensing of floods. Chapter 15 by Schumann et al. discusses the use of EO from space for disaster response assistance. Chapter 16 by Pinel et al. debates the usefulness of remote sensing data for extreme flood event modeling and gives a relevant example from the Amazonian floodplain. Chapter 17 by Kwak et al. demonstrates the use of remote sensing data from MODIS for large-scale flood monitoring in monsoon Asia for global disaster risk reduction. Chapter 18 by Stathopoulos et al. exhibits a new method for mapping flood susceptibility using remote sensing and GIS and demonstrates its use for a region in Greece. Chapter 19 by Mei et al. delivers a systematic review of the state of the art on the use of EO precipitation data for modeling floods.

Section V focuses on remote sensing of storms. Chapter 20 by Sabareesh et al. demonstrates the application of remote sensing data for post-wind storm damage analysis. Chapter 21 by Dutta et al. furnishes a study that shows how precipitation radar EO data can be used for analyzing tropical cyclones. Chapter 22 by Marra et al. discusses how radar rainfall estimates can be used for debris flow early warning systems.

Section VI focuses on landslides. Chapter 23 by Parsinevelos et al. offers a critical review of the use of the unmanned aerial vehicles (UAVs), citizen science, and interferometry remote sensing in landslide hazards. Chapter 24 by Bathrellos et al. demonstrates the use of remote sensing and GIS in developing a landslide susceptibility map using as a case study in a region of Greece.

The editors hope this preface has successfully furnished some insight on the breadth of the topics covered in this book. Users are encouraged to adapt this book in the best way it fits their needs that would help them in understanding the capabilities and potentials of EO technology in the field.

The users of this book can inform the editor of any errors, suggestions, or other comments at george.petropoulos@aber.ac.uk or tanvir.islam@jpl.nasa.gov.

Acknowledgments

The editors express their deepest thanks to the authors of the different chapters who agreed to contribute to the book despite their already very busy schedules. They also express their sincerest gratitude to the reviewers for their useful and insightful review comments and suggestions that helped improve the book. Last but not least, the editors are also grateful to the publisher and the staff at Taylor & Francis for their patience and support for the collaboration in accomplishing the preparation of this book.

Editors

George P. Petropoulos is a Associate Professor (Reader) in remote sensing and GIS at Aberystwyth University, Aberystwyth, United Kingdom. At present he also holds a Marie Curie fellowship from the European Commission (2013–2017). He completed his graduate studies (MSc, PhD) at the University of London, London, United Kingdom in 2008, specializing in Earth observation (EO) modeling.

Dr. Petropoulos' research focuses on exploiting EO data alone or synergistically with land surface process models for computing key state variables of the Earth's energy and water budget, including energy fluxes and soil surface moisture. He is also conducting research on the application of remote sensing technology to land cover mapping and its changes from either anthropogenic activities or geohazards (mainly floods, wildfires, and frost). In this framework, he contributes to the development of open-source software tools in EO modeling and develops and implements all-inclusive benchmarking approaches to either EO operational algorithms/products or surface process models, including advanced sensitivity analysis.

Dr. Petropoulos serves as a council member and trustee of the Remote Sensing and Photogrammetric Society (RSPSoC); he is an Associate Editor and Editorial Board member on several international peer-reviewed scientific journals in EO and environmental modeling. He also serves as a reviewer for various funding bodies, including the European Commission. He has also convened the organization of several scientific sessions at international conferences and on special issues of scientific journals. He is the editor/coeditor of 3 books, author/coauthor of more than 55 peer-reviewed journal articles, and has presented papers in more than 90 international conferences. He has developed fruitful collaborations with key scientists in his area of specialisation globally, and his research work so far has received international recognition through several noteworthy awards and personal fellowships that he has obtained.

Tanvir Islam is with the NASA Jet Propulsion Laboratory and specializes in remote sensing observations. Presently, he is engaged with the development of advanced microwave calibration and retrieval algorithms for NASA's Earth observing missions.

Prior to joining NASA/JPL in 2015, he was with the NOAA/NESDIS/STAR and worked on the development of satellite remote sensing algorithms, with an emphasis on microwave variational inversion techniques (2013–2015). He also held visiting scientist positions at the University of Tokyo, Japan as part of the NASA/JAXA precipitation measurement missions (PMM) algorithm development team, in 2012 and at the University of Calgary, Alberta, Canada in 2015. He earned his PhD in remote sensing at the University of Bristol, Bristol, United Kingdom in 2012.

Dr. Islam was the recipient of the Faculty of Engineering Commendation from the University of Bristol (nominated for a University Prize for his outstanding PhD thesis) in 2012, the JAXA visiting fellowship award in 2012, the CIRA postdoctoral fellowship award in 2013, the Calgary visiting fellowship award in 2015, and the Caltech postdoctoral scholar award in 2015. He has served as a lead guest editor for a special issue on microwave remote sensing for *Physics and Chemistry of the Earth* (Elsevier) and is currently serving on the editorial board of *Atmospheric Measurement Techniques* (EGU) and *Scientific Reports* (Nature). He has published 2 books and more than 60 peer-reviewed papers in leading international journals. His primary research interests include microwave remote sensing, radiometer calibration, retrieval algorithms, radiative transfer theory, data assimilation, mesoscale modeling, cloud and precipitation system, and artificial intelligence in geosciences.

Contributors

Masoud Abdollahi
Department of Geomatics Engineering
Schulich School of Engineering
University of Calgary
Calgary, Alberta, Canada

Zacharias Agioutantis
Department of Mining Engineering
College of Engineering
University of Kentucky
Lexington, Kentucky

Emmanouil N. Anagnostou
Department of Civil and Environmental
    Engineering
University of Connecticut
Mansfield, Connecticut

George D. Bathrellos
Department of Geography and Climatology
Faculty of Geology and Geoenvironment
National and Kapodistrian University of
    Athens
Athens, Greece

Gustavo Bayma-Silva
Embrapa Satellite Monitoring
Campinas, Brazil

Marie-Paule Bonnet
Mixed Laboratory International, Observatory
    for Environmental Change (LMI-OCE)
Institute of Research for Development (IRD)
University of Brasilia (UnB)
Brasília, Brazil

and

UMR 5563 GET/Institute of Research for
    Development (IRD)
Toulouse, France

M. Borga
Department of Land, Environment, Agriculture
    and Forestry
University of Padova
Padova, Italy

Stephane Calmant
Mixed Laboratory International, Observatory
    for Environmental Change (LMI-OCE)
Institute of Research for Development (IRD)
University of Brasilia (UnB)
Brasília, Brazil

and

Hydraulic Research Institute
Federal University of Rio Grande do Sul
Porto Alegre, Brazil

and

UMR 5566 LEGOS/Institute of Research for
    Development (IRD)
Toulouse, France

Christos Chalkias
Department of Geography
Harokopio University
Athens, Greece

David Chaparro
Universitat Politècnica de Catalunya
IEEC/UPC
Barcelona, Spain

Ehsan H. Chowdhury
Department of Geomatics Engineering
Schulich School of Engineering
University of Calgary
Calgary, Alberta, Canada

Daniel Colson
Department of Geography and Earth Sciences
Aberystwyth University
Aberystwyth, United Kingdom

J. D. Creutin
Institut des Géosciences pour l'Environnement
Université de Grenoble Alpes/CNRS
Saint-Martin-d'Hères, France

Nicolas R. Dalezios
Department of Civil Engineering
University of Thessaly
Volos, Greece

and

Department of Natural Resources and
    Agricultural Engineering
Agricultural University of Athens
Athens, Greece

Antônio H. de C. Teixeira
Embrapa Satellite Monitoring
Campinas, Brazil

Meixia Deng
George Mason University
Center for Spatial Information Science and
    Systems
Fairfax, Virginia

Liping Di
George Mason University
Center for Spatial Information Science and
    Systems
Fairfax, Virginia

Jinyang Du
Numerical Terradynamic Simulation Group
College of Forestry and Conservation
University of Montana
Missoula, Montana

Devajyoti Dutta
National Centre for Medium Range Weather
    Forecasting (NCMRWF)
Ministry of Earth Sciences
Noida, India

Efthymios
Department of Civil and Environmental
    Engineering
University of Connecticut
Mansfield, Connecticut

Markus Enenkel
International Research Institute for Climate
    and Society
Columbia University
Palisades, New York

C. R. Fragoso Jr.
Center for Technology
Federal University of Alagoas
Maceió, Brazil

Jeremie Garnier
Institute of Geosciences (LAGEQ)
University of Brasília (UnB)
Brasília, Brazil

Ángel González-Zamora
Instituto Hispano Luso de Investigaciones
    Agrarias
University of Salamanca
Salamanca, Spain

Anil Gupta
Alberta Environmental Monitoring, Evaluation,
    and Reporting Agency
Calgary, Alberta, Canada

Mukesh Gupta
Institut de Ciències del Mar – CSIC
Barcelona Expert Center on Remote Sensing
    (BEC-RS)
Passeig Marítim de la Barceloneta 37–49
Barcelona, Spain

Quazi K. Hassan
Department of Geomatics Engineering
Schulich School of Engineering
University of Calgary
Calgary, Alberta, Canada

Kostas Kalabokidis
Department of Geography
University of the Aegean
Mytilene, Greece

Dionissios P. Kalivas
Soil Science Laboratory
Department of Natural Resources Management
    and Agricultural Engineering
Agricultural University of Athens
Athens, Greece

Kleomenis Kalogeropoulos
Department of Geography
Harokopio University
Athens, Greece

Efthimios Karymbalis
Department of Geography
Harokopio University
Athens, Greece

Youngwook Kim
Numerical Terradynamic Simulation Group
College of Forestry and Conservation
University of Montana
Missoula, Montana

John S. Kimball
Numerical Terradynamic Simulation Group
College of Forestry and Conservation
University of Montana
Missoula, Montana

Nikos Koutsias
Department of Environmental and Natural
    Resources Management
University of Patras
Agrinio, Greece

Youngjoo Kwak
International Centre for Water Hazard and
    Risk Management (ICHARM), under the
    auspices of UNESCO
Tsukuba, Japan

Janice F. Leivas
Embrapa Satellite Monitoring
Campinas, Brazil

Panagiota Louka
Department of Natural Resources Development
    and Agricultural Engineering
Agricultural University of Athens
Athens, Greece

Bristol Powell
International Research Institute for Climate
    and Society
Columbia University
Palisades, New York

F. Marra
Institute of Earth Sciences
Hebrew University of Jerusalem
Jerusalem, Israel

José Martínez-Fernández
Instituto Hispano Luso de Investigaciones
    Agrarias
University of Salamanca
Salamanca, Spain

Masahiro Matsui
Tokyo Polytechnic University
Tokyo, Japan

Yiwen Mei
Department of Civil and Environmental
    Engineering
University of Connecticut
Mansfield, Connecticut

Ran Meng
Environmental and Climate Sciences
    Department
Brookhaven National Laboratory
Upton, New York

Aaron Mills
Department of Geography and Earth Sciences
Aberystwyth University
Aberystwyth, United Kingdom

David Motta Marques
UMR 228 ESPACE-DEV/Institute of Research
    for Development (IRD)
Montpellier, France

E. I. Nikolopoulos
Department of Civil and Environmental
    Engineering
University of Connecticut
Mansfield, Connecticut

Daniel Osgood
International Research Institute for Climate
    and Society
Columbia University
Palisades, New York

Ioannis Papanikolaou
Department of Natural Resources Development
    and Agricultural Engineering
Agricultural University of Athens
Athens, Greece

Panagiotis Partsinevelos
Spatial Informatics Research Group
Laboratory of Geodesy and Geomatics
School of Mineral Resources Engineering
Technical University of Crete
Chania, Greece

George P. Petropoulos
Department of Geography and Earth Sciences
Aberystwyth University
Aberystwyth, United Kingdom

Maria Piles
Barcelona Expert Centre
Institute of Marine Sciences (ICM)
Barcelona, Spain

Sebastien Pinel
RHASA/State of Amazonas University
    (UEA)
Manaus, Brazil

and

Mixed Laboratory International, Observatory
    for Environmental Change (LMI-OCE)
Institute of Research for Development (IRD)
University of Brasilia (UnB)
Brasília, Brazil

Christos Polykretis
Department of Geography
Harokopio University
Athens, Greece

Sudha Radhika
Geethanjali College of Engineering and
    Technology
Hyderabad, India

J. Rafael Cavalcanti
Federal University of Rio Grande do Sul
Hydraulic Research Institute
Porto Alegre, Brazil

Matthew Rodell
NASA Goddard Space Flight Center
Hydrological Sciences Laboratory (617)
Greenbelt, Maryland

A. Routray
National Centre for Medium Range Weather
    Forecasting (NCMRWF)
Ministry of Earth Sciences
Noida, India

Nilda Sánchez
Instituto Hispano Luso de Investigaciones
    Agrarias
University of Salamanca
Salamanca, Spain

Joseph A. Santanello
NASA Goddard Space Flight Center
Hydrological Sciences Laboratory (617)
Greenbelt, Maryland

Joecila Santos Da Silva
RHASA/State of Amazonas University (UEA)
Manaus, Brazil

and

Mixed Laboratory International, Observatory
    for Environmental Change (LMI-OCE)
Institute of Research for Development (IRD)
University of Brasilia (UnB)
Brasília, Brazil

Nathaniel Schaefer
Department of Mining Engineering
College of Engineering
University of Kentucky
Lexington, Kentucky

Guy J.-P. Schumann
Remote Sensing Solutions, Inc.
Monrovia, California

and

School of Geographical Sciences
University of Bristol
Bristol, United Kingdom

Frederique Seyler
Mixed Laboratory International, Observatory
  for Environmental Change (LMI-OCE)
Institute of Research for Development (IRD)
University of Brasilia (UnB)
Brasília, Brazil

and

UMR 228 ESPACE-DEV/Institute of Research
  for Development (IRD)
Montpellier, France

Hariklia D. Skilodimou
Department of Geography and Climatology
Faculty of Geology and Geoenvironment
National and Kapodistrian University of
  Athens
Athens, Greece

Panagiotis Skrimizeas
Hellenic National Meteorological Service,
  Forecasting and Research Division
Athens, Greece

Prashant K. Srivastava
Institute of Environment and Sustainable
  Development
Banaras Hindu University
Varanasi, India

Nikolaos Stathopoulos
Sector of Geological Sciences
Laboratory of Technical Geology and
  Hydrogeology
School of Mining Engineering
National Technical University of Athens
and
Department of Geography
Harokopio University
Athens, Greece

Yukio Tamura
Beijing Jiaotong University
Beijing, China

Achilleas Tripolitsiotis
Spatial Informatics Research Group
Laboratory of Geodesy and Geomatics
School of Mineral Resources Engineering
Technical University of Crete
and
Space Geomatica Ltd.
Chania, Greece

Ioannis X. Tsiros
Meteorology Laboratory
Crop Science Department
Agricultural University of Athens
Athens, Greece

Mercè Vall-llossera
Universitat Politècnica de Catalunya
IEEC/UPC
Barcelona, Spain

Christos Vasilakos
Department of Geography
University of the Aegean
Mytilene, Greece

Ali Levent Yagci
NASA Goddard Space Flight Center
Hydrological Sciences Laboratory (617)
Greenbelt, Maryland

Feng Zhao
Department of Geographical Sciences
University of Maryland
College Park, Maryland

Section I

Remote Sensing of Drought

1 Drought and Remote Sensing
An Overview

Nicolas R. Dalezios

CONTENTS

1.1   INTRODUCTION

Drought is considered as a natural phenomenon recurring at a regional scale throughout history. Essentially, droughts originate from a deficiency or lack of precipitation in a region over an extended period of time and can be regarded as an extreme climatic event associated with water resources deficit (Dalezios et al., 2017a). This is why droughts are also referred to as *nonevents*. Droughts occur in both high and low rainfall areas and virtually all climate regimes. It is recognized that drought is characterized as one of the major natural hazards with significant impact to the environment, society, agriculture, and economy, among others (Dalezios et al., 2017b). Indeed, there are several regions around the world, which are characterized as vulnerable areas due to the combined effect of increased temperature and reduced precipitation in areas already coping with water scarcity (IPCC, 2012). As a result, agricultural production risks could become an issue in these regions as mainly droughts are likely to increase the incidence of crop failure. As yield variability increases, the food supply is at increasing risk (Sivakumar et al., 2005; Dalezios et al., 2017c). Moreover, the impacts of droughts may be severe and are neither immediate nor easily measured. It is difficult to determine the effects of drought as it constitutes a complicated phenomenon, evolving gradually in any single region. In particular, drought impacts are very critical and especially costly affecting more people than any other type of natural disaster universally (Keyantash and Dracup, 2002). All the above may accumulate difficulties in drought assessment and response, which may result into slow progress on drought preparedness plans and mitigation actions. Thus, there is a need to establish the context in which the drought phenomenon and its associated impacts are being described leading to a better definition.

Drought quantification is usually accomplished through indicators and indices. There are several commonly used drought indices based on ground (conventional) and/or remotely sensed data (Du Pissani et al., 1998; McVicar and Jupp, 1998; Kanellou et al., 2009a, 2009b; Mishra and Singh, 2010; Zargar et al., 2011). Traditional drought quantification methods rely on conventional meteorological data, which are limited in a region, often inaccurate, and usually unavailable in near real-time (NRT) (Thenkabail et al., 2004). On the other hand, satellite-based data are consistently available and can be used to detect several drought features and characteristics. Indeed, the growing number and effectiveness of pertinent earth observation satellite systems present a wide range of new capabilities, which can be used to assess and monitor drought hazard and its effects, such as the activities for droughts of the United Nations International Strategy for Disaster Reduction (UNISDR) (UNISDR, 2005, 2015; Dalezios et al., 2017b, 2017d). Remote sensing data and methods can delineate the quantitative spatial and temporal variability of several drought features (Kanellou et al., 2012; Dalezios et al., 2012, 2014). Thus, there is a need for proper remotely sensed quantification of drought and drought impacts. Moreover, drought monitoring is of critical importance in economically and environmentally sensitive regions and is a very significant input in any drought preparedness and mitigation plan.

This chapter focuses on discussing the remote sensing potential and capabilities in drought analysis. Drought definitions and concepts, including types, factors, and features, are initially presented. Then, remote sensing capabilities, in terms of data and methods, are explored in drought analysis and assessment. Specifically, the adjustment of existing drought indices to use remotely sensed data and techniques is reviewed followed by an examination of the remote sensing strategies. Moreover, the reliability of remote sensing data and methods is steadily increasing year by year. Then, remotely sensed drought quantification and assessment are considered, including composite drought indices (CDIs). Representative remotely sensed drought indices are presented, namely the Reconnaissance Drought Index (RDI) (Tsakiris and Vagelis, 2005; Dalezios et al., 2012) and the Vegetation Health Index (VHI) (Kogan, 1995; Dalezios et al., 2014). Applications of both indices, RDI and VHI, are considered in drought early warning systems (DEWS) and monitoring. Finally, remotely sensed drought assessment and management are briefly presented.

1.2 DROUGHT CONCEPTS AND FEATURES

Drought indicators are variables, which describe drought features. Several indicators can also be combined into a single quantitative indicator, namely a drought index (Wilhite et al., 2000). For monitoring drought, drought indices are used based on several drought features, such as severity, duration, onset, end time, areal extent, and periodicity (Dalezios et al., 2000). Moreover, for drought assessment through drought indices, the focus is on the estimation of precipitation shortage and water supply deficit; however, evapotranspiration or temperature may also be included (Tsakiris and Vagelis, 2005; Vicente-Serrano et al., 2010).

Since the last decade, a web service-based environment is being developed for integration of regional and continental drought monitors; for computation and display of spatially consistent systems, such as *in situ* Standardized Precipitation Index (SPI), satellite-based indices, and modeled soil moisture; and for drill-down capacity to regional, national, and local drought products. This research effort has indicated, among others, the research need for CDIs toward a global drought risk modeling system (Brown et al., 2008; Zargar et al., 2011) is also based on remote sensing data and methods. Indeed, due to the complexity of drought, the scientific trend and research need are to consider multiple indicators or CDIs for assessing and monitoring droughts (Hao and AghaKouchak, 2013; Svoboda, 2015). There are indices for all types of drought, although there is no *one-size-fits-all* drought index or indicator. In summary, the approaches to drought assessment are essentially three (Svoboda et al., 2002): (1) single indicator or index (parameter); (2) multiple indicators or indices; and (3) composite or hybrid indicators, which integrate several indicators or indices and converge an evidence approach.

In terms of climate variability, there is medium confidence that since the 1950s, some regions of the world have experienced more intense and longer droughts (IPCC, 2012). Land-use changes have potential impacts on droughts (Arneth et al., 2014), and anthropogenic forcing has contributed to the global trend toward increased drought in the second half of the twentieth century. Extreme climate variables and climate extremes, such as droughts, are projected to experience significant changes over the twenty-first century, just as they have during the past century, in many areas, including Southern Europe, among others (Tarquis et al., 2013; Nastos et al., 2016). There is also medium confidence that the duration and intensity of hydrological droughts will increase in the twenty-first century in some seasons and areas, due to reduced precipitation and/or increased evapotranspiration, although other factors, such as changes in agricultural land cover and upstream interventions, lead to a reduction in river flows or groundwater recharge. Moreover, climate variability and change may affect drought preparedness planning and mitigation measures (Salinger et al., 2005; IPCC, 2012). Thus, climate change has to be considered in all the aspects of drought analysis.

Drought preparedness and mitigation planning based on remote sensing data and methods are considered as essential components of integrated water resources management. It is recognized that there is an international research need for drought preparedness plans through the development of decision support system (DSS) (Wilhite, 2005; Dalezios, 2017). It is also recognized that the drought policy principle has to consider the implementation of preparedness and mitigation measures (Wilhite, 2009; Arneth et al., 2014).

1.2.1 Drought Definitions and Types

Drought is not just a physical phenomenon, because it results from the interplay between a natural event and demands placed on water supply by human-use systems. Drought clearly involves a shortage of water, but realistically it can be defined only in terms of a particular need. It is difficult to find a generally accepted definition of drought. Indeed, there is no universally accepted definition of drought, because there is a wide variety of sectors affected by drought and because of its diverse spatial and temporal distribution (Heim, 2002). More than 150 published definitions of drought have been identified (Niemeyer, 2008). If drought is considered as a phenomenon, then it is certainly an atmospheric phenomenon. Studies in several areas around the world have shown that drought periods are often characterized by a large decrease in the amount of rainfall per rainy day, by an increase in the continentality of clouds, and by lack of rain-producing clouds (Dalezios et al., 2009). In general, droughts have been shown to be associated with persistence of ridges or centers of high-pressure systems at middle level in the troposphere. Furthermore, the corresponding reduced cloud cover results in positive temperature anomalies in the lower atmosphere, which produces the middle-level pressure anomaly and favors subsidence at high level, keeping the atmosphere significantly drier and more stable than normal. Nevertheless, by considering drought as a hazard, there is a tendency to define and classify droughts into different types; however, the relationship between the different types of drought is complex. In the international literature, three operational definitions are considered, namely meteorological or climatological, agrometeorological or agricultural, and hydrological drought (Wilhite et al., 2000). As a fourth type of drought, the socioeconomic impacts of drought can also be considered.

All droughts begin with a deficiency of precipitation in a region over a period of time. These early stages of accumulated departure of precipitation from normal or expected are usually considered as meteorological drought (Dalezios et al., 2017a). A continuation of these dry conditions over a longer period of time, sometimes in association with above normal temperatures, high winds, and low relative humidity quickly results into impacts in the agricultural and hydrological sectors. Specifically, with the exception of meteorological drought, the other types of drought, such as agricultural and hydrological, emphasize on the human or social aspects of drought, in terms of the interaction between the natural characteristics of meteorological drought and human activities that depend on precipitation, to provide adequate water supplies to meet societal and environmental demands. Drought concepts refer to conditions of precipitation deficit, soil moisture, streamflow, plant wilting, wild fires, famine, and other components. Moreover, drought monitoring involves climate data, soil moisture, streamflow, groundwater, reservoir and lake levels, snow pack, short-, medium-, and long-range forecasts, as well as vegetation health/stress and fire danger. A brief description of the aforementioned drought types is as follows:

Meteorological or climatological drought is a region-specific natural event, due to the regional nature of atmospheric phenomena, resulting from multiple causes. It is defined as the degree of dryness specified by precipitation deficiencies and the dry period duration. Meteorological drought is generally characterized by a precipitation anomaly being lower than the average in a region for some period of time and by prolonged and abnormal moisture deficiency.

Agricultural or agrometeorological drought refers to the agricultural impacts resulting from deficiencies in the water availability for agricultural use. Indeed, agricultural drought is described in terms of crop failure and exists when soil moisture is depleted so that crop yield is reduced considerably. Specifically, agricultural drought is defined by the availability of soil water to support crop and forage growth, and there is no direct relationship between precipitation and infiltration of precipitation into the soil. Soils with low water holding capacity are typical of drought-prone areas, which are more vulnerable to agricultural drought.

Hydrological drought is normally defined by the departure of surface and subsurface water from some average conditions over a long time period resulting from meteorological drought. Hydrological drought is considered to be a period during which the actual water supply, either surface water or groundwater, is less than the minimum water supply that is necessary for normal operations in a particular region (watershed). Similar to agricultural drought, there is no direct relationship between precipitation amounts and the status of surface and subsurface water supplies. There is also a significant time lag between departures of precipitation and the appearance of these deficiencies in surface and subsurface of the hydrological system.

Finally, *socioeconomic* drought is defined in terms of loss from an average or expected return and can be measured by both social and economic indicators (Gobron et al., 2007). Indeed, socioeconomic drought refers to the gap between supply and demand of economic goods brought on by the three other types of drought described earlier, such as water, food, raw materials, transportation, hydroelectric power, as a result of a weather-related shortfall in water supply. Socioeconomic drought is different from other types of drought, because its occurrence depends on the spatiotemporal distribution and processes of supply and demand.

1.2.2 FACTORS AND FEATURES OF DROUGHT

Several factors may be implicated as potential causes of drought: El Nino Southern Oscillation (ENSO), abnormal sea surface temperature (SST) patterns in areas other than the equatorial eastern Pacific, soil moisture desiccation, and nonlinear behavior of the climate system. Frequent droughts around the world, and interest in their possible links with phenomena, such as El Nino, keep the hazard in evidence even for the casual observer. For example, in areas such as the Sahel in Africa, where nomadism and intermittent grazing have been prevalent and more or less in balance with environmental conditions, more intensive exploitation has had disastrous results for social systems and ecosystems when drought has struck.

For assessing and monitoring droughts, several drought features are usually detected (Mishra and Singh, 2010; Dalezios et al., 2017a). A description of some key features is as follows:

Severity: Severity or intensity of drought is defined as an escalation of the phenomenon into classes from mild, moderate, severe, and extreme. The severity is usually determined through drought indicators and indices, which include the aforementioned classes. The regions affected by severe drought evolve gradually, and there is a seasonal and annual shift of the so-called epicenter, which is the area of maximum severity.

Periodicity: Periodicity is considered as the recurrence interval of drought.

Duration: Duration of a drought episode is defined as the time interval between the start and end time usually in months.

Onset: The beginning of a drought is determined by the occurrence of a drought episode. The beginning of a drought is assessed through indicators or indices reaching certain threshold value.

End time: End time of a drought episode signifies the termination of drought based again on threshold values of indicators or indices. As drought is a complex phenomenon, it is often difficult to determine the onset and the ending of a drought and on what criteria these determinations should be made.

Areal extent: Areal extent of the drought is considered as the spatial coverage of the phenomenon as is quantified in classes by indicators or indices. Areal extent varies in time, and remote sensing has contributed significantly in the delineation of this parameter by counting the number of pixels in each class.

1.3   REMOTE SENSING IN DROUGHT ANALYSIS

Over the last decades, there is a gradually increasing trend for the use of remote sensing in drought analysis and assessment, and specifically for the detection of several spatial and temporal drought features at different scales (Dalezios et al., 2017e). Moreover, a major consideration for remote sensing use in drought analysis is the extent to which operational users can rely on a continued supply of data (Thenkabail et al., 2004). Indeed, satellite systems provide temporally and spatially continuous data over the globe, and, thus, they are potentially better and relatively inexpensive tools for regional applications, such as drought quantification, monitoring, and assessment, than conventional environmental and weather data. For these types of applications, appropriate remote sensing systems are weather radars and satellites that provide low spatial and high temporal resolution data, because daily coverage and data acquisition are necessary. The series of geosynchronous, polar-orbiting meteorological satellites fulfil the aforementioned requirements, and there are already a long series of datasets. This section presents a brief description of remote sensing systems and their potential in drought analysis and assessment.

1.3.1   REMOTE SENSING SYSTEMS FOR DROUGHT ANALYSIS

The classification of the satellite systems can be based on several criteria. A basic criterion is certainly the wavelength of the electromagnetic radiation, which classifies the systems as being sensitive to visible, infrared, and microwave radiation regions of the spectrum. Another criterion consists of the classification into active and passive satellite systems. Specifically, the active satellite systems transmit energy and record the returned signal. Such systems are weather radars and synthetic aperture radars (SARs), which operate in the microwave portion of the electromagnetic spectrum and are considered as all-weather systems, because they can penetrate clouds without signal attenuation. Active satellite systems are very useful in drought analysis, because precipitation from weather radar is a key parameter, as well as soil moisture can be detected from SAR images. On the other hand, passive satellite systems just record the naturally reflected or transmitted radiation. In drought quantification and assessment, two types of passive remote sensing systems are considered, namely meteorological and environmental or resource satellites. The main differences between the two types of satellites are their spatial and temporal resolutions, which affect their applications and uses. Specifically, meteorological satellites have a rather coarse spatial resolution but high temporal reoccurrence, thus, being suitable mainly for operational applications, such as monitoring drought through changes in the index values. On the other hand, environmental satellites have usually a fine spatial resolution but low temporal reoccurrence, being basically used in land-use classification, such as quantitative classification of drought severity.

The advantage to using remotely sensed data is that they allow for a high-resolution spatial coverage and are updated frequently to allow for NRT analyses, whereas the main drawbacks are the relatively short period of record. Indeed, the number of satellite systems is steadily increasing year by year with a continuous improvement of the spatial resolution. Moreover, there is a recent tendency to increase the number of available bands in these satellites resulting in new and valuable information. New types of remote sensing systems offer online open information for web platforms and are also utilized for monitoring and detecting drought. Such systems are the European Copernicus system with six sentinel satellites (2014–2021) to monitor land, ocean, emergency response, atmosphere, security, and climate change (ESA, 2014) or National Aeronautics and Space Administration's (NASA's) new online satellites for climate change, Global Precipitation Measurement Core Observatory, Orbiting Carbon Observatory-2, and active–passive Soil Moisture. Moreover, massive cloud computing resources and analytical tools for working with big datasets make it possible to extract new information from environmental satellites' imagery with varying spatial resolution, such as Landsat-8 imagery (15 m), RapidEye (5 m), Worldview-3 (0.31 m), or

Pleiades (0.5 m). Thus, digital data processing for CDIs, including satellite imagery, monitoring and preparedness planning, including DSS, could be incorporated into a dynamic web drought platform.

1.3.2 REMOTE SENSING CAPABILITIES IN DROUGHT ANALYSIS

Remote sensing capabilities provide a viable method to offset any loss of information. However, there are dissimilarities in temporal and spatial averages as envisioned by modeling efforts, as existing in the real world and as measured by remote sensing systems. Thus, remotely sensed data in order to be useful for monitoring and assessing droughts must be compatible with mathematical modeling of the corresponding quantification schemes. Moreover, new sensors have higher spatial resolution, a current shortcoming in drought indices products. Novel noise reduction algorithms and other atmosphere correction algorithms improve the thematic accuracy of remote sensing datasets.

1.3.2.1 Drought Quantification and Monitoring

Traditional methods of drought assessment and monitoring rely on conventional rainfall data, which are usually limited in a region, often inaccurate, and, most importantly, difficult to obtain in NRT (Thenkabail et al., 2004). Indeed, some indicators may not have a sufficient record length, and this is usually the case of remotely sensed data. Thus, it is best to consider multiple indicators for DEWS and for detection and verification of the onset, existence, and severity of drought. At the present time, it is possible to have direct estimation of environmental parameters, such as temperature, precipitation, evapotranspiration, soil moisture, snow cover and snow depth, as well as water and energy balance, all of which are involved in drought quantification and assessment. Indeed, remote sensing data and techniques provide, among others, direct measurements of land characteristics, vegetative cover, and components of the hydrological cycle.

Moreover, monitoring the extent of drought is best achieved in near arid areas by the vegetation coverage. This can be achieved through multispectral visible imagery from polar-orbiting satellites for monitoring vegetation conditions and agricultural drought than conventional weather data (Dalezios et al., 2014). For these types of applications, appropriate remote sensing systems are those that provide optimal spatial and temporal resolution data, because daily coverage and data acquisition are necessary. The series of geosynchronous, polar-orbiting meteorological satellites fulfil the aforementioned requirements, and there are already a long series of databases. For example, the Normalized Difference Vegetation Index (NDVI) and the VHI of several satellites, among other indices, are effectively used (Kogan, 1995) and provide good guidance on monitoring the areal extent of drought affected regions.

1.3.2.2 Drought Assessment and Management

Remote sensing methodologies and data can be employed in vulnerability and damage assessment, as well as relief, which involve assistance and/or intervention during or after drought. The possible contribution of remote sensing could be focused on relief and, possibly, preparedness although in many cases remote sensing can make a valuable contribution to disaster prevention in which frequency of observation is not a prohibitive limitation. Moreover, remote sensing is a useful tool to analyze the vegetation dynamics on local, regional, or global scales (Keyantash and Dracup, 2002); to assess the vegetative stress; and to determine the impact of climate on vegetation. Satellite-derived vegetation indices have been extensively used for identifying periods of vegetative stress in crops, which represent an indication of agricultural drought, or generally vegetation. Moreover, soil moisture can be directly measured in the microwave region of the electromagnetic spectrum through satellites, and interpretation of SAR data may also provide additional information on soil moisture (Petropoulos et al., 2015).

1.4 REMOTELY SENSED DROUGHT QUANTIFICATION AND INDICES

Drought quantification is mainly implemented through drought indices. Remotely sensed drought indices use information from remote sensing sensors to map the condition of the land and to detect several drought features based on parameters, such as precipitation or temperature.

1.4.1 REMOTELY SENSED COMPOSITE DROUGHT INDICES

Combined drought indices, which are also termed *hybrid* or *aggregate* or *composite* (Waseem et al., 2015), are derived by incorporating existing drought indicators and indices into a single measure. At the present time, there are several regional/continental drought monitor models, leading to Global Drought Monitor (GDM), which coordinate and exchange information toward a Global Drought Information System (GDIS). Specifically, the four major regional/continental models are the North American Drought Monitor (NADM), which consists of U.S. Drought Monitor (USDM), Canada and Mexico; the European Drought Observatory (EDO) model; the African Drought Monitor (ADM); and the Australian Drought Monitor model (Dalezios et al., 2017b). There is an international need to continue working toward improving drought indices that can take climate change impacts into consideration.

> *USDM*: The USDM system uses a composite of multiple indicators, such as SPI and Palmer Drought Severity Index (PDSI), as well as indicators, such as vegetation and hydrological conditions, into a national weekly map of drought, covering various short- and long-term time frames, to develop a percent ranking methodology for drought monitoring and DEWS, leading to a single product (Svoboda et al., 2002; Wilhite, 2009; Svoboda, 2015). The short-term composite drought index operates on time scales ranging from a few days to a few months and consists of 35% Palmer Z-Index; 25% 3-month precipitation, 20% 1-month precipitation, 13% Climate Prediction Center Soil Moisture Model, and 7% Palmer (Modified) Drought Index. Similarly, the long-term composite drought index operates on time scales ranging from several months to a few years and consists of 25% Palmer Hydrologic Drought Index (PHDI), 20% 12-month precipitation, 20% 24-month precipitation, 15% 6-month precipitation, 10% 60-month precipitation, and 10% Climate Prediction Center Soil Moisture Model (Huang et al., 1996; Wood, 2012).
>
> *EDO model*: When precipitation reduction produces a decrease in soil moisture enough to not satisfy the water demand of the plants, this means the starting of agricultural drought, which is essentially based on satellite data and methods. Specifically, the EDO model combines different drought indices, namely SPI-n (McKee et al., 1993), satellite-based soil moisture anomalies, and the satellite-based fraction of absorbed photosynthetically active radiation (FAPAR) anomalies, which are absorbed by vegetation. This EDO model is proposed as drought indicator due to its sensitivity to vegetation stress (Gobron et al., 2007; Sepulcre-Canto et al., 2012). The method involves a classification scheme based on three drought impact levels: Watch, Warning, and Alert. Two additional levels, Partial recovery and Recovery, identify the stages of the vegetation recovery process. The development of a composite index by combining and integrating meteorological and remote sensing indicators helps to reduce false alarms, such as in vegetation indices, in which a biomass reduction can be generated by causes different from a drought induced by water stress.

1.4.1.1 Remotely Sensed Meteorological Drought Indices

As a general rule, any meteorological drought index can be converted to a remotely sensed index provided that meteorological variables, such as precipitation and/or temperature, are computed by remotely sensed algorithms or methods. Moreover, the use of remotely sensed drought indices is expected to increase drastically in the forthcoming years due to the availability of precipitation and temperature data platforms at a global scale, besides the increasing technological and computational advances. Meteorological drought indices can be used in the context of DEWS in order to provide timely information on drought for decision-making. It should be stated that a meteorological drought index value is essentially considered far more useful than raw data, especially in the case of drought monitoring for NRT decision-making. Moreover, a meteorological drought index can also be used as ground-truth information for modeling efforts or remotely sensed detection of several drought features.

There are precipitation-only drought indices, such as Rainfall Anomaly Index (RAI) (Van-Rooy, 1965), Bhalme–Mooley Drought Index (BMDI) (Bhalme and Mooley, 1980), Drought Severity Index (DSI) (Bryant et al., 1992), National Rainfall Index (NRI) (Gommes and Petrassi, 1994), Effective Drought Index (EDI) (Byun and Wilhite, 1999), and Drought Frequency Index (DFI) (González and Valdez, 2006). For better correlation with drought impacts and accounting for temporal temperature trends, additional meteorological variables have been considered. These include modifications to SPI (McKee et al., 1993) to develop the more comprehensive RDI (Tsakiris and Vagelis, 2005) that incorporates evapotranspiration resulting in better association with impacts from agricultural and hydrological droughts. Moreover, there is a similar index, namely Standardized Precipitation Evapotranspiration Index (SPEI) (Vicente-Serrano et al., 2010), which is sensitive to long-term trends in temperature change. If such trends are absent, SPEI performs similarly to SPI. Indeed, Keetch–Byram Drought Index (KBDI) (1968) has considered temperature and has had wide application to wildfire monitoring. In addition to temperature and evapotranspiration, PDSI (Palmer, 1965) also considers streamflow and soil moisture to give a more complete picture of the water balance. PDSI is categorized as a *comprehensive* drought index (Niemeyer, 2008) and remains a popular index. Improvements include self-calibration capacity and modifications in the estimation of evapotranspiration by replacing the original Thornthwaite method (1948) with other formulations and/or remotely sensed estimation (Dalezios et al., 2012).

1.4.1.1.1 Classification of Meteorological Drought Indices

Table 1.1 presents an indicative list of available and commonly used meteorological drought indices in different classes (Farago et al., 1989; Dalezios et al., 2017a). A brief description of the classes of meteorological drought indices is as follows:

> *Indices of atmospheric drought*: Low humidity is considered as the standard signal of dry spell. Indeed, atmospheric drought is usually described by the water vapor saturation deficit (Selyaninov, 1958).
>
> *Indices of precipitation anomaly*: Several existing precipitation anomaly indices are listed in Table 1.1, such as the Precipitation Index, the Relative Precipitation Sum, the Relative Anomaly, the Standardized Anomaly Index, and the Average Standard Anomaly (WMO, 1975).
>
> *Aridity indices*: The aridity index is based on the evapotranspiration/precipitation ratio (Budyko, 1952). There are several types of aridity indices, some of which are listed in Table 1.1, such as Lang's Rainfall Index, de Martone Aridity Index, Ped's Drought Index (PDI1) (1975), Selyaninov's Hydrothermal Coefficient (1958), Thornthwaite Index (1948), Potential Water Deficit, Potential Evaporation Ratio, Aridity Index: Moisture Available Index, Relative Evaporation, Surface Energy Balance, and Bowen Ratio (Skvortsov, 1950).
>
> *Recursive drought indices*: There are several recursive drought indices consisting of the family of PDSI, which are listed in Table 1.1. Such indices are Fooley Anomaly Index (FAI) (Fensham and Holman, 1999), BMDI, the family of PDSI (Palmer, 1965), SPI (McKee et al., 1993), Surface Water Supply Index (SWSI), Reclamation Drought Index (RDI), Palmer Drought Index (PDI), Crop Moisture Index (CMI), KBDI, EDI, and RDI (Tsakiris and Vagelis, 2005).

TABLE 1.1

Indicative List of Meteorological Drought Classes and Indices

Classification of Drought Indices

1. *Atmospheric Drought Indices*
    1.1 Saturation Deficit
2. *Precipitation Anomaly Indices*
    2.1 Precipitation Index
    2.2 Relative Precipitation Sum
    2.3 Relative Anomaly
    2.4 Standardized Anomaly Index (SAI)
    2.5 Average Standard Anomaly
3. *Aridity Indices*
    3.1 Lang's Rainfall Index
    3.2 De Martone Aridity Index
    3.3 Ped's Drought Index (PDI1)
    3.4 Selyaninov's Hydrothermal Coefficient
    3.5 Thornthwaite Index
    3.6 Potential Water Deficit
    3.7 Potential Evaporation Ratio
    3.8 Aridity Index: Moisture Available Index
    3.9 Relative Evaporation
    3.10 Surface Energy Balance
    3.11 Bowen Ratio

4. *Recursive Indices*
    4.1 Fooley Anomaly Index (FAI)
    4.2 Bhalme–Mooley Drought Index (BMDI)
    4.3 Palmer Drought Severity Index (PDSI)
    4.4 Standardized Precipitation Index
    4.5 Surface Water Supply Index (SWSI)
    4.6 Reclamation Drought Index (RDI)
    4.7 Palmer Drought Index (PDI)
    4.8 Palmer Crop Moisture Index (CMI)
    4.9 Keetch–Byram Drought Index (KBDI)
    4.10 Effective Drought Index
    4.11 Reconnaissance Drought Index (RDI)
5. *Remotely Sensed Information*
    5.1 Crop Water Stress Index (CWSI)
    5.2 Vegetation Index
    5.3 Normalized Difference Vegetation Index
    5.4 Stress Degree Days

Source: Dalezios, N.R. et al., Meteorological drought indices: Definitions, In Eslamian, S. (Ed.), *Handbook of Drought and Water Scarcity (HDWS)*, Vol. 1, Taylor & Francis Group, Abingdon, UK, 2017a.

1.4.1.2   Remotely Sensed Agricultural Drought Indices

Most of the existing and widely used remotely sensed drought indices are based on spectral reflectance of vegetation and, thus, are mainly used as agricultural drought indices, also known as vegetation indices. There are several indices based on remotely sensed information (Wagner et al., 1996), some of which are listed in Table 1.1, such as Crop Water Stress Index (CWSI) (Jackson et al., 1981), Vegetation Index, NDVI (Tsiros et al., 2006) and Stress Degree Days (Idso et al., 1980). Moreover, agricultural drought mainly involves monitoring of soil water balance and the subsequent deficit if a drought occurs (Table 1.2). Indeed, Table 1.2 presents an indicative list of internationally used conventional and remotely sensed agricultural drought indices (Dalezios et al., 2017c, 2017e). Specifically, there are models, such as Relative Soil Moisture (RSM) (Thornthwaite and Mather, 1955), CMI (Palmer, 1965), which is similar to PDSI, however, models short-term agriculture by considering moisture deficit only in the top 5 ft of soil column (Narasimhan and Srinivasan, 2005) and Crop Specific Drought Index (CSDI) (Meyer et al., 1993). Agricultural Drought Index (DTx) (Matera et al., 2007) calculates the daily transpiration deficit (DT) for x days. DTx uses the CRITeRIA soil moisture balance model (Zinoni and Marletto, 2003) with inputs including soil, crop, and weather conditions in addition to temperature anomalies, which affect evapotranspiration. Increased spatial and temporal resolutions are sought in developing Soil Moisture Deficit Index (SMDI) and Evapotranspiration Deficit Index (ETDI) (Narasimhan and Srinivasan, 2005), which are the soil components of the soil and water assessment tool (SWAT) hydrological model.

Remote sensing indices are diverse, and new indices are frequently proposed. Although NDVI has remained popular (Dalezios et al., 2017e), other indices such as the Normalized Difference Water Index (NDWI) (Gao, 1996), Temperature Vegetation Dryness Index (TVDI) (Domenikiotis et al., 2008), Vegetation Drought Response Index (VegDRI) (Brown et al., 2008), Vegetation

TABLE 1.2

Conventional and Satellite-Based Agricultural Drought Indices

Conventional Drought Indices	Satellite-Based Drought Indices
1. Agricultural Drought Index (DT*x*)	1. Normalized Difference Vegetation Index
2. Bhalme–Mooley Drought Index (BMDI)	2. Deviation NDVI index
3. Corn Drought Index	3. Enhanced Vegetation Index (EVI)
4. Crop Moisture Index (CMI)	4. Vegetation Condition Index (VCI)
5. Crop Specific Drought Index	5. Monthly Vegetation Condition Index
6. Evapotranspiration Deficit Index (ETDI)	6. Temperature Condition Index (TCI)
7. Global Vegetation Water Moisture Index	7. Vegetation Health Index (VHI)
8. Leaf Water Content Index (LWCI)	8. Normalized Difference Temperature Index
9. Moisture Availability Index (MAI)	9. Crop Water Stress Index (CWSI)
10. Reclamation Drought Index (RDI)	10. Drought Severity Index (DSI)
11. Soil Moisture Anomaly Index (SMAI)	11. Temperature-Vegetation Dryness Index
12. Soil Moisture Deficit Index (SMDI)	12. Normalized Difference Water Index
13. Soil Moisture Drought Index (SMDI)	13. Remote Sensing Drought Risk Index
14. Standardized Vegetation Index (SVI)	14. Vegetation Drought Response Index
15. Computed Soil Moisture	
16. Agro-Hydro Potential	

Source: Dalezios, N.R. et al., Agricultural drought indices: Combining crop, climate and soil factors, In Eslamian, S. (Ed.), *Handbook of Drought and Water Scarcity (HDWS)*, Vol. 1., Taylor & Francis Group, Abingdon, UK, 2017c.

Condition Index (VCI) (Kogan, 1995), Temperature Condition Index (TCI), and VHI (Kogan, 1995) are currently operationally used (NDMC, 2011; NOAA, 2011). Moreover, Enhanced Vegetation Index (EVI) (Liu and Huete, 1995) and NDWI (Foley, 1957) are also used to monitor vegetation state and health (Sivakumar et al., 2011). Traditionally used bands include near-infrared (NIR), red and shortwave infrared (SWIR). The land surface temperature (LST) has been used as an additional source along with NDVI to improve drought quantification accuracy.

1.4.1.3   Remotely Sensed Hydrological Drought Indices

The advanced PHDI (Palmer, 1965) model considered precipitation, evapotranspiration, runoff, recharge, and soil moisture. The PDSI shows ever lacked the snow component accumulation, which led to the development of SWSI (Shafer and Dezman, 1982), probably the most popular of this group. Reclamation Drought Index (RDI) (Weghorst, 1996) has improved SWSI by incorporating temperature and hence calculates a variable water demand as input. Remote Sensing Drought Index (RSDI) (Stahl, 2001) bases its model on homogeneous drought-stricken regions that comprise several neighboring low-flow gauging stations. In addition, there have been numerous attempts to estimate low-flow indices, such as the estimation of the low-flow index (7Q10), the 7-day, 10-year low-flow, using principal component regression (PCR) based on physiographic and hydrological variables (Eslamian et al., 2010). Two recent indices consider a water balance model: Groundwater Resource Index (GRI) (Mendicino et al., 2008) and Water Balance Derived Drought Index (Vasiliades et al., 2011). The former focuses on groundwater resources and uses geolithological information in a distributed water balance model, whereas the latter uses a model that artificially simulates runoff for ungauged watersheds. Moreover, Palfai Aridity Index (PAI) (Palfai, 1991) considered groundwater, along with temperature and has mainly been applied to basins.

1.4.1.4   Aggregation of Remotely Sensed Drought Indices

Combining drought indices has been increasingly discussed as a means to incorporate and more effectively exploit information that is readily available and proven to be useful in field-specific drought indices (Niemeyer, 2008). In follow-up to the Lincoln Declaration (WMO, 2009), the creation of a new composite hydrological drought index is recommended that would cover stream-flow, precipitation, reservoir levels, snowpack, and groundwater levels (Sivakumar et al., 2011). In general, hybrid drought indices can provide a stronger correlation with actual impacts sustained in the ground. Most hybrid drought indices are comparatively recent, including the USDM and VegDRI (Brown et al., 2008). VegDRI combines SPI and PDSI in addition to two NDVI-based indicators: Percent Average Seasonal Greenness (PASG) and Start of Season Anomaly (SOSA). Moreover, a combination of SPI, SWSI, and PDSI was accomplished to develop the integrated HDI (Karamouz et al., 2009). However, the combinations are mainly based on subjective local experience and are not considered objective. The predicted nonstationarity in future climates (IPCC, 2012) has instigated research for including future temporal patterns in drought characterization. Moreover, the SPEI accounts for the increase in the duration and magnitude of droughts resultant from higher temperatures. Additional research has been conducted for specific regions including Australia and the Czech Republic (Dalezios et al., 2017b).

1.4.2 Description of Representative Remotely Sensed Drought Indices

For drought quantification, two remotely sensed drought indices are briefly presented, namely the meteorological RDI (Tsakiris and Vagelis, 2005) and the agricultural VHI (Kogan, 1995), respectively. The presentation includes the composite use of the two indices for drought quantification. In addition, applications of both indices, RDI and VHI, are considered in DEWS and drought monitoring.

RDI is a new physically based general meteorological index, which can be used in a variety of climatic conditions. Moreover, RDI provides information for the water deficit in a region as it is based not only on precipitation, but also on potential evapotranspiration (PET). In the computation of RDI, the innovation consists of employing the Blaney–Criddle method for PET instead of the Thornthwaite method (1948), because it is more appropriate for the Mediterranean region with dry and hot summers (Blaney and Criddle, 1950). Furthermore, in this application, Blaney–Criddle method is based on brightness temperature (BT) and LST, which are derived from satellite data and constitute an innovative approach (Kanellou et al., 2009a; Dalezios et al., 2012).

Similarly, for the quantitative assessment of agricultural drought, as well as the computation of spatiotemporal features, one of the most reliable and widely used indices is applied, namely the VHI, which combines VCI and TCI. The VCI and TCI, as well as the adjusted VHI, have proven to be useful tools for agricultural drought and for monitoring agricultural crops in vulnerable agroecosystems internationally (Domenikiotis et al., 2004; Dalezios et al., 2014).

1.4.2.1 Study Area and Datasets

1.4.2.1.1 Study Area

A drought-prone study area is selected, namely the Thessaly region (about 14,400 km^2) in central Greece (Figure 1.1). The area has been selected, because it has long records of environmental data and is subject of diachronic research. The Thessaly plain constitutes the main agricultural area of the country. Thessaly is characterized by vulnerable agriculture. Extreme hydrometeorological events, such as hail and droughts, are quite common in the catchment due to the existing water deficit. Droughts are caused mainly by reduced precipitation resulting into lack of soil moisture, increased evapotranspiration, runoff reduction, decrease in streamflow levels in rivers, lakes and dams, lowering of groundwater table, and resulting in water deficit for agriculture (Dalezios et al., 2012).

1.4.2.1.2 Datasets

For the RDI estimation, the following data are utilized: daily precipitation of Thessaly water district in 50 × 50 km^2 spatial analysis derived by ground measurements provided by the Joint

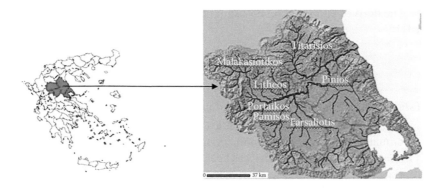

FIGURE 1.1 Geophysical map of Thessaly region. (From Dalezios, N.R. et al., *Nat. Hazard. Earth Syst.*, 14, 2435–2448, 2014. With permission.)

Research Center (JRC) of European commission (EC), Ispra, Italy. The main area is flat with no complex terrain, and the interpolation is effective. Crop coefficients maps are extracted by Corine Hellas 2000 for each month of the year (12 maps). Monthly maps of daytime sunshine duration for 39.39° Middle North Latitude of Thessaly (12 maps) are produced. A time series of 10-day BT images is extracted from Channels 4 and 5 of national oceanic and atmospheric administration (NOAA) satellite for 20 consecutive hydrological years (October 1981–September 2001) with a resolution of 8 × 8 km² provided by NOAA. Similarly, a time series of 10-day NDVI is extracted from Channels 1 and 2 of NOAA satellite for 20 consecutive hydrological years (October 1981–September 2001) with a resolution of 8 × 8 km² provided by NOAA.

For the VHI estimation, the following data are utilized: a time series of 10-day BT images is extracted from Channels 4 and 5 for 20 consecutive hydrological years (October 1981–September 2001) with a resolution of 8 × 8 km² provided by NOAA. Similarly, a time series of 10-day NDVI is extracted from Channels 1 and 2 for 20 consecutive hydrological years (October 1981–September 2001) with a resolution of 8 × 8 km² provided by NOAA.

1.4.2.2 Remotely Sensed Meteorological Drought Index: Reconnaissance Drought Index

The estimation of RDI includes prepossessing of satellite data, calculation of air temperature, estimation of PET with the use of satellite data, rain map extraction, and remotely sensed estimation of RDI (Dalezios et al., 2012). A brief description is as follows:

1.4.2.2.1 Preprocessing

Satellite data that are used originate from 10-day NOAA/advanced very high resolution radiometer (AVHRR) images with a spatial resolution of 8 × 8 km². The initial variables, which are extracted from the aforementioned satellite data, are BT and NDVI on a monthly basis. Then, geometric correction of all images is conducted.

1.4.2.2.2 Computation of Air Temperature

This method is analytically presented in Chapter 6. The monthly BT and NDVI values are processed based on an established algorithm (Becker and Li, 1990) to derive LST values on a pixel basis (Kanellou et al., 2009b). Monthly air temperature maps are then derived from LST satellite images based on a regression analysis between LST values and air temperature measurements from the meteorological station of Larissa, which is located in the region, for the whole period

(1981–2001). The derived empirical relationship between LST and air temperature (T_{air}) is given by Equation 1.1:

$$T_{air} = 0.6143 - LST + 7.3674 \quad R^2 \approx 0.82 \tag{1.1}$$

1.4.2.2.3 Estimation of Potential Evapotranspiration

The estimation of PET is based on Blaney–Criddle (1950) method, which is considered appropriate for subtropical climates with dry and hot summers, such as the Mediterranean region, because it has been applied in California. Originally, PET has been estimated by the Thornthwaite method, which is more appropriate for climates with wet and hot summers (e.g., East United States). The monthly potential evapotranspiration (ET_m), in mm, is estimated from Equation 1.2:

$$ET_m = k * [0.46T + 8.16] * p \tag{1.2}$$

where:

T is the mean monthly air temperature
p is the monthly daytime sunshine duration, which depends on the latitude of the area
k is the crop coefficient, different for each cultivation, vegetation type, and land use

Mean monthly crop coefficients for each vegetation type and land use in 500×500 m^2 pixel size and daytime sunshine duration (p) for each monthly value for the Thessaly water district (39,39° North Latitude) are mapped in a GIS environment (Kanellou et al., 2012). Then, the monthly crop coefficient maps and the maps of daytime sunshine duration are combined with the air temperature maps in order to derive Blaney–Criddle PET on a monthly basis for the whole time period (1981–2001).

1.4.2.2.4 Rain Map Extraction

Rain maps over Thessaly on a monthly basis, which are used for RDI estimation, are provided by JRC, Ispra. These data cover Greece from 1975 to 2005 per 50×50 km^2. From daily rainfall time series, the monthly cumulative rain of each hydrological year from 1975 to 2005 is calculated. Then rain maps are produced every month using linear interpolation.

1.4.2.2.5 Estimation of Remotely Sensed Reconnaissance Drought Index

Estimation of RDI is based on monthly temperature maps, crop coefficient (K_c) maps, sunlight duration maps (p), PET maps based on Blaney–Criddle, and rain maps (P). In this study, RDI is calculated on a monthly and annual basis. At first, the a_k coefficient is estimated (Tsakiris and Vagelis, 2005) from Equation 1.3:

$$a_k = \frac{\sum\limits_{j=1}^{j=k} P_j}{\sum\limits_{j=1}^{j=k} PET_j} \tag{1.3}$$

where P_j and PET_j are the precipitation and potential evapotranspiration, respectively, of the jth month of the hydrological year. The hydrological year for the Mediterranean region starts in October; hence for October, $k = 1$. RDI_n is the Normalized RDI, which is given by

$$RDI_n(k) = \frac{a_k}{\overline{a_k}} - 1 \tag{1.4}$$

TABLE 1.3
RDI Drought Classification Scheme

Drought Categories	RDI Values
Extremely wet	>2.00
Very wet	1.50 to 1.99
Moderately wet	1.00 to 1.49
Near normal	0.99 to −0.99
Moderately dry	−1.00 to −1.49
Severely dry	−1.50 to −1.99
Extremely dry	<−2.00

Source: Tsakiris, G. and Vangelis, H., *Eur. Water.*, 9, 3–11, 2005.

The Standardized RDI (RDI_{st}), which is used in this study, is given by

$$RDI_{st}(k) = \frac{y_k - \overline{y_k}}{\hat{\sigma}_k} \qquad (1.5)$$

where:

y_k is the ln a_k
$\overline{y_k}$ is its arithmetic mean
$\hat{\sigma}_k$ is its standard deviation

The drought categories based on RDI are shown in Table 1.3.

1.4.2.3 Remotely Sensed Agricultural Drought Index: Vegetation Health Index

Remotely sensed monthly VHI images are produced. Preprocessing of the initial satellite images is conducted, involving geometric and atmospheric correction of all images. Specifically, an innovative procedure is used based on certain filters for smoothing the data, leading to improved VHI values (Dalezios et al., 2014). Then, the computation of the VHI is conducted, and monthly VHI images are produced on a pixel basis.

Preprocessing: Ten-day NDVI maps are produced from the corresponding maximum value composite (MVC) images from the original visible (CH1) and NIR (CH2) images, respectively, of NOAA/AVHRR. BT images are produced from CH4 and CH5 images using the formula provided by the info file of the dataset. Then, fluctuations induced by noise are removed. Indeed, the combination of filtering and the MVC can significantly reduce the noise from residual clouds, satellite orbital drift, target/sensor geometry, and fluctuating transparency of the atmosphere. There is also additional noise, which can be related to processing, data errors, or just random noise. Specifically, a *4253 compound twice* median filter is applied to NDVI images, whereas a *conditional* statistical mean spatial filter (window size ranging from 3 × 3 to 7 × 7, according to image needs) is used for smoothing the BT series (Dalezios et al., 2014). Indeed, the BT series present continuous spatial fluctuations, and thus, a spatial filter (statistical mean) is preferred for smoothing CH4 and CH5 BTs. The term *conditional* means that the filter is applied only to the pixels that present errors. Finally, the 10-day NDVI and BT images are integrated into monthly values from the MVC and the mean pixel value, respectively.

TABLE 1.4
VHI Drought Classification Scheme

VHI Values	Vegetative Drought Class	Drought Class Numbers
<10	Extreme drought	1
<20	Severe drought	2
<30	Moderate drought	3
<40	Mild drought	4
>40	No drought	5

Source: Kogan, F.N., *Adv. Space Res.*, 15, 91–100, 1995.

VHI: The computation of monthly VHI images is based on BT and NDVI values. The VHI is the weighted summation of VCI and TCI, both derived from NOAA/AVHRR satellite data (Kogan, 1995). The VCI is based on NDVI and is expressed by Equation 1.6:

$$VCI = 100 * \frac{NDVI - NDVI_{min}}{NDVI_{max} - NDVI_{min}} \tag{1.6}$$

where NDVI, $NDVI_{max}$, and $NDVI_{min}$ are the smoothed 10-day NDVI, its multiyear maximum, and its multiyear minimum, respectively, for each pixel, in a given area. Following the same concept as with VCI, TCI is based on BT values and is given by Equation 1.7:

$$TCI = 100 * \frac{BT_{max} - BT}{BT_{max} - BT_{min}} \tag{1.7}$$

where BT, BT_{max}, and BT_{min} are the smoothed 10-day radiant temperature, its multiyear maximum, and its multiyear minimum, respectively, for each pixel, in a given area. From Equations 1.6 and 1.7, it is evident that VCI and TCI characterize the moisture and thermal conditions of vegetation, respectively. Specifically, thermal conditions are especially important when moisture shortage is accompanied by high temperature, increasing the severity of agricultural drought and having direct impact to vegetation health.

The VHI represents overall vegetation health and is used for drought mapping and crop yield assessment (Kogan, 1995). Table 1.4 presents the four VHI classes of agricultural drought severity, as well as no drought conditions. VHI is expressed by Equation 1.8:

$$VHI = 0.5 * (VCI) + 0.5 * (TCI) \tag{1.8}$$

VCI and TCI vary from zero, for extremely unfavorable conditions, to 100, for optimal conditions.

1.5 REMOTELY SENSED DROUGHT EARLY WARNING SYSTEM AND MONITORING: A CASE STUDY

DEWS focuses on monitoring drought conditions (Wilhite et al., 2000) through the use of drought indicators and indices. Depending on the data availability and quality for any particular area, it may be possible to utilize many drought indices that are available, such as the USDM, and to determine the most suitable for any particular area or season for drought monitoring and DEWS. There is an international need to continue working toward newer and potentially better drought indices that can also account for a changing climate in which there may be a shift in both temperature and precipitation regimes. For illustrative purposes, two case studies using empirical models and leading to DEWS, one based on RDI (Dalezios et al., 2012) and the other based on VHI (Dalezios et al., 2014), respectively, are briefly presented.

1.5.1 Remotely Sensed Meteorological DEWS: Reconnaissance Drought Index

By plotting the cumulative monthly areal extent values of the extreme RDI drought class, that is, class 4 of Table 1.3 with values lower than −2, for all the drought episodes, two figures are produced, namely Figure 1.2 for droughts of large areal extent and Figure 1.3 for droughts of small areal extent, respectively. Furthermore, curve fitting is conducted for each of these figures resulting in the following polynomials, namely Equation 1.9 for droughts of large areal extent and Equation 1.10 for droughts of small areal extent, respectively, both with high coefficient of determination.

$$y = 0.477x^3 - 9.7934x^2 + 78.221x - 36.078 \quad (R^2 = 0.9676) \tag{1.9}$$

$$y = 0.4868x^2 - 3.3415x + 4.78 \quad (R^2 = 0.9618) \tag{1.10}$$

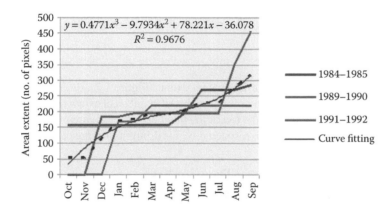

FIGURE 1.2 Cumulative large areal extent (no. of pixels 8 × 8 km²) of extreme drought (>2.0) during drought years based on remotely sensed RDI. (From Dalezios, N.R. et al., *Nat. Hazard. Earth Syst.*, 12, 3139–3150, 2012. With permission.)

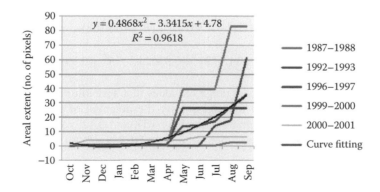

FIGURE 1.3 Cumulative small areal extent (no. of pixels 8 × 8 km²) of extreme drought (>2.0) during drought years based on remotely sensed RDI. (From Dalezios, N.R. et al., *Nat. Hazard. Earth Syst.*, 12, 3139–3150, 2012. With permission.)

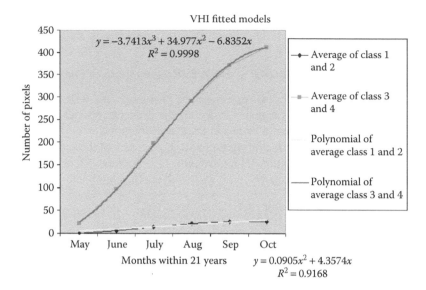

FIGURE 1.4 Fitted models of cumulative areal extent (no. of pixels) of Average Monthly Drought VHI values for the two sums of severity classes. (From Dalezios, N.R. et al., *Nat. Hazard. Earth Syst.*, 14, 2435–2448, 2014. With permission.)

It is worth noticing that for droughts of large areal extent (Figure 1.2), drought starts during the first 3 months of the hydrological year, whereas for droughts of small areal extent (Figure 1.3), drought starts in spring (April). This finding justifies the use of the fitted curves of Figures 1.2 and 1.3 along with the corresponding Equations 1.9 and 1.10, respectively, for drought prognostic assessment or DEWS.

1.5.2 REMOTELY SENSED AGRICULTURAL DROUGHT EARLY WARNING SYSTEMS: VEGETATION HEALTH INDEX

In Figure 1.4, the cumulative monthly areal extent curves of the two merged classes are shown, which correspond to the four VHI severity classes of agricultural drought (Table 1.4). Furthermore, curve fitting is conducted for each of these curves resulting in the following polynomials, namely Equation 1.11 for high severity areal extent drought and Equation 1.12 for low severity areal extent drought, respectively, both with high coefficient of determination.

$$y = 0.0905x^2 + 4.3574x \quad (R^2 = 0.9168) \tag{1.11}$$

$$y = -3.7413x^3 + 34.977x^2 - 6.8352x \quad (R^2 = 0.9998) \tag{1.12}$$

The two curves of Figure 1.4 present the range of values that agricultural drought may take every year during the warm season. These findings signify the potential of using the fitted curves of Figure 1.4, along with the corresponding Equations 1.11 and 1.12, respectively, for first-guess drought prognostic and monitoring assessment leading to DEWS.

1.5.3 RESULTS AND DISCUSSION

This section summarizes the results for drought quantification and monitoring based on two remotely sensed indices, namely RDI and VHI, respectively. For an extensive analysis of results, additional information and references are provided for RDI (Dalezios et al., 2012) and

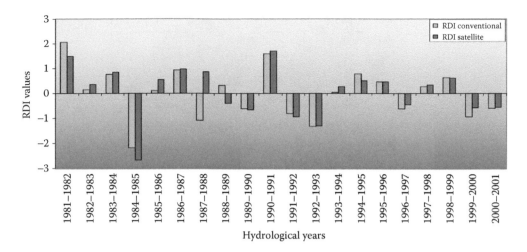

FIGURE 1.5 Annual RDI of Larissa for 1981–2001 (October–September). (From Dalezios, N.R. et al., *Nat. Hazard. Earth Syst.*, 12, 3139–3150, 2012. With permission.)

VHI (Dalezios et al., 2014), respectively. The results of RDI analysis are summarized in Figures 1.2, 1.3, and 1.5. Specifically, Figure 1.5 is a plot of the computed time series of annual RDI for Larisa (1981–2001) in which the negative RDI values represent drought years, and the positive RDI values refer to nondrought years based on the RDI classification scheme of Table 1.3. It should be clarified that the nondrought years also include the near-normal conditions. In Figure 1.5, the term conventional RDI means computation of the RDI from conventional meteorological data at Larisa station, whereas satellite RDI means the corresponding remotely sensed RDI values, as computed for a 3 × 3 pixel area above the Larisa station. The results of Figure 1.5 indicate that there are eight (8) drought episodes during this 20-year period in the study area of Thessaly, Greece. Moreover, the drought periods coincide with the hydrological year in most of the cases, that is, 12 or 13 months duration.

Similarly, VHI is estimated on a monthly basis for a period of 20 years (1981–2001) using satellite data, and the results are summarized in Figures 1.4 and 1.6, respectively. Moreover, several drought features are extracted from VHI images, leading to useful inferences. The results are presented in the upper curve of Figure 1.4 (for high drought severity classes 1 and 2) and in the lower curve of Figure 1.4 (for low drought severity classes 3 and 4), respectively, based on the VHI classification scheme of Table 1.4. Indeed, the initial four VHI severity classes are merged into two in Figure 1.4, namely extreme (class 1) and severe (class 2) drought into one class (upper curve of Figure 1.4). Similarly, moderate (class 3) and mild (class 4) drought are merged into another class (lower curve of Figure 1.4). The reason for merging classes is the relatively small number of pixels in each class in order to develop a sizeable dataset for performing a reliable analysis and fitting models. The analysis of agricultural drought based on VHI results and for the same period shows that drought occurs every year during the warm season, namely from April till October. The majority of pixels are accumulated between mild to moderate drought severity classes indicating a significant decrease in the number of pixels from mild to extreme drought classes for all the months. Finally, for illustrative purposes, Figure 1.6 presents VHI drought severity mapping of Thessaly for 6 months of a drought year, namely from April to September 1985 (Kanellou et al., 2009b). In Figure 1.6, it is evident that drought starts occurring in May with increasing severity and areal extent throughout the warm season with the maximum occurring toward the end of the summer, as expected. Figure 1.6 also shows the spatial variability of agricultural drought severity and extent within Thessaly, as well as the areas of drought persistence.

In summary, the findings justify the composite use of RDI and VHI, as well as other drought indices of different drought types, for drought monitoring and assessment.

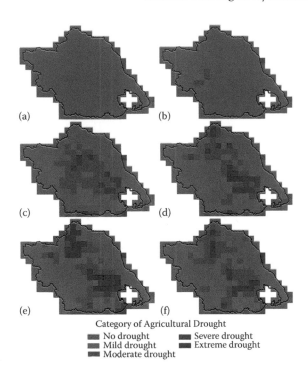

Category of Agricultural Drought

■ No drought	■ Severe drought
■ Mild drought	■ Extreme drought
■ Moderate drought	

FIGURE 1.6 VHI map of Thessaly for 6 months (April–September 1985): (a) April 1985, (b) May 1985, (c) June 1985, (d) July 1985, (e) August 1985, and (f) September 1985. (From Kanellou, E. et al., *Eur. Water J.*, 23, 111–122, 2009b. With permission.)

1.6 DROUGHT RISK ASSESSMENT AND MANAGEMENT

1.6.1 PRINCIPLES AND CONCEPTS

1.6.1.1 Drought Risk Assessment

Drought risk assessment involves risk estimation, which means the risk of such drought events, that is, event probabilities, as well as magnitude–duration–frequency and areal extent relationships for drought assessment (Wilhite et al., 2000; Dalezios et al., 2017d). The risk estimation also involves vulnerability assessment and its uncertainty. The combination of all the specific risks sums up to the total drought risk for all the severities (intensities), return periods, and elements at risk leading to the quantitative risk assessment (QRA) (Wilhite, 2009). Remote sensing methods and data can be employed in all aspects of drought risk assessment, vulnerability, and damage assessment, as well as relief, which involve assistance and/or intervention during or after drought (Dalezios et al., 2017e). Moreover, the areal extent of specific common drought episodes is another important feature of droughts, which could also be detected from remote sensing data and methods.

1.6.1.2 Drought Risk Management

The main principle of drought risk management policy is based on preparedness and mitigation measures (Wilhite, 2005; Arneth et al., 2014). Indeed, preparedness means predisaster activities with the objective to improve the institutional and operational efficiency, as well as to upgrade the level of readiness in order to respond to a drought episode. Moreover, mitigation involves short- and long-term programs, actions, or policies, which are applied in advance or during a drought in order to reduce the degree of risk to human life, productive capacity, and property. However, emergency response is always expected to be a part of drought management, because it is unlikely to consider, avoid, or reduce all potential impacts through mitigation actions. Nevertheless, at the present time, the adopted plans follow a more risk management approach to drought management, becoming more proactive (Dalezios et al., 2009).

The possible contribution of remote sensing could focus on relief and, possibly, preparedness, although in many cases remote sensing can make a valuable contribution to disaster prevention in which frequency of observation is not a prohibitive limitation. Moreover, remote sensing is a useful tool to analyze the vegetation dynamics on local, regional, or global scales (Keyantash and Dracup, 2002) to assess the vegetative stress and to determine the impact of climate on vegetation. Satellite-derived vegetation indices, such as VHI, have been extensively used for identifying periods of vegetative stress in crops, which represent an indication of agricultural drought, or generally vegetation. Moreover, soil moisture can be directly measured in the microwave region of the electromagnetic spectrum through satellites, and interpretation of SAR data may also provide additional information on soil moisture, which is a key factor in agricultural drought (Gobron et al., 2007; Sepulcre-Canto et al., 2012; Petropoulos et al., 2015).

1.6.1.3 Drought Impacts

Drought impacts refer to a multitude of drivers that may turn physical drought causes, such as reduced average precipitation, deficient soil moisture, and low water levels, into disaster events for vulnerable populations and economies. Drought impacts can be classified into direct and indirect impacts (Dalezios et al., 2009). Direct drought impacts may include agricultural production losses, food security problems, reduced cropland, forest and rangeland productivity, reduced water levels, increased fire hazards, livestock and wildlife mortality rates, damage to wildlife and fish habitat, rural livelihoods, as well as urban and economic sectors, among others. In addition, droughts may contribute to ecosystem decline, migration, and conflict. The consequences of these direct impacts are usually considered as indirect impacts. Moreover, drought impacts can also be classified by the affected sector, leading to environmental, economic, or social types of impact.

1.6.1.4 Drought Mitigation

Drought mitigation plans are based mainly on three fundamental components applied either to provincial, national, or regional level (Dalezios et al., 2009). First, an early warning system serves as the basis for decision-making during the development of a drought period. Second, it is important to undertake risk assessment in order to determine the subject and the causes of risk, which are accomplished through impact studies of drought events. Third, it is necessary to specify appropriate mitigation actions in order to reduce the risk of each impact for future drought events (Wilhite et al., 2000; Arneth et al., 2014). Assessment programs include the development of criteria or triggers for specific mitigation actions in response to drought, new data collection networks, early warning and monitoring systems, monitoring climate and water supply conditions, and drought contingency plans.

1.6.2 EXAMPLES OF DROUGHT RISK ASSESSMENT AND MANAGEMENT

Indicative examples of drought risk assessment and management, based on remote sensing data and methods, are presented.

1.6.2.1 Drought Severity–Duration–Frequency Relationships

Statistical frequency analysis of climatic extremes, such as droughts, has been extensively used internationally. However, droughts are not universally quantified phenomena, and frequency analysis of droughts is not easily accomplished. Frequency of drought occurrence cannot sufficiently and fully cover the study of droughts, unless it is quantitatively related to other aspects and terms, such as severity and duration of droughts. This has led to the development of drought severity–duration–frequency (SDF) relationships (Dalezios et al., 2000). Indeed, the frequency of an extreme event is usually expressed by its return period or recurrence interval, which may be defined as the average

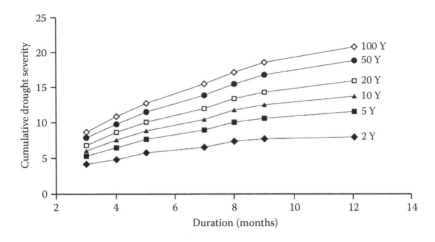

FIGURE 1.7 Severity–duration–frequency curves of several return periods (in years, Y) of drought severity for Bolos (BL) station in Greece. (From Dalezios, N.R. et al., *Hydrolog. Sci. J.*, 45, 751–770, 2000. With permission.)

interval of time within which the magnitude of the event is equated or exceeded once. Moreover, the magnitude of an extreme event is given by the total depth occurring in a particular duration, and data for extreme events can be usually presented by depth–duration–frequency graphs for several points throughout the region of interest.

In this example, the PDSI (Palmer, 1965) is employed, which, in addition to temperature and evapotranspiration, also considers streamflow and soil moisture to give a more complete picture of the water balance. The PDSI can be converted to a remotely sensed index provided that parameters, such as precipitation, temperature, evapotranspiration, or soil moisture, are computed by remotely sensing data and methods. For this application, the severity of drought is defined as the cumulative sum of successive negative values of the Z-index (ΣZ) of PDSI. Similarly, drought duration is considered as the number of successive months with negative Z-index values, whereas drought frequency is considered as the return period of a specific cumulative Z-index (ΣZ) value for successive months. For the development of drought SDF relationships, the computed monthly Z-index time series from 28 stations over Greece are used (Dalezios et al., 2000). A common period from 1957 to 1983 is used for the 28 stations. The analysis is conducted for each station for the identified drought periods, that is, successive negative Z-index values. In Figure 1.7, the developed drought SDF relationships are presented for Volos (BL) station in which each curve represents to one return period. Chapter 6 presents an analytical description of the SDF methodology.

1.6.2.2 Drought Occurrence and Severity Assessment Based on Moderate Resolution Imaging Spectroradiometer Normalized Difference Vegetation Index and Vegetation Condition Index

This application refers to drought occurrence and severity assessment, where the study area is the territory of Kenya. Specifically, the Moderate Resolution Imaging Spectroradiometer (MODIS) NDVI at 250 m ground resolution is used, and the database consists of images from 2000 till now (Klisch et al., 2015). The methodology involves data preprocessing of NDVI values using a modified Whittaker smoother in order to produce weekly NDVI images in NRT (Figure 1.8). Moreover, the data processing includes modeling of the uncertainty range for each pixel and time step, where the uncertainties are computed by a hindcast analysis of the NRT products against an *optimum* filtering (Klisch et al., 2015). In addition, for drought severity assessment, the VCI is estimated on a pixel basis. Then, the weekly VCI values are spatially integrated into administrative units (counties) and also temporally into 1- and 3-monthly time steps incorporating also uncertainty data.

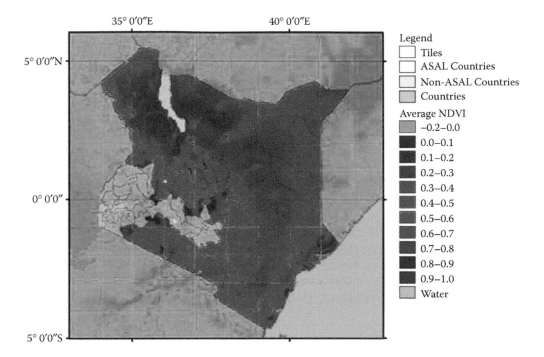

FIGURE 1.8 Average annual NDVI for ASAL counties of Kenya. Non-ASAL counties are shown in grey. The map also shows the 1° × 1 tiling system in which MODIS data is processed. (From Klisch A. et al., Satellite-based drought monitoring in Kenya in an operational setting. The International Archives of the Photogrammetry, RS and Spatial Information Sciences, Volume XL-7/W3, 36th ISRSE, May 11–15, Berlin, Germany, 8 p, 2015. With permission.)

This information is then combined with several socioeconomic indicators in order to reach decisions about drought mitigation and relief in affected counties. A validation of the methodology has also been conducted for drought periods in 3 years, namely 2006, 2009, and 2011, with positive results. There have also been comparisons with existing networks, such as the Famine Early Warning Systems Network (FEWS NET), and the main differences have been recorded.

1.6.2.3 Retrieval of Regional Drought Monitoring by the Modified Normalized Difference Water Index

The objective of this application is the effective regional drought monitoring by improving the remotely sensed soil moisture retrieval and, at the same time, by reducing the effect of the vegetation coverage variation on the accuracy. Indeed, it is well known that the vegetation coverage is a key factor, which affects the soil moisture accuracy. To achieve this objective, the Leaf Area Index (LAI) is used and incorporated into the NDWI (Zhang and Chen, 2015). Specifically, the Relative LAI (RLAI) is used as an indicator, and the application is implemented in two sites, which are located in the north and south of Henan Province, China, respectively. At first, the correlation is computed between the RLAI difference values of the days after turning green for the whole province. In this way, the RLAI distribution is obtained in the province's wheat-producing area. The next step is to use the remotely sensed Modified NDWI based on the following equation: $MNDWI = NDWI \times RLAI$, which leads to the soil moisture distribution of the wheat producing area. The analysis results indicate that the MNDWI mapping can improve the retrieval accuracy of soil moisture, leading to effective drought monitoring under different vegetation coverage, especially vegetation stress.

1.6.2.4 Drought Impacts on Amazon Forest Canopies Using Enhanced Vegetation Index

The objective of this application is to show that the seasonality in greenness of the Amazon rainforests signifies the variations in sun-sensor geometry throughout the year. Needless to say, the Amazon rainforests are the largest forested area in the tropics and constitute a major factor in the global carbon cycle. In this type of application, the MODIS MCD43 is employed, which models the bidirectional reflectance distribution function (BRDF), whereas, at the same time, the sun-sensor geometry remains constant (Brede et al., 2015). For this analysis, the EVI is used to characterize the seasonality in greenness. Specifically, the dataset consists of the 2000–2013 period, and the BRDF-adjusted EVI is computed. The results indicate that the developed time series of EVI demonstrate distinct seasonal patterns with peak values at the beginning of the dry season, whereas, at the same time, it maintains the same pattern of sun geometry expressed as Solar Zenith Angle (SZA). Moreover, in this application, precipitation anomalies are explored through the sensitivity of EVI for the Amazonia area and for the whole period (2000–2013). Specifically, BRDF-adjusted EVI dry season anomalies are compared to two drought indices, namely Maximum Cumulative Water Deficit and SPI. The results have shown that there is no significant relationship between EVI anomalies and drought, which is not compatible with other studies that investigate the drought impact on EVI and forest photosynthetic capacity. Moreover, the predictive potential of EVI for applications in tropical forests has also been examined, which has indicated a high level of uncertainty.

1.7 SUMMARY

In this chapter, an overview of the remote sensing potential in terms of data and methods in all aspects of drought analysis has been explored. For drought quantification, the remote sensing potential focuses mainly on drought indices, which use information from remote sensing sensors to map the condition of the land, to detect several drought features, and to estimate several environmental parameters. Indeed, remote sensing data and techniques provide, among others, direct measurements of land characteristics, vegetative cover, and components of the hydrological cycle, such as precipitation, temperature, evapotranspiration, or soil moisture. Moreover, the scientific trend in drought quantification consists of remotely sensed CDIs at different scales using DSS also toward global web drought platforms, which in certain cases are used operationally. Similarly, drought monitoring has been considered based mainly on DEWS using the aforementioned remotely sensed CDIs. Moreover, drought monitoring could also be addressed through remotely sensed vegetation dynamics and mapping of vegetation stress on local, regional, or global scales, because satellite-derived vegetation indices have been extensively used for identifying periods of agricultural drought, such as NDVI or VHI. In general, drought monitoring is considered semioperational; however, there are also several operational applications internationally. Furthermore, remote sensing methods and data can be effectively employed in all aspects of drought risk assessment, which include risk estimation, that is, drought event probabilities, magnitude–duration–frequency relationships, vulnerability assessment, areal extent mapping, and damage assessment. Finally, the possible contribution of remote sensing in drought management policy could focus on relief, which involves assistance and/or intervention during or after drought. Other contributions could possibly involve drought preparedness, drought impacts, and mitigation measures, although in many cases remote sensing can make a valuable contribution to disaster prevention.

At the present time, it is recognized that there is significant and steadily increasing reliability of remote sensing data and methods throughout the years, mainly due to computational and technological advancements. Moreover, the number of satellite systems is drastically increasing year by year with a continuous improvement of the spatial resolution. Similarly, there is a recent tendency to increase the number of available bands in these satellites resulting in new and valuable information. In addition, new types of remote sensing systems offer online open information for web platforms and are also utilized for monitoring and detecting drought. Indeed, during the last decade, a web service-based environment is being developed for integration of regional and continental drought monitors for computation and display of spatially consistent multiple drought indicators on a global scale.

There are scientific challenges and future perspectives in drought analysis, whereas, at the same time, there are needs and requirements to be addressed. The following list is indicative:

1. Assessment of drought impacts requires the design of a comprehensive database in accordance with the users' needs. Indeed, effective management of, and preparedness for, droughts requires free and unlimited access to relevant databases that allow prediction, monitoring, and assessment.
2. There is a need to assess the forecasting skills for droughts. For example, lack of good forecasting skills in drought is a constraint to improved adaptation, management, and mitigation.
3. The concept of drought monitor map product has to be promoted as a tool for all drought-prone regions to better understand drought severity using CDIs.
4. A methodology for preparation of a drought monitor map needs to be developed with recommendations for minimum, maximum, and optimum data layers.
5. An integration of geographic information system (GIS), remote sensing, simulation models, and other computational methods has to be considered in order to develop more effective DEWS alerts.
6. There is an urgent need for a more risk-based proactive drought management, which would include a timely user-oriented DEWS.
7. There is a need and opportunity to supply design requirements for new satellite sensors, in particular, to drought mitigation.
8. There is a need for regular joint training workshops on national and regional drought monitor products.

REFERENCES

Arneth, A., Brown, C., and Rounsevell, M.D.A. (2014). Global models of human decision-making for land-based mitigation and adaptation assessment, *Nat. Clim. Change.*, 4(7), 550–557.

Becker, F. and Li, Z.L. (1990). Towards a local "split window" method over land surface, *Int. J. Remote Sens.*, 3, 369–393.

Bhalme, H.N. and Mooley, D.A. (1980). Large-scale drought / floods and monsoon circulation, *Mon. Weather Rev.*, 108(8), 1197–1211.

Blaney, H.F. and Criddle, W.D. (1950). Determining water requirements in irrigated areas from climatological and irrigation data. USDA Soil Conservation Service, Technical Paper, No. 96, 48 p.

Brede, B., Verbesselt, J., Dutrieux, L.P., and Herold, M. (2015). Performance of the Enhanced Vegetation Index to detect inner-annual dry season and drought impacts on Amazon forest canopies. In *The International Archives of the Photogrammetry, RS and Spatial Information Sciences, 36th ISRSE*, Vol. XL-7/W3, Berlin, Germany, May 11–15, 8 p.

Brown, J.F., Wardlow, B.D., Tadesse, T., Hayes, J.J., and Reed, B.C. (2008). The Vegetation Drought Response Index (VegDri): An integrated approach for monitoring drought stress in vegetation, *GISci. Remote. Sens.*, 45(1), 16–46.

Bryant, S., Arnell, N.W., and Law, F.M. (1992). The long-term context for the current hydrological drought. *Institute of Water and Environmental Management (IWEM), Conference on the Management of Scarce Water Resources.* October 13–14.

Budyko, M.I. (1952). *Climate Change and National Plan of Environment Modification of Arid USSR Areas* (in Russian). Gidrometeoizdat, Leningrad.

Byun, H.R. and Wilhite, D.A. (1999). Objective quantification of drought severity and duration, *J. Clim.*, 12(9), 2747–2756.

Dalezios, N.R. (Ed.) (2017). *Environmental Hazards Methodologies for Risk Assessment and Management.* London, UK: IWA (book just completed, in press).

Dalezios, N.R., Dunkel, Z., and Eslamian, S. (2017a). Meteorological drought indices: Definitions. In S. Eslamian (Ed.) *Handbook of Drought and Water Scarcity (HDWS)*, Vol. 1, Abingdon, UK: Taylor & Francis Group (accepted, in press).

Dalezios, N.R., Dercas, N., and Eslamian, S. (2017b). Water scarcity management: Part 2: Satellite-based composite drought analysis, *Int. J. Hydrol. Sci. Technol.*, (accepted, in press).

Dalezios, N.R., Gobin, A., Tarquis, A.M., and Eslamian, S. (2017c). Agricultural drought indices: Combining crop, climate and soil factors. In S. Eslamian (Ed.) *Handbook of Drought and Water Scarcity (HDWS)*, Vol. 1. Abingdon, UK: Taylor & Francis Group (accepted, in press).

Dalezios, N.R., Tarquis, A.M., and Eslamian, S. (2017d). Drought assessment and risk analysis. In S. Eslamian (Ed.) *Handbook of Drought and Water Scarcity (HDWS)*. Abingdon, UK: Taylor & Francis Group (accepted, in press).

Dalezios, N.R., Spyropoulos, N.V., and Eslamian, S. (2017e). Remote Sensing in drought quantification and assessment (Chapter 21). In S. Eslamian (Ed.) *Handbook of Drought and Water Scarcity (HDWS)*, Vol. 1. Abingdon, UK: Taylor & Francis Group (accepted, in press).

Dalezios, N.R., Blanta, A., Spyropoulos N.V., and Tarquis, A.M. (2014). Risk identification of agricultural drought for sustainable agroecosystems, *Nat. Hazard. Earth Syst.*, 14, 2435–2448.

Dalezios, N.R., Blanta, A., and Spyropoulos, N.V. (2012). Assessment of remotely sensed drought features in vulnerable agriculture, *Nat. Hazard. Earth Syst.*, 12, 3139–3150.

Dalezios, N.R., Bampzelis, D., and Domenikiotis, C. (2009). An integrated methodological procedure for alternative drought mitigation in Greece, *European Water*, 27(28), 53–73.

Dalezios, N.R., Loukas, A., Vasiliades, L., and Liakopoulos, H. (2000). Severity-duration-frequency analysis of droughts and wet periods in Greece, *Hydrolog. Sci. J.*, 45(5), 751–770.

Domenikiotis, C., Dimakis, E., Kanellou, E., and Dalezios, N.R. (2008). Use of the remotely sensed derived TVDI for drought monitoring in Thessaly watersheds, *AgEng2008: Agricultural and Biosystems Engineering for a Sustainable World*, Crete, Greece, June 23–25.

Domenikiotis, C., Spiliotopoulos, M., Tsiros, E., and Dalezios, N.R. (2004). Early cotton yield assessment by the use of the NOAA/AVHRR derived drought vegetation condition index in Greece, *Int. J. Remote Sens.*, 25(14), 2807–2819.

Du Pissani, C.G., Fouche, H.J., and Venter, J.C. (1998). Assessing rangeland drought in South Africa, *Agri. Syst.*, 57, 367–380.

ESA. (2014). Sentinel, Earth online—ESA, https://earth.esa.int/web/guest/missions/esa-future-missions/sentinel-1, June 4, 2014.

Eslamian, S., Ghasemizadeh, M., Biabanaki, M., and Talebizadeh, M. (2010). A principal component regression method for estimating low flow index, *Water Resour. Manage.*, 24, 2553–2566.

Farago, T., Kozma, E., and Nemes, C. (1989). Drought indices in meteorology, *Időjárás*, 93, 45–59.

Fensham, R.J. and Holman, J.E. (1999). Temporal and spatial patterns in drought related tree dieback in Australian savanna, *J. Appl. Ecol.*, 36, 1035–1050.

Foley, J.C. (1957). *Droughts in Australia: Review of Records from Earliest Years of Settlement to 1955.* Melbourne, Australia: Commonwealth of Australia, Bureau of Meteorology.

Gao, B.C. (1996). NDWI–A normalized difference water index for remote sensing of vegetation liquid water from space, *Remote Sens. Environ.*, 58(3), 257–266. doi:10.1016/S0034-4257(96)00067-3.

Gobron, N., Pinty, B., Melin, F., Taberner, M., Verstraete, M.M., Robustelli, M., Widlowski, J.-L. (2007). Evaluation of the MERIS/ENVISAT fAPAR Product, *Adv. Space Res.*, 39, 105–115.

Gommes, R.A. and Petrassi, F. (1994). Rainfall variability and drought in sub-Saharan Africa since 1960. Rome, Italy: FAO, 100 p.

González, J. and Valdés, J. (2006). New drought frequency index: Definition and comparative performance analysis, *Water Resour. Res.*, 42(11), W11421. doi:10.1029/2005WR004308.

Hao, Z. and AghaKouchak, A. (2013). Multivariate standardized drought index: A parametric multi-index model, *Adv. Water Res.*, 57, 12–18.

Heim, R.R. Jr. (2002). A review of twentieth-century drought indices used in the United States, *Bull. Am. Meteorol. Soc.*, 83(8), 1149–1165.

Huang, J., Van den Dool, H.M., and Georgakakos, K.P. (1996). Analysis of model calculated soil moisture over the United States (1931–1999) and applications to long-range temperature forecasts, *J. Climate*, 9, 1350–1362.

Idso, S.B., Jackson, R.D., Pinter, P.J., Jr., Reginato, R.J., and Hatfield, J.L. (1981). Normalizing the stress-degree-day concept for environmental variability, *Agr. Forest. Meteorol.*, 32, 249–256.

IPCC. (2012). Managing the risks of extreme events and disasters to advance climate change adaptation, Special Report of IPCC, 582 p.

Jackson, R.D., Idso, S.B., Reginato, R.J., and Pinter, P.J., Jr. (1981). Canopy temperature as a crop water stress indicator, *Water Resour. Res.*, 17, 1133–1138.

Kanellou, E., Domenikiotis, C., Tsiros, E., and Dalezios, N.R. (2009b). Satellite-based drought estimation in Thessaly, *Eur. Water J.*, 23(24), 111–122.

Kanellou, E., Spyropoulos, N., Dalezios, N.R. (2012). Geoinformatic intelligence methodologies for drought spatiotemporal variability in Greece, *Water Resour. Manage.*, 26(5), 1089–1106.

Kanellou, E.C., Domenikiotis, C., and Dalezios, N.R. (2009a). Description of conventional and satellite drought indices. pp. 23–59. In G. Tsakiris (Ed.) PRODIM Final Report, EC, 448 p.

Karamouz, M., Rasouli, K., and Nazif, S. (2009). Development of a hybrid index for drought prediction: Case study, *J. Hydrol. Eng.*, 14(6), 617–627. doi:10.1061/(ASCE)HE.1943-5584.0000022.

Keetch, J.J. and Byram, G.M. (1968). A drought index for forest fire control. NOAA, Climatology Division, Asheville, NC, 32 p.

Keyantash, J. and Dracup, J.A. (2002). The quantification of drought: An evaluation of drought indices, *Bull. Am. Meteorol. Soc.*, 83(8), 1167–1180.

Klisch, A., Atzberger, C., and Luminari, L. (2015). Satellite-based drought monitoring in Kenya in an operational setting. The International Archives of the Photogrammetry, RS and Spatial Information Sciences, Volume XL-7/W3, 36th ISRSE, May 11–15, Berlin, Germany, 8 p.

Kogan, F.N. (1995). Application of vegetation index and brightness temperature for drought detection, *Adv. Space Res.*, 15, 91–100.

Liu, H.Q. and Huete, A. (1995). A feedback based modification of the NDVI to minimize canopy background and atmospheric noise, *IEEE Trans. Geosci. Remote Sens.*, 33(2), 457–465. doi:10.1109/36.377946.

Matera, A., Fontana, G., and Marletto, V. (2007). Use of a new agricultural drought index within a regional drought observatory. In G. Rossi (Ed.) *Methods and Tools for Drought Analysis and Management.* Dordrecht, the Netherlands: Springer. pp. 103–124.

McKee, T.B., Doesken, N.J., and Kleist, J. (1993). The relationship of drought frequency and duration to timescales. Preprints, *Eighth Conference on Applied Climatology*, Anaheim, CA, American Meteorological Society, pp. 179–184.

McVicar, T.R., and Jupp, D.L.B. (1998). The current and potential operational uses of remote sensing to aid decisions on drought exceptional circumstances in Australia: A review, *Agr. Syst.*, 57(3), 399–468.

Mendicino, G., Senatore, A., and Versace, P. (2008). A Groundwater Resource Index (GRI) for drought monitoring and forecasting in a mediterranean climate, *J. Hydrol.*, 357(3–4), 282–302. doi:10.1016/j.jhydrol.2008.05.005.

Meyer, S.J., Hubbard, K.G., and Wilhite, D.A. (1993). A crop-specific drought index for corn. I: Model development and validation, *Agron. J.*, 85(2), 388–395.

Mishra, A.K. and Singh, V.P. (2010). A review of drought concepts, *J. Hydrology.*, 39(1–2), 202–216.

Narasimhan, B. and Srinivasan, R. (2005). Development and evaluation of Soil Moisture Deficit Index (SMDI) and Evapotranspiration Deficit Index (ETDI) for agricultural drought monitoring, *Agric. For. Meteorol.*, 133(1–4), 69–88. doi:10.1016/j.agrformet.2005.07.012.

Nastos, P., Dalezios, N.R., and Ulbrich U. (Eds.) (2016). Advances in meteorological hazards and extreme events. Special Issue of NHESS.

NDMC. (2011). Vegetation drought response index. National Climatic Data Center. http://drought.unl.edu/vegdri/VegDRI_Main.htm.

Niemeyer, S. (2008). New drought indices. Options Méditerranéennes. *Série A: Séminaires Méditerranéens*, 80, 267–274.

NOAA. (2011). STAR—Global Vegetation Health Products. NOAA. http://www. star.nesdis.noaa.gov/smcd/emb/vci/VH/vh_browse.php.

Palfai, I. (1991). Az 1990 évi aszály Magyarországon. Vízügyi Közlemények, 2, 117–132.

Palmer, W.C. (1965). Meteorological drought. Weather Bureau Research Paper No. 45, Washington, DC: US Department of Commerce. 58 pp.

Ped, L.A. (1975). On the new drought and over-moistening index (in Russian), *Trans. USSR Hydrometeorol. Center*, 156, 19–39.

Petropoulos, G.P., Ireland, G., and Barrett, B. (2015). Surface soil moisture retrievals from remote sensing: Current status, products and future trends, *Phys. Chem. Earth.*, 83, 36–56. doi:10.1016/j.pce.2015.02.009.

Salinger, J., Sivakumar, M.V.K., and Motha, R.P. (Eds.) (2005). *Increasing Climate Variability and Change: Reducing the Vulnerability of Agriculture and Forestry.* Dordrecht, the Netherlands: Springer. 362 p.

Selyaninov, G.T. (1958). The nature and dynamics of the droughts. In Droughts in the USSR, their Nature, Recurrences and Impact on Crops Yields (in Russian). Gidrometeoizdat, Leningrad.

Sepulcre-Canto, G., Horion, S., Singleton, A., Carrao, H., Vogt, J. (2012). Development of a combined drought indicator to detect agricultural drought in Europe, *Nat. Hazard. Earth Syst.*, 12, 3519–3531.

Shafer, B. and Dezman, L. (1982). Development of a Surface Water Supply Index (SWSI) to assess the severity of drought conditions in snowpack runoff areas, *Proceedings of the Western Snow Conference*, pp. 164–175.

Sivakumar, M.V.K., Motha, R.P., Wilhite, D.A., and Wood, D.A. (2011). Agricultural drought indices, *Proceedings of an Expert Meeting.* Murcia, Spain: WMO. p. 219.

Sivakumar, M.V.K., Motha, R.P., and Das, H.P. (Eds.) (2005). *Natural Disaster and Extreme Events in Agriculture.* Berlin, Germany: Springer.

Skvortsov, A.A. (1950). On the question of heat and water exchange in the surface air (in Russian), *Transactions of Middle Asian State University*, 22(6).

Stahl, K. (2001). Hydrological drought-a study across Europe. Universitätsbibliothek Freiburg.

Svoboda, M. (2015). Overview of drought indicators and their application in the context of a drought early warning and information system, *NENA Regional Water/Drought Platform Stakeholders Workshop*, October 25–27, Cairo, Egypt.

Svoboda, M., LeComte, D., Hayes, M., Heim, R., Gleason, K., Angel, J., B. Rippey, B. et al. (2002). The drought monitor, *Bull. Am. Meteorol. Soc.*, 83(8), 1181–1190.

Tarquis, A.M., Gobin, A., Ulbrich, U., and Dalezios, N.R. (Eds.) (2013). Weather related hazards and risks in agriculture, *Nat. Hazards Earth Syst. Sci.*, 13, 2599–2603.

Thenkabail, P.S., Gamage, M.S.D.N., and Smakhtin, V.U. (2004). The use of remote sensing data for drought assessment and monitoring in southwest Asia, *Research Report, International Water Management Institute*, No. 85, pp. 1–25.

Thornthwaite, C.W. (1948). An approach toward a rational classification of climate, *Geog. Rev.*, 38(1), 55–94.

Thornthwaite, C.W. and Mather, J.R. (1955). The water balance, *Climatology*, 8, 1–104.

Tsakiris, G. and Vangelis, H. (2005). Establishing a drought index incorporating evapotraspiration, *Eur. Water.*, 9(10), 3–11.

Tsiros, E., Domenikitios, C., and Dalezios, N.R. (2006). Aridity mapping with the use of NDVI and satellite derived degree days. In N.R. Dalezios and S. Tzortzios (Eds.) *3rd HAICTA Int. Conference.* COST-University of Thessaly, Volos, Greece, pp. 853–865.

UNISDR. (2015). Reading the Sendai framework for disaster risk reduction 2015–2030. Geneva, Switzerland: United Nations, International Strategy for Disaster Reduction. 34 p.

UNISDR. (2005). Hyogo framework for Action 2005–2015. Building the resilience of nations and communities to disasters. Geneva, Switzerland: United Nations, International Strategy for Disaster Reduction. http://www.unisdr.org/eng/hfa/hfa.htm.

Van-Rooy, M.P. (1965). A rainfall anomaly index (RAI) independent of time and space, *Notos*, 14, 43–48.

Vasiliades, L., Loukas, A., and Liberis, N. (2011). A water balance derived drought index for Pinios River Basin, Greece, *Water Resour. Mgmt.*, 25(4), 1087–1101.

Vicente-Serrano, S.M., Begueria, S., and Lopez-Moreno, J.I. (2010). A multiscalar drought index sensitive to global warming: The standardized precipitation evaportranspiration index, *J. Climate*, 23, 1696–1718.

Wagner, W., Borgeaud, M., and Noll, J. (1996). Soil moisture mapping with the ERS scatterometer, *Earth Obs. Quart.*, 54, 4–7.

Waseem, M., Ajimal, M., and Kim, T.W. (2015). Development of a new composite drought index for multivariate drought assessment, *J. Hydrol.*, 527, 30–37.

Weghorst, K. (1996). The reclamation drought index: Guidelines and practical applications. cedb.asce.org, ASCE, Denver, CO.

Wilhite, D.A. (2005). The role of disaster preparedness in national planning with specific reference to droughts. In M.V.K. Sivakumar, R.P. Motha, and H.P. Das (Eds.) *Natural Disasters and Extreme Events in Agriculture.* Springer, pp. 23–37.

Wilhite, D.A. (2009). The role of monitoring as a component of preparedness planning: Delivery of information and decision support tools. In A. Iglesias, A. Cancelliere, F. Cubillo, L. Garrote, and D. Wilhite (Eds.) *Coping with Drought Risk in Agriculture and Water Supply Systems.* Dordrecht, the Netherlands: Springer Publishers.

Wilhite, D.A., Hayes, M.J., Kinutson, C., and Smith, K.H. (2000). Planning for drought: Moving from crisis to risk management, *J. Amer. Water Res. Assoc.*, 36(4), 697–710.

WMO. (1975). Drought and agriculture. WMO Technical Note 138, Geneva, Switzerland.

WMO. (2009). Lincoln declaration on drought indices. World Meteorological Organization (WMO)http://www. wmo.int/pages/prog/wcp/agm/meetings/wies09/documents/Lincoln_Declaration_Drought_Indices.pdf (accessed April 22, 2011).

Wood, A.W. (2012). The University of Washington Surface Water Monitor: An experimental platform for national hydrologic assessment and prediction. Civil and Environmental Engineering Department, University of Washington, Seattle, WA.

Zargar, A., Sadiq, R., Naser, B., and Khan, F.I. (2011). A review of drought indices, *Environ. Rev.*, 19, 333–349.

Zhang, H. and Huai-liang, C. (2015). The application of modified normalized difference water index by leaf area index in the retrieval of regional drought monitoring. In *The International Archives of the Photogrammetry, RS and Spatial Information Sciences, 36th ISRSE*, Vol. XL-7/W3, Berlin, Germany, May 11–15, 7 p.

Zinoni, F. and Marletto, V. (2003). Prime valutazioni di un nuovo indice di siccità agricola. Atti convegno Aiam 2003, 24–25 maggio 2003, Bologna, Italy, pp. 232–238.

2 Agricultural Drought Monitoring Using Satellite Soil Moisture and Other Remote Sensing Data over the Iberian Peninsula

José Martínez-Fernández, Nilda Sánchez,
and Ángel González-Zamora

CONTENTS

2.1 INTRODUCTION

A natural hazard is a threat of a naturally occurring event that will have a negative effect on people or the environment, and drought is a type of natural hazard that is further aggravated by growing water demand (Mishra and Singh, 2010). Droughts rank first among all natural hazards when measured in terms of the number of people affected (Wilhite, 2000).

    The assessment and monitoring of droughts are of primary importance for freshwater planning and management. Drought is the most important threat that is facing the management of water resources and is one of the most critical issues of the water policy in any country (von Christierson et al., 2011; Walker et al., 1991). Among the three categories of drought that are commonly recognized, that is, meteorological, hydrological, and agricultural drought, agricultural drought has the most direct and immediate impact (Cooley et al., 2015). Meteorological drought is defined as a lack of precipitation over a region for a period of time. Hydrological drought is related to a period with inadequate surface and subsurface water resources for the established water uses of a given water resources management system (Mishra and Singh, 2010). An agricultural drought is considered to begin when the soil moisture available to plants drops to such a level that it adversely affects the crop yield and, hence, agricultural production (Panu and Sharma, 2002). Consequently, drought is a major cause of limited agricultural productivity throughout the world, accounting for a large proportion of crop losses and annual yield variations (Boyer, 1982). Currently, agricultural drought is a key issue in the global change analysis due to its economic implications (Cook and Wolkovich, 2016; Hogg et al., 2013) and is considered as a direct factor of social and political conflicts in developing countries because it can have a catalytic effect, contributing to political unrest (Kelley et al., 2015).

Drought indicators and indices are variables that are used to describe the physical characteristics of drought severity, spatial extent, and duration (Steinemann et al., 2005). Although it has been clearly distinguished from meteorological drought, agricultural drought has usually been studied from a climatological perspective. Most drought assessment methods are based on long-term atmospheric data, such as rainfall and temperature, or on precipitation indices, but they typically do not consider site-specific soil properties (Torres et al., 2013), such as soil water content.

The most commonly used agricultural drought index, the Crop Moisture Index (CMI) (Palmer, 1968), is based on a subset of the calculations required for the Palmer Drought Severity Index (PDSI) (Palmer, 1965), which is primarily a meteorological drought index. The Climatic Moisture Index, although it was first used for forestry applications (Hogg, 1994, 1997), is another agricultural index, and it is calculated by subtracting potential evapotranspiration from precipitation. Purcell et al. (2003) developed the Atmospheric Water Deficit (AWD) index by assuming a simple soil water balance using precipitation and evapotranspiration. These indices have shown good results, and the use of some, as in the case of the CMI, is widespread.

In other cases, soil moisture has been incorporated into agricultural drought analysis, but it was calculated or estimated through water balance or hydrological modeling with the use of climatic variables. Narasimhan and Srinivasan (2005) used the Soil Moisture Deficit Index and the Evapotranspiration Deficit Index (ETDI) for agricultural drought monitoring from simulated soil moisture and evapotranspiration derived from the Soil and Water Assessment Tool (SWAT) model. ETDI uses a concept of water stress from the reference crop evapotranspiration and the actual evapotranspiration, both SWAT model outputs. The Agricultural Reference Index for Drought (ARID) (Woli et al., 2012) is based on a reference crop, which is actively growing grass that completely covers the soil surface and uses a simple soil water balance. Ceppi et al. (2014) used meteorological forecasts and hydrological modeling to simulate soil moisture as a component of a real-time agricultural drought forecasting system. Recently, Qin et al. (2015) applied the Soil Moisture Drought Severity (SMDS) index, calculated from the Community Land Model, in a comparison study with a precipitation index in North China.

Another approach is based on developed indices that directly use soil moisture as a tool to identify and, in some cases, assess agricultural drought. Sridhar et al. (2008) proposed the Soil Moisture Index (SMI), using the soil water content as a quantitative indicator of drought. Although it is not defined as a specifically agricultural drought index, the Multivariate Standardized Drought Index (MDSI) (Hao and AghaKouchak, 2013) integrates a classical climatic approach and soil moisture data. Martínez-Fernández et al. (2015a) proposed the Soil Water Deficit Index (SWDI), which is based on soil moisture measurements and has a hydrological basis and a specific agricultural interpretation.

2.2 REMOTE SENSING OF AGRICULTURAL DROUGHT

Drought monitoring involves the continuous assessment of the natural indicators of drought severity, spatial extent, and impacts (Wilhite and Buchanan-Smith, 2005). Remote sensing can play an important role in drought monitoring strategies because it provides synoptic, rapidly repeating, and spatially continuous information about drought conditions (Hayes et al., 2012). Following these authors, remote sensing products are particularly useful in this field because they provide information required for local-scale monitoring and decision-making that cannot be adequately supported from information derived from traditional, point-based data sources; fill in gaps in information on drought conditions for locations between *in situ* observations and in areas that lack ground-based observational networks; enable earlier drought detection in comparison to traditional climatic indices; and collectively provide a suite of tools and datasets geared to meet the observational needs for a broad range of decision-support activities related to drought.

2.2.1 AGRICULTURAL DROUGHT ANALYSIS THROUGH REMOTE VEGETATION OBSERVATION

A set of agricultural drought monitoring strategies is based on remote sensed products, many of them based on vegetation indices. In fact, the value of satellite remote sensing for drought monitoring was first realized more than two decades ago with the application of the Normalized Difference Vegetation Index (NDVI) from the advanced very high resolution radiometer (AVHRR) for assessing the effect of drought on vegetation (Anyamba and Tucker, 2012). For example, Brown et al. (2008) proposed the Vegetation Drought Response Index (VegDRI), integrating traditional climate-based drought indicators and satellite-derived vegetation index metrics. Otkin et al. (2013) proposed the Evaporative Stress Index (ESI), which uses remotely sensed thermal infrared imagery to estimate evapotranspiration, and demonstrated that ESI anomalies can provide early warning of incipient drought impacts on agricultural systems. Keshavarz et al. (2014) introduced the Soil Wetness Deficit Index, which is calculated from the land surface temperature (LST), and the NDVI is derived from the Moderate Resolution Imaging Spectroradiometer (MODIS) satellite. From the same satellite, Li et al. (2014) obtained and compared the NDVI anomaly with the CMI to assess agricultural drought in the northeast of China.

In general, vegetative drought indices based on the NDVI have been widely and successfully used to identify and monitor areas affected by drought at regional and local scales (Bayarjargal et al., 2006; Bhuiyan et al., 2006; Hayes and Decker, 1998; Kogan, 1997; Tucker and Choudhury, 1987). However, in some cases, the collection of only vegetation data was not sufficient for accurate drought analysis. Therefore, thermal channels were studied to retrieve additional drought information (Kogan, 1995) based on the LST. The temperature of vegetation is as a proxy for the plant stress caused by both scarce and excessive wetness. Furthermore, if LST and NDVI are jointly considered as a surface condition descriptor, surface properties, such as soil water content and evapotranspiration, can be inferred (Carlson, 2013). There is a remarkable inverse relationship between LST and vegetation condition, which in turn is related to the soil moisture content and therefore can be used as an agricultural drought indicator. The LST/NDVI slope retrieved from remote sensing data has been used to assess information related to spatially averaged soil moisture conditions (Goetz, 1997; Sandholt et al., 2002) and in climate and drought monitoring (Karnieli et al., 2010; McVicar and Bierwirth, 2001; Sánchez et al., 2016).

2.2.2 SATELLITE SOIL MOISTURE FOR AGRICULTURAL DROUGHT MONITORING

The great progress in remote sensing during recent decades (Fernández-Prieto et al., 2012) has allowed the scientific community to obtain precise and frequent soil moisture maps anywhere in the world. Global satellite-based soil moisture data are becoming increasingly available (Rebel et al., 2012), and the number of potential applications of that information has increased accordingly (Mecklenburg et al., 2016).

Recently, the global Soil Moisture Climate Change Initiative (SMCCI project) soil moisture dataset from the European Space Agency (ESA) has been generated using active and passive microwave spaceborne instruments since 1978 (Dorigo et al., 2015). In November 2009, the first mission dedicated to soil moisture, Soil Moisture and Ocean Salinity (SMOS, ESA), was launched (Kerr et al., 2010). Since January 2015, the second soil moisture satellite, Soil Moisture Active Passive (SMAP), National Aeronautics and Space Administration (NASA) has been in orbit (Entekhabi et al., 2010). All these initiatives make it possible to generate a global soil moisture map every 1–2 days, with constantly increasing accuracy and a variety of spatial scales. This has enabled certain uses of soil moisture measurements derived from satellite remote sensing data for enhancing drought monitoring systems (Nghiem et al., 2012).

Due to this new perspective and by taking advantage of the availability of soil moisture data, some researchers have recently started to propose new approaches to drought analysis. For example, Scaini et al. (2015) demonstrated that the SMOS-derived soil moisture anomalies can be properly used for drought monitoring assessment. Chakrabarti et al. (2014) used SMOS soil moisture downscaled at 1 km to investigate the effects of agricultural drought on crop yields. Carrão et al. (2016) used the Essential Climate Variable Soil Moisture (ECV SM) product (Liu et al., 2012), derived from the SMCCI project, to produce a single homogenized global dataset for assessing the impacts of agricultural drought throughout the whole South-Central American region. Martínez-Fernández et al. (2016) applied the SWDI using the SMOS soil moisture to assess the agricultural drought dynamics over an agricultural area in Spain. Sánchez et al. (2016) proposed the Soil Moisture Agricultural Drought Index (SMADI) by integrating MODIS products and SMOS soil moisture and assessed its applicability over the Iberian Peninsula at a spatial resolution of 500 m.

2.3 METHODOLOGY

Two approaches have been used to assess the feasibility of using satellite soil moisture data for agricultural drought monitoring over the Iberian Peninsula at different spatial scales. On the one hand, the SWDI, which uses SMOS surface soil moisture (SSM) and soil water parameters, has been used over an agricultural area in the Duero basin (Spain). On the other hand, SMADI, which also uses SMOS SSM together with LST and NDVI from MODIS, has been applied in the same agricultural area as well as in diverse rainfed areas along the Iberian Peninsula.

2.3.1 THE SOIL WATER DEFICIT INDEX

The SWDI was proposed to characterize the agricultural drought based on a soil moisture series and basic soil water parameters (Martínez-Fernández et al., 2015a). SWDI is able to adequately identify the main attributes that define a drought event (i.e., beginning/end, duration, and intensity) and has been formulated with a specific agricultural meaning and interpretation. The SWDI is calculated as follows:

$$\text{SWDI} = \left(\frac{\theta - \theta_{\text{FC}}}{\theta_{\text{AWC}}} \right) 10 \tag{2.1}$$

where:
  θ is the soil water content
  FC is the field capacity
  AWC is the available water content, which is the difference between FC and WP (wilting
    point)

TABLE 2.1
SWDI Severity Categories

SWDI Value	Drought Level
SWDI > 0	No drought
0 > SWDI > −2	Mild
−2 > SWDI > −5	Moderate
−5 > SWDI > −10	Severe
−10 > SWDI	Extreme

When the SWDI is positive, the soil has excess water; when it equals zero, the soil is at the field capacity of water content (i.e., no water deficit). Negative values indicate a soil drought, and the water deficit is absolute (wilting point) when the SWDI reaches $\leq −10$.

The interpretation of the SWDI is close to the concept of the readily available soil water (RAW) of the Food and Agriculture Organization (FAO) guidelines for the determination of the crop water requirements (Allen et al., 1998). The p factor of the RAW definition is the average fraction of total available soil water (TAW) that can be depleted from the root zone before moisture stress occurs. The p factor varies for the main crops from 0.2 (SWDI of −2, close to field capacity) to 0.8. Of the crops considered by Allen et al. (1998), 50% have a p factor below 0.5. Using these considerations, the severity of a drought is categorized as shown in Table 2.1.

In this work, the soil water parameters needed to calculate the SWDI have been obtained from *in situ* analytical determinations. There are other approaches to obtain these parameters based on the statistics of the whole soil moisture series or restricted to the growing season (Hunt et al., 2009). The long-term available satellite soil moisture series (Dorigo et al., 2015) together with the statistical approach could be used to obtain reliable estimates of the soil water parameters (Martínez-Fernández et al., 2015a) and to implement that methodology through strictly earth observation resources. To assess the performance of the SWDI, a comparison analysis was made with the CMI and AWD in the REMEDHUS area for the same period.

2.3.2 The Soil Moisture Agricultural Drought Index

SMADI is a synergistic integration of SMOS-SSM with MODIS-derived LST and water/vegetation indices for agricultural drought monitoring (Sánchez et al., 2016) (Figure 2.1). It focuses on short-term agricultural droughts, and its rationale is based on the inverse relationship between LST and vegetation conditions, which in turn is related to soil moisture content. As different surface types may have different LST/NDVI slopes and intercepts under the same atmospheric and surface moisture conditions, the choice of scale may influence the shape of the relationship (Sandholt et al., 2002). Thus, scaling the temperature and the vegetation index is recommended to minimize the effects of atmospheric temperature changes from one day to the next (Carlson, 2013) and to avoid site and time dependence in the series. The proposed drought index is based on the LST/NDVI slope but uses a normalized version of these variables, following the rationale of the Vegetation Condition Index (VCI) (Kogan, 1990) and the Temperature Condition Index (TCI) (Kogan, 1995).

The VCI is an indicator of environmental stress based on the NDVI normalized with the maximum and minimum range for each pixel over the available imagery (Equation 2.2):

$$\text{VCI} = \frac{(\text{NDVI}_i - \text{NDVI}_{\min})}{(\text{NDVI}_{\max} - \text{NDVI}_{\min})} \qquad (2.2)$$

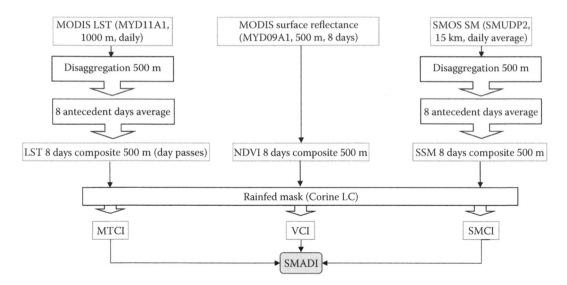

FIGURE 2.1 Conceptual flowchart of the SMADI index retrieval through a multisensor downscaling approach.

where:

NDVI$_i$ is the NDVI for the given day or time interval considered

NDVI$_{min}$ and NDVI$_{max}$ are the absolute 5-year minimum and maximum NDVI, respectively, for each pixel

The TCI also has a normalized formulation but uses the thermal bands to determine temperature-related vegetation stress. Kogan (1995) designed the TCI to be combined with the VCI in an additive manner similar to ($a \times$ VCI $+ b \times$ TCI). However, we aim to highlight the inverse behavior of LST and NDVI; thus, we proposed a modified version of the TCI, the MTCI (Equation 2.3):

$$\text{MTCI} = \frac{(\text{LST}_i - \text{LST}_{min})}{(\text{LST}_{max} - \text{LST}_{min})} \tag{2.3}$$

where:

LST$_i$ is the satellite-derived temperature for the given day or time interval considered

LST$_{max}$ and LST$_{min}$ are the multiyear maximum and minimum, respectively

In this form, high values of MTCI represent dry conditions, whereas low values represent wet or nonstressed conditions, as opposed to the VCI. Therefore, high values of the ratio MTCI/VCI correspond to dry conditions (high values of MTCI and low values of VCI), whereas low values correspond to wet conditions (low MTCI and high VCI).

To incorporate the SSM into the new index, the Soil Moisture Condition Index (SMCI) was defined, similarly to VCI and MTCI, as a normalization of soil moisture values relative to the absolute maximum (SSM$_{max}$) and the absolute minimum (SSM$_{min}$) of the 5-year series (Equation 2.4) in order to obtain normalized SSM ranging from 1 (very dry conditions) to 0 (wet, favorable conditions).

$$\text{SMCI} = \frac{(\text{SSM}_{max} - \text{SSM}_i)}{(\text{SSM}_{max} - \text{SSM}_{min})} \tag{2.4}$$

Finally, the proposed agricultural drought index, SMADI, incorporates the SMCI as a multiplicative factor with the slope of MTCI/VCI (Equation 2.5)

$$SMADI_i = SMCI_i \frac{MTCI_i}{VCI_{i+1}} \tag{2.5}$$

where i corresponds to the given day or time interval considered. The SMADI values are normalized between 0 and 1 to make it comparable to the other datasets. VCI, MTCI, and SMCI are calculated on an 8-day basis due to the MODIS, the 8-day NDVI composite data used in VCI. Note that the VCI selected for a given i corresponds to the ensuing 8-day period in order to consider the time lag between the plant response and the soil moisture conditions. This lag is variable in the literature. Schnur et al. (2010), using MODIS over grass/herbaceous mixed with shrubby rangeland covers in a semiarid climatic area, showed that the correlation reaches a maximum value when the vegetation index lags soil moisture by 5–10 days. In addition, Li et al. (2014) found that the NDVI anomaly responds to CMI with a lag of 10 days. Here, a lag of 8 days was considered, taking into account the time resolution of the MODIS-NDVI composite, the crop type considered, and the purpose of detecting short-term effects.

2.3.3 STUDY AREAS IN THE IBERIAN PENINSULA

2.3.3.1 The REMEDHUS Area

The Soil Moisture Measurement Stations Network of the University of Salamanca (REMEDHUS) is located in Spain (41.1°N–41.5°N; 5.1°W–5.7°W) and includes an agricultural area of approximately 1300 km² in a central semiarid zone of the Duero Basin (Figure 2.2). This area is nearly flat (less than 10% slope on average), and it ranges from 700 to 900 m.a.s.l. The climate is continental–Mediterranean, with approximately 400 mm of average annual precipitation. Mean temperature is 12°C, and the region experiences long, cold winters and hot summers. The average annual reference evapotranspiration is 1025 mm. The land uses are mainly agricultural: winter and spring rainfed cereals and small areas of irrigated crops and vineyards (Sánchez et al., 2012).

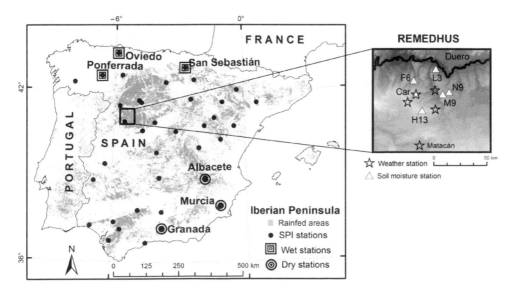

FIGURE 2.2 Iberian Peninsula, REMEDHUS area, and soil moisture/weather station locations.

2.3.3.2   Rainfed Agricultural Areas on the Iberian Peninsula

The Iberian Peninsula is a complex climate system, including dry, temperate, and cold climate categories. Annual average rainfall in the Iberian Peninsula shows a high spatial variability, ranging from less than 200 mm to greater than 1300 mm on average. In general, the highest values are located in the northern and northwestern areas, and the lowest values are recorded in southeastern Spain. The monthly average rainfall varies between years, with notable seasonality, which is stronger in the southern half of the Peninsula, and less notable in northeastern Spain, and with a clear reduction in rainfall in summer (AEMet, 2011). The month with the most rainfall across the Peninsula is December, and the driest month is July, coinciding with the months with the lowest and highest temperature values, respectively.

Based on the Joint Research Center (JRC) European Drought Observatory (EDO) monitoring of the occurrence and evolution of the drought events over Europe, as well as the Spanish Meteorological Agency (AEMet), two periods were especially dry in several areas on the Iberian Peninsula during the period 2010–2014. The most severe drought took place between October 2013 and July 2014 and was located in the southeast (the Murcia and Valencia regions and eastern Andalucía) and in the center of the Iberian Peninsula. These regions were affected by mean and long-term precipitation deficits, leading to significant soil moisture deficits. This was considered by the JRC to be an *exceptional drought*. The persistent lack of rain in central and southern Spain resulted in scarce soil moisture during the grain-filling phase of winter and spring crops. The situation worsened at the beginning of June when maximum daily temperatures were increased above seasonal values, leading to a critical early senescence (EC-JCR, 2014).

The second severe period of drought occurred during 2012. This was a dry and warm year for the entire Iberian Peninsula but was particularly strong in the central and northwestern parts, where the precipitation deficit produced a very dry year. Furthermore, the summer heat wave wiped out 80% of the Spanish crop and raised the price of several products, such as olive oil. Although less marked, 2011 was also a dry year in which the drought was particularly intense in areas that are usually wet, such as the northern part of Spain.

As opposed to drought periods, 2010 was very wet (even extremely wet in some areas) in the southern part (30% more water than usual), leading to a surplus in the water reservoirs. In addition, the first trimester of 2013 was especially wet in the northern and western parts of the Iberian Peninsula.

The rainfed agriculture domain of the Iberian Peninsula (Figure 2.2) is critically exposed to the scarcity of water. In these areas, the sole supply of water comes from precipitation, strongly limiting vegetation growth. Sparse rainfall and warm temperatures during the development stages or the growing season can impact the flowering and grain filling of most cereals, limiting the yield. Thus, the rainfed areas were selected to analyze the performance of SMADI. To mask these areas, the procedure of the crop monitoring and yield forecasting activities of the JRC of the European Commission was followed. The land use–land cover vectorial map from the CORINE 2006 Land Cover project (CLC, version 16, updated 2012) was used to identify and clip the rainfed fields over the MODIS products (Figure 2.2). Next, 39 locations within these areas belonging to the AEMet meteorological network were selected to test SMADI against another drought index in order to assess the spatial and temporal occurrence of drought over the Iberian Peninsula.

2.3.4   Databases

2.3.4.1   Soil Moisture and Climate Databases

2.3.4.1.1   REMEDHUS Area

The REMEDHUS monitoring network performs continuous hourly soil moisture measurements using Hydra probes (Stevens® Water Monitoring System, Inc.) at 5 cm depth and two Envirosmart probes (Sentek Pty. Ltd.) at 25 and 50 cm depths. The soil column of 0–50 cm is considered as the root zone because the main crops are cereals (90% of the area) (Sánchez et al., 2010). In this work, six representative stations (F6, H13, L3, M9, N9, and CAR) were selected to calculate the drought index, and the index data were then averaged for both the surface layer (SWDI-RSSM) and the root zone (SWDI-RRZSM).

The meteorological variables used in the REMEDHUS area were measured at four automatic weather stations located along the network and at one long-term weather station (Matacán) from the AEMet, located close to REMEDHUS (Figure 2.2). The basic variables were daily rainfall and daily potential evapotranspiration (ETo), which were estimated using the Penman–Monteith methodology (Allen et al., 1998). The Matacán station was selected due to the need for long climatic series in the CMI calculation. For the CMI calculation, mean temperature and total precipitation from more than 60 years were provided by the AEMet, together with mean soil water parameters from the REMEDHUS area.

2.3.4.1.2 Iberian Peninsula

Even though drought is a relative condition that differs widely between locations and climates, standardized indices, such as the Standardized Precipitation Index (SPI) (McKee et al., 1993), allow the user to confidently compare historical and current droughts between different climatic and geographic locations. SPI is based only on precipitation records and was designed to quantify the precipitation deficit at multiple timescales in order to reflect the impact of drought on the availability of the different water resources (WMO, 2012). Here, the shortest possible timescale (one month) was chosen. The 1-month SPI may approximate conditions represented by the CMI (WMO, 2012) and is generally used to reflect short-term conditions related to short-term soil moisture and crop stress, especially during the growing season. Following the classification system to define drought thresholds (McKee et al., 1993), a drought event occurs any time; the SPI is continuously negative and reaches an intensity of −1.0 or less and ends when the SPI becomes positive. The SPI data from 39 locations from the AEMet stations (Figure 2.2) were provided from 2001 to 2014 and were compared to SMADI to assess the spatial occurrence and time spans of drought events across the Iberian Peninsula.

2.3.4.2 Satellite Databases

The length of the period of analysis was 5 years (2010–2014), coinciding with the SMOS soil moisture series length. It is a very representative period of the climatic conditions of this area, with contrasting years in terms of the amount of rainfall. However, for the calculations of the drought indices, the series start in June 2010 to avoid the first period of SMOS observations, when more data gaps occurred.

2.3.4.2.1 Moderate Resolution Imaging Spectroradiometer Satellite

MODIS data from the Aqua satellite were chosen for the VCI and MTCI calculation. Aqua is an orbiting satellite with local equatorial crossing times of approximately 1:30 p.m./1:30 a.m. Among the wide variety of MODIS products, the 8-day composite MYD09A1 surface reflectance product at 500 m resolution and the daily LST product (MYD11A1) at 1 km resolution were used in the present study. The reflectance values in different bands were tested to retrieve different alternatives for the VCI (Sánchez et al., 2016), as well as day/night passes. Finally, only day passes and red/near-infrared bands were selected for the LST and the NDVI retrieval and further SMADI calculation.

Owing to the different spatial resolutions of the LST and reflectance products, the LST at 1 km was assigned to the four pixels at the 500 m resolution. Granules h17v04, h17v05, h18v04, and h18v05 covering the Iberian Peninsula were mosaicked and reprojected to geographic coordinates. Regarding their different temporal interval, the daily MYD11A1 product was transformed into an 8-day product similar to the MYD09A1, using the average of the 8 antecedent days. During the period of analysis, no data gaps were found for the study area. This calculation provided 46 composites for each year.

2.3.4.2.2   Soil Moisture and Ocean Salinity Satellite

The SMOS Soil Moisture Level 2 User Data Product (SMUDP2 file) version 5.51 has been used. This product is delivered through ESA over the Icosahedral Snyder Equal Area Earth (ISEA-4H9) grid with equally spaced nodes at ~15 km, known as the Discrete Global Grid (DGG). This product comprises the retrieved surface geophysical parameters and quality indicators. The soil moisture retrieval is associated with a Data Quality Index (DQX), which represents the uncertainty of the retrieval. In this research, a DQX threshold filter of 0.04 m^3 m^{-3} was used to select the best quality SMOS retrievals. In addition, data corrupted by radio frequency interferences (RFIs) have been discarded using the RFI probability flag available from the L2 product. These filters were applied to the ascending and descending orbits separately, and later, the average of the two orbits for each day was calculated. This procedure ensured a more complete series, eliminating gaps due to retrieval failures or corrupted data (González-Zamora et al., 2015).

In the case of the SWDI calculation, daily L2 SSM data from June 2010 to December 2014 of the 11 DGGs overlapping the selected stations in the REMEDHUS area were used to calculate the daily index at each DGG. The index data were averaged over the whole area. Finally, a weekly SWDI average was obtained unless there were less than four daily data during the week. This temporal scale is usually used for agricultural drought monitoring because farmers commonly use a weekly period for irrigation schedules (Purcell et al., 2003). As the monthly scale can be helpful for other water management applications, the SWDI was also calculated and analyzed monthly.

For the SMADI calculation, data of L2 SSM from January 2010 to December 2014 over the Iberian Peninsula were used and combined with the MODIS products. Owing to their different spatial resolutions, each MODIS pixel at 500 m was allocated a corresponding DGG, and similarly to the LST product, SMOS SSM was averaged into an 8 antecedent-day composite, the same interval used for the MODIS products.

2.4   RESULTS AND DISCUSSION

2.4.1   SOIL WATER DEFICIT INDEX ASSESSMENT

Previous works performed in REMEDHUS (Sánchez et al., 2012), on the Iberian Peninsula (Parinussa et al., 2014), and in other areas around the world (Schlenz et al., 2012) have revealed that the SMOS soil moisture estimate provides good and reliable results and that its accuracy has been increasing as new reprocessed data and various scales of analysis have been introduced (González-Zamora et al., 2015). The use of different downscaling strategies (Piles et al., 2014) has considerably increased the spatial resolution of the SMOS soil moisture and, therefore, expanded the fields of application, including agriculture. Taking advantage of all these benefits, the SMOS SSM has been used in the agricultural area of REMEDHUS to characterize and assess the agricultural drought using the SWDI.

The indices calculated with *in situ* SSM (SWDI-RSSM) and with the SMOS L2 product (SWDI-SSSM) show a very high correlation (Table 2.2, Figure 2.3) and good temporal overlap of the series (Figure 2.3b). The SWDI is able to detect drought periods of special relevance in terms of intensity and duration. For instance, the effect of two consecutive dry years (2011–2012) is clearly shown, as the SWDI was almost continuously negative during that period (Figure 2.3). After a very dry winter season, the scarce spring rains were not enough to recharge the soil, and the drought event was extended to the middle of October. In the winter of 2011–2012, the Iberian Peninsula was hit by one of the most severe droughts that were ever recorded (Trigo et al., 2013).

One of the limitations that *a priori* may question the use of soil moisture estimates from satellite is the fact that the observation scope is restricted to the uppermost centimeters of the soil (Njoku and Entekhabi, 1996). However, several authors have demonstrated that there is a very close correlation between the surface and root zone soil moisture (Albergel et al., 2008; Hirschi et al., 2014). In the present work, the SWDI-RSSM and the product calculated for the first 50 cm

TABLE 2.2

Correlation Analysis (Pearson) between SWDI Calculated with REMEDHUS *in situ* Data (0–5 cm, SWDI-RSSM; 0–50 cm, SWDI-RRZSM), SMOS L2 Data (SWDI-SSSM), CMI, and AWD

	SWDI-RSSM	SWDI-RRZSM	SWDI-SSSM
SWDI-RRZSM	0.90		
SWDI-SSSM	0.86	0.73	
CMI	0.77	0.67	0.71
AWD	0.73	0.56	0.83

Note: All the correlations are significant ($p < 0.001$).

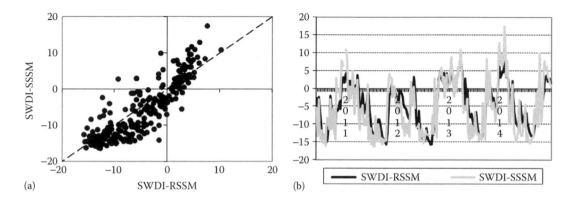

(a)   SWDI-RSSM   (b)

SWDI-RSSM   SWDI-SSSM

FIGURE 2.3   (a) Comparison between SWDI-RSSM and SWDI-SSSM (dashed line 1:1) and (b) temporal evolution of both indices (June 2010, December 2014) in REMEDHUS. (Adapted from Martínez-Fernández, J. et al., *Remote Sens. Environ.*, 177, 277–286, 2016. With permission.)

depth (SWDI-RRZSM) show a very high correlation coefficient ($R = 0.9$). At the same time, the comparison between SWDI-SSSM and SWDI-RRZSM shows a very acceptable relationship (Table 2.2). These results show that the water content measured *in situ* or remotely sensed from the most superficial soil layer would be a good indicator of the water content located in the soil root zone. In relation to the agricultural drought analysis, the reference soil depth has to be the one explored by the roots of the crops.

To assess the SWDI and its feasibility to monitor agricultural drought, it was compared with the CMI and the AWD (Table 2.2). In all the cases, the correlation was good and statistically significant; however, some differences were detected. The comparison with the AWD is always better, especially with the SWDI obtained with SMOS data (Figure 2.4, right). The AWD was proposed as an indicator that expresses the soil water balance from the difference between precipitation and ETo (Purcell et al., 2003). It is a very simple approach but, owing the obtained results, seems to reflect the real soil water dynamics. However, the comparison with CMI shows a less satisfactory result and, to some extent, a less coherent behavior. The correlation coefficient with CMI is lower (Table 2.2), and the comparison reveals that CMI shows a recurring number of zero values (Figure 2.4, left). These particular results are associated with rain events that momentarily interrupt periods of meteorological drought but, as evidenced by the SWDI, are not enough to sufficiently recharge the soil and to soothe or interrupt a period of agricultural drought.

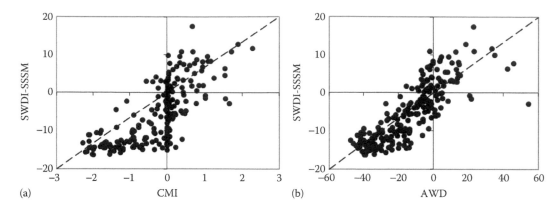

FIGURE 2.4   Comparison between SWDI-SSSM and CMI (a) and AWD (b) at a weekly scale. Dashed line 1:1. (Adapted from Martínez-Fernández, J. et al., *Remote Sens. Environ.*, 177, 277–286, 2016. With permission.)

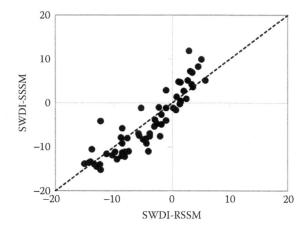

FIGURE 2.5   Comparison between SWDI-RSSM and SWDI-SSSM at the monthly scale. Dashed line 1:1. (Adapted from Martínez-Fernández, J. et al., *Estudios en la Zona no Saturada del Suelo*. Universidad de Alcalá de Henares. Alcalá de Henares, 13, 49–54, 2015b. With permission.)

Although the monthly scale may not be appropriate for certain applications in which the agricultural drought monitoring is needed (e.g., irrigation schedule), it can be useful in relation to water management applications or for the comparison with other indicators of water deficit. For that reason, in this work, the monthly SWDI calculated with *in situ* data and satellite data were compared. The correlation coefficient between SWDI-RSSM and SWDI-SSSM monthly scale is 0.9 in the case of SSM (Figure 2.5) and 0.84 with series of soil moisture in the root zone (not shown). Taking into account the good relationship obtained between the index measured *in situ* and that obtained with SMOS data at different timescales, it is reasonable to think that such a methodology could be useful for the combined analysis of the different types of drought. The data globally supplied by the satellite allow researchers to have information that, until recently, was not affordable and that, therefore, greatly opens the scope of applications.

The results obtained suggest that the index calculated with SMOS data is capable of conveniently capturing, or at least in a way comparable to other indices, the situations of water shortage and identifying periods of drought, as well as the attributes that define a drought period (beginning/end, duration, and intensity).

2.4.2 SOIL MOISTURE AGRICULTURAL DROUGHT INDEX PERFORMANCE

The assessment of SMADI was performed at two scales. First, in the REMEDHUS area, the temporal evolution of SMADI was compared to SWDI at the area-averaged scale and to CMI and SPI at the Matacán weather station (point scale) using the Pearson correlation coefficient (R). In addition, a second temporal analysis consisted of the quantitative comparison of SMADI with R calculated at each Iberian Peninsula AEMet station with available SPI data ($n = 39$).

The results of the correlation with the SWDI for the REMEDHUS average resulted in a statistically significant correlation of -0.75. The negative sense is due to the different description of drought using SWDI (negative values indicate drought conditions) and SMADI (positive values indicate drought conditions). Both indices showed a marked seasonality and revealed the period from winter 2011 to spring 2012 to be the driest of the series (Figure 2.6) (Trigo et al., 2013).

The results of the comparison of CMI and SMADI (Figure 2.7) showed a good inverse correlation ($R = -0.71$). Taking into account that the CMI threshold depicting drought condition goes below -1, only years 2011 and 2012 can be labeled as moderately dry years, with values of CMI between -1 and -2 during the summer period (Figure 2.7a). On the contrary, for the SPI, the correlation worsened to $R = -0.37$, although the relationship was statistically significant. Moreover, for the SPI evolution, it is more difficult to track seasonal cycles, and there was no clear pattern of drought along the whole cycle (Figure 2.7). Note that for the SPI, the abnormally dry conditions began below -1.

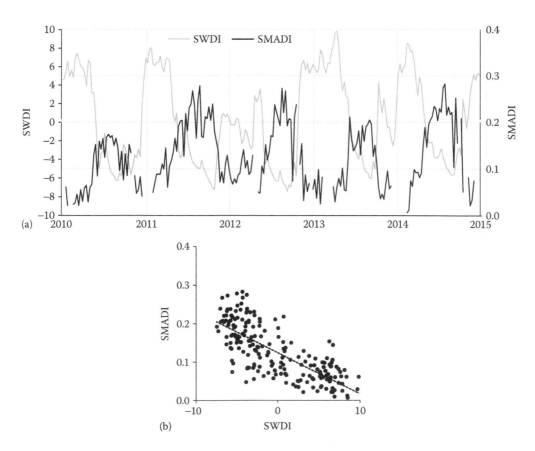

FIGURE 2.6 (a) Temporal evolution of SMADI and SWDI at REMEDHUS and (b) scatter plot of both series. (Adapted from Sánchez, N. et al., *Remote Sens.*, 8, 287, 2016. With permission.)

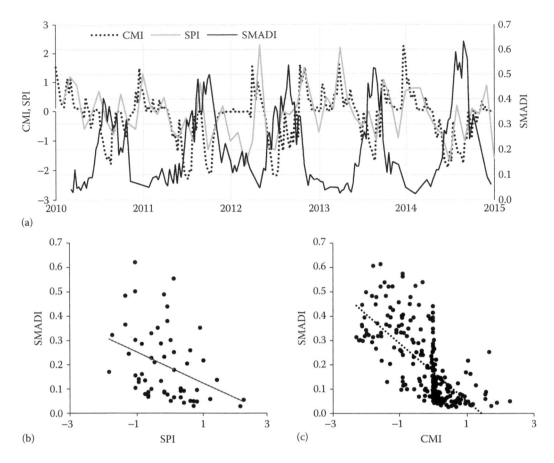

FIGURE 2.7 Temporal evolution of SMADI, CMI, and SPI at the Matacán station (a). Scatter plots of SMADI vs. CMI (b) and SPI (c) are also shown. (Adapted from Sánchez, N. et al., *Remote Sens.*, 8, 287, 2016. With permission.)

The quantitative comparison of SPI with SMADI at the 39 AEMet stations on the Iberian Peninsula led to weak results. Only 9 of the 39 stations showed significant correlations beyond −0.40. These results are in line with the previous analysis in the REMEDHUS area and can be explained through the different timing of both SPI and SMADI and their different approach. SMADI used a shorter, 5-year record of soil moisture instead of the long-term precipitation series used for the SPI (more than 40 years). Due to its nature of short-term deviation from a normal value, the SPI correlation with products, such as anomalies of soil moisture, should be higher, as found in Scaini et al. (2015).

To qualitatively verify the behavior of SMADI and its input parameters under other climatic conditions than that of the REMEDHUS area, three dry areas and three wet areas were selected in the northwestern and southeastern parts of the Iberian Peninsula, respectively (Figure 2.8). The different behavior between dry/wet areas is remarkable. For the wet stations, SMADI rarely exceeded 0.1, indicating a total absence of drought. On the contrary, the three dry stations exhibited values beyond 0.4, indicating drought periods, especially during 2012 and 2014. In addition, a different temporal trend can be tracked over the selected dry/wet areas. A clear increasing SMADI trend along the period 2010–2014 was found for the dry stations (Figure 2.8a through c), coinciding with a decreasing SPI trend (both indices have an opposite sense). This trend was also found in half of the stations located in the dry domain (not shown). On the contrary, all the stations under wet conditions on the Iberian Peninsula showed no trend (Figure 2.8d through f), meaning that no increase or decrease in the drought conditions was described by SMADI through time. It seems that, in the dry

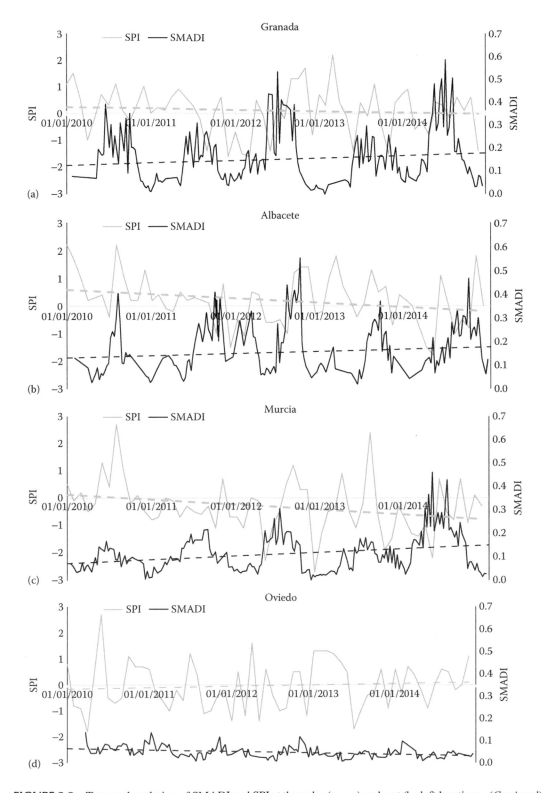

FIGURE 2.8 Temporal evolution of SMADI and SPI at three dry (a, c, e) and wet (b, d, f) locations. (*Continued*)

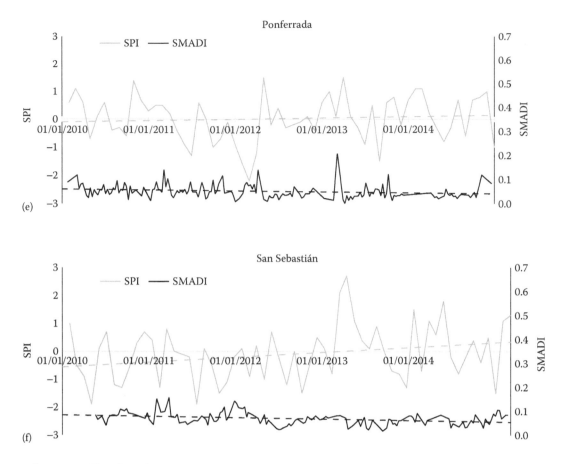

FIGURE 2.8 (Continued) Temporal evolution of SMADI and SPI at three dry (a, c, e) and wet (b, d, f) locations. (Adapted from Sánchez, N. et al., *Remote Sens.*, 8, 287, 2016. With permission.)

areas, the drought tendency is increasing with time. However, these trends must be further analyzed with a longer dataset and additional systematic criteria.

In contrast, no seasonal patterns or differences between dry/wet areas were noticeable in the SPI series. In addition, no annual/seasonal patterns were found, and abnormally dry/wet months ($-2 > SPI > 2$) were found throughout the period. Typically, the shorter the timescale of the SPI (one, three or 12 months), the more the SPI value moves above and below zero, as the deviation with respect to the mean precipitation of a given month is expected to be more variable than that of an annual average (McKee et al., 1993). For example, in regions where rainfall is usually low during a month, as occurs on the Iberian Peninsula during the dry season, SPI results can be misleading due to large negative or positive values resulting from relatively small departures from the median values.

2.5 CONCLUSIONS

The results of this work clearly show that the use of soil moisture and other satellite data products can be a suitable option for agricultural drought monitoring. On the one hand, the soil moisture—and its availability for plants—is the variable that defines agricultural drought. On the other hand, vegetation is the element directly affected by the agricultural drought and, at the same time, is the most reliable indicator of the existence of water stress. Finally, remote sensing offers the ability to measure and monitor all of the variables involved, without restrictions in spatial or temporal scales.

The cases studied in the Iberian Peninsula show the reliability of the different approaches used to properly assess and monitor the agricultural drought at different scales and under different bio-climatic conditions. The specific use of soil moisture or soil moisture combined with vegetation indices and other related variables, all obtained from remote sensing, has been demonstrated to be a feasible methodology for studying agricultural drought. This approach has shown good results and is a step beyond the classical methods based on climatological variables because it is based on the most important variables that define agricultural drought.

Current soil moisture satellites as SMOS and SMAP are providing a unique opportunity to incorporate remote sensing tools into agricultural drought monitoring. In the next few years, even more sophisticated earth observation instruments as those from the Sentinel missions or from the global navigation satellite system (GNSS) reflectometry could address key challenges that exist at the moment in regard to drought early warning and monitoring. The new approaches may face critical issues as, for example, the optimization of the binomial spatial–temporal resolution, in order to increase the accuracy and to have the information in a more immediate way. Therefore, it is expected that earth observation technology will provide precise information in the future years early enough, which helps to prevent or mitigate the drought impacts before it actually hits.

After floods, drought, and specifically agricultural drought, has the greatest impact of all water-related disasters, and the consensus among the projections of future climates is that drought frequency will increase. In a scenario of such climate uncertainty, the new advances and methodologies will be especially relevant for near-future agriculture and for critical issues, such as food security and water scarcity management. The use of new methods for drought monitoring, specifically those from remote sensing, which were unthinkable until very recently, offers a new opportunity for assessment, monitoring, and early alert strategies at any temporal or spatial scale, even at the global scale.

ACKNOWLEDGMENTS

This study was supported by the Spanish Ministry of Economy and Competitiveness (Project AYA2012-39356-C05-05 and ESP2015-67549-C3-3-R) and the European Regional Development Fund (ERDF). SMOS data were provided by the ESA (Project AO-3230). SPI and Matacán data were provided by the Spanish Meteorological Agency, AEMet. MODIS products were retrieved from the NASA EOSDIS Land Processes Distributed Active Archive Center (LP DAAC), USGS/ Earth Resources Observation and Science (EROS) Center.

REFERENCES

AEMet. 2011. Iberian Climate Atlas, 1971–2000. Madrid, Spain: Agencia Estatal de Meteorología, Ministerio de Medio Ambiente y Medio Rural y Marino. Instituto de Meteorologia de Portugal.

Allen, R.G., L.S. Pereira, D. Raes, and M. Smith. 1998. Crop evapotranspiration: Guidelines for computing crop water requirements. Irrigation and Drainage Paper. 56. Rome, Italy: FAO.

Albergel, C., C. Rüdiger, T. Pellarin, J.C. Calvet, N. Fritz, F. Froissard, D. Suquia, A. Petitpa, B. Piguet, and E. Martin. 2008. From near-surface to root-zone soil moisture using an exponential filter: An assessment of the method based on in-situ observations and model simulations. *Hydrology and Earth System Sciences* 12: 1323–1337.

Anyamba, A. and C.J. Tucker. 2012. Historical perspectives on AVHRR NDVI and vegetation drought monitoring. In B.D. Wardlow, M.C. Anderson, J.P. Verdin (Eds.) *Remote Sensing of Drought: Innovative Monitoring Approaches*. Boca Raton, FL: CRC Press. pp. 23–49.

Bayarjargal, Y., A. Karnieli, M. Bayasgalan, S. Khudulmur, C. Gandush, and C.J. Tucker. 2006. A comparative study of NOAA–AVHRR derived drought indices using change vector analysis. *Remote Sensing of Environment* 105: 9–22.

Bhuiyan, C., R.P. Singh, and F.N. Kogan. 2006. Monitoring drought dynamics in the Aravalli region (India) using different indices based on ground and remote sensing data. *International Journal of Applied Earth Observation and Geoinformation* 8: 289–302.

Boyer, J.S. 1982. Plant productivity and environment. *Science* 218: 443–448. doi:10.1126/science.218.4571.443.

Brown, J.F., B.D. Wardlow, T. Tadesse, M.J. Hayes, and B.C. Reed. 2008. The vegetation drought response index (VegDRI): A new integrated approach for monitoring drought stress in vegetation. *GISciences & Remote Sensing* 45: 16–46.

Carlson, T. 2013. Triangle models and misconceptions. *International Journal of Remote Sensing Applications* 3(3): 155–158.

Carrão, H., S. Russo, G. Sepulcre-Canto, and P. Barbosa. 2016. An empirical standardized soil moisture index for agricultural drought assessment from remotely sensed data. *International Journal of Applied Earth Observation and Geoinformation* 48: 74–84.

Ceppi, A., G. Ravazzani, C. Corbari, R. Salerno, S. Meucci, and M. Mancini. 2014. Real-time drought forecasting system for irrigation management. *Hydrology and Earth System Sciences* 18: 3353–3366.

Chakrabarti, S., T. Bongiovanni, J. Judge, L. Zotarelli, and C. Bayer. 2014. Assimilation of SMOS soil moisture for quantifying drought impacts on crop yield in agricultural regions. *IEEE Journal of Selected Topics in Applied Earth Observations and Remote Sensing* 7: 3867–3879.

Cook, B.I. and E.M. Wolkovich. 2016. Climate change decouples drought from early wine grape harvests in France. *Nature Climate Change* 6: 715–719.

Cooley, H., K. Donnelly, R. Phurisamban, and M. Subramanian. 2015. *Impacts of California's Ongoing Drought: Agriculture*. Oakland, CA: Pacific Institute. pp. 24.

Dorigo, W.A., A. Gruber, R.A.M. De Jeu, W. Wagner, T. Stacke, A. Loew, C. Albergel et al. 2015. Evaluation of the ESA CCI soil moisture product using ground-based observations. *Remote Sensing of Environment* 162: 380–395.

EC-JCR. 2014. Crop monitoring in Europe. *MARS Bulletin*. Luxembourg, Luxembourg: AGRI4CAST-JRC/IES MARS Unit.

Entekhabi, D., E. Njoku, P. O'Neill, K. Kellogg, W. Crow, W. Edelstein, J. Entin et al. 2010. The soil moisture active passive (SMAP) mission. *Proceedings of the IEEE* 98. pp. 704–716.

Fernández-Prieto, D., P. van Oevelen, Z. Su, and W. Wagner. 2012. Advances in Earth observation for water cycle science. *Hydrology and Earth System Sciences* 16: 543–549.

Goetz, S.J. 1997. Multisensor analysis of NDVI, surface temperature and biophysical variables at a mixed grassland site. *International Journal of Remote Sensing* 18: 71–94.

González-Zamora, A., N. Sánchez, J. Martínez-Fernández, A. Gumuzzio, M. Piles, and E. Olmedo. 2015. Long-term SMOS soil moisture products: A comprehensive evaluation across scales and methods in the Duero Basin (Spain). *Physics and Chemistry of the Earth* 83–84: 123–136.

Hao, Z. and A. AghaKouchak. 2013. Multivariate standardized drought index: A parametric multi-index model. *Advances in Water Resources* 57: 12–18.

Hayes, M.J. and W.L. Decker. 1998. Using satellite and real-time weather data to predict maize production. *International Journal of Biometeorology* 42(1): 10–15.

Hayes, M.J., M.D. Svodoba, B.D. Wardlow, M.C. Anderson, and F. Kogan. 2012. Drought monitoring: Historical and current perspectives. In B.D. Wardlow, M.C. Anderson, J.P. Verdin (Eds.) *Remote Sensing of Drought: Innovative Monitoring Approaches*. Boca Raton, FL: CRC Press. pp. 1–19.

Hirschi, M., B. Mueller, W. Dorigo, and S.I. Seneviratne. 2014. Using remotely sensed soil moisture for land–atmosphere coupling diagnostics: The role of surface vs. root-zone soil moisture variability. *Remote Sensing of Environment* 154: 246–252.

Hogg, E.H. 1994. Climate and the southern limit of the western Canadian boreal forest. *Canadian Journal of Forest Research* 24: 1835–1845.

Hogg, E.H. 1997. Temporal scaling of moisture and the forest-grassland boundary in western Canada. *Agricultural and Forest Meteorology* 84: 115–122.

Hogg, E.H., A.G. Barr, and T.A. Black. 2013. A simple soil moisture index for representing multi-year drought impacts on aspen productivity in the western Canadian interior. *Agricultural and Forest Meteorology* 178–179: 173–182.

Hunt, E.D., K.G. Hubbard, D.A. Wilhite, T.J. Arkebauer, and A.L. Dutcher. 2009. The development and evaluation of a soil moisture index. *International Journal of Climatology* 29: 747–759.

Karnieli, A., N. Agam, R.T. Pinker, M. Anderson, M.L. Imhoff, G.G. Gutman, N. Panov, and A. Goldberg. 2010. Use of NDVI and land surface temperature for drought assessment: Merits and limitations. *Journal of Climate* 23: 618–633.

Kelley, C.P., S. Mohtadi, M.A. Cane, R. Seager, and Y. Kushnir. 2015. Climate change in the Fertile Crescent and implications of the recent Syrian drought. *PENAS* 112: 3241–3246.

Kerr, Y., P. Waldteufel, J.P. Wigneron, S. Delwart, F. Cabot, J. Boutin, M.J. Escorihuela et al. 2010. The SMOS Mission: New tool for monitoring key elements of the global water cycle. *Proceedings of the IEEE* 98: 666–687.

Keshavarz, M.R., M. Vazifedoust, and A. Alizadeh. 2014. Drought monitoring using a soil wetness deficit index (SWDI) derived from MODIS satellite data. *Agricultural Water Management* 132: 37–45.

Kogan, F.N. 1990. Remote sensing of weather impacts on vegetation in nonhomogeneous areas. *International Journal of Remote Sensing* 11: 1405–1419.

Kogan, F.N. 1995. Application of vegetation index and brightness temperature for drought detection. *Advances in Space Research* 11: 91–100.

Kogan, F.N. 1997. Global drought watch from space. *Bulletin of the American Meteorological Society* 78: 621–636.

Li, R., A. Tsunekawa, and M. Tsubo. 2014. Index-based assessment of agricultural drought in a semi-arid region of Inner Mongolia, China. *Journal of Arid Land* 6(1): 3–15.

Liu, Y.Y., W.A. Dorigo, R.M. Parinussa, R.A.M.D. Jeu, W. Wagner, M.F. McCabe, J.P. Evans, and A.I.J.M. van Dijk. 2012. Trend-preserving blending of passive and active microwave soil moisture retrievals. *Remote Sensing of Environment* 123: 280–297.

Martínez-Fernández, J., A. González-Zamora, N. Sánchez, and A. Gumuzzio. 2015a. A soil water based index as a suitable agricultural drought indicator. *Journal of Hydrology* 522: 265–273.

Martínez-Fernández, J., Sánchez, N., González-Zamora, A., Gumuzzio-Such, A., Herrero-Jiménez, C.M. 2015b. Uso de la humedad del suelo para la monitorización de la sequía agrícola: análisis con mediciones in situ y teledetección. In S. Martínez Pérez, Antonio Sastre Merlín (Eds.) *Estudios en la Zona no Saturada del Suelo.* Vol. 13. Universidad de Alcalá de Henares. Alcalá de Henares. pp. 49–54.

Martínez-Fernández, J., A. González-Zamora, N. Sánchez, A. Gumuzzio, and C.M. Herrero-Jiménez. 2016. Satellite soil moisture for agricultural drought monitoring: Assessment of the SMOS derived soil water deficit index. *Remote Sensing of Environment* 177: 277–286.

McKee, T.B., N.J. Doesken, and J. Kleist. 1993. The relationship of drought frequency and duration of time scales. *Paper presented at Eighth Conference on Applied Climatology*, Anaheim, CA, January 17–23.

Mecklenburg, S., M. Drusch, L. Kaleschke, N. Rodríguez-Fernández, N. Reul, Y. Kerr, J. Font et al. 2016. ESA's Soil Moisture and Ocean Salinity mission: From science to operational applications. *Remote Sensing of Environment.* 180: 3–18. doi:10.1016/j.rse.2015.12.025.

McVicar, T.R. and P.N. Bierwirth. 2001. Rapidly assessing 1997 drought in Papua New Guinea using composite AHVRR imagery. *International Journal of Remote Sensing* 22: 2109–2128.

Mishra, A.K. and V.P. Singh. 2010. A review of drought concepts. *Journal of Hydrology* 391: 202–216.

Narasimhan, B. and R. Srinivasan. 2005. Development and evaluation of Soil Moisture Deficit Index (SMDI) and Evapotranspiration Deficit Index (ETDI) for agricultural drought monitoring. *Agricultural and Forest Meteorology* 133: 69–88.

Nghiem, S.V., B.D. Wardlow, D. Allured, M.D. Svoboda, D. LeComte, M. Rosencrans, S.K. Chan, and G. Neumann. 2012. Microwave remote sensing of soil moisture. Science and applications. In B.D. Wardlow, M.C. Anderson, J.P. Verdin (Eds.) *Remote Sensing of Drought: Innovative Monitoring Approaches.* Boca Raton, FL: CRC Press. pp. 197–226.

Njoku, E.G. and D. Entekhabi. 1996. Passive microwave remote sensing of soil moisture. *Journal of Hydrology* 184: 101–129.

Otkin, J.A., M.C. Anderson, C. Hain, I.E. Mladenova, J.B. Basara, and M. Svoboda. 2013. Examining rapid onset drought development using the thermal infrared–based evaporative stress index. *Journal of Hydrometeorology* 14: 1057–1074.

Palmer, W.C. 1965. Meteorological drought. U.S. Weather Research Paper 45. Washington, DC: U.S. Weather Bureau.

Palmer, W.C. 1968. Keeping track of crop moisture conditions, nationwide: The new crop moisture index. *Weatherwise* 21: 156–161.

Panu, U.S. and T.C. Sharma. 2002. Challenges in drought research: Some perspectives and future directions. *Hydrological Sciences Journal* 47(S): S19–S30.

Parinussa, R.M., M.T. Yilmaz, M.C. Anderson, C.R. Hain, and R.A.M. de Jeu. 2014. An intercomparison of remotely sensed soil moisture products at various spatial scales over the Iberian Peninsula. *Hydrological Processes* 28: 4865–4876.

Piles, M., N. Sánchez, M. Vall-llossera, A. Camps, J. Martínez-Fernández, J. Martínez, and V. González-Gambau. 2014. A dowscaling approach for SMOS land observations: Evaluation of high resolution soil moisture maps over the Iberian Peninsula. *IEEE Journal of Selected Topics in Applied Earth Observations and Remote Sensing* 7: 3845–3857.

Purcell, L.C., T.R. Sinclair, and R.W. McNew. 2003. Drought avoidance assessment for summer annual crops using long-term weather data. *Agronomy Journal* 95: 1566–1576.

Qin, Y., D. Yang, H. Lei, K. Xu, and X. Xu. 2015. Comparative analysis of drought based on precipitation and soil moisture indices in Haihe basin of North China during the period of 1960–2010. *Journal of Hydrology* 526: 55–67.

Rebel, K.T., R.A.M. de Jeu, P. Ciais, N. Viovy, S.L. Piao, G. Kiely, and A.J. Dolman. 2012. A global analysis of soil moisture derived from satellite observations and a land surface model. *Hydrology and Earth System Sciences* 16: 833–847.

Sánchez, N., J. Martínez-Fernández, A. Calera, E. Torres, and C. Pérez-Gutiérrez. 2010. Combining remote sensing and *in situ* soil moisture data for the application and validation of a distributed water balance model (HIDROMORE). *Agricultural Water Management* 98: 69–78.

Sánchez, N., J. Martínez-Fernández, A. Scaini, and C. Pérez-Gutiérrez. 2012. Validation of the SMOS L2 soil moisture data in the REMEDHUS network (Spain). *IEEE Transactions on Geoscience and Remote Sensing* 50(5): 1602–1611.

Sánchez, N., Á. González-Zamora, M. Piles, and J. Martínez-Fernández. 2016. A new Soil Moisture Agricultural Drought Index (SMADI) integrating MODIS and SMOS products: A case of study over the Iberian Peninsula. *Remote Sensing* 8(4): 287.

Sandholt, I., K. Rasmussen, and J. Andersen. 2002. A simple interpretation of the surface temperature/vegetation index space for assessment of surface moisture status. *Remote Sensing of Environment* 79: 213–224.

Scaini, A., N. Sánchez, S.M. Vicente-Serrano, and J. Martínez-Fernández. 2015. SMOS-derived soil moisture anomalies and drought indices: A comparative analysis using in situ measurements. *Hydrological Processes* 29: 373–383.

Schlenz, F., J.T. dall'Amico, A. Loew, and W. Mauser. 2012. Uncertainty assessment of the SMOS validation in the upper Danube catchment. *IEEE Transactions on Geoscience and Remote Sensing* 50: 1517–1529.

Schnur, M.T., H. Xie, and X. Wang. 2010. Estimating root zone soil moisture at distant sites using MODIS NDVI and EVI in a semi-arid region of southwestern USA. *Ecological Informatics* 5(5): 400–409.

Sridhar, V., K.G. Hubbard, J. You, and E.D. Hunt. 2008. Development of the soil moisture index to quantify agricultural drought and its user friendliness in severity-area-duration assessment. *Journal of Hydrometeorology* 9: 660–676.

Steinemann, A.C., M.J. Hayes, and L.F.N. Cavalcanti. 2005. Drought indicators and triggers. In D.A. Wilhite (Ed.) *Drought and Water Crises: Science, Technology, and Management Issues*. Boca Raton, FL: Taylor & Francis Group. pp. 71–92.

Torres, G.M., P.R. Lollato, and T.E. Ochsner. 2013. Comparison of drought probability assessments based on atmospheric water deficit and soil water deficit. *Agronomy Journal* 105: 428–436.

Trigo, R.M., J. Añel, D. Barriopedro, R. García-Herrera, L. Gimeno, R. Nieto, R. Castillo, M.R. Allen, and N. Massey. 2013. The record winter drought of 2011–12 in the Iberian Peninsula. *Bulletin of the American Meteorological Society* 94(9): S41–S45.

Tucker, C.J. and B.J. Choudhury. 1987. Satellite remote sensing of drought conditions. *Remote Sensing of Environment* 23: 243–251.

von Christierson, B., Hannaford, J., Lonsdale, K., Parry, S., Rance, J., Wade, S., and Jones, P. 2011. Impact of long droughts on water resources. Report: SC070079/R5. Environment Agency, Bristol, UK. 129 pp.

Walker, W.R., Hrezo, M.S., and Haley, C.J. 1991. Management of water resources for drought conditions. In Paulson, R.W., Chase, E.B., Roberts, R.S., and Moody, D.W., comp. *National Water Summary 1988–89: Hydrologic Events and Floods and Droughts*. U.S. Geological Survey Water-Supply Paper 2375. pp. 147–156.

Wilhite, D.A. 2000. Drought as a natural hazard: Concepts and definitions. In D.A. Wilhite (Ed.) *Drought: A Global Assessment*, vol. 1. New York: Routledge. pp. 1–18.

Wilhite, D.A. and M. Buchanan-Smith. 2005. Drought as hazard: Understanding the natural and social context. In D.A. Wilhite (Ed.) *Drought and Water Crises: Science, Technology, and Management Issues*. Boca Raton, FL: Taylor & Francis Group. pp. 3–29.

Woli, P., J.W. Jones, K.T. Ingram, and C.W. Fraisse. 2012. Agricultural reference index for drought (ARID). *Agronomy Journal* 104: 287–300.

WMO. 2012. *Standardized Precipitation Index User Guide. Weather-Climate-Water*. Geneva, Switzerland: World Meteorological Organization.

3 Drought Assessments by Coupling Moderate Resolution Imaging Spectroradiometer Images and Weather Data
A Case Study in the Minas Gerais State, Brazil

Antônio H. de C. Teixeira, Janice F. Leivas,
and Gustavo Bayma-Silva

CONTENTS

3.1 INTRODUCTION

Droughts can occur in any climate regime around the world and can arise from a range of hydrometeorological processes that suppress precipitation and/or limit surface water or groundwater availability, creating conditions that are significantly drier than normal or otherwise limiting moisture availability to a potentially damaging extent. Indicators are often used to help track droughts, and tools for their elaboration and application depend on the spatial and timescale (WMO and GWP, 2016).

Drought impacts are significant and widespread in many hydrological basins, increasing disputes over water resources. More conflicts are expected as populations expand, economies grow, and the competition for scarce water supplies during drought events intensifies. Under these circumstances, agreements among different water users, including local communities, to negotiate and agree on the water resources allocation are required. The success of any dialogue depends on the knowledge base and the general trust on water data sources (Teixeira, 2012).

Aiming a more sustainable exploration of the water resources during drought events, water managers must consider the large-scale water balance conditions of the mixed agroecosystems. Drought indicators are important to subsidize policies under water scarcity conditions, aiming to minimize the water use by agriculture, resulting in larger availability for the ecosystems maintenance (Cai et al., 2002).

For drought assessments, besides precipitation, it is also necessary to quantify evapotranspiration, because it is essential for crop yield, and its increase means less water is available for ecological and human uses. Considering the drought indicators used in this chapter, distinctions are important between reference evapotranspiration (ET_0) and actual evapotranspiration (ET). ET_0 is the water flux from a reference surface, not a shortage of water, which may be considered as a hypothetical grass surface with specific characteristics, whereas ET is the real water flux occurring from the surfaces in a specific situation involving all environmental conditions (Allen et al., 1998). Remote sensing algorithms for ET estimations need to be biophysically realistic but should be simple enough for implementation on large scales (Cleugh et al., 2007).

Droughts can adversely affect agriculture and food security and as they evolve, the impacts can vary by region and by season. Agriculture in the Minas Gerais state, Southwest Brazil, has been highlighted in the last decades, with increasing water demands for irrigation in some growing regions. On the one hand, the largest part of the dynamic agricultural products is in general grains and fruits for the external markets because of high investments on technologies and intensive use of monetary resources. On the other hand, the main impact between the use of the water resources and the environment is the pollution caused by the agricultural drainage, which is becoming worse together with several drought events during the recent years (Teixeira et al., 2015a).

In Minas Gerais state, agriculture has diversity as its main characteristic, coexisting regions with intensive technologies and high productivities, with others for subsistence production. The main agricultural growing regions of Minas Triangle and Northwest concentrate large parts of crops, as a consequence of the general favorable both climate and soil. There are disparities, comparing the intensive investments in the Minas Triangle with the technological delay of the North agricultural growing region. As sometimes the river water is also becoming limited as consequences of drought events, these disparities are increasing, even with favorable climatic conditions in the North for irrigated agriculture (Teixeira et al., 2015a).

Although the annual rainfall long-term values are high in some areas of the Minas Gerais state, there are strong natural water deficits in the semiarid region along the years, mainly during the drought events, bringing the need of studies on how to improve the large-scale water productivity, insuring the suitable river flows while contributing to the environmental preservation.

The actual scenario of the Minas Gerais state reveals that even the water being used productively in some agricultural growing regions, proportioning rural development, besides water deficits that affect crop development, the excessive agricultural drainage can adversely influence the water quality and the river flows; both locally and downstream. In addition, the state has experienced severe drought events in some areas, demanding the development and application of drought indicators to subsidize water policies.

Applications of drought indicators at various timescales allow identification of short-term wet periods within long-term droughts or short-term dry spells within long-term wet periods. These indicators are used to provide quantitative assessment of the severity, location, timing, and duration of drought events. Severity refers to the departure from normal of an index. Location refers to the geographic area experiencing drought conditions. The timing and duration are determined by the approximate dates of onset and cessation. A short, relatively low severity, intraseason drought, if it occurs during the moisture sensitive period of a stable crop, can have a more devastating impact on crop yield than a long, more severe drought occurring at a less critical time during the agricultural cycle (WMO and GWP, 2016).

The objective of this chapter is to combine geotechnologies for modeling drought indicators on large scales with satellite images and weather data involving 4 years. Precipitation (P) data were interpolated; the Simple Algorithm for Evapotranspiration Retrieving (SAFER) algorithm was used with the Moderate Resolution Imaging Spectroradiometer (MODIS) reflectance product to estimate ET. The net radiation (R_n) was estimated by the Slob equation, and the ground heat flux (G) was retrieved as a fraction of R_n (Teixeira et al., 2014, 2015b). To consider the climatic water balance and the surface moisture conditions, these parameters were combined to generate two drought indicators for large-scale moisture analyses, emphasizing the main agricultural growing regions in the Minas Gerais state, Southeast Brazil. The results can be used for advancing monitoring in support of risk-based drought management policies and preparedness plans (WMO and CWP, 2016).

This chapter is structured as follows: after the introduction, the study region, dataset, and the steps for modeling are described. Then applications of the drought indicators by using remote sensing methods with MODIS images and weather data are shown in the Minas Gerais state, with emphasis in the main growing agricultural regions.

3.2   STUDY AREA AND DATASET

Figure 3.1 shows the location of the Minas Gerais (MG) state, Southeast Brazil, together with the net of weather stations from the National Meteorological Institute (INMET), and the cropland mask inside the main agricultural growing regions of the state.

Despite being close to the southeastern Brazilian coast, Minas Gerais has no contact with the Atlantic Ocean. More than 90% of the state is at altitudes over 300 m, and about 25% are between 600 and 1500 m. Average daily air temperature (T_a) is in the range from 17°C to 23°C, whereas the annual precipitation (P) values are between 750 and 1800 mm year^{-1}. The eastern side of the state used to be covered by the Atlantic Forest; however, this natural ecosystem has been removed for wood exploration by the development of the cities and by the establishments of the farm. In some areas of the North side, are the *Caatinga* natural species, under the Brazilian semiarid conditions. However, considering the whole state, most areas are occupied by the *Cerrado* ecosystem. The hydrological basins are constituted by the São Francisco and Paraná rivers (Golfari, 1978).

The weather data obtained from 36 weather stations of INMET (www.inmet.gov.br) were incident global radiation (R_G), air temperature (T_a), air humidity (RH), wind speed (u), and precipitation (P). They were used for the ET$_0$ large-scale calculations by the Penman–Monteith (PM) method (Allen et al., 1998) and for the drought indicators. R_G, T_a, ET$_0$, and P were upscaled for the 16-day composing periods of the MODIS reflectance product (spatial resolution of 250 m) and were interpolated by using the moving average method, creating grids with the same spatial resolution of the

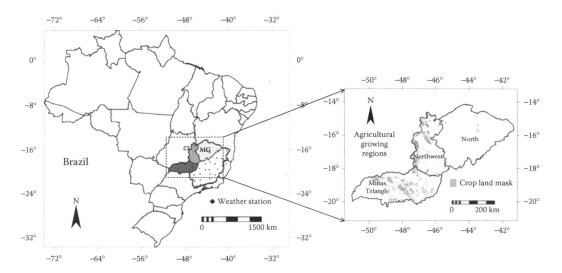

FIGURE 3.1   Location of the of the Minas Gerais (MG) state, Southeast Brazil, the net of weather stations, and the crop land mask inside the main agricultural growing regions.

satellite images. Similar upscaling process for large-scale evapotranspiration estimations was done in Australia (Cleugh et al., 2007).

The cropland mask was acquired from the National Institute of Spatial Researches (INPE) to help understanding the drought events in crops inside the North (N), Northwest (NW), and Minas Triangle (MT) agricultural growing regions of the Minas Gerais state.

3.3   LARGE-SCALE MODELING

Figure 3.2 shows the steps for modeling the drought indicators throughout SAFER algorithm and interpolated precipitation (P) data with MODIS images without their thermal bands.

The surface temperature (T_0) MODIS product was not used because with a lower spatial resolution (1000 km), there were cloud contaminations in some parts of the Minas Gerais state along the years. Instead, T_0 was retrieved by residue in the radiation balance (residual method) after having estimated the atmospheric and surface emissivities (Teixeira et al., 2016a, 2016b).

Cleugh et al. (2007) pointed out that the use of instantaneous measurements of the radiometric surface temperature to calculate time-averaged fluxes led to errors. They emphasized uncertainties in models, which use the MODIS 8-day that is a composite of once-daily overpass at ~10:30 h local time. In this case, the radiometric temperature is determined under a view angle at the satellite overpass time, using emissivities based on vegetation classes at a 1 km grid that differs from that of the MODIS pixel location.

The parameterizations involved for acquiring the radiation and energy balance components, including ET, by using the SAFER algorithm in Figure 3.2 were done in Brazil with simultaneous satellite and field measurements, under strong contrasting agroecosystems and thermohydrological conditions throughout different years (Teixeira et al., 2008, 2014). In addition, acquiring T_0 as

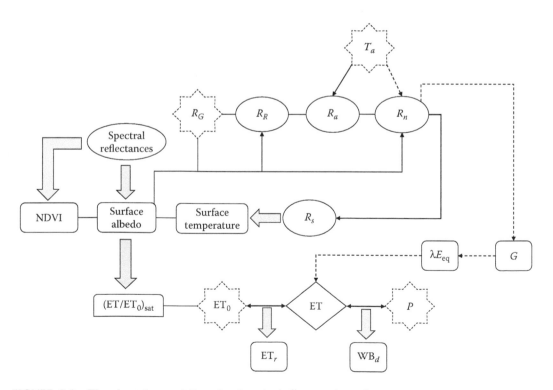

FIGURE 3.2 Flowchart for modeling the drought indicators throughout the application of the Simple Algorithm for Evapotranspiration Retrieving (SAFER) algorithm and interpolated precipitation (P) data to MODIS images without their thermal bands.

residue in the radiation balance gives mutual compensation, reducing possible errors in this model input parameter, as they are self-canceling in the upward and downward long-wave fluxes. Moreover, a correction factor is applied to ET/ET_0 ratio in the algorithm to take into account the atmospheric demand for the current study, in relation to the original modeling conditions.

The SAFER algorithm, based on the PM equation, was elaborated and validated with Landsat images (Teixeira, 2010), when it was called PM2. Later, it was also calibrated and validated with MODIS images in the same original modeling study region (Teixeira et al., 2013). Field data used for these validations involved irrigated crops and natural vegetation (*Caatinga*) from 2001 to 2007, being described in detail in Teixeira et al. (2008). Table grapes were drip irrigated and conducted by an overhead trellis system. Wine grapes conducted by a vertical trellis system and mango orchard were microsprinkler irrigated. The experimental period for *Caatinga* involved different natural species and rainfall conditions above and below the local long-term value.

Thus, with all these strongly contrasting conditions and considerations, one expects sufficient accuracy to obtain the input parameters to achieve dynamic analyses of the drought indicators for the different ecosystems inside the Minas Gerais state, Southeast Brazil.

According to Figure 3.2, the reflectances for bands 1 (α_1) and 2 (α_2), in the red and near infrared of the solar spectrum, respectively, were extracted from the MOD13Q1 product, which provides cloud-free temporal composed images, at 16-day period, totaling 23 images for each band along a year from 2012 to 2015.

For the surface albedo (α_0) calculations, the following equation was applied (Valiente et al., 1995):

$$\alpha_0 = a + b\alpha_1 + c\alpha_2 \tag{3.1}$$

where a, b, and c are regression coefficients, considered as 0.08, 0.41, and 0.14, obtained with different Brazilian vegetation types and thermohydrological conditions (Teixeira et al., 2014, 2015b).

The Normalized Difference Vegetation Index (NDVI) is a measure of the amount of vegetation at the surface:

$$\text{NDVI} = \frac{\alpha_{p(2)} - \alpha_{p(1)}}{\alpha_{p(2)} + \alpha_{p(1)}} \tag{3.2}$$

The reflected solar radiation (R_R) was estimated as follows:

$$R_R = \alpha_0 R_G \tag{3.3}$$

The long-wave atmospheric radiation (R_a) was calculated by applying the Stefan–Boltzmann law:

$$R_a = \sigma \varepsilon_A T_a^4 \tag{3.4}$$

where:
ε_A is the atmospheric emissivity
σ is the Stefan–Boltzmann constant (5.67×10^{-8} W m^{-2} K^{-4})

The radiation balance parameter ε_A was calculated as follows (Teixeira et al., 2016a, 2016b):

$$\varepsilon_A = a_A(-\ln \tau)^{b_A} \tag{3.5}$$

where:
τ is the short-wave atmospheric transmissivity calculated as the ratio of R_G to the incident solar radiation at the top of the atmosphere
a_A and b_A are regression coefficients 0.94 and 0.10, respectively

The regression coefficients of Equation 3.5 in this chapter are in between those obtained for Idaho (Allen et al., 2000; $a_A = 0.85$ and $b_A = 0.09$) and for Egypt (Bastiaanssen et al., 1998; $a_A = 1.08$ and $b_A = 0.26$).

R_n can be described by the 24 h values of net short-wave radiation, with a correction term for net long-wave radiation (Teixeira et al., 2014, 2015b, 2016a, 2016b):

$$R_n = (1 - \alpha_0)R_G - a_L \tau \tag{3.6}$$

where a_L is the regression coefficient of the relationship between net long-wave radiation and τ on a daily scale.

Because of the thermal influence on long-wave radiation via the Stephan–Boltzmann equation, a previous study investigated whether the variations of the a_L coefficient from Equation 3.6 could be explained by variations in 24 h T_a (Teixeira et al., 2008):

$$a_L = dT_a - e \tag{3.7}$$

where d and e are regression coefficients that are found to be 6.99 and 39.93, respectively. A constant value of $a_L = 110$ was previously applied by Bastiaanssen et al. (1998) without considering the thermal spatial differences.

Having estimated R_R, R_a, and R_n, the emitted surface long-wave radiation (R_s) was acquired as residue in the radiation balance equation:

$$R_s = R_G - R_R + R_a - R_n \tag{3.8}$$

Then, the surface temperature (T_0) was estimated as follows (Teixeira et al., 2016a, 2016b):

$$T_0 = \sqrt[4]{\frac{R_s}{\sigma \varepsilon_s}} \tag{3.9}$$

where the surface emissivity (ε_S) was estimated as follows (Teixeira et al., 2016a, 2016b):

$$\varepsilon_s = a_s \ln NDVI + b_s \tag{3.10}$$

a_S and b_S are regression coefficients 0.06 and 1.00, respectively.

The original coefficients of Equation 3.10 are $a_S = 0.047$ and $b_S = 1.009$ (Bastiaanssen et al., 1998), being slightly different from those for Brazil.

Even with small differences on both ε_A and ε_S, when comparing the Brazilian values with those from other environments, estimate errors from these emissivities in the Minas Gerais state should be self-cancelled on the accounting of the upward and downward radiation balance components.

The SAFER algorithm is used to model the ratio of the actual to the reference evapotranspiration based on the input remote sensing parameters (ET/ET$_0$)$_{sat}$, which is then multiplied by ET$_0$ 24 h values from the weather stations to estimate the daily ET large-scale values (Teixeira et al., 2016a, 2016b):

$$(ET/ET_0)_{sat} = \left\{ \exp\left[f + g\left(\frac{T_0}{\alpha_0 NDVI} \right) \right] \right\} \frac{ET_{0\,year}}{5} \tag{3.11}$$

where f and g are the original regressions coefficients, 1.8 and -0.008, respectively. The correction factor (ET$_{0and}$/5) was applied, where ET$_{0year}$ is the annual grid of reference evapotranspiration for Minas Gerais state for the years 2012, 2013, 2014, and 2015, and 5 mm is the ET$_{0year}$ value for the period of the original modeling in the Brazilian Northeast (Teixeira et al., 2016b).

Equation 3.11 does not work for water bodies, that is, when NDVI < 0. Thus, as sometimes some areas are mixtures of land and water in the Minas Gerais state, the equilibrium (eq) evapotranspiration (Raupach, 2001) was considered under these conditions in the SAFER algorithm, and λE_{eq} was retrieved throughout conditional functions and was transformed into ET_{eq}:

$$\lambda E_{eq} = \frac{s(R_n - G)}{s + \gamma} \qquad (3.12)$$

where:
s is the slope of the curve relating saturation water vapor pressure to T_a
G is the ground heat flux
γ is the psychometric constant

For the daily G values, the following equation was used (Teixeira et al., 2016b):

$$\frac{G}{R_n} = a_G \exp(b_G \alpha_0) \qquad (3.13)$$

where a_G and b_G are regression coefficients that are found to be 3.98 and −25.47, respectively.

After considering the results for ET taking into account both Equations 3.11 and 3.12 according to the conditional functions for the NDVI pixel values, a drought indicator, related to soil moisture conditions, the evapotranspiration ratio (ET_r) was applied:

$$ET_r = \frac{ET}{ET_0} \qquad (3.14)$$

High ET_r values indicate that vegetation is well supplied with water, whereas low values mean water stress (Lu et al., 2011), and it can also be used for determining crop water requirements (Teixeira et al., 2015b, 2016c).

Similarly to what was done in Australia (Cleugh et al., 2007) and in the Brazilian Northeast (Teixeira et al., 2016c), another drought indicator was applied for the large-scale analyses, the water balance deficit (WD_d):

$$WB_d = P - ET \qquad (3.15)$$

The indicator represented by Equation 3.15 enables the characterization of the large-scale climatic water balance, taking into account the water consumption and rainfall. Low positive WD_d values may imply the feasibility of rainfed crops, whereas the higher ones indicate unsuitable conditions, due to moisture excess problems and also the possibility of water storage for subsequent drier periods. Negative WD_d values are related to climatic water deficiencies for vegetation and the degree of irrigation needs in crops. Thus, WD_d quantifies the degree of water deficits or excess in terms of mm of water (Teixeira et al., 2016c).

3.4   DROUGHT INDICATORS

3.4.1   RAINFALL AND EVAPOTRANSPIRATION

The trends of P, ET, and ET_0 for each 16-day periods of the MODIS images, during the years from 2012 to 2015, were firstly analyzed in Figure 3.3 in terms of day of the year (DOY), considering the average pixel values for the entire Minas Gerais state, Southeast Brazil.

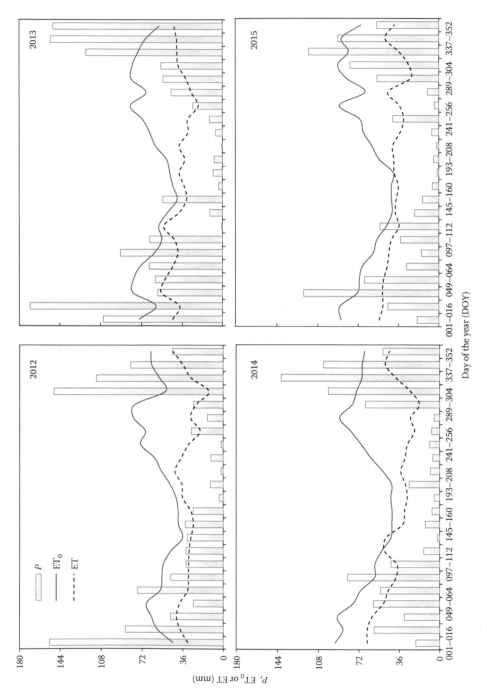

FIGURE 3.3 Average MODIS 16-day pixel values for the totals of precipitation (P), reference evapotranspiration (ET_0), and actual evapotranspiration (ET) during the years from 2012 to 2015 in the Minas Gerais (MG) state, Southwest Brazil.

Concerning P, during the years from 2012 to 2015, one can see two periods with rainfall concentrations from January to April (DOY 001–112) and from the second half of October to the end of December (DOY 289–365). The naturally driest conditions were from the second half of June to the end of September (DOY 177–272), being in 2014 the water scarcity period of longest duration. Taking into account the annual values for the entire state, the year 2015 was the one with the lowest rainfall amounts, presenting average pixel totals of 891 mm year^{-1}, and the year 2013 was the moister year with 1318 mm year^{-1}.

In relation to the atmospheric demand, the ET_0 values had fewer oscillations along the years, when compared to those for rainfall. The periods with the lowest ET_0 pixel averages were after the rainy seasons. At the end of these periods, ET_0 started to rise continuously from the second half of June to the end of October (DOY 177–304), whereas P values decline to increase again only in the second half of October (DOY 289). The driest year 2015 corresponded to the highest ET_0 (1533 mm year^{-1}), whereas the year 2012, with 1304 mm year^{-1}, presented the lowest atmospheric demand.

The largest ET values occurred inside the rainy periods, from January to May (DOY 001–144) and from November to December (DOY 321–365), when the 16-day water fluxes were higher than 40 mm. The lowest rates, in that timescale, below 30 mm, happened from the end of August (DOY 241) to the first half of November (DOY 320). Considering the annual scale and the whole Minas Gerais state, the year with the highest ET (928 mm year^{-1}) was 2015, whereas that with the lowest one was 2012 (737 mm year^{-1}).

ET_0 and ET represented ranges from 122% to 68% (2013) and 165% to 99% (2014), respectively, of P. From the scenarios presented in Figure 3.3, it is clear that the most critical period in relation to drought conditions was from the end of April (DOY 113) to the first half of October (DOY 288) during the year 2014. For minimizing this problem, water storage techniques in areas previously presenting water excess should be encouraged in the water public policies.

After extracting the quarterly (Q) and annual average pixel values of the totals for P, ET_0, and ET inside the main agricultural growing regions of the Minas Gerais state, North (N), Northwest (NW), and Minas Triangle (MT), during the years from 2012 to 2015, they are presented in Table 3.1.

In general, the Q largest rainfall values in the three agricultural growing regions were the first (Q1) and the third (Q3), highlighting the ones for MT. The lowest P happened from May to August (Q2), mainly in the N agricultural growing region during the year 2015.

Regarding the atmosphere demand, Q3 was the quarter period with the highest ET_0 values, with highlights for the year 2015 in the N agricultural growing region, when its total was above 700 mm. The smallest demands occurred in Q2, below 400 mm, with those for TM presenting the lower ET_0. Taking into account the annual values, 2015 and 2012 showed the highest and the smallest ET_0 totals, respectively.

The ET trends were similar to those for ET_0 under good soil moisture conditions; however, for all agricultural growing regions, low rainfall amounts in Q2 promoted strong decline of the ratios ET/ET_0 in the subsequent Q3. Although this ratio was above 70% in Q1 and Q2 along the studied years, in the last quarter it reached to only 30%, with the lowest values happening in NW during the year 2012. The decline of ET/ET_0 in Q3, even under good rainfall conditions, evidenced a gap between P and ET related to the time for the soil that need to recover its good moisture status. In Australia, ET reflected the conditions of rainfall, available energy, and air temperature (Cleugh et al., 2007).

3.4.2 Large-Scale Drought Assessments

Figure 3.4 presents the spatial distribution of the MODIS 16-day average pixel values for the ET_r drought indicator during specific periods of the years from 2012 to 2015 in the Minas Gerais state, Southeast Brazil.

The ET_r spatial and seasonal variations along the years are clear, confirming the sensibility of the SAFER algorithm. The soil moisture differences are mainly noticed when comparing the wettest period of DOY 145–160 (end of May to first half of June), soon after the rainy period, with the one

TABLE 3.1

Quarterly (Q) and Annual Averages for the Drought Indicators, during the Years from 2012 to 2015, in the Agricultural Growing Regions North (N), Northwest (NW) and Minas Triangle (MT) of the Minas Gerais State, Southeast Brazil. Average Values for Totals of Precipitation (P), Reference Evapotranspiration (ET_0) and Actual Evapotranspiration (ET)

Year	2012			2013			2014			2015		
Q	P (mm)	ET_0 (mm)	ET (mm)	P (mm)	ET_0 (mm)	ET (mm)	P (mm)	ET_0 (mm)	ET (mm)	P (mm)	ET_0 (mm)	ET (mm)
						North (N)						
1	577	441	266	449	609	371	307	608	395	259	609	365
2	153	327	172	27	405	188	41	374	184	25	395	201
3	459	548	184	618	647	236	440	643	216	271	735	234
Total	1189	1316	622	1093	1660	796	788	1625	795	555	1740	800
						Northwest (NW)						
1	489	455	247	683	552	371	457	592	385	423	563	366
2	90	346	192	49	397	229	50	386	224	38	395	267
3	428	579	165	595	656	255	526	636	185	412	684	261
Total	1007	1381	604	1327	1605	855	1033	1613	794	872	1643	894
						Minas Triangle (MT)						
1	507	441	266	734	511	379	452	590	408	491	538	368
2	118	319	231	125	383	264	77	370	234	59	382	302
3	504	542	191	565	645	289	540	622	226	596	639	301
Total	1129	1301	688	1423	1539	932	1068	1581	868	1146	1559	972

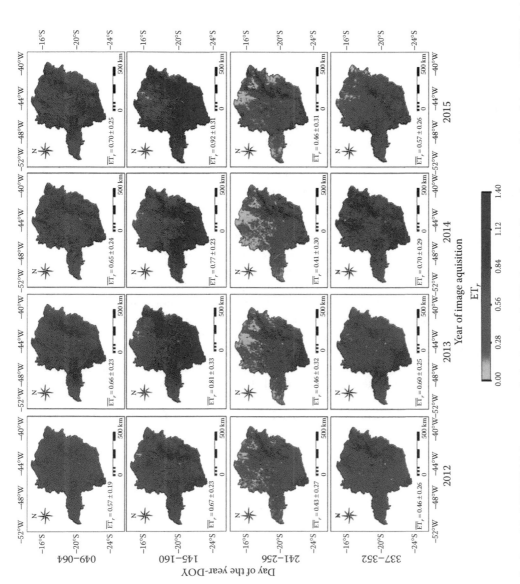

FIGURE 3.4 Spatial distribution of the MODIS 16-day average pixel values for the ET_r drought indicator in terms of day of the year (DOY), during specific periods of the years from 2012 to 2015, in the Minas Gerais state, Southeast Brazil. The bars mean averages, showed together with the standard deviations.

from DOY 241–256 (end of August to the first half of September) at the end of the driest periods (Figure 3.3).

The highest 16-day ET_r values, with average above 0.90 from DOY 145 to 160 (end of May to first half of June) happened in the year 2015, whereas the lowest ones happened in the year 2014 (end of May to first half of June). The strong ET_r variation with the thermohydrological conditions evidenced that this drought indicator can be useful for monitoring the large-scale soil moisture conditions (Teixeira et al., 2016b).

Figure 3.5 shows the spatial distribution of the MODIS 16-day average pixel values for the WB_d drought indicator during specific periods of the years from 2012 to 2015 in the Minas Gerais state, Southeast Brazil.

The WB_d spatial and temporal variations are also clear along the years from 2012 to 2015. The maximum positive values ($P > ET$) occurred in December (DOY 337–352) with 16-day pixel values above 150 mm in the year 2013. The smallest and negative 16-day pixel values ($P < ET$) happened more often from August to September (DOY 241–256), lower than –45 mm, in a large area of the state. By the spatial and temporal trends of the WB_d drought indicator, one can see possibilities of rainfall water storage in wet periods and places with high WB_d for later use during the climatically driest conditions of the years.

Figure 3.6 presents the seasonal variations of the mean 16-day pixel values for the ET_r and WB_d drought indicators in terms of DOY, during the years from 2012 to 2015, in the main agricultural growing regions North (N), Northwest (NW), and Minas Triangle (MT) of the Minas Gerais state, Southeast Brazil.

In general, the highest ET_r values were from DOY 081 (March) to 128 (May), with ET reaching, in some cases to 70% of ET_0, indicating good soil moisture conditions inside the rainy period (Figure 3.3). The largest values for the MT agricultural growing region along the years, besides the climatic effects, could also be attributed to larger irrigated areas, clearly observed during the naturally driest periods of the years. On the one hand, ET_r values (well known as crop coefficient—K_c) can be used for estimating the water requirements at different spatial scales in well-irrigated crops (Teixeira et al., 2015b, 2016c). On the other hand, in natural vegetation, this ratio characterizes the degree of the water status in the root zones (Lu et al., 2011).

Zhang et al. (2012), studying a temperate desert steppe in the Inner Mongolia, China, reported seasonal ET_r variations from mean daily values of 0.16 to maximum of 0.75, similar to several situations of the current study. However, Lu et al. (2011), in the same Chinese region, found ET_r values higher than 1.00 for six different ecosystems, whereas Sumner and Jacobs (2005) reported ET_r values between 0.47 and 0.92 in a nonirrigated pasture site in Florida, United States.

Zhou and Zhou (2009) concluded that air temperature, air humidity, and the available energy were the most important variables for the ET_r variations in a reed marsh in the Northeast China. In the current study, the most important reason for the highest ET_r values was the previous rainy seasons making the soil moister in the subsequent period. However, the ET_r values in natural ecosystems also depend on the stomatal regulation and plant adaptation to water scarcity conditions (Mata-González, 2005).

The large positive WB_d values indicate high probability of water excess with subsequent percolation and runoff, depending on the soil retention capacity, whereas the negative WB_d values may be related to natural climatic water deficiency. Along the studied period, there were concentrations of positive values at the start and end of the years in all agricultural growing regions. However, in 2014 and 2015, some periods presented negative WB_d from January to February, what may have affected the rainfed agriculture, while demanded much irrigation water in irrigated crops. According to Cleugh et al. (2007), low WB_d values reflect reductions of rainfall and/or increases on ET. From the three agricultural regions studied, the North (N) with WB_d annual average value of –245 mm year^{-1} in 2015 was the one with the driest natural conditions.

Within a specific year, similar trends of ET_r and WB_d among the agricultural growing regions were observed; however, they were different among the years, mainly varying with the amount and

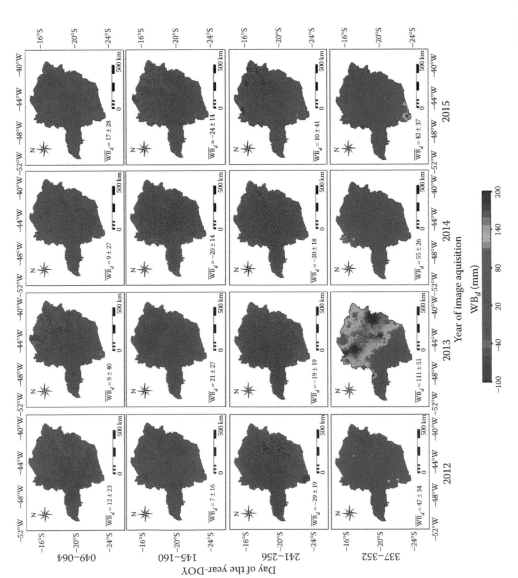

FIGURE 3.5 Spatial distribution of the MODIS 16-day average pixel values for the WB_d drought indicator in terms of day of the year (DOY), during specific periods of the years from 2012 to 2015, in the Minas Gerais state, Southeast Brazil. The bars mean averages, showed together with the standard deviations.

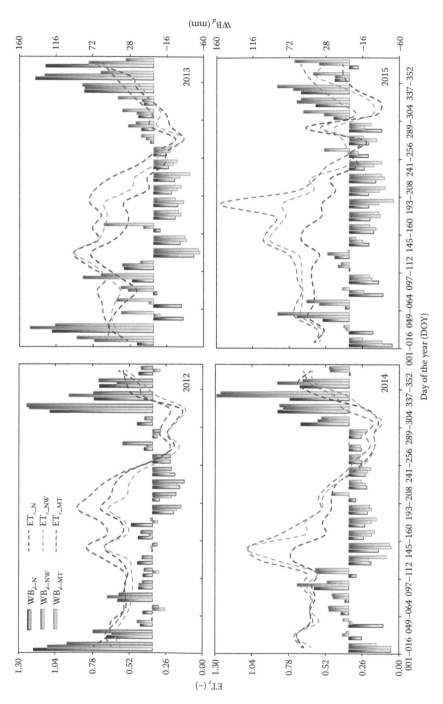

FIGURE 3.6 Seasonal variations of the 16-day mean pixel values for the WB_d and ET_r drought indicators in terms of day of the year (DOY), during the years from 2012 to 2015, in the main agricultural growing regions North (N), Northwest (NW), and Minas Triangle (MT) of the Minas Gerais state, Southeast Brazil.

distribution of rainfalls (Figures 3.3 and 3.6). By the seasonal and spatial behaviors of the drought indicators, one can see that, in general, after the rainy periods, even under situations of negative WB_d values, there is no need of irrigation to satisfy the crop water requirements. However, supplementary irrigation should be beneficial for agriculture from the end of August (DOY 241) to the end of the first half of November (DOY 320), when ET_r dropped below 0.50. In this last case, situations with previous positive WB_d values indicate possibilities of rainfall water storage, as the latter support the crop water requirements throughout irrigation. Considering all the studied years, ET attended 50%, 53%, and 61% of the atmospheric demand, respectively in the agricultural growing regions North (N), Northwest (NW), and Minas Triangle (MT).

Due to the importance for the rural development and under the actual water scarcity conditions, the rainfed agriculture with supplementary irrigation to mitigate the drought impacts should be encouraged in the Minas Gerais state, mainly in the Minas Triangle, which is the wettest agricultural growing region. For the success of this activity, the use of the tools tested here is important for monitoring the drought conditions on large scales.

3.5 CONCLUSIONS AND POLICY IMPLICATIONS

The coupled use of remote sensing parameters from MODIS images and weather data allowed the application of drought indicators along the years from 2012 to 2015 in the Minas Gerais state, Brazil. These indicators can subsidize a better understanding of the water balance dynamics, important issue for mitigating the drought impacts upon agriculture.

Analyzing a series of 4-year data for drought indicators in the agricultural growing regions North, Northwest, and Minas Triangle, it could be concluded that Minas Triangle present the highest moist conditions, whereas North has shown the lowest natural soil moisture levels, with the critical period from July to November. Besides the climatic conditions, the high soil moisture in Minas Triangle could also be attributed to larger irrigated areas.

It was demonstrated that the drought conditions could be analyzed from instantaneous measurements of only the red and infrared spectral radiations from MODIS bands 1 and 2, by modeling the ratio of actual to reference evapotranspiration at the satellite overpass time and the available energy on a 16-day timescale. The combination of these images with weather stations proved to be useful for monitoring water parameters, contributing to the mitigation policies for facing the drought problems in some areas and periods in the Southeast Brazil.

The limitations of the MODIS result because the spatial resolution of the low 1 km thermal resolution was minimized, by retrieving the surface temperature from the residual that emitted long-wave radiation from the surfaces together with weather data, allowing the downscaling of the drought indicators to a 250 m spatial resolution.

REFERENCES

Allen, R.G., Pereira, L.S., Raes, D. and Smith, M. 1998. *Crop Evapotranspiration: Guidelines for Computing Crop Water Requirements*. Rome, Italy: Food and Agriculture Organization of the United Nations.

Allen, R.G., Hartogensis, O. and de Bruin, H.A.R. 2000. Long-Wave Radiation over Alfafa during the RAPID Field Campaign in Southern Idaho; Research Report; Kimberly, ID: University of Idaho.

Bastiaanssen, W.G.M., Menenti, M., Feddes, R.A., Roerink, G.J. and Holtslag, A.A.M. 1998. A remote sensing surface energy balance algorithm for land (SEBAL) 1. Formulation. *Journal of Hydrology* 212–213: 198–212.

Cai, X., Mckinney, C. and Lasdon, S. 2002. A framework for sustainable analysis in water resources management and application to the Syr Darya Basin. *Water Resources Research* 38: 21–24.

Cleugh, H.A., Leuning, R., Mu, Q. and Running, S.W. 2007. Regional evaporation estimates from flux tower and MODIS satellite data. *Remote Sensing of Environment* 106: 285–304.

Golfari, L. 1978. Zoning for reforestation in Brazil and trials with tropical Eucalyptus and Pines in Central–Region. Technical Report. PNUD/FAO: Project BRA/76/027.

Lu, N., Chen, S., Wilske, B., Sun, G. and Chen, J. 2011. Evapotranspiration and soil water relationships in a range of disturbed and undisturbed ecosystems in the semi-arid Inner Mongolia, China. *Journal of Plant Ecolology* 4: 49–60.

Mata-González, R., McLendon, T. and Martin, D.W. 2005. The inappropriate use of crop transpiration coefficients (K_c) to estimate evapotranspiration in arid ecosystems: A review. *Arid Land Research and Management* 19: 285–295.

Raupach, M.R. 2001. Combination theory and equilibrium evaporation. *Quarterly Journal of Royal Meteorological Society* 127: 1149–1181.

Sumner, D.M. and Jacobs, J. 2005. Utility of Penman-Monteith, Priestley-Taylor, reference evapotranspiration, and pan evaporation methods to estimate pasture evapotranspiration. *Journal of Hydrology* 308: 81–104.

Teixeira, A.H. de C., Bastiaanssen, W.G.M., Ahmad, M.D. and Bos, M.G. 2008. Analysis of energy fluxes and vegetation—atmosphere parameters in irrigated and natural ecosystems of semi-arid Brazil. *Journal of Hydrology* 362: 110–127.

Teixeira, A.H. de C. 2010. Determining regional actual evapotranspiration of irrigated and natural vegetation in the São Francisco river basin (Brazil) using remote sensing and Penman-Monteith equation. *Remote Sensing* 2: 1287–1319.

Teixeira, A.H. de C. 2012. Modelling water productivity components in the Low-Middle São Francisco River basin, Brazil. In C. Bilibio, O. Hensel, J. Selbach (Eds.) *Sustainable Water Management in the Tropics and Subtropics and Case Studies in Brazil*. Kassel, Germany: University of Kassel.

Teixeira, A.H. de C., Scherer-Warren, M., Hernandez, F.B.T., Andrade, R.G. and Leivas, J.F. 2013. Large-scale water productivity assessments with MODIS images in a changing semi-arid environment: A Brazilian case study. *Remote Sensing* 11: 5783–5804.

Teixeira, A.H. de C., Hernandez, F.B.T., Lopes, H.L., Scherer-Warren, M. and Bassoi, L.H. 2014. A comparative study of techniques for modeling the spatiotemporal distribution of heat and moisture fluxes in different agroecosystems in Brazil. In G.G. Petropoulos (Ed.) *Remote Sensing of Energy Fluxes and Soil Moisture Content*. Boca Raton, FL: Taylor & Francis Group.

Teixeira, A.H. de C., Leivas, J.F., Andrade, R.G., Victoria, D. de C., Bolfe, E.L. and Silva, G.B. 2015a. Water balance indicators from MODIS images and agrometeorological data in Minas Gerais state, Brazil. *Proceedings of SPIE* 9637: 96370O-1–96370O-14.

Teixeira, A.H. de C., Hernandez, F.B.T., Scherer-Warren, M., Andrade, R.G., Victoria, D. de C., Bolfe, E.L., Thenkabail, P.S. and Franco, R.A.M. 2015b. Water productivity studies from earth observation data: Characterization, modeling, and mapping water use and water productivity. In P.S. Thenkabail (Ed.) *Remote Sensing of Water Resources, Disasters, and Urban Studies*. Boca Raton, FL: Taylor & Francis Group.

Teixeira, A.H. de C., Leivas, J.F. and Silva, G.B. 2016a. Options for using Landsat and RapidEye satellite images aiming the water productivity assessments in mixed agro-ecosystems. *Proceedings of SPIE* 9998: 99980A-1–99980A-11.

Teixeira, A.H. de C., Leivas, J.F., Ronquim, C.C. and Victoria, D. de C. 2016b. Sugarcane water productivity assessments in the São Paulo state, Brazil. *International Journal of Remote Sensing Applications* 6: 84–95.

Teixeira, A. H. de C., Tonietto, J. and Leivas, J.F. 2016c. Large-scale water balance indicators for different pruning dates of tropical wine grape. *Pesquisa Agropecuária Brasileira* 51: 849–857.

Valiente, J.A., Nunez, M., Lopez-Baeza, E. and Moreno, J.F. 1995. Narrow-band to broad-band conversion for Meteosat visible channel and broad-band albedo using both AVHRR-1 and–2 channels. *International Journal of Remote Sensing* 16: 1147–1166.

World Meteorological Organization (WMO) and Global Water Partnership (GWP). 2016. In M. Svoboda and B.A. Fuchs (Eds.) *Handbook of Drought Indicators and Indicators*. Integrated Drought Management Programme (IDMP), Integrated Drought Management Tools and Guidelines Series 2. Geneva, Switzerland.

Zhang, F., Zhou, G., Wang, Y., Yan, F. and Nilsson, C. 2012. Evapotranspiration and crop coefficient for a temperate desert steppe ecosystem using eddy covariance in Inner Mongolia, China. *Hydrological Processes* 26: 379–386.

Zhou, L. and Zhou, G. 2009. Measurement and modeling of evapotranspiration over a reed (*Phragmitesaustralis*) marsh in Northeast China. *Journal of Hydrology* 372: 41–47.

4 The Added Value of Satellite Soil Moisture for Agricultural Index Insurance

Markus Enenkel, Daniel Osgood, and Bristol Powell

CONTENTS

4.1 INTRODUCTION

Weather shocks affect farmers, whose livelihoods depend on agricultural production, both directly and indirectly. Direct impacts are linked to the impact of anomalous weather conditions on the health of crops. Indirect impacts, however, often limit advances in agricultural productivity (e.g., the investment in drought resistant but more expensive seeds) due to the threat of a weather shock, whether it actually occurs or not (Hellmuth et al., 2009). In contrast to conventional loss-based insurance, weather index insurance relies on objective indicators that are agreed upon before the start of the season (Brown et al., 2011; Hellmuth et al., 2009; International Fund for Agricultural Development, 2011). Since *in situ* observations are often not reliably quality controlled and are not available in large regions of the world, satellite data are often the only source of information (International Research Institute for Climate and Society, 2013). If independent satellite-derived key variables of the hydrologic water cycle and indicators of vegetation health agree on specific drought events, this information can be used to decrease uncertainties in the design of insurance indices. Currently, only parts of the hydrologic cycle are considered via satellite data in index insurance due to a strong focus on satellite-derived precipitation and vegetation products. Satellite-derived soil moisture holds a great potential to close the temporal gap between rainfall deficits and the corresponding response of crops. However, to our knowledge, studies focusing on its added value for index insurance are virtually nonexistent. In order for soil moisture to add value beyond existing rainfall estimates, there would need to be evidence that the soil moisture is not only reflecting substantial agreement with rainfall estimates, but also that it is more effectively reflecting biophysical processes related to crop loss.

As an initial test of the potential utility of soil moisture in index insurance, this study evaluates the agreement of satellite-derived rainfall, soil moisture, and vegetation vigor over a study area in central Senegal. Because satellite estimates of vegetative vigor do not directly indicate crop loss, this analysis does not prove the effectiveness of the soil moisture or rainfall as an insurance product. Instead, it is a comparison of how well soil moisture correlates with near-future vegetation health, as compared to satellite rainfall estimates. If soil moisture more significantly predicts vegetative stress behavior, that would provide evidence that soil moisture is likely to be a useful tool in improving index insurance.

Following the introduction, we discuss the role of soil moisture in the hydrologic cycle and the use of space-based microwave sensors for soil moisture estimation. Section 4.3 concentrates on the basics of weather index insurance as well as the necessity to understand the benefits and limitations of satellite-based datasets, which are often the only source of information. Section 4.4 focuses on the region of interest, the Tambacounda region in central Senegal (West Africa). The datasets and methods are presented in Section 4.5. We use satellite-derived rainfall, soil moisture, and vegetation health to identify their spatial and temporal agreement. Section 4.6 presents the results of this study. We conclude with summary and conclusions (Section 4.7) in which we discuss the implications of our findings for weather index insurance. In addition, we highlight current scientific challenges and potential solutions with regard to the exploitation of satellite-derived soil moisture for weather index insurance. Finally, we suggest the combination of well-established and new soil moisture sensors to improve both the spatial and temporal resolution.

4.2 SATELLITE-DERIVED SOIL MOISTURE

The global hydrologic water cycle is a complex and dynamic process that is largely defined by energy- and moisture-related feedback loops (Trenberth et al., 2007). Soil moisture, the water content in the topmost layer of soil, may seem negligible compared to the global water resources. However, it plays a crucial role in these feedback loops with regard to land surface temperature (Miralles et al., 2012), evapotranspiration (Anderson et al., 2011), groundwater recharge (Abelen et al., 2015), moisture supply for boundary layer cloud development (Ek and Holtslag, 2004), and water supply for agriculture (Enenkel et al., 2015; Engman, 1991; Martínez-Fernández et al., 2016). A look at the International Soil Moisture Network (Dorigo et al., 2011), which is hosted by Vienna University of Technology, reveals a very unequal distribution of *in situ* stations on a global scale—many available observations for instance in the United States or Europe and very little on the African continent. Therefore, and because of their ability to provide comparable information on a global scale, satellite-based soil moisture observations are often the only source of information.

Satellite sensors that operate in the microwave domain (wavelengths of around 1 mm to 1 m) are among the most promising technologies for space-based soil moisture estimation (Petropoulos et al., 2015). The use of microwave sensors for soil moisture detection was discovered in the early 1990s (Engman, 1991; Jackson, 1993). Microwaves react sensitively to the dielectric properties of water (the characteristics of a water molecule to align when exposed to an electromagnetic field). Nevertheless, the first operational surface soil moisture dataset from the Advanced Scatterometer (ASCAT) onboard the MetOp satellites only became available in 2007 via the European Organisation for the Exploitation of Meteorological Satellites (EUMETSAT). Soon after, different applications, such as numerical weather forecasting (Dharssi et al., 2011; Drusch, 2007) or flood prediction (Brocca et al., 2012, 2010) started to assimilate ASCAT-derived soil moisture into their models. Although research communities such as remote sensing and weather or climate prediction are naturally relatively close, the gap between remote sensing and socioeconomic applications is far larger. As a consequence, research organizations similar to the International Research Institute for Climate and Society (IRI) at Columbia University have started to focus on both earth observation and socio-economic applications, such as health monitoring, climate change impacts, or social safety nets. This study concentrates on the latter with a particular focus on index-based agricultural insurance for smallholder farmers. In addition to the IRI report to the UN International Labour Organisation (International Research Institute for Climate and Society, 2013), the research group's website provides a comprehensive overview about the use of satellite data to improve index insurance: http://iri. columbia.edu/our-expertise/financial-instruments/using-satellite-data-to-improve-index-insurance/.

4.3   WEATHER INDEX-BASED INSURANCE AND EARTH OBSERVATION

Weather index-based insurance aims at distributing covariate risks (weather shocks that affect large geographical regions) among the people affected and at transferring the local/regional climatic risk to global markets. Traditionally, weather risk index insurance relied on rainfall data to link payouts to abnormal rainfall deficits (drought) or surplus (flood). An example of an element of a typical index used for coverage of hundreds of thousands of people against drought is as follows (Figure 4.1): If the cumulative rainfall in a predefined temporal window is less than a certain threshold (the *trigger* value), the payout rises linearly up to 100% (the *exit* value). This exit value represents a rainfall amount below which agricultural production is severely threatened.

The advantage of index insurance lies mainly in four areas (Hellmuth et al., 2009; International Fund for Agricultural Development, 2011):

- Costly, postdisaster loss assessment is not required, leading to lower premiums and faster payouts.
- The index is an objective indicator that insures the risk of a weather shock, not crop failure. As a consequence, farmers have no incentive to let the crops fail (*moral hazard*).
- Key parameters (e.g., the envisaged payout frequency, the hazard and crops insured) are discussed and agreed upon in collaboration with users.
- Index insurance can help to unlock agricultural potential, for instance by allowing farmers to access credit, which can be used to increase their productivity (drought-resistant seeds, irrigation, etc.), or by generating more income in good years to compensate bad ones.

However, apart from issues related to user-tailored knowledge transfer, a limited number of brokers to promote demand (Brown et al., 2011), and the harmonization with other risk management strategies, the largest source of uncertainty with regard to the overall index design lies in the input data. The design team must try to minimize the basis risk, the mismatch between the index, and the actual impact of a weather shock by using *in situ* and/or satellite data. Several sophisticated satellite

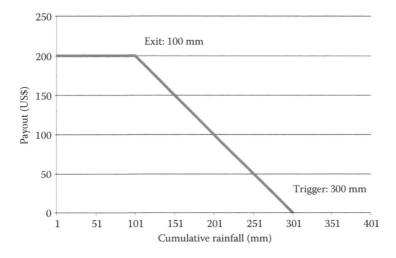

FIGURE 4.1   Schematic example of index-based insurance.

rainfall datasets already assimilate thousands of quality-controlled *in situ* observations in their satellite retrieval algorithms (Funk et al., 2015; Novella and Thiaw, 2013); others use local measurements for historical calibration (e.g., Tarnavsky et al., 2014). Nevertheless, the fact that rainfall is not detected directly, but via estimating the top temperature of clouds via sensors in the infrared domain and a low number of *in situ* observations in large parts of the world, leads to uncertainties with regard to the estimated rainfall quantity. For instance, over complex topography or coastal regions satellite rainfall products often do not perform well (Dinku et al., 2007, 2008, 2010).

While rainfall-based indices can potentially be used throughout the season, indices that are based on satellite-derived indicators of vegetation greenness/health are currently tested to cover the later parts of the growing seasons in which the crops are already developed. It has to be noted that satellite-based vegetation indicators, such as the Normalized Difference Vegetation Index (NDVI) or the Enhanced Vegetation Index (EVI), cannot be directly used as an indicator for yield (Lopresti et al., 2015; Ren et al., 2008). Both satellite rainfall estimates and NDVI/EVI indices have been widely utilized in index insurance, sometimes in combination (Greatrex et al., 2014).

4.4 REGION OF INTEREST

The study area is located in central Senegal (Figure 4.2), one of the countries that participate in the R4 Rural Resilience Initiative of the United Nations world food programme (WFP) and OXFAM, covering many tens of thousands of farms (http://www.wfp.org/climate-change/initiatives/r4-rural-resilience-initiative). R4 links four risk management strategies: (1) improved natural resource management, (2) agricultural insurance, (3) access to microcredit, and (4) savings. The IRI is a partner with the WFP and OXFAM to support the agricultural insurance component.

According to the European Space Agency (ESA) Climate Change Initiative (CCI) land cover dataset (version v 1.6.1), the region of interest (red rectangle in Figure 4.2) is characterized mainly by rain-fed cropland. The staple/cash crops are maize millet, rice, and groundnut. The Food and Agriculture Organization of the United Nations (UN FAO) crop calendar (www.fao.org/agriculture/seed/cropcalendar/) lists May/June as the planting/sowing months for maize. The maize harvest usually starts in August/September. There is a high interannual and interdecadal rainfall variability in key agricultural production areas (Fall et al., 2006; McSweeney et al., 2010), leading to an increased drought risk.

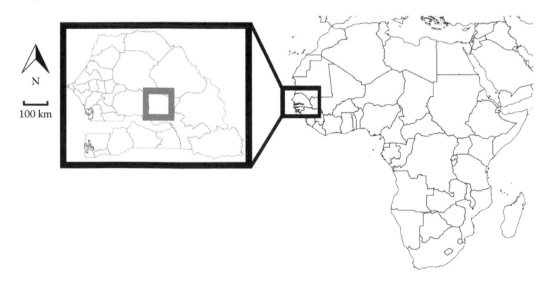

FIGURE 4.2   Illustration of the study area (grey rectangle) in central Senegal (Tambacounda region). The dimension of the study area is 1 × 1 degree (roughly 109 × 109 km).

According to Reliefweb (2002), a major drought with severe agricultural impacts was recorded in 2002. Furthermore, Senegal was affected by droughts in 2006, 2007, and 2011 and affected by flood events in 2009 and 2012. Between 2008 and 2011, the food security-related vulnerability was aggravated by economic shocks, which were partly related to the global financial crisis (UN WFP, 2014). The last severe drought, which resulted in socioeconomic consequences, was recorded in 2014/2015 (Famine Early Warning Systems Network, 2014). One of the latest vulnerabilities and food security assessments of the UN WFP (2014) stated that 42% of all households were suffering from food insecurity in 2014.

4.5   DATASETS AND METHODS

4.5.1   Precipitation

According to Funk et al. (2015), the Climate Hazards Group Infrared Precipitation with Stations (CHIRPS) dataset is designed to support the agricultural drought monitoring activities of the U.S. Agency for International Development, Famine Early Warning Systems Network (FEWS NET). It is a quasi-global (50°S–50°N), high resolution (0.05°; ~4.5 km at the equator), daily, pentadal (5-daily), and monthly precipitation dataset with a climatology dating back to 1981. A preliminary product, which includes the Global Telecommunication System (GTS) of the UN World Meteorological Organization gauge observations (around 14,000 stations in 2014), is available 2 days after the end of the pentad for several regions, such as Mexico. The final, global product is distributed during the third week of the following month. All observations are based on infrared cold cloud duration (CCD) observations. The algorithm relies on a high-resolution climatology and a novel blending scheme, whose method to assign interpolation weights considers the spatial correlation structure of CCD estimates.

4.5.2   Soil Moisture

Within the CCI of the ESA different active (radar) and passive (radiometer) sensors are combined to generate a daily surface soil moisture product (Liu et al., 2011, 2012; Wagner et al., 2012). In contrast to active microwave systems, which record the backscatter of the emitted microwave beam, radiometers detect emitted brightness temperatures, a product of surface temperature and surface emissivity. Although higher soil moisture content leads to higher backscatter values, the brightness temperature decreases (Mulder and Jeu, 2003). The ESA CCI soil moisture product uses the passive component in regions with low vegetation cover, the active component in regions with moderate, and an average of both in regions where both components agree well ($P > 0.65$). The dataset covers the years 1978 to 2014 (v02.2) with large gaps before 1992 (McNally et al., 2016). For the purpose of this study, we use an in-house product of Vienna University of Technology, which is based on the methods developed by Liu et al. (2011, 2012) and covers one additional year (2015).

Maybe the biggest advantage of soil moisture retrieval via microwave sensors is the independence from weather conditions. In particular for agricultural applications, it is an added value if the observations are not impeded by clouds, which are present during large parts of the rainy season. Although physical limitations allow only the detection of surface soil moisture in the topmost (less than 2 cm) soil layer, the application of an infiltration model allows estimations of soil moisture in the root zone of plants (Albergel et al., 2008; Wagner et al., 1999). However, the major drawback is the comparably low resolution of the ESA CCI datasets. It provides volumetric soil moisture at a spatial resolution of 0.25° (Liu et al., 2012). In addition, retrieval is not possible or flawed over frozen/snow-covered soils, over complex topography (e.g., over mountain ranges) or in the presence of dense vegetation (tropical forests) (Dorigo et al., 2012).

4.5.3 VEGETATION GREENNESS/HEALTH

The EVI is an advanced version of the NDVI (Tucker, 1979), which is one of the most widely used vegetation indicators for vegetation, agricultural, and drought monitoring (Anyamba and Tucker, 2005; Jiang et al., 2008; Karnieli et al., 2010; Klisch and Atzberger, 2016). Both NDVI and EVI correspond to the vegetative landscape changes caused by factors such as moisture supply and evapotranspiration. Their retrieval is based on observations in visible and near-infrared channels, which react sensitively to the chlorophyll content of leaves. However, the EVI exploits additional information from the blue channel, which has a wavelength of 470 nm to reduce atmospheric distortions caused by particles in the air and ground cover below the vegetation (Huete et al., 2002). In general, the EVI is less sensitive to the chlorophyll content of plants than the NDVI but more sensitive to structural variations in the canopy cover, as expressed by the Leaf Area Index (LAI).

For the purpose of this study, we use the EVI that is operationally distributed by the U.S. Geological Survey (USGS) Land Processes Distributed Active Archive Center (LP DAAC). It is based on observations from the two Moderate Resolution Imaging Spectroradiometer (MODIS) sensors Terra/Aqua. The spatial resolution is 500 m/1 km at 16-day compositing periods with a temporal coverage since 2000.

4.5.4 METHODS

Figure 4.3 illustrates the methodological approach underlying this study. First, we extract CHIRPS, ESA CCI soil moisture, and EVI for the region of interest based on a geographic bounding box (14.5°W–13.5°W and 13.5°N–14.5°N). After temporal matching to get monthly averages and spatial matching of datasets at different spatial resolutions, we calculate standardized anomalies for all variables (Equation 4.1).

$$z = \frac{\text{Monthly average} - \text{Longterm monthly mean for the same month}}{\text{Standard Deviation}} \qquad (4.1)$$

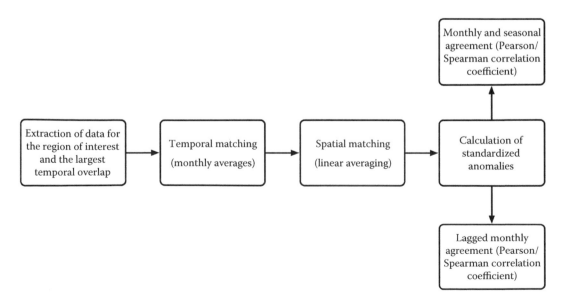

FIGURE 4.3   Method—Schematic illustration.

The time series analysis concentrates on the analysis of

- Monthly and seasonal agreement of precipitation, soil moisture, and vegetation via Pearson (*P*) and Spearman (*S*) correlation coefficient.
- Lagged monthly agreement via Pearson and Spearman correlation coefficient, whereas we shift the data at time steps of 1 month to account for the delayed response of vegetation greenness to changes in moisture supply.

4.6 RESULTS

Table 4.1 shows the agreement of precipitation/soil moisture, precipitation/vegetation greenness, and soil moisture/vegetation greenness. For all combinations, the highest *P*-correlation and comparable *S*-correlations are found for the standardized anomalies calculated for the June–July averages. We find the highest agreement between soil moisture and vegetation greenness/health anomalies (*P* = 0.76, *S* = 0.76). The correlations for rainfall and vegetation greenness/anomalies in the same period are 0.54 (Pearson) and 0.51 (Spearman).

In addition, with regard to anomalies over the entire season (June to August/September), we observe the highest correlations for soil moisture and vegetation anomalies. The agreement between rainfall and soil moisture anomalies reaches its maximum over June–July (*P* = 0.72, *S* = 0.69), whereas a lagged response of soil moisture was not considered. Overall, the agreement of rainfall/soil moisture and vegetation greenness decreases as the season proceeds. As the response of vegetation greenness to anomalies in moisture is usually delayed, we also calculate the correlation between rainfall/soil moisture and vegetation with a 1-month delay.

TABLE 4.1
Correlations of Precipitation (CHIRPS), Soil Moisture (ESA CCI), and Vegetation (EVI) for the Entire Growing Period of Central Senegal and 2-Month Averages (June/July, July/August, and August/September)

Variable *A*	Variable *B*	Coverage	Months Averaged	Correlation (Pearson)	Correlation (Spearman)
Precipitation	Soil moisture	1992–2015	June–August	0.63	**0.71**
Precipitation	Soil moisture	1992–2015	June–September	0.66	0.70
Precipitation	Soil moisture	1992–2015	June–July	**0.72**	0.69
Precipitation	Soil moisture	1992–2015	July–August	0.38	0.52
Precipitation	Soil moisture	1992–2015	August–September	0.48	0.53
Soil moisture	Vegetation	2000–2015	June–August	0.74	0.74
Soil moisture	Vegetation	2000–2015	June–September	0.72	0.69
Soil moisture	Vegetation	2000–2015	June–July	**0.76**	**0.76**
Soil moisture	Vegetation	2000–2015	July–August	0.63	0.56
Soil moisture	Vegetation	2000–2015	August–September	0.51	0.40
Precipitation	Vegetation	2000–2015	June–August	0.49	0.43
Precipitation	Vegetation	2000–2015	June–September	0.52	0.43
Precipitation	Vegetation	2000–2015	June–July	**0.54**	0.51
Precipitation	Vegetation	2000–2015	July–August	0.40	**0.56**
Precipitation	Vegetation	2000–2015	August–September	0.31	0.26

Note: The bold numbers highlight the highest correlation for each pair.

TABLE 4.2

Correlations of Precipitation (CHIRPS)/Soil Moisture (ESA CCI) and Vegetation (EVI) Considering a Temporal Lag of 1 Month for Vegetation for 2000–2015

Variable *A*	Variable *B*	Month (Variable *A*)	Month (Variable *B*)	Correlation (Pearson)	Correlation (Spearman)
Precipitation	Vegetation	June	July	0.36	0.42
Precipitation	Vegetation	July	August	**0.64**	**0.72**
Precipitation	Vegetation	August	September	0.33	0.24
Soil moisture	Vegetation	June	July	0.50	0.55
Soil moisture	Vegetation	July	August	**0.82**	**0.83**
Soil moisture	Vegetation	August	September	0.64	0.64

Note: The bold numbers highlight the highest correlation for each pair.

Table 4.2 illustrates the lagged correlations for standardized precipitation/soil moisture anomalies and the standardized vegetation anomaly. The agreement of soil moisture and vegetation greenness is generally higher than for precipitation. We find the best correlation for soil moisture anomalies in July and for vegetation greenness anomalies in August ($P = 0.82$, $S = 0.83$).

The following illustrations demonstrate the seasonal agreement of standardized anomalies for the precipitation/soil moisture and vegetation between the years 2000 and 2015. With regard to the Tambacounda region, in particular, the years 2001, 2002, 2014, and 2015 had been mentioned by local farmers as severe drought years (IRI/R4 project report, unpublished). The green line in Figure 4.4 represents vegetation greenness/health. Both the 2002 and the 2014/2015 events are well captured in the EVI time series. However, the CHIRPS rainfall anomaly misses the largest event in 2002.

In Figure 4.5, we replaced the rainfall component with seasonal standardized soil moisture. With regard to the drought in 2002, soil moisture and EVI anomalies match very well. Both time series exhibit an anomaly of around two standard deviations below the mean. In addition, the 2014 and 2015 events are well reflected in both datasets. Over the 2001 agricultural season, the agreement between EVI anomalies and soil moisture anomalies is higher than for rainfall. However, neither EVI nor soil moisture indicates a severe drought.

The maps in Figure 4.6 illustrate the spatial agreement between precipitation and soil moisture with vegetation health throughout all seasons (June–September) from 2000 to 2015. The results for the entire country are comparable to the time series analysis in the central Tambacounda region. The correlation of the coarser resolution soil moisture dataset and vegetation health is generally higher than the correlation of precipitation and vegetation health. Only in the Southeast, which is characterized by higher altitudes, the correlations are negative if no temporal lags are considered.

Figure 4.7 shows the agreement of precipitation and soil moisture with vegetation health, whereas a time lag of 1 month is considered. The top left image, for instance, shows the correlation of precipitation in June with vegetation health in July. Again, the findings are in line with the time series analysis for the Tambacounda region. We observe the highest correlation for soil moisture conditions during July and the response of vegetation health in August.

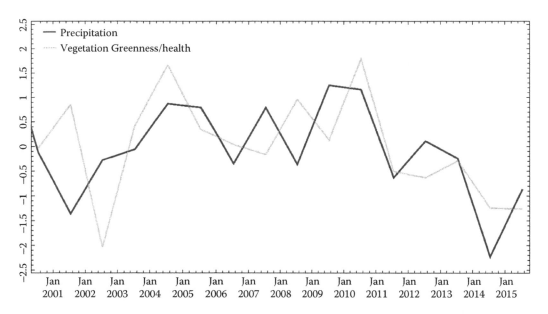

FIGURE 4.4 Time series of standardized June–August anomalies for precipitation and vegetation health/greenness, expressed in standard deviations ($P = 0.49$, $S = 0.43$).

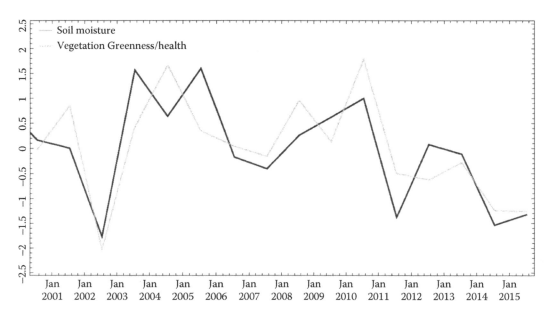

FIGURE 4.5 Time series of standardized June–August anomalies for soil moisture and vegetation health/greenness, expressed in standard deviations ($P = 0.74$, $S = 0.74$).

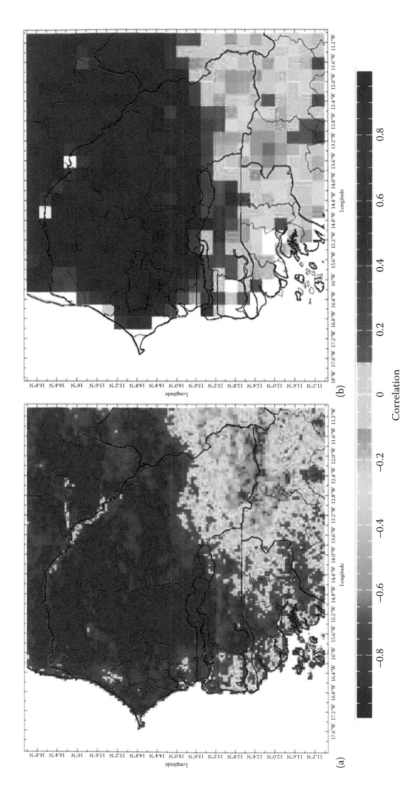

FIGURE 4.6 Seasonal (June–September) Pearson correlation (a) between precipitation and vegetation health and (b) between soil moisture and vegetation health.

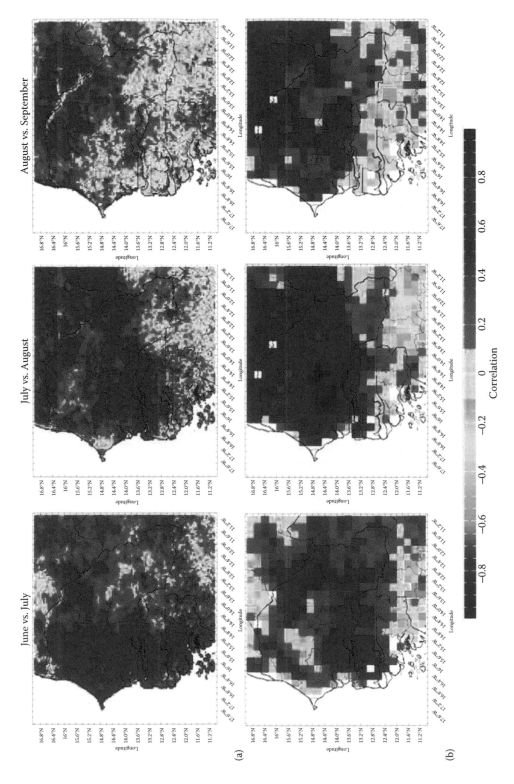

FIGURE 4.7   Lagged Pearson correlation for (a) precipitation and vegetation health as well as (b) soil moisture and vegetation health.

4.7   SUMMARY AND CONCLUSIONS

If integrated with complementary risk management strategies, weather index-based insurance can help to unlock the agricultural potential of smallholder farmers. Satellite data play an important role in the design of the insurance indices. One main task of researchers is to select, analyze, and combine different datasets to maximize the match between satellite-derived information and the impact of the insured peril (e.g., drought) on the ground. Current agricultural index insurance programs are mainly driven by satellite-derived rainfall or estimations of vegetation greenness/health. Satellite soil moisture holds the potential to indicate both drought and near-future vegetation growth (Qiu et al., 2014; Zribi et al., 2010). However, studies that analyze the added value of satellite soil moisture, a relatively new variable in earth observation, are very scarce. Consequently, we focus on an initial analysis of the agreement between three satellite-derived datasets over a rain-fed agricultural area in central Senegal: precipitation (CHIRPS), soil moisture (ESA CCI), and vegetation greenness/health (EVI).

Our findings indicate that in the study region the agreement of standardized seasonal (June–August) anomalies of rainfall and vegetation greenness is lower than the agreement of soil moisture and vegetation greenness (Table 4.1). As vegetation usually takes time to respond to changes in moisture, we also investigate the lagged (1 month) agreement. Again, the correlation between soil moisture and vegetation is higher than for rainfall and vegetation, whereas we observe the best *forecasting* capability of vegetation anomalies in July via soil moisture anomalies ($P = 0.82$, $S = 0.83$) in July (Table 4.2). Figures 4.4 and 4.5 illustrate the time series of standardized seasonal anomalies for the Tambacounda region in central Senegal. Although the seasonal rainfall anomaly fails to capture the severe drought event in 2002, both the soil moisture and vegetation anomalies agree with regard to severity and timing. A moisture deficit in 2014 and 2015 is more or less captured by all datasets. The 2001 season, which had been mentioned by local farmers as a *bad year* during IRI field visits, is reflected in the rainfall anomaly but not in soil moisture or vegetation greenness. Possible explanations are a preseason moisture deficit that affected the start of the season and a different (e.g., flood) or nonclimatic (e.g., locust) impact on the crops.

The findings in this study, provide strong evidence for the consideration of satellite soil moisture to complement other datasets. Therefore, we suggest five approaches to better exploit the added value of satellite soil moisture for weather index insurance and to overcome challenges with regard to the development of operational systems: First, following the example of satellite-based rainfall and vegetation health estimations, interdisciplinary research must identify if the coarse resolution (0.25°, roughly 26 km at the equator) or satellite soil moisture reflects local conditions (drought impacts, detection of the start of season, etc.) sufficiently well for a specific region of interest. Second, specific methods need to be developed for the purpose of index insurance design that uses soil moisture as a validation for one or more satellite-derived rainfall datasets. The SM2Rain algorithm of Brocca et al. (2014, 2013), which basically inverts the water balance equation to estimate rainfall from satellite soil moisture, is a promising approach. Third, the reliance on polar-orbiting satellites results in data gaps. Several methods (e.g., data filling via a z-score transformations of nearby data to fill the missing data) need to be tested to avoid the distortion of historical payout simulations. Fourth, the ESA CCI dataset is updated approximately every year. Currently, methods are developed to extend the historical reference with near real-time (NRT) observations from radars and/or radiometers. Studies will have to show the consistency of NRT and historical data, which is crucial for the calculation of anomalies or insurance indices. *In situ* soil moisture observations, which are scarce on the African continent, are therefore an indispensable source of information. Fifth, data continuity needs to be guaranteed for operational weather index insurance. The Copernicus portal of the European Commission, for instance, provides a satellite-derived root-zone soil moisture (Soil Water Index) as a daily product on global scale and free of charge.

With regard to Senegal high uncertainties, which are intrinsic to long-range climate projections, make it hard to draw conclusions about future changes in mean annual rainfall. However, most climate projections indicate a clear increase in the frequency of hot days and nights (McSweeney et al., 2010). Naturally, this might have negative impacts on the vulnerability of smallholder farmers. In combination with user-tailored knowledge transfer and institutional partnerships, scaling up index insurance can become an even more important component of long-term poverty alleviation (Greatrex et al., 2014). Earth observation from satellites will inevitably play an important role in the development and scaling up of weather risk indices. Satellite-derived soil moisture holds the potential to optimize this process.

Future research will have to focus on the latest generation of soil moisture missions, such as the Sentinel mission of the European Commission and ESA. In contrast to National Aeronautics and Space Administration's (NASA's) Soil Moisture Active Passive (SMAP) mission, whose key radar sensor failed in 2015, Sentinel 1A and B provides global soil moisture estimates every 6 to 12 days (Malenovský et al., 2012). Operational products are available at a spatial resolution of 1 km. Methods that combine the high temporal resolution of the two ASCAT sensors onboard the MetOp satellites, which individually cover more around 82% of the global land surface every day (Wagner et al., 2013), and the high spatial resolution of Sentinel 1A/B are therefore very promising.

ACKNOWLEDGMENTS

This study was partly funded within the NASA Interdisciplinary Science Program (award number #NNX14AD63G).

REFERENCES

Abelen, S., Seitz, F., Abarca-del-Rio, R., Güntner, A., 2015. Droughts and floods in the La Plata basin in soil moisture data and GRACE. *Remote Sens.*, 7, 7324–7349. doi:10.3390/rs70607324

Albergel, C., Rüdiger, C., Pellarin, T., Calvet, J.-C., Fritz, N., Froissard, F., Suquia, D., Petitpa, A., Piguet, B., Martin, E., 2008. From near-surface to root-zone soil moisture using an exponential filter: An assessment of the method based on in-situ observations and model simulations. *Hydrol. Earth Syst. Sci.*, 12, 1323–1337. doi:10.5194/hess-12-1323-2008

Anderson, M.C., Kustas, W.P., Norman, J.M., Hain, C.R., Mecikalski, J.R., Schultz, L., González-Dugo, M.P. et al., 2011. Mapping daily evapotranspiration at field to continental scales using geostationary and polar orbiting satellite imagery. *Hydrol. Earth Syst. Sci.*, 15, 223–239. doi:10.5194/hess-15-223-2011

Anyamba, A., Tucker, C.J., 2005. Analysis of Sahelian vegetation dynamics using NOAA-AVHRR NDVI data from 1981–2003. *J. Arid Environ.*, 63, 596–614. doi:10.1016/j.jaridenv.2005.03.007

Brocca, L., Ciabatta, L., Massari, C., Moramarco, T., Hahn, S., Hasenauer, S., Kidd, R., Dorigo, W., Wagner, W., Levizzani, V., 2014. Soil as a natural rain gauge: Estimating global rainfall from satellite soil moisture data. *J. Geophys. Res. Atmos.*, 119, 5128–5141. doi:10.1002/2014JD021489

Brocca, L., Melone, F., Moramarco, T., Wagner, W., Naeimi, V., Bartalis, Z., Hasenauer, S., 2010. Improving runoff prediction through the assimilation of the ASCAT soil moisture product. *Hydrol. Earth Syst. Sci.*, 14, 1881–1893. doi:10.5194/hess-14-1881-2010

Brocca, L., Moramarco, T., Melone, F., Wagner, W., 2013. A new method for rainfall estimation through soil moisture observations. *Geophys. Res. Lett.*, 40, 853–858. doi:10.1002/grl.50173

Brocca, L., Moramarco, T., Melone, F., Wagner, W., Hasenauer, S., Hahn, S., 2012. Assimilation of surface- and root-zone ASCAT Soil Moisture Products Into Rainfall and Runoff Modeling. *IEEE Trans. Geosci. Remote Sens.*, 50, 2542–2555. doi:10.1109/TGRS.2011.2177468

Brown, M.E., Osgood, D.E., Carriquiry, M.A., 2011. Science-based insurance. *Nat. Geosci.*, 4, 213–214. doi:10.1038/ngeo1117

Dharssi, I., Bovis, K.J., Macpherson, B., Jones, C.P., 2011. Operational assimilation of ASCAT surface soil wetness at the Met Office. *Hydrol. Earth Syst. Sci.*, 15, 2729–2746. doi:10.5194/hess-15-2729-2011

Dinku, T., Ceccato, P., Grover-Kopec, E., Lemma, M., Connor, S.J., Ropelewski, C.F., 2007. Validation of satellite rainfall products over East Africa's complex topography. *Int. J. Remote Sens.*, 28, 1503–1526. doi:10.1080/01431160600954688

Dinku, T., Chidzambwa, S., Ceccato, P., Connor, S.J., Ropelewski, C.F., 2008. Validation of high-resolution satellite rainfall products over complex terrain. *Int. J. Remote Sens.*, 29, 4097–4110.

Dinku, T., Ruiz, F., Connor, S.J., Ceccato, P., 2010. Validation and intercomparison of satellite rainfall estimates over Colombia. *J. Appl. Meteorol. Climatol.*, 49, 1004–1014. doi:10.1175/2009JAMC2260.1

Dorigo, W.A., Wagner, W., Hohensinn, R., Hahn, S., Paulik, C., Drusch, M., Mecklenburg, S., van Oevelen, P., Robock, A., Jackson, T., 2011. The international soil moisture network: A data hosting facility for global in situ soil moisture measurements. *Hydrol. Earth Syst. Sci.*, 15, 1609–1663. doi:10.5194/hessd-8-1609-2011

Dorigo, W., de Jeu, R., Chung, D., Parinussa, R., Liu, Y., Wagner, W., Fernández-Prieto, D., 2012. Evaluating global trends (1988–2010) in harmonized multi-satellite surface soil moisture. *Geophys. Res. Lett.*, 39. doi:10.1029/2012GL052988

Drusch, M., 2007. Initializing numerical weather prediction models with satellite-derived surface soil moisture: Data assimilation experiments with ECMWF's Integrated Forecast System and the TMI soil moisture data set. *J. Geophys. Res. Atmos.*, 112, D03102. doi:10.1029/2006JD007478

Ek, M.B., Holtslag, A.A.M., 2004. Influence of soil moisture on boundary layer cloud development. *J. Hydrometeorol.* 5, 86–99.

Enenkel, M., Reimer, C., Dorigo, W., Wagner, W., Pfeil, I., Parinussa, R., De Jeu, R., 2015. Combining satellite observations to develop a daily global soil moisture product for a wide range of applications. *Hydrol. Earth Syst. Sci.*, 11549–11589. doi:10.5194/hessd-12-11549-2015

Engman, E.T., 1991. Applications of microwave remote sensing of soil moisture for water resources and agriculture. *Remote Sens. Environ.*, 35, 213–226. doi:10.1016/0034-4257(91)90013-V

Fall, S., Semazzi, F.H.M., Niyogi, D.D.S., Anyah, R.O., Bowden, J., 2006. The spatiotemporal climate variability over Senegal and its relationship to global climate. *Int. J. Climatol.*, 26, 2057–2076. doi:10.1002/joc.1355

Famine Early Warning Systems Network, 2014. Significantly below-average harvests in Senegal contribute to increasing food insecurity, at: http://www.fews.net/west-africa/senegal/alert/december-3-2014 (accessed April 5, 2016).

Funk, C., Peterson, P., Landsfeld, M., Pedreros, D., Verdin, J., Shukla, S., Husak, G. et al., 2015. The climate hazards infrared precipitation with stations—A new environmental record for monitoring extremes. *Sci. Data*, 2, 150066. doi:10.1038/sdata.2015.66

Greatrex, H., Hansen, J.W., Garvin, S., Diro, R., Blakeley, S., Le Guen, M., Rao, K.N., Osgood, D.E., 2014. Scaling up index insurance for smallholder farmers: Recent evidence and insights, CCAFS Report No. 14, Copenhagen: CGIAR Research Program on Climate Change, Agriculture and Food Security (CCAFS). Available: www.ccafs.cgiar.org.

Hellmuth, M.E., Osgood, D.E., Hess, U., Moorhead, A., Bhojwani, H., (Eds.) 2009. *Index Insurance and Climate Risk: Prospects for Development and Disaster Management.* Palisades, NY: International Research Institute for Climate and Society.

Huete, A., Didan, K., Miura, T., Rodriguez, E.P., Gao, X., Ferreira, L.G., 2002. Overview of the radiometric and biophysical performance of the MODIS vegetation indices. *Remote Sens. Environ.*, 83, 195–213. doi:10.1016/S0034-4257(02)00096-2

International Fund for Agricultural Development, 2011. Weather index-based insurance in agricultural development–A technical guide. Available: https://www.ifad.org/documents/10180/2a2cf0b9-3ff9-4875-90ab-3f37c2218a90.

International Research Institute for Climate and Society, 2013. Using satellites to make index insurance scalable: Final IRI Report to the International Labour Organisation - Microinsurance Innovation Facility, Columbia University. Available: http://iri.columbia.edu/resources/publications/Using-Satellites- Scalable-Index-Insurance-IRI-ILO-report/.

Jackson, T.J., 1993. III. Measuring surface soil moisture using passive microwave remote sensing. *Hydrol. Process.*, 7, 139–152. doi:10.1002/hyp.3360070205

Jiang, Z., Huete, A., Didan, K., Miura, T., 2008. Development of a two-band enhanced vegetation index without a blue band. *Remote Sens. Environ.*, 112, 3833–3845. doi:10.1016/j.rse.2008.06.006

Karnieli, A., Agam, N., Pinker, R.T., Anderson, M., Imhoff, M.L., Gutman, G.G., Panov, N., Goldberg, A., 2010. Use of NDVI and land surface temperature for drought assessment: Merits and limitations. *J. Clim.*, 23, 618–633. doi:10.1175/2009JCLI2900.1

Klisch, A., Atzberger, C., 2016. Operational drought monitoring in Kenya using MODIS NDVI time series. *Remote Sens.*, 8, 267. doi:10.3390/rs8040267

Liu, Y., Parinussa, R.M., Dorigo, W.A., De Jeu, R.A.M., Wagner, W., van Dijk, A.I.J.M., McCabe, M.F., Evans, J.P., 2011. Developing an improved soil moisture dataset by blending passive and active microwave satellite-based retrievals. *Hydrol. Earth Syst. Sci.*, 15, 425–436. doi:10.5194/hess-15-425-2011

Liu, Y.Y., Dorigo, W.A., Parinussa, R.M., de Jeu, R.A.M., Wagner, W., McCabe, M.F., Evans, J.P., van Dijk, A.I.J.M., 2012. Trend-preserving blending of passive and active microwave soil moisture retrievals. *Remote Sens. Environ.*, 123, 280–297. doi:10.1016/j.rse.2012.03.014

Lopresti, M.F., Di Bella, C.M., Degioanni, A.J., 2015. Relationship between MODIS-NDVI data and wheat yield: A case study in Northern Buenos Aires province, Argentina. *Inf. Process. Agric.*, 2, 73–84. doi:10.1016/j.inpa.2015.06.001

Malenovský, Z., Rott, H., Cihlar, J., Schaepman, M.E., García-Santos, G., Fernandes, R., Berger, M., 2012. Sentinels for science: Potential of Sentinel-1, -2, and -3 missions for scientific observations of ocean, cryosphere, and land. *Remote Sens. Environ.*, 120, 91–101. doi:10.1016/j.rse.2011.09.026

Martínez-Fernández, J., González-Zamora, A., Sánchez, N., Gumuzzio, A., Herrero-Jiménez, C.M., 2016. Satellite soil moisture for agricultural drought monitoring: Assessment of the SMOS derived soil water deficit index. *Remote Sens. Environ.*, 177, 277–286. doi:10.1016/j.rse.2016.02.064

McNally, A., Shukla, S., Arsenault, K.R., Wang, S., Peters-Lidard, C.D., Verdin, J.P., 2016. Evaluating ESA CCI soil moisture in East Africa. *Int. J. Appl. Earth Obs. Geoinf.*, 48, 96–109. doi:10.1016/j.jag.2016.01.001

McSweeney, C., Lizcano, G., New, M., Lu, X., 2010. The UNDP climate change country profiles: Improving the accessibility of observed and projected climate information for studies of climate change in developing countries. *Bull. Am. Meteorol. Soc.*, 91, 157–166. doi:10.1175/2009BAMS2826.1

Miralles, D.G., van den Berg, M.J., Teuling, A.J., de Jeu, R.A.M., 2012. Soil moisture-temperature coupling: A multiscale observational analysis. *Geophys. Res. Lett.*, 39, L21707. doi:10.1029/2012GL053703

Mulder, E.W.A.M., de Jeu, R.A.M., 2003. *Retrieval of Land Surface Parameters Using Passive Microwave Remote Sensing.* Stichting Natuurpublicaties Limburg.

Novella, N.S., Thiaw, W.M., 2013. African rainfall climatology version 2 for famine early warning systems. *J. Appl. Meteorol. Climatol.*, 52, 588–606. doi:10.1175/JAMC-D-11-0238.1

Petropoulos, G.P., Ireland, G., Barrett, B., 2015. Surface soil moisture retrievals from remote sensing: Current status, products & future trends. *Phys. Chem. Earth Parts A/B/C*, 83–84, 36–56. doi:10.1016/j.pce.2015.02.009

Qiu, J., Crow, W.T., Nearing, G.S., Mo, X., Liu, S., 2014. The impact of vertical measurement depth on the information content of soil moisture times series data. *Geophys. Res. Lett.*, 41, 4997–5004. doi:10.1002/2014GL060017

Reliefweb, 2002. Cattle dying, stench of death: Mauritania, Senegal face drought. Available: http://reliefweb.int/report/mauritania/cattle-dying-stench-death-mauritania-senegal-face-drought (accessed May 10, 2016).

Ren, J., Chen, Z., Zhou, Q., Tang, H., 2008. Regional yield estimation for winter wheat with MODIS-NDVI data in Shandong, China. *Int. J. Appl. Earth Obs. Geoinf.*, 10, 403–413. doi:10.1016/j.jag.2007.11.003

Tarnavsky, E., Grimes, D., Maidment, R., Black, E., Allan, R.P., Stringer, M., Chadwick, R., Kayitakire, F., 2014. Extension of the TAMSAT satellite-based rainfall monitoring over Africa and from 1983 to present. *J. Appl. Meteorol. Climatol.*, 53, 2805–2822. doi:10.1175/JAMC-D-14-0016.1

Trenberth, K.E., Smith, L., Qian, T., Dai, A., Fasullo, J., 2007. Estimates of the global water budget and its annual cycle using observational and model data. *J. Hydrometeorol.*, 8, 758–769. doi:10.1175/JHM600.1

Tucker, C.J., 1979. Red and photographic infrared linear combinations for monitoring vegetation. *Remote Sens. Environ.*, 8, 127–150. doi:10.1016/0034-4257(79)90013-0

UN WFP, 2014. Juillet 2014 Analyse Globale de la Vulnérabilité, de la Sécurité Alimentaire et de la Nutrition (AGVSAN) - Sénégal.

Wagner, W., Dorigo, W., de Jeu, R., Fernandez, D., Benveniste, J., Haas, E., Ertl, M., 2012. Fusion of active and passive microwave observations to create an essential climate variable data record on soil moisture. *ISPRS Ann. Photogramm, Remote Sens. Spat. Inf. Sci.* 7, 315–321. doi:10.5194/isprsannals-I-7-315-2012

Wagner, W., Hahn, S., Kidd, R., Melzer, T., Bartalis, Z., Hasenauer, S., Figa-Saldaña, J. et al., 2013. The ASCAT soil moisture product: A review of its specifications, validation results, and emerging applications. *Meteorol. Z.*, 22, 5–33. doi:10.1127/0941-2948/2013/0399

Wagner, W., Lemoine, G., Rott, H., 1999. A method for estimating soil moisture from ERS scatterometer and soil data. *Remote Sens. Environ.*, 70, 191–207. doi:10.1016/S0034-4257(99)00036-X

Zribi, M., Paris Anguela, T., Duchemin, B., Lili, Z., Wagner, W., Hasenauer, S., Chehbouni, A., 2010. Relationship between soil moisture and vegetation in the Kairouan plain region of Tunisia using low spatial resolution satellite data. *Water Resour. Res.*, 46. doi:10.1029/2009WR008196

5 Detecting the 2012 Drought in the Southeastern United States with Moderate Resolution Imaging Spectroradiometer- and Gravity Recovery and Climate Experiment-Based Drought Indicators[*]

Ali Levent Yagci, Joseph A. Santanello, Matthew Rodell, Meixia Deng, and Liping Di

CONTENTS

---

[*] We declare that this manuscript (in full or part) has not been published elsewhere.

5.1 INTRODUCTION

Drought is one of the devastating natural hazards, which often recurs when plants cannot sustain their growth as a result of water deficit. Its occurrence interferes with agricultural production by significantly reducing crop yields, in turn damaging the global economy. As the world population has been steadily growing, food supply must keep up with this increasing demand.

In this regard, several drought monitoring tools such as United States Drought Monitor (USDM) (Svoboda et al. 2002) and Global Agricultural Drought Monitoring and Forecasting System (GADMFS) (Deng et al. 2013) have been developed to detect onset, duration, extent, and severity of drought and to timely inform the state and government agencies, stakeholders, farmers, and public so that its devastating effects can be mitigated.

Daily observations obtained by satellites orbiting in space are indispensible to routinely track the Earth's ground and surface water resources and natural hazards such as droughts and floods because they provide spatially continuous synoptic view of the Earth. In the last decade, many efforts have been devoted to drought monitoring. Drought is relatively defined as natural phenomenon, generally identified by the deviations of precipitation (e.g., meteorological drought), soil water (e.g., agricultural drought), and groundwater and streamflow (e.g., hydrological drought) from their long-term average condition (Wilhite 2000).

Remotely sensed vegetation indices such as the Normalized Difference Vegetation Index (NDVI) have been extensively used to track droughts (Kogan 2001), especially from the NOAA's advanced very high resolution radiometer (AVHRR) because of its long record (e.g., ≈ 30 years). Vegetation indices are good surrogate measures of photosynthetically functioning vegetation (Tucker and Choudhury 1987). Because drought hinders the photosynthetic activity of plants, large-scale reduction in NDVI over a region (e.g., statewide) can be associated with droughts. After completing 10 years in orbit, the products of National Aeronautics and Space Administration's (NASA's) Moderate Resolution Imaging Spectroradiometer (MODIS) have also been used to monitor droughts (Yagci et al. 2012; Deng et al. 2013; Yagci et al. 2013). MODIS acquires observations in narrower bands than the AVHRR instrument, successfully avoiding the water vapor absorption in the visible-red and near-infrared (NIR) region of the electromagnetic spectrum. Therefore, MODIS-NDVI products attain relatively larger values and better accuracy in exhibiting temporal profiles of forests than the AVHRR-NDVI data (Huete et al. 2002).

In addition to NDVI, ability of surface brightness temperature (T_b) or land surface temperature (LST) to track drought has been successfully tested and validated against the crop yields in the state of Texas, United States (Yagci et al. 2011) and around the globe (Kogan 2001). LST is a better indicator of surface temperature conditions than T_b because it is corrected for surface emissivity and estimated from surface radiance, that is, atmospherically corrected surface radiance reaching the sensor. LST is a proxy for moisture availability and evapotranspiration conditions such that water depletion in the plant root zone leads to stomatal closure, reduced transpiration, and subsequently elevated canopy temperatures (Anderson and Kustas 2008). Drought detected by NDVI and LST products is referred to as vegetative drought or agricultural drought.

In recent years, a new way has surfaced to monitor drought through analysis of the terrestrial water storage (TWS) anomalies. The monthly variations in the Earth's gravitational signal measured by twin satellites of the Gravity Recovery and Climate Experiment (GRACE) have been shown to relate to monthly TWS changes with roughly 1.5 cm accuracy at regional scales (Wahr et al. 2004). GRACE-derived TWS is coarsely resolved and contains vertically integrated information about surface and subsurface water conditions; therefore, its spatial, temporal, and vertical decomposition into soil moisture and groundwater components achieved through data assimilation into the Catchment Land Surface Model (CLSM) aids in its interpretation and application to drought monitoring (Houborg et al. 2012; Rodell 2012). The resulting groundwater and soil moisture wetness fields are appropriate for hydrological and agricultural drought monitoring applications, respectively.

USDM is a collaborative effort by the National Drought Mitigation Center at the University of Nebraska–Lincoln, the Departments of Commerce and Agriculture and outside experts to summarize weekly drought conditions across the United States (Svoboda et al. 2002). Despite the fact that USDM is the premier drought product for the United States, it does have certain shortcomings such as a tendency toward overestimation of drought areal coverage and difficulty in representing the local-scale (e.g., county-scale) conditions, which have been highlighted by several studies (Brown et al. 2008; Tadesse et al. 2005).

The conterminous United States experienced a vast costly drought in 2012, which caused disastrous impacts on agriculture and livestock industries, totaling nearly $30 billion losses (Rippey 2015). The drought of 2012 was similar to the drought of 1988 in terms of cost and the megadrought of the 1950s in terms of areal coverage (Rippey 2015). In this study, characteristics of the 2012 drought are examined using the drought maps derived from the aforementioned approaches. Each method is rather distinct in terms of input type and source, theoretical background, and level of complexity. Their results are intercompared in 2012, and their similarities and discrepancies are also highlighted in Southeastern United States.

5.2 DATA AND METHODS

5.2.1 NORMALIZED DIFFERENCE VEGETATION INDEX (NDVI)

NDVI is a measure of vegetation greenness, ranging from −1 to 1. Presence of chlorophyll pigments in plant leaves causes visible sunlight in the *red* region of the spectrum to be absorbed for photosynthesis, and sunlight in the NIR region of the spectrum is substantially reflected due to the cell structure of the leaves. Therefore, green healthy functioning vegetation always attains larger NDVI value than brown stressed vegetation. Swain et al. (2011) demonstrated that NDVI in the drought year of 2002 was considerably smaller than NDVI during the nondrought year, 2007 over the croplands and grasslands of Nebraska, United States. The 16-day composite MODIS-NDVI products (collection 5) were retrieved from the NASA's Land Processes Distributed Active Archive Center (LP DAAC). The level 3 NDVI products, abbreviated as MOD13A2.005, are compiled from radiometrically-, geometrically-, and atmospherically-corrected surface reflectances and have 1 km spatial resolution. The compositing algorithm, the constrained view angle maximum value composite (CV-MVC), picks the best available NDVI observation that is noncloudy and closest to nadir view to represent the vegetation conditions during the 16-day period (Solano et al. 2010).

5.2.2 LAND SURFACE TEMPERATURE (LST)

LST is a proxy variable for moisture availability and evapotranspiration conditions (Anderson and Kustas 2008). Elevated LSTs are typical during drought years as opposed to LSTs observed in normal or wet years because plants are not transpiring to cool off the canopy. Similarly, Swain et al. (2011) demonstrated that LST increased during the 2002 drought year in comparison to the 2007 normal year in the croplands (corn) and grasslands of Nebraska. The collection 5 daytime MODIS-LST products were retrieved from the NASA's LP DAAC. The level 3 LST products, abbreviated as MYD11A2.005, are composited over a 8-day period with 1 km spatial resolution and are calculated from radiometrically, geometrically, and atmospherically corrected surface radiances. Unlike 16-day NDVI composites, the 8-day LST composite is the average of all noncloudy LSTs during the 8-day period (Wan 2007).

5.2.3 VEGETATION CONDITION INDEX (VCI)

The Vegetation Condition Index (VCI) was introduced to separate the annually varying NDVI component due to prevailing weather conditions from long-term component of NDVI (e.g., climate, soil,

TABLE 5.1
USDM Drought Classification Scheme

Category	Description	Percentiles
D0	Abnormally dry	21–30
D1	Moderate drought	11–20
D2	Severe drought	6–10
D3	Extreme drought	3–5
D4	Exceptional drought	0–2

and land cover type) (Kogan 1997). The index ranges from 0 to 100 and can be calculated with the following formula:

$$VCI_c = 100 \times \frac{NDVI_c - NDVI_{min}}{NDVI_{max} - NDVI_{min}} \qquad (5.1)$$

where:

$NDVI_{min}$ and $NDVI_{max}$ are the multiyear minimum and maximum NDVI values, respectively
$NDVI_c$ is the NDVI value of the compositing period of interest

For instance, if VCI of the 177th day of 2012 is the interest, then $NDVI_c$ is the NDVI value of the 177th day of 2012. VCI values of 0 and 100 indicate the worst and best vegetation conditions, respectively. Prior to VCI calculation, low-quality NDVI pixels that are covered with cloud, cloud shadows, and adjacent to clouds were removed based on quality flags in the corresponding quality assurance (QA) layers that come with the NDVI products. The resulting gaps in NDVI products were filled by interpolation. NDVI observations from two preceding and following 16-day periods along with their corresponding day of year (DOY) information were used to interpolate gaps and downscale to 8-day temporal resolution. The VCI-based drought maps were compiled by the percentile-based classification scheme given in Table 5.1.

5.2.4 TEMPERATURE CONDITION INDEX (TCI)

Similar to VCI, TCI was designed to highlight LST changes due to prevailing weather conditions (Kogan 1997). It ranges from 0 to 100 and can be calculated with the following formula:

$$TCI_c = 100 \times \frac{LST_{max} - LST_c}{LST_{max} - LST_{min}} \qquad (5.2)$$

where:

LST_{min} and LST_{max} are the multiyear minimum and maximum LST values, respectively
LST_c is the LST value of the compositing period of interest

For instance, if TCI of the 177th day of 2012 is the interest, then LST_c is the LST value of the 177th day of 2012. Minimum and maximum TCI values (e.g., 0 and 100) indicate the worst and best vegetation conditions, respectively. Prior to TCI calculation, LST products underwent a masking process in which all cloudy LST observations were removed. The incomplete LST time series were filled by temporal interpolation using LST observations from two preceding and following 8-day compositing periods. The TCI-based drought maps were categorized by the drought classification scheme in Table 5.1 to identify drought-affected areas.

5.2.5 United States Drought Monitor (USDM)

The team of roughly 15 authors of the USDM combines meteorological, agricultural, and hydrological drought indicators such as Palmer Drought Severity Index (PDSI), Climate Prediction Center (CPC) soil moisture model, United States Geological Survey (USGS) weekly streamflow, Standardized Precipitation Index (SPI), and other drought indices to produce weekly drought maps, by focusing on broadscale conditions (e.g., state-level). In turn, it may not be used to infer local-scale (e.g., county-level) conditions. Drought is classified by percentiles into five different severities: abnormally dry, moderate, severe, extreme, and exceptional drought, as outlined in Table 5.1 (The National Drought Mitigation Center 2016). In the end, a blend of drought indicators with different weights determined subjectively by the experts contributes to the final drought map (Svoboda et al. 2002), and this map is updated weekly and is disseminated via the USDM website (http://droughtmonitor.unl.edu/Home.aspx).

5.2.6 Gravity Recovery and Climate Experiment-Based (GRACE) Drought Indicators

Earth's gravity field varies in space and time as a result of heterogeneities and movements of mass at the surface, including redistribution of TWS. GRACE detects these gravitational variations as they perturb the orbits of its twin satellites (Tapley et al. 2004; Wahr et al. 2004) and uses them to infer monthly changes in TWS at regional scales (>150,000 km^2) (Swenson et al. 2006). In addition to its coarse spatial and temporal resolutions, GRACE alone cannot separate changes in groundwater, soil moisture, surface waters, and snow/ice (Rodell and Famiglietti 1999). Zaitchik et al. (2008) proposed a data assimilation method based on the CLSM (Koster et al. 2000) to downscale and vertically decompose GRACE-based TWS. Later, Houborg et al. (2012) applied this data assimilation approach to GRACE-derived TWS and produced drought indicators for surface soil moisture (SFSM), root-zone soil moisture (RTZSM), and groundwater storage (GWS), which conformed to the percentile ranges proposed by the USDM (Table 5.1). SFSM and RTZSM are indicative of agricultural drought, whereas GWS can be used to map the extent and severity of hydrological drought. These experimental GRACE-based drought/wetness maps are now produced weekly at 0.125° resolution for the continental United States. They are used as an input to the USDM and are disseminated weekly via this website, http://drought.unl.edu/monitoringtools/nasagracedataassimilation.aspx.

5.2.7 Study Area

The study area is the Southeastern United States, where a humid warm temperate climate is prevalent according to Köppen–Geiger climate classification (Kottek et al. 2006). The land cover is mainly dominated by forests (mostly deciduous), cultivated crops, and hay/pasture according to the National Land Cover Database 2011 (NLCD 2011). Summers are characteristically hot and wet with frequent thundershowers. Evaporative demand is high during summers, which makes the region very susceptible to drought when seasonal rainfall is delayed.

Basins in the study area (Figures 5.1 and 5.2) were retrieved from the website of the Watershed Boundary Dataset (WBD) (http://nhd.usgs.gov/wbd.html) to compare the drought indicators on the basin level. The WBD contains boundaries of drainage areas developed by the collaborative effort among the U.S. federal agencies in consistent with national federal standards, and topographic and hydrologic features across the United States and territories (U.S. Geological Survey and the U.S. Department of Agriculture, Natural Resources Conservation Service 2013). Each basin in the WBD is defined as the level 3 hydrological unit and assigned a unique identifier, hydrological unit code (HUC). In this chapter, we follow the naming conventions of hydrological units established in the WBD, region (level 1), basin (Level 3), and watershed (level 5), in the descending order with respect to areal size.

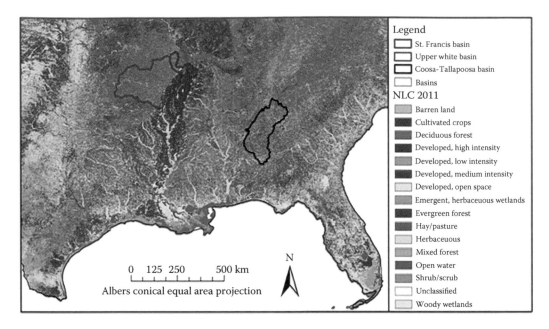

FIGURE 5.1 Study area and boundaries of basins defined in the Watershed Boundaries Dataset (WBD). The background image is the land cover/land use subset from the National Land Cover Database 2011. (Homer, C. et al., *Photogramm. Eng. Remote Sen.*, 81, 345–354, 2015.)

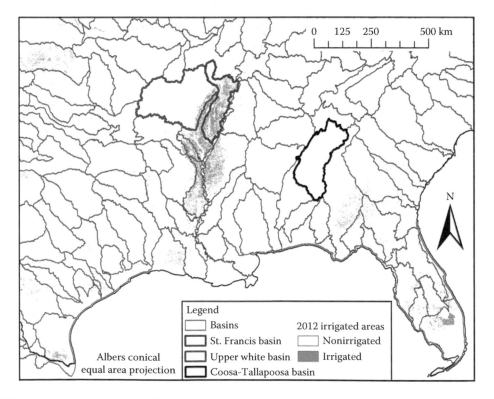

FIGURE 5.2 Irrigated areas in 2012 with respect to basins in the study area. Irrigation map is the subset of the MODIS-based Irrigated Areas Database. (Brown, J.F. and Pervez, M.S., *Agric. Syst.*, 127, 28–40, 2014.)

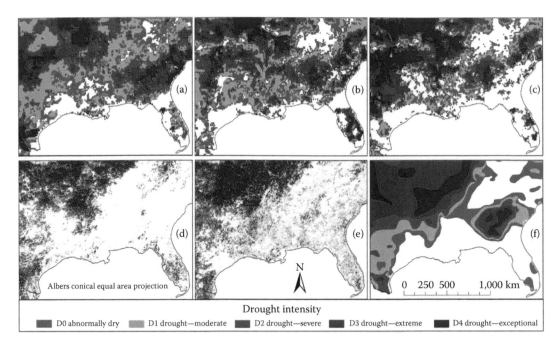

Drought intensity

■ D0 abnormally dry   ■ D1 drought—moderate   ■ D2 drought—severe   ■ D3 drought—extreme   ■ D4 drought—exceptional

FIGURE 5.3 Drought maps of (a) GRACE-based GWS, (b) RTZSM, (c) SFSM, (d) MODIS-based TCI, (e) VCI, and (f) USDM. The USDM drought map is valid from August 7 to August 13, 2012, and all other maps are valid between August 6 and August 12, 2012.

Various crops such as corn, soybeans, rice, winter wheat, sorghum, cotton, and peanuts are grown in the study area, particularly in the Lower Mississippi region along the Mississippi river (Figure 5.1). During hot seasons, crops are irrigated to support crop growth and ensure high crop yields, and irrigation is primarily concentrated over the Lower Mississippi region (Figure 5.3) according to the irrigation map, extracted from the MODIS Irrigated Agriculture Dataset for the United States (MIrAD-US). Pervez and Brown (2010) developed a geospatial model by combining remote sensing inputs such as MODIS-NDVI and NLCD products with United States Department of Agriculture (USDA) Census of Agriculture irrigated area statistics to produce 2012 irrigated agriculture areas dataset at 250 m resolution.

5.3   RESULTS

The spatial extent and severity of the 2012 drought are mapped by all drought indicators as described in Section 5.2. The identical classification scheme (Table 5.1) is employed to identify drought-affected regions and quantify severity of drought, ensuring that they are all in the same units. Therefore, percentile-based classification allows us to visually and quantitatively analyze the drought results and draw meaningful conclusions. Visual comparison is necessary to analyze the spatial extent of drought reported by all drought indicators, whereas quantitative examination enables to intercompare results with respect to drought onset, end, and intensity. It is crucial to reemphasize that drought maps based on GWS percentiles are an indicator of hydrological drought, whereas VCI-, TCI-, RTZSM- and SFSM-based drought maps provide agricultural drought conditions. On the other hand, USDM-based drought maps collectively contain information about hydrological, meteorological, and agricultural drought.

5.3.1 Spatial Representation of Drought

GRACE- and MODIS-based maps are shown side-by-side in Figure 5.3 along with the USDM map on August 6, 2012. These maps are valid for the week of August 6–12, 2012, except that USDM map is valid for the week of August 7–13, 2012. Good correspondence between TCI- and VCI-based maps was observed, although VCI indicated relatively large drought extent. Both maps were also generally in good agreement with the USDM map and GRACE-SFSM, although they displayed more extensive drought extent than MODIS-based drought indices. One stark discrepancy among all indicators was seen in Georgia where both GRACE-derived indices and USDM suggested severe-to-exceptional agricultural drought, whereas VCI and TCI did not indicate any drought. Over Central United States, drought extent reported by all indicators was in complete agreement. Of all the indicators, the largest drought extent was reported by GRACE-GWS and GRACE-RTZSM on August 6, 2012 (Figure 5.3).

Another disagreement in indices was observed over the Lower Mississippi region where the land is cultivated for agricultural production. Crops in this region were irrigated in 2012 according to irrigated agriculture map (Figure 5.2). Over this region, VCI did not report widespread reduced vegetation activity (Figure 5.4), and TCI did not indicate elevated LST in comparison to other years, both indicating a response of the respective index to the irrigation signal. On the other hand, severe-to-exceptional drought was reported in the USDM- and GRACE-derived SFSM over the St. Francis basin (Figure 5.4), which is more representative of the natural hydrometeorological conditions in 2012 in nonirrigated locations.

According to GRACE-based maps, groundwater, root zone, and surface soil moisture all deviated negatively from their historical averages throughout the study area, further signaling both agricultural and hydrological drought throughout Southeastern United States. In Georgia where VCI and TCI did not detect drought on August 6, 2012, both USDM- and GRACE-based drought indicators detected severe-to-exceptional drought. Over irrigated agriculture of the Lower Mississippi region, GRACE-based drought indicators were in agreement with USDM but not with the MODIS-based indicators (Figures 5.3 and 5.4). Disagreements between MODIS and GRACE indices were generally situated along the Appalachian Mountains (e.g., Blue Ridge mountains, and Ridge and Valley), Piedmont Plateau, and Atlantic Coastal Plains. Over these regions, GRACE drought indicators reported severe-to-exceptional groundwater and soil moisture depletion in 2012. Drought reported

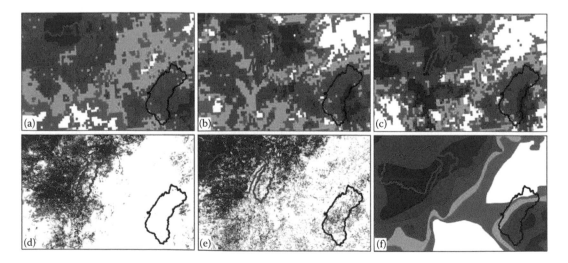

FIGURE 5.4 Close-up view of the drought maps of (a) GRACE-based GWS, (b) RTZSM, (c) SFSM, (d) MODIS-based TCI, (e) VCI, and (f) USDM over three basins on August 6, 2012 (USDM map is on August 7, 2012). Basin names are given in Figures 5.1 and 5.2.

Precipitation percent of average
June–August 2012
Averate period: Twentieth Century

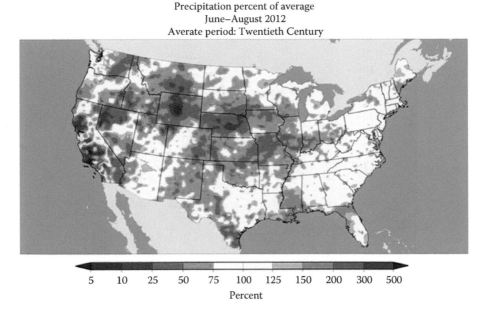

5 10 25 50 75 100 125 150 200 300 500
Percent

FIGURE 5.5 Three-month percent of normal precipitation for the time period of June–August, 2012. (From NOAA-National Climatic Data Center, 3-month percent of normal precipitation. National temperature and precipitation maps, http://www.ncdc.noaa.gov/monitoring-content/sotc/national/grid-prcp/prcp-pon-201206-201208.gif., 2012. With permission.)

by GRACE-SFSM was not seen in VCI and TCI maps along the Appalachian Mountains. Broadly, discrepancies between GRACE-SFSM and MODIS indices seemed to be concentrated over highly elevated areas along the Appalachian Mountains (i.e., Blue Ridge Mountains).

There is a well-known lagged response of vegetation (i.e., NDVI) to precipitation (Di et al. 1994), and Ji and Peters (2003) suggested 3-month lag of NDVI to precipitation deficit. For this reason, 3-month percent of normal precipitation for the time period of June–August of 2012 (Figure 5.5) was retrieved from the NOAA's National Climatic Data Center (NCDC) (http://www.ncdc.noaa. gov/temp-and-precip/). This precipitation deficit map broadly matched drought extent indicated by VCI on August 6, 2012, whereas smaller drought extent was reported by TCI. Both USDM and GRACE-SFSM indicated comparatively larger drought extent.

5.3.2 DROUGHT INTENSITY

Apart from analysis of spatial extent of drought, quantitative examination of drought intensity is essential to reveal similarities and differences across indices. The comparison is conducted based on the basin-level averages of drought indicators. Time series of these averages for each drought indicator are formed and plotted together in Figures 5.6 and 5.7, and their results are discussed. The location of three basins in the study area: Coosa-Tallapoosa (HUC6 = 031501), St. Francis (HUC6 = 080202), and Upper White (HUC6 = 110100) can be seen in Figures 5.1 and 5.2. Coosa-Tallapoosa basin was selected for analysis because MODIS-based drought indicators did not indicate any drought on August 6, 2012, in contrast to USDM- and GRACE-derived indicators (Figure 5.4). St. Francis basin was impacted by the irrigation signal seen only in VCI and TCI, and all drought indicators were in good agreement in Upper White basin. Using basin boundaries, time series of VCI, TCI, RTZSM, SFSM, and GWS were constructed between April 30, 2012 and October 1, 2012 on a weekly basis (Figures 5.6 and 5.7).

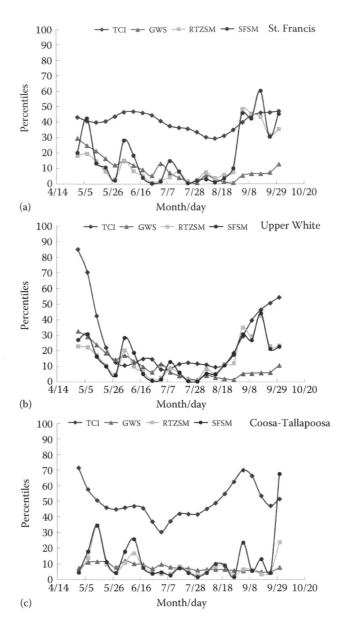

FIGURE 5.6 Basin averages of Vegetation Condition Index (VCI), groundwater storage (GWS), root-zone soil moisture (RTZSM), and surface soil moisture (SFSM) percentiles in (a) St. Francis, (b) Upper White, and (c) Coosa-Tallapoosa between April 30 and October 1, 2012.

The results (Figure 5.6a) show that VCI was relatively constant above the drought threshold (>30, Table 5.1) in St. Francis basin throughout 2012 where agriculture is irrigated (Figure 5.2). Similarly, VCI did not report any drought throughout the 2012 growing season in Coosa-Tallapoosa basin where precipitation deficit was not seen between June and August of 2012 (Figure 5.5). However, TCI fluctuated substantially around the drought threshold throughout 2012 in St. Francis basin (Figure 5.7a) unlike Upper White (Figure 5.7b), indicating drought from May 14 to May 27, no drought from May 28 to June 17, drought from June 18 to July 8, and no drought from July 9 to July 15. Moreover, TCI reported drought during the late June and early July of 2012 (Figure 5.7c) and at other times, no drought was indicated by TCI in Coosa-Tallapoosa basin. From early June to

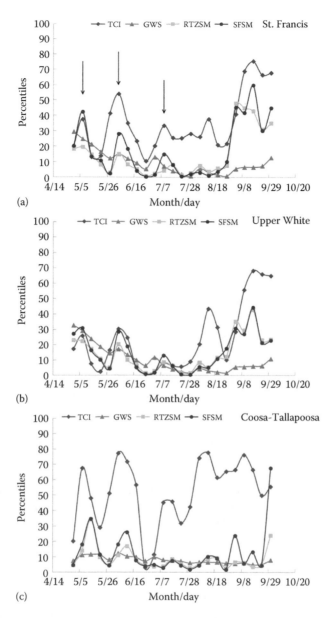

FIGURE 5.7 Basin averages of Temperature Condition Index (TCI), groundwater storage (GWS), root-zone soil moisture (RTZSM), and surface soil moisture (SFSM) percentiles in (a) St. Francis, (b) Upper White, and (c) Coosa-Tallapoosa between April 30 and October 1, 2012.

late August in 2012, good correspondence was observed between all GRACE-based and MODIS-based drought indicators in Upper White basin (Figures 5.6b and 5.7b), identifying drought conditions. GRACE-derived indicators implied that all three basins experienced severe-to-exceptional drought during the 2012 growing season.

Correlation analysis was conducted using the time series of drought indicators in 2012. Each time series is composed of 23 weekly observations spanning from April 30 to October 1, 2012. The results revealed that TCI had higher statistically significant relationship at 0.01 significance level with both SFSM and RTZSM than GWS in St. Francis and Upper White basins (Table 5.2). TCI did not display any relation to groundwater variations in all basins. On the other hand, VCI exhibited

TABLE 5.2

Correlation Coefficients (r) between VCI, TCI, SFSM, RTZSM, and GWS in St. Francis, Upper White, and Coosa-Tallapoosa Basins

N = 23	St. Francis		Upper White		Coosa-Tallapoosa	
r = 0.53	VCI	TCI	VCI	TCI	VCI	TCI
GWS	0.46	−0.08	0.64	−0.13	−0.26	−0.08
RTZSM	0.43	0.75	0.67	0.71	0.07	0.17
SFSM	0.52	0.78	0.67	0.64	0.17	0.23
TCI	0.44		0.44		0.40	

Note: Time series are composed of observations between April 30 and October 1, 2012. Statistically significant *r* at 0.01 significance level ($\alpha = 0.01$) are underlined. The critical *r* value is 0.53 at 0.01 significance level.

statistically significant relationship with GWS, RTZSM, and SFSM only in Upper White basin. Finally, there was no statistically significant correlation among any MODIS- and GRACE-based indicators in Coosa-Tallapoosa basin.

Lagged response of NDVI and NDVI-based drought indices to soil moisture at various depths up to 100 cm was reported by other studies (Adegoke and Carleton 2002; Peng et al. 2014) such that response of plants to soil moisture changes is not concurrent, rather exhibits some time lag. Time lags up to 7 weeks are considered, and additional basin averages of GRACE-derived GWS, RTZSM, and SFSM are computed starting from January 16 until October 1, 2012, ensuring that correlation coefficients are always computed from 23 weekly observations of all drought indicators, and the time period matches the growing season when vegetation is not dormant (i.e., April 30 to October 1). The results (Table 5.3) show that correlations among drought indicators improved considerably, thus suggesting that VCI exhibited lagged response to changes in surface and root-zone soil moisture in St. Francis and Upper White basins. On the other hand, no lag was found between TCI- and GRACE-based RTZSM and SFSM, thus suggesting that LST varies simultaneously with SFSM and RTZSM during dry years. Again, there was no significantly lagged relationship among all indicators in Coosa-Tallapoosa basin. Overall, VCI lagged behind RTZSM and SFSM about 2 weeks in St. Francis and Upper White basins. Therefore, TCI responded to changes in SFSM and RTZSM more quickly than VCI in St. Francis and Upper White basins. Furthermore, the results pointed out that VCI and TCI had a positive relationship in all basins, yet only statistically significant at 0.01

TABLE 5.3

The Lags and Their Correlation Coefficients (r) between VCI, TCI, SFSM, RTZSM, and GWS in St. Francis, Upper White, and Coosa-Tallapoosa Basins

N = 23	St. Francis				Upper White				Coosa-Tallapoosa			
r = 0.53	VCI		TCI		VCI		TCI		VCI		TCI	
	Lag	r	lag	r	lag	r	lag	r	lag	r	lag	r
GWS	0	0.46	0	−0.08	0	0.64	0	−0.13	7	0.14	0	−0.08
RTZSM	2	0.55	0	0.75	2	0.87	0	0.71	0	0.07	0	0.17
SFSM	1	0.57	0	0.78	2	0.83	0	0.64	0	0.17	0	0.23
TCI	1	0.46			3	0.84			1	0.50		

Note: Statistically significant *r* and lag at 0.01 significance level ($\alpha = 0.01$) are underlined. The critical *r* value is 0.53 at 0.01 significance level.

TABLE 5.4

The Lags and Their Correlation Coefficients (r) among GRACE-Derived Drought Indicators in St. Francis, Upper White, and Coosa-Tallapoosa Basins

$N = 31$	SFSM					
$r = 0.46$	St. Francis		Upper White		Coosa-Tallapoosa	
	lag	*r*	lag	*r*	lag	*r*
RTZSM	0	0.93	0	0.97	0	0.75
GWS	4	0.89	4	0.89	4	0.69
	RTZSM					
GWS	5	0.94	5	0.93	3	0.81

Note: The critical r value is 0.46 at 0.01 significance level.

level in Upper White basin (Table 5.3). Time delay of 3 weeks between VCI and TCI was observed in Upper White basin.

The correlation analysis among GRACE-derived SFSM, RTZSM, and GWS revealed that SFSM was strongly correlated with RTZSM and GWS in all basins (Table 5.4), although relationship was relatively less strong in Coosa-Tallapoosa basin in 2012. SFSM relation to RTZSM was concurrent, whereas a time lag of 4 weeks was observed between SFSM and GWS in all basins (Table 5.4). The results also suggested that there was a strong lagged relationship between RTZSM and GWS in all basins, and the lag was 5 weeks in St. Francis and Upper White basin and 3 weeks in Coosa-Tallapoosa basin.

5.4 DISCUSSION

Over irrigated agriculture in the Lower Mississippi region, VCI did not report any drought although USDM clearly indicated drought in 2012. Especially in St. Francis basin, VCI provided more consistent results as opposed to TCI because LST responds more rapidly to prevailing weather conditions and irrigation events than NDVI. Furthermore, there was no discernible variation in SFSM, RTZSM, and GWS unlike that observed in TCI over irrigated fields of St. Francis basin. It can be concluded that when agricultural fields were irrigated in 2012, LST decreased rapidly, and subsequently TCI signaled no drought. When the surface became dry before the next irrigation event, TCI reported drought after the sudden increase in LST (Figure 5.7a). In conclusion, discrepancy between MODIS- and GRACE-based results in St. Francis can be easily explained by irrigation, where irrigation is not considered in the decomposition of GRACE-based TWS into SFSM, RTZSM, and GWS (Houborg et al. 2012).

Correlation analysis revealed that the relationship between VCI- and GRACE-based SFSM and RTZSM is not concurrent, but rather lagged in St. Francis and Upper White basins, whereas TCI had concurrent positive relationships with both GRACE-derived SFSM and RTZSM. VCI exhibited roughly a 2-week lag to surface and root-zone soil moisture in 2012. Such conclusions with NDVI-based indices were achieved by other studies (Adegoke and Carleton 2002; Peng et al. 2014), as well. Correlations between VCI and other drought indicators were statistically significant at 99% confidence level and improved considerably when the lag effect is taken into consideration in St. Francis and Upper White basins. However, the results of the correlation analysis in St. Francis basin should be interpreted with caution because the transfer of groundwater to surface through irrigation and subsequently infiltration of that water down to root zone are not explicitly handled in CLSM. Besides, the land is heavily subject to anthropogenic effects (e.g., irrigation, harvesting of crops, and farming practices), and timing of these events can vary annually. Therefore, such drivers could be partly responsible for poorer correlation of VCI to SFSM, RTZSM, and TCI in St. Francis basin in comparison to Upper White basin. In Coosa-Tallapoosa, no statistically significant relationship observed between VCI and TCI could be attributed to frequent thundershowers, a common weather activity in summer across this region. We demonstrated that TCI fluctuated substantially throughout the 2012 growing season as opposed to VCI because LST responds to wetting events (e.g., irrigation and thundershowers) more quickly than NDVI.

We theorize that the timing of irrigation events can be detected by LST or TCI in which LST responds rapidly to irrigation event as sharp changes were seen in TCI time series in St. Francis as opposed to Upper White basin. The methodology developed by Pervez and Brown (2010) only decides whether or not a pixel is irrigated, but it does not supply any information about the timing of watering events. We suspect that sudden changes in the time series could be a sign of irrigation as depicted with arrows in Figure 5.7a. However, LST products must be combined with MIrAD irrigation dataset to eliminate likely errors because sharp fluctuations observed in Coosa-Tallapoosa (Figure 5.7c) could lead to false positives (i.e., Type I error). More research is needed to validate our claim.

Utility of VCI to monitor meteorological drought was investigated by Quiring and Ganesh (2010); however, we demonstrated that although USDM indicated drought conditions (i.e., meteorological drought) over irrigated agriculture in the Lower Mississippi region, drought was not reported by VCI during the 2012 growing season (Figure 5.6a). Therefore, VCI may not be a reliable indicator of meteorological drought, but agricultural drought.

Our analysis of the 2012 drought in the Southeastern United States demonstrated that the agreements and disagreements over the extent and intensity of the 2012 drought exist among USDM-, GRACE- and MODIS-based drought indicators. We demonstrated that precipitation between June and August (Figure 5.5) was at normal levels in which disagreements between MODIS, GRACE, and USDM were seen over Georgia. In addition, two principal factors, irrigation and lagged response of vegetation to variations in soil moisture, could be partially responsible for these disagreements. Another factor that may contribute to these disagreements is the type of drought reported by these indicators such that GRACE-GWS is a measure of hydrological drought indicator, whereas the rest could be more suitable in depicting agricultural drought conditions.

5.5 CONCLUSIONS

USDM-, GRACE-, and MODIS-based drought maps were successful in depicting the drought of 2012 despite disagreements over its extent and intensity, and they all indicated that Southeastern United States experienced severe-to-exceptional drought in 2012. Both MODIS-based and GRACE-SFSM drought maps closely mimicked the surface conditions depicted in the USDM maps except over irrigated areas, Georgia, and along the Appalachian Mountains (e.g., Blue Ridge mountains, and Ridge and Valley). However, short-term precipitation deficit map agreed with MODIS indices in these regions, indicating normal precipitation conditions compared to long-term average conditions. GRACE-based GWS implied that majority of the Southeastern United States experienced moderate-to-extreme hydrological drought, thus suggesting that groundwater sources severely depleted during the drought of 2012. We demonstrated that disagreements over the extent and intensity of the 2012 drought across all drought indicators could result from irrigation, complex lagged response of vegetation to precipitation and soil moisture, and the type of drought these indicators report (e.g., meteorological, agricultural, and hydrological drought).

At present, the main challenge is to extend the length of MODIS-NDVI and MODIS-LST observations with historical records from the AVHRR sensors. However, AVHRR products still suffer from inherent problems attributed to sensor changes and degradation, volcanic eruptions, and satellite orbital drift (Huete et al. 2002; Gallo et al. 2005) despite numerous efforts to correct these issues (Tucker et al. 2005). In addition, wider bandwidths and coarser spatial resolution of AVHRR spectral bands than MODIS and difficulties in applying atmospheric correction on visible and thermal bands further complicate such endeavor. The follow-on sensor to AVHRR and MODIS instruments is the Visible Infrared Imaging Radiometer Suite (VIIRS) onboard the Suomi National Polar-orbiting Partnership (Suomi NPP) spacecraft. Fortunately, the VIIRS instrument ensures the data continuity in NDVI and LST records because the specification of VIIRS bands closely resembles MODIS bands (Gallo et al. 2005).

At the time of writing, the GRACE mission has survived well beyond its original design lifetime of 5 years. Due to degradation of the onboard battery cells, the instruments are powered down for roughly 2 out of every 5 months, resulting in a loss of data. During those months, the GRACE-based drought indicators rely more heavily on standard meteorological inputs to the model. GRACE's orbit is slowly decaying such that mission operations are expected to end no later than late 2017. Fortunately, the GRACE follow-on mission is scheduled for launch in February 2018. That mission will provide gravimetric measurements and derived TWS data that are similar to those of GRACE but with some small to moderate improvement in spatial resolution and accuracy. Hence, production of the GRACE-based drought indicators will be able to continue. In addition, ongoing research projects include the goals of expanding the drought indicators to the global scale, using them as the starting point for 30–90-day drought forecasts, and integrating other satellite observations in addition to those of GRACE through multivariate data assimilation.

Future research in satellite drought monitoring will focus on the products of two new missions, Soil Moisture Active Passive (SMAP), and Global Precipitation Measurement (GPM). SMAP and GPM missions provide spatially continuous, but coarse, soil moisture and precipitation estimates across the Earth's surface, which can be downscaled or assimilated into Land Surface Models (LSM) to derive finer resolution products. Later, in a similar fashion to VCI and TCI, drought indicators can be calculated from these products. In this case, 6 or more years of data as well as one drought and wet year within that time period are required to generate reliable drought information.

ACKNOWLEDGMENTS

This research was supported by an appointment to the NASA Postdoctoral Program at the Goddard Space Flight Center (GSFC), administered by Universities Space Research Association (USRA) under a contract, NNH15C048B.

REFERENCES

Adegoke, J. O. and A. M. Carleton. 2002. Relations between soil moisture and satellite vegetation indices in the U.S. Corn Belt. *Journal of Hydrometeorology* 3(4): 395–405.

Anderson, M. C. and W. Kustas. 2008. Thermal remote sensing of drought and evapotranspiration. *Eos, Transactions American Geophysical Union* 89(26): 233. doi:10.1029/2008EO260001.

Brown, J. F., B. Wardlow, T. Tadesse, M. J. Hayes, and B. C. Reed. 2008. The vegetation drought response index (VegDRI): A new integrated approach for monitoring drought stress in vegetation. *GIScience & Remote Sensing* 45(1): 16–46. doi:10.2747/1548-1603.45.1.16.

Brown, J. F. and M. S. Pervez. (2014). Merging remote sensing data and national agricultural statistics to model change in irrigated agriculture. *Agricultural Systems* 127: 28–40. https://doi.org/10.1016/j.agsy.2014.01.004.

Deng, M., L. Di, W. Han, A. L. Yagci, C. Peng, and G. Heo. 2013. Web-service-based monitoring and analysis of global agricultural drought. *Photogrammetric Engineering & Remote Sensing* 79(10): 929–943. doi:10.14358/PERS.79.10.929.

Di, L., D. C. Rundquist, and L. Han. 1994. Modelling relationships between NDVI and precipitation during vegetative growth cycles. *International Journal of Remote Sensing* 15(10): 2121–2136. doi:10.1080/01431169408954231.

Gallo, K., L. Ji, B. Reed, J. Eidenshink, and J. Dwyer. 2005. Multi-platform comparisons of MODIS and AVHRR normalized difference vegetation index data. *Remote Sensing of Environment* 99(3): 221–231. doi:10.1016/j.rse.2005.08.014.

Homer, C., J. Dewitz, L. Yang, S. Jin, P. Danielson, G. Xian, J. Coulston, N. Herold, J. Wickham, and K. Megown. 2015. Completion of the 2011 National Land Cover Database for the Conterminous United States – Representing a Decade of Land Cover Change Information. *Photogrammetric Engineering & Remote Sensing* 81(5): 345–354.

Houborg, R., M. Rodell, B. Li, R. Reichle, and B. F. Zaitchik. 2012. Drought indicators based on model-assimilated Gravity Recovery and Climate Experiment (GRACE) terrestrial water storage observations. *Water Resources Research* 48(7): W07525. doi:10.1029/2011WR011291.

Huete, A. R., K. Didan, T. Miura, E. P. Rodriguez, X. Gao, and L. G. Ferreira. 2002. Overview of the radio-metric and biophysical performance of the MODIS vegetation indices. *Remote Sensing of Environment* 83(1–2): 195–213. doi:10.1016/S0034-4257(02)00096-2.

Ji, L. and A. J. Peters. 2003. Assessing vegetation response to drought in the northern great plains using vegetation and drought indices. *Remote Sensing of Environment* 87(1): 85–98. doi:10.1016/S0034-4257(03)00174-3.

Kogan, F. 1997. Global drought watch from space. *Bulletin of the American Meteorological Society* 78(4): 621–636.

Kogan, F. 2001. Operational space technology for global vegetation assessment. *Bulletin of the American Meteorological Society* 82(9): 1949–1964.

Koster, R. D., M. J. Suarez, A. Ducharne, M. Stieglitz, and P. Kumar. 2000. A catchment-based approach to modeling land surface processes in a general circulation model: 1. Model structure. *Journal of Geophysical Research: Atmospheres* 105(D20): 24809–24822. doi:10.1029/2000JD900327.

Kottek, M., J. Grieser, C. Beck, B. Rudolf, and F. Rubel. 2006. World map of the Köppen-Geiger climate classification updated. *Meteorologische Zeitschrift* 15(3): 259–263. doi:10.1127/0941-2948/2006/0130.

The National Drought Mitigation Center. 2016. U.S. drought monitor classification scheme. United States Drought Monitor. http://droughtmonitor.unl.edu/AboutUs/ClassificationScheme.aspx.

NOAA-National Climatic Data Center. 2012. 3-month percent of normal precipitation. National temperature and precipitation maps. http://www.ncdc.noaa.gov/monitoring-content/sotc/national/grid-prcp/prcp-pon-201206-201208.gif.

Peng, C., M. Deng, and L. Di. 2014. Relationships between remote-sensing-based agricultural drought indicators and root zone soil moisture: A comparative study of Iowa. *IEEE Journal of Selected Topics in Applied Earth Observations and Remote Sensing* 7(11): 4572–4580. doi:10.1109/JSTARS.2014.2344115.

Pervez, M. S. and J. F. Brown. 2010. Mapping irrigated lands at 250-M scale by merging MODIS data and national agricultural statistics. *Remote Sensing* 2(10): 2388–2412. doi:10.3390/rs2102388.

Quiring, S. M. and S. Ganesh. 2010. Evaluating the utility of the Vegetation Condition Index (VCI) for monitoring meteorological drought in Texas. *Agricultural and Forest Meteorology* 150(3): 330–339. doi:10.1016/j.agrformet.2009.11.015.

Rippey, B. R. 2015. The U.S. drought of 2012. *Weather and Climate Extremes*, 10: 57–64. doi:10.1016/j.wace.2015.10.004.

Rodell, M. 2012. Satellite gravimetry applied to drought monitoring. In B. D. Wardlow, M. C. Anderson, and J. P. Verdin (Eds.) Remote Sensing of Drought, *Drought and Water Crises*, pp. 261–278. Boca Raton, FL: CRC Press.

Rodell, M. and J. S. Famiglietti. 1999. Detectability of variations in continental water storage from satellite observations of the time dependent gravity field. *Water Resources Research* 35(9): 2705–2723. doi:10.1029/1999WR900141.

Solano, R., K. Didan, A. Jacobson, and A. R. Huete. 2010. MODIS vegetation index (MOD13) C5 user's guide version 2. Vegetation Index and Phenology Lab, The University of Arizona, Tucson, AZ. http://vip.arizona.edu/documents/MODIS/MODIS_VI_UsersGuide_01_2012.pdf.

Svoboda, M., D. Lecomte, M. Hayes, R. Heim, K. Gleason, J. Angel, B. Rippey et al. 2002. The drought monitor. *Bulletin of the American Meteorological Society* 83(8): 1181–1190.

Swain, S., B. D. Wardlow, S. Narumalani, T. Tadesse, and K. Callahan. 2011. Assessment of vegetation response to drought in Nebraska using Terra-MODIS land surface temperature and normalized difference vegetation index. *GIScience & Remote Sensing* 48(3): 432–455. doi:10.2747/1548-1603.48.3.432.

Swenson, S., P. J.-F. Yeh, J. Wahr, and J. Famiglietti. 2006. A comparison of terrestrial water storage variations from GRACE with in situ measurements from Illinois. *Geophysical Research Letters* 33(16): L16401. doi:10.1029/2006GL026962.

Tadesse, T., J. Brown, and M. Hayes. 2005. A new approach for predicting drought-related vegetation stress: Integrating satellite, climate, and biophysical data over the U.S. Central Plains. *ISPRS Journal of Photogrammetry and Remote Sensing* 59(4): 244–253. doi:10.1016/j.isprsjprs.2005.02.003.

Tapley, B. D., S. Bettadpur, M. Watkins, and C. Reigber. 2004. The gravity recovery and climate experiment: Mission overview and early results. *Geophysical Research Letters* 31(9): L09607. doi:10.1029/2004GL019920.

Tucker, C. J. and B. J. Choudhury. 1987. Satellite remote sensing of drought conditions. *Remote Sensing of Environment* 23(2): 243–251. doi:10.1016/0034-4257(87)90040-X.

Tucker, C. J., J. Pinzon, M. Brown, D. Slayback, E. Pak, R. Mahoney, E. Vermote, and N. El Saleous. 2005. An extended AVHRR 8-Km NDVI dataset compatible with MODIS and SPOT vegetation NDVI data. *International Journal of Remote Sensing* 26(20): 4485–4498. doi:10.1080/01431160500168686.

U.S. Geological Survey and the U.S. Department of Agriculture, Natural Resources Conservation Service. 2013. Techniques and methods 11–A3. In *Federal Standards and Procedures for the National Watershed Boundary Dataset (WBD)*, 4th ed., 63. Reston, VA: U.S. Geological Survey and the U.S. Department of Agriculture, Natural Resources Conservation Service. http://pubs.usgs.gov/tm/11/a3/.

Wahr, J., S. Swenson, V. Zlotnicki, and I. Velicogna. 2004. Time-variable gravity from GRACE: First results. *Geophysical Research Letters* 31(11): L11501. doi:10.1029/2004GL019779.

Wan, Z. 2007. *Collection-5 MODIS Land Surface Temperature Products Users' Guide*. Santa Barbara, CA: ICESS, University of California. http://www.icess.ucsb.edu/modis/LstUsrGuide/MODIS_LST_products_Users_guide_C5.pdf.

Wilhite, D. A. 2000. Drought as a natural hazard: Concepts and definitions. In D. A. Wilhite (Ed.) *Drought: A Global Assessment, 1st ed. Hazards and Disasters*. pp. 3–18. London, UK: Routledge.

Yagci, A. L., L. Di, and M. Deng. 2013. The effect of land-cover change on vegetation greenness-based satellite agricultural drought indicators: A case study in the southwest climate division of Indiana, USA. *International Journal of Remote Sensing* 34(20): 6947–6968. doi:10.1080/01431161.2013.810824.

Yagci, A. L., L. Di, M. Deng, W. Han, and C. Peng. 2011. Agricultural drought monitoring from space using freely available MODIS data. In *Proceedings of 18th William T. Pecora Memorial Remote Sensing Symposium*. Herndon, VA: ASPRS.

Yagci, A. L., L. Di, M. Deng, G. Yu, and C. Peng. 2012. Global agricultural drought mapping: Results for the year 2011. In *2012 IEEE International Geoscience and Remote Sensing Symposium*, 3764–3767. Munich, Germany: IEEE. doi:10.1109/IGARSS.2012.6350498.

Zaitchik, B. F., M. Rodell, and R. H. Reichle. 2008. Assimilation of GRACE terrestrial water storage data into a land surface model: Results for the Mississippi river basin. *Journal of Hydrometeorology* 9(3): 535–548. doi:10.1175/2007JHM951.1.

Section II

---

*Remote Sensing of Frost and
Sea Ice Hazards*

6 Frost and Remote Sensing
An Overview of Capabilities

Nicolas R. Dalezios and George P. Petropoulos

CONTENTS

6.1 INTRODUCTION

Frost is a natural environmental risk, which occurs when the air temperature at the surface becomes equal to 0°C or below (Webb and Snyder, 2013). One of the areas directly affected by frost with disastrous results is agriculture (WMO, 2010; IPCC, 2013). Indeed, the occurrence of frost in active growth may cause major damage, even the total loss of production. In addition, frost may cause injuries to sensitive crops in which the extent of injuries depends on frost type, severity, frequency, and duration. Frost marks the end of growth for the plant, because it can freeze plant tissue; thus, it is considered significant, because it can occur at either end of the growing season (Figure 6.1). Thus, frost crop damages result in significant economic losses and are one of the most important threats to farmers (Louka et al., 2015). Moreover, the impact of a severe winter extends beyond economic loss: indeed, production losses in industry, crop losses, transportation losses in revenue, losses in retail sales, and losses resulting from increased energy consumption during cold winters. Frost forecasting and monitoring help farmers to reduce any possible crop injuries. In addition, frost protection methods help farmers to combat frost. At present, there is clearly a global requirement for more and better spatial and temporal information on frost occurrence and its associated risk (Dalezios, 2017). Moreover, there is an increasing recognition of the potential value of meteorological information in decision-making relevant to frost.

FIGURE 6.1   Frost damage on apple trees in the stage of (a) swollen buds, (b) bud burst, (c) green cluster, (d) bloom, (e) petal fall, and (f) fruit set. (From ELGA–GAIO, Estimation manual of frost damage in apple trees, Available: www.elga.grv, 2003. With permission.)

Earth observation (EO) or remote sensing is the acquisition and interpretation of spectral measurements made at a distant location to obtain information about the Earth's surface. During the last decades, the technological advances in the field of remote sensing have resulted in the gradual improvement of the level of accuracy in quantitative assessment of several environmental parameters and variables including frost. At present, EO data and techniques provide direct measurements of land characteristics, which are relevant to frost risk assessment, monitoring, and mitigation, such as vegetative cover, meteorological, environmental, and hydrological parameters (e.g., temperature and evapotranspiration) (EM-DAT, 2012).

The objective of this chapter is to present an overview on the relationship between frost hazard and EO and to discuss the potential and capabilities of this technology in that respect. At first, frost concepts are presented followed by a classification of frost into different types based on several criteria. Then, remote sensing capabilities in frost hazard are discussed. This is followed by a presentation of frost forecasting (before the event), frost monitoring (during the event), and frost assessment (after the event), along with the corresponding remote sensing methods. Specifically, frost quantification is described, which includes frost modeling and forecasting, where remote sensing methods are also presented. Next, frost early warning systems (FEWSs) and the potential of remote sensing in this context are discussed. Subsequently, frost frequency analysis methods with reference to remote sensing are covered. Finally, a discussion on frost assessment in terms of impacts is furnished; mitigation methods are also summarized providing some case studies on the use of EO in frost-related studies.

6.2   FROST CONCEPTS

Frost may be considered either as a climatic condition or a form of mineral, or even a hazard with several human impacts (Webb and Snyder, 2013). At first, frost as a climatic condition refers mainly to hoar frost or white frost that accumulates on the surfaces in places with appropriate temperatures and lack of sunshine. Indeed, the *hoar* implies a gray tone observed on objects covered with frost. Hoar frost is indicative of the following three conditions:

1. Surfaces on which the frost forms must be 0°C or below.
2. The surrounding air is saturated at 0°C or slightly below.
3. Nuclei are present so that the process of sublimation can take place.

Second, frost as a form of mineral refers to a solid phase of water, because it crystallizes according to the hexagonal system featuring six-sided plates, needles, clusters, and columns. Finally, frost as a hazard implies near 0°C temperatures. Thus, a plant may suffer *chilling* injury but not as a result of frost or freezing. In general, frost is a symptom of a climatic condition in which temperatures have been reduced through radiation or advection to the freezing stage. It is thus clear that, in quantitative terms, frost is a function of temperature.

The term *frost* is used by the public to describe the condition when plants experience freezing injury (again see Figure 6.1 as an example). Specifically, the word *frost* refers to the formation of ice crystals on surfaces, either by freezing of dew or a phase change from vapor to ice (Blanc et al., 1963). Indeed, there are several definitions of frost, such as the following:

1. When the surface temperature drops below 0°C and the existence of low air temperature that causes damage or death to the plants (Ventskevich, 1958).
2. The occurrence of a temperature less than or equal to 0°C measured in a *Stevenson-screen* shelter at a height between 1.25 and 2.0 m (Hogg, 1971).
3. Frost is defined as the condition, which exists when the air temperature near the Earth's surface drops below 0°C (Kalma et al., 1992). A freeze exists when over a widespread region the air remains below freezing (0°C) for a sufficient period of time, which is at least 1 or 2 days.
4. Snyder and Paulo de Melo-Abreu (2005) define frost mainly for tropical areas as follows: "A freeze exists when over a widespread region the air remains below freezing."

Water within plants may or may not freeze during a frost event. Plants freeze that result into intracellular or extracellular freezing. Indeed, intracellular freezing is immediately fatal, whereas extracellular freezing may cause injuries. Moreover, the injury extent depends on plant resistance to frost, which is associated with a critical temperature below which a malfunction or death of the cell takes place.

6.3   FROST CLASSIFICATION SCHEME

The term *remotely sensed frost classification* is used when temperature is measured or estimated through remote sensing data and methods. The frost phenomenon is characterized by varying severity and extent. The scale of the atmospheric mechanisms, which constitutes the driving force of frost occurrence, varies considerably. Frost, as a function of temperature, is generally classified into different types (Kalma et al., 1992), based on certain criteria, and is summarized as follows:

1. *The criterion of frost genesis*: Frost is classified based on the genesis criterion into radiation frost and advection frost, respectively, which are caused by synoptic meteorological conditions. Radiation frost is the result of intense radiation during the night, which causes cooling of the land surface, and is associated with high pressure systems during calm and cloudless nights (Figure 6.2a). Specifically, radiation frosts originate from intense long-wave radiation, which causes cooling of the surface of the earth due to energy loss during calm and clear nights. These types of frost are usually characterized by temperature inversion, clear and calm nights, and usually air temperature higher than

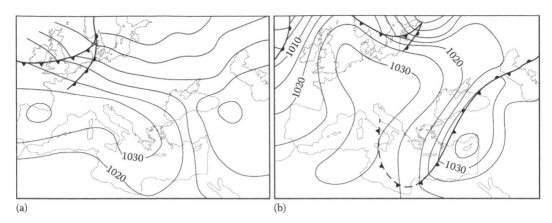

(a)   (b)

FIGURE 6.2 (a) Synoptic surface map with predominantly anticyclonic circulation over Europe favoring radiation frost over Greece. (b) Synoptic surface map showing strong invasion of cold air masses, following the cold front passage and causing advection frost. (Dalezios, N.R. et al., Remote sensing in drought quantification and assessment, In Eslamian, S. (Ed.), *Handbook of Drought and Water Scarcity (HDWS)*, Vol. 1 of 3-Volume, Chapter 21, Taylor & Francis Group, Boca Raton, FL, 2017. With permission.)

0°C during daytime. Indeed, a feature in radiation frosts is usually temperature inversion, which means that the temperature at an average height of about 15 m above the ground is usually higher about 4°C–5°C than the corresponding surface temperature. An example of temperature inversion is presented in Figure 6.3, which shows a valley with trees planted on the valley slopes and deep in the valley, where, during the night, due to radiation the cool air reaches the bottom of the valley, while forcing the warm air to rise upward. On the other hand, advection frost is produced by a sudden and strong invasion of cold air masses often from polar region, which usually follows the passage of a cold front from a region (Figure 6.2b). Advection frosts usually develop during the day or night, and they are also related to high winds and a well-mixed atmosphere with air temperatures below 0°C during daytime. Specifically, these frosts occur in valleys and surface depressions, when radiation frost conditions exist in a region, although invasion of cold air on a local scale leads to sudden temperature drop within the radiation frost area. Needless to say, the most damaging frosts are the advection frosts followed by radiation frosts.

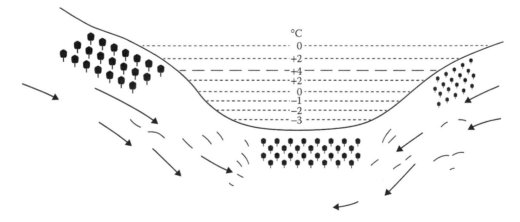

FIGURE 6.3 Formation of frost pocket in a valley. (Dalezios, N.R. et al., Remote sensing in drought quantification and assessment, In Eslamian, S. (Ed.), *Handbook of Drought and Water Scarcity (HDWS)*, Vol. 1 of 3-Volume, Chapter 21, Taylor & Francis Group, Boca Raton, FL, 2017. With permission.)

2. *The criterion of total or partial frost*: Based on duration, frost is characterized as total frost when the maximum daily temperature is equal to 0°C or below and partial frost when the minimum daily temperature is less than or equal to 0°C. If the total frost lasts for several consecutive days, it can have catastrophic consequences, because in such cases frost penetrates into the soil, resulting in the destruction of seeds and roots. In general, the most common type of frost is partial frost, which is equally devastating, especially when it occurs during spring, where the plants are in active growth stage.

3. *The frost criterion of humidity*: Atmospheric humidity is another frost criterion, which may cause condensation of water droplets creating ice crystals (frost), along with temperature drop to negative values. This type of frost is called white frost. On the other hand, when there is no condensation along with the drop in temperature below 0°C, then this type of frost is called black frost.

4. *The criterion of frost severity*: Frost is classified based on the severity criterion into mild frost when the temperature ranges between 0°C and −4°C, moderate frost when the temperature ranges between −4.1°C and −10°C, and severe frost when the temperature ranges from −10.1°C and below.

6.4 REMOTE SENSING IN FROST ANALYSIS: AN OVERVIEW

6.4.1 REMOTE SENSING SYSTEMS AND CAPABILITIES IN FROST ASSESSMENT

In environmental analysis and, in particular, in frost quantification and assessment, two types of (passive) remote sensing systems are considered, namely *meteorological* and *environmental* or *resource* satellites. The main differences between these two types of satellites are their spatial and temporal resolutions, which affect their applications and uses. Specifically, meteorological satellites have a rather coarse spatial resolution, but high temporal reoccurrence, thus, being suitable mainly for operational monitoring applications of frost. On the other hand, environmental satellites are generally characterized by fine spatial resolution but low temporal reoccurrence, being basically used in applications, which do not change dramatically over time, for example, qualitative features of frost, such as the identification of land use/cover types and detection of several frost features (e.g., frost areal extent, frost assessment), FEWSs, and monitoring.

The majority of meteorological satellites operate at heights ranging between 800 and 1,500 km, called low orbit satellites, whereas the rest operate at approximately 36,000 km, called high or geostationary orbit satellites, respectively. The first class consists of the polar or near-polar orbit satellites with a sun-synchronous orbit, because they cross the equator at the same time, such as the series of NOAA-N. Similarly, the second class consists of the so-called geostationary satellites, because they appear to be *stationary* at specific locations over the equator and move in the same direction and at the same rate that Earth is spinning. There are several such satellites, for example, Meteosat and geostationary operational environmental satellite (GOES), that cover the Earth and provide images every 30 min with a spatial resolution of 2.5 km in the visible part of the electromagnetic spectrum, and they are mainly used in operational meteorology, weather forecasting and monitoring.

The first environmental satellite was the Landsat series, with the first one launched in 1971. This satellite is considered as one of the most representative and successful satellites of this type that is still in orbit (with Landsat 8) providing valuable data and information. Essentially, the main feature of the series of environmental satellites is the gradual improvement of their spatial and spectral resolution throughout the years. There are many such satellites in orbit at present, such as ASTER, SPOT, or the recent Sentinel-2 (ESA, 2014), just to mention a few. The trend of further improving the spatial resolution continues reaching the level of microremote sensing in the order of 1 m or smaller with new satellites, such as Quick bird, Ikonos, and WorldView-2/-3.

6.4.2 REMOTE SENSING CAPABILITIES IN FROST ASSESSMENT

The application and utility of EO technology to frost assessment are growing rapidly, mainly due to the increasing number of satellite systems that are launched and due to their continuously improved technical capabilities. Indeed, satellite images and data are consistently available and can be used to detect several frost features. Remote sensing methodologies and techniques can be employed in several aspects of frost, such as vulnerability and damage assessment and warning. The possible contribution of remote sensing could be focused on relief and, possibly, preparedness or warning (e.g., Foot, 1993). However, in many cases remote sensing can make a valuable contribution to disaster prevention in which frequency of observation is not such a prohibitive limitation. In fact, EO data and methods can delineate the spatial and temporal variability of several frost features in quantitative terms (Louka et al., 2015).

A major consideration for development of EO technology for frost assessment and disaster reduction is the extent to which operational users can rely on a continued supply of data. Remote sensing capabilities provide a viable method to offset any loss of information (Jupp et al., 1998). However, remote sensing has been of increasing value for improving the ability to delineate and simulate the spatial features of frost. Furthermore, there are considerations to be accounted for, related to the dissimilarities in temporal and spatial averages as envisioned by modeling efforts, as that exist in the real world and as measured by remote sensing systems. Thus, EO data to be useful for monitoring and assessing frosts must be compatible with mathematical modeling of the corresponding quantification schemes.

6.5 REMOTELY SENSED FROST QUANTIFICATION

Remotely sensed frost quantification involves temperature measurement or estimation through remote sensing data and methods. In particular, Table 6.1 presents a list of extreme air temperature indices for frost or heat waves assessment and monitoring (http://cccma.seos.uvic.ca/ETCCDMI/list_27_indices.html.—see in Dalezios, 2016). The contribution of meteorology to the frost phenomenon can be summarized in the early detection of the phenomenon, as well as the consideration of the climatic conditions in order to estimate the frequency of probability of frost occurrence in a certain area (Dalezios, 2016). Then the annual temperature variability can be described, along with frost forecasting and modern frost assessment methods.

Annual temperature variability: The annual temperature variability is approximated with sufficient accuracy by sinusoidal functions of the form:

$$T_i(D) = A_i + B_i \sin\left(\tfrac{360}{365} D - F_i\right) \tag{6.1}$$

where:

The quantities A_i (in °C), B_i (in °C), and F_i (in degrees) are constants for each location
The variable D represents the day of the year ($D = 1, \ldots, 365$)
The index i denotes the mean, that is, mean maximum (mean–max), mean minimum (mean–min), absolute maximum (max–max), and absolute minimum (min–min) daily temperature

6.5.1 FROST MODELING AND FORECASTING

Frost modeling and forecasting are very important especially for the prevention of disasters in agriculture and particularly in crops. Frost modeling and forecasting are conducted through the analysis of daily synoptic weather charts, as well as the application of empirical models based on remote sensing data and methods. Theoretical and applied methods are implemented, which combine air temperature, dew point temperature, and land surface temperature (LST) through heat transfer. There are two levels of frost forecasting, namely general and local forecast. General forecast consists of the identification

TABLE 6.1

Extreme Air Temperature Indices Recommended by the ETCCDMI

ID	Indicator Name	Indicator Definitions	Units
TXx	Max Tmax	Let Tx_{kj} be the daily maximum temperatures in month k, period j. The maximum daily maximum temperature each month is then $TXx_{kj} = \max(Tx_{kj})$	°C
TNx	Max Tmin	Let Tn_{kj} be the daily minimum temperatures in month k, period j. The maximum daily minimum temperature each month is then $TNn_{kj} = \max(Tx_{kj})$	°C
TXn	Min Tmax	Let Tx_{kj} be the daily maximum temperatures in month k, period j. The minimum daily maximum temperature each month is then $TXn_{kj} = \min(Tx_{kj})$	°C
TNn	Min Tmin	Let Tn_{kj} be the daily minimum temperatures in month k, period j. The minimum daily minimum temperature each month is then $TNn_{kj} = \min(Tn_{kj})$	°C
TN10p	Cold nights	Let Tn_{ij} be the daily minimum temperature on day i in period j and let $Tn_{in}10$ be the calendar day 10th percentile centered on a 5-day window (Zhang et al., 2005b). The percentage of time is determined where $Tn_{ij} < Tn_{in}10$	Days
TX10p	Cold days	Let Tx_{ij} be the daily maximum temperature on day i in period j and let $Tx_{in}10$ be the calendar day 10th percentile centered on a 5-day window (Zhang et al., 2005b). The percentage of time is determined where $Tx_{ij} < Tx_{in}10$	Days
TN90p	Warm nights	Let Tn_{ij} be the daily minimum temperature on day i in period j and let $Tx_{in}90$ be the calendar day 90th percentile centered on a 5-day window (Zhang et al., 2005b). The percentage of time is determined where $Tn_{ij} > Tn_{in}90$	Days
TX90p	Warm days	Let TX_{ij} be the daily maximum temperature on day i in period j and let $Tx_{in}90$ be the calendar day 90th percentile centered on a 5-day window (Zhang et al., 2005b). The percentage of time is determined where $Tx_{ij} > Tx_{in}90$	Days
DTR	Diurnal temperature range	Let Tx_{ij} and Tn_{ij} be the daily maximum and minimum temperature, respectively on day i in period j. If I represents the number of days in j, then $$DTR_j = \frac{\sum_{i=1}^{I}(Tx_{ij} - Tn_{ij})}{I}$$	°C
FDO	Frost days	Let Tn_{ij} be the daily minimum temperature on day i in period j. Count the number of days where $Tn_{ij} < 0°C$	Days
SU25	Summer days	Let Tx_{ij} be the daily maximum temperature on day i in period j. Count the number of days where $Txy_{ij} > 25°C$	Days
IDO	Ice days	Let Tx_{ij} be the daily maximum temperature on day i in period j. Count the number of days where $TX_{ij} < 0°C$	Days
TR20	Tropical nights	Let Tn_{ij} be the daily minimum temperature on day i in period j. Count the number of days where $Tn_{ij} > 20°C$	Days
GSL	Growing season Length	Let T_{ij} be the mean temperature on day i in period j. Count the number of days between the first occurrence of at least 6 consecutive days with $T_{ij} > 5°C$ and the first occurrence after July 1 (January 1 in SH) of at least 6 consecutive days with $T_{ij} < 5°C$	Days
WSDI*	Warm spell duration indicator	Let Tx_{ij} be the daily maximum temperature on day i in period j and let $Tx_{in}90$ be the calendar day 90th percentile centered on a 5-day window (Zhang et al., 2005b). Then the number of days per period is summed where, in intervals of at least 6 consecutive days: $Tx_{ij} > Tx_{in}90$	Days

Note: See also 1130 http://cccma.seos.uvic.ca/ETCCDMI/list_27_indices.html.

of the properties and characteristics of the air mass, which prevails over the area during the night and early morning. In the local forecast, some representative sites of the area are selected, such as a valley, or the slopes or the top of a hill. The drop in temperature during the night depends on the duration of the night, meaning that the longer the duration of the night, the greater the drop in temperature is and the heat lost by the ground. The amount of heat lost by the land surface is a function of soil moisture and temperature, cloud cover, vegetation cover, air temperature, and other factors.

Some frost forecasting methods are presented as follows (Bagdonas et al., 1978):

1. *Rule of wet bulb temperature*: This method takes into account the effect of atmospheric water vapor, which prevents heat loss from the land surface.

$$T_{min} = aT_w - bT_d - c \tag{6.2}$$

   where:
       T_w is the wet bulb temperature at the time of sunset of day N
       T_d is the dry bulb temperature at the same day and time
       a, b, c are the constants, depending on the location

   Many times a and b are very small, so Equation 6.2 can be simplified to $T_{min} = T_w - c$ and is called Kammerman formula.

2. *Gold formula*: This formula computes the minimum temperature T_{min} as follows:

$$T_{min} = aT_{1500Z} + bT_{dew1500Z} - c \tag{6.3}$$

   where:
       T_{1500Z} is the temperature at 15:00Z
       $T_{dew1500Z}$ is the dew point temperature at 15:00Z
       a, b, and c are constants

3. *Multiple linear regression*: For the forecast of the minimum temperature Y of the day, a multiple linear regression can be used as follows:

$$Y = a_1X_1 + a_2X_2 + a_3X_3 + a_5X_5 + a_5X_5 \tag{6.4}$$

   where:
       X_1 is the dry bulb temperature
       X_2 is a function of the cloud cover: $X_2 = 0.9\,m + 0.5\,k$, where m is the amount of low cloud, and k is the amount of medium clouds into eighths
       X_3 is a function of the wind speed and direction
       X_4 is the dew point temperature
       X_5 is a function dependent on whether frost has occurred the previous day. If frost has occurred the previous day, then it takes the value 1 and if frost does not occur, then it takes the value 0.

   The variables a_1–a_5 are calculated using the least squares method in a specified area.

4. *Model ANGELA*: The physical model of ANGELA system (WMO, 2010) refers to the drop in temperature at night. In this model, the LST is a function of temperature at the sunset time and several hours after the sunset, given by

$$T_n = T_s - K \times n^{1/2} \tag{6.5}$$

   where:
       T_n is the temperature of n hours after sunset in °C
       T_s is the temperature at the time of sunset in °C

K is the coefficient of temperature drop

n is the number of hours after sunset

5. *Rule of maximum–minimum*: It has been shown that the minimum temperature (T_{min}) on the next day ($N + 1$) follows a linear relationship with the maximum temperature (T_{max}) of day (N), when the forecast is issued, namely:

$$T_{min} = aT_{max} - b, \quad \text{where } a, b \text{ are constants} \tag{6.6}$$

6. *Craddock formula*: It is an empirical formula for the minimum temperature T_{min} based on the dry bulb temperature T_d at 12:00Z and dew point temperature T_{dew} also at 12:00Z.

$$T_{min} = aT_d + bT_{dew} + 2.12 + c \tag{6.7}$$

The values of the temperatures are given in degrees °F. The parameters a and b are constants. The constant c is given as a function of the average cloud on prognostic hours 18:00Z, 24:00Z, and 06:00Z and the average wind speed (in knots) at the same hours. The formula is not valid when there is fog at night.

7. *Faust formula for soil frost*: Faust has provided an empirical formula in which if the amount of cloud at night is less than 2/8, and if the average wind speed is less than 2 knots, then soil frost occurs when the sum $[T + [1/(2T_{dew})]]$ at 14:00 local time is less than 79°F.

6.5.2 Remote Sensing Methods

Frost assessment using EO-based methods is based on temperature observations from infrared (IR) bands of meteorological satellites, such as Meteosat or MODIS. From these satellites, brightness temperature is usually observed from thermal IR channels on a pixel basis from which LST can be computed. Meteorological satellites are characterized by high temporal resolution, for example, for Meteosat every 30 min, but with rather coarse spatial resolution ranging between 1 and 2.5 km or a few hundred meters for MODIS. As a result, temperature monitoring can be conducted through meteorological satellites leading to estimation and assessment of frost, especially for radiation frost in cloudless nights during spring season in frost-prone valleys or land surface patches (Domenikiotis et al., 2004, 2006). For illustrative purposes Table 6.2a delineates the surface 549 temperatures from a Meteosat IR image in the area of Katerini in Northern Greece consisting of 26 pixels (Dalezios and Lavrediadou, 1995). Moreover, Table 6.2b presents the number of pixels for each corresponding temperature from Meteosat IR 552 during the night of March 31, 1994 in the same area of Katerini in Northern Greece 553 (Dalezios and Lavrediadou, 1995).

Remotely sensed LST and air temperature: The extraction of LST includes the use of empirical equation such as the one shown in the following equation (Dalezios et al., 2012):

$$T = (\text{image pixel} + 31,990) * 0.005 \tag{6.8}$$

where:

image pixel is the pixel value from the thermal band

T is temperature in Kelvin (°K), which is then converted to values of degrees Celsius (°C)

For the elimination of the water vapor effect in IR radiation and the transmitted radiation from the surface, the algorithm of *split window* (Becker and Li, 1990) is employed. This method achieves

TABLE 6.2

(a) Observed Temperatures from Meteosat IR (2.5 × 2.5 km Resolution) at 04:00 a.m. of March 31, 1994 in the Area of Katerini in Northern Greece. (b) Number of Pixels for Each Corresponding Temperature during the Night of March 31, 1994 in the Area of Katerini from Meteosat IR

(a)

		1	1	2
		1	0	1
		0	0	1
2	1	-1	0	1
1	0	-2	-1	0
0	-1	-2	-1	0
			0	-1

(b)

Time	Area of Katerini in Northern Greece					
(March 31, 1994)	3°C	2°C	1°C	0°C	-1°C	-2°C
3:30	–	5	8	8	4	1
4:00	–	2	8	9	5	2
5:00	–	10	9	6	1	–
6:00	1	12	12	5	1	–
6:30	2	9	9	6	–	–
7:00	6	10	6	4	–	–

Source: Dalezios, N.R. and Lavrediadou, E.E., *Adv. Space Res.*, 15, 123–126, 1995.

atmospheric correction of the satellite data including water vapor absorption. The equation of Becker and Li is given by Equation 6.9:

$$T = 1.274 + (T_4 + T_5)/2[1 + 0.15616\{(1-e)/e\} - 0.482\,de/e**2]$$
$$+ (T_4 - T_5)/2[6.26 + 3.989\{(1-e)/e\} - 38.33\,de/e**2] \tag{6.9}$$

where:

   T is the LST in °C,
   T_4 and T_5 are the values of thermal bands 4 and 5 of the satellite, respectively

The variables e and de of Equation 6.9 are defined by

$$e = \frac{(e_4 + e_5)}{2} \tag{6.10}$$

$$de = e_4 - e_5 \tag{6.11}$$

where e_4 and e_5 are the reflection values of bands 4 and 5, respectively, which estimate the transmission of IR radiation from the surface and are given by Van de Griend and Owe (1993) empirical equations:

$$e_4 = 1.0094 + 0.047\ln(\text{NDVI}) \tag{6.12}$$

$$e_5 = e_4 + 0.01 \tag{6.13}$$

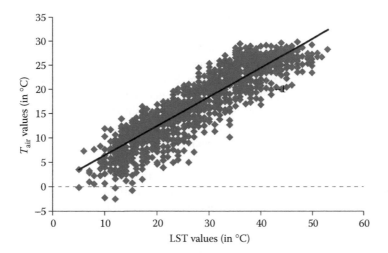

FIGURE 6.4 LST and air temperature (T_{air}) for Larisa (°C). (From Dalezios, N.R. et al., *Nat. Hazard Earth Syst. Sci.*, 12, 3139–3150, 2012.)

where NDVI are the values of the index. A regression analysis is conducted between the derived monthly LST values (derived from Equation 6.9) and the corresponding monthly air temperature (T_{air}) values, which are shown in Figure 6.4. This equation was developed from Larisa meteorological station in Greece using data from a 20-year period (1981–2001).

$$T_{air} = 0.6143 - LST + 7.3674 \qquad R^2 \approx 0.82 \qquad (6.14)$$

Finally, from Equation 6.14 monthly air temperature maps (images) of Thessaly may be produced on a pixel basis (for more details on the method see Dalezios et al., 2012).

6.6 REMOTELY SENSED FROST EARLY WARNING SYSTEMS

Frost monitoring is based on the development of FEWSs. Indeed, frost occurrence and warning are based on temperature and its spatiotemporal variability (Moeletsi et al., 2013; Shen et al., 2015). Quantification of frost hazard uses a methodological approach based on the minimum temperature consideration through EO data and methods. Indicative examples of FEWS and frost monitoring are presented based on remote sensing data and methods.

1. *Kalman filtering approach*: For monitoring and forecasting or nowcasting frost, a so-called phenomenological approach is used based on Kalman filtering, which belongs to estimation and control theory (see e.g., Dalezios, 1987). Specifically, a one step-ahead forecasting on a pixel basis using 2D satellite temperature images is considered. The database consists of a series of satellite records (e.g., Landsat, Meteosat, national oceanic and atmospheric administration (NOAA)/advanced very high resolution radiometer (AVHRR)) from which temperature is extracted on a pixel basis. In this way, temperature time series are developed for each pixel, and then the one step-ahead forecasting is attempted. The adopted approach comes from the optimal estimation theory, and in the current application the adaptive Kalman filter is employed (Dalezios, 1987). The system model is the so-called phenomenological temperature model, which is based on the assumption that the daily temporal variability of temperature follows a sinusoidal function.
2. *Georgia's Extreme-Weather Neural-Network Informed Expert (GENIE) system*: This application refers to an expert system, entitled *GENIE*. GENIE incorporates the knowledge of experts, such as agrometeorologists, and additional information on air temperature, dew point temperature, and wind speed into a fuzzy expert system. GENIE is designed to be used by

Georgia producers in order to present warning levels of frost and freeze mainly for blueberries and peaches (Chevalier et al., 2012). Forecasting of air temperature and dew point temperature across the state of Georgia for 1–12 h ahead is conducted through artificial neural network (ANN). Moreover, observed wind speed, along with the aforementioned forecasts, are used as input variables for this fuzzy expert system. Specifically, five levels of frost and freeze are considered by experts to describe the prevailing meteorological conditions. Then, this classification is used to develop fuzzy logic rules and membership functions for GENIE. Additional scenarios are presented to GENIE for evaluation. GENIE is available to Georgia producers through a web-based interface (www.georgiaweather.net). Specifically, the five general warnings are (1) no frost or freeze, (2) possible frost, (3) mild frost, (4) severe frost, and (5) hard freeze. The crisp output of the fuzzy expert system, which represents the relative level of damage associated with the given weather conditions, is arbitrarily restricted to the continuous set [0, 1000]. Warnings are generated based on the value of this output as follows:

Output: [0, 200] = No frost or freeze
Output: [201, 400] = Possible frost
Output: [401, 600] = Mild frost
Output: [601, 800] = Severe frost
Output: [801, 1000] = Hard freeze

A user by receiving the warning level and the continuous output value is able to differentiate between conditions, such as a possibly less threatening mild frost (output = 405) and a mild frost that is dangerously close to a severe frost (output = 595). A temperature of 4°C is the upper limit, because no frost occurs at that temperature. Similarly, a lower limit of −5°C is chosen, because, at this temperature, severe damage due to freezing conditions is certain. The current observed wind speed is used as part of the expert system. A wind speed of 16 km/h is used as the upper threshold, and half that value (8 km/h) is used as the lower threshold. The possible weather scenarios, thus, consist of the various combinations of integer air temperature and dew point temperature values, along with the three ranges of wind speed (less than 8 km/h, between 8 and 16 km/h, and greater than 16 km/h). Each of these scenarios is labeled using one of the previously described five warning levels. As an example, the following rule, which is related to severe frost conditions, can be inferred based on the provided and classified information: "If the air temperature is greater than −2.5°C and less than 4.5°C, and the dew point temperature is greater than −2.5°C and less than −1.5°C, and the wind speed is over 16 km/h, then severe frost conditions exist."

3. *Frost risk mapping model*: This approach is based on developing a deterministic model to predict frost hazard in agricultural land utilizing remotely sensed imagery, and GIS is developed (see, e.g., study by Louka et al., 2015—see Figure 6.5). The model is based on the main factors that govern frost risk including environmental parameters, such as LST and geomorphology. Its implementation is based primarily using EO data from polar-orbiting sensors, supported—in some instances—by ancillary ground observation data. Topographical parameters required in the model include the altitude, slope, steepness, aspect, topographic curvature, and extent of the area influenced by water bodies. Additional data required include land use and vegetation classification (i.e., type and density). In general, the adopted methodology consists of three basic stages:

Step 1: Development of a frost hazard model in a GIS environment. The proposed frost hazard model combines the following parameters and takes the generic form:

$$\text{Frost hazard} = 0.3 \times \text{elevation} + 0.10 \times \text{aspect} + 0.10 \times \text{slope} + 0.15 \times \text{CTI}$$
$$+ 0.15 \times \text{dist.water} + 0.10 \times \text{curvature} + 0.10 \times \text{landuse}$$

(6.15)

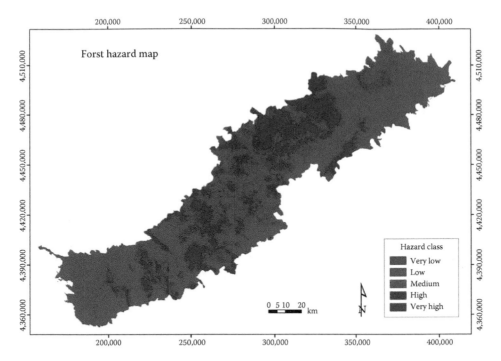

FIGURE 6.5 Frost hazard map of the study area produced by the frost hazard model (projection system: Greek Grid). The map highlights areas where the estimated frost hazard level is high (red colored) in which the terrain factors might be more favorable to frost incidents than in other areas (yellow and green colored). (Adapted from Louka, P. et al., Frost risk assessment tool in agriculture utilizing remote sensing and GIS techniques, In Srivastava, P.K., Pandey, P., Kumar P. et al. (Eds.), *Geospatial Technology for Water Resource Applications*, CRC Press, Taylor & Francis Group, Boca Raton, FL, 2015. With permission.)

Step 2: Development of a frost frequency map product based on multitemporal analysis of MODIS LST data. The first processing stage consists of the selection of areas, where LST was below 0°C, therefore, indicating frost conditions, and it includes the following steps: the images were processed with the Band Math Tool, using the equation:

$$T_s < 273°K \quad \text{(which equals to 0°C)} \tag{6.16}$$

where T_s represents the LST. The objective of the second processing stage is to separate the areas, where temperature information is recorded from areas with no data due to cloud cover or other cause. Similar processing steps are followed for each image with the Band Math Tool, using the equation:

$$T_s < 0°K \tag{6.17}$$

Step 3: Mapping of frost damage distribution on agricultural land, based on ground observations, is conducted in order to study the spatial and temporal distribution of frost risk. Such data that record frost damages are important, because they offer an independent data source, which can also help to validate the model. The processed data consist of 10-year frost incidents (2000–2010) or longer.

6.7   FROST FREQUENCY ANALYSIS

Frost is characterized by several features, such as the recurrence period, the space–time–frequency analysis, and the frost persistence (IPCC, 2007; Dalezios, 2017). The recurrence period is extremely useful, because it allows the determination of the expected extreme value of a quantity, based on the average of the extreme values of the same magnitude in a given period. If frequency of frost occurrence is defined, the likelihood of partial frost occurrence for several consecutive days, whereas frost severity is defined as the absolute minimum (negative) temperature, which occurs on successive days of frost. Moreover, frost duration is defined as the number of consecutive days in which there is partial frost. Frost has also been related to the average length of frost-free period (or growing season in days) during the year and the frost severity–duration–frequency relationships. Needless to say, the frost-free period is a very important parameter in several parts of the world. A frost-free period is defined as the period between the latest spring and the earliest autumn frosts. For example, in Greece, analysis of measurements from 86 meteorological stations showed that the frost-free period reaches more than 240 days per year mainly in the southern regions and decreases from south to north (Dalezios, 2015). Information on the occurrence of frost has an economic effect on high-value crops, although crops can be protected. Frost risk maps and dates of first and last frost are simple but useful applications to agriculture. These maps are made at the macro- to mesoscale and are useful for specifying general planting dates for crops and for the assessment of crop damage when combined with phenological data (WMO, 2010).

A methodology for frost frequency analysis is presented next (Dalezios et al., 2000). The daily minimum temperature below 0°C is used as a threshold, and the number of days with such a temperature, or partial frost, is identified for each station. The number of successive days with minimum temperature below 0°C defines the duration of a frost episode, whereas the recorded absolute minimum temperature below 0°C identifies the intensity or severity of the frost episode. Several episodes are grouped according to their duration, for the whole period of study, and are ranked according to their severity for each station. Similarly, the frequency of an extreme event is usually expressed by its return period or recurrence interval, which may be defined as the average interval of time within which the magnitude of the event is equated or exceeded once. The analysis of extreme events is usually presented by severity–duration–frequency relationships for several stations throughout the region of interest. The results of frost frequency analysis are shown in Table 6.3. In this table, column 1 shows the ranking numbers; column 2 shows the absolute minimum temperature values in ascending order; column 3 shows the corresponding probability (P) of occurrence using the Weibull plotting position equation, where m is the current ranking number and n is the total number of data points; and column 4 shows the corresponding return period T duration using the equation, where P was previously defined.

For the estimation of extreme events, such as frosts, in which the return periods are required, when the severities and duration are given, it is necessary to assume a particular mathematical form of the frequency distribution. Several theoretical distributions have been tested against the cumulative severities of extreme phenomena of various durations. These include the Extreme Value Index (EVI, Gumbel), the generalized extreme value (GEV), the three parameter lognormal (LN3), and the log-Pearson (LP3) distributions (Dalezios et al., 2000). Application of the nonparametric Kolmogorov–Smirnov two sample tests at 95% confidence level and visual inspection of the fitting of the aforementioned theoretical frequency distributions to cumulative intensity values indicate that the EVI provides overall, a reasonable and acceptable approximation of the frequency of the calculated severity values (Dalezios and Lavrediadou, 1994). Furthermore, the EVI has been used in numerous studies of extreme phenomena. Data used in this study include daily series of minimum

TABLE 6.3

Frost Absolute Minimum Temperatures of 2-Day Duration with the Corresponding Probabilities (*P*) and Return Periods (*T*) for Agrinio Station

Rank	Absolute Minimum Temperature	Probability P = m/(n + 1)	Return Period T = 1/P	Rank	Absolute Minimum Temperature	Probability P = m/(n + 1)	Return Period T = 1/P
1	0.2	0.025	40.00	21	2.4	0.525	1.90
2	0.5	0.05	20.00	22	2.6	0.55	1.82
3	0.6	0.075	13.33	23	2.6	0.575	1.74
4	0.8	0.1	10.00	24	2.6	0.6	1.67
5	1	0.125	8.00	25	2.8	0.625	1.60
6	1	0.15	6.67	26	2.8	0.65	1.54
7	1.2	0.175	5.71	27	3	0.675	1.48
8	1.3	0.2	5.00	28	3	0.7	1.43
9	1.3	0.225	4.44	29	3	0.725	1.38
10	1.4	0.25	4.00	30	3.2	0.75	1.33
11	1.4	0.275	3.64	31	3.5	0.775	1.29
12	1.6	0.3	3.33	32	3.6	0.8	1.25
13	1.6	0.325	3.08	33	3.6	0.825	1.21
14	1.8	0.35	2.86	34	3.8	0.85	1.18
15	1.8	0.375	2.67	35	4	0.875	1.14
16	1.9	0.4	2.50	36	4.8	0.9	1.11
17	2	0.425	2.35	37	4.8	0.925	1.08
18	2.2	0.45	2.22	38	5	0.95	1.05
19	2.2	0.475	2.11	39	7	0.975	1.03
20	2.3	0.5	2.00				

Source: Dalezios, N.R. and Lavrediadou, E.E., Frost severity-duration-frequency relationships, In *Proceedings, 2nd Greek Scientific Conference on Meteorology, Climatology and Atmospheric Physics*, Thessaloniki, Greece, September 29–30, pp. 27–34, 1994.

temperature from 15 meteorological stations in Greece. The threshold is temperature below 0°C. A brief description of the steps, which are followed to develop the *severity–duration–frequency* (SDF) relationships, is presented as follows:

Step 1: Probability Tables. The frost episodes for each station are identified, when the minimum temperature for successive days is below 0°C. Absolute minimum temperatures of partial frost below 0°C of each episode are used in order to rank the episode's severity. In this way, multiple episodes for the whole period are calculated for several durations (Table 6.3).

Step 2: Fitting Gumbel Distribution. For each frost episode, the identified absolute minimum temperatures versus the corresponding return period are plotted, and the EVI distribution (Dalezios and Lavrediadou, 1994) is fitted to the plotted data points, which has the following cumulative distribution function (Equation 6.18):

$$F(x) = \exp\left[-\exp(-A \cdot (x - U))\right] \tag{6.18}$$

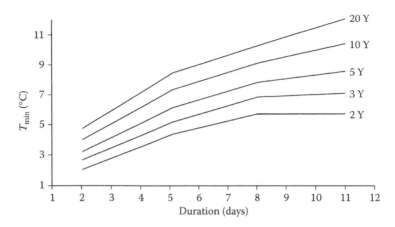

FIGURE 6.6 Frost severity–duration–frequency curves for Kavala station in Northern Greece. (From Dalezios, N.R. and Lavrediadou, E.E., Frost severity-duration-frequency relationships, In *Proceedings, 2nd Greek Scientific Conference on Meteorology, Climatology and Atmospheric Physics*, Thessaloniki, Greece, September 29–30, pp. 27–34, 1994. With permission.)

where *A* and *U* are the fitted parameters, which are computed for each duration from the data. The procedure of fitting EVI distribution is applied to all the identified episodes for each station.

Step 3: *SDF Curves*. Finally, using the Gumbel distribution, cumulative periods are computed, which correspond to return periods of 2, 5, 10, 25, 50, and 100 years, respectively, for each identified frost duration. The SDF curves appear to be as expected, because for decreasing frequencies, there is a corresponding increase in severities or intensities. There is a corresponding increase in frost severities, which tend to become asymptotic to the *x* axis, as seen in Figure 6.6.

6.8   FROST IMPACTS AND MITIGATION

Frost damage is responsible for more economic losses than any other weather-related phenomenon in many regions across the globe (Jonsson et al., 2004; Pulatov et al., 2015). With sufficient warning, producers can minimize the potential damages caused by frost and freeze events. However, the severity of these events is dependent on several factors including air temperature, dew point temperature, and wind speed. Assessing this risk is not easily quantifiable and requires methodologies and decision support systems (DSSs) to address the process. Moreover, EO data and methods can assist to assess the extent of frost damages in agriculture. Specifically, satellite-based methods to estimate LST or air temperature, such as the methods described earlier (see Section 6.5.2), or methods to monitor vegetation through the Normalized Difference Vegetation Index (NDVI) (e.g., see Lin and Lv, 2010; Prabhakara et al., 2015), as well as frost risk mapping methods (e.g., Louka et al., 2015) can certainly contribute to frost damage assessment. Indicative examples and case studies are presented later in this chapter (see Section 6.8.3).

6.8.1   FROST IMPACTS AND PREVENTION

Agriculture is a field of economic activity, which is directly affected by frost, usually with disastrous results. The frost losses in agriculture are too large and sometimes reach the total loss of production. The critical temperature below which there is no damage to the plants depends mainly on the development stage of the plant tissue. The damaging effect of frost on plants is related to cellular scale activity. Factors, which determine the amount of damage, for example, in horticulture, are the rate and speed of temperature drop, freezing conditions, the rate of temperature increase, the time of freezing, the botanical species and variety, the vegetative part and stage of the plant, and the age of the plant. Winter frosts contribute the most to total frost injuries. The most damaging events are the successive frosts, that is, advection frosts followed by severe radiation frosts. Frost damages account forto more than half of the total damages due to weather-related phenomena in Greece (Dalezios, 2015). Contribution refers to total frost injuries. Fruit trees damages account to more than 50% of the total frost injuries. Special horticulture plantation in the Mediterranean region, such as vines, may account for 20% of the losses, although damages are infrequent. Further studies are needed to relate reliable forecasting with economic impact on both local and regional scale. In addition, more studies are needed to indicate the economic benefits of accurate frost forecasting for different crop categories.

The accumulated experience of frost as a hazard leads to techniques for frost or freeze prevention. These methods aim at developing plants that mature in shorter periods of time, or in the field of climatology, which may define the areal extent of frost incidence for any particular area. Indeed, the greatest emphasis in frost prevention is in the marginal or fringe areas, where the farmer is willing to risk planting in the hope to produce a crop. The frost prevention techniques applied in agriculture include both treatment of the phenomenon and crop resistance. It is observed that the best time to protect an area from frost is even before planting. This means the need of careful selection of plant species, the area, and the time of planting. Long before, agriculture is considered for a particular area or crop; a climatic record or history would reveal the probabilities of frost or freeze incidence and the chance of success if the proper plans and methods are used. Additional terrain analysis helps to identify local microclimatic conditions in basins, alluvial fans, slopes, plains, or other topographic conditions. In the competition for the early harvest–high price crop, farmers for many years have tried to protect young plants with plastic or paper shields. However, the shield protects within limited temperature ranges.

6.8.2 Frost Protection and Mitigation Methods

One of the main issues of this chapter remains whether EO data and methods can help farmers and producers to plan and implement frost protection measures effectively, such as passive and active methods, which are described in Sections 6.8.2.1 and 6.8.2.2. At the present time, it is recognized that the increasing reliability of remote sensing, along with the continuously increasing technological and computational advancements, signals a positive impact on the field of agriculture. Specifically, the previously presented methodologies and examples of FEWS and frost monitoring systems can certainly assist farmers to develop and use appropriate protection measures at the farm level. Moreover, if these systems are incorporated into DSS, then integrated farm systems are developed for optimal farm and production management.

The aforementioned FEWS and DSS should be linked to the development and establishment of the so-called agroclimatic classification (Tsiros et al., 2009). Indeed, the agroclimatic potential of agricultural areas has to be assessed in order to achieve sustainable and efficient use of natural resources in combination with production maximization. Specifically, zones indicating water availability are combined with topographic features and soil types in order to identify areas for sustainable production. First, Aridity Index (AI) and Vegetation Health Index (VHI) are used in order to define zones adequate for sustainable farming according to water limitations. As crop growth is affected by water supply, these zones are named as water limited growth environment (WLGE) zones. VHI is derived from NOAA/AVHRR data, whereas in AI computations both satellite and conventional field data are used. Then, WLGE zones are combined with soil maps and a digital elevation model (DEM) of the area under investigation in order to define zones appropriate for sustainable production, namely noncrop-specific agroclimatic zones. By incorporating the additional remotely sensed indices, such as growing degree days (GDD) and direct solar radiation (DSR), crop-specific agroclimatic zones are developed leading to the final identification of sustainable production classes, namely high, medium, and low productivity. These zones can be further used for agroclimatic classification.

To limit the damage to agriculture from frosts, various protection methods are used around the globe (Monzon et al., 2007; Ghaemi et al., 2009; Smyth and Skates, 2009). These methods are classified as either passive or active and are outlined in Sections 6.8.2.1 and 6.8.2.2. Most of the practical techniques listed in these sections to combat the phenomenon of frost are only effective for radiation frosts, although some can be applied to advection frosts.

6.8.2.1 Passive Methods

The passive protection measures include microclimatological research prior to any use of a field and selection of the most suitable crop for each region. Passive methods are utilized prior to the frost event in order to avoid or minimize injuries. Passive methods include the following:

1. *Proper site selection*: It is probably the most important passive protection method. Hilltops and middle portions of hillsides are most volatile to advection frosts, where the temperatures observed are usually higher than down-wind sides and low spots that are sheltered from the wind. Exactly the opposite is observed during a radiation frost event.
2. *Cold air drainage management*: Cold air drained downhill can be diverted by using methods such as building a solid wall or using a wooden fence. This diversion can effectively provide protection.
3. Plant selection.
4. Avoiding soil cultivation.
5. Plant covers.
6. Canopy trees provide protection from radiation frosts because there is an enhanced longwave radiation downward from the trees. The effect is that temperatures are higher under these trees than in the open space.

Passive methods of frost protections are used more widely in agricultural regions basically because they are more cost-effective and more beneficial. For example, in Greece, passive protection methods are widely utilized to help farmers combat freezing injuries (Dalezios, 2015). Low prices and overproduction are responsible for the unwillingness of farmers to invest in active protection methods.

6.8.2.2 Active Methods

Active frost protection measures focus to modify the microclimate of the field in order to avoid low temperatures leading to frost in crops. In other words, they cover the development of physical equipment, such as fans, heaters, brushes, sprinklers, and plant shields to modify temperatures or reduce radiation. Active protection methods are costly, because most of them are fuel dependent. Active methods are deployed during a frost night. Whatever method or combination of methods is chosen to prevent destruction by frost, the choice is usually to modify temperatures a few degrees, usually not more than four or five. The techniques involve reducing radiation, improving wind circulation, discouraging sublimation, or creating a fog or smoke cover. The physical modification of the environment in the immediate condition of a frost or freeze hazard includes the deployment of the following: wind machines, sprinklers, heaters, surface irrigation, or foggers. Specifically, wind machines and microsprinklers are used as active protection methods only in a small scale. For example, horizontal (conventional) blowing machines are deployed in the Argolic plain in Greece to protect citrus, namely sweet oranges and mandarins (Dalezios, 2015).

6.8.3 REMOTELY SENSED FROST DAMAGE: EXAMPLES AND CASE STUDIES

A number of indicative examples and case studies of frost damage assessment are presented based on remote sensing data and methods.

1. *Frost damage risk*: This application refers to a comprehensive method to quantify frost damage risk in different sweet cherry production areas of South Patagonia and to estimate the potential impact of frost control systems on risk reduction (Cittadini et al., 2006). A theoretical–empirical approach is considered due to lack of historical weather

data. At first, it is assumed that frost damage for any specific day of the growing season occurs when the minimum temperature on that day is below the specific lethal temperature for the phenological stage. Indeed, phenological models are used for the prediction of the phenological stage. Moreover, remote sensing data and methods are used for the estimation of areal minimum temperature. Then, for each production location of South Patagonia, frost damage probability is estimated as the frequency of seasons in which at least one damaging frost occurs, which means damage is greater than or equal to 90% of the reproductive organs during the growing season. Finally, frost damage risk is reduced based on active frost control methods. The analysis has indicated that the frequency of years with at least one killing frost decreases dramatically when the minimum temperature increases by 3°C, using active frost control systems (Cittadini et al., 2006). This methodology appears useful to identify the main and secondary variables affecting frost damage risk. Thus, this type of quantitative analysis can support growers in decision-making on required investments and operational costs of the equipment for frost control, on the basis of potential impact of a particular control system on mean yields and yield stability.

2. *Frost risk mapping for agroclimatic suitability*: This application refers to the development of an agroclimatic suitability library for crop production. The database consists of climatic data from 20 to 33 years for 41 meteorological stations in the Bolivian Altiplano (Geerts et al., 2006). For validation purposes, four agroclimatic indicators are used, namely the reference evapotranspiration, the length of the rainy season, the severity of intraseasonal dry spells, and the monthly frost risks for each station. Indeed, monthly frost risks are identified for temperatures below which crop frost damage occurs. It is clear that temperature thresholds differ between crop types and within each crop type during the growing season. Specifically, three temperature thresholds are considered, namely −8°C, −6°C, and −4°C, respectively. As temperatures at crop canopy height are generally up to 1°C lower than those recorded at screen height, the monthly probabilities of frost occurrence at least once are equal to or lower than −7°C, −5°C, and −3°C, which are computed for the period from September to May for 39 meteorological stations. Indeed, the input data used are either daily minimum temperatures or monthly absolute minimum temperatures. Alternatively, remote sensing data could be used for the required regional estimation of LST. Moreover, the point data are entered in a GIS environment and interpolated using kriging in order to obtain regional estimates. In this application, quinoa is considered, an important crop in the region that is cultivated during the short and irregular rainfall season and that is well adapted to the frequent occurrence of drought and frost. The GIS library is used to mark agroclimatic zones, where irrigation could improve quinoa production. Specifically, irrigation requirements are used to assess the vulnerability of the delineated zones. In this application, two regions with a high vulnerability are selected, namely a severe drought risk and an acceptable frost risk region.

3. *Assessment of frost occurrence and severity*: In this application, remotely sensed assessment of frost occurrence and severity is considered. Specifically, frost occurrence is monitored by remote sensing based on the difference of vegetation index values and on the differences in canopy temperature (Lin and Lv, 2010). Needless to say, remote sensing has proven to be a feasible tool in monitoring crop growth, especially after stresses. Indeed, based on the difference of NDVI and canopy temperature (CT), it is possible to combine data of frost damage and crop development. For this application, the remote sensing potential is considered for temporal Landsat ETM images in order to monitor the frost occurrence on cotton fields in specified regions. In particular, in the cotton zone without frost occurrence, NDVI values are apparently higher, and the corresponding canopy temperature is 26.4°C. Moreover, when mild frost occurs in the area, the corresponding canopy temperature becomes 27.6°C. Finally, in the case of severe frost

occurrence in the area, the corresponding canopy temperature becomes 29.3°C. In summary, based on the difference of NDVI and canopy temperature, frost occurrence and the degree of frost severity can be monitored.

4. *Frost risk assessment*: This application covers the assessment of frost risk in tropical highlands such as the Andes, where there are human activities at altitudes up to 4200 m, and night frost may occur throughout the year. Specifically, in these semiarid and cold regions with sparse meteorological networks, remote sensing and topographic modeling are considered for potentially delineating the effect of geomorphology on topoclimatology. In this application, the integration of night MODIS LST and the extraction of physiographic descriptors from a DEM contribute to explore how regional and landscape-scale features influence frost occurrence in the southern Altiplano of Bolivia (Pouteau et al., 2011). Specifically, based on the high correlation between night LST and minimum air temperature, frost occurrence in early, middle, and late summer periods is computed from satellite observations and mapped at a 1 km resolution over a 45,000 km² area. Moreover, physiographic modeling of frost occurrence is then conducted comparing multiple regression (MR) and boosted regression trees (BRT) in which physiographic predictors are latitude, elevation, distance from salt lakes, slope steepness, potential insolation, and topographic convergence. In particular, insolation impact on night frost is tested assuming that ground surface warming in the daytime reduces frost occurrence in the following night.

   The results indicated that BRT models explain 74%–90% of frost occurrence variation, thus, showing better performance than the MR method. Moreover, inverted BRT models allow the downscaling of frost occurrence maps at 100 m resolution, illustrating local processes, such as cold air drainage. In addition, minimum temperature lapse rates show seasonal variation and mean values higher than those reported for temperate mountains. Indeed, in successive application at regional and subregional scales, BRT models reveal noticeable impacts of latitude, elevation, and distance to salt lakes at large scales, whereas topographic convergence, slope, and insolation show effects at local scales (Figure 6.7). The daytime insolation on night frost occurrence at local scale is considered significant, particularly in the early summer and midsummer periods, when solar astronomic forcing is maximum, thus, allowing a prognostic potential. Finally, there are also seasonal variations and interactions in physiographic effects.

5. *Spring frost damage day* (*SFDD*): This application considers spring frost damage. At first, it is well known that spring temperatures affect plant phenology. Moreover, recent spring warming coincides with earlier and longer nonfrozen season trends, as well as, earlier spring canopy onset resulting, in general, in increased vegetation productivity. However, the frost damage risk increases due to earlier spring onset, with potential negative impacts to productivity. Indeed, the occurrence, severity, and regional impact of frost events are difficult to monitor from sparse weather stations. This application deals with the development of spring frost day (SFD) and SFDD metrics from a long-term (>30 year) record. Specifically, the database consists of a satellite microwave remote sensing record of daily landscape freeze–thaw (FT) status and optical-IR sensor-based phenology record of start of season (SOS) and day of peak (DOP) canopy cover (Kim et al., 2014). The analysis has shown a decreasing regional SFD trend coincident with spring warming, whereas the SFDD is generally increasing. Moreover, although spring warming reduces frost occurrence, an earlier SOS trend increases vegetation frost damage risk. In addition, satellite-derived vegetation gross primary production (GPP) and vegetation greenness (EVI2) anomalies are used to assess the environmental impacts of the SFD and SFDD changes. The results indicate that higher SFD and SFDD levels coincide with reduced vegetation growth in spring, although only the SFDD shows significant correlation with EVI2 summer growth anomalies.

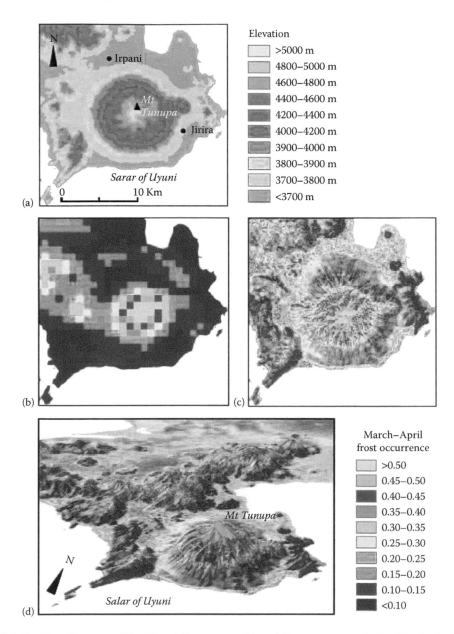

FIGURE 6.7   Elevation map of the Mount Tunupa area (a), and frost occurrence in the March–April period mapped at 1 km resolution from MODIS observations (b), at 100 m resolution (c), and in 3D view using regional BRT (d). Frost occurrence is scaled between 0 and 1 as the probability of daily occurrence of negative T_s values in the March–April period.

6.9   SUMMARY

In this chapter, an overview of the remote sensing data and methods in frost hazard analysis has been presented, along with its potential. At first, basic concepts and several characteristics of frost have been described, thus developing an understanding of temperature extremes. Then, risk identification of frost has been examined, including remotely sensed frost quantification. For frost quantification, the remote sensing potential focuses mainly on temperature extremes and frost indices, which use information from remote sensing sensors to map the condition of the land, detect several frost features and estimate several environmental parameters. Indeed, remote sensing data and methods provide, among others, direct measurements of land characteristics, vegetative cover, and components of the hydrological cycle, namely temperature, precipitation, soil moisture, or evapotranspiration, among others. Moreover, the scientific trend in frost quantification consists of remotely sensed temperature extremes and frost indices at different scales also using DSS toward web frost hazard platforms, which in certain cases have become operational. Furthermore, frost monitoring has been examined based mainly on modeling and forecasting methods, as well as FEWS. In general, frost monitoring is considered semi-operational; however, there are also several operational applications internationally. Moreover, remote sensing data and methods can be effectively used in frost risk assessment, which includes frost risk estimation, that is, frost event probabilities, SDF relationships, vulnerability assessment, areal extent mapping, and damage assessment. Finally, the possible contribution of remote sensing in frost risk management policy could focus on frost preparedness, frost impacts, prevention, and mitigation measures.

The future outlook of remotely sensed frost hazard analysis is promising due to significant and steadily increasing reliability of remote sensing data and methods throughout the years, mainly due to computational and technological, as well as scientific advancements. Indeed, the number of satellite systems is increasing year by year with a continuous improvement of their spatial and temporal resolution. Moreover, there is a current trend to increase the number of available bands in these satellites resulting in new and valuable information. In addition, there is a new challenge, because new types of remote sensing systems offer online open information for web platforms and are also utilized for monitoring and detecting frost.

ACKNOWLEDGMENTS

Dr. Petropoulos gratefully acknowledges the financial support provided by the Marie Curie Career Re-Integration Grant *TRANSFORM-EO* project supporting his participation to this research work. Authors would like to thank the anonymous reviewers for their comments that resulted in the improvement of the manuscript.

REFERENCES

Bagdonas, A., J. C. Georg, and J. F. Gerber (1978). Techniques of frost prediction and methods of frost and cold protection. Technical Note 157, 487, WMO.

Becker, F. and Z. L. Li (1990). Towards a local "split window" method over land surface. *International Journal of Remote Sensing* 11(3), 369–393.

Blanc, M. L., H. Geslin, I. A. Holzberg, and B. Mason (1963). Protection against frost damage. Technical Note 51, WMO.

Chevalier, R. F., G. Hoogenboom, R. W. McClendon, and J. O. Paz (2012). A web-based fuzzy expert system for frost warnings in horticultural crops. *Environmental Modelling & Software* 35, 84–91.

Cittadini, E. D., N. de Ridder, P. L. Peri, and H. van Keulen (2006). A method for assessing frost damage risk in sweet cherry orchards of South Patagonia. *Agricultural and Forest Meteorology* 14, 235–243.

Dalezios, N. R. (1987). Development of a watershed system using estimation theory. In *Proceedings 3rd Greek Hydrotechnical Conference*, Greek Hydrotechnical Union, Thessaloniki, Greece. October 7–9, pp. 621–630.

Dalezios, N. R. (2015). Agrometeorology: Analysis and Simulation (in Greek). Kallipos: Libraries of Hellenic Universities (also e-book) 481 pages, November 2015.

Dalezios, N. R. (Ed) (2017). *Environmental Hazards Methodologies for Risk Assessment and Management.* 525 p. London, UK: IWA.

Dalezios, N. R., A. Blanta, and N. V. Spyropoulos (2012). Assessment of remotely sensed drought features in vulnerable agriculture. *Natural Hazards and Earth System Sciences* 12, 3139–3150.

Dalezios, N. R., A. Loukas, L. Vasiliades, and H. Liakopoulos (2000). Severity-duration-frequency analysis of droughts and wet periods in Greece. *Hydrological Science Journal* 45(5), 751–770.

Dalezios, N. R. and E. E. Lavrediadou (1995). Features of frost-affected areas from digital METEOSAT IR images. *Advances in Space Research* 15(11), 123–126.

Dalezios, N. R. and E. E. Lavrediadou (1994). Frost severity-duration-frequency relationships. In *Proceedings, 2nd Greek Scientific Conference on Meteorology, Climatology and Atmospheric Physics*, Thessaloniki, Greece. September 29–30, pp. 27–34.

Dalezios, N. R., N. V. Spyropoulos, and S. Eslamian (2017). Remote sensing in drought quantification and assessment. In S. Eslamian (Ed.) *Handbook of Drought and Water Scarcity (HDWS)*, Vol. 1 of 3-Volume, Chapter 21 Boca Raton, FL: Taylor & Francis Group (accepted, in press).

Domenikiotis, C., M. Spiliotopoulos, E. Kanellou, and N. R. Dalezios (2004). Mapping of temperature—Related areas in Greece for the study of radiation Frost. In *7th Panhellenic Geographical Conference*, Hellenic Geographical Society, Mytilini, Greece. October 14–17, 2004, pp. 74–81.

Domenikiotis, C., M. Spiliotopoulos, E. Kanellou, and N. R. Dalezios (2006). Classification of NOAA/AVHRR Images for Mapping of Frost Affected Areas in Thessaly, Central Greece. *International Symposium GIS and Remote Sensing: Environmental Applications*, University of Thessaly (UTH), Volos. November 7–9, 2003, pp. 25–32.

ELGA–GAIO (2003). Estimation Manual of Frost Damage in Apple Trees. Available: www.elga.grv [accessed January 13, 2015].

EM-DAT (2012). The OFDA/CRED International data base. Université catholique de Louvain, Brussels, Belgium. Available: http://www.emdat.be.

ESA (2014). Sentinel, earth online–ESA. Available: https://earth.esa.int/web/guest/missions/esa-future-missions/sentinel-1 [April 6, 2014].

Foot, J. S. (1993). Hazard warning in meteorology: The importance of remote sensing. In *Proceedings of IDNDR Conference on Natural Hazards and Remote Sensing*, London, UK, March 8–9, pp. 13–16.

Geerts, S., D. Raes, M. Garcia, C. Del Castillo, and W. Buytaert (2006). Agro-climatic suitability mapping for crop production in the Bolivian Altiplano: A case study for quinoa. *Agricultural and Forest Meteorology* 139, 399–412.

Ghaemi, A. A., M. R. Rafiee, and A. R. Sepaskhah (2009). Tree-temperature monitoring for frost protection of orchards in semi-arid regions using sprinkler irrigation. *Agricultural Sciences in China* 8(1), 98–107.

Hogg, W. H. (1971). Spring frosts. *Agriculture* 78(1) 28–31.

IPCC (2013). Climate change 2013: The physical science basis. In T. F. Stocker, D. Qin, G.-K. Plattner, M. Tignor, S. K. Allen, J. Boschung, A. Nauels, Y. Xia, V. Bex, and P. M. Midgley (Eds.) *Contribution of Working Group I to the Fifth Assessment Report of the Intergovernmental Panel on Climate Change*. Cambridge, UK: Cambridge University Press, 1535 pp. doi:10.1017/CBO9781107415324.

IPCC (2007). Climate change 2007: Impacts, adaptation and vulnerability. In M. L. Parry, O. F. Canziani, J. P. Palutikof, P. J. Van Der Linden, and C. E. Hanson (Eds.) *The Intergovernmental Panel on Climate Change*. Cambridge, UK: Cambridge University Press. 976 pp.

Jonsson, A. M., M. L. Linderson, I. Stjernquist, P. Schlyter, L. Barring (2004). Climate change and the effect of temperature backlashes causing frost damage in Picea abies. *Global and Planetary Change* 44, 195–207.

Jupp, D. L. B., G. Tian, T. R. McVicar, Y. Qin, and F. Li. (1998). Monitoring soil moisture and drought using AVHRR satellite data I: Theory. CSIRO Earth Observation Centre Technical Report, 98.1, Canberra, ACT.

Kalma, J. D., G. P. Laughlin, J. M. Caprio, and P. J. C. Hamer (1992). *Advances in Bioclimatology*. The Bio Climatology of Frost, Vol. 2. Berlin, Germany: Springer.

Kim, Y., J. S. Kimball, K. Didan, and G. M. Henebry (2014). Response of vegetation growth and productivity to spring climate indicators in the conterminous United States derived from satellite remote sensing data fusion. *Agricultural and Forest Meteorology* 194, 132–143.

Lin, H. and X. Lv (2010). Monitoring frost disaster of cotton based on difference of vegetation index and canopy temperature by remote sensing. In *2nd Conference on Environmental Science and Information Application Technology (ESIAT 2010)*, IEEE, pp. 552–557.

Louka, P., I. Papanikolaou, G. P. Petropoulos, and N. Stathopoulos (2015). Frost risk assessment tool in agriculture utilizing remote sensing and GIS techniques. In P. K. Srivastava, P. Pandey, P. Kumar, D. Han, and A. S. Raghubanshi (Eds.) *Geospatial Technology for Water Resource Applications*. Boca Raton, FL: CRC Press, Taylor & Francis Group.

Moeletsi, M. E., S. G. Moopisa, S. Walker, M. Tsubo (2013). Development of an agroclimatological risk tool for dryland maize production in the free state province of South Africa. *Computers and Electronics in Agriculture* 95, 108–121.

Monzon, J. P., V. O. Sadras, P. A. Abbate, and O. P. Caviglia (2007). Modelling management strategies for wheat–soybean double crops in the south-eastern Pampas. *Field Crops Research* 101, 44–52.

Pouteau R., S. Rambal, J. P Ratte, F. Gogι, R. Joffre, and T. Winkel (2011). Downscaling MODIS-derived maps using GIS and boosted regression trees: The case of frost occurrence over the arid Andean highlands of Bolivia. *Remote Sensing of Environment* 115, 117–129.

Prabhakara K., W. D. Hively, and G. W. McCarty (2015). Evaluating the relationship between biomass, percent groundcover and remote sensing indices across six winter cover crop fields in Maryland, United States. *International Journal of Applied Earth Observation and Geoinformation* 39, 88–102.

Pulatov, B., M. L. Linderson, K. Hall, and A. M. Jönsson (2015). Modeling climate change impact on potato crop phenology, and risk of frost damage and heat stress in northern Europe. *Agricultural and Forest Meteorology* 214–215, 281–292.

Shen, W., C. Zou, D. Liua, Y. Ouyang, H. Zhang, C. Yang, S. Bai, and N. Lin (2015). Climate-forced ecological changes over the Tibetan Plateau. *Cold Regions Science and Technology* 114, 27–35.

Smyth, M. and H. Skates (2009). A passive solar water heating system for vineyard frost protection. *Solar Energy* 83, 400–408.

Snyder, R. L. and J. Paulo de Melo-Abreu (2005). *Frost Protection: Fundamentals, Practice, and Economics*, Vol. 1. Rome: FAO Environment and Natural Resources Series. http://www.fao. org/docrep/008/y7223e/y7223e00.htm.

Tsiros, E., C. Domenikiotis, and N. R. Dalezios (2009). Sustainable production zoning for agroclimatic classification using GIS and remote sensing. *IDŐJÁRÁS* 113(1–2), 55–68.

Van de Griend, A. A. and M. Owe (1993). On the relationship between thermal emissivity and the normalized difference vegetation index for natural surfaces. *International Journal of Remote Sensing* 14(6), 1119–1137.

Ventskevich, G. Z. (1958). Agrometeorology. Translated from the Russian by the Israel Programme for Scientific Translation, Jerusalem, 1961.

Webb, L. and R. L. Snyder (2013). Frost hazards. In P. T. Bobrowsky (Ed.) *Encyclopedia of Natural Hazards*, Dordrecht, the Netherlands: Springer, pp. 363–366.

WMO (2010). Guide to agricultural meteorological practices. WMO–No. 134, 799 p.

7 Remote Sensing of Sea Ice Hazards
An Overview

Mukesh Gupta

CONTENTS

7.1 INTRODUCTION

A natural process or event, which becomes a potential threat to human life and property, is known as a natural hazard (Keller and DeVecchio 2012). Sea ice covers about 7.3% of Earth's surface, or about 11.8% of the surface of the World Ocean (Weeks 2010). During summer, it reflects sunlight back into the space and during winter, it radiates heat into the space (Perovich 1996). Sea ice acts as a large lid on the polar oceans, controlling the exchange of heat and mass across the ocean-sea ice-atmosphere (OSA) interface, thus playing a significant role in the Earth's climate system (Serreze and Barry 2005). The accelerated melting of Arctic sea ice in recent decades may have long-term impact on the Earth's climate likely to affect the lives and properties of billions of humans (Stroeve et al. 2007; IPCC 2014). This chapter identifies the hazards related to sea ice, factors that are responsible for inducing it or that directly play a role in causing the hazard, and the remote sensing-based investigation, detection, and possible mitigation of the ice hazard.

Sea ice, being a complex substance, behaves differently under different atmospheric and oceanic conditions at a range of temporal (seconds to decades) and spatial (mm to thousands of km) scales (Wadhams 2000). The hazardous nature of sea ice stems from its physical, rheological, morphological, thermodynamic, and hydrodynamic properties that vary with prevailing environmental conditions (Laxon 1994). This makes sea ice as one of the most complex geophysical substances for making predictions of its behavior. Sea ice also grows vertically undersea (called *keel* growth) (Steiner et al. 1999), which represents a major hazard for under ice navigation/operation of automated underwater vehicles, and is dangerous for scuba diving. Keels that extend up to the ocean floor (called *stamukhi*) pose a significant danger to the running oil pipelines or cables (Astafiev et al. 1991).

With the advent of spaceborne sensors in 1970s for monitoring polar oceans, it became possible to study sea ice in greater details than ever before covering all aspects of its behavior; thus, spaceborne sensors helped in identifying associated ice hazards and its possible prediction (Carsey 1992). Altimeters (Laxon et al. 2013), passive microwave radiometers (Tian-Kunze et al. 2014), and multispectral optical sensors (Liu et al. 2015a) have enabled the detection of ice thickness from space. However, better accuracy is required to have greater confidence in retrieving ice thickness from spaceborne observations, which can be assimilated in ice forecast models (Yang et al. 2014). Synthetic aperture radars (SARs) have been used to monitor the movement of sea ice, which is always in mobile state except the landfast ice (Kwok et al. 1998). SAR images are, in fact, being used commercially for navigation (Johannessen et al. 1992), ice tracking (Kwok et al. 1990), and high-resolution science experiments over polar regions (Maslowski and Lipscomb 2003). Multispectral radiometers provide wealth of information, however, limited to cloud-free conditions, on the sea ice leads (Onana et al. 2013), polynyas (Meier et al. 2013), and hazardous thaw holes (Digby 1984). Spaceborne scatterometers (Haarpaintner et al. 2004) and passive microwave radiometers (Meier and Stroeve 2008) have shown promise in identifying sea ice edge, thus greatly helping in the detection of extent of sea ice in a given area of the ocean where commercial or other activities involving life and property are conducted.

The focus of this chapter will be on the remote sensing detection of the potentially hazardous sea ice behavior and its mitigation measures. There are four sections in this chapter. Section 7.2 identifies the potential role of sea ice and its geophysical properties as a natural hazard. It covers sea ice thickness, ice motion, ice leads, ice roughness, melt ponds and thaw holes, ice edge, and oil spill in sea ice. Section 7.3 suggests mitigation measures to understand and circumvent the severity of ice hazard. Section 7.4 concludes the chapter with suggestions for future avenues and techniques required to make reasonable predictions related to sea ice hazards. In each section, we first describe the important physical processes related to ice hazard followed by how Earth observation (EO) technology can be helpful in a better understanding of those processes.

7.2 REMOTE SENSING OF SEA ICE AS NATURAL HAZARD

7.2.1 ICE THICKNESS

Sea ice is frozen seawater. During its formation at various temporal and spatial scales, it exhibits different properties depending on prevalent atmospheric conditions (Untersteiner 1986). Although the thickness of sea ice is a major hazard for ship navigation, the navigability of ships depends on the internal stresses of ice (Thorndike 1986). Generally, the icebreakers can easily cut through ~1 m (or more) thick ice if the ice is porous and has very little strength. Ice rheology plays a big role in response to different stresses acting on sea ice (Feltham 2008).

Sea ice is polycrystalline and behaves as a viscoelastic solid. However, sea ice in the marginal ice zone (MIZ) can be explained through different ice rheology representations, for example, plastic (Hibler 1989) and a rheology based on ice floe mechanics (Shen et al. 1987). Ice has near-zero tensile strength, which means if we apply divergent stresses in two dimensions, it will dilate and easily break apart (Figure 7.1). Ice has very high compressive strength, that is, it is difficult to crush the ice under compression (Timco and Frederking 1990). Ice also has significant shear strength (Saeki et al. 1985), which implies that when shear stress is applied on ice, it is more likely to slip and deform. This shear property facilitates the main cause of rafting and piling up of ice and the formation of rubble, pressure ridges, and hummocks under the influence of differential forces at the ice edge and in the MIZ due to wind and waves (Leppäranta 2011). The severity of sea ice hazard is thus governed by its rheological properties to a great extent.

The thickness of sea ice is not uniform everywhere. The ratio of the part of ice above mean sea level (sail height) and the part extending into the ocean below mean sea level (keel draft) is about 1:4.5 (Wadhams et al. 1992). An ice distribution is required for operational and theoretical

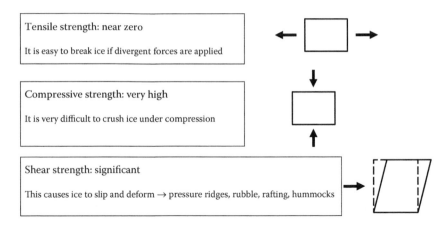

FIGURE 7.1   Rheological properties of sea ice.

computations, which can help reduce the impact of the ice hazard (Bourke and Garrett 1987). In a recent study, the sea ice distribution, $g(h)$ has been theoretically determined (Toppaladoddi and Wettlaufer 2015) as (Equation 7.1),

$$g(h) = N(q)h^q e^{-h/H} \tag{7.1}$$

where q and H are expressible in terms of moments over the transition probabilities between thickness categories. Due to harsh polar environment, technological limitations, and poor spatial and temporal coverage, the direct measurement of ice thickness distribution is still a challenge for the scientific community and technologists (Rothrock 1986). Over the past several decades, continuous satellite-based observations (see following paragraphs) in the polar areas have helped enormously in dealing with aforementioned issues.

Ice thickness is one of the most underestimated geophysical variables that are important to climate modeling (Battisti et al. 1997). The climate models include sea ice as it plays a significant role in controlling the heat budget (therefore global warming) and global ocean circulation of the Earth system (Vinnikov et al. 1999). However, the volume of sea ice is inaccurately represented due to uncertainty in ice thickness and its distribution given that the surface and bottom of sea ice is nonuniform, anisotropic, and heterogeneous. Most climate models use thermodynamic model of the growth of sea ice, which is unverified and thus underrepresents the reality. There are nil or scarce direct observations of ice thickness measured in the field. Satellite observations are close proxies, but not accurate, of the thickness variable. Altimeters (CryoSat-2; Ice, Cloud, and land Elevation Satellite [ICESat]), passive microwave (Soil Moisture and Ocean Salinity [SMOS]), SAR systems (e.g., Radarsat-2) have been used to estimate sea ice thickness to a variable degree of accuracy, which is essentially limited to thin ice. No satellite observations are able to provide accurate thickness of full ice draft due to uncertainties in snow depth, snow and sea ice density, and challenges in backscattering and waveform interpretation in altimetry (Laxon et al. 2013; Ricker et al. 2014) (Table 7.1). However, submarine-based sonar measurements provide good thickness observations, but these are spatially insignificant. Various thermodynamic model estimates suffer from complete lack or inadequate representation of ice thickness distribution. Thus, the problem of sea ice thickness and its adequate representation in various regional and global climate models is still a topic of ongoing research.

Kaleschke et al. (2012) provided an algorithm for estimating ice thickness up to 0.5 m from the SMOS brightness temperature. High penetration depth at L-band (1.4 GHz) and high brightness temperature contrast of over 100 K between ice and open water reflect in the L-band emission as increasing sea ice thickness (Huntemann et al. 2014). Therefore, it is imperative to assess the potential of retrieving sea ice thickness with SMOS (spatial resolution 35 km at nadir). The estimation of

TABLE 7.1

A Summary of Available EO Technological Resources, Their Strengths, Limitation/Challenges, and Level of Confidence for Different Geophysical Variables

Geophysical Variable Related to Sea Ice Hazard	EO Technology (Ground-Based, Airborne, or Spaceborne)	Strengths	Limitation/Challenges	Level of Confidence in EO technology
Ice thickness	CryoSat-2, ICESat, SMOS, Radarsat-2, MODIS	Good polar coverage	Limited to thin ice, waveform issues, lack of snow depth and density, no ice thickness distribution observations	Low to moderate
Sea ice motion	Radarsat-2, Envisat ASAR, SSM/I, AMSR-E	All weather	Requires good coherency between images	Moderate to high
Sea ice leads	SPOT, MODIS, AMSR-E, CryoSat-2, Envisat ASAR	Continuous, near real-time	Cloud artifacts, inaccurate ice concentration	High
Sea ice roughness	All passive microwave and SAR sensors	All weather	Lack of wind over ice, anisotropy, spatial and temporal challenges	Low to moderate
Melt ponds	MODIS, TerraSAR-X, Radarsat-2	All weather, multi-spectral	Pond interconnections and water classification	Moderate to high
Ice edge	QuikSCAT, AMSR-E, SSM/I	All weather	Uncertainty in ice concentration, coarse spatial resolution	Low to moderate
Oil spill in sea ice	SAR sensors, hyperspectral sensors, acoustic, passive microwave sensors, UV sensors	Multisensor	Oil weathering, lack of physical basis of oil–ice interaction	Low

ice thickness from passive microwave brightness temperature involves a number of approximations and uncertainties, which also depend on the type of emissivity model of sea ice. Kaleschke et al. (2012) assumed the bulk ice temperature and bulk ice salinity, which turns out to be a major shortcoming of this algorithm. Tian-Kunze et al. (2014) overcame this issue by considering the profiles of ice temperature and salinity varying with depth in the ice. Figure 7.2 provides a comparison of ice thickness retrieval using two algorithms: Kaleschke et al. (2012) (Algorithm I) and Tian-Kunze et al. (2014) (Algorithm II). Algorithm II enables ice thickness estimates up to 1.5 m. However, the coarse resolution (35 km grid size) of SMOS prevents detection of smaller leads and polynyas and ignores the effect of snow cover in thin ice thickness estimation.

Although microwave instruments (e.g., SMOS) are preferred due to their all-weather capability, other remote sensing instruments (optical and thermal), for example, Moderate Resolution Imaging Spectroradiometer (MODIS), have also demonstrated capability in estimating ice thickness.

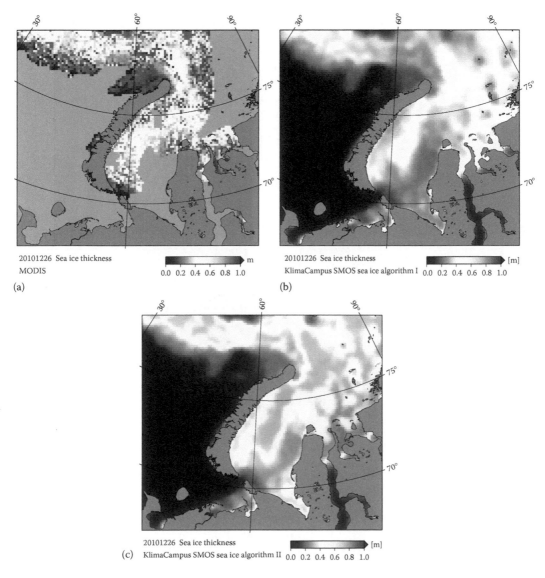

FIGURE 7.2 (a) MODIS ice thickness with 12.5 km grid resolution. (b) Ice thickness derived using Kaleschke et al. (2012) (Algorithm I). (c) Ice thickness obtained using Tian-Kunze et al. (2014) (Algorithm II). (From Tian-Kunze, X. et al., *Cryosphere*, 8, 997–1018, 2014. With permission.)

The MODIS ice thickness of the same area is shown in Figure 7.2a for comparison with SMOS ice thickness. MODIS ice thickness is calculated using thermal ice temperature and High-Resolution Limited Area Model (HIRLAM) (Mäkynen et al. 2013). This MODIS algorithm provides ice thickness up to 0.99 m. It is shown that MODIS ice thickness values larger than 0.5 m are areas of thick ice without accurate ice thickness estimates. This uncertainty is due to exclusion of solar shortwave radiation and surface albedo, ambiguity in thin cloud detection, and inaccurate numerical weather prediction forcing data.

Apart from passive microwave and optical satellite imageries, satellite radar altimeters such as CryoSat-2 (Ku-band) have also been used to estimate sea ice volume and thickness (Laxon et al. 2013). Figure 7.3 shows a recent near real-time (NRT) sea ice thickness product (at 5 km grid; also

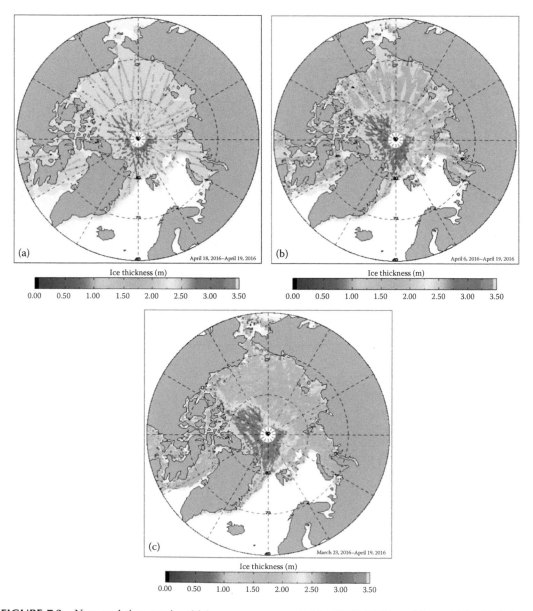

FIGURE 7.3 Near real-time sea ice thickness maps generated by CPOM/ESA at 5 km grid for (a) 2 days, (b) 14 days, and (c) 28 days during March–April 2016. (Courtesy of Centre for Polar Observation and Modelling Data Portal (CPOM): http://www.cpom.ucl.ac.uk/csopr/seaice.html)

available at 1 km for individual sectors) generated by Centre for Polar Observation and Modelling/ European Space Agency (CPOM/ESA) for 2-, 14-, and 28-day time period. Ice thickness is calculated by combining the freeboard measurements and estimates of snow depth and density derived from climatology (Warren et al. 1999; Laxon et al. 2013). NRT data are preliminary fast access products, which ignore all the precise corrections that come with the final products made available 30 days later. As the ice thickness is estimated using the freeboard and ice/snow density, it is able to provide full thickness of the ice (from keel-depth to sail-height). Similar to CryoSat-2 ice thickness, Geoscience Laser Altimeter System (GLAS) instrument (wavelength 1064 nm) onboard ICESat provided ice thickness calculated using freeboard and snow density (Kwok and Cunningham 2008). However, these products were available till 2008. Thus, CryoSat-2 provides one of the best available datasets of ice thickness. This can be tremendously useful in dealing with ice hazards and mitigation plans.

7.2.2 SEA ICE MOTION

Moving ice can be more hazardous than the landfast ice that remains attached to the shore. The continuous motion of sea ice further aggravates the hazard caused by it. This motion is governed by a number of forces that act on sea ice. Under various forces, the ice moves in the ocean in a divergent or convergent manner to create surface roughness. The forces given in the following act as a total force (F) that controls the sea ice motion (Equation 7.2) (Leppäranta 2011):

$$F = \tau_a + \tau_w + F_C + F_i + F_t \tag{7.2}$$

where:

τ_a and τ_w are wind and water drag, respectively
F_C is Coriolis force
F_i is internal stress
F_t is sea surface tilt

Four major external forces working on sea ice control the deformation and movement of an ice floe: (1) wind force; (2) water drag—it is the frictional force between ice and seawater, which is controlled by water density, temperature, and underwater currents; (3) Coriolis force—this is one of the precisely calculated forces on sea ice unlike wind and water drag, which are computed using semiempirical formulae (Steele et al. 1997). The Coriolis force arises due to Earth's rotation. An ice floe experiences acceleration caused by rotation of the Earth deflecting the original trajectory of the floe to move clockwise in the northern hemisphere. The magnitude of Coriolis force is given by Equation 7.3:

$$F_C = 2m\omega U \sin\phi \tag{7.3}$$

where:

m is the mass of the ice floe
ω is angular velocity of Earth $= 7.272 \times 10^{-5}$ rad s^{-1}
U is ice velocity
ϕ is latitude

The ice floe appears to move toward the right of its velocity direction in the northern hemisphere. The Coriolis force is also observed in the atmosphere. It is zero at the equator and maximum toward the poles; therefore, big ice floes due to their large mass experience greater Coriolis force in the Arctic. Other forces are, for example, internal ice stress—it acts on a unit area of ice in an ice floe due to transmitted stress within the ice floe. Winds or ocean currents play a greater role in generating internal stresses in addition to other forces acting on ice (Leppäranta 2011). The net stress can be the result of all the stress vectors on a given ice floe. The sustainability of stress within the ice is dependent on ice thickness distribution, ice rheology, and the strength of ice cover. Multiyear ice, which is thicker than first-year ice, is more likely to sustain greater internal ice stresses. All the aforementioned forces control the motion of a sea ice floe, which can be monitored using the microwave satellite images.

Sea ice motion is calculated from SAR images acquired under similar conditions in two or three consecutive passes (i.e., the repeat cycles of SAR) using different algorithms including the most popular cross-correlation method (e.g., Thomas et al. 2011). The ice motion algorithms produce satisfactory results with the C-band SAR images commonly used in sea ice remote sensing. In a recent study, Lehtiranta et al. (2015) have found that L-band images produce better results than C-band images for ice tracking. There are, however, limitations in estimating ice motion from consecutive satellite images (Table 7.1). The ice surface requires coherency, which means that it must not change during the time span between two images. The microwave signatures (active or passive) of sea ice are highly sensitive to meteorological conditions such as ice surface temperature, wind, and air temperature (Gupta 2014). This alters the ice surface characteristics (surface roughness and dielectric constant), thereby changing its microwave response. Second, only average ice velocity can be determined. Generally, two images from the satellite repeat pass that are 2–3 days apart are used for ice motion estimation, but this is difficult to achieve as most radar satellites have a repeat cycle of more than 2–3 days for the same site. Image pairs from two different satellites provide an ideal dataset required for ice motion estimation, but this suffers from the two images being of different characters and requires additional processing. Figure 7.4 shows the ice motion vectors obtained in the Gulf of Bothnia using Radarsat SAR and Envisat ASAR images that are 2 days apart (Lehtiranta et al. 2015).

Ice displacement vectors can also be obtained from the passive microwave imageries, Special Sensor Microwave/Imager (SSM/I, 85 GHz), and Scanning Multichannel Microwave Radiometer (SMMR, 37 GHz) (Kwok et al. 1998). The magnitude of the normalized cross-correlation coefficient of brightness temperature is used as a measure of similarity between features in the two passive microwave images. Daily ice motion vectors are derived from the

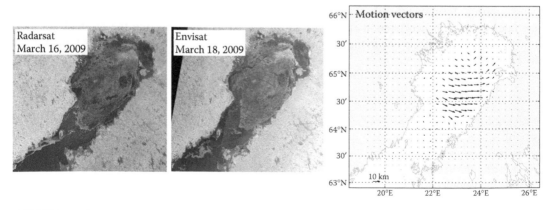

FIGURE 7.4 Sea ice motion vectors calculated from combining C-band Radarsat SAR (March 16, 2009) and Envisat ASAR (March 18, 2009) images in the Gulf of Bothnia of Baltic Sea. (From Lehtiranta, J. et al., *Cryosphere*, 9, 357–366, 2015. With permission.)

89 GHz channel of Advanced Microwave Scanning Radiometer for Earth Observing System (AMSR-E) at 6.25 km spatial grid (Kimura et al. 2013).

7.2.3 Sea Ice Leads

Leads are narrow openings (1–10^4 m wide and several km long) in the ice, which can turn into a hazard for people working nearby, for example, scientists set up an instrument near a lead (Esau 2007). Leads are of utmost importance as they provide an easy path for navigating ships. Second, leads are the hot spots of exchange of mass, momentum, and heat at the OSA interface. The leads cover 1%–5% of the central Arctic (Miles and Barry 1998) and up to 20% of the Arctic marginal seas, accounting for more than 70% of the upward heat fluxes (Lindsay and Rothrock 1995; Inoue et al. 2005). Refrozen leads can camouflage the ice field, thus turning hazardous. Narrow leads are about twice more efficient in transmitting turbulent heat fluxes than large leads. In a recent study, Marcq and Weiss (2012) have shown that lead widths are power law distributed. Turbulent heat fluxes over leads depend on the distribution of lead width and decrease with increasing lead width (Esau 2007; Marcq and Weiss 2012). Leads have been successfully identified using optical and microwave EO technology to be discussed later.

Sea ice leads can be easily detected on optical satellite imageries. Figure 7.5a shows a grayscale Satellite Pour l'Observation de la Terre (SPOT; Earth-observing satellite) image of sea ice cover with image center at 80°11′N, 108°33′W. The peaks represent different states of refreezing over the three largest leads (Figure 7.5b). The peaks clearly discriminate leads from the ice. Adequate lead detection algorithms (e.g., AMSR-E lead detection algorithm by Röhrs and Kaleschke (2014); CryoSat-2 lead detection algorithm by Wernecke and Kaleschke (2015)) are used for automatic identification of leads in an image. Onana et al. (2013) have developed a new algorithm (sea-ice lead detection algorithm using minimal signal [SILDAMS]) that extracts leads and classifies ice types within the lead from an airborne visible imagery.

The fraction of leads in a sea ice cover can be seen as a parameter reflecting the loss in mechanical strength of ice pack controlling the degree of mobility of pack ice. Figure 7.6 shows lead area fraction obtained from the AMSR-E passive microwave imagery (ICDC 2003; Röhrs and Kaleschke 2012); and CryoSat-2 image (ICDC 2013; Wernecke and Kaleschke 2015). In a recent study, Ivanova et al. (2016) have investigated the lead fraction estimates from AMSR-E and

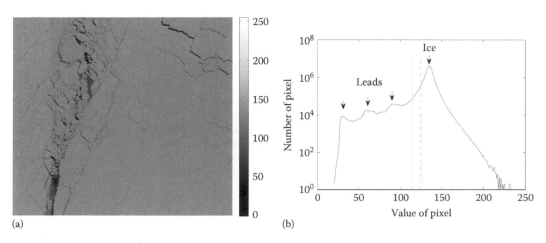

(a)   (b)

FIGURE 7.5 (a) Grayscale SPOT image of sea ice cover and (b) pixel histogram of the SPOT image; the dashed lines are thresholds of 115 and 125. (From Marcq, S. and Weiss, J., *Cryosphere*, 6, 143–156, 2012. With permission.)

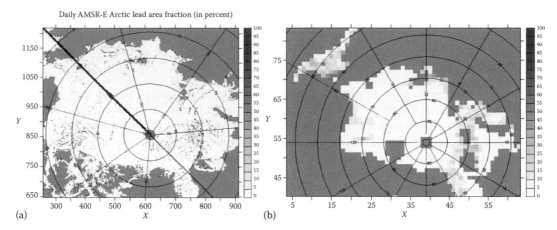

(a)   (b)

FIGURE 7.6 (a) Lead area fraction in percentage (%) from AMSR-E passive microwave imagery of March 4, 2003. (From ICDC [Integrated Climate Data Center], AMSR-E lead area fraction for the Arctic, (March 4, 2003), Integrated Climate Data Center [ICDC, http://icdc.zmaw.de/], University of Hamburg, Hamburg, Germany, 2003. With permission.); (b) lead area fraction (%) from CryoSat-2 image of February 1, 2013. (From ICDC [Integrated Climate Data Center], CryoSat-2 lead area fraction for the Arctic, [February 1, 2013], [ICDC, http://icdc.zmaw.de], University of Hamburg, Hamburg, Germany, 2013. With permission.)

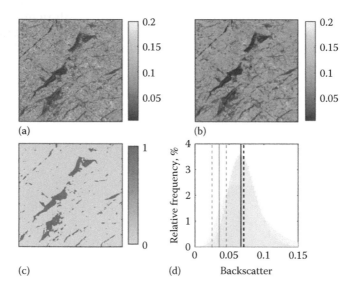

FIGURE 7.7 Lead area fraction calculated using threshold technique using Envisat ASAR HH polarization imagery of March 1, 2009: (a) backscatter, (b) median-filtered backscatter, (c) lead detected (lead-1, ice-0), (d) histogram of the image. Gray and black lines are the thresholds and 1.5 standard deviation as dashed line. The beige bars are the unfiltered signal. Gray dashed line is the mean. (From Ivanova, N. et al., *Cryosphere*, 10, 585–595, 2016. With permission.)

Envisat ASAR images (Figure 7.7). Sea ice leads are warmer in comparison to surrounding sea ice cover during wintertime. This property of leads can be utilized to detect them using a thermal infrared imagery such as MODIS (Figure 7.8) (Willmes and Heinemann 2015, 2016). Some of the challenges in ice lead detection are cloud artifact filtering and improved ice concentration estimation (Table 7.1).

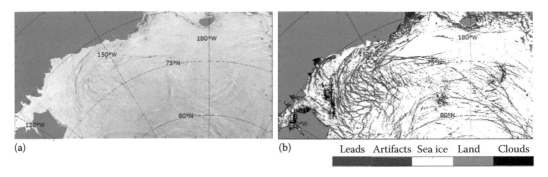

FIGURE 7.8   (a) MODIS band 2 (841–876 nm), April 4, 2009, swath normalized and daily averaged reflectance. (b) Fuzzy cloud artifact filter (FCAF) applied to extract leads. (From Willmes, S. and Heinemann, G., *Remote Sens.*, 8, 4, 2016. With permission.)

7.2.4   SEA ICE ROUGHNESS

Sea ice roughness (aerodynamic or physical) is the manifestation of turbulent processes at the OSA interface (McPhee 2002). With a few exceptions, sea ice always turns into a rough field (ice surface as well as ice bottom) to a varying degree of spatial scales under the influence of atmospheric and oceanic forcing (Andreas 1987). Sea ice hazard is more difficult to mitigate under rough ice conditions. The rough ice terrain is almost inaccessible, difficult for ships to navigate, and unpredictable. The rough ice (rubble ice, pressure ridges) dramatically alters the heat and momentum fluxes in the atmospheric boundary layer (McPhee 1992). The regional and global climate models exclude the atmospheric boundary layer above sea ice due to poor understanding of the boundary layer physics, inaccurate parameterizations, and difficulty in acquiring relevant *in situ* data over sea ice for validation (Ebert and Curry 1993; Gettelman and Rood 2016). Since the advent of satellite remote sensing of polar ice in 1970s, one of the major research objectives of scientists has been to incorporate sea ice boundary layer into the regional numerical models for a better understanding of the effects of ice roughness on the regional climate and associated processes (Herzfeld et al. 2015).

Efforts have been put to derive information on ice roughness from ship-based, airborne, and satellite remote sensing techniques (Gupta 2015). This requires parameterization of boundary layer processes of momentum and heat fluxes at the OSA interface (Andreas 2002). New parameterizations of air–ice and ice–water drag coefficients have been developed as a result of advances in instrumentation, field experiments, and laboratory work (Lu et al. 2016). The aerodynamic roughness length is determined from wind flow characteristics measured from a meteorological tower (Zippel and Thomson 2016). It is possible to estimate the aerodynamic roughness length from the measurements of surface characteristics. Such surface features are the size and distribution of roughness elements (rms height) (Kustas and Brutsaert 1986; Vries et al. 2003). Parameterization of ice drag coefficient using EO technology (e.g., laser altimeter) is currently of much interest to the scientific community and needs to be explored further (Elvidge et al. 2016).

Satellite or airborne altimetry is extremely useful for measuring physical roughness of the ice-covered ocean (Rivas et al. 2006) and is the only known way of parameterizing the drag coefficient using remote sensing (Petty et al. 2016). Sea ice roughness distribution is acquired from a helicopter-based laser altimeter and an electromagnetic induction system in an area dominated by first-year ice and MIZ (Gupta 2014, 2015). However, altimeters provide point/profile measurements and do not cover large swaths, which are important requirements for measuring ice roughness (Table 7.1). The altimeter data can be resampled to match the spatial resolution of other microwave satellite data such as SSM/I. Some of the satellite altimeters used in Arctic applications are Seasat, European Remote Sensing (ERS-1/2) satellite, Envisat Radar Altimeter-2 (past); CryoSat-2, Satellite with ARgos and ALtika (SARAL) (current) (Kwok and Rothrock 2009); and Sentinel-3, ICESat-2 (future).

Although parameterization of aerodynamic roughness using remote sensing is still an emerging science, different EO techniques (e.g., polarimetric coherences and ratios) have been utilized to estimate the physical roughness of various sea ice surfaces (Wakabayashi et al. 2004). Polarimetric coherences and ratios (ground-based or spaceborne) are helpful in discriminating ice types and roughness categories (Gupta 2014). Ship-based observations of co- (linear), cross-polarized, and circular polarimetric coherences (ρ_{VVHH}, ρ_{HHVH}, and ρ_{RRLL}) are used to evaluate ice surface discrimination using polarimetric radar operating in C-band (Equations 7.4 through 7.8) (Gupta 2015)

$$\rho_{VVHH} = \frac{\left\langle \left| S_{VV} S_{HH}^* \right| \right\rangle}{\sqrt{\left\langle \left| S_{VV} \right|^2 \right\rangle \left\langle \left| S_{HH} \right|^2 \right\rangle}} \tag{7.4}$$

$$\rho_{HHVH} = \frac{\left\langle \left| S_{HH} S_{VH}^* \right| \right\rangle}{\sqrt{\left\langle \left| S_{HH} \right|^2 \right\rangle \left\langle \left| S_{VH} \right|^2 \right\rangle}} \tag{7.5}$$

$$\rho_{RRLL} = \frac{\left\langle \left| S_{HH} - S_{VV} \right|^2 \right\rangle - 4 \left\langle \left| S_{HV} \right|^2 \right\rangle}{\left\langle \left| S_{HH} - S_{VV} \right|^2 \right\rangle + 4 \left\langle \left| S_{HV} \right|^2 \right\rangle} \tag{7.6}$$

$$\gamma_{co} = \frac{S_{VV}}{S_{HH}} \tag{7.7}$$

$$\gamma_{cross} = \frac{S_{HV}}{S_{HH}} \tag{7.8}$$

where S is the complex scattering matrix. The elements of S are complex numbers containing magnitude and phases of transformed electric field. An asterisk (*) represents the complex conjugate. The brackets $\langle . \rangle$ represent ensemble averages of the observed data. The field-based acquisition of C-band signatures of ice physical roughness forms the basis of improving EO algorithms using similar sensors from space platform. In a recent study, Fors et al. (2016) have utilized polarimetric coherences and ratios using Radarsat-2 (C-band) SAR images for estimating sea ice surface physical roughness.

A SAR provides detailed information on the surface and volume scattering from the sea ice surface (Onstott 1992). SAR scattering of sea ice is a major and very important tool for improving our knowledge of surface roughness in the ice-covered ocean as the microwave signatures of ice vary with the changing surface dielectric and physical properties (Drinkwater 1989). SAR sensors are highly sensitive to the changes occurring at the ice surface. SAR sensors have been operated on airborne (Nakamura et al. 2005) and satellite platforms (Dierking 2010). A summary of successful/planned SAR missions is given in Table 7.2.

Satellite-based passive microwave sensors (e.g., AMSR-E) provide nearly complete temporal (1–2 day) data coverage (especially for polar regions; e.g., sea ice concentration, extent), which SARs and altimeters do not provide (Stroeve et al. 2006). The spatial resolution of passive microwave sensors at the most desired frequencies is, however, approximately limited to 30 km (Spreen et al. 2008). This restricts subpixel evaluation (higher resolution) of sea ice passive microwave emission signatures using satellite-based sensors (Gupta 2014). Stroeve et al. (2006) have investigated the

TABLE 7.2

A Summary of Past, Current, and Future SAR Missions Suitable for Sea Ice Hazard Studies

Period	Mission	Reference
Past	Spaceborne Imaging Radar-C (SIR-C)	Lang et al. (2014)
	ERS SAR	Johannessen et al. (1992)
	Envisat ASAR	Lehtiranta et al. (2015)
	Radarsat-1 SAR	Lehtiranta et al. (2015)
	Japanese Earth Resources Satellite (JERS-1) SAR	Dierking and Busche (2006)
	Advanced Land Observing Satellite—Phased Array type L-band Synthetic Aperture Radar (ALOS PALSAR)	Dierking (2010)
Current	Radar Imaging Satellite (RISAT-1) SAR	Srisudha et al. (2013)
	RADARSAT-2	El-Hilo et al. (2013)
	Sentinel-1	Muckenhuber et al. (2016)
	ALOS-2	—
Future	Sentinel-3	—
	ALOS-3	—
	Radarsat Constellation Mission (RCM)	—
	NASA-Indian Space Research Organisation (ISRO) SAR (NISAR)	—

TABLE 7.3

A Summary of Past, Current, and Future Passive Microwwave Satellite Missions Suitable for Sea Ice Hazard Studies

Period	Mission	Reference
Past	SSM/I	Kwok et al. (1998)
	AMSR-E	Kimura et al. (2013)
	SMMR	Kwok et al. (1998)
Current	AMSR-2	Karvonen (2014)
	SSMI/S	Fetterer et al. (2002)
	SMOS	Huntemann et al. (2014)
	Soil Moisture Active Passive (SMAP)	Ludwig (2016)
Future	SSMI/S	—

effect of surface roughness on passive microwave emissions from sea ice. The passive microwave emissions from the ocean surface increase with increasing surface roughness due to increased surface area (Hollinger 1971). Table 7.3 provides a summary of some of the past, current, and future passive microwave missions suitable for sea ice hazard studies. Figure 7.9 shows monthly sea ice concentration derived from Defense Meteorological Satellite Program (DMSP)-F17/Special Sensor Microwave Imager/Sounder (SSMIS) passive microwave brightness temperature for March 2016 (Fetterer et al. 2002). These types of satellite observations have become an ideal source of information in mitigating sea ice-related hazards and for proper planning and management of resources in the polar regions.

FIGURE 7.9   Monthly sea ice concentration for March 2016 derived from Defense Meteorological Satellite Program (DMSP)-F17/Special Sensor Microwave Imager/Sounder (SSMIS) passive microwave brightness temperature. (Courtesy of National Snow and Ice Data Center/NASA Earth Observatory; From Fetterer, F. et al., Updated daily. Sea Ice Index, Version 1. [Sea ice concentration], National Snow and Ice Data Center [NSIDC], Boulder, CO, 2002, accessed April 26, 2016. With permission.)

7.2.5   MELT PONDS

Melt ponds are puddles of meltwater from the surface of first-year or multiyear sea ice, usually occurring during the melt onset and lasting till they refreeze or drain into the ocean (Fetterer and Untersteiner 1998). A significant change in the surface roughness, ice morphology, and topography is observed after the melt onset in mid-June in the Arctic (Scharien and Yackel 2005). The melt-water forms a network of ponds over the surface. The depth and area of initially small and shallow melt ponds increase as the summer progresses (Maykut 1986). Eventually, the meltwater drains into the sea through holes called *thaw holes*. The melt ponds that do not drain through thaw holes refreeze in the fall (Taylor and Feltham 2004). It is the thaw holes that are hazardous and can lead to loss of instruments installed usually during summer on sea ice or accidents during hunting (Laidler et al. 2009). In addition to thaw holes being hazardous, melt ponds also affect the variability of surface albedo of the sea ice cover, thus impacting the heat budget of the atmospheric boundary layer (Morassutti and LeDrew 1996). It should also be mentioned that thaw holes are the sources of sunlight for the flora and fauna underneath sea ice, and Arctic seals usually come out of these holes for sunlight (Digby 1984). Thaw holes serve as one of the favorite spots for polar bears to look for food. The next paragraph discusses how optical and microwave remote sensing techniques can be used to detect and estimate melt pond coverage.

As the melt ponds can significantly alter the surface albedo of the sea ice cover, it is possible to study melt pond coverage using images from optical satellites such as MODIS (Tschudi et al. 2008). Estimates of melt pond coverage are made as melt pond fraction, which is the ponded area relative to the sea ice cover (Schröder et al. 2014). Figure 7.10 shows the surface albedo curves for different sea ice surface types including various stages of melt pond development (Rösel et al. 2012). Schroder et al. (2014) have observed an increase in melt pond fraction in the Arctic since 1979 correlating with

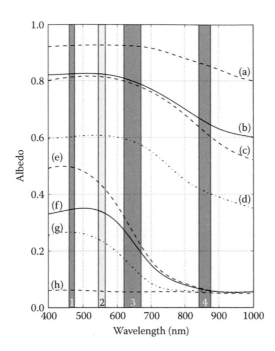

FIGURE 7.10 Spectral albedo for different sea ice surfaces. (a) Snow-covered ice (dry snow), (b) cold bare ice, (c) wet snow, (d) melting first-year ice, (e) young melt pond, (f) and (g) two types of older melt ponds, and (h) open water. MODIS bands are shown as grey vertical strips: 1 (band3, 459–479 nm), 2 (band4, 545–565 nm), 3 (band1, 620–670 nm), and 4 (band2, 841–876 nm). (From Rösel, A. et al., *Cryosphere*, 6, 431–446, 2012. With permission.)

the Arctic sea ice minimum extent, and the increase of melt ponds (4%/decade in July) results in the observed decrease in summer albedo (3%/decade in July/August). In a recent study, Liu et al. (2015b) have observed an increase in melt pond fraction during 2000–2011 showing that the Arctic sea ice minimum extent can be predicted using the melt pond fraction estimated from MODIS (Figure 7.11).

Optical satellites such as MODIS are constrained by limited solar illumination and persistent cloud cover in the polar regions. The potential of high-resolution airborne SAR (0.3 m) and TerraSAR-X images has been explored in mapping the melt ponds on first-year ice (Kim et al. 2013) and on multiyear

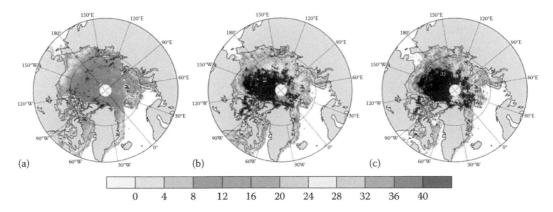

FIGURE 7.11 Melt pond fraction obtained from MODIS from May 9 to (a) June 2, (b) June 26, and (c) July 20 during 2000–2011. Color represents the averaged pond fraction for the given day. Black areas are the statistically significant correlations (integrated from May 9 to the given day) between the pond fraction and the extent of sea ice in September. (Liu, J. et al., *Environ. Res. Lett.*, 10, 054017, 2015b. With permission.)

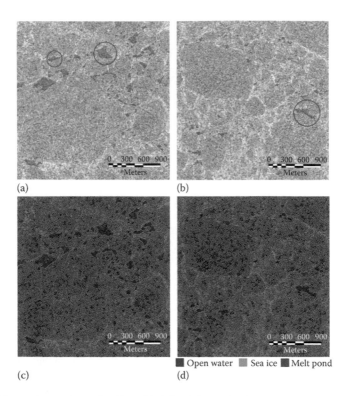

FIGURE 7.12 Melt pond mapping using TerraSAR-X HH-polarized amplitude image (a, b), and rule-based machine learning approach (random forest) for melt pond mapping of the same scene (c, d). Yellow circles indicate well-identified melt ponds. (From Han, H. et al., *Remote Sens.*, 8, 57, 2016. With permission.)

ice (Han et al. 2016). The shape and size of melt ponds derived from high-resolution TerraSAR-X provided greater details and better accuracy than that obtained from aerial photographs. However, pond fraction estimation, pond interconnections, and open water classification using SAR still require further improvements (Figure 7.12). Scharien et al. (2012) have assessed, using high-resolution surface-based C-band radar scatterometry, the role of various geophysical parameters that control the melt pond behavior in the high Arctic and MIZ. Figure 7.13 shows Radarsat-2-retrieved melt pond fraction using a

FIGURE 7.13 Melt pond fraction retrieved using a Radarsat-2 C-band copolarized ratio (VV/HH)-based model (called CV model) in the Parry Channel, Canadian Arctic Archipelago. Land is masked as white. Scene R1—May 12, R2—June 13, R3—June 20, R4—June 24, and R5—June 26, 2012. (From Scharien et al., 2014. With permission.)

model (called cross-validation (CV) model) that uses copolarized ratio (VV/HH) (Scharien et al. 2014). As suggested, the retrievals using Radarsat-2 still require more work to ensure accurate pond detection (Table 7.1); however, the transition from ice-covered regime to the formation of melt ponds (melt onset) is well detected by the Radarsat-2 images (R1 to R2, Figure 7.13). A better understanding of the formation, development, and disappearance of melt ponds can tremendously assist in planning to mitigate melt pond hazards.

7.2.6 Ice Edge

According to World Meteorological Organization (WMO), ice edge is defined as "the demarcation at any given time between the open sea and sea ice of any kind, whether fast or drifting. It may be termed compacted or diffuse (cf. ice boundary)." Accurate location and extent of the boundary between open water and sea ice are important for various activities including mitigation of sea ice hazards (Meier and Stroeve 2008). Ice edge detection is essential for secure navigation, protection of oil rigs from ice hazards, and for high-resolution geophysical modeling (Bitz et al. 2005). Optical identification of ice edge is relatively quick and easy, only if the optical satellite images are cloud-free, which is uncommon in the polar regions. Satellite images and airborne reconnaissance surveys provide valuable information on the ice edge location, extent, and likely future shifts. Microwave remote sensing is preferable due to its all-weather capability, for example, day/night or under cloud cover. Active microwave scatterometer, for example, QuikSCAT has shown potential for ice edge detection in the polar regions. The algorithm uses active polarization ratio (APR) defined by Tonboe and Ezraty (2002) as Equation 7.9,

$$\text{APR} = \frac{(\sigma_{0H} - \sigma_{0V})}{(\sigma_{0H} + \sigma_{0V})} \tag{7.9}$$

The skill to detect ice edge also relies upon the sea ice type to be appropriately interpreted using the satellite image (active or passive). For example, frazil ice can be visually interpreted on SAR images; however, a similar signature from QuikSCAT may be interpreted as open water (Haarpaintner et al. 2004). Polarization and gradient ratios have been widely used for estimating sea ice concentration and ice extent using AMSR-E and SSM/I brightness temperature at different frequencies (Swift and Cavalieri 1985). Sea ice concentration calculations are based on certain algorithms, for example, National Aeronautics and Space Administration (NASA) Team and Bootstrap algorithm, which use a percentage (e.g., 10%) of ice concentration thresholds in a pixel area (Comiso et al. 1997). This defines the ice edge; however, this estimate may render spurious ice edge identification due to atmospheric moisture and wind-roughened ocean surface (Table 7.1).

In recent years, ice edge prediction has drawn utmost attention of researchers and stakeholders (e.g., oil companies) who require accurate location of sea ice edge for hazard-free ship navigation, conducting scientific experiments, and planning of infrastructure development for installation of new offshore oil platforms. U.S. National Ice Center (NIC) generates daily ice edge products using multiple sources of NRT satellite data (visible, infrared, passive microwave, scatterometer, and SAR images), buoy data, satellite-derived products, and meteorological data (Posey et al. 2015). These products define ice edge as areas of less than 10% sea ice concentration (Figure 7.14) and are used for navigational purposes to avoid ice hazards. Ice edge detection techniques combining different kinds of satellite sensors are also in use, for example, QuikSCAT, AMSR-E, and SSM/I (Meier and Stroeve 2008). In recent decades, the volume of data from various kinds of satellite sensors, airborne sensors, and *in situ* observations have grown substantially. This useful data is assimilated in thermodynamic/numerical models of ice edge detection to get the most accurate sea ice–water edge discrimination, and the products are continually updated with new observations at high temporal and spatial resolution (Posey et al. 2015).

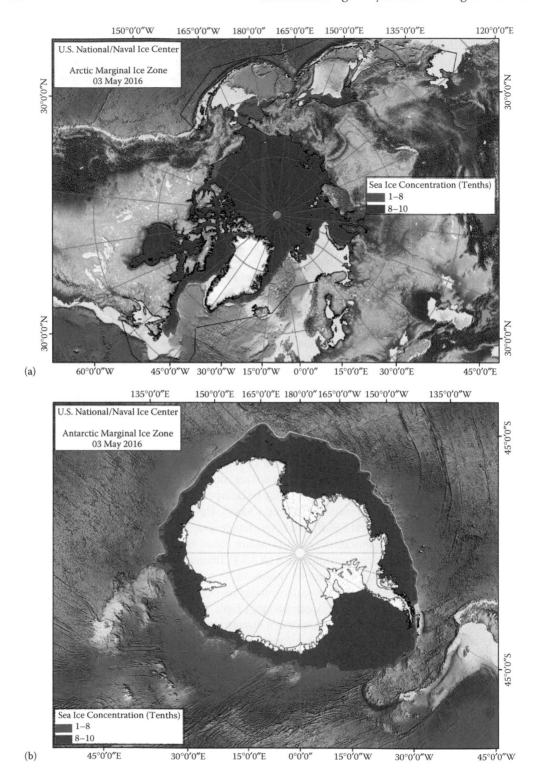

FIGURE 7.14   United States National Ice Center (NIC) ice edge products showing the ice edge and the marginal ice zone in yellow (a) Arctic, and (b) Antarctic. (From USNIC (United States National Ice Center/Naval Ice Center), Daily ice analysis products, http://www.natice.noaa.gov/daily_graphics.htm (Accessed on May 3, 2016), 2016. With permission.)

7.2.7 Oil Spill in Sea Ice

This section describes the science behind the remote sensing detection of oil spill in sea ice. The mitigation of oil spill in sea ice is discussed in Section 7.3. Of all the hazards related to sea ice, oil spill is one of the most devastating hazards from the viewpoint of biodiversity of polar regions (Owens et al. 1998). Due to inaccessibility and extreme cold temperatures, the hydrocarbon exploration in the Arctic, in case of an accident, can turn into an irrecoverable ecological hazard (Potter et al. 2012). There is a need to understand the oil spill in sea ice, resulting advection and dispersion of oil-smeared ice, and its impact on the vulnerability of the coastlines, landfast ice, and biodiversity at the ice bottom (Wilkinson et al. 2013). Satellite remote sensing is near ideal for monitoring such disasters on timely and regular basis. Optical and microwave sensors have been efficiently used for oil spill management, detection, and mitigation purposes for open ocean and ice-free seas (Brekke et al. 2014; Liu et al. 2016). Oil spill in sea ice poses greater challenges in terms of spatial and temporal frequency of observations from three different regions affected by the oil spill: (1) open water, (2) MIZ, and (3) ice-covered ocean (Dickins and Buist 1999; Afenyo et al. 2016). The optical and microwave signatures of these three distinct regions are difficult to characterize with available scientific understanding of oil spill in sea ice (Table 7.1).

Active microwave: Electromagnetic response of oil in sea ice is expected to have a lower relative permittivity than newly formed sea ice (Brekke et al. 2014). The SAR signatures of oil-smeared sea ice also pose difficulty due to changing nature of physical and chemical properties of oil over time. This is called *weathering* of oil in sea ice. Weathering process includes spreading, drift, evaporation, dissolution, dispersion, emulsification, flocculation, biodegradation, and oxidation. The dielectric constant, a parameter that radar is highly sensitive to, is dependent on the type of oil and its characteristics. It is possible to identify oil spill in sea ice using the radar sensors; however, it is hindered by the time varying spatial variability of the target. Brekke et al. (2014) have shown that the copolarization ratio (VV/HH) using SAR can be used for discriminating oil in sea ice from freshly frozen ice-covered Arctic. The dielectric properties of oil-smeared sea ice under varying wind conditions are less understood, indicating a need for dielectric models of a mixture of water and different types of oil and sea ice. The field, lab, and modeling experiments conducted by Bradford et al. (2008) indicate that ground penetrating radar (GPR) methods can detect oil films under sea ice as long as adequate energy reaches the ice–water interface (Bradford et al. 2010).

Hyperspectral: Near-infrared and optical remote sensing (multi- and hyperspectral) have been shown to have greater potential than SAR in detecting oil in sea ice. Hyperspectral sensors can provide a continuous reflectance spectrum from the mixture of different types of sea ice and oil. Liu et al. (2016) have measured the reflectance of seawater, pack ice, crude oil, and its mixture. The oil-contaminated ice has lower reflectance than that of compact ice for wavelengths smaller than 490 nm and higher beyond 510 nm (Figure 7.15c). Figure 7.15d shows the reflectance spectra of the pack ice contaminated with a thin film of oil (Liu et al. 2016). The reflectance of the polluted pack ice is lower than that of clean ice between 560 and 710 nm, which clearly discriminates the two types of ice (Liu et al. 2015c). Although oil can be distinctively identified in an oil-polluted sea ice environment using the visible and near-infrared hyperspectral channels, the time-varying behavior of oil-infested thick ice cover is unknown.

Acoustic: Acoustic methods show a great promise of detecting oil in sea ice (Fingas and Brown 2013). Acoustic waves propagate in different media (types of oil and ice) with different speeds and attenuation factors. This property of acoustic waves forms the basis of oil detection in, on, and under sea ice (Collis et al. 2016). Oil also behaves similar to a

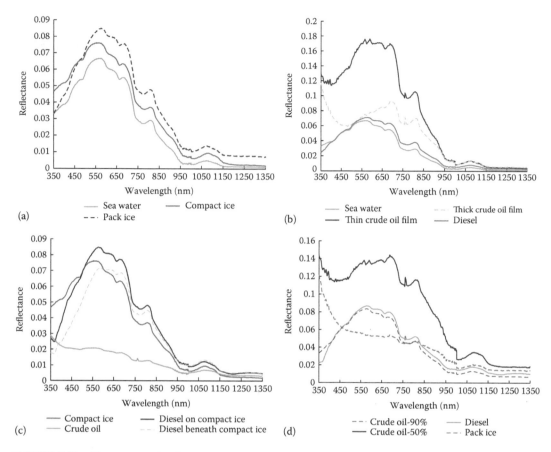

FIGURE 7.15 The measured reflectance spectra of oil in sea ice. (a) Seawater, pack ice, and compact ice, (b) Seawater contaminated with light diesel oil film, (c) Compact ice with oil film, and (d) Pack ice contaminated with the oil film. (From Liu, B. et al., *J. Spectrosc.*, Article ID 6584314, 2016. With permission.)

solid that allows the propagation of secondary waves (S-waves or shear waves) in addition to transverse and compressive waves. S-waves travel much slower than the primary waves, and their arrival is delayed at the acoustic sensor (Wilkinson et al. 2007). If there are no sediments in sea ice, it is much easier to disseminate oil signal from that of sea ice (Fingas and Brown 2000). The operation of underwater/ice vehicles with acoustic sensors can be laborious and expensive enterprise.

Passive microwave: It is possible to detect oil in sea ice using a passive microwave sensor operated from airborne or spaceborne platform (Fingas 2015). Crude oil has higher emissivity than the sea ice and seawater (Fingas and Brown 2000). This property of crude oil can easily identify areas of oil spill in the MIZ or the pack ice region. The detection of oil in sea ice using passive microwave sensors has its own limitations. The atmospheric correction of emitted radiation is critical for such investigation; second, the spatial resolution of passive microwave sensors is too coarse for an effective oil spill response. Sea ice thickness and oil layer thickness can significantly alter the observations (Brekke and Solberg 2005). However, passive microwave radiometers have a unique potential of operating under all-weather conditions (Jha et al. 2008).

Ultraviolet: Ultraviolet (UV) remote sensing is very useful for detecting oil sheen on the ocean surface due to high reflectivity of UV radiation. UV method can provide a positive indication of oil-mixed sea ice in a MIZ; however, UV sensors suffer from spurious signals from sun glint, biomass, and wind disturbances to the sea surface (Fingas 2015). Similarly, laser fluorescence sensors identify areas of UV light absorption, indicating oil presence; however, the signal can also come from chlorophyll a, yellow substance, and colored dissolved organic matter.

7.3 MITIGATION MEASURES

With the exception of oil in sea ice hazard, other ice-related hazards can be effectively mitigated using sophisticated EO technology that is already in use and is being practiced by polar nations (Jha et al. 2008; Fingas 2015). Strategic planning and adequate protocols for operations in cold environment of polar regions can effectively mitigate many of the challenges of ice hazards related to melt ponds, ice edge, ice roughness and thickness, leads, and ice motion. The development of new technologies for the detection and better understanding of the hazard in a sea ice-covered ocean and related strategic plans are required for the mitigation of consequences of the hazard (Owens et al. 1998). The mitigation can benefit by having the proximity to infrastructure in the extreme environmental conditions in polar regions.

As oil spill in sea ice is the most devastating geohazards among other ice hazards, we keep our focus on the mitigation strategies of this aspect of ice hazard. Oil in sea ice hazard cleanup and mitigation measures are currently insufficient, and there exists a little understanding of the science behind the interaction of oil with sea ice during winter and summer seasons (Kauranen 2016).

The following can be the approaches to mitigation plans for oil hazard in sea ice:

- *In-situ* burning (transfer from ocean to atmosphere): *In-situ* burning of oil is the most common practice today; however, it only transfers the pollutants from the ocean to the atmosphere, which can have impact on global climate change (Benner et al. 1990).
- A wide range of oil types—knowledge of physical and chemical properties: There can be numerous kinds of oil types in the oil spill hazard in sea ice; a prior knowledge of physical and chemical properties of each oil type may help the mitigation (Fingas 2012).
- Toxicity testing of chemical herding agents before *in situ* burning: Before practicing dispersant use and spraying of herding agents in the Arctic/Antarctic, the toxicity testing of herding agents is required to contain the hazard from expanding further (Buist et al. 2011). Limited understanding exists on the use of dispersants for oil spill treatment in Arctic.
- Develop capacity for large-scale direct source injection of dispersant for subsea and surface oil spill.
- Knowledge of wave energy in ice for effective application of dispersants (Ardhuin et al. 2016): The understanding of ice behavior upon interaction with waves is still in infancy. The addition of dispersants to the oil–ice system makes it a complex media to study and prevents from making accurate model predictions.
- Mechanical recovery of oil in the presence of ice—extremely limited expertise.
- Combat interference of ice in mechanical recovery.
- Infrastructure for storage and disposal of recovered oil.
- Coastal vulnerability mapping is required (Barnhart et al. 2014).
- Shoreline cleanup and prespill assessment should be completed.
- Mitigation tactics should be based on geomorphological processes and biological parameters.
- Emphasis should be placed on natural recovery processes for shorelines.
- Detection of buried oil on the shorelines and landfast ice indicates a need of thorough understanding of coastal processes (Eicken and Mahoney 2014).

7.4   CONCLUSIONS

This chapter provided an overview of available and future remote sensing techniques that can be used to (1) understand the scientific nature of the ice hazard, (2) make possible predictions, and (3) take effective mitigation measures. Despite our inability to confidently predict sea ice behavior due to lack of complete understanding of sea ice processes (physical, biological, chemical, geomorphological), the microwave satellites (e.g., Radarsat, AMSR-2, SMAP, Sentinel-1) have boosted our confidence manifold in the level of maturity, readiness, and NRT response for planning and execution of mitigation measures related to sea ice hazards in the Arctic or Antarctic. NRT MODIS rapid response Earth data and Radarsat imageries are already in use onboard various icebreakers in the Arctic for enhanced navigability. Theoretical approaches and parameterization of sea ice-related processes are continuously being rectified with the availability of large volumes of *in situ* and remote sensing observations. Microwave remote sensing has been and continues to be extremely useful in monitoring sea ice in the polar regions. The first decade of the twenty-first century has witnessed unprecedented surge in ship- and space-based observations resulting in better understanding of ice hazards, for example, ice edge, thaw holes, rubble ice and ridges, ice leads and polynyas, and oil spill in sea ice. However, sea ice thickness and oil hazard in sea ice are still not well understood due to the complex nature of the processes involved and extreme polar environment. The main challenges that we face today are the poor understanding of the physics of turbulent (ice) boundary layer and inadequate parameterization of sea ice processes (at varying spatial and temporal scales) that are much needed for inclusion in regional and global climate and ice prediction models. Future avenues of work are open on the integrated use of various kinds of data (*in situ* and remote sensing) in conjunction with the theoretical models and data assimilation. Our ability to predict sea ice behavior and associated geohazards is slated for improvement, as we grow with the development and integrated utilization of the EO technology as the way forward.

REFERENCES

Afenyo, M., Khan, F., Veitch, B., and Yang, M. 2016. Modeling oil weathering and transport in sea ice. *Marine Poll. Bull.*, 107(1), 206–215, doi:10.1016/j.marpolbul.2016.03.070.

Andreas, E. L. 1987. A theory for the scalar roughness and the scalar transfer coefficients over snow and sea ice. *Boundary Layer Meteorol.*, 38(1–2), 159–184. doi:10.1007/BF00121562.

Andreas, E. L. 2002. Parameterizing scalar transfer over snow and ice: A review. *J. Hydrometeorol.*, 3(4), 417–432.

Ardhuin, F., Sutherland, P., Doble, M., and Wadhams, P. 2016. Ocean waves across the arctic: Attenuation due to dissipation dominates over scattering for periods longer than 19 s. *Geophys. Res. Lett.*, 43, 5775–5783. doi:10.1002/2016GL068204.

Astafiev, V. N., Polomoshnov, A. M., and Truskov, P. A. 1991. Stamukhi on the Northern Sakhalin Offshore. In *The First International Offshore and Polar Engineering Conference*. International Society of Offshore and Polar Engineers, Document ID: ISOPE-I-91-127.

Barnhart, K. R., Overeem, I., and Anderson, R. S. 2014. The effect of changing sea ice on the physical vulnerability of Arctic coasts. *Cryosphere*, 8(5), 1777–1799. doi:10.5194/tc-8-1777-2014.

Battisti, D. S., Bitz, C. M., and Moritz, R. E. 1997. Do general circulation models underestimate the natural variability in the arctic climate? *J. Clim.*, 10(8), 1909–1920.

Benner Jr, B. A., Bryner, N. P., Wise, S. A., Mulholland, G. W., Lao, R. C., and Fingas, M. F. 1990. Polycyclic aromatic hydrocarbon emissions from the combustion of crude oil on water. *Environ. Sci. Technol.*, 24(9), 1418–1427. doi:10.1021/es00079a018.

Bitz, C. M., Holland, M. M., Hunke, E. C., and Moritz, R. E. 2005. Maintenance of the sea-ice edge. *J. Clim.*, 18(15), 2903–2921. doi:10.1175/JCLI3428.1.

Bourke, R. H. and Garrett, R. P. 1987. Sea ice thickness distribution in the Arctic Ocean. *Cold Reg. Sci. Technol.*, 13(3), 259–280. doi:10.1016/0165-232X(87)90007-3.

Bradford, J. H., Dickins, D. F., and Brandvik, P. J. 2010. Assessing the potential to detect oil spills in and under snow using airborne ground-penetrating radar. *Geophysics*, 75(2), G1–G12. doi:10.1190/1.3312184.

Bradford, J. H., Liberty, L. M., and Dickins, D.F. 2008. Locating oil spills under sea ice using ground-penetrating radar. *The Leading Edge*, 27(11), 1424–1435. doi:10.1190/1.3011014.

Brekke, C., Holt, B., Jones, C., and Skrunes, S. 2014. Discrimination of oil spills from newly formed sea ice by synthetic aperture radar. *Remote Sens. Environ.*, 145, 1–14. doi:10.1016/j.rse.2014.01.015.

Brekke, C. and Solberg, A. H. 2005. Oil spill detection by satellite remote sensing. *Remote Sens. Environ.*, 95(1), 1–13. doi:10.1016/j.rse.2004.11.015.

Buist, I., Potter, S., Nedwed, T., and Mullin, J. 2011. Herding surfactants to contract and thicken oil spills in pack ice for in situ burning. *Cold Reg. Sci. Technol.*, 67(1), 3–23. doi:10.1016/j.coldregions.2011.02.004.

Carsey, F. D. (Ed.) 1992. *Microwave Remote Sensing of Sea Ice*. Washington, DC: American Geophysical Union.

Collis, J. M., Frank, S. D., Metzler, A. M., and Preston, K. S. 2016. Elastic parabolic equation and normal mode solutions for seismo-acoustic propagation in underwater environments with ice covers. *J. Acoust. Soc. Am.*, 139(5), 2672–2682. doi:10.1121/1.4946991.

Comiso, J. C., Cavalieri, D. J., Parkinson, C. L., and Gloersen, P. 1997. Passive microwave algorithms for sea ice concentration: A comparison of two techniques. *Remote Sens. Environ.*, 60(3), 357–384. doi:10.1016/S0034-4257(96)00220-9.

Dickins, D. F. and I. Buist. 1999. Oil spill countermeasures for ice covered waters. *J. Pure Appl. Chem.*, 71(1), 173–191, doi:10.1351/pac199971010173.

Dierking, W. 2010. Mapping of different sea ice regimes using images from Sentinel-1 and ALOS synthetic aperture radar. *IEEE Trans. Geosci. Remote Sens.*, 48(3), 1045–1058. doi:10.1109/TGRS.2009.2031806.

Dierking, W. and Busche, T. 2006. Sea ice monitoring by L-band SAR: An assessment based on literature and comparisons of JERS-1 and ERS-1 imagery. *IEEE Trans. Geosci. Remote Sens.*, 44(4), 957–970, doi:10.1109/TGRS.2005.861745.

Digby, S. A. 1984. Remote sensing of drained ice areas around the breathing holes of ice-inhabiting seals. *Can. J. Zool.*, 62(6), 1011–1014, doi:10.1139/z84-143.

Drinkwater, M. R. 1989. LIMEX'87 ice surface characteristics: Implications for C-band SAR backscatter signatures. *IEEE Trans. Geosci. Remote Sens.*, 27(5), 501–513, doi:10.1109/TGRS.1989.35933.

Ebert, E. E. and Curry, J. A. 1993. An intermediate one-dimensional thermodynamic sea ice model for investigating ice-atmosphere interactions. *J. Geophys. Res. Oceans*, 98(C6), 10085–10109. doi:10.1029/93JC00656.

Eicken, H. and Mahoney, A. R. 2014. Sea ice: Hazards, risks, and implications for disasters (Chapter 13). In Ellis, J. and Sherman, D. J. (Eds.) *Coastal and Marine Hazards, Risks, and Disasters*. Waltham, MA: Academic Press.

El-Hilo, S., Staples, G., Zagon, T., Isaacs, D., Arkett, M., Clausi, D., and Nadeau, C. 2013. Classification and concentration estimation of Arctic sea ice using RADARSAT-2. In *ESA Special Publication*, 709, p. 37.

Elvidge, A. D., Renfrew, I. A., Weiss, A. I., Brooks, I. M., Lachlan-Cope, T. A., and King, J. C. 2015. Observations of surface momentum exchange over the marginal-ice-zone and recommendations for its parameterization. *Atmos. Chem. Phys.*, 16. 1545–1563. doi:10.5194/acp-16-1545-2016.

Esau, I. N. 2007. Amplification of turbulent exchange over wide Arctic leads: Large-eddy simulation study. *J. Geophys. Res.*, 112, D08109. doi:10.1029/2006JD007225.

Feltham, D. L. 2008. Sea ice rheology. *Annu. Rev. Fluid Mech.*, 40, 91–112. doi:10.1146/annurev.fluid.40.111406.102151.

Fetterer, F., K. Knowles, W. Meier, and M. Savoie. 2002. Updated daily. Sea Ice Index, Version 1. [Sea ice concentration]. Boulder, CO: National Snow and Ice Data Center (NSIDC). doi:10.7265/N5QJ7F7W (Accessed: April 26, 2016).

Fetterer, F. and Untersteiner, N. 1998. Observations of melt ponds on Arctic sea ice. *J. Geophys. Res.*, 103(24), 821–824. doi:10.1029/98JC02034.

Fingas, M. 2012. *The Basics of Oil Spill Cleanup*. Boca Raton, FL: CRC Press.

Fingas, M. 2015. Detection of oil in, with, and under ice and snow (Chapter 14). In *Handbook of Oil Spill Science and Technology*. Hoboken, NJ: John Wiley & Sons.

Fingas, M. and Brown, C. 2000. A review of the status of advanced technologies for the detection of oil in and with ice. *Spill Sci. Technol. Bull.*, 6(5), 295–302. doi:10.1016/S1353-2561(01)00056-1.

Fingas, M. and Brown, C. E. 2013. Detection of oil in ice and snow. *J. Marine Sci. Eng.*, 1(1), 10–20. doi:10.3390/jmse1010010.

Fors, A. S., Brekke, C., Gerland, S., Doulgeris, A. P., and Beckers, J. F. 2016. Late summer Arctic sea ice surface roughness signatures in C-band SAR data. *IEEE J. Sel. Topics Appl. Earth Obs. Remote Sens.*, 9(3), 1199–1215. doi:10.1109/JSTARS.2015.2504384.

Gettelman, A. and Rood, R. B. 2016. Simulating the ocean and sea ice (Chapter 6). In *Demystifying Climate Models*. Berlin, Germany: Springer.

Gupta, M. 2014. On the estimation of physical roughness of a marginal sea ice zone using remote sensing. PhD diss., University of Manitoba. http://hdl.handle.net/1993/23836.

Gupta, M. 2015. Various remote sensing approaches to understanding roughness in the marginal ice zone. *Phys. Chem. Earth, Parts A/B/C*, 83, 75–83. doi:10.1016/j.pce.2015.05.003.

Han, H., Im, J., Kim, M., Sim, S., Kim, J., Kim, D. J., and Kang, S. H. 2016. Retrieval of melt ponds on Arctic multiyear sea ice in summer from TerraSAR-X dual-polarization data using machine learning approaches: a case study in the Chukchi Sea with mid-incidence angle data. *Remote Sens.*, 8(1), 57. doi:10.3390/rs8010057.

Haarpaintner, J., Tonboe, R. T., Long, D. G., and Van Woert, M. L. 2004. Automatic detection and validity of the sea-ice edge: An application of enhanced-resolution QuikScat/SeaWinds data. *IEEE Trans. Geosci. Remote Sens.*, 42(7), 1433–1443. doi:10.1109/TGRS.2004.828195.

Herzfeld, U. C., Hunke, E. C., McDonald, B. W., and Wallin, B. F. 2015. Sea ice deformation in Fram Strait–Comparison of CICE simulations with analysis and classification of airborne remote-sensing data. *Cold Reg. Sci. Technol.*, 117, 19–33. doi:10.1016/j.coldregions.2015.05.001.

Hibler, W. D. III. 1989. Arctic ice-ocean dynamics. In Y. Herman (Ed.) *The Arctic Seas. Climatology, Oceanography, Geology, and Biology*, Van Nostrand Reiohold. New York: Springer, pp. 47–91.

Hollinger, J. 1971. Passive microwave measurements of sea surface roughness. *IEEE Trans. Geosci. Electron.*, 3(9), 165–169. doi:10.1109/TGE.1971.271489.

Huntemann, M., Heygster, G., Kaleschke, L., Krumpen, T., Mäkynen, M. and Drusch, M. 2014. Empirical sea ice thickness retrieval during the freeze up period from SMOS high incident angle observations. *Cryosphere*, 8(2), 439–451. doi:10.5194/tc-8-439-2014.

ICDC (Integrated Climate Data Center) 2003. AMSR-E lead area fraction for the Arctic, (04 March 2003), Integrated Climate Data Center (ICDC, http://icdc.zmaw.de/), University of Hamburg, Hamburg, Germany (Accessed on April 15, 2016).

ICDC (Integrated Climate Data Center) 2013. CryoSat-2 lead area fraction for the Arctic, (February 1, 2013), (ICDC, http://icdc.zmaw.de), University of Hamburg, Hamburg, Germany (Accessed on April 15, 2016).

Inoue, J., M. Kawashima, Y. Fujiyoshi, and M. Wakatsuchi. 2005. Aircraft observations of air-mass modification over the sea of Okhotsk during sea- ice growth. *Bound.-Lay. Meteorol.*, 117, 111–129. doi:10.1007/s10546-004-3407-y.

IPCC (Intergovernmental Panel on Climate Change) 2014. Observations: Cryosphere. In: *Climate Change 2013–The Physical Science Basis*. Cambridge, UK: Cambridge University Press, pp. 317–382.

Ivanova, N., Rampal, P., and Bouillon, S. 2016. Error assessment of satellite-derived lead fraction in the Arctic. *Cryosphere*, 10(2), 585–595. doi:10.5194/tc-10-585-2016.

Jha, M. N., Levy, J., and Gao, Y. 2008. Advances in remote sensing for oil spill disaster management: State-of-the-art sensors technology for oil spill surveillance. *Sensors*, 8(1), 236–255. doi:10.3390/s8010236.

Johannessen, O. M., Sandven, S., Skagseth, O., Kloster, K., Kovacs, Z., Sauvadet, P., Geli, L., Weeks, W., and Louet, J. 1992. ERS-1 SAR ice routing of L'Astrolabe through the Northeast Passage. In ESA, Environment Observation and Climate Modelling Through International Space Projects. Volume 3: Earth Observation Space Programmes, SAFISY Activities, Strategies of International Organisations, Legal Aspects, pp. 997–1002.

Kaleschke, L., X. Tian-Kunze, N. Maaß, M. Mäkynen, and M. Drusch. 2012. Sea ice thickness retrieval from SMOS brightness temperatures during the Arctic freeze-up period. *Geophys. Res. Lett.*, 39, L05501. doi:10.1029/2012GL050916.

Karvonen, J. 2014. A sea ice concentration estimation algorithm utilizing radiometer and SAR data. *The Cryosphere*, 8(5), 1639–1650. doi:10.5194/tc-8-1639-2014.

Kauranen, A. 2016. Oil spill tests on ice prove Arctic quests risky. PHYSORG, April 3. Accessed April 3, 2016. https://phys.org/news/2016-04-oil-ice-arctic-quests-risky.html

Keller, E. A. and D. E. DeVecchio. 2012. *Natural Hazards: Earth's Processes as Hazards, Disasters, and Catastrophes*. Upper Saddle River, NJ: Prentice Hall.

Kim, D. J., Hwang, B., Chung, K. H., Lee, S. H., Jung, H. S., and Moon, W. M. 2013. Melt pond mapping with high-resolution SAR: The first view. *Proc. IEEE*, 101(3), 748–758. doi:10.1109/JPROC.2012.2226411.

Kimura, N., Nishimura, A., Tanaka, Y., and Yamaguchi, H. 2013. Influence of winter sea-ice motion on summer ice cover in the Arctic. *Polar Res.*, 32, 20193. doi:10.3402/polar.v32i0.20193.

Kustas, W. P. and Brutsaert, W. 1986. Wind profile constants in a neutral atmospheric boundary layer over complex terrain. *Bound.-Lay. Meteorol.*, 34, 35–54. doi:10.1007/BF00120907.

Kwok, R. and Cunningham, G. F. 2008. ICESat over Arctic sea ice: Estimation of snow depth and ice thickness. *J. Geophys. Res. Oceans*, 113(C8), 1–17. doi:10.1029/2008JC004753.

Kwok, R., Curlander, J. C., McConnell, R., and Pang, S.S. 1990. An ice-motion tracking system at the Alaska SAR facility. *IEEE J. Ocean. Engg.*, 15(1), 44–54. doi:10.1109/48.46835.

Kwok, R. and Rothrock, D. A. 2009. Decline in Arctic sea ice thickness from submarine and ICESat records: 1958–2008. *Geophys. Res. Lett.*, 36(L15501), 1–5. doi:10.1029/2009GL039035.

Kwok, R., Schweiger, A., Rothrock, D. A., Pang, S., and Kottmeier, C. 1998. Sea ice motion from satellite passive microwave imagery assessed with ERS SAR and buoy motions. *J. Geophys. Res. Oceans*, 103(C4), 8191–8214. doi:10.1029/97JC03334.

Laidler, G. J., Ford, J. D., Gough, W. A., Ikummaq, T., Gagnon, A. S., Kowal, S., Qrunnut, K., and Irngaut, C. 2009. Travelling and hunting in a changing Arctic: Assessing Inuit vulnerability to sea ice change in Igloolik, Nunavut. *Clim. Change*, 94(3–4), 363–397. doi:10.1007/s10584-008-9512-z.

Lang, W., Wu, J., Zhang, X., Yang, X., and Meng, J. 2014. Detection of ice types in the Eastern Weddell Sea by fusing L-and C-band SIR-C polarimetric quantities. *Int. J. Remote Sens.*, 35(19), 6874–6893. doi:10.1080/01431161.2014.960615.

Leppäranta, M. 2011. *The Drift of Sea Ice*. Chichester, UK: Springer.

Laxon, S. 1994. Sea ice altimeter processing scheme at the EODC. *Int. J. Remote Sens.*, 15(4), 915–924. doi:10.1080/01431169408954124.

Laxon, S. W., Giles, K. A., Ridout, A. L., Wingham, D. J., Willatt, R., Cullen, R., Kwok, R. et al. 2013. CryoSat-2 estimates of Arctic sea ice thickness and volume. *Geophys. Res. Lett.*, 40(4), 732–737. doi:10.1002/grl.50193.

Lehtiranta, J., Siiriä, S., and Karvonen, J. 2015. Comparing C-and L-band SAR images for sea ice motion estimation. *Cryosphere*, 9(1), 357–366. doi:10.5194/tc-9-357-2015.

Lindsay, R. W. and Rothrock, D. A. 1995. Arctic sea ice leads from advanced very high resolution radiometer images. *J. Geophys. Res.*, 100(C3), 4533–4544. doi:10.1029/94JC02393.

Liu, C., Chao, J., Gu, W., Xu, Y., and Xie, F. 2015a. Estimation of sea ice thickness in the Bohai Sea using a combination of VIS/NIR and SAR images. *GISci. Remote Sens.*, 52(2), 115–130. doi:10.1080/15481603.2015.1007777.

Liu, B., Li, Y., Zhang, Q., and Han, L. 2016. Assessing sensitivity of hyperspectral sensor to detect oils with sea ice. *J. Spectrosc.*, Article ID 6584314. doi:10.1155/2016/6584314.

Liu, B., Zhu, W., and Li, Y. 2015c. Spectral characteristics analysis of oil among sea ice. *Proc. IEEE Int. Geosci. Remote Sens. Symp. (IGARSS'15)*, pp. 3450–3453, doi:10.1109/IGARSS.2015.7326562.

Liu, J., Song, M., Horton, R. M., and Hu, Y. 2015b. Revisiting the potential of melt pond fraction as a predictor for the seasonal Arctic sea ice extent minimum. *Environ. Res. Lett.*, 10(5), 054017. doi:10.1088/1748-9326/10/5/054017.

Lu, P., Li, Z., and Han, H. 2016. Introduction of parameterized sea ice drag coefficients into ice free-drift modeling. *Acta Oceanol. Sin.*, 35(1), 53–59. doi:10.1007/s13131-016-0796-y.

Ludwig, V. 2016. Comparison of SMOS and SMAP satellite sea ice concentrations derived from 1.4 GHz brightness temperatures. Master Thesis, Universität Hamburg, School of Integrated Climate System Science (SICSS), Hamburg. http://hdl.handle.net/11858/00-001M-0000-0029-73DF-8.

Mäkynen, M., Cheng, B., and Similä, M. 2013. On the accuracy of thin-ice thickness retrieval using MODIS thermal imagery over Arctic first-year ice. *Ann. Glaciol.*, 54(62), 87–96. doi:10.3189/2013AoG62A166.

Marcq, S. and Weiss, J. 2012. Influence of sea ice lead-width distribution on turbulent heat transfer between the ocean and the atmosphere. *Cryosphere*, 6, 143–156. doi:10.5194/tc-6-143-2012.

Maslowski, W. and Lipscomb, W.H. 2003. High resolution simulations of Arctic sea ice, 1979–1993. *Polar Res.*, 22(1), 67–74. doi:10.1111/j.1751-8369.2003.tb00097.x.

Maykut, G. A. 1986. The surface heat and mass balance. In N. Untersteiner (Ed.) *The Geophysics of Sea Ice*, New York: Plenum Press, pp. 395–464.

McPhee, M. G. 1992. Turbulent heat flux in the upper ocean under sea ice. *J. Geophys. Res.: Oceans*, 97(C4), 5365–5379. doi:10.1029/92JC00239.

McPhee, M. G. 2002. Turbulent stress at the ice/ocean interface and bottom surface hydraulic roughness during the SHEBA drift. *J. Geophys. Res. Oceans*, 107(C10), 1–15. doi:10.1029/2000JC000633.

Meier, W. N., Gallaher, D., and Campbell, G. G. 2013. New estimates of Arctic and Antarctic sea ice extent during September 1964 from recovered Nimbus I satellite imagery. *Cryosphere*, 7(2), 699–705. oi:10.5194/tc-7-699-2013.

Meier, W. N. and Stroeve, J. 2008. Comparison of sea-ice extent and ice-edge location estimates from passive microwave and enhanced-resolution scatterometer data. *Ann. Glaciol.*, 48(1), 65–70. doi:10.3189/172756408784700743.

Miles, M. W. and Barry, R. G. 1998. A 5-year satellite climatology of winter sea ice leads in the western Arctic. *J. Geophys. Res.*, 103(C10), 21,723–21,734. doi:10.1029/98JC01997.

Morassutti, M. P. and LeDrew, E. F. 1996. Albedo and depth of melt ponds on sea-ice. *Int. J. Climatol.*, 16(7), 817–838.

Muckenhuber, S., Korosov, A. A., and Sandven, S. 2016. Open-source feature-tracking algorithm for sea ice drift retrieval from Sentinel-1 SAR imagery. *Cryosphere*, 10(2), 913–925. doi:10.5194/tc-10-913-2016.

Nakamura, K., Wakabayashi, H., Naoki, K., Nishio, F., Moriyama, T., and Uratsuka, S. 2005. Observation of sea-ice thickness in the Sea of Okhotsk by using dual-frequency and fully polarimetric airborne SAR (Pi-SAR) data. *IEEE Trans. Geosci. Remote Sens.*, 43(11), 2460–2469. doi:10.1109/TGRS.2005.853928.

Onana, V., Kurtz, N. T., Farrell, S. L., Koenig, L. S., Studinger, M., and Harbeck, J. P. 2013. A sea-ice lead detection algorithm for use with high-resolution airborne visible imagery. *IEEE Trans. Geosci. Remote Sens.*, 51(1), 38–56. doi:10.1109/TGRS.2012.2202666.

Onstott, R. G. 1992. SAR and scatterometer signatures of sea ice (Chapter 5). In F. Carsey (Ed) *Microwave Remote Sensing of Sea Ice*, Washington, DC: American Geophysical Union.

Owens, E. H., Solsberg, L. B., West, M. R., and McGrath, M. 1998. Field guide for oil spill response in Arctic waters. Emergency Prevention, Preparedness and Response Working Group (EPPR). Yellowknife, NT: Environment Canada, 348 p.

Perovich, D. K. 1996. *The Optical Properties of Sea Ice* (No. MONO-96-1). Hanover, NH: Cold Regions Research and Engineering Lab.

Petty, A. A., Tsamados, M. C., Kurtz, N. T., Farrell, S. L., Harbeck, J. P., Feltham, D. L., and Richter-Menge, J. A. 2016. Characterizing Arctic sea ice topography using high-resolution IceBridge data. *Cryosphere*, 10(3), 1161–1179. doi:10.5194/tc-10-1161-2016.

Posey, P. G., Metzger, E. J., Wallcraft, A. J., Hebert, D. A., Allard, R. A., Smedstad, O. M., Phelps, M. W. et al. 2015. Improving Arctic sea ice edge forecasts by assimilating high horizontal resolution sea ice concentration data into the US Navy's ice forecast systems. *Cryosphere*, 9(4), 1735–1745. doi:10.5194/tc-9-1735-2015.

Potter, S., Buist, I., Trudel, K., Dickins, D., and Owens, E. 2012. Spill response in the Arctic offshore. Prepared for the American Petroleum Institute and the Joint Industry Programme on Oil Spill Recovery in Ice. Washington, DC: American Petroleum Institute.

Ricker, R., Hendricks, S., Helm, V., Skourup, H., and Davidson, M. 2014. Sensitivity of CryoSat-2 Arctic sea-ice freeboard and thickness on radar-waveform interpretation. *Cryosphere*, 8(4), 1607–1622. doi:10.5194/tc-8-1607-2014.

Rivas, M. B., Maslanik, J. A., Sonntag, J. G., and Axelrad, P. 2006. Sea ice roughness from airborne LIDAR profiles. *IEEE Trans. Geosci. Remote Sens.*, 44(11), 3032–3037. doi:10.1109/TGRS.2006.875775.

Röhrs, J. and Kaleschke, L. 2012. An algorithm to detect sea ice leads by using AMSR-E passive microwave imagery. *Cryosphere*, 6(2), 343–352. doi:10.5194/tc-6-343-2012.

Rösel, A., Kaleschke, L., and Birnbaum, G. 2012. Melt ponds on Arctic sea ice determined from MODIS satellite data using an artificial neural network. *Cryosphere*, 6(2), 431–446. doi:10.5194/tc-6-431-2012.

Rothrock, D. A. 1986. Ice thickness distribution–Measurement and theory. In N. Untersteiner (Ed.) *The Geophysics of Sea Ice*. New York: Plenum Press, pp. 551–575.

Saeki, H., Ono, T., Zong, N. E., and Nakazawa, N. 1985. Experimental study on direct shear strength of sea ice. *Ann. Glaciol.*, 6, 218–221.

Scharien, R. K. and Yackel, J. J. 2005. Analysis of surface roughness and morphology of first-year sea ice melt ponds: Implications for microwave scattering. *IEEE Trans. Geosci. Remote Sens.*, 43(12), 2927–2939. doi:10.1109/TGRS.2005.857896.

Scharien, R. K., Yackel, J. J., Barber, D. G., Asplin, M., Gupta, M., and Isleifson, D. 2012. Geophysical controls on C band polarimetric backscatter from melt pond covered Arctic first year sea ice: Assessment using high-resolution scatterometry. *J. Geophys. Res.: Oceans*, 117(C9), C00G18. doi:10.1029/2011JC007353.

Scharien, R. K., Hochheim, K., Landy, J., and Barber, D. G. 2014. First-year sea ice melt pond fraction estimation from dual-polarisation C-band SAR - Part 2: Scaling in situ to Radarsat-2. *Cryosphere*, 8(6), 2163–2176. doi:10.5194/tc-8-2163-2014.

Schröder, D., Feltham, D. L., Flocco, D., and Tsamados, M. 2014. September Arctic sea-ice minimum predicted by spring melt-pond fraction. *Nat. Clim. Change*, 4(5), 353–357. doi:10.1038/nclimate2203.

Serreze, M. C. and Barry, R. G., 2005. *The Arctic Climate System* (Vol. 22). Cambridge, UK: Cambridge University Press.

Shen, H. H., Hibler, W. D., and Lepparanta, M. 1987. The role of ice floe collisions in sea ice rheology. *J. Geophys. Res.*, 92(C7), 7085–7096. doi: 10.1029/JC092iC07p07085.

Spreen, G., Kaleschke, L., and Heygster, G. 2008. Sea ice remote sensing using AMSR-E 89-GHz channels. *J. Geophys. Res. Oceans*, 113(C2), 1–14. doi:10.1029/2005JC003384.

Srisudha, S., Kumar, A. S., Jain, D. S., and Dadhwal, V. K. 2013. Detection and size distribution analysis of ice floes near Antarctica using RISAT-1 imagery. *Curr. Sci.*, 105(10), 1400–1403.

Steele, M., Zhang, J., Rothrock, D., and Stern, H. 1997. The force balance of sea ice in a numerical model of the Arctic Ocean. *J. Geophys. Res. Oceans*, 102(C9), 21061–21079. doi:10.1029/97JC01454.

Steiner N., Harder M., and Lemke P. 1999. Sea-ice roughness and drag coefficients in a dynamic-thermodynamic sea-ice model for the Arctic. *Tellus A* 51, 964–978, doi:10.1034/j.1600-0870.1999.00029.x.

Stroeve, J., Holland, M. M., Meier, W., Scambos, T., and Serreze, M. 2007. Arctic sea ice decline: Faster than forecast. *Geophys. Res. Lett.*, 34(9), 1–5. doi:10.1029/2007GL029703.

Stroeve, J. C., Markus, T., Maslanik, J. A., Cavalieri, D. J., Gasiewski, A. J., Heinrichs, J. F., Holmgren, J., Perovich, D. K., and Sturm, M. 2006. Impact of surface roughness on AMSR-E sea ice products. *IEEE Trans. Geosci. Rem. Sens.*, 44, 3103–3117. doi:10.1109/TGRS.2006.880619.

Swift, C. T. and Cavalieri, D. J. 1985. Passive microwave remote sensing for sea ice research. *Eos, Trans. Amer. Geophys. Union*, 66(49), 1210–1212. doi:10.1029/EO066i049p01210.

Taylor, P. D. and Feltham, D. L. 2004. A model of melt pond evolution on sea ice. *J. Geophys. Res. Oceans*, 109(C12), 1–19. doi:10.1029/2004JC002361.

Tian-Kunze, X., Kaleschke, L., Maaß, N., Mäkynen, M., Serra, N., Drusch, M., and Krumpen, T. 2014. SMOS-derived thin sea ice thickness: Algorithm baseline, product specifications and initial verification. *Cryosphere*, 8(3), 997–1018. doi:10.5194/tc-8-997-2014.

Timco, G. W. and Frederking, R. M. W. 1990. Compressive strength of sea ice sheets. *Cold Reg. Sci. Technol.*, 17(3), 227–240. doi:10.1016/S0165-232X(05)80003-5.

Thomas, M., Kambhamettu, C. and Geiger, C. A. 2011. Motion tracking of discontinuous sea ice. *IEEE Trans. Geosci. Remote Sens.*, 49(12), 5064–5079. doi:10.1109/TGRS.2011.2158005.

Thorndike, A. S. 1986. Kinematics of sea ice. In N. Untersteiner (Ed.) *The Geophysics of Sea Ice*. New York: Plenum Press, pp. 489–549.

Tonboe, R. and Ezraty, R. 2002. Monitoring of new-ice in Greenland waters. *IEEE Int. Geosci. Remote Sens. Symp.*, (IGARSS'02), 3, 1932–1934. doi:10.1109/IGARSS.2002.1026304.

Toppaladoddi, S. and Wettlaufer, J. S. 2015. Theory of the sea ice thickness distribution. *Phys. Rev. Lett.*, 115(14), 148501. doi:10.1103/PhysRevLett.115.148501.

Tschudi, M. A., Maslanik, J. A., and Perovich, D.K. 2008. Derivation of melt pond coverage on Arctic sea ice using MODIS observations. *Remote Sens. Environ.*, 112(5), 2605–2614. doi:10.1016/j.rse.2007.12.009.

Untersteiner, N. (Ed.) 1986. *The Geophysics of Sea Ice*. New York: Plenum Press.

USNIC (United States National Ice Center/Naval Ice Center) 2016. Daily ice analysis products. http://www.natice.noaa.gov/daily_graphics.htm (Accessed on May 3, 2016).

Vinnikov, K. Y., Robock, A., Stouffer, R. J., Walsh, J. E., Parkinson, C. L., Cavalieri, D. J., Mitchell, J. F., Garrett, D., and Zakharov, V. F. 1999. Global warming and Northern Hemisphere sea ice extent. *Science*, 286(5446), 1934–1937. doi:10.1126/science.286.5446.1934.

Vries, A. C. D., Kustas, W. P., Ritchie, J. C., Klaassen, W., Menenti, M., Rango, A., and Prueger, J. H. 2003. Effective aerodynamic roughness estimated from airborne laser altimeter measurements of surface features. *Int. J. Remote Sens.*, 24(7), 1545–1558. doi:10.1080/01431160110115997.

Wadhams, P., Tucker III, W. B., Krabill, W. B., Swift, R. N., Comiso, J. C., and Davis, N.R. 1992. Relationship between sea ice freeboard and draft in the Arctic Basin, and implications for ice thickness monitoring. *J. Geophys. Res.*, 97(C12), 20325–20334. doi:10.1029/92JC02014.

Wadhams, P. 2000. *Ice in the Ocean*. London, UK: CRC Press.

Wakabayashi, H., Matsuoka, T., Nakamura, K., and Nishio, F. 2004. Polarimetric characteristics of sea ice in the Sea of Okhotsk observed by airborne L-band SAR. *IEEE Trans. Geosci. Remote Sens.*, 42(11), 2412–2425. doi:10.1109/TGRS.2004.836259.

Warren, S. G., Rigor, I. G., Untersteiner, N., Radionov, V. F., Bryazgin, N. N., Aleksandrov, Y. I., and Colony, R. 1999. Snow depth on Arctic sea ice. *J. Clim.*, 12(6), 1814–1829.

Weeks, W. F. 2010. *On Sea Ice*. Fairbanks, AK: University of Alaska Press.

Wernecke, A. and Kaleschke, L. 2015. Lead detection in Arctic sea ice from CryoSat-2: quality assessment, lead area fraction and width distribution. *Cryosphere*, 9(5), 1955–1968. doi:10.5194/tc-9-1955-2015.

Wilkinson, J., Maksym, T., and Singh, H. 2013. Capabilities for detection of oil spills under sea ice from autonomous underwater vehicles. Polar Ocean Services and the Woods Hole Oceanographic Institution for the Arctic Response Technology Joint Industry Programme, Woods Hole, MA, 104 pp. http://www.arcticresponsetechnology.org/wp-content/uploads/2013/10/Report%205.2%20-%20CAPABILITIES%20FOR%20DETECTION%20OF%20OIL%20SPILLS%20UNDER%20SEA%20ICE.pdf.

Wilkinson, J. P., Wadhams, P., and Hughes, N. E. 2007. Modelling the spread of oil under fast sea ice using three-dimensional multibeam sonar data. *Geophys. Res. Lett.*, 34(22), 1–5. doi:10.1029/2007GL031754.

Willmes, S. and Heinemann, G. 2015. Pan-Arctic lead detection from MODIS thermal infrared imagery. *Ann. Glaciol.*, 56(69), 29–37. doi:10.3189/2015AoG69A615.

Willmes, S. and Heinemann, G. 2016. Sea-ice wintertime lead frequencies and regional characteristics in the Arctic, 2003–2015. *Remote Sens.*, 8(1), 4. doi:10.3390/rs8010004.

Yang, Q., Losa, S.N., Losch, M., Tian-Kunze, X., Nerger, L., Liu, J., Kaleschke, L., and Zhang, Z. 2014. Assimilating SMOS sea ice thickness into a coupled ice-ocean model using a local SEIK filter. *J. Geophys. Res. Oceans*, 119(10), 6680–6692. doi:10.1002/2014JC009963.

Zippel, S. and Thomson, J. 2016. Air-sea interactions in the marginal ice zone. *Elem. Sci. Anth.*, 4(1), 000095. doi:10.12952/journal.elementa.000095.

8 Satellite Microwave Remote Sensing of Landscape Freeze–Thaw Status Related to Frost Hazard Monitoring

Youngwook Kim, John S. Kimball, and Jinyang Du

CONTENTS

8.1 INTRODUCTION

An increasing frequency and intensity of extreme weather events have been reported under recent climate change (Solomon et al., 2007; Marino et al., 2011) and are associated with hydrometeorological hazards, including frost, drought, flooding, and wildfires (Bello and Aina, 2014; Yaodong, 2005; Wang et al., 2012; Westerling, 2016). These natural disasters result in significant loss of life and property and negatively impact regional and national economies (Martino et al., 2009). Frost hazard monitoring is critical in countries where natural, agricultural, and man-made environments are directly influenced by the frost season. Spring freeze–thaw (FT) conditions influence processes involved in surface and groundwater storages, including soil permeability and water infiltration, the timing of seasonal snowmelt, river ice breakup, and the spring flood pulse (Kim et al., 2015; Park et al., 2016a). Warmer springs promote snowpack melting, which lowers the land surface albedo, allows greater absorption of incoming solar radiation, and further intensifies the melting process; relatively rapid and extensive thawing, and snowmelt increases the risk of flooding (Todhunter, 2001; Whitfield, 2012) and landslides (Gauthier and Hutchinson, 2012). The timing and duration of the seasonal FT transition between predominantly frozen winter conditions and summer nonfrozen conditions are closely related to civil and transportation infrastructure damage risk (Thomachot et al., 2005; Larsen et al., 2008; Li et al., 2014). Ground surface deformations caused by seasonal surface uplift (frost heave) and subsidence (thaw settlement), and longer-term surface subsidence associated with degradation of ice-rich permafrost, pose major threats to the safety and stability of human infrastructure (Nelson et al., 2001; Kääb, 2008; Chang and Hanssen, 2015). Although it may be paradoxical that global warming leads to an increase in the frequency of vegetation frost damage, several studies have reported that unusual winter thawing and earlier springs have led to tree dieback (Braathe, 1957, 1995; Bourque et al., 2005) and that the combination of earlier spring phenology onset followed by a sudden frost resulted in vegetation damage and productivity declines (Rigby and Porporato, 2008; Martin et al., 2010; Hufkens et al., 2012).

Satellite remote sensing has been widely used for monitoring frost season length (Kim et al., 2014a) and for detecting frost-related occurrence and damage, including crop losses and freeze injury (Feng et al., 2009; Papagiannaki et al., 2014), vegetation productivity decline (Gu et al., 2008; Kim et al., 2014b), and frost heave (Kääb, 2008). Satellite microwave remote sensing technology has been used to develop efficient tools for regional and global retrieval, and monitoring of frost-related metrics including landscape FT status (Kim et al., 2011, 2012; Du et al., 2015), Arctic ice phenology (Kang et al., 2012; Park et al., 2016a), snowmelt dynamics (Rawlins et al., 2005; Kim et al., 2015), and thaw–refreeze events (Derksen et al., 2009; Bartsch et al., 2010; Wilson et al., 2013). The landscape FT status detected from microwave remote sensing is sensitive to weather and climate conditions and provides a useful metric for determining frozen temperature constraints on ecohydrology, surface energy, and permafrost processes (Kim et al., 2014a; Zhang et al., 2011; Park et al., 2016b). The frost season indicated from satellite microwave sensor observations influences the chilling requirements for vegetation dormancy and frost resistance, plant bud break, and blossoming (Bennie et al., 2010; Yu et al., 2010). The timing and duration of seasonal frozen temperatures also effectively bound the potential growing season for crops and natural vegetation, thereby influencing annual productivity (Buermann et al., 2013; Eccel et al., 2009; Fengjin and Lianchun, 2011).

In this chapter, we provide an overview of satellite active and passive microwave sensor characteristics, pertaining to the retrieval of landscape FT status. We summarize some of the major FT classification algorithms and introduce a global daily FT Earth System Data Record (FT-ESDR) recently developed from similar long-term satellite passive microwave sensor observations. The FT-ESDR is used to illustrate several frost-related metrics over the Northern Hemisphere, including the number of spring frost days (SFDs), which have been linked to anomalous declines in vegetation productivity and changing crop planting practices. Finally, recommendations are made for improving the detection and monitoring of frost-related environmental impacts, including developing finer resolution and multifrequency satellite FT data records, and for improving the integration of multisensor observations with other data for more comprehensive assessments.

8.2 SATELLITE MICROWAVE SENSOR CHARACTERISTICS RELATED TO THE FREEZE–THAW RETRIEVAL

Many operational satellite sensors provide global coverage and frequent temporal revisit capabilities suitable for natural hazards assessment and monitoring (Chen et al., 2012; Gähler, 2016). For frost hazard monitoring, the satellite detection of landscape FT status can provide greater spatial coverage and accuracy than *in situ* measurements and spatially interpolated temperatures from sparse weather station networks, or regional weather model predictions (Gisnas et al., 2014; Kollas et al., 2014). Moreover, the limited temperature measurements available in weather station data-sparse areas, including boreal-Arctic and mountainous regions, constrain capabilities for regional assessment and monitoring of frost hazards (Andre et al., 2015; Kim et al., 2011).

Terrestrial remote sensing applications generally exploit up to three spectral wavelength (or frequency) regions, including visible and near-infrared (VNIR; 0.4–1 µm), thermal-infrared (TIR; 3.5–20 µm), and microwave bands (frequencies between 0.3 and 300 GHz). Remote sensing using lower frequency microwave bands has certain advantages for FT detection and monitoring, including greater vegetation transparency and surface penetration ability, and less sensitivity to atmosphere cloud and aerosol contamination relative to VNIR and TIR bands. The global coverage and overlapping orbital geometry of polar-orbiting satellites, coupled with relative microwave insensitivity to solar illumination and atmosphere effects, allow for potential FT monitoring day and night throughout the year, especially at higher latitudes where frozen temperatures are more common and have generally greater ecosystem impact. Microwave sensors detect natural radiation emitted by the land surface (passive system) or the surface radar backscatter energy pulse originating from the sensor (active system). The sensitivity of passive and active microwave sensors to land surface FT is influenced by several factors, including microwave frequency and polarization, sensor incidence angle, surface moisture content, atmospheric parameters, and landscape dielectric properties.

Passive microwave sensors (i.e., radiometers) provide brightness temperature (T_b) retrievals sensitive to surface temperature and emissivity. The emissivity depends on the surface moisture content, roughness, and dielectric properties (Bateni et al., 2013). The T_b retrieval represents the mean emitted radiation from the integrated land surface within the sensor footprint, including bare soil, vegetation, open water bodies, and snow cover (when present) elements (Holmes et al., 2015; Kim et al., 2011). The microwave emission and sensing depth are frequency dependent and generally shallower for wet surface conditions, and deeper for dry conditions due to the low emissivity and high attenuation of liquid water relative to dry soil (Ulaby and Long, 2014). Frozen conditions, whereby liquid water in the landscape is immobilized as ice, exhibit microwave emissivity characteristics similar to dry soil. The emissivity of dry soil and pure water ice is relatively high at microwave frequencies, which correspond to their low dielectric constants. Upon melting, the new release of liquid water in the landscape results in an increase in the surface dielectric constant and corresponding decrease in surface emissivity. Large characteristic changes in surface dielectric properties and emissivity occur between predominantly frozen and nonfrozen conditions during FT transitions, resulting in a large T_b temporal shift and associated FT signal relative to background noise effects. Potential noise effects influencing FT retrieval performance include sensor calibration accuracy, non-FT-related changes in surface moisture conditions including seasonal phenology or disturbance-related variations in vegetation structure and water content, and rainfall events and associated surface wetting and drying. The emissivity of wet soil decreases at lower (e.g., L-band) microwave frequencies (Raytheon, 2000), leading to a larger T_b response to surface moisture and FT variability relative to higher frequency observations (McDonald and Kimball, 2005; Das et al., 2014). Microwave polarization is another important property, which refers to the predominant horizontal (H) or vertical (V) orientation of the electrical field of an electromagnetic wave. The frequency-dependent difference between H- and V-polarized T_b retrievals increases with surface moisture content, whereas lower sensor incidence angles show less T_b polarization difference (Raytheon, 2000). The T_b sensitivity to vegetation is strongly influenced by vegetation opacity, which increases with higher microwave frequency and biomass density. Increase in vegetation opacity generally leads to reduced FT sensitivity to underlying soil conditions, but greater sensitivity to FT conditions within the vegetation canopy. The T_b sensitivity to deeper vegetation and soil layers is generally proportional to the landscape moisture content and sensor frequency; higher moisture levels in soil and vegetation layers reduce the effective depth of T_b sensitivity, whereas lower frequency (C-, L-band) sensor observations generally have greater depth of sensitivity than higher frequency (e.g., X-, K-band) observations. The T_b sensitivity to snow cover depends on the snowpack wetness, density, depth, and ice crystal structure, and it is more significant at higher frequencies (e.g., X-, K-band) due to more snowpack volume scattering (Ulaby et al., 1986; Lemmetyinen et al., 2011; Ulaby and Long, 2014). A constant sensor look angle and sampling time are generally optimal for the FT retrieval, although the Soil Moisture and Ocean Salinity (SMOS) and Aquarius radiometers have successfully used multitemporal T_b observations at variable sensor look angles for FT classification (Roy et al., 2015; Rautiainen et al., 2016).

The underlying characteristics and physical basis for the FT retrieval are similar between passive and active microwave sensors. Active microwave sensors (e.g., radars, scatterometers) emit pulses of electromagnetic energy that are backscattered from the land surface, whereas a portion of the returned backscatter is measured by the sensor detector at a defined frequency, polarization, and incidence angle. The received backscatter (σ^0) signal represents the complex microwave frequency and polarization-dependent interactions with aggregate landscape dielectric and structural characteristics. The σ^0 signal from natural snow cover is affected by three general parameters (Du et al., 2010a) relating to sensor characteristics (e.g., frequency, polarization, and sensing geometry), snow cover (e.g., snow density, particle size distribution and stratification, and free liquid water content), and subsurface conditions (e.g., dielectric and roughness at the snow-ground interface). For vegetated land, radar σ^0 is also sensitive to vegetation dielectric properties and canopy structure, including biomass water content, leaf and branch scatterer size, shape, and orientation (Du et al., 2010b). The radar σ^0 sensitivity to vegetation is proportional to microwave frequency and polarization, and the size and distribution of the vegetation scattering elements (Ulaby et al., 1986; Elachi, 1987). The vegetation scattering effects are generally more significant at higher frequencies (McDonald and Kimball, 2005) and for radar relative to passive microwave remote sensing. Active microwave sensors generally enable measurements with finer spatial resolution than passive sensors due to stronger radar energy emissions and the use of synthetic aperture radar (SAR) processing techniques to enhance signal-to-noise relative to passive microwave sensors that rely on detecting low levels of natural microwave emissions. Compared with available satellite radiometers and scatterometers, which generally have high temporal fidelity (~1–2 days) but coarse (~ 25 km) spatial resolution, currently available satellite SARs provide finer (on the order of 10 m to 100 m) spatial resolution, but coarse (on the order of weeks) temporal repeat observations. Interferometric synthetic aperture radar (InSAR) processing measures the phase differences between SAR images of the same location taken at different times and is capable of deriving FT-induced surface deformations with centimeters precision (Bürgmann et al., 2000; Lu and Dzurisin, 2014; Schaefer et al., 2015). InSAR has also been used for detecting frost hazards (e.g., permafrost deformation) related to seasonal FT transitions (Chang and Hanssen, 2015; Chen et al., 2012; Kääb, 2008; Liu et al., 2010).

8.3   LANDSCAPE FREEZE–THAW CLASSIFICATION ALGORITHMS

Satellite VNIR and TIR sensors provide snow-covered area, surface albedo, and land surface temperature (LST) observations that can be used to infer landscape FT status. However, for most of the cryosphere, frequent cloud cover, low solar elevation angles, shadowing, and low solar illumination limit regional monitoring from these sensors to relatively coarse 8–16-day temporal composites necessary to mitigate atmospheric effects (Cihlar et al., 1997). Satellite microwave remote sensing has unique capabilities that allow near real-time monitoring of landscape FT state without many of the limitations of VNIR and TIR sensors (McDonald and Kimball, 2005; Kim et al., 2011, 2012). The basic physical principle applied in FT classification is the relatively strong sensitivity of T_b and σ^0 to large temporal shifts in landscape dielectric properties as the landscape transitions between predominantly frozen and nonfrozen conditions. The relative magnitude of the T_b or σ^0 response to the FT state transition is dependent upon microwave frequency, surface moisture content, and aboveground land elements, including vegetation and snow cover (Ulaby et al., 1986). Despite the complex interactions that occur between microwave signals and land elements, the large dielectric change and the associated T_b and σ^0 response that occur from thawing and freezing of water in vegetation, snow, and soil allow for the FT classification (McDonald and Kimball, 2005; Kim et al., 2011, 2012). Low spatial resolution and high temporal revisit passive microwave sensors are well suited for detecting frequent FT state transitions from T_b time series, whereas finer resolution, but low-to-moderate temporal revisit active microwave sensors are better able to resolve spatial heterogeneity in landscape FT state transitions over complex terrain, heterogeneous land–water boundaries, and land cover conditions (Podest et al. 2014). However, radar imaging of regions with high topographic relief is subject to geometric distortions, including foreshortening and layover, depending on terrain slope-aspect variations and sensor viewing geometry. Radar shadow may also occur over mountainous areas when the radar beam is unable to reach the ground surface target due to terrain obstructions, which result in no backscattered signal and FT information loss for the terrain shadowed areas (Henderson and Lewis, 1998).

General approaches for classifying landscape FT status from satellite active and passive microwave sensors include seasonal threshold, moving window, temporal edge detection, multichannel combinations, decision tree, and probabilistic model algorithms (Table 8.1). Temporal edge detection algorithms are suitable for the identification of dominant FT state transitions from T_b (Liu et al., 2005; Kimball et al., 2006) or σ^0 observations (Park et al., 2011; Mortin et al., 2012). Moving window algorithms are useful for temporally consistent T_b or σ^0 observations and determine FT state transitions by comparing microwave retrievals for a selected location and period with a temporal moving window average T_b or σ^0 condition defined from a preceding period (Frolking et al., 1999; Rawlins et al., 2005; Wang et al., 2008). Seasonal threshold algorithm (STA) approaches examine the temporal progression of microwave T_b (Smith et al., 2004; Kim et al., 2011, 2012; Podest et al., 2014) or σ^0 (Du et al., 2015; Naeimi et al., 2012; Podest et al., 2014; Wilson et al., 2013) relative to seasonal reference frozen or nonfrozen conditions. The STA approach is well suited for determining daily FT conditions and for identifying multiple FT transition events, although algorithm performance and classification accuracy depend on the quality and stability of FT reference conditions. Multichannel combination algorithms exploit microwave signal differences between ice and liquid water from two or more frequencies or polarizations (e.g., ratio, differences, and normalization) for both σ^0 (Bartsch et al., 2007) and T_b (Zhao et al., 2011; Wang et al., 2013; Rautiainen et al., 2014; Guo et al., 2015; Roy et al., 2015). Decision tree algorithms are similar to multichannel

TABLE 8.1
Summary of the Major FT Classification Approaches and Their Relative Advantages and Disadvantages

Classification Approach	Advantages	Disadvantages
Temporal Edge Detection Algorithms	Suitable for identification of dominant FT state transitions	Less efficient for detecting smaller FT events, including daily or multiple FT transitions
Moving Window Algorithms	Suitable for identifying FT state transitions for a selected location and period on the basis of temporal anomalies computed relative to a preceding period	Sensitive to potential errors from non-FT-related, short-term microwave fluctuations
Seasonal Threshold Algorithms	Suitable for determining daily and multiple FT transition events	Classification accuracy depends on the quality and stability FT reference conditions
Multichannel combinations	Suitable for deriving composite classifications that distinguish FT conditions from different landscape elements	Potential errors from microwave frequency-dependent differences in sensing depths and sampling footprints
Decision Tree Algorithms	Useful for distinguishing FT conditions from precipitation events and in sparse vegetation and dry climate zones constrained by lower FT signal-to-noise	Less efficient for global domain due to larger computational costs and ancillary data requirements
Probabilistic Model Algorithm	Unsupervised classification from FT state transition probabilities that require no training data	Limitations from inherent assumptions of statistical models

combination algorithms; in that, multiple microwave channels are used to classify FT status. The decision tree algorithm approach has been found to effectively distinguish FT conditions from precipitation events and in sparse vegetation and drier climate zones in which other approaches are constrained by lower FT signal-to-noise (Jin et al., 2009; Chai et al., 2014; Han et al., 2015). The probabilistic model algorithm approach computes FT state transition probabilities using statistical models (e.g., statistical Markov model) for recognizing temporal patterns in T_b or σ^0 time series (Zwieback et al., 2012). All of these algorithms exploit the dynamic temporal response of T_b or σ^0 to the large characteristic changes in surface moisture and dielectric properties that occur as the landscape transitions between predominantly frozen and nonfrozen conditions. A major algorithm assumption is that the microwave temporal response to FT transitions is larger than other potential factors influencing T_b or σ^0 variability, including seasonal variations in snow cover, surface wetness, and vegetation cover. This assumption generally holds for northern temperate, boreal, Arctic, and alpine biomes with relatively well-defined frozen seasons and FT transition periods. However, large rainfall events can cause transient T_b or σ^0 shifts similar to FT transitions, whereas a smaller temporal dielectric response to FT transitions under dry surface or soil conditions can degrade FT signal-to-noise. Sensor footprint temporal geolocation instability and mixed land and water heterogeneity in coastal or open water body dominant areas can also degrade FT classification accuracy.

Global and regional FT classifications have been generated using a variety of satellite microwave active and passive systems and frequencies as summarized in Table 8.2. Sensor frequencies used for FT retrievals have included L-band (1–2 GHz), C-band (4–8 GHz), X-band (8–12 GHz), Ku-band (12–18 GHz), and Ka-band (27–40 GHz) observations. Satellite FT detection using radiometers has generally low-to-moderate spatial resolutions (>25 km) ranging from shallower characteristic

---

TABLE 8.2

Overview of Satellite Active and Passive Microwave Systems Used for FT Classification

Mission (Operation Period)	Frequency (GHz)	Native Footprint Size	Relevant Studies
Passive Radiometer			
SMMR (1979–1987)	18	55 km × 41 km	Smith et al. (2004)
	37	27 km × 18 km	Kim et al. (2011, 2012) and Smith et al., 2004
SSM/I(S)(1987–present)	19	70 km × 45 km	Chai et al. (2014), Jin et al. (2009), Smith et al. (2004), and Podest et al. (2014)
	37	38 km × 30 km	Chai et al. (2014), Jin et al. (2009), Kim et al. (2011, 2012), and Smith et al. (2004)
AMSR-E (2002–2011)	18.7	27 km × 16 km	Chai et al. (2014), Han et al. (2015), and Zhao et al. (2011)
	36.5	14 km × 8 km	Chai et al. (2014), Han et al. (2015), Kim et al. (2011, 2012), and Zhao et al. (2011)
AMSR2 (2012–present)	36.5	12 km × 7 km	Kim et al. (2017)
SMOS (2009–present)	1.4	~42 km	Rautiainen et al. (2014, 2016), Roy et al. (2015), and Zwieback et al. (2012)
SMAP (2015–present)	1.41	39 km × 47 km	Entekhabi et al. (2010) and Dunbar et al. (2015)
Aquarius (2011–2015)	1.413	62 km × 68 km–75km × 100 km	Roy et al. (2015)

(Continued)

TABLE 8.2 (*Continued*)

Overview of Satellite Active and Passive Microwave Systems Used for FT Classification

Mission (Operation Period)	Frequency (GHz)	Native Footprint Size	Relevant Studies
		Active Scatterometer	
NSCAT (1996–1997)	14	9 km × 32 km	Frolking et al. (1999) and Kimball et al. (2001)
QuikSCAT/Seawinds (1999–2009)	13.4	25 km × 37 km	Bartsch et al. (2007), Colliander et al. (2010, 2012), Podest et al. (2014), Rawlins et al. (2005), Wilson et al. (2013), and Zwieback et al. (2012)
ASCAT (2006–present)	5.255	25–50 km	Bartsch et al. (2012), Naeimi et al. (2012), and Zwieback et al. (2012)
Aquarius (2011–2015)	1.26	76–94 km × 96–156 km	Xu et al. (2016)
		Active SAR	
ERS1/SAR (1991–2000)	5.30	6–30 m	Rignot et al. (1994) and Podest et al. (2014)
ERS2/SAR (1995–2011)	5.30	6–30 m	Rignot et al. (1994)
Radarsat-1 (1995–2008)	5.30	8–100 m	Murphy et al. (2001)
Radarsat-2 (2007–present)	5.405	1–100 m	Jagdhuber et al. (2014)
JERS-1 SAR (1992–1998)	1.27	18 m	Podest et al. (2014)
Envisat/ASAR (2002–2012)	5.331	30–1000 m	Park et al. (2011)
ALOS/PALSAR (2006–2011)	1.27	7–100 m	Colliander et al. (2011) and Du et al. (2015)
ALOS2/PALSAR2 (2014–present)	1.20	1–100 m	Rosenqvist et al. (2014)
TerraSAR-X (2007–present)	9.65	1–16 m	Antonova et al. (2016)
Sentinel-1A (2014–present)	5.405	5–40 m	Malenovský et al. (2012)
Sentinel-1B (2016–present)	5.405	5–40 m	Malenovský et al. (2012)
SMAP (2015)	1.26	1–3 km	Entekhabi et al. (2010) and Dunbar et al. (2015)

sensing depths at Ka-band to deeper sensing depths at L-band. Similar multifrequency passive microwave radiometer measurements from overlapping sensor records (e.g., Scanning Multichannel Microwave Radiometer [SMMR], Special Sensor Microwave Imager [SSMIS], Advanced Microwave Scanning Radiometer for Earth Observing System [AMSR-E], Advanced Microwave Scanning Radiometer 2 [AMSR2]) operating on polar orbiting operational environmental satellites have enabled the development of long-term (>35years) global data records that can track FT daily, seasonal and interannual variability, and multidecadal trends (Kim et al., 2011, 2012). Satellite Ku-band and C-band scatterometers (e.g., quick scattermeter (QuikSCAT) and advanced scatterometer (ASCAT)) have moderate resolution (~25 km) to retrieve surface FT state, but with greater sensitivity to vegetation and snow cover conditions. Lower frequency (<10 GHz) SAR provides relatively finer resolution (10–100 m), but coarse (on the order of weeks) temporal repeat FT observations. The lower frequency L-band retrievals from SAR and passive microwave radiometers (e.g., SMOS, Soil Moisture Active Passive [SMAP]) also provide potentially enhanced sensitivity to soil FT conditions (Entekhabi et al. 2010).

8.3.1 A SATELLITE PASSIVE MICROWAVE GLOBAL FREEZE–THAW DATA RECORD

A long-term global satellite data record of daily landscape FT conditions (i.e., the FT-ESDR) is used in this chapter to illustrate FT-related applications for frost hazard assessment and monitoring. The FT-ESDR is a publicly accessible database developed from calibrated overlapping satellite microwave T_b time series extending over more than 35 years of continuous observations (Kim et al., 2014c, 2017). The FT-ESDR was derived using a STA temporal change classification of similar 37 GHz V-polarized (pol) daily T_b records from the SMMR, Special Sensor

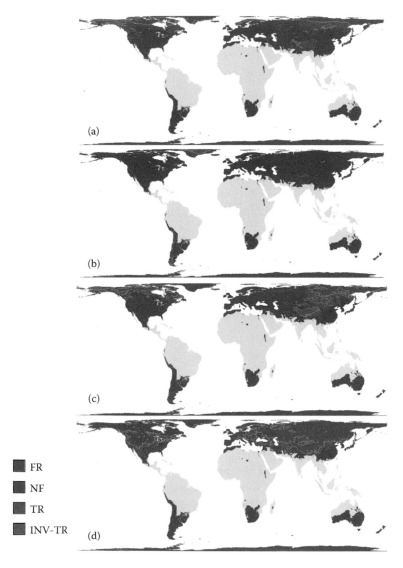

FIGURE 8.1 Selected daily combined (CO) FT-ESDR classification results for 2007, representing: (a) Apr 19 (DOY 109), (b) Jul 20 (DOY 201), (c) Oct 30 (DOY 303), and (d) Dec 26 (DOY 360). White and grey colors denote respective open water bodies and land areas outside of the FT-ESDR domain. FR denotes frozen conditions (AM and PM frozen); NF denotes non-frozen (AM and PM thawed); TR denotes transitional (AM frozen and PM thawed) and INV-TR denotes Inverse transitional (AM thawed and PM frozen) conditions.

Microwave Imager (SSM/I), and SSMIS. The FT-ESDR provides a daily classification of the predominant frozen or nonfrozen status of the land surface within a satellite sensor footprint. The 37 GHz frequency T_b retrievals have relatively high atmospheric transmittance and are sensitive to surface layer conditions (Holmes et al., 2013; Andre et al., 2015). The V-pol T_b retrieval is less sensitive to surface moisture variations and snow, and ice stratifications than H-pol T_b retrievals (Owe et al., 2008; Green et al., 2012). The FT-ESDR distinguishes twice daily (AM and PM) FT status from ascending and descending satellite overpass observations, posted to a 25 km EASE-Grid geographic projection (Brodzik and Knowles, 2002). The FT-ESDR also includes similar global daily FT data records derived from the National Aeronautics and Space Administration's (NASA's) AMSR-E daily (AM and PM overpass) 36.5 GHz (V-pol) T_b retrievals (June 2002–September 2011); the AMSR-E sensor ceased operations in 2011, whereas the associated FT record has been extended to more recent years using compatible T_b measurements from the Japan Aerospace Exploration Agency's (JAXA's) AMSR2 sensor record. Satellite ascending and descending orbital data time series are processed separately to produce information on morning (AM), afternoon (PM) and composite daily FT conditions (CO). Four categorical daily FT classification levels are provided, including frozen (AM and PM frozen), nonfrozen (AM and PM thawed), transitional (AM frozen and PM thawed), and inverse transitional (AM thawed and PM frozen) status. The FT-ESDR provides a daily FT classification across the vegetation–snow–soil continuum within a global domain (Figure 8.1). The global FT-ESDR domain encompasses all FT affected areas where seasonal frozen temperatures are a significant constraint to surface water mobility and ecosystem processes (Kim et al., 2017).

8.3.2 FT DETECTION FROM THE PHASED ARRAY L-BAND SYNTHETIC APERTURE RADAR ACTIVE SENSOR

The phased array L-band synthetic aperture radar (PALSAR) sensor onboard the JAXA's Advanced Land Observing Satellite (ALOS) and its successor PALSAR2 are among the few L-band spaceborne radars available for studying L-band SAR-based landscape FT detection. The PALSAR ScanSAR mode was capable of acquiring radar σ^0 data over a swath as large as 350 km, resulting in variable incident angle radar σ^0 retrievals. Different from the coarse resolution (~25 km) and high temporal repeat observations (~1–2 days revisit time) of satellite radiometers and scatterometers, PALSAR ScanSAR had a much finer (~100 m) spatial resolution, but much lower temporal fidelity (46 day orbit revisit). A similar STA approach as the FT-ESDR was used with PALSAR L-band σ^0 to classify the FT pattern over Alaska (Du et al. 2015); these results are used in this chapter to illustrate spatial scale differences between the relatively coarse FT-ESDR and finer resolution SAR classification. A similar STA approach was used for production of a global FT product derived from NASA's SMAP L-band radar σ^0 retrievals (Entekhabi et al., 2010; Dunbar et al., 2015). The STA relies on a pixel-wise FT temporal change classification of radar σ^0 differences from reference frozen or nonfrozen conditions.

An example FT state map derived from 100 m resolution PALSAR retrievals over a subregion of Alaska (67.6°N–68.1°N, 155.6° W–157.2°W) for April 29, 2007 is shown in Figure 8.2. The resulting FT classification shows the general spring FT transition for the subregion where frozen conditions are more prevalent at higher elevations, whereas the lower valley areas and south-facing slopes show predominately thawed conditions. The relatively fine-scale SAR retrievals distinguish large characteristic FT spatial heterogeneity congruent with the influence of terrain slope and aspect, and vegetation cover on local microclimate variability. This level of FT spatial heterogeneity is lost in the coarser scale FT-ESDR retrievals (Figure 8.1), though the FT-ESDR distinguishes daily FT variability and has global coverage.

FIGURE 8.2 Spring FT pattern over a sub-region of Alaska (67.6° N-68.1°N, 155.6° W-157.2°W) derived from 100 m resolution PALSAR L-band radar backscatter retrievals for April 29, 2007, following Du et al. (2015). The resulting image is draped over a digital terrain map. Open water bodies are denoted in black.

8.3.3 Freeze–Thaw Metrics Related to Frost Hazards

Various FT-related metrics including the start, end, and length of the frost season have been used as climate indicators (Easterling et al., 2002; Moonen et al., 2002; Rawlins et al., 2016). Spatial and temporal changes in FT-related climate indicators have been linked with terrestrial ecosystem impacts, including the timing of spring bud burst and leaf out, photosynthetic activity, and the timing and pathways of animal migrations (Wolfe et al., 2005; Hufkens et al., 2012; Schwartz et al., 2013). Previous studies have documented a variety of FT-related climate indicators using *in situ* field measurements (Bourque et al., 2005; Augspurger, 2013; McCabe et al., 2015), including icing day ($T_{max} < 0°C$), frost day ($T_{min} < 0°C$), and cumulative plant growing degree day metrics (Alexander et al., 2006; Richardson et al., 2006). SFDs determined from surface freezing air temperatures have been developed for studying temporal variability of the spring frost season (Heino et al., 1999; Bonsal et al., 2001; Frich et al., 2002; Kunkel et al., 2004) and for analyzing frost impacts on spring phenology (Linkosalo et al., 2000; Augspurger, 2013; Lenz et al., 2016) and crop planting dates (Parker et al., 2016). Frost events have also been considered as significant hazardous frost processes for improving ecosystem models (Poirier et al., 2010; Rammig et al., 2010) and phenological models (Cannell and Smith, 1986; Cittadini et al., 2006).

The frost season derived from the FT-ESDR is defined as an accumulation of frozen days during a year. The frost seasons were derived for each EASE-Grid cell within the global domain on an annual basis over the 36-year (1979–2014) FT-ESDR. The resulting 36 annual counts were averaged for each grid cell to obtain a mean annual frost season map. The resulting annual frost season and frost probability maps are shown in Figure 8.3. These results show a global mean annual frost season of 126.9 ± 119.0 (spatial SD) days for the 1979–2014 satellite record, and a general increase in the average frost season and probability of frost occurrence at higher latitudes and elevations. Longer frost season duration and higher probability of frost are found in mountainous areas, including the Rocky Mountains, Alps, Andes, and the Tibetan plateau. The frost season determines the timing and effective duration of the growing season, and the

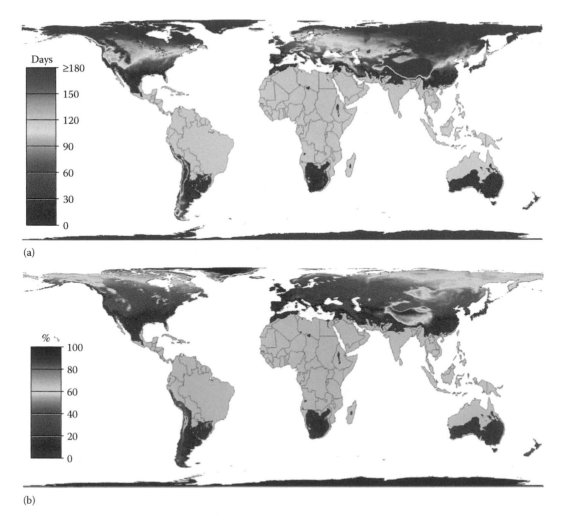

(a)

(b)

FIGURE 8.3 (a) Average annual frost season (frozen and transitional days) and (b) mean annual frost occurrence probability (%) derived from the FT-ESDR 36-year (1979–2014) record.

seasonal progression of land–atmosphere water, energy, and carbon exchange in which frozen temperatures are a major constraint to ecosystem processes and land surface water mobility (Penuelas et al., 2009; Kim et al., 2014b).

The timing and duration of freezing temperatures in spring have been linked to large variations in spring phenology and annual vegetation productivity in temperate and northern climate areas (Tubiello et al., 2007; Vitasse et al., 2009; Schwartz et al., 2013). In this chapter, we define a SFD metric as the total number of FT-ESDR derived frozen or transitional frost days in spring (March–May) (Kim et al. 2014b) in which the resulting SFD metric ranges from 0 (no spring frost events) to 92 (complete spring frozen period) days. The SFD results show a general latitudinal gradient and a mean of 44.0 ± 29.3 (spatial SD) days over the Northern Hemisphere (Figure 8.4a), with large interannual (SD) variability (Figure 8.4b). The SFD variability is more extreme along the boundaries of major climate zones and air masses, including the interior continental United States, Central Canada, Europe, Southern Greenland, Central Asia, and Southern China.

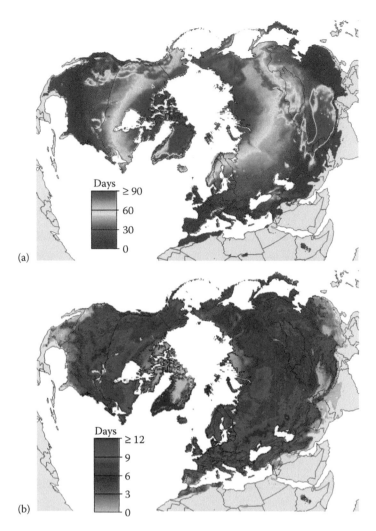

FIGURE 8.4 (a) Average annual spring frost day (SFD) metric (days) and (b) SFD temporal standard deviation (SD) derived from the FT-ESDR 36-year (1979–2014) record. Grey denotes land areas outside of the FT-ESDR domain.

8.4 SATELLITE DETECTION OF FROST DAMAGE-RELATED IMPACTS ON VEGETATION GROWTH

The end of frost in spring is a prerequisite for bud-burst and canopy onset, whereas earlier plant leaf out and late spring frost events can cause frost damage to vegetation, including leaf pigment injuries, withering and loss of canopy photosynthetic capability (Ensminger et al., 2004; Lazarus et al., 2006; Oksanen et al., 2006). Frost damage can also affect a wide range of vegetation (Awaya et al., 2009; Hufkens et al., 2012; Kreyling et al., 2012), including crops (Rodrigo, 2000; Carter et al., 2006; Fengjin and Lianchun, 2011). For example, extraordinary warm spring temperatures in May and June of 2007 were recorded in the northeastern United States relative to the long-term (1971–2000) climatology; the anomalous warm temperatures were followed by a late spring freezing event extending across several states (Blunden et al., 2011; Guirguis et al., 2011). The effects of this event were widespread and ranged from mild leaf damage to complete canopy defoliation. The anomalous spring warming in 2007 led to early canopy leaf out and plant growth across eastern and central North America. The relatively early spring leaf out coupled with the subsequent late spring frost event resulted in widespread frost damage to vegetation growth, including forests and managed cropland (Gu et al., 2008; Kim et al., 2014b).

Several studies have documented frost damage-related patterns and severity to vegetation using satellite VNIR remote sensing ranging from relatively fine resolution Landsat imagery (Olthof et al., 2004; King et al., 2005; Wang et al., 2012) to Moderate Resolution Imaging Spectroradiometer (MODIS) and Satellite Pour l'Observation de la Terre (SPOT; Earth-observing satellite) imagery (Silleos et al., 2002; Feng et al., 2009; Currit and Clair, 2010). These studies have focused on the use of image classification techniques and change detection for analyzing frost-related impacts on vegetation greenness, photosynthetic canopy structure, and productivity. LST from advanced very high resolution radiometer (AVHRR) and MODIS TIR remote sensing has also been used for detecting and mapping frost occurrence and extent (Kerdiles et al., 1996; Tait and Zheng, 2003; Pouteau et al., 2011). However, LST monitoring of transient frost occurrences from global satellite VNIR and TIR sensors is constrained by data loss and temporal compositing requirements necessary to mitigate cloud and atmosphere contamination effects. VNIR signals are sensitive to plant photosynthetic canopy phenology and productivity, whereas passive microwave remote sensing is well suited for monitoring high temporal (e.g., daily) variations in landscape FT status. LST retrievals from TIR sensors may also provide effective information for FT monitoring (Hachem et al., 2009), although capabilities for global monitoring of transient FT events may be constrained by cloud and atmosphere contamination effects and information loss.

A satellite data fusion approach combining vegetation greenness and FT information from VNIR and passive microwave sensors was recently applied for assessing frost damage-related impacts to ecosystem productivity (Kim et al., 2014b). VNIR sensor-derived land surface phenology from a vegetation greenness index (VI) was combined with FT-ESDR-derived frost season metrics to define the number of damaging frost days (NFD) over the conterminous United States. The NFD was defined as the sum of FT-ESDR-observed spring frost events following VI-defined canopy onset when vegetation is more vulnerable to frost injury. In this chapter, the NFD record is extended over the entire Northern Hemisphere and a longer period (1981–2014). The annual spring start of season (SOS) and day of peak (DOP) VI greenness for the study period were obtained from a globally consistent and continuous 5.6 km resolution (0.05° × 0.05°) satellite VNIR sensor-based vegetation phenology record (Barreto-Munoz, 2013; Didan et al., 2016a). The 5.6 km SOS and DOP data were projected to a consistent 25 km global EASE-Grid format and used with the FT-ESDR to determine the number of frost days occurring between spring canopy onset and peak seasonal canopy development for each year of record and grid cell over the Northern Hemisphere domain.

The NFD metric defines the accumulation of FT-ESDR-classified frost (frozen or transitional) days between SOS and DOP as determined from the VNIR phenology record. The resulting NFD pattern varies according to different locations, terrains, climates, vegetations, and land cover conditions, ranging up to a maximum period defined by the SOS and DOP difference. The Northern Hemisphere shows a mean NFD count of 7.7 ± 9.4 (spatial SD) days (Figure 8.5a), with large year-to-year variability (Figure 8.5b). Unlike the SFD metric representing the sum of all frost days in spring (Figure 8.4), the NFD metric only accounts for frost days occurring during the period of active canopy development, when vegetation is more sensitive to frost damage and impaired growth (Kim et al., 2014b).

Early spring leaf development has the potential to increase exposure to potential frost damage (Strimbeck et al., 1995; Inouye, 2008), including loss of stored carbon and nutrients, and reduction of photosynthetic carbon uptake by natural and cropland vegetation (Gu et al., 2008; Martin et al., 2010). Previous studies indicate an increasing NFD regional trend over the continental United States (Kim et al., 2014b). Similarly, the spatial mean NFD trend over the Northern Hemisphere domain is 0.05 ± 0.62 day year^{-1} (spatial SD) for the 1981–2014 records. The temporal NFD trends are defined using prewhitened Kendall's tau statistics screened for outliers (≥ ±2 temporal SD). The regional

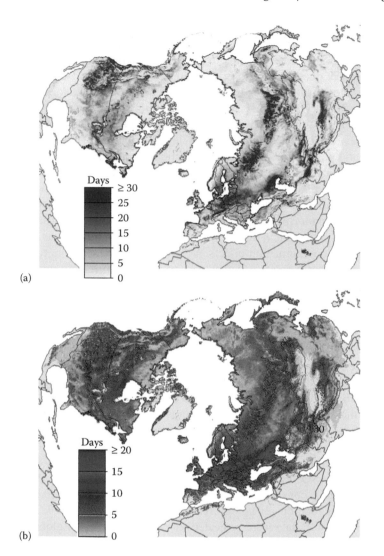

FIGURE 8.5 (a) Average annual number of NFD damaging frost days and (b) NFD temporal standard deviation (SD) derived from the FT-ESDR 34-year (1981–2014) record. Grey denotes land areas outside of the FT-ESDR domain or no availability of VI defined SOS and DOP vegetation phenology metrics.

distribution of the NFD trend is heterogeneous over the Northern Hemisphere (Figure 8.6), indicating relatively large increases in damaging frost events over western North America, northern Europe, and central Eurasia. The NFD trend is increasing for 34.5% of the Northern Hemisphere domain. As indicated in a previous study (Kim et al., 2014b), a positive NFD trend indicates greater risk of damaging frost events leading to loss of vegetation productivity. The general increase in NFD occurrence is associated with an earlier spring vegetation greening trend. Regional trends toward earlier onset of the growing season have been attributed to global warming and the relaxation of cold temperature constraints to vegetation growth, which paradoxically increases frost damage risk in many areas. The positive NFD trend implies that the danger of leaf damage from spring frosts may be greater than the potential gains to annual vegetation growth from early leaf out (Kim et al., 2014b).

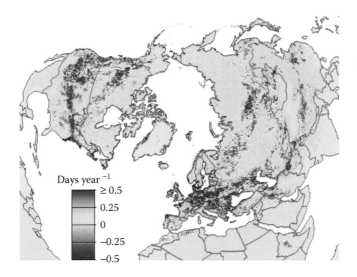

FIGURE 8.6   Regional Kendall's tau trend patterns (days yr^{-1}) derived from the number of NFD damaging frost days (1981–2014) over the Northern Hemisphere. Grey denotes land areas outside of the FT-ESDR domain or no availability of VI defined SOS and DOP vegetation phenology metrics.

8.5   APPLICATIONS USING ENHANCED RESOLUTION SATELLITE FREEZE–THAW DATA RECORDS

Similar daily 36.5 GHz, V-pol T_b records from the AMSR-E and AMSR2 sensors were integrated to produce a consistent FT daily data record over the Northern Hemisphere. The AMSR-E sensor was operational on the NASA Aqua satellite from June 2002 to October 2011 and provided twice-daily global T_b observations with sensor descending/ascending orbital equatorial crossings at 1:30 AM/PM local time. After the loss of AMSR-E normal operations on October 4, 2011, the successor AMSR2 sensor was launched on May 18, 2012 on the JAXA global change observation mission – water satellite 1 (GCOM-W1) satellite. AMSR2 is similar to AMSR-E in sensor configuration, including microwave frequencies, sensor incidence angles, and orbital equatorial crossing times. The AMSR-E 36.5 GHz orbital swath T_b data have a native footprint resolution of 14 km × 8 km (Kawanishi et al., 2003); these data were merged with similar frequency T_b orbital swath (L1R) data from AMSR2 with a native 12 km × 7 km footprint resolution (Imaoka et al., 2010) by resampling the T_b records to a consistent 6 km polar EASE-Grid projection format (Brodzik et al., 2012, 2014). The AMSR FT retrieval is obtained using the STA approach, which classifies daily T_b variations in relation to grid cell-wise FT thresholds calibrated using 6 km resolution surface air temperature (SAT) maps downscaled from coarser (0.25°) spatial resolution European reanalysis (ERA)-Interim global reanalysis daily surface meteorological data using a digital elevation map (DEM) and environmental lapse rates (ELR). The daily grid cell-wise ELR was derived from the linear regression relationship between DEM elevations and Aqua MODIS LST retrievals. The resulting FT record is mapped to a 6 km resolution polar EASE-Grid for the Northern Hemisphere domain and spans more than 14 years of observations (2002–2016) from both AMSR-E and AMSR2 sensors. The resulting FT record shows relatively enhanced delineation of the spring thaw pattern over Alaska relative to the coarser (25 km) resolution global FT-ESDR (Figure 8.7). Here the primary spring thaw date is estimated from the daily FT records for each grid cell as the first day for which 12 out of 15 consecutive days from January

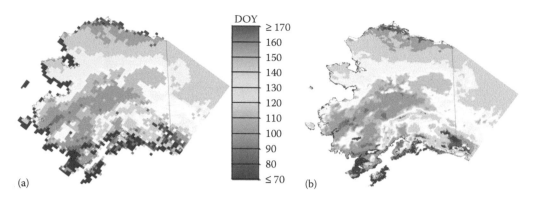

FIGURE 8.7   Spring primary thaw pattern over Alaska derived from AMSR-E 36GHz V-pol T_b retrievals at 25-km (a) and 6 km (b) resolution for 2007.

to June were classified as nonfrozen (Kim et al., 2014a). Other spatial enhancement techniques are available and may provide greater accuracy, including Backus–Gilbert, truncated singular value decomposition (TSVD), smoothing filter-based intensity modulation (SFIM) techniques (Migliaccio and Gambardella 2005; Santi 2010; Lenti et al., 2014), machine learning algorithm, and data assimilation-based fusion of coarser passive microwave and finer scale VNIR, TIR, and radar data (Zhan et al., 2006; Song et al., 2014; Kou et al., 2016). Other fine-scale auxiliary data (e.g., topography) can also be used for downscaling coarser resolution data using statistical correlations (Reichle et al., 2001; Chauhan et al., 2003) or distributed physically based models (Pellenq et al., 2003; Merlin et al., 2013).

For the frost hazard monitoring case study, the AMSR-E 6 km daily FT data records were used for analyzing the documented anomalous spring freezing event that occurred in April 2007 over the continental United States (Gu et al. 2008). The AMSR-E daily FT state maps of frost affected areas within nine states (MO, IL, IN, KY, TN, GA, AL, MS, and AR) reporting unusual spring freezing events in April 2007 are shown in Figure 8.8. The frost maps are also presented with corresponding VI anomaly maps indicated from the satellite VNIR Enhanced Vegetation Index (EVI2) (Barreto-Munoz, 2013; Didan et al., 2016b). The EVI2 metric is derived at 5.6 km resolution and is used as a proxy for changes in vegetation photosynthetic canopy cover and productivity (Kim et al., 2010; Zhang et al., 2014). The EVI2 anomalies represent the difference between the daily VI retrieval and the corresponding daily mean climatological value derived from the 2000 to 2014 satellite records. These results indicate that the extent of frost-affected areas is considerably larger on April 8, 2007 and extends over nine states. The enhanced 6 km resolution AMSR-E frost sequence maps are closer in scale to the 5.6 km resolution EVI2 record than the coarser (25 km) global FT-ESDR, providing improved delineation of frost patterns and associated impacts to vegetation growth over the affected regions. Frost occurrence is primarily captured as nighttime freezing followed by subsequent daytime thawing indicated by the AMSR-E T_b orbital acquisitions at 1:30 AM/PM local time, and recorded as transitional (AM frozen, PM thaw) frost events by the AMSR-E FT daily composite. The anomalous frost occurrence is followed by a widespread decrease in EVI2-derived photosynthetic canopy cover and vegetation productivity. The frost-related damage to vegetation growth persists even after the cold snap has ended. Vegetation recovery from damaging spring frost events may extend up to several weeks or more following frost occurrence and can also result in plant mortality and annual productivity decline (Gu et al. 2007, Kim et al. 2014).

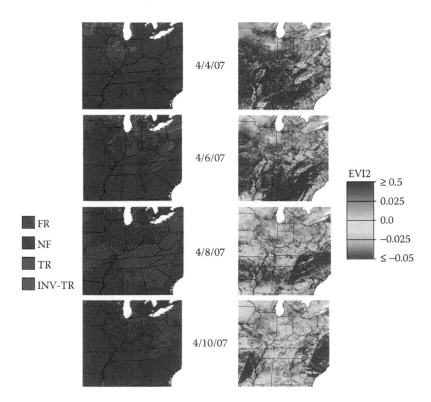

FIGURE 8.8 (a) Anomalous spring frost and (b) vegetation greenness (EVI2) patterns for April 2007 over the central United States including: MO, IL, IN, KY, TN, GA, AL, MS, and AR. The frost pattern is derived from 6-km resolution AMSR-E daily T_b retrievals, and includes frozen (FR; AM and PM frozen), non-frozen (NF; AM and PM thawed), transitional (TR; AM frozen and PM thawed) and Inverse transitional (INV-TR; AM thawed and PM frozen) FT categories. The EVI2 anomaly maps (5.6-km resolution) indicate higher (in red) and lower (in blue) canopy greenness and productivity differences from the EVI2 climatological (2000–2014) mean for each grid cell.

8.6　CONCLUSIONS

Accurate and timely assessment of frost risk and occurrence is critical in countries where natural and man-made environments are directly influenced by freezing temperatures. Satellite remote sensing provides an effective means for global monitoring of frost occurrence and associated ecological impacts. The FT-ESDR introduced in this chapter is one of the most consistent and longest global satellite environmental data records, spanning all significant frost-affected land areas and providing daily delineation of FT status that can be used to inform frost risk and damage assessments. Satellite radar sensors (e.g., SAR) are capable of finer spatial delineation of frost events than passive microwave sensors, which may have particular value over heterogeneous land cover and mountainous regions, but at the expense of degraded temporal fidelity for regional monitoring. The increasing cost and risk of extreme weather events and natural disasters have led to a growing demand for near real-time geospatial detection and monitoring capabilities, including suddenly evolving frost events. Capabilities for near-real time FT monitoring are enabled by currently operational global environmental satellites, whereas further investments in rapid data processing and distribution services would facilitate disaster early warning, assessment, and rapid response planning.

Warmer springs associated with global climate change are promoting earlier onset of vegetation growth and increased risk of frost damage in many areas that are affected by seasonal frozen temperatures. These environmental trends not only impact vegetation productivity but may also alter plant community composition, ecosystem function, and biosphere–atmosphere interactions. A shift in the extent and frequency of thawing and freezing with climate warming may influence the structure and duration of seasonal snow cover, adversely affecting animal habitats, migration and foraging success, and human infrastructure (Jorgenson et al., 2001; Li et al., 2014; Kim et al., 2015; Leblond et al., 2016; Riseth et al., 2016).

The NASA's SMAP mission began effective operations in April 2015, producing new global observations of landscape FT status and associated impacts on water mobility and ecosystem productivity with enhanced L-band microwave sensitivity to surface soil conditions underlying vegetation and snow cover, and with near daily temporal monitoring capabilities (Dunbar et al., 2015). However, wet snow and dense vegetation can still obscure the soil FT signal. The use of model data assimilation and data integration techniques utilizing a diversity of environmental observations from satellite VNIR and microwave sensors, active and passive microwave frequencies and polarizations, and other geospatial data may enable enhanced delineations of snow, soil, and vegetation elements from the aggregate FT retrieval (Farhadi et al., 2015; McColl et al., 2016; Podest et al., 2014). Despite loss of the SMAP radar, which ceased functioning after approximately 2.5 months of operations, similar operational FT products are being developed using the SMAP radiometer, which continues normal operations. Enhanced resolution FT products are also being developed from SMAP L-band radiometer retrievals with 9 km spatial gridding through postprocessing of overlapping T_b retrievals (Das et al., 2016). Better understanding of relations between surface conditions and microwave signals at finer spatial resolution and various frequencies would enhance satellite microwave remote sensing capabilities for frost hazard monitoring over complex terrain and land cover areas. Other satellite microwave sensor records are available and may provide additional FT information, including the Tropical Rainfall Measuring Mission's (TRMM) Microwave Imager (TMI) and the precipitation radar (PR); these sensors potentially provide higher temporal fidelity FT observations in lower latitude and high elevation regions (e.g., Himalaya and Rocky Mountain ranges), and relatively finer spatial resolution retrievals (~14 km for TMI and ~4.5 km for PR; Viltard et al., 2006). Continuing satellite microwave observations from existing global operational satellites allow for the development and extension of global environmental data records capable of distinguishing transient weather events from periodic climate anomalies and long-term climate trends in frost events, risk probabilities, and hazards. The recent launch of the next-generation satellite sensors (e.g., SMAP) designed for FT detection and producing operational FT data products provide for new frost hazard applications. All of these recent advances offer capabilities for new investigations and regional monitoring of frost hazards and environmental impacts.

ACKNOWLEDGMENTS

This work was conducted at the University of Montana with support from the NASA Making Earth Science Data Records for Use in Research Environments (MEaSUREs) program (NNX14AB20A).

REFERENCES

Alexander, L. V., X. Zhang, T. C. Peterson, J. Caesar, B. Gleason, A. M. G. Klein Tank, M. Haylock et al. 2006. Global observed changes in daily climate extremes of temperature and precipitation. *Journal of Geophysical Research*, 111, D05109.

Andre, C., C. Ottle, A. Royer, and F. Maignan. 2015. Land surface temperature retrieval over circumpolar Arctic using SSM/I-SSMIS and MODIS data. *Remote Sensing of Environment*, 162, 1–10.

Augspurger, C. K., 2013. Reconstructing patterns of temperature, phenology, and frost damage over 124 years: Spring damage risk is increasing. *Ecology*, 94 (1), 41–50.

Awaya, Y., K. Tanaka, E. Kodani, and T. Nishizono. 2009. Responses of a beech (*Fagus crenata* Blume) stand to late spring frost damage in Morioka, Japan. *Forest Ecology and Management*, 257, 2359–2369.

Antonova, S. et al. 2016. Spatio-temporal variability of X-band radar backscatter and coherence over the Lena River Delta, Siberia. *Remote Sensing of Environment*, 182, 69–191.

Barreto-Munoz, A. 2013. Multi-sensor vegetation index and land surface phenology Earth science data records in support of global change studies: Data quality challenges and data explorer system. Available: http://arizona.openrepository.com/arizona/handle/10150/301661 (Accessed: September 21, 2016).

Bartsch, A., R. A. Kidd, W. Wagner, and Z. Bartalis. 2007. Temporal and spatial variability of the beginning and end of daily spring freeze/thaw cycles derived from scatterometer data. *Remote Sensing Environment*, 106 (3), 360–374.

Bartsch, A., T. Kumpula, B. C. Forbes, and F. Stammler. 2010. Detection of snow surface thawing and refreezing in the Eurasian Arctic with QuickSCAT: Implications for reindeer herding. *Ecological Applications*, 20, 234–2358.

Bateni, S. M., C. Huang, S. A. Margulis, E. Podest, and K. McDonald. 2013. Feasibility of characterizing snowpack and the freeze-thaw state of underlying soil using multifrequency active/passive microwave data. *IEEE Transactions on Geoscience and Remote Sensing*, 51 (7), 4085–4102.

Bello, O. M. and Y. A. Aina. 2014. Satellite remote sensing as a tool in disaster management and sustainable development: Towards a synergistic approach. *Procedia – Social and Behavioral Sciences*, 120, 365–373.

Bennie, J., E. Kubin, A. Wiltshire, B. Huntley, and R. Baxter. 2010. Predicting spatial and temporal patterns of bud-burst and spring frost risk in north-west Europe: The implications of local adaptation to climate. *Global Change Biology*, 16, 1503–1514.

Blunden, J., D. S. Arndt, and M. O. Baringer.2011. State of the climate in 2010. *Bulletin of the American Meteorological Society*, 92, 1–266.

Bonsal, B. R., X. Zhang, L. A. Vincent, and W. D. Hogg, 2001. Characteristics of daily and extreme temperatures over Canada. *Journal of Climate*, 14, 1959–1976.

Bourque, C. P. A., R. M. Cox, D. J. Allen, P. A. Arp, and F. Meng. 2005. Spatial extent of winter thaw events in eastern North America: Historical weather records in relation to yellow birch decline. *Global Change Biology*, 11, 1477–1492.

Braathe, P. (1957). Is there a connection between the birch dieback and the March thaw of 1936? *The Forestry Chronicle*, 33, 358–363.

Braathe, P. (1995). Birch dieback caused by prolonged early spring thaws and subsequent frost. *Norwegian Journal of Agriculture Science Supplement*, 20, 50.

Brodzik, M. J., B. Billingsley, T. Haran, B. Raup and M. H. Savoie. 2012. EASE-Grid 2.0: Incremental but significant improvements for Earth-gridded data sets. *ISPRS International Journal of Geo-Information*, 1, 32–45.

Brodzik, M. J., B. Billingsley, T. Haran, B. Raup and M. H. Savorie. 2014. EASE-Grid 2.0: Incremental but significant improvements for Earth-gridded data sets. *ISPRS International Journal of Geo-Information*, 3 (3), 1154–1156.

Brodzik, M. J. and K. W. Knowles. 2002. EASE-Grid: A versatile set of equal area projections and grids. In M. Goodchild (Ed.) *Discrete Global Grids*. Santa Barbara, CA: National Center for Geographic Information and Analysis.

Buermann, W., P. R. Bikash, M. Jung, D. H. Burn, and M. Reichstein. (2013). Earlier springs decrease peak summer productivity in North American boreal forest. *Environmental Research Letters*, 8, 024027.

Bürgmann, R., G. Hilley, A. Ferretti and Novali, F., 2006. Resolving vertical tectonics in the San Francisco bay area from permanent scatterer InSAR and GPS analysis. *Geology*, 34 (3), 221–224.

Cannell, M. G. R. and R. I. Smith. 1986. Climatic warming, spring budburst and frost damage on trees. *Journal of Applied Ecology*, 23, 177–191.

Carter, P. and D. Wiersma. 2006. Early season frost damage on Corn. *Crop Insights*, 16 (4), 1–6.

Chai L., L. Zhang, Y. Zhang, Z. Hao, L. Jiang, and S. Zhao. 2014. Comparison of the classification accuracy of three soil freeze–Thaw discrimination algorithms in China using SSMIS and AMSR-E passive microwave imagery, *International Journal of Remote Sensing*, 35 (22), 7631–7649.

Chang, L. and R.F., Hanssen. 2015. Detection of permafrost sensitivity of the Qinghai–Tibet railway using satellite radar interferometry. *International Journal of Remote Sensing*, 36(3), 691–700.

Chauhan, N. S., S. Miller and P. Ardanuy. 2003. Spaceborne soil moisture estimation at high resolution: A microwave-optical/IR synergistic approach. *International Journal of Remote Sensing*, 24 (22), 4599–4622.

Chen, K., S. B. Serpico, and J. A. Smith. 2012. Remote sensing of natural disasters. *Proceedings of the IEEE*, 100 (10), 2794–2797.

Cihlar, J. H. Ly, Z. Li, J. Chen, H. Pokrant, and F. Huang. 1997. Multitemporal, multichannel AVHRR data sets for land biosphere studies-Artifacts and corrections. *Remote Sensing of Environment*, 60, 35–57.

Cittadini, E. D., N. de Ridder, P. L. Peri, and H. ver Keulen. 2006. A method for assessing frost damage risk in sweet cherry orchards of South Patagonia. *Agricultural and Forest Meteorology*, 141, 235–243.

Colliander, A. K. McDonald, R. Zimmerman, E. Podest, R. Schroeder, J. Kimball, and E. Njoku. 2011. Active and Passive multi-scale microwave remote sensing of the Alaska Ecological Transect: Application to SMAP freeze/thaw state validation planning, *Proceedings of the International Geoscience and Remote Sensing Symposium (IGARSS)*, Vancouver, Canada, 24–29 July 2011, 3156–3159.

Colliander, A. K. McDonald, R. Zimmerman, R. Schroder, J. S. Kimball, and E. G. Njoku. 2012. Application of QuikSCAT Backscatter to SMAP Validation planning: Freeze/Thaw State Over ALECTRA Sites in Alaska From 2000 to 2007. *IEEE Transaction on Geoscience and Remote Sensing*, 50 (2), 461–468.

Colliander, A., K. McDonald, R. Zimmermann, T.Linke, R.Schroeder, J. Kimball, and E. Njoku. 2010. QuikSCAT backscatter sensitivity to landscape freeze/thaw state over ALECTRA sites in Alaska from 2000 to 2007 application to SMAP validation planning. *Proceedings of the International Geoscience and Remote Sensing Symposium (IGARSS)*, Honolulu, HI, 25–30 July 2010, 1269–1272.

Currit, N. and S. B. St Clair. 2010. Assessing the impact of extreme climatic events on aspen defoliation using MODIS imagery. *Geocarto International*, 25 (2), 133–147.

Das, N. N., D. Entekhabi, R. S. Dunbar, E. G. Njoku, and S. H. Yueh. 2016. Uncertainties estimates in the SMAP combined active-passive downscaled brightness temperature. *IEEE Transactions on Geoscience and Remote Sensing*, 54 (2), 640–650.

Das, N. N., D. Entekhabi, E. G. Njoku, J. J. C. Shi, J. T. Johnson, and A. Colliander. 2014. Tests of the SMAP combined radar and radiometer algorithm using airborne field campaign observations and simulated data. *IEEE Transactions on Geoscience and Remote Sensing*, 52 (4), 2018–2028.

Derksen, C., M. Sturm, G. E. Liston, J. Holmgren, H. Huntington, A. Silis, and D. Solie. 2009. Northwest Territories and Nunavut snow characteristics from a subarctic traverse: Implications for passive micro-wave remote sensing. *Journal of Hydrometeorology*, 10, 448–463.

Didan K. and A. B. 2016a. NASA MEaSUREs Vegetation Index and Phenology (VIP) Phenology NDVI Yearly Global 0.05Deg CMG. NASA EOSDIS Land Processes DAAC. http://doi.org/10.5067/MEaSUREs/VIP/VIPPHEN_NDVI.004 (Accessed: September 23, 2016).

Didan K. and A. B. 2016b. NASA MEaSUREs Vegetation Index and Phenology (VIP) Vegetation Indices Daily Global 0.05Deg CMG. NASA EOSDIS Land Processes DAAC. http://doi.org/10.5067/MEaSUREs/VIP/VIP01.004 (Accessed: September 23, 2016).

Du, J., J. S. Kimball, M. Azarderakhsh, R. S. Dunbar, M. Moghaddam, and K. C. McDonald. 2015. Classification of Alaska spring thaw characteristics using L-band radar remote sensing. *IEEE Transactions on Geoscience and Remote Sensing*, 53 (1), 542–556.

Du, J., J. Shi, and H. Rott. 2010a. Comparison between a multi-scattering and multi-layer snow scattering model and its parameterized snow backscattering model. *Remote Sensing of Environment*, 114 (5), 1089–1098.

Du, J., J. Shi, and R. Sun. 2010b. The development of HJ SAR soil moisture retrieval algorithm. *International Journal of Remote Sensing*, 31(14), 3691–3705.

Dunbar, S., X. Xu, A. Colliander, C. Derksen, K. McDonald, E. Podest, E. Njoku, J. Kimball, and Y. Kim. 2015. SMAP L3 radar northern hemisphere daily 3 km EASE-grid freeze/thaw state. Version 2. [indicate subset used]. Boulder, Colorado USA: NASA DAAC at the National Snow and Ice Data Center. doi:10.5067/GO4QPNEM0BKF.

Easterling, D. R. 2002. Recent changes in frost days and the frost-free season in the United States. *Bulletin of American Meteorological Society*, 83, 1327–1332.

Eccel, E., R. Rea, A. Caffarra, and A. Crisci. 2009. Risk of spring frost to apple production under future climate scenarios: The role of phenological acclimation. *International Journal of Biometeorology*, 53, 273–286.

Elachi, C. 1987. *Introduction to the Physics and Techniques of Remote Sensing*, New York: John Wiley & Sons, p. 413.

Ensminger, I., D. Sveshnikov, D. A. Campbell, C. Funk, S. Jansson, J. Lloyd, O. Shibistova, and G. Oquist. 2004. Intermittent low temperatures constrain spring recovery of photosynthesis in boreal Scots pine forests. *Global Change Biology*, 10, 995–1008.

Entekhabi, D. et al. 2010. The soil moisture active passive (SMAP) mission. *Proceedings of the IEEE*, 98 (5), 704–716.

Farhadi, L., R. H. Reichle, G. J. M. De Lannoy, and J. S. Kimball, 2015. Assimilation of freeze/thaw observations into the NASA catchment land surface model. *Journal of Hydrometeorology*, 16, 730–743.

Feng, M., W. Yang, L. Cao, and G. Ding. 2009. Monitoring winter what freeze injury using multi-temporal MODIS data. *Agricultural Sciences in China*, 8 (9), 1053–1062.

Fengjin, X. and S. Lianchun. 2011. Analysis of extreme low-temperature events during the warm season in Northeast China. *Natural Hazards*, 58, 1333–1344.

Frich, P., L. V. Alexander, P. Della-Marta, B. Gleason, M. Haylock, A. M. G. Klein Tank, and T. Peterson. 2002. Observed coherent changes in climatic extremes during the second half of the twentieth century. *Climate Research*, 19, 193–212.

Frolking, S., K. C. McDonald, J. S. Kimball, J. B. Way, R. Zimmermann, and S. W. Running. 1999. Using the space-borne NASA scatterometer (NSCAT) to determine the frozen and thawed seasons. *Journal of Geophysical Research*, 104 (D22), 27895–27907.

Gähler, M. 2016. Remote sensing for natural or man-made disasters and environmental changes. In M. Marghany. (Ed.). *Environmental Applications of Remote Sensing*, InTech. doi:10.5772/62183. Available: http://www.intechopen.com/books/environmental-applications-of-remote-sensing/remote-sensing-for-natural-or-man-made-disasters-and-environmental-changes.

Gauthier, D. and D. J. Hutchinson. 2012. Evaluation of potential meteorological triggers of large landslides in sensitive glaciomarine clay, eastern Canada. *Natural Hazards and Earth System Sciences*, 12, 3359–3375.

Gisnas, K., S. Westermann, T. V. Schuler, T. Litherland, K. Isaksen, J. Boike, and B. Etzelmuller. 2014. A statistical approach to represent small-scale variability of permafrost temperatures due to snow cover. *The Cryosphere*, 8, 2063–2074.

Green, J. C. Kongoli, A. Prakash, M. Sturm, C. Duguay, and S. Li. 2012. Quantifying the relationships between lake fraction, snow water equivalent and snow depth, and microwave brightness temperatures in an arctic tundra landscape. *Remote Sensing of Environment*, 127, 329–340.

Gu, L., P. J. Hanson, W. M. Post, D. P. Kaiser, B. Yang, R. Nemani, S. G. Pallardy, T. Meyers. 2008. The 2007 Eastern US spring freeze: Increased cold damage in a warming world? *BioScience*, 58 (3), 253–262.

Guirguis, K. A. Gershunov, R. Schwartz, and S. Bennett. 2011. Recent warm and cold daily winter temperature extremes in the Northern Hemisphere. *Geophysical Research Letters*, 38, 2–7.

Guo, B., Y. Zhou, J. Zhu, W. Li, F. Wang, L. Wang, and L. Jiang. 2015. An estimation method of soil freeze-thaw erosion in the Qinghai-Tibet Plateau. *Natural Hazards*, 78, 1843–1857.

Hachem, S., Allard, M., and Duguay, C. 2009. Using the MODIS land surface temperature product for mapping permafrost: An application to northern Québec and Labrador, Canada. *Permafrost and Periglacial Processes*, 20(4), 407–416.

Han, M., K. Yang, J. Qin, R. Jin, Y. Ma, J. Wen, Y. Chen, L. Zhao, L. Zhu, and W. Tang. 2015. An algorithm, based on the standard deviation of passive microwave brightness temperatures for monitoring soil surface freeze/thaw state on the Tibetan Plateau. *IEEE Transactions on Geoscience and Remote Sensing*, 53 (5), 2775–2783.

Heino, R., R. Brazdil, E. Forland, H. Tuomenvirta, H. Alexandersson, M. Beniston, C. Pfister et al.,1999. Progress in the study of climatic extremes in northern and central Europe. *Climate Change*, 42, 151–181.

Henderson, F. M. and A. J. Lewis. 1998. In: F.M. Henderson, A.J. Lewis (Eds.), *Principles and Applications of Imaging Radar* (Volume 2) *of Manual of Remote Sensing* (3rd ed). New York: John Wiley & Sons.

Holmes, T. R. H., W. T. Crow, C. Hain, M. C. Anderson, and W. P. Kustas. 2015. Diurnal temperature cycle as observed by thermal infrared and microwave radiometers. *Remote Sensing of Environment*, 158, 110–125.

Holmes, T. R. H., W. T. Crow, M. T. Yilmaz, T. J. Jackson, and J. B. Basara. 2013. Enhancing model-based land surface temperature estimates using multiplatform microwave observations. *Journal of Geophysical Research: Atmosphere*, 118, 577–591.

Hufkens, K., M. A. Friedl, T. F. Keenan, O. Sonnentag, A. Bailey, J. O'Keefe, and A. D. Richardson. 2012. Ecological impacts of a widespread frost event following early spring leaf-out. *Global Change Biology*, 18, 2365–2377.

Imaoka, K. M. Kachi, M. Kasahara, N. Ito, K. Nakagawa, and T. Oki. 2010. Instrument performance and calibration of AMSR-E and AMSR2. *International Archives of the Photogrammetry, Remote Sensing and Spatial Information Science*, 38 (8), 13–18.

Inouye, D.W. 2008. Effects of climate change on phenology, frost damage, and floral abundance of montane wildflowers. *Ecology*, 89 (2), 353–362.

Jagdhuber, T. et al. 2014. Identification of soil freezing and thawing states using SAR polarimetry at C-Band. *Remote Sensing*, 6 (3), 2008–2023.

Jin, R., X. Li, and T. Che. 2009. A decision tree algorithm for surface soil freeze/thaw classification over China using SSM/I brightness temperature. *Remote Sensing of Environment*, 113, 2651–2660.

Jorgenson, M. T., C. H. Racine, J. C. Walters, and T. E. Osterkamp. 2001. Permafrost degradation and ecological changes associated with a warming climate in central Alaska. *Climate Change*, 48, 551–579.

Kääb, A. 2008. Remote Sensing of permafrost-related problems and hazards. *Permafrost and Periglacial Processes*, 19, 107–136.

Kang, K. K., C. R. Duguay, and S. E. Howell. 2012. Estimating ice phenology on large northern lakes from AMSR-E algorithm development and application to Great Bear Lake and Great Slave Lake, Canada. *The Cryosphere*, 6, 235–254

Kawanishi, T. J., T. Sezai, Y. Ito, K. Imaoka, T. Takashima, Y. Ishido, A. Shibata, M. Miura, H. Inahata, and R. W. Spencer. 2003. The advanced scanning microwave radiometer for the Earth Observing System (AMSR-E): NASDA's contribution to the EOS for global energy and water cycle studies. *IEEE Transactions on Geoscience and Remote Sensing*, 41 (2), 184–194.

Kerdiles, H., M. Grondona, R. Rodriguez, and B. Seguin. 1996. Frost mapping using NOAA AVHRR data in the Pampean region, Argentina. *Agricultural and Forest Meteorology*, 79, 157–182.

Kim, Y., J. S. Kimball, J. Glassy, and J. Du. 2017. An extended global Earth system data record on daily landscape freeze-thaw determined from satellite passive microwave remote sensing. *Earth System Science Data*, 9 (1), 133–147. doi:10.5194/essd-9-133-2017.

Kim, Y., J. S. Kimball, D. A. Robinson, and C. Derksen. 2015. New satellite climate data records indicate strong coupling between recent frozen season changes and snow cover over high northern latitudes. *Environmental Research Letters*, 10, 084004.

Kim, Y., J. S. Kimball, K. Zhang, K. Didan, I. Velicogna, and K. C. McDonald. 2014a. Attribution of divergent northern vegetation growth responses to lengthening non-frozen seasons using satellite optical-NIR and microwave remote sensing. *International Journal of Remote Sensing*, 35 (10), 3700–3721.

Kim, Y., J. S. Kimball, K. Didan, and G. M. Henebry. 2014b. Responses of vegetation growth and productivity to spring climate indicators in the conterminous United States derived from satellite remote sensing data fusion. *Agricultural and Forest Meteorology*, 194, 132–143.

Kim, Y., J. S. Kimball, J. Glassy, and K. C. McDonald. 2014c. MEaSUREs Global Record of Daily Landscape Freeze/Thaw Status. Version 3. [1979–2011]. Boulder, Colorado USA: NASA DAAC at the National Snow and Ice Data Center. http://dx.doi.org/10.5067/MEASURES/CRYOSPHERE/nsidc-0477.003

Kim, Y., J.S. Kimball, K. Zhang, and K.C. McDonald, 2012. Satellite detection of increasing Northern Hemisphere non-frozen seasons from 1979 to 2008: Implications for regional vegetation growth. *Remote Sensing of Environment*, 121, 472–487.

Kim, Y., J. S. Kimball, K. C. McDonald, and J. Glassy. 2011. Developing a global data record of daily landscape freeze/thaw status using satellite microwave remote sensing. *IEEE Transactions on Geoscience and Remote Sensing*, 49 (3), 949–960.

Kim, Y., A. R. Huete, T. Miura, Z. Jiang. 2010. Spectral compatibility of vegetation indices across sensors: Band decomposition analysis with Hyperion data. *Journal of Applied Remote Sensing*, 4, 043520.

Kimball, J. S., K. C. McDonald, K. C., and M. Zhao. 2006. Spring thaw and its effect on terrestrial vegetation productivity in the western arctic observed from satellite microwave and optical remote sensing. *Earth Interactions*, 10, 1–22.

Kimball, J. S., K. C. McDonald, A. R. Keyser, S. Frolking, S. W. Running, 2001. Application of the NASA Scatterometer (NSCAT) for determining the daily frozen and nonfrozen landscape of Alaska. *Remote Sensing of Environment*, 75 (1), 113–126.

King, D. J., I. Olthof, P. K. E. Pellikka, E. D. Seed, and C. Butson. 2005. Modelling and mapping damage to forests from an ice storm using remote sensing and environmental data. *Natural Hazards*, 35, 321–342.

Kollas, C., C. F. Randin, Y. Vitasse, and C. Korner. 2014. How accurately can minimum temperatures at the cold limits of tree species be extrapolated from weather station data? *Agricultural and Forest Meteorology*, 184, 257–266.

Kou, X., L. Jiang, Y. Bo, S. Yan, and L. Chai. 2016. Estimating of land surface temperature through blending MODIS and AMSR-E data with the Bayesian maximum entropy method. *Remote Sensing*, 8 (2), 105.

Kreyling, J., D. Thiel, K. Simmnacher, E. Willner, A. Jentsch, and C. Beierkuhnlein. 2012. Geographic origin and past climatic experience influence the response to late spring frost in four common grass species in central Europe. *Ecography*, 35, 268–275.

Kunkel, K. E., D. R. Easterling, K. Hubbard, and K. Redmond. 2004. Temporal variations in frost-free season in the United States: 1895–2000. *Geophysical Research Letters*, 31, L03201.

Lazarus, B. E., P. G. Schaberg, G. J. Hawley, and D. H. DeHayes. 2006. Landscape-scale spatial patterns of winter injury to red spruce foliage in a year of heavy region-wide injury. *Canadian Journal of Forest Research*, 36, 142–152.

Larsen, P. H., S. Goldsmith, O. Smith, M. L. Wilson, K. Strzepek, P. Chinowsky, and B. Saylor. 2008. Estimating future costs for Alaska public infrastructure at risk from climate change. *Global Environmental Change-Human and Policy Dimensions*, 18 (3), 442–457.

Leblond, M., M. St-Laurent, and S. D. Cote. 2016. Caribou, water, and ice—Fine-scale movements of a migratory arctic ungulate in the context of climate change. *Movement Ecology*, 4, 14.

Lemmetyinen, J., A. Kontu, J. Karna, J. Vehvilainen, M. Takala, and J. Pullianinen. 2011. Correcting of the influence of frozen lakes in satellite microwave radiometer observations through applications of a microwave emission model. *Remote Sensing of Environment*, 115, 3695–3706.

Lenti, F., F. Numziata, M. Migliaccio, and G. Rodriguez. 2014. Two-dimensional TSVD to enhance the spatial resolution of radiometer data. *IEEE Transactions on Geoscience and Remote Sensing*, 52 (5), 2450–2458.

Lenz, A., G. Hoch, C. Korner, and Y. Vitasse. 2016. Convergence of leaf-out towards minimum risk of freezing damage in temperate trees. *Functional Ecology*, 30 (9), 1480–1490. doi:10.1111/1365-2435.12623.

Li, S., Y. Lai, W. Pei, S. Zhang, and H. Zhong. 2014. Moisture-temperature changes and freeze-thaw hazards on a canal in seasonally frozen regions. *Nature Hazards*, 72, 287–308.

Linkosalo, T., T. R. Carter, R. Hakkinen, and P. Hari. 2000. Predicting spring phenology and frost damage risk of Betula spp. under climate warming: A comparison of two models. *Tree Physiology*, 20, 1175–1182.

Liu, L., T. Zhang, and J. Wahr. 2010. InSAR measurements of surface deformation over permafrost on the North Slope of Alaska. *Journal of Geophysical Research: Earth Surface*, 115(F3).

Liu, H., L. Wang, and K. C. Jezek. 2005. Wavelet-transform based edge detection approach to derivation of snowmelt onset, end and duration from satellite passive microwave measurements. *International Journal of Remote Sensing*, 26 (21), 4639–4660.

Lu, Z. and D. Dzurisin. 2014. InSAR imaging of Aleutian volcanoes. In *InSAR Imaging of Aleutian Volcanoes*. Berlin, Germany: Springer, pp. 87–345.

Malenovský, Z., H. Rott, J. Cihlar, M. E. Schaepman, G. Garcia-Santos, R. Fernandes, and M. Berger. 2012. Sentinels for science: Potential of Sentinel-1,-2, and-3 missions for scientific observations of ocean, cryosphere, and land. *Remote Sensing of Environment*, 120, 91–101.

Marino, G. P., D.P. Kaiser, L. Gu, D. M. Ricciuto. 2011. Reconstruction of false spring occurrences over the southeastern United States, 1901–2007: An increasing risk of spring freeze damage? *Environmental Research Letters*, 6, 024015–024023.

Martin, M., K. Gavazov,C. Korner,S. Hattenschwiler,C. Rixen. 2010. Reduced early growing season freezing resistance in alpine treeline plants under elevated atmospheric CO_2. *Global Change Biology*, 16, 1057–1070.

Martino, L., C. Ulivieri, M. Jahjah, and E. Loret. 2009. Remote sensing and GIS techniques for natural disaster monitoring. In P. Olla. (Ed.). *Space Technologies for the Benefit of Human Society and Earth*. Dordrecht , the Netherlands: Springer, pp. 331–382. doi:10.1007/978-1-4020-9573-3_14.

McCabe, G. J., J. L. Betancourt, and S. Feng. 2015. Variability in the start, end, and length of frost-free periods across the conterminous United States during the past century. *International Journal of Climatology*, 35 (15), 4673–4680.

McColl, K. A., A. Roy, C. Derksen, A. G. Konings, S. H. Alemohammed, and D. Entekhabi. 2016. Triple collocation for binary and categorical variables: Application to validating landscape freeze/thaw retrievals. *Remote Sensing of Environment*, 176, 31–42.

McDonald, K. C. and J. S. Kimball, 2005. Hydrological application of remote sensing: Freeze-thaw states using both active and passive microwave sensors. In M.G. Anderson and J.J. McDonnell (Eds.) *Encyclopedia of Hydrological Sciences*. Part 5. Remote Sensing. Chichester, UK: John Wiley & Sons. doi:10.1002/0470848944.hsa059a.

Merlin, O., M. J. Escorihuela, M. A. Mayoral, O. Hagolle, A. A. Bitar, and Y. Kerr. 2013. Self-calibrated evaporation-based disaggregation of SMOS soil moisture: An evaluation study at 3 km and 100 m resolution in Catalunya, Spain. 2013. *Remote Sensing of Environment*, 130, 25–38.

Migliaccio, M. and A. Gambardella. 2005. Microwave radiometer spatial resolution enhancement. *IEEE Transactions on Geoscience and Remote Sensing*, 43 (5), 1159–1169.

Moonen, A. C., L. Ercoli, M. Mariotti, and A. Masoni. 2002. Climate change in Italy indicated be agrometeorological indices over 122 years. *Agricultural and Forest Meteorology*, 111, 13–27.

Mortin, J., T. M. Schroder, A. W. Hansen, B. Holt, ad K. C. McDonald. 2012. Mapping of seasonal freeze-thaw transitions across the pan-Arctic land and sea ice domains with satellite radar. *Journal of Geophysical Research*, 117, C08004.

Murphy, M. A., I. P. Martini, and R. Protz. 2001. Seasonal changes in subarctic wetlands and river ice breakup detectable on RADARSAT images, southern Hudson Bay Lowland, Ontario, Canada. *Canadian Journal of Remote Sensing*, 27 (2), 143–158.

Naeimi, V., C. Paulik, A. Bartsch, W. Wagner, R. Kidd, S. Park, K. Elger, and J. Bioke. 2012. ASCAT surface state flag (SSF): Extracting information on surface freeze/thaw conditions from backscatter data using an empirical threshold-analysis algorithm. *IEEE Transactions on Geoscience and Remote Sensing*, 50 (7), 2566–2582.

Nelson, F. E., O. A. Anisimov, and N. I. Shiklomanov. 2001. Subsidence risk from thawing permafrost. *Nature*, 410, 889.

Oksanen, E., V. Freiwald, N. Prozherina, and M. Rousi. 2006. Photosynthesis of birch (Betual pendula) is sensitive to springtime frost and ozone. *Canadian Journal of Forest Research*, 35, 703–712.

Olthof, I., D. J. King, and R. A. Lautenschlager. 2004. Mapping deciduous forest ice storm damage using Landsat and environmental data. *Remote Sensing of Environment*, 89, 484–496.

Owe, M., R. de Jeu, and T. Holmes. 2008. Multisensor historical climatology of satellite-derived global land surface moisture. *Journal of Geophysical Research*, 113, F01002.

Papagiannaki, K., K. Lagouvardos, V. Kotroni, and G. Papagiannakis. 2014. Agricultural losses related to frost events: Use of the 850 hPa level temperature as an explanatory variable of the damage cost. *Natural Hazards and Earth System Sciences*, 14, 2375–2386.

Park, H., Y. Kim, and J. S. Kimball. (2016b). Widespread permafrost vulnerability and soil active layer increases over the high northern latitudes inferred from satellite remote sensing and process model assessments. *Remote Sensing of Environment*, 175, 349–358.

Park, H., Y., Yoshikawa, K. Oshima, Y. Kim, T. Ngo-Duc, and J. S. Kimball, and D. Yang. (2016a). Quantification of warming climate-induced changes in terrestrial arctic river ice thickness and phenology. *Journal of Climate*, 29, 1733–1754.

Park, S., A. Bartsch, D. Sabel, W. Wagner, V. Naeimi, and Y. Yamaguchi. 2011. Monitoring freeze/thaw cycles using ENVISAT ASAR Global Mode. *Remote Sensing of Environment*, 115, 3457–3467.

Parker, P. S., J. S. Shonkwiler, and J. Aurbacher. 2016. Cause and consequence in Maize planting dates in Germany. *Journal of Agronomy and Crop Science*, 203 (3), 227–240. doi:10.1111/jac.12182.

Pellenq, J., J. Kalma, G. Boulet, G. M. Saulnier, S. Wooldridge, Y. Kerr, and A. Chehbouni. 2003. A disaggregation scheme for soil moisture based topography and soil depth. *Journal of Hydrology*, 276, 112–127.

Penuelas, J., T. Ruishauser, and I. Filella. 2009. Phenology feedbacks on climate change. *Science*, 324, 887–888.

Podest, E., K. C., McDonald, and J. S. Kimball. 2014. Multisensor microwave sensitivity to freeze/thaw dynamics across a complex boreal landscape. *IEEE Transactions on Geoscience and Remote Sensing*, 52 (11), 6818–6828.

Poirier, M., A. Lacointe, and T. Ameglio. 2010. A semi-physiological model of cold hardening and dehardening in walnut stem. *Tree Physiology*, 30, 1555–1569.

Pouteau, R. S. Rambal, J. Ratte, F. Goge, R. Joffre, and T. Winkel. 2011. Downscaling MODIS-derived maps using GIS and boosted regression trees: The case of frost occurrence over the arid Andrean highlands of Bolivia. *Remote Sensing of Environment*, 115, 117–129.

Rammig, A., A. M. Jonsson, T. Hickler, B. Smith, L. Barring, and M. T. Sykes. 2010. Impacts of changing frost regimes on Swedish forests: Incorporating cold hardiness in a regional ecosystem model. *Ecological Modelling*, 221, 303–313.

Rautiainen, K., J. Lemmetyinen, M. Schwank, A. Kontu, C. B. Menard, C. Matzler, M. Drusch, A. Wiesmann, J. Ikonen, and J. Pulliainen. 2014. Detection of soil freezing from L-band passive microwave observations. *Remote Sensing of Environment*, 147, 206–218.

Rautiainen, K., T. Parkkinen, J. Lemmetyinen, M. Schwank, A. Wiesmann, J. Ikonen, C. Derksen et al. 2016. SMOS prototype algorithm for detecting autumn soil freezing. *Remote Sensing of Environment*, 180, 346–360.

Rawlins, M. A., R. S. Bradley, H. F. Diaz, J. S. Kimball, and D. A. Robinson. 2016. Future decreases in freezing days across North America. *Journal of Climate*, 29, 6923–6935. doi:10.1175/JCLI-D-15-0802.1.

Rawlins, M. A., K. C. McDonlad, S. Frolking, R. B. Lammers, M. Fahnestock, J. S. Kimball, and C. J. Vorosmarty. 2005. Remote sensing of snow thaw at the pan-Arctic scale using the SeaWinds scatterometer, *Journal of Hydrology*, 312 (1–4), 294–311.

Raytheon. 2000. Special Sensor Microwave/Image (SSM/I) User's Interpretation Guide.

Reichle, R. H., D. Entekhabi, and D. B. McLaughlin. 2001. Downscaling of radio brightness measurements for soil moisture estimation: A four-dimensional variational data assimilation approach. *Water Resources Research*, 37 (9), 2353–2364.

Richardson, A. D., A. S. Bailey, E. G. Denny, C. W. Martin, and J. O'keefe. 2006. Phenology of a northern hardwood forest canopy. *Global Change Biology*, 12 (7), 1174–1188.

Rigby, J. R. and Porporato, A., 2008. Spring frost risk in a changing climate. *Geophysical Research Letters*, 35, L12703.

Rignot, E. and J. B. Way. 1994. Monitoring freeze—thaw cycles along North—South Alaskan transects using ERS-1 SAR. *Remote Sensing of Environment*, 49 (2), 131–137.

Rodrigo, J. 2000. Spring frosts in deciduous fruit trees-morphological damage and flower hardiness. *Scientia Horticulturae*, 85, 155–173.

Riseth, J. A., H. Tommervik, and J. W. Bjerke. 2016. 175 years of adaptation: North Scandinavian Sami reindeer herding between government policies and winter climate variability (1835–2010). *Journal of Forest Economics*, 24, 186–204.

Rosenqvist, A. et al. 2014. Operational performance of the ALOS global systematic acquisition strategy and observation plans for ALOS-2 PALSAR-2. *Remote Sensing of Environment*, 155, 3–12.

Roy, A., A. Royer, C. Derksen, L. Brucker, A. Langlois, A. Mialon, and Y. H. Kerr. 2015. Evaluation of space-borne L-band radiometer measurements for terrestrial freeze/thaw retrievals in Canada. *IEEE Journal of Selected Topics in Applied Earth Observations and Remote Sensing*, 8 (9), 4442–4459.

Santi, E. 2010. An application of the SFIM technique to enhance the spatial resolution of spaceborne microwave radiometer. *International Journal of Remote Sensing*, 31 (9), 2419–2428.

Schaefer, K., L. Liu, A. Parsekian, E. Jafarov, A. Chen, T. Zhang, A. Gusmeroli, S., Panda, H. A. Zebker, and T. Schaefer. 2015. Remotely sensed active layer thickness (ReSALT) at Barrow, Alaska using Interferometric Synthetic Aperture Radar. *Remote Sensing*, 7 (4), 3735–3759.

Schwartz, M. D., T. R. Ault, and J. L. Betancourt. 2013. Spring onset variations and trends in the continental United States: Past and regional assessment using temperature-based indices. *International Journal of Climatology*, 33 (13), 2917–2922.

Silleos, N., K. Perakis, and G. Petsanis. 2002. Assessment of crop damage using space remote sensing and GIS. *International Journal of Remote Sensing*, 23 (3), 417–427.

Smith, N. V., S. S. Saatchi, and J. T. Randerson. 2004. Trends in high northern latitude soil freeze and thaw cycles from 1988 to 2002. *Journal of Geophysical Research*, 106, D12101.

Solomon, S., D. Qin, M. Manning et al. (Eds.) 2007. *Climate Change 2007: The Physical Science Basis.* New York: Cambridge University Press.

Song, C., L. Jia, and M. Menenti. 2014. Retrieving high-resolution surface soil moisture by downscaling AMSR-E brightness temperature using MODIS LST and NDVI data. *IEEE Journal of Selected Topics in Applied Earth Observations and Remote Sensing*, 7 (3), 935–942.

Strimbeck, G. R., P. G. Schaberg, D. H. DeHayes, J. B. Shane, G. J. Hawley, 1995. Mid-winter dehardening of montane red spruce during a natural thaw. *Canadian Journal of Forest Research*, 25, 2040–2044.

Tait, A. and X. Zheng. 2003. Mapping frost occurrence using satellite data. *Journal of Applied Meteorology*, 42, 193–203.

Thomachot, C., N. Matsuoka, N. Kuchitsu, and M. Morii. 2005. Frost damage of bricks composing a railway tunnel monument in Central Japan: Field monitoring and laboratory simulation. *Natural Hazards and Earth System Science*, 5 (4), 465–476.

Todhunter, P. E. 2001. A hydroclimatological analysis of the Red River of the North snowmelt flood catastrophe of 1997. *Journal of the American Water Resources Association*, 37 (5), 1263–1278.

Tubiello, F.N., J. Soussana, S. M. Howden. 2007. Crop and pasture response to climate change. *Proceedings of the National Academy of Sciences*, 104 (50), 19686–19690.

Ulaby, F. T. and D.G. Long. (2014). *Microwave Radar and Radiometric Remote Sensing*. Ann Arbor, MI: University of Michigan Press.

Ulaby, F. T., R. K. Moore, and A. K. Fung. 1986. *Microwave Remote Sensing: Active and Passive*, Vol. 1–3. Dedham, MA: Artech House.

Viltard, N., C. Burlaud, C. D. Kummerow. 2006. Rain retrieval from TMI Brightness temperature measurements using a TRMM PR-based database. *Journal of Applied Meteorology and Climatology*, 45, 455–466.

Vitasse, Y., A. Lenz, and C. Korner. 2014. The interaction between freezing tolerance and phenology in temperate deciduous trees. *Frontiers in Plant Science*, 5, 1–12.

Vitasse, Y., S. Delzon, E. Dufrene, J. Pontailler, J. Louvet, A. Kremer, and R. Michalet, 2009. Leaf phenology sensitivity to temperature in European trees: Do within-species populations exhibit similar responses? *Agricultural Forest Meteorology*, 149, 735–744.

Wang, H., G. Xiaohe, J. Wang, and Y. Dong. 2012. Monitoring winter wheat freeze injury based on multi-temporal data. *Intelligent Automation & Soft Computing*, 18 (8), 1035–1042.

Wang, L., C. Derksen, and R. Brown. 2008. Detection of pan-Arctic terrestrial snowmelt from QuikSCAT, 2000–2005. *Remote Sensing of Environment*, 112 (10), 3794–3805.

Wang, L., C. Derksen, R. Brown, and T. Markus. 2013. Recent changes in pan-Arctic melt onset from satellite passive microwave measurements. *Geophysical Research Letters*, 40, 1–7.

Westerling, A. L. R. 2016. Increasing western US forest wildfire activity: Sensitivity to changes in the timing of spring. *Philosophical Transactions B*, 371: 20150178.

Whitfield, P. H. 2012. Floods in future: A review. *Journal of Flood Risk Management*, 5 (4), 336–365.

Wilson, R. R., A. Bartsch, K. Joly, J. H. Reynolds, A. Orlando, and W. M. Loya. 2013. Frequency, timing, extent, and size of winter thaw-refreeze events in Alaska 2001–2008 detected by remotely sensed microwave backscattered data. *Polar Biology*, 36, 419–426.

Wolfe, D. W., M. D. Schwartz, A. N. Lakso, Y. Otsuki, R. M. Pool, and N. J. Shaulis. 2005. Climate change and shifts in spring phenology of three horticultural woody perennials in northeastern USA. *International Journal of Biometeorology*, 49, 303–309.

Xu, X., C. Derksen, S. Yueh, R. S. Dunbar, and A. Colliander. 2016. Freeze/thaw detection and validation using Aquarius L-band backscattering data. *IEEE Journal of Selected Topics in Applied Earth Observations and Remote Sensing*, 9 (4), 1370–1381.

Yaodong, D. 2005. Frost and high temperature injury in China. In M. V. K. Sivakumar, R. P. Motha, H. P. Das. (Eds.) *Natural Disasters and Extreme Events in Agriculture*. Berlin, Germany: Springer, pp.145–157.

Yu, H., E. Luedeling, and J. Xu. 2010. Winter and spring warming result in delayed spring phenology on the Tibetan Plateau. *Proceedings of the National Academy of Sciences*, 107 (51), 22151–22156.

Zhan, X., P. R. Houser, J. P. Walker, and W. T. Crow. 2006. A method for retrieving high-resolution surface soil moisture from Hydros L-band radiometer and radar observations. *IEEE Transactions on Geoscience and Remote Sensing*, 44 (6), 1534–1544.

Zhang, K., J. S. Kimball, Y. Kim, and K. C. McDonald, 2011. Changing freeze-thaw seasons in northern high latitudes and associated influences on evapotranspiration. *Hydrological Processes*, 25, 4142–4151. doi:10.1002/hyp.8350.

Zhang, X., B. Tan, and Y. Yu. 2014. Interannual variations and trends in global land surface phenology derived from enhanced vegetation index during 1982–2010. *International Journal of Biometeorology*, 58, 547–564.

Zhao, T., L. Zhang, L. Jiang, S. Zhao, L. Chai, and R. Jin. 2011. A new soil freeze/thaw discriminant algorithm using AMSR-E passive microwave imagery. *Hydrological Processes*, 25: 1704–1716.

Zwieback, S., A. Bartsch, T. Melzer, and W. Wagner. 2012. Probabilistic fusion of Ku- and C-band scatterometer data for determining the freeze/thaw state. *IEEE Transactions on Geoscience and Remote Sensing*, 50 (7), 2583–2594.

9 Temperature Fluctuation and Frost Risk Analysis on a Road Network by Coupling Remote Sensing Data, Thermal Mapping, and Geographic Information System Techniques

Panagiota Louka, Ioannis Papanikolaou,
George P. Petropoulos, Nikolaos Stathopoulos,
and Ioannis X. Tsiros

CONTENTS

9.1   INTRODUCTION

Safe transportation during the winter months depends highly on temperatures and on the presence of snow or ice on the road surface. During periods with wintery conditions, there is a higher risk of car accidents due to road slipperiness and delays due to lower commuting speed. In more extreme weather, the road may be inaccessible even for essential traffic such as ambulances, fire engines, and police cars, causing a major issue for civil protection agencies.

Road closures can lead to significant economic losses. According to the Washington State Department of Transportation between 1992 and 2004, one of the main motorways of the state (I-90) had been closed on average 120 h per year, causing an annual loss of at least $17.5 million dollars. In addition, the crash rate on the I-90 highway in the presence of snow has been found to be about five times the rate compared to clear conditions (Federal Highway Administration 2006).

Winter road maintenance costs depend on the type of climate of each country. In Japan, for example, which has the snowiest roads in the world, approximately £1.1 billion is spent annually for clearing and melting the snow from the roads (Chapman et al. 2001). In North America, where the climate conditions vary greatly throughout the country, £1.7 billion is spent annually on winter road maintenance, which accounts for 40% of global spending approximately (Boselly et al. 1993). In the United Kingdom, which is considered as a country with marginal winter environment, £140 million is spent every year on winter road maintenance (Cornford and Thornes 1996). At the same time, the side effects of salt corrosion and damage to vehicles and structures cause an extra annual cost of £100 million (Thornes 2000). It is apparent from the aforementioned references that significant amount of money could be saved annually by optimizing winter maintenance measures.

A precise prediction concerning road surface temperature (RST) in a road network is critical especially during the winter months even in areas with temperate climate, as marginal temperatures may be observed during nighttime. Highway authorities use this type of information as a reference tool for planning precautionary measures, such as salting and gritting of the road. It is essential for highway engineers and meteorologists to plan a network of field meteorological stations, which will assist in the monitoring of weather conditions of roads. Finally, this information is important to be communicated to road users so that they can adjust their driving behavior accordingly.

Prediction of frost daily duration and intensity is very important for roadway maintenance personnel in order for them to decide when, where, and which protection measures should be used (Greenfield and Takle 2006). Winter nighttime road temperatures may vary by over 10°C across a network. Such variation means that some stretches of road may fall below freezing temperatures, whereas other may remain well above freezing. It is vital to identify not only when ice or frost conditions are likely to occur but also to identify which road sections are most susceptible (Shao et al. 1997).

Two main strategies are currently used for winter maintenance: deicing and anti-icing. In the first method, chemicals are applied that melt ice and snow. In the latter, the use of chemicals is preventive as their target is to reduce ice by hindering bonds between ice crystals and road pavement. Anti-icing is preferred to deicing as it reduces the total chemical use, is more environmentally friendly, and allows a higher level of public service (Berrocal et al. 2010).

However, both strategies have important corrosive and environmentally damaging effects on soil, vegetation, streams, road surface, and to the vehicles themselves (Shao et al. 1996; Ramakrishna and Viraraghavan 2005). Thus, understandably, being able to provide a precise prediction of road weather conditions, precautionary measures could be avoided if they are unnecessary, as well as their environmental and economic cost could be reduced.

The aim of the study included in this chapter is to present a methodology that aims at the analysis of different sources of information regarding temperature fluctuation of the road surface, including remote sensing data. In addition, it attempts to predict low temperature and frost risk levels along a stretch of road by applying a geographic information system (GIS) model, which incorporates geographical and environmental parameters.

More specifically, the main objectives of this chapter are

- To predict frost hazard on a stretch of a mountainous motorway with a GIS model based primarily on remote sensing data.
- To analyze the temperature variation through the process of thermography measurements and Earth observation (EO) data.
- To evaluate the predicted frost risk by the GIS model in comparison to the recorded fluctuation of temperature along the studied stretch of the motorway.

9.2 PREVIOUS WORK ON ROAD SURFACE TEMPERATURE DISTRIBUTION

A plethora of research has been focused on the variation of RST and on methods of predicting road surface conditions.

Numerical models have been developed since the 1970s, which predict RST based on the meteorological data of a weather station (e.g., Shao 1990; Thornes 1991; Jacobs and Raatz 1996; Wood and Clark 1999 Crevier and Delage 2001; Bouilloud and Martin 2006; Yahia 2006; Bouilloud et al. 2009). In order to extrapolate the forecast throughout the road network, the technique of thermal mapping was employed (e.g., Thornes 1991; Shao et al. 1997; Gustavsson 1999; Postgard and Lindqvist 2001; Shao 2000). A different approach was the use of empirical methods and the development of local climatological models (e.g., Bogren et al. 1992; Gustavsson et al. 1998; Rayer 1987).

The most recent methods for the prediction of road weather conditions are based on the numerical modeling of geographical parameters that affect RST variation (Chapman et al. 2001; Bradley et al. 2002; Chapman and Thornes 2006). These models are developed in a GIS environment and exploit the enhanced potential of remote sensing data (Eriksson 2001; Fry 2010; Hammond et al. 2010; Vinter 2015).

9.2.1 THERMOGRAPHY

Infrared thermography has been used in road climatological studies since 1975 as a reliable and effective method to describe and display RST spatial variation (Lindqvist 1975; Rosema and Welleman 1977; Stove et al. 1987). The results from thermal measurements have been used to detect cold spots along a road and possible positions for temperature sensors (Gustavsson and Bogren 1991). They are the basic methods for recording actual surface temperature variation along a road and are the main inputs to forecast models (Fry 2010).

Thermal surveys are organized with an infrared thermometer that can be mounted to a vehicle, such as a car or a helicopter or for stationary measurements; the sensors are mounted on specific road spots (Paumier and Arnal 1998; Chapman and Thornes 2005). The data collection is usually organized during nighttime, before dawn, as minimum daily temperatures occur during this period (Shao et al. 1997).

Infrared systems used in RST surveys usually consist of a scanner unit, a display unit, and a computer that processes and corrects thermal data with special software. Infrared radiation emitted by objects in the wave band of 2–5 μm is converted to electronic video signals. These signals are processed by the video scanner and are presented in gray tone thermal image. The images can be converted to color images through computer processing, presenting temperature values (Gustavsson and Bogren 1991).

It is crucial that thermography measurements are applied under strict quality control in order to eliminate possible sources of error (Shao et al. 1997). These include the determination of external parameters such as the emissivity of road materials, temperature differences due to reflection of surrounding space of the road, and effect of atmospheric temperature variations.

Finally, another drawback on the use of thermal mapping is that it is an expensive method, as it requires the use of special equipment. In addition at least five different nocturnal surveys must be organized to ensure satisfactory coverage of a range of atmospheric stability (Chapman and Thornes 2006).

9.2.2 GEOGRAPHICAL AND ENVIRONMENTAL PARAMETERS OF ROAD SURFACE TEMPERATURE

Temperature variation and likelihood of frost are determined by the energy receipt and loss at the road surface. This energy flow is controlled by a variety of environmental, geographical–geomorphological (exposure, altitude, traffic, etc.), and meteorological (wind speed, cloud cover, etc.) factors. Significant variation, up to 10°C, may be recorded in RST from one location to another because of these parameters (Shao et al. 1997).

The major factors of temperature fluctuation and frost risk are briefly reviewed in Table 9.1, based on previous research. These include geographical parameters, such as elevation, slope gradient and aspect, parameters connected to the surrounding topography, and finally factors connected to the land use and cover, such as vegetation or human activity.

The importance of the effect of frost risk factors is differentiated depending on the prevailing weather conditions. During periods with climatic instability, when cloudy and windy weather is observed, the most important factor is elevation. On calm days, temperature variation is affected greatly by the incoming solar radiation determined. Topography factors such as slope and aspect as well as screening of the road influence importantly RST. During clear and calm nights, the distribution of temperature is influenced by cold air flows and pooling in low lying areas, and an inverse relationship between temperature and elevation is observed (Eriksson 2001).

9.2.3 MODELING ROAD SURFACE TEMPERATURE FLUCTUATION ON GEOGRAPHIC INFORMATION SYSTEM ENVIRONMENT

Nowadays, with growing availability of spatial referenced data and advances of computing power, more accurate and complex GIS environmental models may be produced. The incorporation of geographical parameters in the GIS models assists in the forecasting of the microclimate of the surrounding of the road network and thus improves the produced results.

One of the first attempts of modeling the relative variation of road climate with GIS was implemented by Gustavsson et al. (1998). It was based on topography and land use factors in combination to climate data. The road climate model was further developed by including information sources such as traffic accidents and winter road maintenance data (Norrman et al. 2000; Eriksson 2001).

Another example of a GIS-based model, which provides a dynamic prediction of road conditions, has been developed by Chapman and Thornes (2006). The model combines geographical parameters such as altitude, land use, road construction and the sky view factor, and meteorological data in order to provide spatiotemporal forecast of RST. According to their results in West Midlands, UK, the model can predict the spatial variation of the road network successfully up to 73%. Another model focused on the impact of urban heat island on RSTs has been developed by Bradley et al. (2002), who analyzed the spatial correlation of topography and classified Landsat imagery.

A more recent attempt of developing a deterministic frost risk model in a GIS environment was accomplished by Louka et al. (2016). The model was based on topographic parameters, including altitude, slope, steepness, aspect, topographic curvature, and environmental factors linked to land use, vegetation classification, and the existence of water bodies. The data used in the implementation of the model were primarily remote sensing data from Moderate Resolution Imaging Spectroradiometer (MODIS) and Advanced Spaceborne Thermal Emission and Reflection Radiometer (ASTER) sensors and ancillary ground observation data used mainly for the validation of the model's results. It was applied on agricultural land (tree crops and arable farms), and the model's results were found

TABLE 9.1

Geographical and Environmental Temperature and Frost Risk Parameters

Parameter	Interaction with Frost Risk	Measurement Technique
Latitude	It is linked to incoming shortwave radiation and as a result daytime RST (Chapman et al. 2001).	GPS
Altitude	RST decreases as altitude rises according to lapse rates, with a nonlinear relationship. The importance of this factor is higher under unstable weather conditions (Shao et al. 1997; Chapman et al. 2001; Pouteau et al. 2011).	Derived from DEM
Hydrographic network	Water bodies through their thermal inertia have a protective role as far as frost risk is considered. Sea, lakes, and rivers reduce frost risk up to a distance of 20 km (Fridley 2009; Pouteau et al. 2011).	Derived from vector data
Slope and aspect	They are related to the amount of incoming shortwave radiation and daytime RST. Areas with a south orientation and a slope gradient of 20% receive double shortwave radiation than a flat area during January (Radcliffe and Lefever 1981; Oliver and Dolph 1992; Fridley 2009; Pouteau et al. 2011).	Derived from DEM
Topography/ geomorphology	Its impact is correlated to catabatic flow of cold air, which creates cold air pools in hollows and valley bottoms (Oke 1987; Soderstrom and Magnusson 1995; Pouteau et al. 2011; Bogren and Gustavsson 1991).	Derived from DEM
Screening/sky view factor	It is closely related to incoming shortwave radiation and radiation losses (Gustavsson and Bogren 1991; Chapman et al. 2001; Grimmond et al. 2001; Chapman and Thornes 2004; Bogren et al. 2000).	Calculated from fish-eye photographs
Land use	The thermal properties of surfaces vary with land use. Forested areas act as a protective barrier to cold air masses leading to temperature differences up to 3°C (Gustavsson et al. 1998). In addition, anthropogenic activities reduce frost risk due to urban heat island impact (Rouse and Wilson 1969; Chapman and Thornes 2006).	Derived from remote sensing data
Road construction	RST is affected by the road bed materials and their specific thermal properties, such as heat capacity and thermal conductivity (Gustavsson and Bogren 1991; Eriksson 2001).	Derived from remote sensing data
Traffic	RST increases with traffic due to limitation of long-wave radiation by surfaces, tyre friction, increased turbulence, and heat radiation by vehicles' engines (e.g., Gustavsson and Bogren 1991; Thornes 1991)	Traffic counters/ empirical formulae
Special features/ thermal singularities	RST is usually lower in bridges, and frost is more often reported on them than on adjoining roads. There is usually a lack of compensating heat flow from the ground and higher loss of long-wave radiation during night. RST differentiation depends on the type of the surrounding of the bridges and on whether they cross water (Takle 1990; Gustavsson and Bogren 1991; Greenfield and Takle 2006).	Aerophotography/ remote sensing data/field surveys

to correlate well or fairly well with the remotely sensed land surface temperature (LST) data over the vast majority (83.39%) of the study area (Louka et al. 2016).

Winter road maintenance can also be optimized by combining GIS with weather data. The minimization of gritting costs has been a major subject of research, which is based on a heuristic algorithm devised in a GIS environment by Li and Eglese (1996). Optimization models and solution algorithms have been developed for the routing of vehicles for spreading operations, ploughing roadways, and loading snow and for transporting it to disposal sites (Muyldermans et al. 2003; Perrier et al. 2007; Tagmouti et al. 2007).

GIS models can significantly reduce costly field surveys, including thermal mapping, and can produce accurate forecasts covering a wide area of a road network (Fry 2010). Field surveys are often work, time, and money consuming, and their results depend on the accessibility of an area. In addition thermal mapping results may be representative of the temperature variation for the specific climate conditions of the day the survey was conducted. On the contrary, modeling road conditions with GIS model can produce forecasts that are not focused on a fixed climatic scenario, can incorporate a multitude of parameters, and can reduce costs at the same time.

9.2.4 REMOTE SENSING OF ROAD SURFACE TEMPERATURE FLUCTUATION

With the development of EO technologies, remote sensing data of higher accuracy or temporal availability has become a major part of current applications. Most of these data have the advantage of being free of charge and easily accessible online.

EO data can be used as an input for GIS models focused on RST variation and frost risk and to improve their ability to produce accurate predictions. The geographical and geomorphological factors that are incorporated in models developed in a GIS environment are derived from digital surface model (DSM). The higher spatial resolution data offer higher accuracy in the models' output. For example, ASTER Global Digital Elevation Model (GDEM) data with a mean spatial resolution of 30 m can be replaced by light detection and ranging (LIDAR) data with grid spacing of 1 m or less (Fry 2010).

Additionally, remote sensing data that provide ready to use products with information on day and nighttime LST are available. The main sources of nighttime LST information are imagery by MODIS and ASTER sensors. The values of remotely sensed LST are determined from Planck's Law using the emissivity of thermal infrared (TIR) bands.

LST information derived by the MODIS (MOD11A1) has a course spatial resolution of 1 km but a very high temporal resolution with more than one revisit per day for some areas of the world. On the other hand, ASTER LST product (AST-08–L.2) contains information on surface kinetic temperature at 90 m resolution for land areas only. It has a 16-day revisit, but due to cloud cover data it may be available in bigger time intervals. Its relative accuracy is calculated to be 0.3 K (https://lpdaac.usgs.gov/).

Additionally, Landsat 8 Thermal Infrared Sensor (TIRS) is the very recent thermal infrared sensor, which provides two adjacent thermal bands from which LST can be derived. Several approaches have been used for LST inversion from TIRS, including the radiative transfer equation (RTE)-based method, the split-window (SW) algorithm, and the single channel (SC) method. The estimated accuracy is lower than 1 K for RTE, whereas the SW algorithm has moderate accuracy and the SC method has the lowest accuracy (Yu et al. 2014).

A new development in the acquirement of remotely sensed temperature information is the latest mission of the European Space Agency, Sentinel-3. Data concerning sea and land surface temperature (SLST) data will be provided, among others, on a daily basis with a spatial resolution of 1 km, according to the user guide of Sentinel-3 products. The produced sea surface temperature (SST) has an accuracy of better than 0.3 K under certain cloud-free conditions (i.e., >20% cloud-free samples within each area). The possible accuracy of the derived LST is estimated to be 1 K, especially at night when differential surface heating is absent. The sea and surface temperature radiometer (SLSTR) instrument also has a temporal stability of 0.1 K/decade (https://earth.esa.int).

Currently, there are no remote sensing temperature data, which combine high spatial resolution and revisit capabilities. However, in temperature fluctuation studies, it is often essential to have datasets with high spatial and temporal resolution. During the data selection for these studies, a critical choice has to be made, which creates significant limitations: either using data with high spatial but low temporal resolution or data with high temporal but low spatial resolution. Effectively synthesizing high temporal resolution imagery with high spatial resolution imagery can potentially ease this limitation (Liu and Weng 2012). Data fusion models have been developed, such as spatial and temporal adaptive reflectance fusion Model, that provide the ability to produce ASTER-like daily LST with promising results (Gao et al. 2006; Coll et al. 2007; Liu and Weng 2012; Semmens et al. 2016).

An emerging technology that may overcome this restriction of remote sensing data is based on the unmanned aerial vehicles (UAVs), which can monitor frost conditions in near real-time. As several demonstrations organized by National Aeronautics and Space Administration (NASA) have shown, UAVs can be used to monitor frost conditions and plan emergency response measures (Anand 2007). UAVs are expected to replace satellites in several applications. They have the advantage of being easily retrieved, upgraded, and retasked. They can also be moved to a different location from their automated global positioning system (GPS)-generated flight pattern at little cost. On the other hand, satellites have relatively fewer respective options. However, there are still issues to be resolved with the use of UAVs, such as their integration with the civilian airspace and the setting of regulations so as to avoid risk to other airspace users and the development of algorithms for frequency allocation and avoidance of collisions (Odido and Madara 2013).

9.3 EXPERIMENTAL SETUP

This study is focused on the analysis of a period of frost conditions in northwestern Greece during January 2008, when thermography measurements were conducted on a part of a major motorway of northern Greece, Egnatia motorway.

The selection of the days of the thermography was based on the forecast of the weather conditions for the nearest meteorological station of Kozani, which is presented in Table 9.2. According to reliable national weather forecasts, temperatures below 0°C were forecasted along with clear sky, conditions very favorable for frost to occur (Bolam Model Weather Forecast 2008; POSEIDON System 2008).

9.3.1 Study Area

The study area is a 26 km portion of a major highway in northern Greece, Egnatia motorway, a dual carriage way with two traffic lanes per direction, a central reserve and an emergency lane throughout its length. It is the main way for the East–West crossing of the South Balkans.

Egnatia motorway has a total length of 670 km and runs across Greece with an east–west direction from Igoumenitsa port in the prefecture of Thesprotia to Kipi in the prefecture of Evros. It is linked to the borders of all Greece's northern bordering countries, that is, Albania, FYROM, Bulgaria, and Turkey, through nine major axes. The motorway can be characterized as one of the most mountainous roads of Europe as more than 200 km is situated in altitudes over 500 m, whereas its highest altitude is over 1000 m in the area of Metsovo. Its climatic conditions have a significant variation and, especially during the winter months, extreme weather events may occur unexpectedly causing difficulty to road users, as the motorway crosses through rough landscape (Figure 9.1). (www.egnatia.eu).

The studied motorway is situated in a mountainous region of northern Greece where, during winter and early spring, frost and snow conditions often occur. This part of Egnatia motorway is one of the most demanding in terms of weather extremes and terrain complexity. It includes 15 tunnels from which the longest two bore tunnel has a total length of 2.225 m. In addition, it includes six bridges, none of which cross over water, with the longest one being 456 m long. The area under

TABLE 9.2
Forecasted Climatic Conditions according to the National Weather Forecast for the Meteorological Station of Kozani

Date	January 25–26 (02:00)	January 26–27 (02:00)	January 29–30 (02:00)
Temperature (°C)	−1	−2	−5
Relative humidity (%)	76	50	43
Wind direction—intensity (Beaufort force)	N − 1	SW − 2	N − 2
Sky	Clear	Clear	Clear

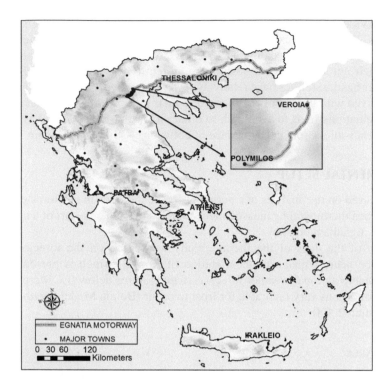

FIGURE 9.1   View of study area.

study is subject to an altitude range of 650 m, as elevation varies between 800 m in Polymilos and 150 m in Veroia. The sky view factor along the road is relatively constant with major differentiations that are observed only close to tunnel entrances and exits (Bouris et al. 2010).

9.3.2   DATASETS

This research is based on the combined use of EO and GIS datasets, either derived directly through internet open sources or derived by the process of thermography measurements, as presented in Table 9.3.

First of all, ASTER GDEM V002 data were retrieved from the online data pool, courtesy of the NASA Land Processes Distributed Active Archive Center (LP DAAC), United States geological survey (USGS)/Earth Resources Observation and Science (EROS) Center, Sioux Falls, South Dakota (https://lpdaac.usgs.gov/data_access/data_pool). ASTER GDEM is a product of NASA and

TABLE 9.3
Datasets of Present Study

Data	Source
ASTER GDEM, V. 2	http://reverb.earthdata.nasa.gov (date of access: September 16, 2016)
CORINE CLC2012 (V 18.5.1)	http://land.copernicus.eu/pan-european/corine-land-cover/clc-2012 (date of access: September 16, 2016)
MODIS LST (MYD11A1-L3)	http://reverb.earthdata.nasa.gov (date of access: October 15, 2016)
HYDROGRAPHIC NETWORK	www.geodata.gov.gr (date of access: September 16, 2016)
ROAD NET WORK	EGNATIA S.A.
THERMOGRAPHY	EGNATIA S.A.

the Japan Ministry of Economy, Trade and Industry (METI). The dataset comprised of the granules N40E21-22 and has a spatial resolution of 30 m and the mean accuracy is estimated to be 12.41–13.6 m for the area of Greece (Chrysoulakis et al. 2004; Miliaresis and Paraschou 2011)

Land use and cover data were included in this research by Corine Land Cover (CLC) inventory. The selected spatial data belong to CLC2012, Version 18.5.1 data with a positional accuracy of 100 m (http://land.copernicus.eu/pan-european/corine-land-cover/clc-2012).

Datasets by MODIS with information on LST were also obtained from the online Data Pool, courtesy of NASA's Earth Observing System Data and Information System (EOSDIS) EOSDIS (2009), Land Processes Distributed Active Archive Center (LP DAAC), USGS/ EROS Center, Sioux Falls, South Dakota (https://lpdaac.usgs.gov/data_access/data_pool). The product MYD11A was selected, which provides information on nighttime LST with a spatial resolution of 1 km on a daily basis. It is a level L3 product, available in tiles of 1.113 km^2 and its estimated accuracy is 1 K in land and under clear sky conditions, according to MODIS LST ATBD, V. 3.3 (https://modis.gsfc.nasa.gov/data/ atbd/ atbd_mod11.pdf). These data were captured by Aqua satellite and were preferred to MODIS satellite, because the predicted overpass time of Aqua is approximately at 11:30–01:00 for the period of study (https://cloudsgate2.larc.nasa.gov/cgi-bin/predict/predict.cgi). A dataset of seven images was created corresponding to the week the thermography was conducted (January 25–31, 2008). The principle of this wider date selection was to assess a set of dates closest to the thermography measurements with low cloud cover. This dataset was considered to be representative of LST fluctuation for the period of study.

Ancillary data concerning the hydrographic network of the study area, including lakes, rivers, and the sea were obtained from the Greek national open data catalogue of website www.geodata.gov.gr.

Finally, thermography measurements as well as the vector file of the Egnatia motorway were kindly granted by Egnatia SA. Thermography measurements provided information on air temperature (T_{air}) and RST with an estimated accuracy of $\pm 2°C$ (Bouris et al. 2010).

9.4 METHODOLOGY

The methodology, which was adopted in the present study consisted of four basic stages and are outlined as follows:

1. Assessment of frost risk levels along the selected section of the Egnatia motorway through the application of GIS modeling
2. Analysis of thermography results in order to extract temperature distribution along the motorway
3. Processing of MODIS images for the period of time the thermography was conducted
4. Overlay of all results from the GIS frost risk model and temperature variation maps derived from MODIS and thermography data

9.4.1 MAPPING AREAS OF HIGH FROST RISK

The levels of frost hazard along the selected stretch of the Egnatia motorway were assessed by applying a variant of the GIS model developed by Louka et al. (2016). The model is based on geographical and environmental factors related to temperature fluctuation and frost risk. The selection of the frost model parameters was based on the relevant literature review (Table 9.1). More specifically, the model incorporates the following input parameters: elevation (*E*), slope aspect (*A*) and gradient (*S*), curvature (*C*), Compound Topographic Index (CTI), Euclidean distance from water bodies (*H*), and land use (LU).

It is based on the multiattribute decision-making methods and on the application of weighted linear combination (WLC) in a GIS environment. These multicriteria evaluation procedures are considered as major support tools with applications in many different spatial decision fields (e.g. Drobne and Lisec 2009; Malczewski 2011).

For each frost risk factor, a corresponding thematic map was created with a spatial resolution of 100 m. Each thematic map was reclassified using Natural Breaks method in GIS, one of the most common grouping data methods for decision-making processes (Mitchell 1999; Simpson and Human 2008; Sunbury 2013). The initial raw data were converted into classes with values ranging from 1 to 5. The classes were assigned according to previous research on the relevance of each factor with frost (Radcliffe and Lefever 1981; Soderstrom and Magnusson 1995; Gustavsson et al. 1998; Gessler et al. 2000; Geiger et al. 2003; Chapman and Thornes 2006; Fridley 2009; Pouteau et al. 2011).

In the maps contained in Table 9.4, the reclassification of each frost parameter is presented. The background of the maps is the reclassification of the surrounding area of the motorway and the reclassification of the pixels of the Egnatia motorway is presented in colours ranging from light grey to black. So, there are two color ramps in the maps' legends.

By analyzing the reclassification of the frost risk parameters along the motorway and the corresponding maps, the importance of each frost risk factor can be derived. The factors that are ranked mainly in the high-risk zone are curvature (Table 9.4d) and Euclidean distance from water bodies (Table 9.4g). Elevation (Table 9.4a) and slope gradient (Table 9.4c) are ranked in the moderate class. Finally, slope aspect (Table 9.4b), land use (Table 9.4f), and CTI (Table 9.4e) in the low to very low frost risk class are ranked. In conclusion, frost hazard in the study area is mainly attributed to the rough terrain of the motorway's surroundings, whereas there is no significant proximity to water bodies to act as a protective layer to low temperatures.

According to the parametrization followed for the GIS models, the reclassified attribute layers were aggregated with their corresponding assigned weight with multiplication and addition overlay operations in GIS. The values extracted from the model formed the basis of the final frost hazard map after a normalization of the raster values.

9.4.1.1 Weighted Linear Combination Model

Weights were assigned to reclassify attribute maps corresponding to each factor's relative importance and effect on the risk of frost. The factor with the most prominent effect on frost was assigned with the highest weight and opposite. WLC was introduced in the assignment of the weights for the criteria of the map layers.

The levels of frost hazard (FH) values were computed for the study area according to the following formula:

$$FH = 0.3 \times E + 0.10 \times A + 0.10 \times S + 0.15 \times CTI + 0.15 \times H + 0.10 \times C + 0.10 \times LU \qquad (9.1)$$

9.4.1.2 Analytical Hierarchy Model

In order to assess the consistency of the factors rating, the method of analytical hierarchy (AH) was also applied in assigning the weights after pairwise comparisons between the frost factors. The new factors' weights were generated by a pairwise comparison matrix. The Consistency Index (CI) of the procedure was calculated to be 0.0388, which is considered satisfactory, as it is lower than the threshold of 0.10 appointed by Saaty (2008).

The frost hazard parameter combination, which was produced by AH is the following:

$$FH = 0.39 \times E + 0.07 \times A + 0.10 \times S + 0.21 \times CTI + 0.15 \times H + 0.05 \times C + 0.03 LU \qquad (9.2)$$

9.4.2 Thermography Measurements

As already mentioned, thermography measurements were conducted on a 26 km stretch of the Egnatia motorway between the intersections of Polymilos and Veroia. Measurements were conducted during the nights of 25th, 26th, and 30th of January 2008 between 01:00 and 05:00 h. The scope of the measurements was the identification of areas that are more prone to low temperatures and ice formation (Bouris et al. 2010).

TABLE 9.4

Reclassification of Thematic Maps

Reclassified Thematic Map	Reclassification of Risk Factor

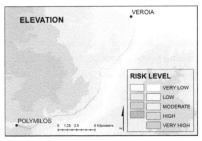

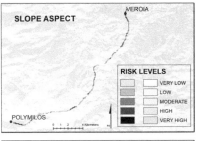

a. Elevation

Risk Level	Classes
Very low	0–250 m
Low	250–500
Moderate	500–850 m
High	850–1,300 m
Very high	1,300–2,396

b. Slope Aspect

Risk Level	Classes
Very low	157.5–202.5
Low	112.5–157.5 and 202.5–247.5
Moderate	67.5–112.5 and 247.5–292.5
High	22.5–67.5 and 292.5–337.5
Very high	337.5–360.0 and 0.0–22.5

c. Slope Gradient (°)

Risk Level	Classes
Very low	26.2°–56.1°
Low	17.6°–26.2°
Moderate	10.6°–17.6°
High	4.4°–10.6°
Very high	0°–4.4°

d. Curvature

Risk Level	Classes
Very low	0.5–3.6
Low	0.1–0.5
Moderate	−0.1–0.1
High	−0.5–−0.1
Very high	−3.28–−0.5

e. CTI

Risk Level	Classes
Very low	11.9–21.0
Low	9.8–11.9
Moderate	8.2–9.8
High	6.8–9.2
Very high	4.1–6.8

(Continued)

TABLE 9.4 (*Continued*)
Reclassification of Thematic Maps

Reclassified Thematic Map   **Reclassification of Risk Factor**

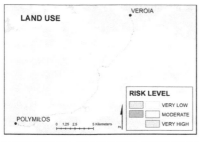

f. Land Use

Risk Level	Classes
Very low	Wetlands, forests, artificial surfaces, and water bodies
Moderate	Cultivated land, mixed vegetation, and open fields
Very high	Narrow vegetation and pastures

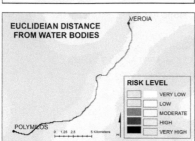

g. Distance from Water Bodies

Risk Level	Classes
Very low	$D_s < 10$ km and $D_{lr} < 1.5$ km
Low	$D_s < 10$ km and $1.5 < D_{lr} < 3$ km *or* $10 < D_s < 20$ km and $D_{lr} < 1.5$ km
Mode Rate	$D_s < 10$km and $D_{lr} > 3$ km *or* $10 < D_s < 20$ km and $1.5 < D_{lr} < 3$ km *or* $D_s > 20$ km and $D_{lr} < 1.5$ km
High	$10 < D_s < 20$ km and $D_{lr} > 3$ km *or* $D_s > 20$ km and $1.5 < D_{lr} < 3$ km
Very high	$D_s > 20$km and $D_{lr} > 3$ km

Note: D_s: Distance from sea shore
D_{lr}:   Distance from lakes and rivers

Over 12,000 thermal images were recorded and processed in total, so that RST profiles were produced for the selected stretch of road for the three dates by the research team of Bouris et al. 2010. These temperature profiles of the Egnatia motorway were kindly granted by Egnatia SA for processing within the framework of the study presented in this chapter. More specifically, road surface and air temperature values were extracted from the initial temperature profiles for points distributed every 100 m along the motorway.

The fluctuation of the average temperature of the road surface (black line) and the air (blue line) was calculated for the three dates of the measurements, and the temperature profiles of Figure 9.2 were produced. Average RST values are depicted with black lines, and air temperature variation is represented

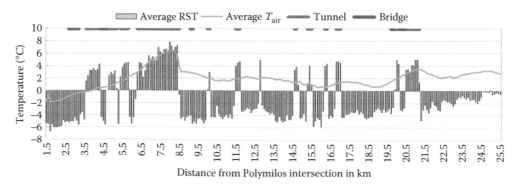

FIGURE 9.2 Average road surface temperature and air temperature profile produced by extracting thermography measurement values per 100 m for the 3 days of measurements.

with a blue line. The areas when a bridge or a tunnel exists along the motorway, which artificially alter the temperature conditions, are marked with a green or red line in the top part of the diagram.

According to the thermography measurements, average RST ranged from 7.76°C to −6.67°C, with an average value of −1.37°C for the portion of the Egnatia motorway. Abrupt changes in the RST were recorded in positions in which tunnels or bridges are located. Differences of more than 10°C were measured even within a few meters' distance. Especially when the motorway enters tunnels, the RST is about 10°C–12°C higher than the surrounding parts before and after the tunnel, which are exposed to weather conditions.

Air temperature values ranged between 6.5°C and −1.9°C with an average value of 1.89°C. The surface of the road was colder than the air with differences up to 8°C for the length of the motorway that does not include tunnels or bridges. On the contrary, inside tunnels, air was colder than the surface of road, with differences reaching up to 3.86°C. In general, the fluctuation of T_{air} was significantly smoother than that of the RST.

In the next phase of processing of the thermography results, a series of transformations were conducted aiming at the overlay and comparison of the produced temperature fluctuation and the results of the frost models.

- First, the points that were positioned inside tunnels or on bridges were extracted from the calculations to avoid any artificially induced temperature variations. In addition, a buffer zone of 100 m on either side of each tunnel and bridge was also extracted from the analysis. As a result, 137 points were selected and included in further analysis.
- Next, temperature information of the remaining points was interpolated, and temperature contours were produced. These contours were rasterized into layers with the same raster analysis of the GIS model results, that is, cell size of 100 m and Greek grid projection system.
- Then, the raster cells that intersect the portion of the motorway were selected, and the values of their centroids were extracted. New charts were produced showing the temperature profile of the road. In addition, maps presenting temperature fluctuation of air and the surface of the road were created.

This procedure facilitated an optimal overlay of thermography measurements and frost model results.

9.4.3 Moderate Resolution Imaging Spectroradiometer Land Surface Temperature

LST data were acquired for this study from MODIS imagery, which was collected from NASA's EOSDIS site and processed through the Modis conversion toolkit (MCT) tool of environment for visualizing images software (ENVI) software.

The 1000 m spatial resolution MODIS daily LST data were resampled to 100 m using the nearest neighbor method. Then, the images were reprojected to Greek grid projection system in order to be compatible to the frost hazard model results.

New raster datasets were produced for each day of study by extracting cells that intersect the selected portion of the motorway, excluding tunnels and bridges. The average LST values were calculated during the week of the thermography measurements along the 26 km stretch of Egnatia motorway.

A map containing LST distribution was produced, as well as the LST profile along the road with values of the centroids of the rater dataset was created.

9.5 RESULTS

9.5.1 Geographic Information System Model Results

Following the procedure described in Section 9.3.1, the predicted spatial distribution of the frost hazard was produced. Frost hazard level was calculated for each pixel with values ranging between 0 and 1, with value "1" representing areas more susceptible to low temperature values and frost conditions and value 0 indicating areas less susceptible to frost.

A map that presents frost hazard along the selected stretch of Egnatia motorway was produced for both GIS models (Figure 9.3a and b). The parts of the motorway that are depicted in red are the most frost-prone parts of the road, whereas green parts are considered as less prone to frost.

In addition, by extracting the centroids' cell values of model results, a profile of frost hazard along the road was produced and is presented in Figure 9.3c.

By analyzing the distribution of GIS model results (Figure 9.4), the studied portion of the Egnatia motorway is situated in a relative low frost risk zone, but it contains parts of high to very high frost risk levels with a total length of 4.65 km according to WLC GIS model and 6.24 km according to AH GIS model. In general, AH model estimates higher levels of frost risk than WLC model.

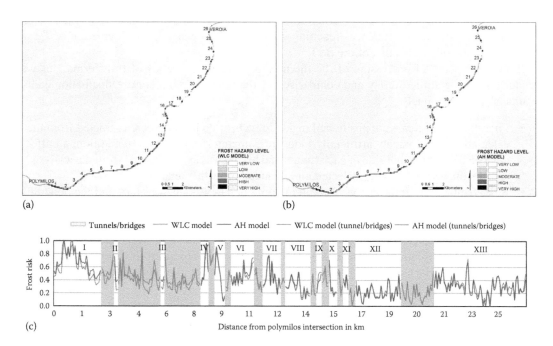

FIGURE 9.3 Spatial distribution of estimated frost hazard levels according to WLC model results (9.3.a) and AH model results (9.3.b) and profile of the models' results along the motorway (9.3.c).

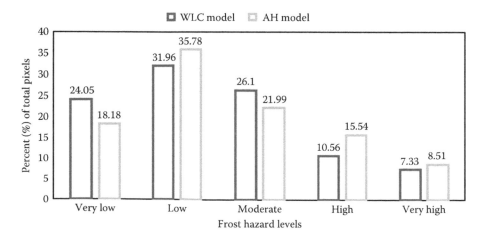

FIGURE 9.4 Distribution of WLC and AH model results for the portion of the Egnatia motorway displayed in five hazard ranks.

In the frost hazard maps of the two GIS models, the areas that are not contained in tunnels or bridges were numbered from "I" to "XIII," and the results of the frost hazard models are analyzed as follows.

The major frost risk areas (both high and very high risk) of the studied motorway are noticed in area I (0–2.5 km from Polymilos Intersection). In this portion of motorway, elevation is comparatively high, and topography characteristics, including aspect and CTI, are favorable to frost, whereas there is no proximity to water bodies that can act as a protective layer.

Moderate estimated frost risk is predicted in areas II (3.5–3.6 km from Polymilos intersection) and III (6.0–6.2 km from Polymilos intersection). They are small-sized areas contained between tunnels. In addition, the areas VI (10.3–11.3 from Polymilos intersection) and VII (11.9–12.7 km from Polymilos intersection) are ranked mainly in the moderate frost risk zone and contain few pixels ranked in the high frost risk zone. In all of these areas, AH frost model provides with a higher frost risk prediction than WLC.

Low-to-moderate frost risk is estimated in areas VIII (13.0–14.5 km from Polymilos intersection) and IX (14.8–15.1 km from Polymilos intersection). A reverse tend among the models is noticed in these areas as the WLC is predicting higher frost risk than AH, contrary to the previously analyzed areas. In addition, the areas XI (16.4–16.6 km from Polymilos intersection) and XII (17.2–19.5 km from Polymilos intersection) are mainly ranked as low-to-moderate frost hazard zones.

A high variability in estimated frost risk is noticed in areas IV (8.6–9.0 km from Polymilos intersection) and V (9.4–9.9 km from Polymilos intersection). The GIS model results in these areas range from low to very high frost risk levels. The predicted frost risk peaks are mainly attributed to curvature and CTI and to a lesser extent to elevation. In addition, there seems to be a tendency of cold air pools to form, as aspect and slope gradient are favorable to cold air drainage. A high degree of variability in the results of the frost risk levels is also contained in area X (15.6–16.1 km from Polymilos intersection). It contains a part in which high frost risk is predicted due to the factors of slope gradient, curvature, and no proximity to water bodies.

Finally, area XIII (21.3–26.0 km from Polymilos intersection) can be divided into two parts: The first part (21.3–23.2 km from Polymilos intersection) is a moderate frost hazard zone with a minor part of high frost hazard levels attributed to the parameters of CTI and slope aspect. The produced frost hazard levels are higher according to WLC model results than AH model. The remaining part of this area, closer to Veroia, is mainly ranked as low-to-moderate frost hazard levels, with the WLC model providing with higher predictions of frost risk.

9.5.2 THERMOGRAPHY RESULTS

Figure 9.5c shows the RST for the three dates of measurements and the average temperature values. The recorded temperature variation is similar for all three dates, which indicate that cold spots of the motorway are consistent, as they exhibit a clear pattern. This observation is in agreement with previous studies, which suggested that general temperature trends remained similar along the motorway. With the exception of extreme nights due to cold air advection, the same locations have the tendency to remain relatively colder due to systematic variation in geographical parameters (Chapman et al. 2001).

The recorded RST values were below freezing temperature for almost the entire length of the studied part of the motorway, when tunnels were excluded. The average RST values have a range of 6.9°C and an average value of −3.6°C (Figure 9.5b).

The most extreme conditions were recorded during the third day, when temperature values up to −9.27°C were recorded. During the first 2 days of measurements, relatively milder weather conditions were prevailed.

In Figure 9.5a, a map is included that presents RST fluctuation along the road in five classes with natural Jenks classification method. The study area is divided into 14 subregions numbered in Latin similar to Section 9.5.1 and the results are analyzed as follows: The lowest RST values were

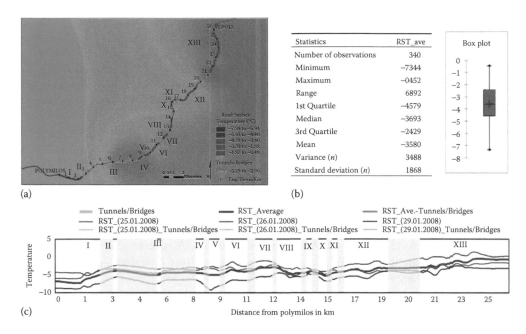

Statistics	RST_ave
Number of observations	340
Minimum	−7344
Maximum	−0452
Range	6892
1st Quartile	−4579
Median	−3693
3rd Quartile	−2429
Mean	−3580
Variance (n)	3488
Standard deviation (n)	1868

FIGURE 9.5 Spatial distribution of road surface temperature along Egnatia motorway, based on the thermography measurements (9.5.a), statistical analysis of the results (9.5.b) and temperature profile of the motorway (9.5.c). (Courtesy of Egnatia SA).

recorded in area I, which is in agreement to frost hazard model results. Relatively lower RST was also observed in regions III, IV, V, part of VIII, X, and XI. Moderate RST was seen in areas VI, part of VIII and XI, and relatively higher RST was seen in areas XII and XIII.

Similar to RST results, the fluctuation of air temperature was produced and is presented in Figure 9.6. T_{air} variation also has a similar pattern for the dates of measurements, and the cold spots

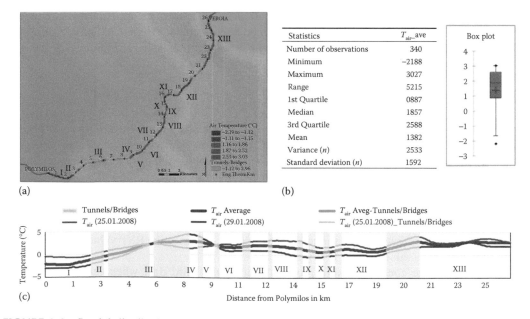

Statistics	T_{air}_ave
Number of observations	340
Minimum	−2188
Maximum	3027
Range	5215
1st Quartile	0887
Median	1857
3rd Quartile	2588
Mean	1382
Variance (n)	2533
Standard deviation (n)	1592

FIGURE 9.6 Spatial distribution of air temperature along Egnatia motorway, based on the thermography measurements (9.6.a), statistical analysis of the results (9.6.b) and temperature profile of the motorway (9.6.c). (Courtesy of Egnatia SA).

TABLE 9.5

Pearson's Correlation Matrix of GIS Model Results and Temperature Measurements

Variables	LSTave	RSTave	T_{air} ave	WLC Norm	AH Norm
LSTave	1	0.659	0.640	−0.386	−0.490
RSTave	0.659	1	0.820	−0.467	−0.561
T_{air} ave	0.640	0.820	1	−0.412	−0.510
WLC norm	−0.386	−0.467	−0.412	1	0.973
AH norm	−0.490	−0.561	−0.510	0.973	1

are recorded in constant positions throughout different weather conditions. The mean T_{air} for the period of study is 1.4°C, which is almost 4°C higher than mean RST for the same dates. In general, air temperature was above freezing temperature for the greater extent of the motorway.

The fluctuation of mean T_{air} is a lot smoother than that of the RST values, with values ranging between 3°C and −2.2°C. In addition, air temperature values had a lower variance compared to RST, which indicates that T_{air} is characterized by lower fluctuation and in general is more constant than the RST (Figure 9.6b).

Figure 9.6a shows a map presenting T_{air} variation along the motorway, presenting the study area classified into five classes depending on air temperature values. Similar to RST distribution, the most severe weather conditions were recorded in area I. In addition, relatively lower temperatures were recorded in areas II, IX, X, and XII. Finally, relatively milder weather conditions were observed in areas IV, V, VII, VIII, and XIII.

By comparing RST and T_{air} distribution, it is interesting to note that the variability between them is striking in some parts of the motorway. For example, in areas IV and V, temperatures below freezing point were observed in the surface of the road, whereas T_{air} was recorded to reach values of 5°C. This is a clear evidence that the recording of air temperature is not a representative indicator of frost conditions on the surface of the motorway.

In addition, in regions such as XII, the variation of T_{air} shows relatively low values contrary to RST variation, which shows moderate to relatively higher values. In this area, according to the reclassification of the frost risk factors (Table 9.5), slope gradient and aspect are favorable to frost conditions. So, it is possible that the microclimatic conditions in this area due to rough topography affect air temperature more intensively than RST. In general, frost spots indicated by air temperature measurements are not necessarily frost spots on the surface of the motorway and do not affect road surface conditions.

9.5.3 Temperature Distribution Based on Moderate Resolution Imaging Spectroradiometer Results

The distribution of LST for the study area was extracted based on satellite-derived LST data by MODIS. Figure 9.7 shows the calculated average LST fluctuation along the studied portion of the Egnatia motorway for the period of the thermography measurements.

According to the LST profile of Figure 9.7c, all dates demonstrated similar variation in LST values. The average values of LST ranged between −5.14°C and −0.03°C, with a mean value of −0.808°C. The recorded LST variance is 1.427°C, which is lower to that of air and RST (Figure 9.7b).

In the map of Figure 9.7a, the motorway is reclassified into five classes with natural Jenks method, and the areas not contained in tunnels or bridges were numbered in Latin. It is not surprising that the lowest LST values were found in area I, which is consistent to thermal mapping and frost hazard model results. In addition, frost spots were observed in areas II, V, VI, VIII, IX, and X. Relatively

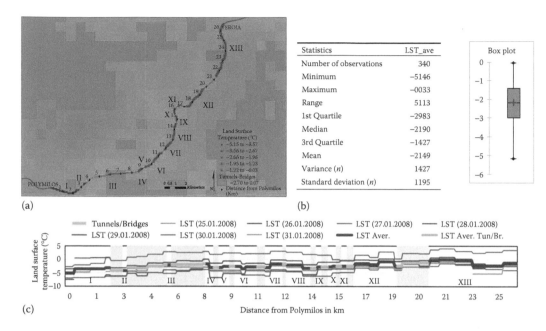

FIGURE 9.7 Spatial distribution of land surface temperature along Egnatia motorway, based on the MODIS imagery (9.7.a), statistical analysis of the results (9.7.b) and temperature profile of the motorway (9.7.c).

milder conditions were recorded in areas III, XI, and the last part of XIII. Finally, higher LST values were observed in areas IV, VII, XII, and XIII.

There is a similarity in the temperature fluctuation of RST and LST in maps of Figures 9.5a and 9.7a. The main exceptions are noted in areas IV and XIII. In area IV, LST presents a variation having relatively high values, followed by a drop of temperature. In area XIII, which according to the thermal mapping and GIS model is expected to have the highest temperatures of the study area, LST shows moderate values.

In particular, the MODIS covers a 1 km pixel that averages the conditions in which a more granular approach is needed so as to capture the impact of microclimatic factors of topography. In addition, these differences could be attributed to the loss of data due to cloud cover, which is a significant limitation in the use of MODIS LST data. In this study, in order to minimize the effect of data loss on the calculated average value of LST, seven MODIS images were processed instead of only the three dates of thermal mapping. This was considered necessary, as during the third day of the thermography measurement, there were missing LST values due to cloud cover, which would artificially affect the calculated average LST. So, although the possible source of error due to data loss has been lessened to a great extent, it has not been eliminated and still causes differentiations in the results.

9.5.4 STATISTICAL ANALYSIS: EVALUATION OF RESULTS

9.5.4.1 Comparative Evaluation of the Results of All Processes

In order to evaluate the correlation of the results produced from all the processes carried out in this project, a composite chart was created and is presented in Figure 9.8. The areas contained in tunnels or bridges are obviously excluded from this chart.

The profiles of road surface, air, and LST are displayed in Figure 9.8a. In the RST and LST temperature profile, error bars are included to indicate their margins of error. These were defined with the strictest error limits, according to the specification of the initial data. More specifically, an error bar of 2°C was applied, which is the estimated accuracy of the thermography measurements

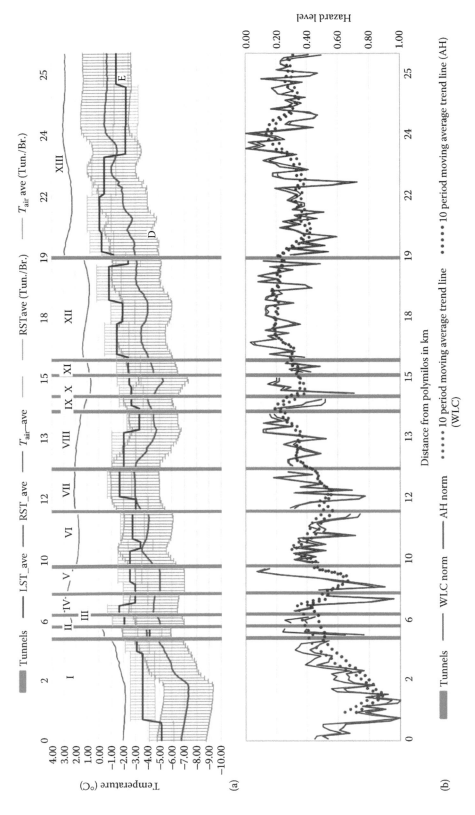

FIGURE 9.8 Road surface, air, and land surface temperature profiles (9.8.a) in juxtaposition to GIS frost hazard model results (9.8.b).

(Bouris et al. 2010). Accordingly, the LST error bar was defined to be 1°C, as this is the estimated accuracy of MODIS LST products, in land and under clear sky conditions (https://modis.gsfc.nasa.gov/data/atbd/atbd_mod11.pdf). The error bars are useful because they signify whether differences between the datasets are statistically important or if the displayed datasets are in agreement.

In Figure 9.8b, the normalized GIS model results are presented. A 10-period moving average trend line is also included in the chart so as to smooth fluctuations of GIS results and enhance their pattern. In order to facilitate the visual interpretation of the outcome of the GIS models and its comparison with measured temperature fluctuation from the thermography and MODIS, the model results are shown in a reverse mode. Higher model result values represent areas of higher frost risk, where lower temperature values are expected. So, the areas where there is a correspondence between the temperature profiles and model trend lines are areas where the GIS models are expected to represent sufficient temperature fluctuation.

By visually analyzing the aforementioned composite chart, it is obvious that along most of length of the portion of the motorway, the RST and LST temperature profiles follow similar patterns, indicating that temperature variation along the road is represented adequately by both the thermography and MODIS. RST is estimated to be lower than remotely sensed LST with a difference of nearly 1.5°C. The difference in temperature values pointed out that this study is based on the temperature fluctuation as it is recorded by thermal mapping and MODIS and not on the absolute temperature values. The two datasets have a different recording method and also the emissivity values that have been assigned to the same area may be differentiated. However, this differentiation is statistically low and within error bars for almost the full length of the road, taking the estimated accuracy of the data into consideration. Only toward the last 3 km of the road, an inversion is noticed, as LST is estimated to be lower than RST. This observation could be attributed to the coarse spatial resolution of MODIS, which have the tendency to average the climatic conditions in areas of 1 km.

The trend lines of GIS model results show a corresponding variation to the RST profile for almost 80% of the length of the motorway that was analyzed. In areas I, VI, XII, and XIII, it is striking that the trend lines of the frost model results follow a similar pattern to that of RST as it was produced from thermal mapping. This fitting is not observed in such an extent between frost model results and LST fluctuation produced by MODIS. This was expected as the spatial analysis of the frost model results is 100 m and is more compatible to thermography measurements in which values were extracted in 100 m intervals, whereas MODIS images have a coarser spatial analysis, which demonstrates more large-scale climatic variations.

In area I where the lowest temperature values were recorded by the thermography and where the frost models predict the highest frost levels, MODIS also recorded the lowest LSTs. In this area, according to the reclassification of the frost risk factors (Table 9.5), the elevation is ranked in the highest risk levels of the study area. Elevation seems to be a dominant frost risk factor, especially in large-scale analysis, which is detected and represented by MODIS.

The specific features of the areas where the predictions of the frost models are not in good agreement with the recorded temperature comprise issues to be further analyzed, so as to enhance the models results. Between the two versions of the models, the AH seems to have a better fit with RST variation, which indicates that AH has improved the model's efficiency in representing frost hazard in the local scale of a motorway. WLC model has shown to produce results that correlate well or fairly well when applied in wider areas, such as agricultural land (Louka et al. 2016).

9.5.4.2 Pearson's Correlation

The recorded temperature values and the results of the two versions of the GIS model were also analyzed to evaluate the correlation of the model results with the recorded temperature fluctuation. This process was executed using the Pearson's correlation coefficient in Excel's statistical analysis software add-on, XLSTAT. The results are shown in Table 9.5 containing the correlation matrix of all datasets.

9.5.4.2.1   Correlation of Temperature Variation from Thermal Mapping and Moderate Resolution Imaging Spectroradiometer

Concerning the correlation of temperature data, which are obtained from different and independent sources, a significant positive relationship is derived. The Pearson's correlation coefficient of road surface and air temperature recorded during the thermography and LST obtained by MODIS imagery is calculated to be 0.659 and 0.640, respectively.

The correlation between RST and air temperature is higher (0.820), as expected, because it is derived by the same measurement method and instruments.

9.5.4.2.2   Correlation of Geographic Information System Model Results and Recorded Temperature Variation

As shown in Table 9.5, the two versions of the GIS models have a negative relationship, with the recorded temperature fluctuation both from thermal mapping and remote sensing data. This outcome is not surprising, as the higher values of frost risk model results are linked to lower temperatures and higher risk of frost.

Between the two versions of the GIS model, it is apparent that the AH version provided a more accurate prediction than the WLC model. The first has a high negative association assessed to be −0.561 with RST and −0.510 with air temperature. In addition, its association with LST is marginally high (−0.49). The corresponding Pearson's correlation coefficient between WLC results and road surface, air, and LST was calculated to be −0.467, −0.412 and −0.386, respectively. So there seems to be moderate to marginally high correlation.

9.5.4.3   Moran's I Autocorrelation Analysis

In order to further investigate the pattern of temperature variation along the motorway and the existence of areas with significantly lower or higher temperatures, the spatial analysis software GeoDA was employed (Anselin et al. 2006). The degree of spatial autocorrelation of the results of the processes of this study was analyzed based on the local Moran's I statistic test. The matrix K4 nearest neighbor was selected for these analyses, as it combined nearly the highest values of Moran's I index while providing with sufficient clustering of the data. Figure 9.9 shows the produced LISA cluster maps, which classify the motorway by type of spatial association. The study area is divided into 14 numbered patches similar to Sections 9.5.1 through 9.5.3.

The dark red locations represent spatial clusters of high values surrounded by high values, forming a high–high region. Similarly, the dark blue locations indicate spatial clusters with low values surrounded by low values, forming a low–low region. The spatial outliers are regions with high values surrounded by low value regions or the opposite and are marked with light red or light blue and are called high–low and low–high regions, respectively.

The overall pattern of the cluster maps of the GIS models seems to have a similarity. Clusters of high risk are observed in areas I, IV, and V. Less susceptible to frost seem to be parts of areas VIII, XII, and XIII. An insignificant autocorrelation was detected in the GIS model results for the remaining motorway.

The RST and LST cluster maps seem to follow a similar pattern, as clusters of low temperatures appear to be in areas I, II, VI, and VIII. High-temperature clusters are observed in areas XII and XIII.

The cluster map of T_{air} is quite differentiated to the RST and LST cluster maps, except from area I where a low-temperature cluster is detected.

T_{air} cluster maps show a similarity to those of LST and RST in the low-temperature regions (areas I and II). On the contrary, high-temperature cluster maps are contained in areas IV, V, and XIII, which are different to LST and RST cluster maps.

According to the autocorrelation analysis results, there seems to be an efficient prediction of the GIS frost models for areas I, XII, and XIII, as in these areas clusters of high levels of frost risk are produced by the models, and clusters of low RST and LST values are observed.

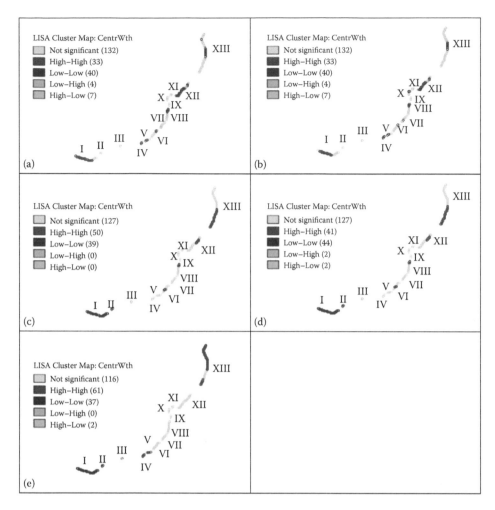

FIGURE 9.9 Cluster maps of GIS model results and RST and LST average values along the studied portion of the Egnatia motorway. (a) AH GIS model, (b) WLC GIS model, (c) RST, (d) LST, and (e) T_{air}.

Further research could be addressed to the remaining study area so as to optimize the output of the GIS models with proper adjustments.

9.6 CONCLUSIONS

This study provides important insights into how remote sensing imagery and thermography measurements might be combined to study climatological hazards such as frost.

Thermography measurements offered a high spatial resolution and enabled the better understanding of the RST patterns as it has the ability to produce a very precise thermal map of a specific area such as a motorway. However, the complementation of more repetitions of the thermography measurements and under more diverse weather conditions are required to produce an even more representative RST profile.

Remotely sensed LST derived by MODIS was processed to produce temperature fluctuation along the motorway. MODIS imagery is freely available data and is provided with ready to use LST information through the internet on a daily basis, provided cloud-free weather conditions prevail. But their low spatial resolution of 1 km and the possible loss of data due to clouds are significant limitations in studies of road weather conditions and could produce misleading results. However, downscaling in areas of diverse landscape and geomorphology should be performed with caution and by employing techniques such as data fusion, which enables the exlpoitation of the advantage of high temporal resolution of MODIS and spatial resolution of Landsat-8 and ASTER sensors. In addition, remotely sensed temperature data will be available in the near future by sources such as Sentinel-3, which will provide an alternative LST data source. The potential offered by these data is a subject of further investigation concerning temperature fluctuation studies.

Finally, the hazard levels of frost have been forecasted by applying a GIS frost model, which incorporates geographical and environmental variables. The use on GIS modeling to predict frost hazard levels on a motorway offers the possibility of forecasting frost risk for an area independent of its accessibility and may reduce costly field surveys. It requires the development and parametrization of a model according to the climatic conditions of an area and the availability of high-resolution remote sensing data to produce more accurate results. Future research should take place to analyze and better explain the impact and the interaction of the factors addressed in the model or even to include additional environmental or climatic variables. The accuracy of risk assessment could be improved, and the resultant hazard maps would be more valuable to road engineers and maintenance personnel for creating safety protocols for specific parts of the Egnatia motorway.

The methodology analyzed herein is applicable to areas with different climatic conditions with the appropriate adjustments. It enables the exploitation of thermography measurements and its coupling with remote sensing data and GIS tecnhiques aiming at the identification of areas along a motorway that runs a high frost and ice risk. The ability to estimate areas more suspectible to low temperatures and frost conditions along a motorway is an issue of great importance. With the development of GIS and EO techniques, it is possible to be achieved with greater accuracy and more timely.

ACKNOWLEDGMENT

The authors are thankful to the company "Egnatia Odos SA" for providing the thermography measurement data, which were used in this study.

REFERENCES

Anand, S. (2007). Domestic use of unmanned aircraft systems: An evaluation of policy constraints and the role of industry consensus standards. *ASTM Standardization News*, 35, 30.

Anselin, L., Syabri, I., and Kho, Y. (2006). GeoDa: an introduction to spatial data analysis. *Geographical Analysis*, 38(1), 5–22.

Berrocal, V. J., Raftery, A. E., Gneiting, T., and Steed, R. C. (2010). Probabilistic weather forecasting for winter road maintenance. *Journal of the American Statistical Association*, 105(490), 522–537.

Bogren, J. and Gustavsson, T. (1991). Nocturnal air and road surface temperature variations in complex terrain, *International Journal of Climatology*, 11, 443–455.

Bogren, J., Gustavsson, T., Karlsson, M., and Postgård, U. (2000). The impact of screening on road surface temperature. *Meteorological Applications*, 7(2), 97–104.

Bogren, J., Gustavsson, T., and Lindqvist, S. (1992). A description of a local climatological model used to predict temperature variations along stretches of road. *Meteorological Magazine*, 121(1440), 157–164.

Bolam Model Weather Forecast. National Observatory of Athens. Available: http://cirrus.meteo.noa.gr/forecast/bolam/index.htm

Boselly, S. E., Doore, G. S., Thornes, J. E., Ulbery, C., and Einst, D. D. (1993). Road weather information systems. Volume 1: Research report. Washington, DC.

Bouilloud, L. and Martin, E. (2006). A coupled model to simulate snow behavior on roads. *Journal of Applied Meteorology and Climatology*, 45(3), 500–516.

Bouilloud, L., Martin, E., Habets, F., Boone, A., Le Moigne, P., Livet, J., and Marchetti, M. (2009). Road surface condition forecasting in France. *Journal of Applied Meteorology and Climatology* 48(12), 2513–2527.

Bouris, D., Theodosiou, T., Rados, K., Makrogianni, M., Koutsoukos, K., and Goulas, A. (2010). Thermographic measurement and numerical weather forecast along a highway road surface. *Meteorological Applications*, 484(03), 474–484.

Bradley, A. V., Thornes, J. E., Chapman, L., Unwin, D., and Roy, M. (2002). Modelling spatial and temporal road thermal climatology in rural and urban areas using a GIS. *Climate Research*, 22(1), 41–55.

Chapman, L. and Thornes, J., E. (2004). Real-time sky-view factor calculation and approximation. *Journal of Atmospheric and Oceanic Technology*, 21(5), 730–741.

Chapman, L. and Thornes, J. E. (2005). The influence of traffic on road surface temperatures: Implications for thermal mapping studies. *Meteorological Applications*, 12, 371–380.

Chapman, L. and Thornes, J. E. (2006). A geomatics-based road surface temperature prediction model. *Science of the Total Environment*, 360(1), 68–80.

Chapman, L., Thornes, J. E., and Bradley, A. V. (2001). Modelling of road surface temperature from a geographical parameter database. Part 2: Numerical. *Meteorological Applications*, 8, 421–436.

Chrysoulakis, N., Abrams, M., Feidas, H., and Velianitis, D. (2004). Analysis of ASTER multispectral stereo imagery to produce DEM and land cover databases for Greek islands: The REALDEMS project. *Proceedings of e-Environment Progress and Challenge*, p. 411–424.

Cornford, D. and Thornes, J. E. (1996). A comparison between spatial winter indices and expenditure on winter road maintenance in Scotland. *International Journal of Climatology*, 16, 339–357.

Coll, C., Caselles, V., Valor, E., Niclòs, R., Sánchez, J. M., Galve, J. M., and Mira, M. (2007). Temperature and emissivity separation from ASTER data for low spectral contrast surfaces. *Remote Sensing of Environment*, 110(2), 162–175.

Crevier, L. P. and Delage, Y. (2001). METRo: A new model for road-condition forecasting in Canada. *Journal of Applied Meteorology*, 40, 2026–2037.

Drobne S. and Lisec A. (2009). Multi-attribute decision analysis in GIS: Weighted linear combination and ordered weighted averaging. *Nature*, 4(26), 28.

Earth Observing System Data and Information System (EOSDIS). 2009. Earth Observing System ClearingHOuse (ECHO)/Reverb, Version 10.X [online application]. Greenbelt, MD: EOSDIS, Goddard Space Flight Center (GSFC) National Aeronautics and Space Administration (NASA). Available: http://reverb.earthdata.nasa.gov.

Eriksson, M. (2001). Winter road climate investigations using GIS. Doctoral Thesis–University of Gothenburg, Gothenburg, Sweden.

Federal Highway Administration (2006). Best practices for road weather management. Version 2.0. Available: http://ops.fhwa.dot.gov/ Weather/best_practices /CaseStudies/029.pdf

Fridley, J. D. (2009). Downscaling climate over complex terrain: high finescale (<1000 m) spatial variation of near-ground temperatures in a montane forested landscape (Great Smoky Mountains). *Journal of Applied Meteorology and Climatology*, 48(5), 1033–1049.

Fry, R. (2010). Improving road ice prediction through the introduction of GIS generated geographical parameters to an ice- forecasting model. PhD thesis. University of Glamorgan Wales, Treforest, Wales.

Gao, F., Masek, J., Schwaller, M., and Hall, F. (2006). On the blending of the Landsat and MODIS surface reflectance: Predicting daily Landsat surface reflectance. IEEE. *Transactions on Geoscience and Remote Sensing*, 44(8), 2207–2218.

Geiger, R. F., Aron, R. H. and Todhunter, P. (2003). *The Climate Near the Ground*. Lanham, MD: Rowman & Littlefield Publishers, p. 584.

Gessler P. E., Chadwick O. A., Chamran F., Althouse L. and Holmes K. (2000). Modeling soil-landscape and ecosystem properties using terrain attributes. *Soil Science Society of America Journal*, 64, 2046–2056.

Greenfield, T. M. and Takle, E. S. (2006). Bridge frost prediction by heat and mass transfer methods. *Journal of Applied Meteorology and Climatology*, 45(3), 517–525.

Grimmond, C. S., Potter, S. K., Zutter, H. N., and Souch, C. (2001). Rapid methods to estimate sky-view factors applied to urban areas. *International Journal of Climatology*, 21(7), 903–913.

Gustavsson, T. and Bogren, J. (1991). Infrared thermography in applied road climatological studies. *International Journal of Remote Sensing*, 12(9), 1811–1828.

Gustavsson, T. (1999). Thermal mapping–A technique for road climatological studies. *Meteorological Applications*, 6, 385–94.

Gustavsson, T., Karlsson, M., Bogren, J., and Lindqvist, S. (1998). Development of temperature patterns during clear nights. *Journal of Applied Meteorology*, 37(6), 559–571.

Hammond, D., Chapman, L., Thornes, J. E. and White, S. P. (2010) Verification of route-based winter road maintenance weather forecasts. *Theoretical and Applied Climatology*, 100, 371–384.

Jacobs, W. and Raatz, W. E. (1996). Forecasting road-surface temperatures for different site characteristics. *Meteororological Applications*, 3, 243–256.

Li, L. Y. and Eglese, R. W. (1996). An interactive algorithm for vehicle routeing for winter—Gritting. *Journal of the Operational Research Society*, 47(2), 217–228.

Liu, H. and Weng, Q. (2012). Enhancing temporal resolution of satellite imagery for public health studies: A case study of West Nile Virus outbreak in Los Angeles in 2007. *Remote Sensing of Environment*, 117, 57–71.

Lindqvist, S. (1975). Infraredtermografiska tillampningar inom vagklimatologin, *The Swedish Geographical Yearbook*, vol. 51, pp. 117–123. (With English abstract.).

Louka, P., Papanikolaou, I., Petropoulos, G., and Stathopoulos, N. (2016). A deterministic model to predict frost hazard in agricultural land utilizing remotely sensed imagery and GIS. *Geospatial Technology for Water Resource Applications*. Portland, OR: CRC Press.

Malczewski, J. (2011). Local weighted linear combination. *Transactions in GIS*, 15(4), 439–455.

Miliaresis, G. C. and Paraschou, C.V. (2011). An evaluation of the accuracy of the ASTER GDEM and the role of stack number: A case study of Nisiros Island, Greece. *Remote Sensing Letters*, 2(2), 127–135.

Mitchell, A. (1999). The ESRI guide to GIS analysis. *Geographic Patterns and Relationships*, vol. 1. Redlands, CA: Environmental Systems Research Institute.

Muyldermans, L., Cattrysse, D., and Van Oudheusden, D. (2003). District design for arc-routing applications. *Journal of the Operational Research Society*, 54, 1209–1221.

Norrman, J. (2000). Slipperiness on roads–An expert system classification. *Meteorological Applications*, 7, 27–36.

Odido, D. and Madara, D. (2013). Emerging technologies: Use of unmanned aerial systems in the realisation of vision 2030 goals in the Counties. *International Journal of Applied Science and Technology* 3(8), 107–27.

Oke, T. R. (1987). *Boundary Layer Climates*. London, UK: Routledge.

Oliver, W. W. and Dolph, K. L. (1992). Mixed-conifer seedling growth varies in response to overstory release. *Forest Ecology and Management*, 48(1–2), 179–183.

Paumier, J. L. and Arnal, M. (1998). Experimentation Previroute sur l'autoroute A75 dans le Cantal (Previroute experiments on A75 highway in Cantal department). *Revue Generale des Routes et Aerodromes*, 758, 44–51.

Perrier, N., Langevin, A., and Campbell, J. F. (2007). A survey of models and algorithms for winter road maintenance. Part IV: Vehicle routing and fleet sizing for plowing and snow disposal. *Computers & Operations Research*, 34(1), 258–294.

POSEIDON System. (2008). Hellenic Center for Marine Research. Available: http://www.poseidon.hcmr.gr

Postgard, U. and Lindqvist, S. (2001). Air and road surface temperature variations during weather change. *Meteorological Applications*, 8(1), 71–84.

Pouteau, R., Rambal, S., Ratte, J. P., Gogé, F., Joffre, R., and Winkel, T. (2011). Downscaling MODIS-derived maps using GIS and boosted regression trees: The case of frost occurrence over the arid Andean highlands of Bolivia. *Remote Sensing of Environment*, 115(1), 117–129.

Radcliffe J. E. and Lefever K. R. (1981). Aspect influences on pasture microclimate at Coopers Creek, North Canterbury. *New Zealand Journal of Agricultural Research*, 24, 55–66.

Ramakrishna, D. M. and Viraraghavan, T. (2005). Environmental impact of chemical deicers–A review. *Water, Air, and Soil Pollution*, 166(1), 49–63.

Rayer, P. J. (1987). The Meteorological Office forecast road surface temperature model. *Meteorological Magazine*, 116, 180–191.

Rosema, A. and Welleman, A. G. (1977). Microclimate and winter slipperiness. A study of factors influencing slipperiness, with application of thermal infrared observation techniques. *Niwars*, p.38.

Rouse, W. R. and Wilson, R. G. (1969). Time and space variations in the radiant energy fluxes over sloping forested terrain and their influence on seasonal heat and water balances at a middle latitude site. *Geografiska Annaler. Series A. Physical Geography*, p.160–175.

Saaty, T. L. (2008). Decision making with the analytic hierarchy process. *International Journal of Services Sciences*, 1(1), 83–98.

Shao, J. (2000). Fuzzy categorization of weather conditions for thermal mapping *Journal of Applied Meteorology and Climatology*, 39, 1784–1790.

Shao, J. C., Swanson, R., Patterson, P., Lister, J., and Mcdonald, A. N. (1997). Variation of winter road surface temperature due to topography and application of thermal mapping. *Meteorological Applications*, 4, 131–137.

Semmens, K. A., Anderson, M. C., Kustas, W. P., Gao, F., Alfieri, J. G., McKee, L., and Xia, T. (2016). Monitoring daily evapotranspiration over two California vineyards using Landsat 8 in a multi-sensor data fusion approach. *Remote Sensing of Environment*, 185, 155–170.

Simpson, D. M. and Human, R. J. (2008). Large-scale vulnerability assessments for natural hazards. *Natural Hazard*, 47(2), 143–155.

Soderstrom, M. and Magnusson, B (1995). Assessment of local agroclimatological conditions—A methodology. *Agricultural and Forest Meteorology*, 72, 243–260.

Stove, G. C., Kennie, T. J., and Harrison, A., (1987). Airborne thermal mapping for winter highway maintenance using the Barr and Stroud IR18 thermal video frame scanner. *International Journal of Remote Sensing*, 8, p.1077–1084.

Sunbury T. M. (2013). The role and challenges of utilizing GIS for public health research and practice. *Technology & Innovation*, 15(2), 91–100.

Tagmouti, M., Gendreau, M., and Potvin, J. Y. (2007). Arc routing problems with time-dependent service costs. *European Journal of Operational Research*, 181(1), 30–39.

Takle, E. S. (1990). Bridge and roadway frost: Occurrence and prediction by use of an expert system. *Journal of Applied Meteorology*, 29(8), 727–734.

Thornes, J. E. (2000). Road salting—An international cost/benefit review. In R. M. Geertman (Ed.) *Proceedings of the 8th World Salt Symposium*. Volume 2, Amsterdam, the Netherlands: Elsevier, pp. 787–790.

Thornes, J. E. (1991). Thermal mapping and road-weather information systems for highway engineers. *Highway Meteorology*, pp. 39–67.

Vinter, R. (2015). Utilization of Lidar data and street view images in road environment monitoring. PhD thesis, Aalto University, Finland.

Wood, N. L. H. and Clark, R. T. (1999). The variation of road-surface temperatures in Devon, UK during cold and occluded front passage. *Meteorological Applications*, 6, 111–118.

Yahia, J. C., (2006). Développement et validation d'un modé le physique de prévision de la tempé rature de surface de re- vê tement de la chaussée (Development and validation of a physical model of road surface temperature forecasting). PhD thesis, University Blaise Pascal, France.

Yu, X., Guo, X., and Wu, Z. (2014). Land surface temperature retrieval from Landsat 8 TIRS—Comparison between radiative transfer equation-based method, split window algorithm and single channel method. *Remote Sensing*, 6(10), 9829–9852.

INTERNET SOURCES

https://lpdaac.usgs.gov/
https://sentinel.esa.int/ web/sentinel/user-guides/sentinel-3-slstr/
www.egnatia.eu
https://lpdaac.usgs.gov/data_access/data_pool
http://land.copernicus.eu/pan-european/corine-land-cover/clc-2012
https://lpdaac.usgs.gov/data_access/data_pool
https://modis.gsfc.nasa.gov/data/ atbd/atbd_mod11.pdf
https://modis.gsfc.nasa.gov/data/ atbd/atbd_mod11.pdf
www.geodata.gov.gr
https://earth.esa.int

Section III

Remote Sensing of Wildfires

10 Wildfires and Remote Sensing
An Overview

Nicolas R. Dalezios, Kostas Kalabokidis,
Nikos Koutsias, and Christos Vasilakos

CONTENTS

10.1 INTRODUCTION

Wildland/forest fires (or merely wildfires) are considered one of the most widespread environmental hazards. This hazard contributes significantly to climate change and soil degradation (NC, 2015; IPCC, 2012). Destruction of vegetation by wildfires can potentially affect the hydrological cycle, as well as land surface, due to the increase in the surface albedo, increase in surface runoff, decrease in evapotranspiration, increase in erosion, and occurrence of floods and deserts. Moreover, the burning of biomass may contribute, along with gases, to the greenhouse effect and can originate the destruction of the ozone layer.

Wildfires have been a natural disturbance of wildland ecosystems, such as in the Mediterranean Basin or in California, United States. Indeed, fire has played an important role in shaping many plant communities that grow in a climatic region characterized by hot and dry summer conditions, with high evapotranspiration rates (Sedaei et al., 2017). During the last decades, human activities have disturbed the delicate natural balance between fire activity and the regeneration processes. It is recognized that among the main reasons that can explain these facts, the following can be emphasized: abandonment of farmlands and forests, urban expansion into forest areas, and the constant increase of tourist development in wooded zones. With focus on the Mediterranean region, wildfires occur in bordering countries of this area during the dry season and mainly affect pine forests, shrublands, and less frequently cultivated fields. For example, on an average, about 1,500 wildfires are recorded annually in Greece, which affect a surface area of roughly 50,000 ha (Kailidis, 1990; Kalabokidis et al., 2013). Although the vast majority are small-scale fires, a small number of them (about 5%) are considered as large-scale fires that burn more than 100 ha, even when the associated burned area reaches 70% of the aforementioned total. In United States, large-scale fires usually burn more than 400 ha.

This chapter focuses on the remote sensing potential in wildfires. At first, wildfires definitions and concepts are presented, along with parameters, causes, and factors affecting wildfires, as well as wildfire mechanisms and protection. Then, remote sensing concepts, as well as capabilities in wildfires, are presented. Specifically, remotely sensed data and methods for wildland fuel modeling (prefire conditions), fire early warning systems (FEWS), wildfire monitoring (during an event activity), and postfire assessment are analyzed. Remote sensing examples and case studies are presented on these topics.

10.2 WILDFIRE DEFINITIONS AND CONCEPTS

There are three types of forest fires, namely ground, surface, and crown fires (Omi, 2015). *Ground* fires, which burn on the ground or below ground vegetation, are often best controlled by excavating trenches or *firelines* down into the mineral soil layer, which cannot burn. Ground fires typically burn by smoldering and can burn slowly for days to years, as exemplified by peat fires. *Surface* fires burn along the surface and tend to move quickly. Crawling or surface fires are fueled by low-lying vegetation, such as leaf and timber litter, debris, grass, and low-level shrubbery. Surface and ground fuel types are especially susceptible to ignition due to *spotting*, which is shedding burning biomass. In Australian bushfires, spot fires are known to occur as far as 20 km from the fire front (VBRC, 2010). *Ladder* fuel is material between low-level vegetation and tree canopies, such as small trees, downed logs, and vines that may be ignited during an advancing fire front and spread flames into the tree canopy/crown. *Crown* fires are most dangerous and spread very fast. They occur on top of the trees, where fire can jump from crown to crown, often over firebreaks. The ignition of a crown fire, termed crowning, depends on the density of the suspended material, canopy height, canopy continuity, and sufficient surface and ladder fuels to reach the tree crowns.

10.2.1 PARAMETERS AND CAUSES OF WILDFIRES

Meteorological parameters, such as rainfall, air temperature, relative humidity, and wind velocity, have a significant effect in the initiation and spread of forest fires. Specifically, drought periods followed by dry and hot winds blowing from arid continental interiors over a period of days create a cumulative heating and drying effect on vegetation (Sedaei et al., 2017). These are atmospheric conditions, which promote dry lightning storms and are a frequent ignition source. Moreover, environmental factors affecting fire danger are topography, type of fuel, and fuel moisture content (FMC) which is one of the most significant factors that affect the ignition and spread of forest fires (Omi, 2005).

> *Rainfall* of long duration is capable of depositing significant amounts of water and drenches the flammable forest matter, thereby, increasing its resistance to the initiation and spread of fire. On the other hand, as expected, short-duration precipitation has a very low effect in increasing the fuel moisture. Besides precipitation duration and amount, seasonality also plays an important role. *Wind* effect on forest fires depends on its speed and direction. At the initial stages, wind provides the necessary oxygen supply and at the maturity stage it affects the propagation velocity through flame tilting. Wind speed that exceeds 2.8 m s^{-1} has been shown to have a 45% slope equivalent (Spanos et al., 1998). The wind direction is also important, because it determines the moisture content of the wind and the spread direction of the fire. Dry winds are more dangerous with respect to fire outbreak and spread. Large-scale fires show a high propagation velocity (2.5 km h^{-1}) and usually evolve to crown fires. High *air temperature* in combination with dry spells is extremely dangerous

for the initiation and spread of wildfires. Wildfires with spatial extension over 500 ha habitually come about when the temperature is greater than 30°C. *Relative humidity* determines the amount of ignition heat and affects the number of incidents and their spatial extension. Dry air results in dry and highly flammable forest biomass, whereas during nighttime the increase of relative humidity results in the increase of the FMC.

Environmental parameters affecting the fire danger are the variety and extent of fuel (vegetation and dead organic material), fuel moisture, and topography (altitude, slope, and aspect). The fuel loading of an area depends on land use, that is, forest or cultivated area, the vegetation species, and the vegetation condition. Moisture content depends both on fuel size and atmospheric conditions. Thin dead materials, for example, needles, respond rapidly to moisture changes in the atmosphere. On the other hand, branches of large size and thick leaf layers maintain their moisture for several days after rain episodes, especially when calm conditions prevail and saturation vapor pressure deficit is low. Topography plays an important role in the initiation and spread of forest fires, as well as in vegetation regrowth following a fire event. Altitude affects the vegetation period through temperature and moisture changes in the atmospheric environment and the groundwater supply of the plants.

Causes of wildfires differ from one area to another due to environmental factors, such as vegetation type, lightning frequency, climatic conditions, and anthropogenic activities. Causes of wildfires can be classified into three broad categories: (1) wildfires caused by natural causes, (2) wildfires caused by human activities; and (3) wildfires caused by dubious–unknown reasons. In Mediterranean type of ecosystems, only the smallest percentage of fires can be considered to be natural, such as caused by lightning, sparks, or volcanic eruptions, compared to human-caused fires. Wildfires attributable to anthropogenic activities are evoked either randomly, such as by negligence or accidents or on purpose, such as arsons.

Factors affecting wildfires: The spread of forest fires varies based on the flammable material volume and its vertical arrangement. For example, fuels uphill from a fire are more readily dried and warmed by the fire than those downhill. Fuel arrangement and density are governed in part by topography, as land shape determines factors, such as available sunlight and water for plant growth.

10.2.2 WILDFIRE MECHANISMS AND PROTECTION

Wildfire mechanisms: A wildfire front is the portion sustaining continuous flaming combustion in which unburned material meets active flames, or the smoldering transition between unburned and burned material (NWCG, 2008). Even before the flames of a wildfire arrive at a particular location, heat transfer from the wildfire front warms the air to more than 800°C, which preheats and dries flammable materials, causing them to ignite faster and allowing the fire to spread. Wildfires have a rapid forward rate of spread when burning through dense, uninterrupted fuels. Wildfires can advance tangential to the main front to form a flanking front, or burn in the opposite direction of the main front by backing motion.

Protection from wildfires: In order to have an integrated wildfire protection system, two kinds of measures are needed, namely prevention and suppression measures. *Prevention* refers to preemptive methods of reducing the wildfire risk, as well as lessening fire severity and spread. The various techniques that can be used for reduction of anthropogenic fires fall into two general categories, namely to reduce danger–risk and to manage it. Other measures to prevent fire ignition and spread include the creation of firebreaks or fuelbreaks or modification of flammable fuels. The use of different types of vegetation cover is considered as a very important measure to prevent large fires. Moreover, wildfire prevention programs around the world may employ techniques, such as wildland fire use and prescribed or controlled burns. Multiple fuel treatments are often needed to influence future fire risk. *Suppression* depends on the technologies available in the area in which the wildfire occurs. The key to controlling and suppressing a wildfire is getting human power and equipment to the scene in the shortest possible time. During the evolution of the fire, as information from the field and data from different sources (e.g., weather and satellite maps) become available, the firefighting strategy can be modified from the first-response plan (Borealforest, 2016). The choice of whether to apply retardants depends on the scale, location, and intensity of the wildfire. Past fire suppression, along with other factors, has resulted in the larger, more intense wildfire events that are seen today.

Assessment of burned area: The delineation of the burned area depends on the type of burned vegetation, the soil type, the time interval after the fire, and the extent of the fire, that is, totally or partially burned (Domenikiotis et al., 2002). In all these cases, the spectral signatures of the land cover objects do not have a characteristic pattern. After the cease of a fire, significant reduction of the vegetation is expected, and values corresponding to complete lack of chlorophyll elements are an indication of the burned area. Vegetation indices are an acceptable technique for identifying vegetation changes. The methods for mapping the burned areas are usually based either on the thermal signal or on the use of an index or on an algorithm utilizing both information.

10.3   REMOTE SENSING CAPABILITIES IN WILDFIRES

Remote sensing is a useful tool for providing information before, during, and after a wildfire through the visible, infrared, and microwave portion of the electromagnetic spectrum (ESA, 2004). Specifically, in the visible and infrared, meteorological satellites, such as the sun-synchronous NOAA-N series, or the geostationary Meteosat and geostationary operational environmental satellite (GOES), and environmental satellites, such as Landsat or satellite pour l'observation de la terre (SPOT), which are polar or near-polar low-orbit satellites, are mainly used. Indeed, meteorological satellites have a rather coarse spatial resolution, but high temporal reoccurrence, thus being suitable mainly for operational monitoring applications of wildfires and enable detecting meteorological parameters quantitatively, such as precipitation, humidity, wind, and temperature, among others. On the other hand, environmental satellites are generally characterized by fine spatial resolution, but low temporal reoccurrence, being basically used in land use/cover types and detection of several wildfire features, such as areal extent, postfire assessment, FEWS, and monitoring. Moreover, in the microwave portion, there are active sensors, such as synthetic aperture radar (SAR) and light detection and ranging (LiDAR), which can provide significant data in wildfire analysis, because they can map the vertical structure of vegetation. Similarly, satellite sensors European remote sensing (ERS-1), Japanese earth resources satellite (JERS-1), and Radarsat, as well as airborne sensors, have been widely used in wildfires (Zhang et al., 2016).

The application and utility of Earth observation (EO) technology to wildfire assessment are growing rapidly, mainly due to the increasing number of satellite systems that are launched, along with the increasing remote sensing reliability and the continuously improving technological advances. The trend of further improving the spatial resolution continues reaching the level of microremote sensing in the order of 1 m or smaller with new satellites. Moreover, there is a very recent tendency to increase the number of available bands in these satellites resulting in new and valuable information in wildfire analysis. New types of remote sensing systems can provide online open information for web platforms. Such systems are NASA's new online satellites for climate change, Orbiting Carbon Observatory-2, Global Precipitation Measurement Core Observatory, or Soil Moisture Active Passive, as well as the European Copernicus system with six Sentinel satellites (2014–2021) to monitor land, ocean, emergency response, atmosphere, security, and climate change (ESA, 2014). Moreover, massive cloud computing resources and analytical tools for working with big datasets make it possible to extract new information from environmental satellites with varying spatial resolution, such as Landsat-8 (15 m), QuickBird, Ikonos, RapidEye (5 m), Pleiades (0.5 m), or Worldview-3 (0.31 m).

A major consideration for development of EO technology for wildfires and disaster reduction is the extent to which operational users can rely on a continued supply of data. Remote sensing capabilities provide a viable method to offset any loss of spatial information. Indeed, satellite images and data are consistently available and have been of increasing value for improving the ability to delineate and simulate the spatial features of wildfires. Moreover, monitoring the extent of wildfire is best achieved in near arid areas by the vegetation coverage (Dalezios [Ed.], 2017). It is recognized that remote sensing methodologies and techniques can be employed in several aspects of wildfire, such as vulnerability and damage assessment, as well as relief, which involves assistance and/or intervention during or after a wildfire event. Similarly, a potential contribution of remote sensing could be focused on wildfire preparedness or warning, although in many cases remote sensing can make a valuable contribution to disaster prevention.

10.4   REMOTE SENSING IN WILDLAND FUEL MODELING

10.4.1   FUEL TYPES AND MODELS

The possibility of using satellite data for land use/land cover classification, vegetation monitoring, and mapping has been a research and operational subject for years. Satellite remote sensing has been proved to effectively assist in fuel type mapping of large areas with low costs. Both passive and active sensors can be used based on various algorithms with high accuracy, but each method presents both advantages and limitations. The main approach that is widely used is the extraction of the vegetation types and their reclassification into surface fuel models, based on fuel characteristics. Supervised classification, unsupervised classification, principal component analysis, and tasseled cap transformation of medium resolution imagery along with ancillary data have been widely used (van Wagtendonk and Root, 2003; Francesetti et al., 2006; Palaiologou et al., 2013) at low or no cost nowadays. Multitemporal images were used to identify the different fuel types based on their phenology (Chuvieco et al., 2003). Sensors with very high resolution, such as QuickBird and Ikonos, have also been used (Giakoumakis et al., 2002; Arroyo et al., 2006; Gitas et al., 2006; Lasaponara and Lanorte, 2007a; Mallinis et al., 2008) based mainly on object-oriented classification algorithms. However, the major drawback of passive sensors is the fact that they cannot see underneath the canopy and under cloudy conditions. Thus, fuel structural characteristics cannot be quantified for all fuel layers.

Sensors with high temporal and low spatial resolution have also been used in fuel models retrieval. For example, Uyeda et al. (2015) used the Moderate Resolution Imaging Spectroradiometer (MODIS) to study postfire vegetation recovery using a pixel-explicit approach to generate maps of postfire biomass recovery and fuel development; they found that the Normalized Difference Vegetation Index (NDVI) time series reveal signals of biomass accumulation including some noise from precipitation and site variability. Other studies are based on the usage of the advanced spaceborne thermal emission and reflection radiometer (ASTER), showing overall accuracies of more than 90% (Guang-xiong et al., 2007; Lasaponara and Lanorte, 2007b). At another study, Fernández-Manso et al. (2014) used fraction images from linear spectral mixture analysis to estimate above ground biomass (AGB) based on ASTER data. The modeling was based on multiple regression between field measurements as the dependent variable and ASTER spectral bands, fraction images, NDVI, or tasseled cap components as the independent variables. According to their results, the model that includes the green vegetation fraction image, the soil fraction image, and the shade fraction image improved AGB estimation. Spectral mixture analysis methods were also applied in hyperspectral data; Jia et al. (2006) used airborne visible infraRed imaging spectrometer (AVIRIS) data to estimate forest canopy cover, discriminating among the dominant canopy types (Douglas-fir and ponderosa pine). They concluded that spectral mixture analysis approaches offer the capacity to assess important fuel attributes including canopy cover, species composition, and burn severity over montane conifer forests. However, airborne hyperspectral data are available only for a small extent upon request with high costs.

Opposed to passive sensors, active sensors such as SAR and LiDAR can provide significant data for mapping stand biomass due to the fact that they can discriminate the vertical structure of the vegetation. Various studies proved that SAR and LiDAR can describe and quantify the fuel potential of the stands, and they may be useful for estimating surface fuel models, crown bulk densities, and canopy dimensions (Keane, 2015).

Satellite sensors European remote sensing (ERS)-1, Japanese earth resources satellite (JERS)-1, and Radarsat, as well as airborne sensors, have been widely used to estimate fuel characteristics, such as foliar biomass, tree volume, tree height, and canopy closure (Toutin and Amaral, 2000; Austin et al., 2003; Li et al., 2007; Saatchi et al., 2007; Garestier et al., 2008; Huang et al., 2009; Ho Tong Minh et al., 2016; Zhang et al., 2016). Both polarimetric and interferometric measurements of SAR sensors can provide valuable information especially at low frequencies in which they are sensitive to crown and stem biomass. Moreover, the interferometric measurements in conjunction with allometric equations can provide forest height parameters (Saatchi and Moghaddam, 2000). It should be noted that fuel retrieval based on SAR signal depends on the bands (i.e. frequencies) and their polarization (Saatchi et al., 2007) (Figure 10.1).

The most promising results of fuel properties mapping are provided by airborne LiDAR. LiDAR data consist of a discrete point measurement of ranges, that is, distance between the sensor and the target. From these measurements, one can calculate elevations coupled with the strength of the return signal; and the fuel strata can be described in three dimensions (Figure 10.2). Many authors explored the usage of LiDAR (Riaño et al., 2003, 2004; Andersen et al., 2005; Popescu and Zhao, 2008; Hermosilla et al., 2014), while the first forest applications of airborne laser have been reported 30 years ago (Nelson et al., 1984). During the last decade, LiDAR data have been extensively used in conjunction with multispectral data. Erdody and Moskal (2010) used LiDAR and near-infrared imagery as explanatory variables in regression models to correlate them with field canopy fuel metrics. According to their results, LiDAR data presented strong relationship with field data, whereas the addition of the near-infrared increased the accuracy only by 3%–4%. Therefore, the cost of very high-resolution imagery is not worthy unless other usage of this imagery is performed at the same time; for example, for vegetation species and health mapping. A small increase of the accuracy was also found by Ruiz et al. (2016), who compared data fusions of low-density LiDAR, WorldView-2, and the new and cost-free Sentinel-2. They used an object-based classification with field data. Four different models were developed based on the different dataset combinations. All models could

FIGURE 10.1   Comparison of the penetration depth for different SAR bands.

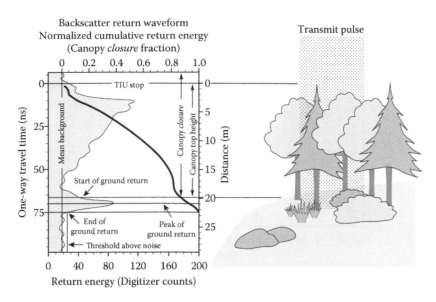

FIGURE 10.2 Forest fuel stratification from LiDAR pulse. (From Harding, D.J. et al., *Remote Sens. Environ.*, 76, 283–297, 2001. With permission.)

discriminate the different fuel types with an overall accuracy of almost 90%. The model that uses only the LiDAR data has shown the lower accuracy, whereas it had been only 3%–4% lower than using all the data together. However, if the user wants to discriminate specific classes with the minimum confidence level, then the fusion of LiDAR and a very high-resolution image of WorldView-2 is essential. It should be mentioned that the usage of only a multispectral imagery presented an overall accuracy level below 70%. Except the cost-free Sentinel-2, another low-cost approach was proposed by Hudak et al. (2016) based on LiDAR and Landsat. Marino et al. (2016) also used low-density LiDAR data (1 pulse/m²) and cost-free Landsat-8 operational land imager (OLI). They applied a random forest classifier and obtained an overall accuracy of 82%, when they used the standard Northern Forest Fire Laboratory fuel modeling scheme as target and obtained an overall accuracy of 70%, when they used site-specific fuel models. Kramer et al. (2016) investigated a new method for predicting ladder fuel levels in the field using photographs with a calibration banner and remotely-sensed data using LiDAR.

10.4.2 Fuel Moisture Content

The significant influence of the FMC in wildfires has been recognized by wildfire managers and scientists (Pollet and Brown, 2007). Ignition and behavior of wildfires have great sensitivity in FMC, which is a key parameter in risk assessment (Vejmelka et al., 2016). In wildfires, it is critical to know both live and dead FMCs (Danson and Bowyer, 2004). Live fuel moisture (LFM) is based on the process of transpiration and soil water dynamics, whereas dead fuel moistures are influenced by the process of evaporation. The most common technique for the estimation of fuel moisture is the gravimetric sampling, that is, the ratio of weight of the water in the sample or material to the dry weight of the sample (Countryman and Dean, 1979).

The drawbacks of fuel moisture estimation based on field sampling or measurement by remote automatic weather stations can be overcome by using satellite remotely sensed data (Chuvieco et al., 2004). Through remote sensing, investigators can estimate vegetation conditions directly in which water stress affects vegetation electromagnetic behavior (Prosper-Laget et al., 1995) (Figure 10.3). Although remote sensing is considered as an advantage on fuel moisture estimation, satellites on the other hand have their own specifications. For example, satellites such as Landsat, with high spatial resolution, provide precise information of vegetation for fuel types, but they have low temporal resolution unlike satellites such as MODIS, which have a revisit cycle that is less than the one of Landsat

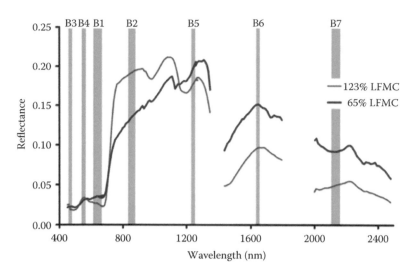

FIGURE 10.3 Reflectance response of different live fuel moisture values from AVIRIS sensor. MODIS bands in gray columns. (From Yebra, M. et al., *Remote Sens. Environ.*, 136, 455–468, 2013. With permission.)

(Verbesselt et al., 2006). As a result, the latter satellites provide regular update of information on vegetation water stress considering that water content temporal frequency can vary significantly (Sow et al., 2013).

Most researchers are based on both optical and thermal parts of the electromagnetic spectrum, using the reflectance properties of vegetation (Sow et al., 2013). Other studies have shown that short-wave infrared (SWIR) is the most sensitive channel to water absorption (Chuvieco, 2009). However, for vegetation analysis, the best approach is the usage of vegetation indices (Sow et al., 2013), and the most common of them is NDVI, when analyzing it at multitemporal series (Alonso et al., 1996). It is assumed that NDVI is sensitive to water content changes because it is based on chlorophyll activity and vegetation vigor (Chuvieco et al., 2004). For example, it is assumed that NDVI and FMC may be related because of the strong correlation between leaf chlorophyll content and leaf moisture content (Ceccato et al., 2001; Pettorelli, 2013). Nevertheless, many researchers disagree about the direct measurement of vegetation water content from NDVI. Although they use vegetation indices as a proxy for indirect water stress detection, the connection between FMC and chlorophyll is not direct because it depends on plant species and their phenological status, atmospheric pollution, nutrient deficiency, toxicity, plant disease, and radiation stress (Ceccato et al., 2001; Dasgupta et al., 2005). Usually, the approach to retrieve FMC though NDVI is done when estimating live FMC; but measuring dead FMC through remote sensing is difficult, because dead fuels do not have variation in chlorophyll of leaves, whereas weather influences water variations (Verbesselt et al., 2006). On the other hand, plant drying causes a decrease in Leaf Area Index and chlorophyll content, so FMC could be indirectly measured as the result of the effects from this procedure (Figure 10.3). Positive linear correlations were reported using real FMC data and satellite-derived NDVI (Jones and Reinke, 2009).

Similar approaches were made from Gao (1996), who used near-infrared (NIR) and SWIR bands to propose NDVI. The basic concept about using NIR is that plants have high reflectance in this part of the spectrum (Jones and Reinke, 2009), and indices using NIR and SWIR estimate water content directly because they follow wavelengths that are related more to water absorption channels. Several other studies (Prosper-Laget et al., 1995; Alonso et al., 1996; Dasgupta et al., 2005; Hong et al., 2007; Sow et al., 2013) extended the use of different indices by analyzing the combination of NDVI and land surface temperature (LST), making temperature measurements significant on vegetation water stress detection. This relationship is based on the spectral reflectance of plant leaves in the red and near-infrared range, combined with the thermal mass of plant leaves relative to soil (Sow et al., 2013). Furthermore, soil moisture and thus NDVI reduction are due to the increase in evapotranspiration caused by temperature rise (Sun and Kafatos, 2007). In addition, there is a negative correlation between NDVI and LST. Water stress can rise the surface temperature rapidly (Wan et al., 2004). A number of studies based on the relation between FMC real data and satellite indices have been proposed to estimate water content and showed good correlations, especially FMC with NDVI and LST (Chuvieco, 2009). For example, Chuvieco et al. (2004) used a regression model combining FMC with NDVI to LST ratio to evaluate the connection between real ground FMC and satellite data. There is a positive correlation, and plant drying reduces chlorophyll activity.

Opposed to optical sensors (infrared and thermal), radars have been rarely used in fuel moisture estimation. SAR backscatter signal is influenced by the moisture in the forest floor, the canopy (including woody parts), and the weather (rainfall) (Leblon, 2005). Most of the SAR applications refer to soil moisture estimation. Initial approaches correlate the backscattering signal with the Canadian Fire Weather Index and weather variables. However, temporal changes of signal are related with foliar moisture content (Leblon et al., 2002). Airborne SAR has also been used for the estimation of LFM. Backscatter intensity and polarimetric decomposition components were linearly correlated to field measurements (Tanase et al., 2015). As a conclusion, forest fuel moisture can be roughly estimated based on SAR backscatter signal because the signal is quite influenced by the biomass, surface roughness, and moisture conditions of the soil and vegetation (Bourgeau-Chavez et al., 2013).

10.5 REMOTE SENSING-BASED FIRE EARLY WARNING SYSTEMS AND MONITORING

10.5.1 Remotely Sensed Fire Early Warning Systems

The development of FEWS is an essential tool in the framework of wildfire prevention and presuppression planning. The type of vegetation is very significant in fire risk mapping and forecasting. For example in Figure 10.4, a fire initiation risk map of Greece based on the vegetation cover is

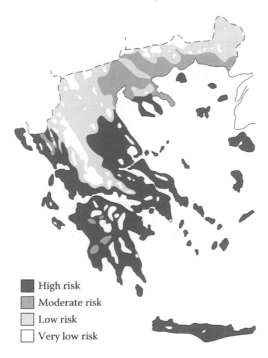

High risk
Moderate risk
Low risk
Very low risk

FIGURE 10.4 Fire initiation risk map for Greece, derived from the vegetation cover. (From Sedaei, L. et al., *Handbook of Drought and Water Scarcity* [*HDWS*], Taylor & Francis Group, Boca Raton, FL, 2017. With permission.)

presented. Moreover, parameters, such as fuel type, fuel moisture, wind, and topography, constitute inputs, among others, to fire danger predicting systems that have been developed for fire prevention and suppression. Two of the most effective operational and widely used systems are the National Fire Danger Rating System (NFDRS) in United States (US Forest Service, 2012) and the Canadian Forest Fire Danger Rating System (CFFDRS) in Canada (Canadian Forest Service, 1992); both rely on remote sensing data. Indeed, the contribution of remote sensing to FEWS in operational systems for use in fire management around the world can be achieved through the estimation of indices used for prefire risk detection models (Leblon et al., 2012). These indices, in combination with meteorological parameters and other fire risk assessment indices, can be employed for forecasting wildfires.

It is recognized that wildfire danger forecasting is one of the most important components of wildfire management (Chowdhury et al., 2015). This subject involves several critical issues, such as (1) current operational wildfire danger forecasting systems and their limitations; (2) remote sensing-based fire danger monitoring systems and operational perspective; (3) remote sensing-based fire danger forecasting systems and their functional implications; and (4) synergy between operational forecasting systems and remote sensing methods. Theoretically, it is possible to overcome the uncertainty associated with the interpolation techniques in point-based measurements of meteorological variables by using remote sensing data. Fire danger condition systems could be broadly classified into two major groups: fire danger forecasting and monitoring systems. Indeed, most of the monitoring systems focus on determining the danger during and/or after the period of image acquisition.

Moreover, fire danger modeling is a key component of such systems. A case study is presented, where eight different spectral vegetation indices are selected to determine the most appropriate index in specific regions of Spain (Bisquert et al., 2014). Specifically, 6 years of MODIS images, along with ground fire data in a 10 × 10 km grid basis, are utilized. The Enhanced Vegetation Index (EVI) provides the best results (Bisquert et al., 2014). Based on these results, a simple fire danger model is established, using logistic regression, by combining the EVI variation with other variables, such as fire history in each cell and period of the year. Another example is the development of a fire risk index in Malaysia, where severe wildfire episodes along with atmospheric pollution and haze, occur as a result of prolonged dry seasons following El-Niňo. Thus, there is a need to develop better systems for effective wildfire management with three objectives: detecting hot spots, developing a fire risk index, and generating spatial analysis for remotely sensed detected fires. In this case, a simple FEWS for wildfire detection in Malaysia is developed (Chowdhury et al., 2015), where thermal bands of MODIS are used to extract hot spot information and to generate a fire risk map, which is also based on MODIS NDVI values.

The Canadian Fire Weather Index (FWI) is one of the most widely known indices and provides numerical rating of relative wildland fire potential in a standard fuel type on level terrains. Essentially, FWI consists of six components that individually and/or collectively account for the effects of fuel moisture and wind on fire behavior. In this example, a procedure is followed that produces operational daily maps of fire danger over Euro-Mediterranean (DaCamara et al., 2014). The maps are based on integrated use of vegetation cover, weather data, and fire activity as detected by satellite remote sensing. Statistical models based on two-parameter generalized Pareto (GP) distributions adequately fit the observed samples of fire duration and are significantly improved when the FWI is integrated as a covariate of scale parameters of GP distributions. Moreover, probabilities of fire duration exceeding specified thresholds are then used to calibrate FWI leading to five classes of fire danger, where fire duration is estimated on the basis of 15 min data provided by Meteosat Second Generation (MSG) satellites and corresponds to the total number of hours in which fire activity is detected in a single MSG pixel in a day. The resulted classes of fire danger provide useful information for wildfire management and are based on the Fire Risk Mapping product that is disseminated on a daily basis by the EUMETSAT Satellite Application Facility on Land Surface Analysis (DaCamara et al., 2014).

Furthermore, a new system is presented, which is called Fire Urgency Estimator in Geosynchronous Orbit (FUEGO) and is based on a small telescope. A small telescope with contemporary detectors and significant computing capacity in geosynchronous orbit can detect small (12 m²) fires on the surface of the earth, cover large areas, such as most of the western United States every few minutes, and attain very good signal-to-noise ratio against Poisson fluctuations in a second. As a result, such a satellite could operate and reject the large number of expected systematic false alarms from a number of sources. It is possible to probe the sensitivity of a fire detection satellite in geosynchronous orbit through a number of algorithms that can help reduce false alarms and show efficacy on a few alarms (Pennypacker et al., 2013). In FUEGO, the framework is developed for a geosynchronous satellite with new imaging detectors, software, and algorithms that can detect heat from early and small fires, and operate on minute-scale detection times.

There is recent research to develop web information systems/platforms for wildfire prevention and management. As an example, the AEGIS platform is based on remote sensing (Kalabokidis et al., 2016). The AEGIS platform assists with early fire warning, fire planning, fire control, and coordination of firefighting forces by providing online access to information that is essential for wildfire management (Figure 10.5). The system uses a number of spatial and nonspatial data sources to support key system functionalities. Specifically, land use/land cover maps are produced by combining field inventory data with high-resolution multispectral satellite images, namely RapidEye. These data support wildfire simulation tools that allow the users to examine potential fire behavior and hazard with the minimum travel time fire spread algorithm. Moreover, the system uses a minimum number of information from end users, such as fire duration, ignition point, and weather information to conduct a fire simulation. Indeed, AEGIS offers three types of simulations, that is, single-fire propagation, point-scale calculation of potential fire behavior, and burn probability analysis, similar to the FlamMap fire behavior modeling software. Furthermore, artificial neural networks (ANNs) are utilized for wildfire ignition risk assessment based on various parameters, training methods, activation functions, preprocessing methods, and network structures. In addition, the combination of ANNs and expected burned area maps are used to generate integrated output map of fire hazard prediction.

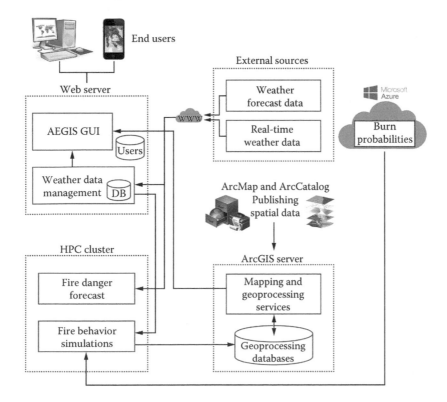

FIGURE 10.5 Architectural components of the AEGIS platform showing the linkages among data and computing resources. (From Kalabokidis, K. et al., *Nat. Hazards Earth Syst. Sci.*, 16, 643–661, 2016. With permission.)

Moreover, the system uses weather information obtained from remote automatic weather stations and weather forecast maps. The system and associated computation algorithms utilize parallel processing methods, such as high-performance computing and cloud computing, which ensure the required computational power for real-time applications. In addition, all AEGIS functionalities are accessible to authorized end users through a web-based graphical user interface. Finally, an innovative smartphone application, AEGIS App, also provides mobile access to the web-based version of the system.

10.5.2 REMOTELY SENSED WILDFIRE DETECTION AND MONITORING

There are a number of automatic and semiautomatic detection and monitoring systems used globally (Alkhatib, 2014). Early detection efforts have focused on early response, accurate results in both daytime and nighttime, and the ability to prioritize fire risk (Ambrosia et al., 1998). Currently, public hotlines, fire lookouts in towers, and ground and aerial patrols can be used for early detection of wildfires. Moreover, near real-time systems have gained ground in recent years as a possible solution to human operator error. Specifically, satellite and aerial monitoring through the use of planes, helicopter, or drones can provide a wider view and may be sufficient to monitor very large and high-risk areas. These more sophisticated systems employ global positioning system (GPS) and infrared or high-resolution visible cameras to identify and target wildfires (ESA, 2006). In addition, satellite-mounted sensors, such as Envisat's Advanced Along Track Scanning Radiometer, can measure infrared radiation emitted by fires, identifying hot spots greater than 39°C. The NOAA's Hazard Mapping System combines remote sensing data from satellite sources, such as GOES, MODIS, and advanced very high resolution radiometer (AVHRR), for detection of fire and smoke plume locations (NOAA, 1998). However, satellite detection is prone to offset errors, anywhere from 2 to 3 km for MODIS and AVHRR data and up to 12 km for GOES data (Ramachandran et al., 2008).

10.5.2.1 Remote Sensing of Wildfire Detection

Remote sensing techniques can be considered fully operational for wildfire detection (Leblon et al., 2012). At local scale, they are mainly based on the use of visible and infrared cameras for the detection of active fires or smoke plumes. Fire detection at this scale is focused on support to wildfire fighting operations. Specifically, active sensors, such as the ERS-1, SAR, and Radarsat, have proven their capacity for monitoring fires under all-weather conditions. Similarly, at large scale, information is provided by geostationary satellite sensors (GOES), spinning enhanced visible and infrared imager (SEVIRI) or sun-synchronous sensors (AVHRR, advanced along-track scanning radiometer (AATSR), MODIS) with operational capabilities (e.g., SEVIRI) for active fire mapping. The high revisit time of the geostationary satellites provides frequent information (15–30 min) that is indicated for monitoring fire processes and fire effects. Moreover, although sun-synchronous satellites provide a lower revisit time (1–2 daily passes), they provide global fire information that is essential for the monitoring of wildfire processes and their effects on ecosystems, the atmosphere, and climate. The fire detection algorithms can be divided into four generic categories: bispectral, threshold, spatial contextual, and multitemporal fire detection methods (Hua and Shao, 2016). At the present time, the Collection 6 active fire detection algorithm data acquired by the MODIS sensor shows improved performance compared to previous version, with reduced omission errors and reduced false alarms (Giglio et al., 2016). Thus, the daily and monthly gridded summaries of fire pixels are used in regional and global climate modeling (Amraoui et al., 2015).

A number of examples and case studies are described on wildfire detection based on remote sensing data and methods. Specifically, a fire identification and detection index is presented, namely the Brightness Temperature Threshold Index (BTTI) (Dalezios [Ed.], 2017), which is given by the following formula:

$$\text{BTTI} = 10 + \left[\frac{(\text{pixelvalue} - 50)}{10} \right] \qquad (10.1)$$

The BTTI is used for the identification of forest-risk areas, or areas, where a forest fire already occurs. The index is based on values of brightness temperature above a critical threshold. The BTTI is computed a few times per day from NOAA/AVHRR satellite digital data with a resolution of 1.1×1.1 km^2. Similarly, the BTTI can be computed from Meteosat satellite digital data for initial fire assessment, because the spatial resolution is 5×5 km^2, although the information is available for the whole Europe every 30 min. Another fire detection index is the Fire Potential Index (FPI), which is based on relative greenness (RG) estimates and measured using the Visible Atmospheric Resistant Index (VARI) derived from MODIS data (Schneider et al., 2008). VARI is selected, instead of NDVI, because it has been shown to have the strongest relationship with LFM out of a wide selection of MODIS-derived indices in shrublands. The results show that VARI-FPI is far better than NDVI-FPI in distinguishing between fire and no-fire events for historical wildfire data, as applied in southern California. In another case study, the Angström's index, along with NDVI and three meteorological parameters, namely temperature, relative humidity, and wind are used for fire risk detection and assessment of a burned area (Spanos et al., 1998). Specifically, the differences in temperature and relative humidity were not significant, although wind speed shows considerable variations. The Angström's index values suggested very favorable and favorable fire conditions. There was a reduction of the NDVI values for the affected pixels by the fire.

The smoke plume produced by the fire can both be detected with satellite imagery and modeled numerically. However, most models designed to study smoke plumes are developed for controlled burns and not wildfires. For wildfire detection, a model is used to compare model simulations with different types of satellite imagery (Kuester et al., 2005). Specifically, application is attempted in the 2003 Aspen Fire in the mountains north of Tucson, Arizona, where analysis of satellite imagery of a wildfire smoke plume is conducted in conjunction with model simulations of this plume. The analysis results indicate that this plume model can be used to adequately simulate the fire plume as depicted in the satellite imagery, when the plume achieves a sufficient altitude. Moreover, the results show that for weak fires and low wind conditions, the plumes often follow the local surface topography.

A comprehensive review of current and potential remote sensing methods is presented, which are used to assess fire behavior and effects and ecological responses to fire, by Lentile et al. (2006). Specifically, remote sensing sensors have been used to estimate features of active fires, to map area burned, and to assess postfire ecological effects. Indeed, uncertainties about fire severity, burn severity, and related aspects can lead to potential misuse of the inferred information for fire management. In addition, the development and interpretation of remote sensing products are assessed, including potentials and limitations of a variety of approaches for monitoring of active fires and their postfire ecological effects, as well as challenges and future perspectives.

Furthermore, it is recognized that thermal remote sensing is widely used in the detection, study, and management of biomass burning occurring in open vegetation fires (Wooster et al., 2013). It is also mentioned that such fires may be planned for land management, may occur as a result of accidental ignition by humans, or may result from lightning or even from other natural phenomena. Indeed, vegetation fires are closely related to high temperatures, which means that thermal remote sensing is the appropriate tool for their identification. The theoretical basis of the key approaches used involves: (1) detection of actively burning fires, (2) characterization of subpixel fires, and (3) estimation of fuel consumption and smoke emissions. Specifically, the types of thermal remote sensing methods indicate how operational fire management has benefited from the resulting information. Moreover, the significance and magnitude of biomass burning, both within global and regional scales, justify the importance of thermal remote sensing to the study and management of open vegetation burning.

10.5.2.2 Remote Sensing of Wildfire Monitoring

The operational monitoring and mapping of the burning areas are very important aspects in dealing with emergency situations and the quantitative estimation of the affected area. By directly observing the plant's radiometric response, it is possible to record the canopy reaction to environmental stresses and constrains directly and in real time. Methods usually applied are based on the thermal signal generated by flaming or smoldering combustion, and the daily fire growth (Chuvieco, 2009).

The use of remotely sensed contextual algorithms can potentially improve the detection of active fires, as compared to simple thresholding algorithms. At the present time, the data acquired by the MODIS sensor have become the standard for fire monitoring at regional to global scales and are used for environmental policy and decision-making.

A program of active fire mapping is briefly presented, namely the design and development of a rapid response system (RRS) for mapping and monitoring wildfires for the entire United States twice daily with MODIS (Quayle, 2002). The system provides an automated, rapid, and reliable approach for (1) daily acquisition and processing of remotely sensed data for active fire detection; (2) processing, analysis, and mapping of active fire detection data, and the production of other related products; and (3) distributing wildfire mapping products via the Internet. Moreover, enhancements for the RRS have included the following: (1) integrating Aqua MODIS data to provide an additional view of fire conditions approximately 2–3 h after each Terra MODIS pass, (2) providing cartographic enhancements to current map products, (3) integrating MODIS fire detections and other geospatial data into an interactive Web map interface, and (4) providing additional wildfire geospatial products and information. It is also mentioned that operational use of remotely sensed fire monitoring has been implemented in five African countries, that is, South Africa, Namibia, Botswana, Senegal, and Ethiopia (Flasse et al., 2004).

Several fire properties for monitoring of active wildland and agricultural fires are considered based on observations by operational polar-orbiting and geostationary satellites. Specifically, simulations of synthetic remote sensing pixels comprised of observed high-resolution fire data, along with ash or vegetation background, demonstrate that fire properties including flame temperature, fractional area, and radiant-energy flux, can best be estimated from concurrent radiance measurements at wavelengths near 1.6, 3.9, and 12 μm (Riggan et al., 2000). Indeed, successful observations at night may be made at scales to at least 1 km for the cluster of fire data simulated. Moreover, during the daytime, uncertainty in the composition of the background and its reflection of solar radiation would limit successful observations to a scale of 100 m or less. Nevertheless, measurements at the three wavelengths in the long-wave infrared would be unaffected by reflected solar radiation and could be applied to separate flame properties in a binary system of flame and background.

Moreover, a project is briefly presented on the fuel treatment effectiveness of wildland–urban interface (WUI). The WUI fuel treatments were designed to increase fire fighter safety, to protect people and property, and to mitigate severe fire effects on natural resources and were proved to be, in general, effective (Hudak et al., 2010). Specifically, a case study was examined for the 2007 East Zone and Cascade wildfires in central Idaho, which were burned through fuel treatments in the WUI surrounding two local communities. Indeed, the analysis results demonstrated that fuel treatment effectiveness can also be assessed remotely, using data from Burned Area Reflectance Classification (BARC) maps that are customarily produced by the United States forest service (USFS) Remote Sensing Applications Center (RSAC) in response to major wildfires in the United States. Moreover, a simple GIS analysis was used to calculate the proportion of high severity burned area and to compare between treatment units and untreated lands in the two study landscapes. It was found that in both landscapes, a higher proportion of untreated lands was burned at high severity than in treated lands. This result was consistent with near real-time postfire BARC mapping indicative of fire severity or after one-year postfire BARC mapping indicative of burn severity.

Two new airborne infrared imaging systems for rapid airborne fire mapping, namely FireMapper and OilMapper, which operate in a *snapshot* mode, are briefly presented. Indeed, both systems delineate the real-time display of single image frames in any selected spectral band (Hoffman et al., 2005). Specifically, these single frames are displayed to operational use with full temperature calibration. Moreover, a rapid *tactical* imaging mode is available for low-level operation, such as lead plane use during wildfire operations. For operational effectiveness, all of the images are tagged with GPS and are recorded on hard drives. Furthermore, in order to support fire and disaster management, the raw image frames are transmitted from the aircraft, via satellite, to a data processing facility, which generates the digital map products and then transmits them electronically back to the incident monitoring in the field (Hoffman et al., 2005). Then, after the individual image frames are mosaicked and orthorectified, they are overlaid on digital map base layers, such as United States geological survey (USGS) Digital Raster Graphics. These two systems show a wide dynamic range (up to 1200°C) without saturating, and the thermal infrared maps are color coded by temperature, resulting in direct interpretation of fire severities and oil spill thicknesses (Stow et al., 2014). Eventually, fully annotated maps are generated and transmitted to the fire incident management for use in the field.

Remote sensing of active fires offers consistent near real-time geospatial information, in terms of wildfire mapping, that has proven useful for strategic operations. However, fire management may require higher spatial resolution products for tactical operations. Indeed, the recently developed 375 m Visible Infrared Imaging Radiometer Suite (VIIRS) active fire detection algorithm improves the current spatial resolution of active fire detection products, showing a higher level of agreement with available airborne data (Schroeder et al., 2014). In addition, the improved spatial sampling of the VIIRS sensor and the 12 h revisiting time produce consistent daily data, allowing multiple observations of fires lasting several days.

10.6   REMOTE SENSING AND POSTFIRE ASSESSMENT

Assessment of the diverse consequences of wildland fires on environment, economy, and society is supported by proper data on fire activity acquired through advanced and powerful monitoring tools. A critical issue that affects fire management is the lack of multiscale spatially explicit information of fire occurrence. Such information summarized in thematic maps of various contents is important in fire management, whereas it is considered as the basis for protecting and restoring fire-affected natural ecosystems worldwide (Koutsias and Karteris, 1998). Spatiotemporal data of fire history help fire scientists and managers to understand the underlying causal factors of fire behavior, to assess the role of climatic anomalies on fire regimes, to interpret postfire vegetation dynamics, and to provide consistent fire statistics (Koutsias et al., 2013).

Available tools to create spatially explicit information on past wildland fire events are restricted to the availability of satellite data, such as the Landsat satellites where available multispectral scanner system (MSS) images exist from 1972 and Thematic Mapper images exist from 1984. Field surveys, aerial photography, and satellite remote sensing, including lately also drones, can be used conceptually for mapping new fire events. Field survey has serious limitations for burned land mapping, because it only provides general statistics due to time and cost restrictions, but it is a highly accurate method at local scales (Koutsias et al., 1999). Aerial photography covers larger geographical areas than field surveys and processes the data with less cost, but still its use in burned land mapping is minimal again mainly due to cost constraints. Remote sensing, especially nowadays when satellites of improved spectral and spatial resolution are in orbit, is an ideal alternative for collecting and processing the required information in a relatively inexpensive and timely fashion (Koutsias and Karteris, 1998). This technology, especially after the free release of Landsat archives from USGS and European space agency (ESA) and the availability also of Sentinel-2 by ESA, can be used to provide data of higher spatial resolutions at global scales, along with periodic spectral data in the visible and infrared part of the electromagnetic spectrum (Koutsias et al., 1999).

During the last 30 years, there was an active period of methods development using advanced techniques that integrate geostatistics and support vector machines and artificial neural networks (Boschetti et al., 2010; Gómez and Martín, 2011; Petropoulos et al., 2011; Mallinis and Koutsias, 2012; Lanorte et al., 2013). Burned area mapping at local, regional, and global scales has achieved very high classification accuracies. Although satellite remote sensing appears to be a suitable approach to map and monitor burned areas compared to others, this method is not free of errors, because there are still various limitations to be resolved.

The postfire spectral signal of burned areas is determined (Pereira et al., 1997) by (1) the deposition of charcoal as the direct result of burning that is a unique consequence of fire burning and

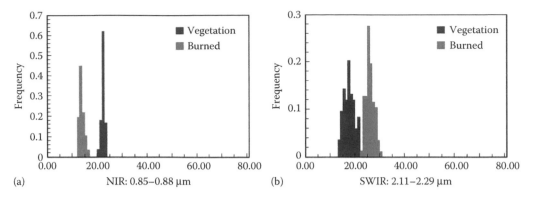

FIGURE 10.6 Histogram data plots of burned and vegetation areas in (a) NIR and (b) SWIR part of the spectrum. (Modified from Koutsias, N. and Pleniou, M., *Int. J. Remote Sens.*, 36, 3714–3732, 2015. With permission.)

(2) the removal of photosynthetic vegetation that may be also caused by other factors than fires (Robinson, 1991). Consequently, the spectral pattern of burned areas is characterized first by a strong decrease in reflectance in NIR region of the spectrum because of the destruction of the leaf cell structure, and second by a strong increase in reflectance in SWIR because of the reduction of water content, which absorbs radiation in this spectral region (Pereira et al., 1997; Koutsias and Karteris, 1998). This particular spectral behavior of burned areas in NIR and SWIR (Figure 10.6) is the basis for the development of the Normalized Burn Ratio (NBR) index (Key and Benson, 2006). NBR is a modification of NDVI by replacing Red with SWIR, which has been proposed by Lopez-Garcia and Caselles (1991) and later verified by Koutsias and Karteris (2000), although the replacement of the Red with SWIR channels has a long history in remote sensing (Ji et al., 2011).

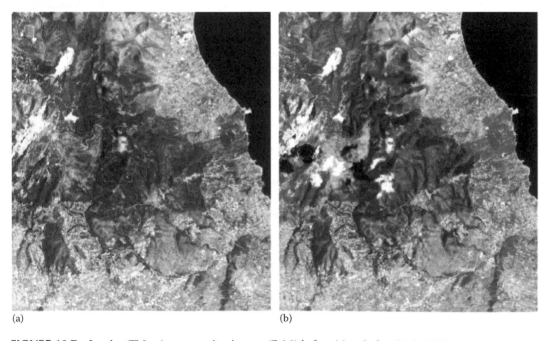

FIGURE 10.7 Landsat TM color composites images (7,4,1) before (a) and after (b) the 1995 wildfire in Penteli, Greece. This band compilation highlights the burned area. In image (a) the Penteli area appears as green brown, and in (b) as light to dark red. (From Domenikiotis, C. et al., *Int. J. Remote Sens.*, 23, 4235–4246, 2002. With permission.)

An additional example is presented for mapping the affected burned forested area of about 63 km² (Domenikiotis et al., 2002). The applied method attempted to assess the agreement of the NDVI, extracted by NOAA/AVHRR and evaluate that by comparing with the NDVI produced with Landsat TM data, delineating the burned areas. Figure 10.7a and b show color composites of the Landsat images before and after the forest fire, respectively. As expected, Landsat TM described the burned area with more details. NDVI abrupt changes before and after the fire were the basis for mapping the extent of burned area and estimating the damage in near real time. The magnitude of such changes depends on the amount of burned area per pixel, the vegetation density, and the dominating species. It should be emphasized that although the overall agreement of both datasets was similar, the Landsat TM was much more accurate when it came to the estimation of the burned areas only.

Spectral overlapping between burned areas and other land cover types exists, and it is responsible for the confusion observed when trying to discriminate the burned areas spectrally. Confusions are observed between the burned areas and (1) water bodies, especially in cases of topographically shadowed areas, recently burned surfaces, mixed land–water, and water–vegetation pixels; (2) urban areas, although this can be eliminated by masking out urban areas; (3) shadows as a result of either irregular terrain, found especially in mountainous areas, or the presence of cloud shadows; (4) slightly burned land and unburned vegetation that is associated mainly with mixed pixels (Koutsias et al., 1999).

Despite any potential limitations, remote sensing technology has been used at global scale to map and monitor burned areas. Global fireproducts, based for example on MODIS (Justice et al., 2002), have been utilized in studies referring mainly to continental scales for characterizing global fire regimes (Chuvieco et al., 2008) or for estimating global biomass burning emissions (Korontzi et al., 2004). Annually resolved fire perimeters based on MODIS data are provided by the European Forest Fire Information System of the European Commission in an effort to provide consistent fire statistics over Europe. Systematic fire products using medium-to-high resolution satellite data (e.g., Landsat) are not common on a global basis, mainly due to cost constraints on gathering and processing medium- or high-resolution satellite data series (Koutsias et al., 2013). The recent developments in informatics technology together with the freely available appropriate

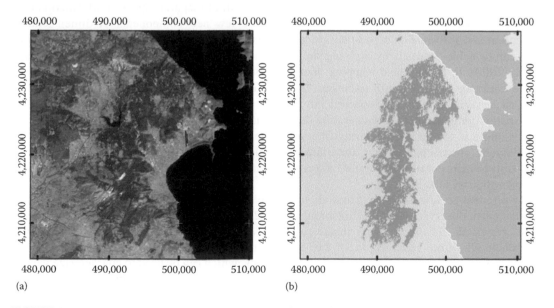

(a)   (b)

FIGURE 10.8   A Landsat image (RGB: 742) showing a fire scar in red color (a) and the result of mapping the burned area (b). (Modified from Pleniou, M. et al., *J. Maps*, 8, 499–506, 2012. With permission.)

satellite data (e.g., Landsat, Sentinel-2) offer new possibilities to develop such global products (Figure 10.8).

Currently, there is an evident redirection of the research to the development of automatically supported methods that are able to process the satellite data without human interference. Semi- or fully automated methods are not very common in studies referring to local or regional scales in which Landsat has been extensively applied to monitor burned areas. A method based on a two-phased algorithm that automatically maps burned areas has been developed by Bastarrika et al. (2014), although a training phase is used to improve classification accuracy. In addition, a rule-based semiautomatic method to map burned areas and to reconstruct recent fire history has been developed by Koutsias et al. (2013). Automatic or semiautomatic techniques minimize human intervention and allow the algorithms to be applied to a series of satellite images in which hundreds of images might be used in the processing chain. Such methods enhance objectivity and reduce the required processing time by eliminating the time-consuming face of the classifier training (Hirschmugl et al., 2017). An example of time series MODIS satellite data (MYD09A1, MODIS/AQUA 8-day L3, RGB-721) from the northwest part of the Peloponnese suffered from the 2007 fires is presented in Figure 10.9.

Koutsias et al. (1999) and Nioti et al. (2011) summarized and discussed the main findings about some strategies followed in burned land mapping studies, such as (1) the use of multitemporal ver- sus single-date satellite data, (2) the use of multispectral transformations with emphasis on princi- pal component analysis, and (3) the application of postclassification processing by a 3 × 3 majority filter. For the multi- versus single-date satellite images, research findings demonstrated that meth- ods using a multitemporal dataset are more effective than those using only a single postfire image. In the multidate approach, the confusions between burned areas and other land cover types with similar spectral behavior that remain unchanged are minimized (Koutsias et al., 1999). However in multitemporal data, radiometric and geometric misregistration may result in under- or overesti- mation of the burned areas. Mallinis and Koutsias (2012) observed that the majority of the errors are distributed in the borders between burned and unburned vegetation in which mixed pixels are found. In that sense, problems can be enhanced when geometric misregistration is considered.

The multispectral character of the satellite data promotes the application of various multi- variate statistical methods, mainly methods dealing with dimensionality reduction. Specifically, principal component analysis and/or vegetation indices aim to separate spectral information (dis- tributed in the original spectral channels) into those few new components. If dimensionality reduction and information separation are accomplished successfully, then the desired informa- tion can be achieved easily by applying simple further processing such as thresholding (Koutsias et al., 2000). In burned land mapping, methods considering the spatial information of the sat- ellite data can be very useful (Koutsias, 2003). Purely pixel-based multispectral classification approaches only consider the spectral information at pixel basis; these approaches do not take the spatial component into consideration defined mainly by the surrounding neighbor pixels, as for example, a simple 3 × 3 majority filter at the postprocessing level or an autocovariate in the autologistic approach. However, there are other more advanced and sophisticated approaches available to incorporate spatial information into digital classification either during postprocessing or prior to pixel labeling.

In a recent review paper on "methods for mapping forest disturbance and degradation from optical earth observation data" (Hirschmugl et al., 2017), two main change detection approaches were reported: (1) the classical image-to-image approach and (2) the time series analysis approach. The time series approach can be very useful to create temporal profiles extracted from the spectral signal of time series satellite images, which can be used to characterize vegeta- tion phenology, and thus to be helpful for monitoring vegetation recovery in fire-affected areas (Figure 10.10). Vegetation phenology is an important element of vegetation characteristics that can be useful in vegetation monitoring, especially when satellite remote sensing observations are used.

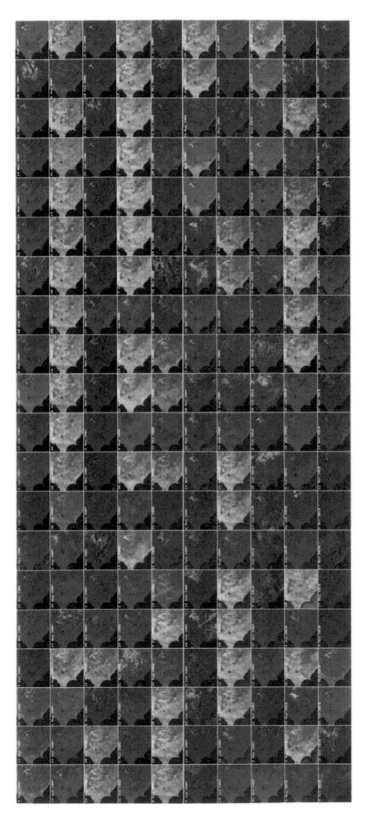

FIGURE 10.9 Time series MODIS satellite data (MYD09A1, MODIS/AQUA 8-day L3, RGB-721) from the northwest part of the Peloponnese, Greece, suffered from the 2007 fires.

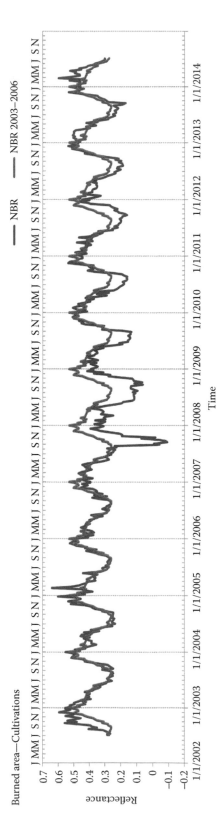

FIGURE 10.10 Time series NBR index of a 2007 fire affected area in the Peloponnese, Greece. Vegetation recovery can be assessed by monitoring vegetation phenology from remote sensing observations.

10.7 SUMMARY

In this chapter, an overview of the remote sensing, along with potentials and capabilities in wildfire hazard analysis, has been presented. Remote sensing data and methods provide direct measurements of land characteristics, vegetative cover, and components of the hydrological cycle, namely temperature, rainfall, evapotranspiration, and soil moisture, just to mention a few. The international contribution of remote sensing to FEWS for use in fire management has been recognized and can be achieved through indices used in prefire risk detection modes that can be considered fully operational. Moreover, the operational monitoring and mapping of the burning areas are very important aspects in dealing with emergency situations and the quantitative estimation of the affected areas. Furthermore, satellite remote sensing proved to effectively assist in fuel type mapping of large areas with low costs. Both passive and active sensors can be used based on various algorithms with high accuracy. In addition, the significant influence of the FMC in wildfires has been recognized. Specifically, LFM is based on the process of transpiration and soil water dynamics, whereas dead fuel moistures are influenced by the process of evaporation. Remote sensing is considered as an advantage on fuel moisture, because vegetation conditions can be directly estimated in which water stress affects vegetation electromagnetic behavior. Burned area mapping at local, regional, and global scales has achieved very high classification accuracies. In addition, temporal profiles extracted from the spectral signal of time series satellite images can be used to characterize vegetation phenology, and thus can be used for monitoring vegetation recovery in fire-affected areas.

The last 30 years was an active period of method development including advanced techniques that integrate geostatistics and support vector machines and artificial neural networks. At the present time, research is being redirected toward the development of automatically supported methods that are able to process the satellite data without human interference. Semi- or fully automated methods are not very common in studies referring to local or regional scales, where Landsat has been extensively applied to monitor burned areas. Nevertheless, the future outlook of remotely sensed wildfire analysis is promising, because there is diachronically significant progress and steadily increasing reliability of remote sensing data and methods that are based on scientific as well as technological and computational advancements year by year. Moreover, the number of satellite systems is continuously increasing with an improvement of their spatial and temporal resolution. The current trend is also to increase the number of available bands in these satellites, leading to additional and useful information. Finally, there is a new challenge due to the fact that new types of remote sensing systems offer online open information for web platforms that are also employed in wildfire monitoring and detecting processes.

REFERENCES

Alkhatib, A.A.A. 2014. A review on forest fire detection techniques. Hindawi Publishing Corporation. *International Journal of Distributed Sensor Networks*, 2014, Article ID 597368, 12p. http://dx.doi.org/10.1155/2014/597368.

Alonso, M., Camarasa, A., Chuvieco, E., Cocero, D., Kyun, I., Martin, M.P., Salas, F.J. 1996. Estimating temporal dynamics of fuel moisture content Mediterranean species from NOAA-AVHRR data. *EARSeL Advances in Remote Sensing*, 4(4), 9–24.

Ambrosia, V.G., Buechel, S.W., Brass, J.A., Peterson, J.R., Davies, R.H., Kane, R.J., Spain, S. 1998. An integration of remote sensing, GIS, and information distribution for wildfire detection and management. October 1998. *Photogrammetric Engineering and Remote Sensing*, 64, 977–985.

Amraoui, M., Pereira, M.G., DaCamara, C.C., Calado, T.J. 2015. Atmospheric conditions associated with extreme fire activity in the western Mediterranean region. *Science of the Total Environment*, 524–525, 32–39.

Andersen, H.E., McGaughey, R.J., Reutebuch, S.E. 2005. Estimating forest canopy fuel parameters using LiDAR data. *Remote Sensing of Environment*, 94, 441–449.

Arroyo, L.A., Healey, S.P., Cohen, W.B., Cocero, D. 2006. Using object-oriented classification and high-resolution imagery to map fuel types in a Mediterranean region. *Journal of Geophysical Research: Biogeosciences*, 111, G04S04. doi:10.1029/2005JG000120.

Austin, J.M., Mackey, B.G., Van Niel, K.P. 2003. Estimating forest biomass using satellite radar: An explor-atory study in a temperate Australian Eucalyptus forest. *Forest Ecology and Management*, 176, 575–583.

Bastarrika, A., Alvarado, M., Artano, K., Martinez, M., Mesanza, A., Torre, L., Ramo, R., and Chuvieco, E. 2014. BAMS: A tool for supervised burned area mapping using landsat data. *Remote Sensing*, 6, 12360–12380.

Bisquert, M., Sánchez J.M., and Caselles, V. 2014. Modeling fire danger in Galicia and Asturias (Spain) from MODIS images. *Remote Sensing*, 6, 540–554. doi:10.3390/rs6010540.

Borealforest.org. 2016. Forest fire suppression. http://www.borealforest.org/world/innova/fire_suppression.htm

Boschetti, M., Stroppiana, D., and Brivio, P.A. 2010. Mapping burned areas in a Mediterranean environment using soft integration of spectral indices from high-resolution satellite images. *Earth Interactions*, 14, 1–20.

Bourgeau-Chavez, L.L., Leblon, B., Charbonneau, F., Buckley, J.R. 2013. Assessment of polarimetric SAR data for discrimination between wet versus dry soil moisture conditions. *International Journal of Remote Sensing*, 34(16), 5709–5730.

Canadian Forest Service. 1992. Development and Structure of the Canadian Forest Fire Behaviour Prediction System. Canadian Forest Service, Information Report ST-X-3, Ottawa, ONT., 63 p.

Ceccato, P., Flasse, S., Tarantola, S., Jacquemoud, S., Gregoirea, J.M. 2001. Detecting vegetation leaf water content using reflectance in the optical domain. *Remote Sensing of Environment*, 77, 22–33.

Chowdhury, E.H., Hassan, Q.K. 2015. Operational perspective of remote sensing-based forest fire danger fore-casting systems. *ISPRS Journal of Photogrammetry and Remote Sensing*, 104, 224–236.

Chuvieco, E. 2009. Earth Observation of Wildland Fires in Mediterranean Ecosystems. Dordrecht, the Netherlands: Springer, 87 p.

Chuvieco, E., Cocero, D., Riaño, D., Martin P., Martinez-Vega, J., De la Riva, J., Perez, F. 2004. Combining NDVI and surface temperature for the estimation of live fuel moisture content in forest fire danger rating. *RS of Environment*, 92, 322–331.

Chuvieco, E., Giglio, L., and Justice, C. 2008. Global characterization of fire activity: Toward defining fire regimes from Earth observation data. *Global Change Biology*, 14, 1488–1502.

Chuvieco, E., Riano, D., van Wagtendonk, J., and Morsdof, F. 2003. Fuel loads and fuel type mapping. In E. Chuvieco (Ed.), *Wildland Fire Danger Estimation and Mapping: The Role of Remote Sensing Data*. Singapore: World Scientific, pp. 119–142.

Countryman, C.M. and Dean, W.A. 1979. Measuring moisture content in living chaparral: a field user's manual. USDA Forest Service General Technical Report PSW-036, 28 p.

DaCamara, C.C., Calado, T.J., ErmidaA, S.L., Trigo, I.F., Amraoui, M., and Turkman, K.F. 2014. Calibration of the fire weather index over Mediterranean Europe based on fire activity retrieved from MSG satellite imagery. *International Journal of Wildland Fire*, 23(7), 945–958.

Dalezios, N.R. (Ed.) 2017. *Environmental Hazards Methodologies for Risk Assessment and Management*. Publisher IWA, London UK, 525 pages.

Danson, F.M. and Bowyer, P. 2004. Estimating live fuel moisture content from remotely sensed reflectance. *Remote Sensing of Environment*, 92(3), 309–321.

Dasgupta, S., Qu, J.J., and Hao, X. 2005. Evaluating remotely sensed live fuel moisture estimations for fire behavior predictions. *EastFIRE Conference 2005*. Fairfax, VA: George Mason University.

Domenikiotis, C., Dalezios, N.R., Loukas, A., and Karteris, M. 2002. Agreement assessment of NOAA/AVHRR NDVI with Landsat TM NDVI for mapping burned forested areas. *International Journal of Remote Sensing*, 23, 4235–4246.

Erdody, T.L. and Moskal, L.M. 2010. Fusion of LiDAR and imagery for estimating forest canopy fuels. *Remote Sensing of Environment*, 114(4), 725–737.

ESA. 2004. Satellites are tracing Europe's forest fire scars. http://www.esa.int/Our_Activities/Observing_the_Earth/Satellites_are_tracing_Europe_s_forest_fire_scars

ESA. 2006. Airborne campaign tests new instrumentation for wildfire detection. http://www.esa.int/Our_Activities/Observing_the_Earth/The_Living_Planet_Programme/Airborne_campaign_tests_new_instrumentation_for_wildfire_detection

ESA. 2014. Sentinel, Earth online–ESA, https://earth.esa.int/web/guest/missions/esa-future-missions/sentinel-1, [Accessed: April 6, 2014].

Fernández-Manso, O., Fernández-Manso, A., and Quintano, C. 2014. Estimation of aboveground biomass in mediterranean forestsby statistical modelling of ASTER fraction images. *International Journal of Applied Earth Observation and Geoinformation*, 31(1), 45–56.

Flasse, S.I., Trigg, S.N., Ceccato, P.N., Perryman, A.H., Hudak, A.T., Thompson, M.W., Brockett, B. H. et al. 2004. Remote sensing of vegetation fires and its contribution to a fire management information system (Chapter 8). *Wildland Fire Management Handbook for Sub-Sahara Africa*, pp. 158–211.

Francesetti, A., Camia, A., and Bovio, G. 2006. Fuel type mapping with Landsat TM images and ancillary data in the Prealpine region of Italy. *Forest Ecology and Management*, 234, S259.

Gao, B.C. 1996. NDWI-A normalized difference water index for remote sensing of vegetation liquid water from space. *Remote Sensing of Environment*, 58, 257–266.

Garestier, F., Dubois-Fernandez, P.C., and Papathanassiou, K.P. 2008. Pine forest height inversion using single-pass X-band PolInSAR data. *IEEE Transactions on Geoscience and Remote Sensing*, 46, 59–68.

Giakoumakis, N.M., Gitas, I.Z., and San-Miguel, J. 2002. Object-oriented classification modelling for fuel type mapping in the Mediterranean, using LANDSAT TM and IKONOS imagery—Preliminary results. In: D.X. Viegas (Eds.), *Forest Fires Research & Wildland Fire Safety*. Rotterdam, the Netherlands: Millpress.

Giglio, L., Schroeder, W., and Justice, C.O. 2016. The collection 6 MODIS active fire detection algorithm and fire products. *Remote Sensing of Environment*, 178, 31–41.

Gitas, I.Z., Mitri, G.H., Kazakis, G., Ghosn, D., and Xanthopoulos, G. 2006. Fuel type mapping in Annapolis, Crete by employing QuickBird imagery and object-based classification. *Forest Ecology and Management* 234(S1), S228. doi:10.1016/j.foreco. 2005.08.255.

Gómez, I. and Martín, M.P. 2011. Prototyping an artificial neural network for burned area mapping on a regional scale in Mediterranean areas using MODIS images. *International Journal of Applied Earth Observation and Geoinformation*, 13, 741–752.

Guang-xiong, P., Jing, L., Yun-hao, C., and Abdul-patah, N. 2007. A forest fire risk assessment using ASTER images in peninsular Malaysia. *Journal of China University of Mining & Technology*, 17, 232–237.

Harding, D.J., Lefsky, M.A., Parker, G.G., and Blair, J.B. 2001. Laser altimeter canopy height profiles methods and validation for closed-canopy, broadleaf forests. *Remote Sensing of Environment*, 76(3), 283–297.

Hermosilla, T., Ruiz, L.A., Kazakova, A.N., Coops, N.C., and Moskal, L.M. 2014. Estimation of forest structure and canopy fuel parameters from small-footprint full-waveform LiDAR data. *International Journal of Wildland Fire*. 23, 224–233.

Hirschmugl, M., Gallaun, H., Dees, M., Datta, P., Deutscher, J., Koutsias, N., and Schardt, M. 2017. Methods for mapping forest disturbance and degradation from optical earth observation data: A review. *Current Forestry Reports*, 3, 1–14.

Ho Tong Minh, D., Le Toan, T., Rocca, F., Tebaldini, S., Villard, L., Réjou-Méchain, M., Phillips, O.L. et al. 2016. SAR tomography for the retrieval of forest biomass and height: Cross-validation at two tropical forest sites in French Guiana. *Remote Sensing of Environment*, 175, 138–147.

Hoffman, J.W., Coulter, L.L., Luciani E.M., and Riggan P.J. 2005. Rapid turn-around mapping of wildfires and disasters with airborne infrared imagery from the new firemapper 2.0 and oilmapper systems. *ASPRS Annual Conference: Geospatial Goes Global: From Your Neighborhood to the Whole Planet*, March 7–11, Baltimore, MD, 8p.

Hong, S., Lakshmi, V., and Small, E.E. 2007. Relationship between vegetation biophysical properties and surface temperature using multisensor satellite data. *Journal of Climate*, 20, 5593–5606.

Hua, L. and Shao, G. 2016. The progress of operational forest fire monitoring with infrared remote sensing. *Journal of Forestry Research*, 1–15. doi:10.1007/s11676-016-0361-8.

Huang, S., Crabtree, R.L., Potter, C., and Gross, P. 2009. Estimating the quantity and quality of coarse woody debris in Yellowstone post-fire forest ecosystem from fusion of SAR and optical data. *Remote Sensing of Environment*, 113(9), 1926–1938.

Hudak, A.T., Bright, B.C., Pokswinski, S.M., Loudermilk, E.L., O'Brien, J.J., Hornsby, B.S., Klauberg, C., and Silva, C.A. 2016, Mapping forest structure and composition from low-density LiDAR for informed forest, fuel, and fire management at Eglin Air Force Base, Florida, USA. *Canadian Journal of Remote Sensing*, 42(5), 411–427.

Hudak, A.T., Jain, T.B., Morgan, P., and Clark J.T. 2010. Remote sensing of WUI fuel treatment effectiveness following the 2007 wildfires in central Idaho. *Proceedings of 3rd Fire Behavior and Fuels Conference*, October 25–29, Spokane, Washington, USA. International Association of Wildland Fire, Birmingham, AL, pp. 1–11.

IPCC. 2012. Managing the risks of extreme events and disasters to advance climate change adaptation. A special report of working groups I and II of the intergovernmental panel on climate change [Field, C.B., V. Barros, T.F. Stocker, D. Qin, D.J. Dokken, K.L. Ebi, M.D. Mastrandrea, K.J. et al. (Eds.)]. Cambridge University Press, Cambridge, UK, and New York, 582 pp.

Ji, L., Zhang, L., Wylie, B.K., and Rover, J. 2011. On the terminology of the spectral vegetation index (NIR-SWIR)/(NIR+SWIR). *International Journal of Remote Sensing*, 32, 6901–6909.

Jia, G.J., Burke, I.C., Kaufmann, M.R., Goetz, A.F.H., Kindel, B.C., and Pu, Y. 2006. Estimates of forest canopy fuel attributes using hyperspectral data. *Forest Ecology and Management*, 229, 27–38.

Jones, S., Reinke, K. 2009. *Innovations in Remote Sensing and Photogrammetry.* Lecture Notes in Geoinformation and Cartography. Berlin, Germany: Springer, 215 p.

Justice, C.O., Giglio, L., Korontzi, S., Owens, J., Morisette, J.T., Roy, D., Descloitres, J., Alleaume, S., Petitcolin, F., and Kaufman, Y. 2002. The MODIS fire products. *Remote Sensing of Environment,* 83, 244–262.

Kailidis, D.S. 1990. *Forest Fires,* 3rd ed. Thessaloniki, Greece: Giahoudi Giapouli Publications, 510 p.

Kalabokidis, K., Ager, A., Finney, M., Athanasis, N., Palaiologou, P., and Vasilakos, C. 2016. AEGIS: A wildfire prevention and management information system. *Natural Hazards Earth System Sciences,* 16, 643–661.

Kalabokidis, K., Iliopoulos, N., Gliglinos, D. 2013. *Fire Meteorology and Forest Fire Behavior in a Changing Climate.* Athens, Greece: ION Publications. 400 p.

Keane, R.E. 2015. *Wildland Fuel Fundamentals and Applications.* Switzerland: Springer International Publishing.

Key, C. H. and Benson, N.C. 2006. Landscape assessment: Ground measure of severity, the Composite Burn Index; and remote sensing of severity, the Normalized Burn Ratio. In D.C. Lutes, R.E. Keane, J.F. Caratti, C.H. Key, N.C. Benson, S. Sutherland, and L.J. Gangi (Eds.) *FIREMON: Fire Effects Monitoring and Inventory System.* Ogden, UT: USDA Forest Service, Rocky Mountain Research Station, pp. 1–51.

Korontzi, S., Roy, D.P., Justice, C.O., and Ward, D.E. 2004. Modeling and sensitivity analysis of fire emissions in southern Africa during SAFARI 2000. *Remote Sensing of Environment,* 92, 255–275.

Koutsias, N. 2003. An autologistic regression model for increasing the accuracy of burned surface mapping using Landsat Thematic Mapper data. *International Journal of Remote Sensing,* 24, 2199–2204.

Koutsias, N. and Karteris, M. 1998. Logistic regression modelling of multitemporal Thematic Mapper data for burned area mapping. *International Journal of Remote Sensing,* 19, 3499–3514.

Koutsias, N. and Karteris, M. 2000. Burned area mapping using logistic regression modeling of a single post-fire Landsat-5 Thematic Mapper image. *International Journal of Remote Sensing,* 21:673–687.

Koutsias, N. and Pleniou, M. 2015. Comparing the spectral signal of burned surfaces between Landsat-7 ETM+ and Landsat-8 OLI sensors. *International Journal of Remote Sensing,* 36, 3714–3732.

Koutsias, N., Karteris, M., Fernandez-Palacios, A., Navarro, C., Jurado, J., Navarro, R., and Lobo, A. 1999. Burned land mapping at local scale. In E. Chuvieco (Ed.) *Remote Sensing of Large Wildfires in the European Mediterranean Basin.* Berlin, Germany: Springer-Verlag, pp. 15–187.

Koutsias, N., Karteris, M., and Chuvieco, E. 2000. The use of Intensity-Hue-Saturation transformation of Landsat-5 Thematic Mapper data for burned land mapping. *Photogrammetric Engineering & Remote Sensing,* 66, 829–839.

Koutsias, N., Pleniou, M., Mallinis, G., Nioti, F., and Sifakis, N.I. 2013. A rule-based semi-automatic method to map burned areas: Exploring the USGS historical Landsat archives to reconstruct recent fire history. *International Journal of Remote Sensing* 34, 7049–7068.

Kramer, H.A., Collins, B.M., Lake, F.K., Jakubowski, M.K., Stephens, S.L., and Kelly, M. 2016. Estimating ladder fuels: A new approach combining field photography with LiDAR. *Remote Sensing,* 8, 766.

Kuester, M.A., Marshall J., and Emery, W.J. 2005. Remote sensing and modeling of wildfires. *IEEE Transactions on Geoscience and Remote Sensing,* 43, 5729–5732.

Lanorte, A., Danese, M., Lasaponara, R., and Murgante, B. 2013. Multiscale mapping of burn area and severity using multisensor satellite data and spatial autocorrelation analysis. *International Journal of Applied Earth Observation and Geoinformation,* 20, 42–51.

Lasaponara, R. and Lanorte, A. 2007a. On the capability of satellite VHR QuickBird data for fuel type characterization in fragmented landscape. *Ecological Modelling,* 204, 79–84.

Lasaponara, R. and Lanorte, A. 2007b. Remotely sensed characterization of forest fuel types by using satellite ASTER data. *International Journal of Applied Earth Observation and Geoinformation,* 9(3), 225–234.

Leblon, B. 2005. Monitoring forest fire danger with remote sensing. *Natural Hazards,* 35(3), 343–359.

Leblon, B., Bourgeau-Chavez, L., and San-Miguel-Ayanz, J. 2012. Use of remote sensing in wildfire management (Chapter 3). *Sustainable Development—Authoritative and Leading Edge Content for Environmental Management.* Croatia, Rijeka: INTECH, 55–82.

Leblon, B., Kasischke, E., Alexander, M., Doyle, M., and Abbott, M. 2002. Fire danger monitoring using ERS-1 SAR images in the case of northern boreal forests. *Natural Hazards,* 27(3), 231–255.

Lentile, L.B., Holden, Z.A., Smith, A.M.S., Falkowski, M.J., Hudak, A.T., Morgan, P., Lewis, S.A., Gessler P.E., and Benson, N.C. 2006. Remote sensing techniques to assess active fire characteristics and post-fire effects. *International Journal of Wildland Fire,* 15, 319–345.

Li, X., Yeh, A.G.O., Wang, S., Liu, K., Liu, X., Qian, J., and Chen, X. 2007. Regression and analytical models for estimating mangrove wetland biomass in South China using Radarsat images. *International Journal of Remote Sensing,* 28, 5567–5582.

Lopez Garcia, M. J. and Caselles, V. 1991. Mapping burns and natural reforestation using Thematic Mapper data. *Geocarto International,* 6, 31–37.

Mallinis, G. and Koutsias, N. 2012. Comparing ten classification methods for burned area mapping in a Mediterranean environment using Landsat TM satellite data. *International Journal of Remote Sensing*, 33, 4408–4433.

Mallinis, G., Mitsopoulos, I.D., Dimitrakopoulos, A.P., Gitas, I.Z., and Karteris, M. 2008. Local-scale fuel-type mapping and fire behavior prediction by employing high-resolution satellite imagery. *IEEE Journal of Selected Topics in Applied Earth Observations and Remote Sensing*, 1(4), 230–239.

Marino, E., Ranz, P., Tomé, J.L., Noriega, M.Á., Esteban, J., and Madrigal, J. 2016. Generation of high-resolution fuel model maps from discrete airborne laser scanner and Landsat-8 OLI: A low-cost and highly updated methodology for large areas. *Remote Sensing of Environment*, 187, 267–280.

National Oceanic and Atmospheric Administration (NOAA). 1998. Hazard mapping system fire and smoke product. Satellite and information service. Available: http://www.ospo.noaa.gov/Products/land/hms.html.

National Wildfire Coordinating Group (NWCG). 2008. Glossary of wildland fire terminology. Available: http://gacc.nifc.gov/nrcc/dc/idgvc/dispatchforms/glossary.pdf.

Nature Conservancy. 2015. Global fire initiative: Fire and climate change. Available: https://www.conservationgateway.org/ConservationPractices/FireLandscapes/Pages/fire-landscapes.aspx.

Nelson, R., Krabill, W., and Maclean, G. 1984. Determining forest canopy characteristics using airborne laser data. *Remote Sensing and Environment*, 15, 201–212.

Nioti, F., Dimopoulos, P. and Koutsias, N. 2011. Correcting the fire scar perimeter of a 1983 wildfire using USGS archived Landsat satellite data. *GIScience & Remote Sensing*, 48, 600–613.

Omi, P.N. 2005. *Forest Fires: A Reference Handbook*. Contemporary world issues. Santa Barbara, CA: ABC-CLIO.

Palaiologou, P., Kalabokidis, K., and Kyriakidis, P. 2013. Forest mapping by geoinformatics for landscape fire behavior modelling in coastal forests, Greece. *International Journal of Remote Sensing*, 34(12), 4466–4490.

Pennypacker, C.R., Jakubowski, M.K., Kelly, M., Lampton, M., Schmidt, C., Stephens, S., and Tripp, R. 2013. FUEGO—Fire urgency estimator in geosynchronous orbit—A proposed early-warning fire detection system. *Remote Sensing*, 5, 5173–5192.

Pereira, J.M.C., Chuvieco, E., Beaudoin, A., and Desbois, N. 1997. Remote sensing of burned areas: A review. In E. Chuvieco (Ed.) *A Review of Remote Sensing Methods for the Study of Large Wildland Fires*. *Alcala de Henares* Spain: Universidad de Alcala.

Petropoulos, G. P., Kontoes, C., and Keramitsoglou, I. 2011. Burnt area delineation from a uni-temporal perspective based on Landsat TM imagery classification using support vector machines. *International Journal of Applied Earth Observation and Geoinformation*, 13, 70–80.

Pettorelli, N. 2013. *The Normalized Difference Vegetation Index*. Oxford University Press, New York. pp. 1–224.

Pleniou, M., Xystrakis, F., Dimopoulos, P., and Koutsias, N. 2012. Maps of fire occurrence—Spatially explicit reconstruction of recent fire history using satellite remote sensing. *Journal of Maps* 8, 499–506.

Pollet, J. and Brown, A. 2007. *Fuel Moisture Sampling Guide*. Ogden, UT: Utah State Office, Bureau of Land Management, pp. 1–30.

Popescu, S.C. and Zhao, K. 2008. A voxel-based LiDAR method for estimating crown base height for deciduous and pine trees. *Remote Sensing of Environment*, 112, 767–781.

Prosper-Laget, V., Douguedroit, A., and Guinot, J.P. 1995. Mapping the risk of forest fire occurrence using NOAA satellite information. *EARSeL Advances in Remote Sensing*, 4(3), 30–38.

Quayle, B. 2002. Rapid mapping of active wildland fires: Integrating satellite remote sensing, GIS, and internet technologies. Joint interim report, College Park, MD: USDA-NASA-Univerasity of Maryland, 7 p.

Ramachandran, C., Misra, S., and Obaidat, M. S. 2008. A probabilistic zonal approach for swarm-inspired wildfire detection using sensor networks. *International Journal of Communication Systems*, 21(10), 1047–1073. doi:10.1002/dac.937.

Riaño, D., Chuvieco, E., Condés, S., González-Matesanz, J., and Ustin, S.L. 2004. Generation of crown bulk density for Pinus sylvestris L. from LiDAR. *Remote Sensing of Environment*, 92, 345–352.

Riaño, D., Meier, E., Allgöwer, B., Chuvieco, E., and Ustin, S.L. 2003. Modeling airborne laser scanning data for the spatial generation of critical forest parameters in fire behavior modeling. *Remote Sensing of Environment*, 86, 177–186.

Riggan, P.J., Hoffman, J.W., and Brass, J.A. 2000. Estimating fire properties by remote sensing. USDA interim report, 7 p.

Robinson, J.M. 1991. Fire from space: Global fire evaluation using infrared remote sensing. *International Journal of Remote Sensing*, 12, 3–24.

Ruiz, L.Á, Recio, J.A., Crespo-Peremarch, P., Sapena, M. 2016. An object-based approach for mapping forest structural types based on low-density LiDAR and multispectral imagery, *Geocarto International*, 31, 1–15.

Saatchi, S., Halligan, K., Despain, D.G., and Crabtree, R.L. 2007. Estimation of forest fuel load from radar remote sensing. *IEEE Transactions on Geoscience and Remote Sensing*, 45(6), 1726–1740.

Saatchi S. and Moghaddam, M. 2000. Estimation of crown and stem water content and biomass of boreal forest using polarimetric SAR imagery. *IEEE Transactions on Geoscience and Remote Sensing*, 38(2), 697–709.

Schneider, P., Roberts, D.A., and Kyriakidis, P.C. 2008. A VARI-based relative greenness from MODIS data for computing the fire potential index. *Remote Sensing of Environment*, 112, 1151–1167.

Schroeder, W., Oliva, P., Giglio, L., and Csiszar, I. 2014. The new VIIRS 375 m active fire detection data product: Algorithm description and initial assessment. *Remote Sensing of Environment*, 143, 85–96.

Sedaei, L., Sedaei, N., Cox, J.P., Dalezios, N.R., and Eslamian, S. 2017. Forest fire mitigation under water shortage (Chapter 27). In Prof S. Eslamian (Ed.) *Handbook of Drought and Water Scarcity (HDWS)*. Boca Raton, FL: Taylor & Francis Group.

Sow, M., Mbow, C., Hély, C., Fensholt, R., and Sambo, B. 2013. Estimation of herbaceous fuel moisture content using vegetation indices and land surface temperature from MODIS data. *Remote Sensing*, 5, 2617–2638.

Spanos, S.I., Zarpas C.D., and Dalezios N.R. 1998. Meteorological and satellite indices for the assessment of forest fires. *International Symposium on Applied Agrometeorology and Agroclimatology*, Volos, Greece, April 24–26, 1996, Luxembourg: European Commission, pp. 583–588.

Stow, D.A., Riggan, P.J., Storey, E.J. and Coulter, L.L. 2014. Measuring fire spread rates from repeat pass airborne thermal infrared imagery. *Remote Sensing Letters*, 5(9), 803–812.

Sun, D. and Kafatos, M. 2007. Note on the NDVI-LST relationship and the use of temperature-related drought indices over North America. *Geophysical Research Letters*, 34, 1–4.

Tanase, M.A., Panciera, R., Lowell, K., and Aponte, C. 2015. Monitoring live fuel moisture in semiarid environments using L-band radar data. *International Journal of Wildland Fire*, 24, 560–572.

Toutin, T. and Amaral, S. 2000. Stereo RADARSAT data for canopy height in Brazilian forest. *Canadian Journal of Remote Sensing*, 26, 189–199.

US Forest Service. 2012. The Wildland Fire Assessment System. Available: http://www.wfas.net/.

Uyeda, K.A., Stow, D.A., Riggan, P.J. 2015. Tracking MODIS NDVI time series to estimate fuel accumulation. *Remote Sensing Letters*, 6(8), 587–596.

van Wagtendonk, J.W. and Root, R.R. 2003. The use of multi-temporal Landsat Normalized Difference Vegetation Index (NDVI) data for mapping fuel models in Yosemite National Park, USA. *International Journal of Remote Sensing*, 8, 1639–1651.

Vejmelka, M., Kochanski, A.K., and Mandel, J. 2016. Data assimilation of dead fuel moisture observations from remote automated weather stations. *International Journal of Wildland Fire*, 25(5), 558–568.

Verbesselt, J., Somers, B., Aardt, J., Jonckheere, I., and Coppin, P. 2006. Monitoring herbaceous biomass and water content with SPOT VEGETATION time-series to improve fire risk assessment in savanna ecosystems. *Remote Sensing of Environment*, 101(3), 399–414.

Victorian Bushfires Royal Commission (VBRC), Australia. 2010. ISBN 978-0-9807408-2-0. http://www.royalcommission.vic.gov.au/finaldocuments/summary/PF/VBRC_Summary_PF.pdf.

Wan, Z., Wang, P., and Li, X. 2004. Using MODIS land surface temperature and normalized difference vegetation index products for monitoring drought in the southern Great Plains, USA. *International Journal of Remote Sensing*, 25(1), 61–72.

Wooster, M.J., Roberts, G., Smith, A.M.S., Johnston, J., Freeborn, P., Amici, S., and Hudak, A.T. 2013. Thermal remote sensing of active vegetation fires and biomass burning events (Chapter 18). In C. Kuenzer and S. Dech (Eds.) *Thermal Infrared Remote Sensing: Sensors, Methods, Applications*. Remote Sensing and Digital Image Processing 17, Berlin, Germany: Springer, pp. 347–390.

Yebra, M., Dennison, P.E., Chuvieco, E., Riaño, D., Zylstra P., Hunt Jr, E.R., Danson, F.M., Qi, Y., and Jurdao, S. 2013. A global review of remote sensing of live fuel moisture content for fire danger assessment: Moving towards operational products. *Remote Sensing of Environment*, 136, 455–468.

Zhang, Y., He, C., Xu, X., and Chen, D. 2016. Forest vertical parameter estimation using PolInSAR imagery based on radiometric correction. *ISPRS International Journal of Geo-Information*, 5(10), 186. doi:10.3390/ijgi5100186.

11 A Review on European Remote Sensing Activities in Wildland Fires Prevention

David Chaparro, Mercè Vall-llossera, and Maria Piles

CONTENTS

11.1 INTRODUCTION

Wildland fires are an issue of major concern in human, economic, and environmental terms. They cause human victims and also produce important economic losses, for instance, by destroying residential structures and agricultural lands. Wildfire impacts on the environment are huge; they produce greenhouse gas (GHG) emissions and affect carbon budgets, vegetation characteristics, and the energy balance (Chuvieco et al. 2016).

Nowadays, Earth observing (EO) satellites provide information of a wide range of environmental variables at different spatiotemporal scales. Managing this information and providing it to end users are expected to lead to a more comprehensive monitoring of the Earth's system and, in particular, of fire risks and impacts. In this respect, the European Commission (EC) established the Copernicus Program to organize and deliver remote sensing and *in situ* observational datasets; it coordinates a set of services related to environmental and security issues and provides up-to-date information of the whole planet.

This chapter provides an overview of the Copernicus program and of its opportunities and challenges in wildfire prevention focusing on (1) the applicability of land cover and land use data from the local to the global scale, (2) the examples of fire risk studies based on fuel loads and vegetation hydric status information, and (3) the importance of soil moisture (SM) status to detect dry conditions posing a risk of fire.

A comprehensive review on the use of EO information for wildfires risk evaluation, fire monitoring, burned area mapping, and the analysis of fire impacts is provided. Particular emphasis is given to the activities of the European Forest Fire Information System (EFFIS). The possibility of complementing fire risk indices using new spaceborne observations acquired at L-band (i.e. the water frequency channel) is discussed. In that sense, the Soil Moisture and Ocean Salinity (SMOS) mission, launched in 2009, and the Soil Moisture Active Passive (SMAP) mission, launched in 2014, are providing for the first time global information of water content in soils and vegetation, which crucially condition fire ignition and propagation.

Section 11.5 is focused on presenting new research showing the applicability of spaceborne-derived SM data in fire risk assessment services. The complementarity between SM and surface temperature information to derive fire risk thresholds is described. In addition, the relationship between moisture–temperature anomalies, drought situations, and wildfire episodes is described. Finally, a new fire risk model based on SM and land surface temperature (LST) conditions is presented. Results from this model including fire risk maps for the Iberian Peninsula are provided.

11.2 MONITORING WILDFIRE-RELATED FACTORS FROM EARTH OBSERVING SATELLITES TO IMPROVE FIRE RISK ASSESSMENT

Wildfires are a multifaceted phenomenon involving interactions among human and environmental factors. A comprehensive framework to assess wildfires causality and risk should focus on four thematic areas: anthropogenic causes, vegetation characteristics, topography, and weather. First and foremost, humans must be considered as a main component affecting wildfires behavior. Human activities (e.g., land clearing, agriculture, resettlements, negligence, or arson, among others) are principal causes of fire ignition in most areas of the world. Moreover, humans change the availability, continuity, and distribution of fuels as a result of land use changes, sprawl of urban areas, and systematic fire extinction. Consequently, most wildfires burn in patchy landscapes where forest, agricultural and urban zones intermingle, representing a risk for human lives and beings. In particular, the expansion of wildland–urban interface (WUI) areas and the fuel accumulation due to systematic suppression of fires increase these risks (Cohen et al. 2008; Syphard et al. 2008). In addition, vegetation, topography, and weather conditions strongly influence fire behavior and extent (Syphard et al. 2008; Verdú et al. 2012). The abundance, distribution, and structure of vegetation are determinant in wildfire ignition and spread (Whelan 1995) and respond to natural and anthropogenic factors. Locally, topography influences climate and vegetation characteristics, fuel moisture, and wind effects (Whelan 1995). Finally, weather determines the fire environment and the ignition conditions (Padilla and Vega-García 2011). In particular, wind is the most important driver of fire spread (Whelan 1995). Nevertheless, temperature and rainfall are also crucial for fire ignition and spread. In that sense, the human-induced climate change leads to more frequent and intense droughts and extreme temperature events. As a result, fire seasons have lengthened, and the area affected by anomalous long exposure to weather situations posing a risk of fire has increased globally (Jolly et al. 2015).

11.2.1 AN OVERVIEW ON THE COPERNICUS PROGRAM: MONITORING THE EARTH

The ability of satellites to monitor the Earth at different spatiotemporal scales and spectral frequencies is of great interest to study wildfires risk. Taking advantage of this potential, the European

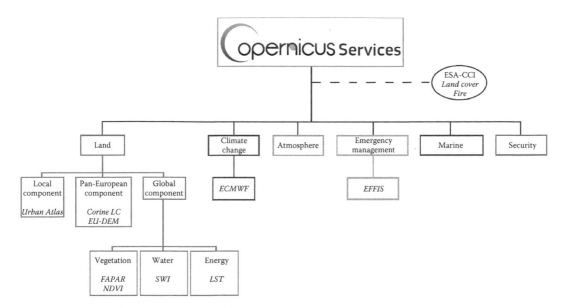

FIGURE 11.1 Copernicus program services. Land, climate change, and emergency management services are detailed as they contribute to fire prediction applications. Datasets and systems mentioned in the text are written in italics. Note that the ESA Climate Change Initiative (ESA–CCI) is a project linked to Copernicus but not included in any particular service (dashed line). (Adapted from Copernicus, 2016a. Copernicus in brief, http://www.copernicus.eu/main/copernicus-brief, Last accessed: July 10, 2016. With Permission.)

Union has developed the Copernicus Program, aiming to monitor the Earth. It supplies a wide range of services among which are satellite, airborne, and *in situ* information. Satellite datasets are obtained from the Sentinel platforms (i.e., the Copernicus specific missions) and from other EO contributing missions. Data are classified into six main services: land, marine, atmosphere, climate change, emergency management, and security. The Copernicus land service permits a multiscalar approach as it is structured in the local, pan-European, and global components. From the point of view of fire management, these components allow studying prefire conditions providing key data such as land cover and land use changes, topography, vegetation conditions, and soil moisture and temperature. In addition, Copernicus accounts with an emergency management system, which facilitates near real-time monitoring of active fires. Figure 11.1 shows the structure of the Copernicus services.

11.2.2 Earth Observing Data Sets Applied to Fire Prevention

11.2.2.1 Mapping Land Cover and Land Use to Understand Fire–Human Interaction and Fire Behavior

Satellite-derived *land cover and land use* data provide information on fuel types and their coexistence with human activities. This is needed to evaluate fire ignition and propagation risks threatening human lives and beings, as well as agricultural and environmental services.

The anticipation to fire danger situations in the WUI areas is crucial to protect human populations and their economic activities. In that sense, the *Urban Atlas* (Copernicus 2016b) provides an accurate picture of urban sprawl in the fringe of urban zones. It contains land cover and land use data for ~700 European cities (>50,000 habitants) and their surroundings. The Urban Atlas is built from very high-resolution satellite imagery (mainly SPOT-5), which enhances its applicability in fire prevention. For instance, it has been used to detect urban structures, which could be endangered under extreme fire conditions (Mitsopoulos et al. 2014).

Beyond urban zones, the density of human population decreases, and agriculture and natural vegetation lands predominate. Consequently, the number of fire ignitions is lower but, in contrast, fires may propagate throughout large areas. They can affect different vegetation covers and land uses, which can be classified from satellite-derived land cover maps. At a global scale, the European space agency-climate change initiative (*ESA–CCI*) *Land Cover* supplies land cover data at 300 m resolution, with accuracy between 70.8% and 74.4% (ESA 2016a). In Europe, the Corine Land Cover (CLC) map contains harmonized data of land cover and its changes since 1990. This information is used to estimate the effect of land covers on fire spread. Particularly, it has been found that larger fires burn coniferous forests and scrublands, whereas smaller ones burn broad-leaved forests, agro-forestry areas, crops, and urban zones (Bajocco and Ricotta 2008, Verdú et al. 2012).

The influence of land covers on fire propagation is also linked to other factors. Actually, coniferous forests are often located in areas where topography complicates the extinction tasks, and where steep slopes facilitate the rapid spread of fire. In addition, topography affects the local vegetation distribution (Whelan 1995). Hence, *topography* is considered in fire risk modeling using digital elevation models (DEM) from national databases (e.g., Verdú et al. 2012). Actually, a European DEM (EU-DEM) is now available (Copernicus 2016b). Its application in fire risk studies should be a matter of future works.

11.2.2.2 Spaceborne-Derived Measurements of Fuel and Soil Conditions

Together with vegetation types and their local determinants, *fuel loads and vegetation hydric status* are main factors in fire risk assessment as they condition the availability and flammability of fuels. Under homogeneous fire-prone meteorological conditions within a region, large fires occur majorly in areas with high fuel loads (i.e., vigorous vegetation). These areas are detected by high values of the fraction of absorbed photosynthetically active radiation (FAPAR), which is related to gross primary productivity (GPP). Hence, high values of FAPAR may favor fire occurrence and propagation (Knorr et al. 2011). Other spaceborne vegetation indices, such as the Normalized Difference Vegetation Index (NDVI), have also been used as an indicator of vegetation water stress in fire risk studies (Wang et al. 2013).

NDVI has also been used to estimate live fuel moisture content (FMC), a crucial variable in fire ignition and propagation (Chuvieco et al. 2014). Initially, the link between NDVI and FMC was studied for grasslands, with good correlation between both variables (Hardy and Burgan 1999). However, weak correlations were obtained in shrublands and forests (Chuvieco et al. 1999). To bridge this gap, the combined use of NDVI and LST was applied to estimate FMC and associated fire risk (Chuvieco et al. 2004). Note that an increase in LST can be interpreted as an increase in vegetation water stress and/or soil surface dryness (Chuvieco et al. 2004, Li et al. 2016).

Consequently, the study of the *soil surface state* is paramount to detect dry conditions increasing the risk of fire. LST and SM are directly related to live FMC and also to the moisture content of dead fuels (e.g., litter), which is as well a main driver of fire propagation (Chuvieco et al. 2014). SM and LST are variables which can be estimated at the global scale using spaceborne sensors. Satellite-derived LST datasets can be obtained from several platforms (see Tomlinson et al. 2011 for a review, and Copernicus 2016b). Concerning SM state, Copernicus provides the Soil Water Index (SWI) as the percentage of water in the soil at eight different time lengths (from 1 to 100 days). These are currently derived from MetOp C-band radar backscatter measurements. These satellites are planned to provide continuous data until at least 2020, complementing the quantitative SM estimates (in m^3/m^{-3}) provided by the European Space Agency (ESA) SMOS mission and the National Aeronautics and Space Administration (NASA) SMAP mission (Naeimi et al. 2009). This chapter provides an overview of the quantitative SM information provided by microwave L-band satellites and of the applicability of SMOS data in forest fires risk assessment (Sections 11.4.1 and 11.5).

11.3   THE EUROPEAN FOREST FIRE INFORMATION SYSTEM

Knowledge and technologies applicable to forest fires are increasing rapidly during the past years as described in Section 11.2. This means that, nowadays, there is a broad range of data available to support fire-related policies and researches. To take profit of this information, it is required to account with well-organized systems devoted to acquire, process, and offer data and products encompassing all fire stages. Built on this scope, the EFFIS was established in 2003 to provide a wide range of fire information at a pan-European level and to support European Union environmental policies (San-Miguel-Ayanz et al. 2013). From 2015, EFFIS is being integrated to the Copernicus program in order to provide new emergency management and risk assessment tools at European level.

EFFIS offers a comprehensive approach to wildfires, encompassing fire danger forecast, fire near real-time monitoring, burnt area mapping, and fire impacts. These issues are described hereafter, and a more detailed information is found in San-Miguel-Ayanz et al. (2012) and is available at forest.jrc.eu/effis:

1. *Fire danger forecast*: it is based on the well-known Canadian Fire Weather Index (FWI; Van Wagner 1987). This index uses *in situ* meteorological measurements (temperature, relative humidity, wind, and rain), which are applied to derive the six components of the FWI (Figure 11.2). Three of these components are fuel moisture codes, which correspond to the moisture contents of surface, medium depth, and deep organic matter (Fine Fuel Moisture Code [FFMC], Duff Moisture Code [DMC], and Drought Code [DC], respectively). The combination of the latter two results on the Build Up Index (BUI), and the joining of the FFMC with wind speed data builds the Initial Spread Index (ISI). Finally, ISI and BUI are grouped to obtain the FWI (Figure 11.2). In the case of EFFIS, the FWI information is derived from meteorological forecast models. Hence, the risk maps are provided for the current day and predicted for the next 8 days (16 km resolution; forest.jrc.eu/effis).

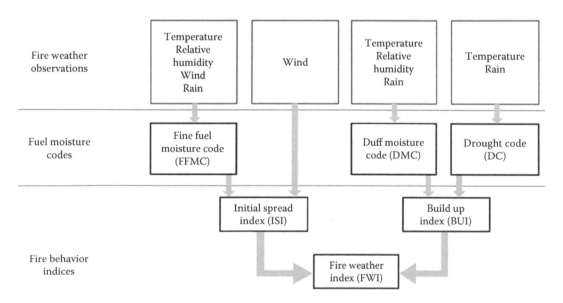

FIGURE 11.2   Diagram showing the six components of the Canadian Fire Weather Index (FWI). (Adapted from Government of Canada 2016, Canadian wildland fire information system, http://cwfis.cfs.nrcan.gc.ca/background/summary/fwi, Last accessed: July 13, 2016. With Permission.)

2. *Fire monitoring*: EFFIS takes advantage of the Moderate Resolution Imaging Spectro-radiometer (MODIS) active fire monitoring product (250 m resolution) in order to provide near real-time fire active hotspots that are useful to emergency management systems. EFFIS filters these products using ancillary data (e.g., CLC map) to reduce false alarms and to classify the type of fire. The final data are published at forest.jrc.eu/effis.

3. *Burned area mapping*: Wildfires data are collected separately by each of the 22 contributing countries and are provided each year to the Joint Research Center of the EC. This center checks and processes the data in order to build the European Fire Database. Fire data contain the time of fire ignition, its size, its location and its causes, among other information. The validation phase includes checking for consistency of time, location, and size (San-Miguel-Ayanz et al. 2012). The European Fire Database has been compared to the Rapid Damage Assessment dataset (derived from MODIS information) with an overall agreement between 40% and 90% depending on the fire size (period 2006–2009; Vilar et al. 2015). The database was officially initiated in 2000, but longer records (>25 years) are available for Mediterranean countries, and are used to model the influence of climate change on forest fires in Southern Europe, as well as to understand spatial, temporal, and seasonal trends (San-Miguel-Ayanz 2012).

4. *Fire impacts*: The estimation of fire impacts is necessary to evaluate fire damages and to assess future derived risks. For instance, wildfires contribute to the emissions of GHG and, actually, these emissions may constitute a large proportion of the GHGs released to the atmosphere. After fire occurrences, EFFIS estimates wildfire emissions. Furthermore, the analysis of the potential soil erosion is planned to be a new information provided by EFFIS in order to estimate landslides and flood risks derived from fire impacts on soils. In addition, it is expected to provide a new EFFIS module devoted to the study of vegetation recovery after large fires. It will be based on the comparison of prefire and postfire remote sensing images (Joint Research Center 2015).

11.4 MEASURING WATER CONTENT IN SOILS AND VEGETATION FROM SPACE: APPLICABILITY TO WILDFIRE PREVENTION

Water content in live and dead fuels influences the probability of ignition and the potential propagation of wildfires (Chuvieco et al. 2014). Normally, fire risk indices estimate the moisture of dead fuels from the interpolation of *in situ* meteorological data, and not from direct measurements with global coverage. This is the case, for instance, of most common fire risk indices such as the FWI (Van Wagner 1987 and Section 11.3), the McArthur Forest Fire Danger Index (FFDI; McArthur 1967), or the National Fire Danger Rating System (NFDRS; Deeming et al. 1977). Such indices afford reliable and widely used fire risk information, but could potentially be complemented with global and direct measurements of fuel moisture variables. In addition, the live vegetation water content (VWC), in turn, is difficult to evaluate, and until now it is not generally included.

In Section 11.2.2, we have introduced some datasets providing direct information on fuel moisture. Previous studies present evidence that the combination of NDVI and LST data allows for a good approximation to live FMC in grasslands and scrublands (Chuvieco et al. 2004). Still, results of this approximation in forests are not satisfactory (Chuvieco et al. 1999). Remotely sensed SM datasets, in turn, provide information of the dead fuel moisture conditions and of the water available to the superficial root layers of vegetation. There are at present two missions specifically dedicated to SM monitoring: the ESA's SMOS and the NASA's SMAP. In addition, recent research has shown that it is possible to retrieve VWC from L-band (Konings et al. 2016, Piles et al. 2016, Konings et al. 2017). Also, Copernicus currently provides the SWI from MetOp satellites as a proxy for SM state (Naeimi et al. 2009).

11.4.1 Spaceborne Quantitative Measurements of Soil Moisture

Global and continental SM data have been obtained from microwave sensors since 1980s. This information is essential for the development of long-term consistent SM series serving the climate research community, which is the main goal of the ESA Soil Moisture Climate Change Initiative program (Dorigo et al. 2016, ESA 2016b).

Nevertheless, the challenge of measuring Earth emissivity at the optimal frequency for SM estimation (1.4 GHz; L-band) has been overcome recently with the launch of SMOS (2009), Aquarius (2011), and SMAP (2016) missions. Theoretical and practical research has evidenced that L-band is the optimal frequency to measure SM. The main advantages of L-band sensors are that emissivity originates from the top ~5 cm of soil, the atmosphere is almost transparent, and measurements are sensitive to SM through vegetation of up to 5 kg·m^{-2} water content (Ulaby et al. 1981).

The ESA SMOS satellite was the first mission specifically dedicated to SM measurements. Its unique payload is an L-band interferometric radiometer, the Microwave Imaging Radiometer with Aperture Synthesis (MIRAS). After more than 6 years in orbit, SMOS continues providing global SM maps every 3 days, with a target accuracy of 0.04 m^3 · m^{-3} (Kerr et al. 2012). The spatial resolution of the SMOS data is ~40 km (L2 product). This meets the needs for global applications, but it is still too coarse to serve regional and local uses. Downscaling techniques allow improving the spatial resolution to 1 km (Merlin et al. 2005, 2008, Piles et al. 2011a, 2014). The NASA SMAP satellite has a real aperture L-band radiometer and an L-band radar. The specific advantages of both instruments allow for a high temporal resolution and enhance the spatial resolution from 36 km to 9 km (Entekhabi et al. 2010). However, the SMAP active–passive operations were ceased abruptly with the failure of the SMAP radar on July 2016. Nonetheless, the SMAP radiometer is continuing to make measurements, and the use of Sentinel-1 data as a replacement of its radar is under evaluation. The Aquarius satellite was an L-band mission shared between NASA and the Comisión Nacional de Actividades Espaciales (CONAE; Argentina). This mission, which was designed for measuring ocean salinity, was proved to be valid for SM retrievals (Bindlish et al. 2015). Nevertheless, the Aquarius mission ended in June 2015.

Particularly, the SMOS-derived SM data has served to different applications related to the assessment of drought and vegetation water stress conditions. SMOS-derived SM anomalies have been correlated with two drought indices: the Standard Precipitation Index (SPI; McKee 1993) and the Standard Evaporation Precipitation Index (SPEI; Vicente-Serrano et al. 2010). The study was conducted at different timescales (10–120 days), and it was found that 30-day anomalies were highly correlated to the drought conditions reported by both indices in the Duero basin (Spain; Scaini et al. 2014). In addition, the SMOS data have been used to develop two agricultural drought indices: the Soil Water Deficit Index (SWDI; Martínez-Fernández et al. 2015, 2016) and the Soil Moisture Agricultural Drought Index (SMADI; Sánchez et al. 2016). Both indices appropriately track the agricultural droughts affecting the growing season and the impact on yield. Finally, SMOS data allowed detecting water stress conditions leading to vegetation dieback in grasslands of Australia (Ross et al. 2014) and in forests of Catalonia (northeastern Spain; Chaparro et al. 2016a). Consequently, the ability of detecting drought situations with SMOS data has suggested its applicability on forest fire risk evaluation studies. Section 11.5 presents an overview of the research conducted in this field and the main results obtained.

11.4.2 Microwave Retrievals of Vegetation Optical Depth from Space

Passive microwave SM inversion techniques need to account for the effect of vegetation optical depth (VOD) and vegetation scattering albedo in surface emissivity in order to retrieve SM. Recent research has shown that these two vegetation parameters can also be retrieved alongside SM (Konings et al. 2016, Piles et al. 2016). VOD is directly proportional to VWC, and the vegetation scattering albedo is related to the canopy structure. An algorithm was proposed to retrieve microwave-based SM and

VOD and first evaluated using 3 years of global observations from Aquarius (Konings et al. 2016). At the global scale, the resulting VOD distribution was coherent with climate gradients, with seasonal precipitation dynamics, and with the expected patterns of canopy biomass. Furthermore, VOD retrievals showed temporal dynamics to be consistent with precipitation and drought periods, and captured both water retention and drying processes in canopies. The same method was applied to the first annual cycle of SMAP data, and the resulting SM, VOD, and albedo measurements were consistent with expected vegetation dynamics and precipitation regimes (Piles et al. 2016; Konings et al. 2017). VOD provides complementary information to the data obtained from indices such as NDVI, LAI, or FAPAR. Future work in microwave vegetation parameter retrievals is expected to generate VWC maps, which could be applied to detect vegetation water stress conditions posing a risk of fire.

11.4.3 VEGETATION WATER CONTENT RETRIEVAL FROM VISIBLE AND INFRARED INDICES

Indices based on the infrared and optic spectral bands are also sensitive to the water content of vegetation. A research from Ullah et al. (2014) compared a variety of indices and concluded that the mid infrared (MIR) and the shortwave infrared (SWIR) bands were the most sensitive spectral regions to changes in VWC. On the contrary, the thermal infrared (TIR) showed limited sensitivity to VWC. A review on the methods and indices for the estimation of VWC is provided by Roberto et al. (2012) showing different products available, such as (1) the Normalized Difference Infrared Index (NDII), from Landsat Thematic Mapper, (2) the Global Vegetation Moisture Index (GVMI), from satellite Pour l'Observation de la Terre (SPOT)-VGT, (3) the Normalized Difference Water Index from MODIS, and (4) the Normalized Multiband Drought Index (NMDI) from MODIS. Finally, note that the widely used NDVI has also been applied as an estimator of the moisture content of live fuels (Chuvieco et al. 2014).

11.5 FIRE RISK ASSESSMENT FROM REMOTELY SENSED SOIL MOISTURE

Studying the relationship between drought and forest fires is feasible due to the availability of satellite estimates of SM, which can be applied to fire risk prediction. First approaches in this research line were conducted in central Siberia by Bartsch et al. (2009) and Forkel et al. (2012). In both studies, dry soil conditions were detected by moisture anomalies (derived from European remote sensing satellite (ERS)-1/2 and advanced microwave scanning radiometer (AMSR)-E satellites) and led to large and frequent fires in the region. These results opened an avenue to the applicability of remotely sensed SM in fires research. It was observed that SMOS-derived SM data at a coarse resolution (L2 product) was well correlated with the FWI in Siberia. However, this relationship was found only in some areas, suggesting that further research was needed (Shvetsov 2013).

The application of SMOS data at a higher resolution was expected to enhance the capability of detecting droughts increasing the risk of fires. This has been explored in the Iberian Peninsula by the Barcelona Expert Center (BEC), applying the downscaling algorithm described in Piles et al. (2011a and 2014). This method merges SMOS observations (L2) with higher spatial resolution MODIS NDVI and LST data into 1 km SM estimates (BEC L4 product). Recently, it has been enhanced with a new version (L4v3 product) complementing MODIS surface temperature (LST) data when it is masked by clouds. Consequently, it provides SM data regardless of weather conditions. The new product is supplied by the BEC (2015). Sections 11.5.1 to 11.5.3 report studies in the Iberian Peninsula, which were based on the L4 and L4v versions of the high-resolution SM product.

11.5.1 DRY AND HEATED SOILS FAVOR WILDFIRE IGNITION
AND SPREAD IN THE IBERIAN PENINSULA

The Iberian Peninsula is a particularly interesting region for the study of SM and its relationship with wildfires due to its high fire activity and due to its contrasting climate and fire patterns. The area influenced by the Mediterranean climate suffers recurrent drought situations in summer.

This limits the fuel availability but, in turn, favors the dryness of the combustible fuel and its accumulation in areas which remain unburned for years. This leads to the burning of few but large fires in the region. In contrast, the northwestern Iberian Peninsula is influenced by the humid Atlantic climate, facilitating the accumulation of large amounts of fuel with high water content. This brings to the occurrence of many fires, which are limited in extent (Moreno et al. 2005; Verdú et al. 2012). Nevertheless, anomalously dry situations bring to large wildfires in Portugal and Galicia (Trigo et al. 2006; Fischer et al. 2007). Then, the study of droughts is essential to prevent fires in the Iberian Peninsula. The analysis of SM patterns from satellite information, and the application of complementary data, should enhance the capability to anticipate droughts and fire risk situations (see Section 11.5.2 and Chaparro et al. 2016b).

In that sense, the relationship among SM, LST, and fires was initially studied in Piles et al. (2011b). This research analyzed the complementarity between the SMOS-derived SM (L4; 1 km) and the MODIS LST (1 km). SM–LST values before fire ignitions showed that soils were drier and warmer in fire-affected areas in comparison to unburned perimeters. Later, the SM–LST complementarity was further explored to assess the risk of wildfires. To this objective, fires burning in all the Iberian Peninsula were obtained from the EFFIS database (European Comission 2010) for the period 2010–2014. These wildfires were grouped in three categories (<500 ha, 500–3,000 ha, and >3,000 ha) and the relationship between SM–LST and wildfires spread was investigated. The L4v.3 product provided SM data at 6 a.m., and the European centre for medium-range weather forecasts (ECMWF) interim reanalysis (ERA)-Interim LST reanalysis models (at noon; ECMWF 2015) were also used. LST data were linearly interpolated from 0.125° to 1 km. Median SM and LST values before burning were calculated and assigned to each fire perimeter. Soil surface conditions prior to fire occurrences were compared to a database containing the unburned conditions of SM and LST in the Iberian Peninsula for each day of the study period (2010–2014) and for each flammable pixel in the region (the only pixels considered were those containing natural or agricultural vegetation in the CLC map; EEA 2006). To exclude burned cells, those pixels affected by fire after 2005 (obtained from the EFFIS database for the period 2006–2014) were not included. The dataset finally contained ~$1.9 \cdot 10^7$ SM–LST values for each unburned cell and day during all the study period. In addition, a single median value for each year and variable was calculated to summarize overall SM and LST for the unburned cells (Chaparro et al. 2015, 2016c, 2016d).

Results showed that 70% of fires <500 ha burned under drier and hotter conditions than the yearly median SM–LST values obtained for the unburned pixels in the region. This percentage increased to 88% for fires between 500 and 3000 ha and to 90% for fires larger than 3000 ha. Normally, most fires occurred in very dry and hot conditions (SM<0.10 $m^3 \cdot m^{-3}$ and LST>300 K), which are frequent during July and August (i.e., summer) in the study area (Chaparro et al. 2015, 2016c). Nevertheless, important fire episodes were detected out of summer season in 2012. This is reported in Figure 11.3, where the accumulated number of fires and burned area during 2012 are plotted. Here, the two main fire patterns, which are common in the region are detected: (1) on summer, few but large fires burned (mainly in the Mediterranean region) and (2) during February and March, as well as in September, many small fires occurred (mainly in the northwest; Figure 11.3 and Chaparro et al. 2016d).

In Figure 11.4, largest fires are plotted beyond the median SM–LST for 2012. The warm climate of the Iberian Peninsula implies that dry and warm conditions are quite usual in the region, as shown by the density plot. This facilitates the ignition and spread of wildfires on summer. Still, in 2012, a high number of fire outbreaks were registered in relatively wet and cold conditions (see Figure 11.4), probably corresponding to fires burned out of summer season. Further analysis was needed to understand why these fires occurred under these conditions.

To this purpose, the relationship between wildfires and SM–LST data was analyzed, but this time including anomalies of both variables. SM and LST anomaly time series at 9-day and 30-day timescales were computed in order to detect drought periods posing a risk of fire. The anomaly calculation comprised three steps: (1) monthly means of each variable were computed (2) a linear

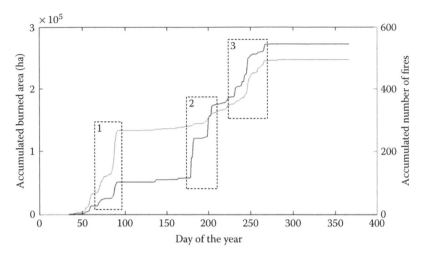

FIGURE 11.3 Accumulated number of wildfires (light grey) and accumulated burned area (dark grey) during 2012 in the Iberian Peninsula. Two patterns are observed: numerous small wildfires burning on February–March (1) and October (3), and few but large fires burning in the Mediterranean (2).

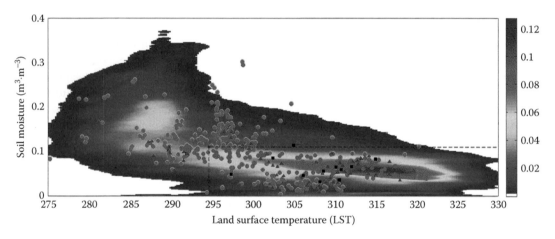

FIGURE 11.4 SM–LST comparison between burned and unburned areas during 2012 in the Iberian Peninsula. Green points, red triangles, and black squares correspond to wildfires <500 ha, 500–3000 ha and >3000 ha respectively. They are plotted in function of moisture and temperature conditions up to 3 days before forest fire occurrences. Black dashed lines show the median SM and LST in 2012 unburned pixels. The distribution for both variables in not burned areas during 2012 is shown as a density plot of pixels and days presenting each pair of moisture–temperature values.

interpolation was performed between mean values of each pair of consecutive months and these time series represented the average moisture and temperature conditions and (3) the difference between moving means and the corresponding average conditions was calculated from the day of interest to 9 and 30 days backward (Chaparro et al. 2016b).

In Figures 11.5 and 11.6, the SM–LST mean conditions and anomalies for 2011 and 2012 are plotted, respectively. The month of occurrence is also included. Fires burning under cold/wet conditions corresponded to those detected in February and March 2012. Fires on September 2012 burned under dry and warm conditions, as those occurred on July and August (Figure 11.6). Similarly, many fires burned in areas with extremely warm and dry soils on October 2011 (Figure 11.5). Note that the two important fire episodes detected in October 2011 and February–March 2012 were linked to negative moisture and positive temperature anomalies (Figures 11.5b and 11.6b). These results showed that

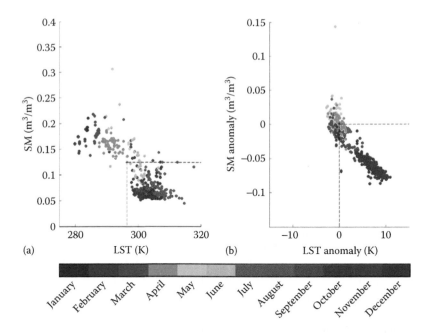

FIGURE 11.5 (a) Mean SM and mean LST before fires occurring in 2011. Black dashed lines show the yearly (2011) median values of temperature and moisture considering the entire Iberian Peninsula. (b) SM and LST anomalies before fires occurring in 2011. In both (a) and (b) plots, fires are represented per months (colorbar). All the variables were calculated at a 30-day time scale. (Adapted from Chaparro, D. et al., *Eur. J. Remote Sens.*, 2016b. With Permission.)

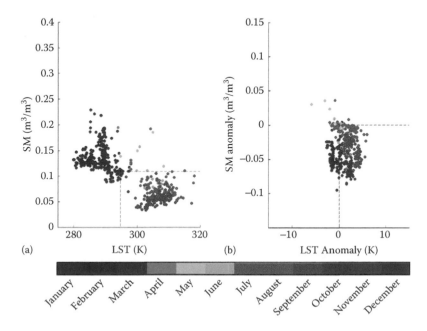

FIGURE 11.6 (a) Mean SM and mean LST before fires occurring in 2012. Black dashed lines show the yearly (2012) median values of temperature and moisture considering the entire Iberian Peninsula. (b) SM and LST anomalies before fires occurring in 2012. In both (a) and (b) plots, fires are represented per months (colorbar). All the variables were calculated at a 30-day time scale. (Adapted from Chaparro, D. et al., *Eur. J. Remote Sens.*, 2016b. With Permission.)

anomalies at 30-day timescale allowed detecting extremely dry conditions leading to a large number of fire occurrences even out of summer season. Similar patterns were found for anomalies at 9-day timescale (Chaparro et al. 2016b).

11.5.2 STUDY OF DROUGHTS INVOLVING RISK OF WILDFIRES IN THE NORTHWESTERN IBERIAN PENINSULA

As it is explained in Section 11.5.1, many fires burned in the region in October 2011 and February–March 2012. These fires mainly affected the northwestern Iberian Peninsula. To study their relationship with the intensity and duration of droughts, Chaparro et al. (2016b) analyzed the moisture, temperature, and fire trends in that region. To delimit the study area, the Spanish ecoregions map (see Padilla and Vega-García, 2011) and the phytogeographic regions of Portugal (Paes do Amaral 2000; Agência Portuguesa do Ambiente 2015) were used. Figure 11.7 shows the demarcated area, which included the northern, humid areas of Portugal and the northern regions of Spain, which are influenced by the Atlantic climate. Moisture and temperature anomalies trends were summarized computing overall median anomalies in the region each day. Only flammable pixels were considered in this calculation (this excluded water and urban pixels from the CLC map; EEA 2006), and pixels burned from 2006 to 2014 were excluded (EFFIS database; European Comission 2010). Finally, time series of median anomalies in the region were plotted using a 9-day and 30-day mean moving windows. These time series were compared with the number of fire occurrences. Figures 11.8 and 11.9 show the resulting plots at 9-day and 30-day timescales, respectively. In these figures, drought situations and wet/cold conditions are drawn and can be compared to the number of fires reported for each period of 9 days (Figure 11.8) and 30 days (Figure 11.9).

Results in Figure 11.9 show how the intense fire activity in October 2011 coincided with the drought situation reported from August to November 2011. In that case, the peak of driest and warmest soil conditions (-0.09 m$^3 \cdot$ m^{-3} and 8 K) matched in time with the large number of fires burned during that month (304 fires; Figure 11.9c). Considering the fire period from February to March 2012 (267 wildfires), SM anomalies reached -0.10 m$^3 \cdot$ m^{-3} (Figure 11.9a), coinciding with the highest values of temperature anomalies (+1.6 K; Figure 11.9b). Soils were continuously dry for

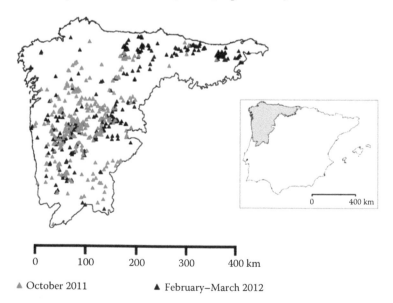

▲ October 2011   ▲ February–March 2012

FIGURE 11.7 The area of interest is sited in the NW Iberian Peninsula (right box), where fires burning in October 2011 (grey triangles) and February–March 2012 (black triangles) were studied. (Adapted from Chaparro, D. et al., *Eur. J. Remote Sens.*, 2016b. With Permission.)

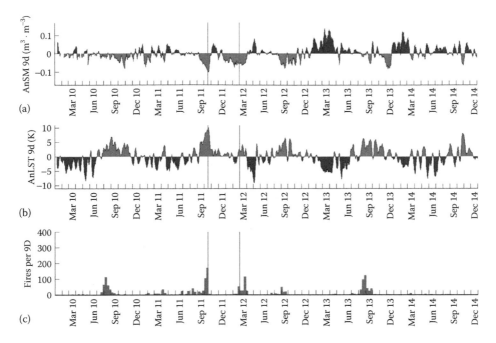

FIGURE 11.8 Dry/warm (light grey) and wet/cold (dark grey) periods compared to the number of fire outbreaks (9-day basis) in the northwestern Iberian Peninsula during the period 2010–2014. (a) Median anomalies of SM at 9-day time scale; (b) median anomalies of LST at 9-day time scale; (c) number of fires per 9-day groups. Dark lines in October 2011 and February–March 2012 correspond to the end of the anticyclone periods.

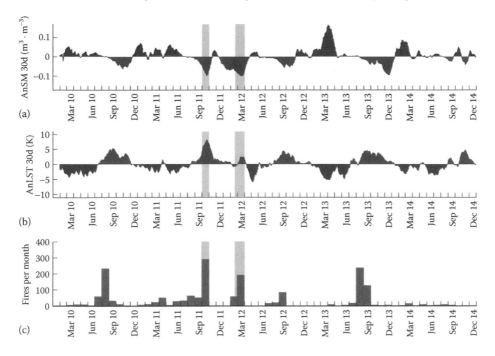

FIGURE 11.9 Dry (red) and wet (blue) periods compared to the monthly number of fire outbreaks in the northwestern Iberian Peninsula during the period 2010–2014. (a) Median anomalies of SM at 30-day time scale; (b) median anomalies of LST at 30-day time scale; (c) number of fires grouped per month. Shaded areas represent the drought-fire periods, October 2011 and February–March 2012. (Adapted from Chaparro, D. et al., *Eur. J. Remote Sens.*, 2016b. With Permission.)

more than 3 months before this episode, although the duration of above average temperatures was only limited to the month of March (Figure 11.9b).

Interestingly, both fire periods coincided with synoptic meteorological anomalous situations. Particularly, abnormal anticyclonic activities in the western and central Iberian Peninsula occurred between October 11 and 18, 2011 and between February 21 and March 2, 2012. These produced above mean air temperatures and low air humidity (Amraoui et al. 2013). In Figure 11.8, gray lines have been plotted on October 18, 2011 and March 2, 2012, showing the moisture–temperature conditions during the anticyclonic situation.

Results show that driest and warmest soils occurred simultaneously to the anticyclonic period on mid-October 2011, which also matched in time to the highest number of fire ignitions during that month (Figure 11.8). The fire period on February and March 2012 began under the anticyclonic situation and during the long and intense drought that had started 3 months ago, according to the SM anomalies (Figures 11.8 and 11.9).

The spatial patterns of moisture and temperature anomalies have also been studied here. Figure 11.10 shows the maps of the anomalies of both variables computed before, during, and after the anticyclone. The spatial distribution of high surface temperatures and low SM between October 11 and 18, 2011 (Figure 11.10b and e) matched the spatial distribution of high air temperatures and low air humidity. These extended from the central Iberian plateau to the Atlantic coast. The meteorological anomalies were especially intense in the northwestern Iberian Peninsula (see Figure 11.2 in

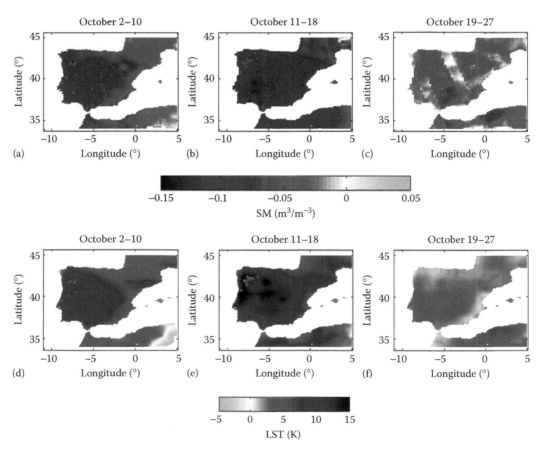

FIGURE 11.10 9-day SM anomalies (up row) on (a) October 10, 2011, (b) October 18, 2011, and (c) October 27, 2011. 9-day land surface temperature anomalies (down row) on (d) October 10, 2011, (e) October 18, 2011, and (f) October 27, 2011. The period October 11–18 (red rectangle) matches in time with the abnormal synoptic anticyclone activity. Green dots correspond to fires larger than 500 ha in the northwestern Iberian Peninsula.

Amraoui et al. 2013). In addition, note that the drought intensity was lower between October 2 and 10 (i.e. before the anticyclonic interval; Figure 11.10a and d), increased between October 10 and 18 (Figure 11.10b and e; i.e. during the anticyclone), and clearly diminished after the anticyclonic period (October 19–27, 2011; Figure 11.10c and f). This pattern matches the evolution of the number of wildfires during that month (Figure 11.8). Particularly, the occurrences of large fires (>500 ha) rose between October 2–10 (8 fires) and October 11–18 (20 fires; Figure 11.10).

In conclusion, the anomaly time series from SMOS-derived data and ERA-Interim models demonstrated a coherent fitting with the atmospheric conditions on October 2011 and February–March 2012. This stresses the capacity of the computed anomalies to detect droughts increasing the risk of forest fires (at least out of summer season). This also suggests that the study of temperature datasets and remotely sensed SM should be explored to enhance fire risk assessment methodologies.

Finally, note that not all the drought periods reported in Figures 11.8 and 11.9 led to high fire activity. From August to November 2010, and from August to December 2013, the most adverse moisture and temperature conditions did not match with time and, possibly due to this reason, fires burned only during the summer months (fire activity was important on August 2013, as shown in Figure 11.9). In September 2012, the drought was intense but shorter and milder than the reported during October 2011 and February–March 2012, and the number of wildfires was lower (109).

11.5.3 DEVELOPMENT AND APPLICATION OF FIRE RISK MODELS IN THE IBERIAN PENINSULA

This section summarizes the results from the researches described in Chaparro et al. (2016c and 2016d), where moisture–temperature datasets are applied to develop fire risk indices and maps in the Iberian Peninsula. Again, fire perimeters in the region were obtained from the EFFIS database (European Comission 2010) for the period 2010–2014. The datasets and the methodology used to obtain moisture and temperature conditions previous to each fire, as well as the dataset studied, are the same as those described in Section 11.5.1.

11.5.3.1 Development and Operational Use of an Empirical Fire Risk Model

Moisture and temperature conditions prior to fire occurrences were compared among different wildfire extents (<500 ha, 500–3000 ha, and >3000 ha). Large fires were normally related to extremely dry and hot soils, whereas small fires could occur, generally, under milder conditions. A linear relationship was found among SM, LST, and the burned area (Chaparro et al. 2016c). The configuration of fire risk thresholds to develop an empirical model of risk was based on this result and was complemented by the study of SM and LST anomalies (Table 11.1). More details concerning to the methodology applied are provided in Chaparro et al. (2016c).

The model was validated with an independent sample, which was classified by burned area categories. The extent of each fire was compared to the risk category predicted by the model. Results

---

TABLE 11.1

Third Quartile for Soil Moisture (SM) and First Quartile for Land Surface Temperature (LST) Determine Each Risk Category

Risk Thresholds	Low Risk (1)	Ignition (<500 ha)	Large Fire (500–3000 ha) (2)	Very Large Fire (>3000 ha)
SM (3rd Q; m$^3 \cdot$ m^{-3})	>0.11	<0.11	<0.09	<0.08
LST (1st Q; K)	<300	>300	>304	>306

Note: Anomalies of both SM and LST variables change the risk categories as follows: (1) Low risk Pixels are finally classified as ignition risk pixels when the SM anomaly is negative; (2) Pixels where risk of large fires (500–3000 ha) is predicted are finally considered under risk of very large fire (>3000 ha) when the LST anomaly is positive.

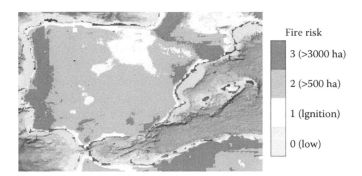

FIGURE 11.11 Fire risk map for July 11, 2016. Four risk levels are presented (from low risk to high risk indicating possibility of fires >3000 ha).

showed that 87% of fires <500 ha, 59% of fires >500 ha, and 56% of fires >3000 ha were correctly predicted (Chaparro et al. 2016c).

Finally, fire risk maps were put on operational application through the Barcelona Expert Centre website (http://cp34-bec.cmima.csic.es/NRT) on July 2015. Maps are produced at 1 km spatial resolution and encompass the Iberian Peninsula, the south of France, and the north of Africa (45°N–34°N, –11°W – 5°E). This fire risk maps (see an example in Figure 11.11), as well as the high resolution SM maps, are routinely included in the fire risk prevention service bulletin prepared by the provincial government of Barcelona (DIBA). This bulletin is delivered daily to the forest rangers during the summer fire prevention campaign.

These results and their operational applicability support the use of surface moisture and temperature information in fire risk prevention services. Still, further research considering other crucial variables on wildfires is needed to develop a more comprehensive fire risk assessment framework. In that sense, Section 11.5.3.2 presents a recently developed fire risk model encompassing SM, temperature, land cover, and ecological regions information (Chaparro et al. 2016d).

11.5.3.2 Modeling Potential Wildfire Spread

The relationship found between SM–LST data and burned area showed that wet and cold soils limited the spread of forest fires that, in contrast, could be propagated under drought conditions (see Section 11.5.3.1). This suggested that the maximum extent that a fire could reach under certain moisture–temperature conditions could be estimated. To this objective, the burned area was logarithmically transformed and was plotted separately as a function of SM and LST. Figure 11.12

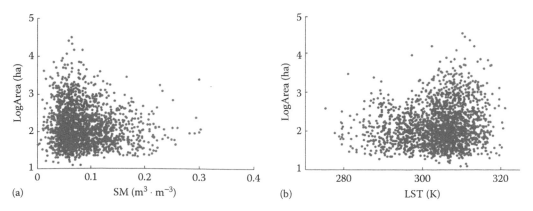

FIGURE 11.12 (a) Decimal logarithm of burned area plotted as a function of soil moisture. (b) Decimal logarithm of burned area plotted as a function of land surface temperature. (Adapted from Chaparro, D. et al., *IEEE J. Sel. Topics Appl. Earth Observ. Remote Sens.*, 9, 2818–2829. With Permission.)

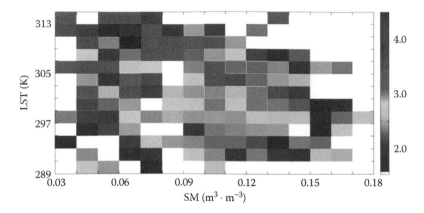

FIGURE 11.13 Decimal logarithm of maximum burned area (colorbar) per SM–LST bins is plotted. Bins of 2K and $0.01 m^3 \cdot m^{-3}$ led to the model in Equation 11.1 with $R^2 = 0.43$.

shows the resulting triangle-shaped plots in which a regression could be fitted along the imaginary hypotenuse of both triangles (Chaparro et al. 2016d).

From these results, a fire risk model was built on the basis of the moisture–temperature complementarity. As no strong redundancy between SM and LST was found ($r = -0.54$; Pearson correlation coefficient), both variables were combined in a single model. To perform the regression analysis, the authors binned SM and LST variables, and included all wildfires larger than the 90th percentile of burned area in the model for each bin. This methodology was similar to that found in other studies based on triangular-shaped relationships between environmental variables (Moran et al. 1994, Sandholt et al. 2002). Forest fires with moisture and temperature at the extremes of the variables' ranges (beyond percentiles 5th and 95th) were eliminated, as the sample of fires is reduced in that conditions not showing a representative maximum according to the SM–LST data. In addition, the effect of the bin width was studied considering several binning possibilities (from $[0.005 m^3 \cdot m^{-3} - 0.5 K]$ to $[0.02 m^3 \cdot m^{-3} - 2 K]$). Finally, the binning employed was $0.01 m^3 \cdot m^{-3}$ for SM and 2K for LST, and the sample was 183 fires. Figure 11.13 shows the resulting plot in which SM–LST paired values condition fire spread. The binning that was chosen permitted the best fitting ($R^2 = 0.43$) for the linear model defined in (Equation 11.1):

$$\text{LogArea} \approx a \cdot \text{SM} + b \cdot \text{LST} \tag{11.1}$$

where:
  LogArea corresponds to the decimal logarithm of burned area
  SM states for soil moisture
  LST states for land surface temperature
  Their corresponding coefficients are a and b

This result was considered as a good basis to improve the model. The set of variables was completed including land cover, the month of occurrence of the fire, and the region of occurrence. Land cover was obtained from CLC Map (EEA 2006) at 250 m and transformed to 1 km grid. Later, these data were reclassified to eight land cover categories: broadleaved forests, coniferous forests, mixed forests, heathlands and moors, sclerophyllous vegetation, natural grasslands, woodland–shrub transitions, and sparse vegetation areas. Considering the geographical classification, it aimed to group wildfires depending on the biogeophysical and climatic characteristics of the region where they burned. The 53 ecoregions defined by the Spanish Forest Service were useful to this proposal (see Padilla and Vega-García 2011). In Portugal, the geographical division was based on the phytogeographic regions from the environmental atlas of Portugal (Paes do Amaral 2000). Finally, to simplify the study, the regions were grouped (Figure 11.14). Three regions were considered in Spain,

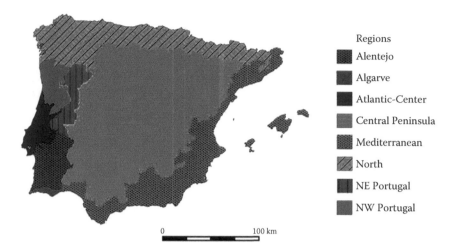

FIGURE 11.14 Geographical division of the Iberian Peninsula. (Adapted from Chaparro, D. et al., *IEEE J. Sel. Topics Appl. Earth Observ. Remote Sens.*, 9, 2818–2829. With Permission.)

TABLE 11.2

Classification of the Model Output (Predicted Potential Burned Area) into Five Risk Categories

	Risk Categories				
Predicted area (ha)	<10	10–100	100–1,000	1,000–10,000	>10,000
Risk category	Low	Moderate	High	Very high	Extreme

that is, Mediterranean, northwestern, and central areas. This permitted to separate the main climate and fire regimes in the country. In Portugal, five regions were considered: northwestern Portugal, northeastern Portugal, Atlantic center, Alentejo, and Algarve (Figure 11.14). Finally, one land cover datum and one region were assigned to each fire in the database, considering the modal category within each wildfire perimeter.

The sample for each variable and category was detailed (see Table 11.2 in Chaparro et al. 2016d). Fires burning on January, May, November, and December were excluded as the sample for these months was lesser than 5. The same criterion was applied to land cover (agricultural areas were excluded) and regions (Alentejo, Algarve, and Central Atlantic regions were excluded). Finally, a linear model was fitted (Equation 11.2):

$$\text{LogArea} \approx c + M + \text{LC} + R + a_{(M,R)} \cdot \text{SM} + b_{(R)} \cdot \text{LST} \tag{11.2}$$

where:

c is the intercept
M, LC, and R correspond to the additive terms of month, land cover, and region, respectively
The term $a_{(M,R)}$ is the slope for SM, which depends on the month and the region
$b_{(R)}$ is the slope for LST, which depends on the region

The model explained 68% of the variance of the potential burned area. In particular, SM and LST explained 33.1% and 19.8%, respectively, land cover explained the 6.6%, regions explained 3.1%, and the month of occurrence explained 2.5%. Interestingly, SM was found to be the most important explanatory variable. Dry soils facilitated fire spread during all the studied months and in four of the studied regions. The effect was remarkable in the Mediterranean, coherently with the climate–fire

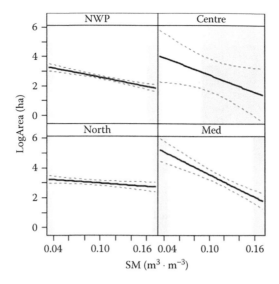

FIGURE 11.15 Modeled effect of soil moisture on the potential extension of wildfires for each of the significant interactions between soil moisture and regions: northwestern Portugal (NWP), central Peninsula (Centre), northern Iberian Peninsula (North), and Mediterranean (Med). Black lines show the modeled effects and dashed lines show 95% confidence intervals. Shaded rectangles show moisture values out of the range for each region, where effects are extrapolated. The effect was significant in the central region ($p < 0.05$) and highly significant ($p < 0.001$) in north, northwestern Portugal, and the Mediterranean. (Adapted from Chaparro, D. et al., *IEEE J. Sel. Topics Appl. Earth Observ. Remote Sens.*, 9, 2818–2829. With Permission.)

relationship in the region (Figure 11.15). High temperatures facilitated fire propagation in some of the northernmost regions (see Figure 11.16). On the contrary, an inverse relationship was found in the Mediterranean, where probably the high importance of moisture conditions and unstudied factors (e.g., wind) could explain this unexpected behavior (Figure 11.16). Finally, the largest spread of fires occurred in coniferous forests and in summer months.

The model was validated, and 83.3% of accuracy was obtained. The remaining 16.7% of fires burned larger areas than the maximum predicted. The maximum excess reported was 44.6 ha. As some months, land covers, and regions were eliminated from the model, the authors provided

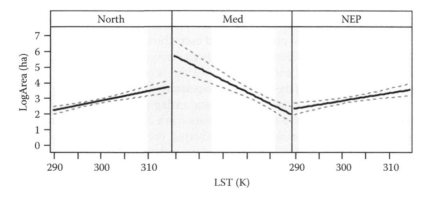

FIGURE 11.16 Modeled effect of the surface temperature on the potential extension of wildfires for each of the significant interactions between land surface temperature and regions: northern Iberian Peninsula (North), Mediterranean (Med) and northeastern Portugal (NEP). Black lines show the modeled effect and dashed lines show 95% confidence intervals. Shaded rectangles show temperature values out of the range for each region, where effects are extrapolated. All the effects were highly significant (p < 0.001). (Adapted from Chaparro, D. et al., *IEEE J. Sel. Topics Appl. Earth Observ. Remote Sens.*, 9, 2818–2829. With Permission.)

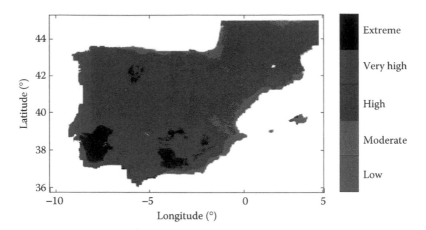

FIGURE 11.17 Fire risk map for June 24, 2010, obtained from the model explained in Chaparro et al. (2016d).

complementary equations excluding these variables one by one. The accuracy and the excess were similar to the main model (see Chaparro et al. 2016d).

Finally, the reclassification of the model outputs (Table 11.2) allows obtaining fire risk maps as the one presented in Figure 11.17.

In conclusion, the model provided a good estimate of the potential spread of fires. It demonstrated the applicability of spaceborne SM data and surface temperature information on fire risk modeling. Chaparro et al. (2016d) suggested that an operational implementation was feasible applying forecasted or observed temperatures to provide near real-time information. The model implementation is restricted to the Iberian Peninsula, and it should be recalibrated in order to apply it in other regions of the world. Finally, the need of a longer dataset was stressed in the article in order to account with a more robust model in all months, land covers, and regions.

11.6 CONCLUSIONS

Wildfires threaten humans and environment, and their occurrence and impacts are expected to increase under present climate changing conditions. In this context, remote sensing technologies offer a great opportunity to assess the potential risk of fires. Satellite missions and their derived products facilitate information for a wide range of environmental variables at different spatial scales. This information is currently processed, organized, and provided to end users through unified service platform initiatives such as the Copernicus program in Europe. Applications derived from Copernicus datasets encompass several thematic areas (e.g., WUIs, land cover and land use, vegetation conditions, or SM) and contribute to further our understanding of the fire risk phenomenon, assessment, and prevention.

In order to improve our capacity of preventing and extinguishing fires, as well as to manage fire impacts, it is essential to deal with fire-related information in a comprehensive manner, considering different fire stages: fire risk evaluation, active fire monitoring, burned areas, and postfire impacts. In that sense, the ESA–CCI Fire project, the Copernicus system, and the EFFIS are three initiatives that effectively contribute to enhance our capacity to deal with each of these fire aspects. In particular, the accurate assessment of fire risk is the crucial step preventing wildfire ignition and its derived impacts. At this moment, fire risk indices are based on meteorological data. The use of new remote sensing information could complement and improve these indices. Particularly, L-band missions launched during the last decade permit retrieving for the first time VOD and surface soil moisture information, which are linked to water content of live and dead fuels, and allow detecting drought situations posing a risk of fire.

The combined use of SM with surface temperature has been shown to provide an effective mean to evaluate fire risk conditions in the Iberian Peninsula. Empirical models in the region have been developed to set risk thresholds depending on moisture and temperature values calculated in absolute and relative terms. In addition, moisture and temperature anomalies with respect to mean climatic values have been related to adverse atmospheric conditions and fire episodes. Finally, a linear predictive model has been proposed to assess the potential spread of fires in near real time using satellite imagery only. Fire risk maps have been obtained using this approach and have been validated showing high accuracy (>80%).

Still, further enhancement and evaluation of remotely sensed data and fire risk models are needed for practical improvements in the field of wildfire prevention. First, the integration of soil and VWC data in fire risk indices should be explored. The interaction of water depletion with fire-related factors (e.g., LST, wind and land cover) should be further studied. Second, groundwater estimations from the Gravity Recovery and Climate Experiment (GRACE) allowed developing fire predictive maps in the United States (Skibba 2015), showing a new research line with high potentiality to improve fire risk assessment.

This review has detailed how recent remote sensing technologies deriving water content from soils and vegetation are opening a new path toward improving fire risk models. It is anticipated that, through their combination with the Copernicus datasets and their application within a comprehensive fire management framework, they will lead to significant progress in wildfire prevention services.

REFERENCES

Agência Portuguesa do Ambiente. 2015. Sistema Nacional de Informação de Ambiente. http://sniamb.apambiente.pt. Last accessed: March 29, 2016.

Amraoui, M., M. L. R. Liberato, T. J. Calado, C. C. DaCamara, L. P. Coelho, R. M. Trigo, and C. M. Gouveia. 2013. Fire activity over Mediterranean Europe based on information from Meteosat-8. *Forest Ecology and Management* 294:62–75. doi:10.1016/j.foreco.2012.08.032.

Bajocco, S. and C. Ricotta. 2008. Evidence of selective burning in Sardinia (Italy): Which land-cover classes do wildfires prefer? *Landscape Ecology* 23(2):241–48. doi:10.1007/s10980-007-9176-5.

Bartsch, A., H. Balzter, and C. George. 2009. The influence of regional surface soil moisture anomalies on forest fires in Siberia observed from satellites. *Environmental Research Letters* 4(4):045021.

BEC (Barcelona Expert Centre). 2015. CP34-BEC: BEC data distribution and visualization services. 2012–2015. http://cp34-bec.cmima.csic.es/data/data-access. Last accessed: June 13, 2015.

Bindlish, R., T. Jackson, M. Cosh, T. Zhao, and P. O'Neill. 2015. Global soil moisture from the Aquarius/SAC-D satellite: Description and initial assessment. *IEEE Geoscience and Remote Sensing Letters* 12(5):923–927. doi:10.1109/LGRS.2014.2364151.

Chaparro, D., M. Vall-llossera, M. Piles, A. Camps, and C. Rüdiger. 2015. Low soil moisture and high temperatures as indicators for forest fire occurrence and extent across the Iberian Peninsula. *Geoscience and Remote Sensing Symposium (IGARSS), 2015 IEEE International*, 3325–3328, July 26–31, 2015.

Chaparro, D., J. Vayreda, J., M. Vall-llossera, M. Banqué, M. Piles, A. Camps, and J. Martínez-Vilalta. 2016a. The role of climatic anomalies and soil moisture in the decline of drought-prone forests. *IEEE Journal of Selected Topics in Applied Earth Observations and Remote Sensing.* 10(2):503–514.

Chaparro, D., M. Piles, M. Vall-llossera, and A. Camps. 2016b. Surface moisture and temperature trends anticipate drought conditions linked to wildfire activity in the Iberian Peninsula. *European Journal of Remote Sensing.* 49:955–971.

Chaparro, D., M. Piles, and M. Vall-llossera. 2016c. Remotely sensed soil moisture as a key variable in wildfires prevention services: Towards new prediction tools using SMOS and SMAP data. In P. Srivastava, G. Petropoulos, Y. H. Kerr (Eds.) *Satellite Soil Moisture Retrieval.* Amsterdam, the Netherlands: Elsevier.

Chaparro, D., M. Vall-llossera, M. Piles, A. Camps, C. Rüdiger, and R. Riera-Tatché. 2016d. Predicting the extent of wildfires using remotely sensed soil moisture and temperature trends. *IEEE Journal of Selected Topics in Applied Earth Observations and Remote Sensing* 9(6):2818–2829.

Chuvieco, E., I. Aguado, S. Jurdao, M. L. Pettinari, M. Yebra, F. J. Salas, S. Hantson et al. 2014. Integrating geospatial information into fire risk assessment. *International Journal of Wildland Fire* 23(5): 606–619. doi:10.1071/WF12052.

Chuvieco, E., D. Cocero, D. Riaño, P. Martin, J. Martínez-Vega, J. De La Riva, and F. Pérez. 2004. Combining NDVI and surface temperature for the estimation of live fuel moisture content in forest fire danger rating. *Remote Sensing of Environment* 92(3):322–331. doi:10.1016/j.rse.2004.01.019.

Chuvieco, E., M. Deshayes, N. Stach, D. Cocero, and D. Riaño. 1999. Short-term fire risk: Foliage moisture content estimation from satellite data. In E. Chuvieco (Ed.) *Remote Sensing of Large Wildfires in the European Mediterranean Basin*. Berlin, Germany: Springer-Verlag, pp. 17–38.

Chuvieco, E., C. Yue, A. Heil, F. Mouillot, I. Alonso-Canas, M. Padilla, J. Miguel Pereira, D. Oom, and K. Tansey. 2016. A new global burned area product for climate assessment of fire impacts. *Global Ecology and Biogeography* 25(5):619–629. doi:10.1111/geb.12440.

Cohen, J. D. 2008. The wildland-urban interface fire problem: A consequence of the fire exclusion paradigm. *Forest History Today* 20–26.

Copernicus. 2016a. Copernicus in brief. http://www.copernicus.eu/main/copernicus-brief. Last accessed: July 10, 2016.

Copernicus. 2016b. Copernicus land monitoring services. http://land.copernicus.eu. Last accessed: July 10, 2016.

Deeming, J. E., R. E. Burgan, and J. D. Cohen. 1977. The national fire-danger rating system–1978. USDA forest service, rocky mountain forest and range experiment station, General Technical Report INT-39. Odgen, UT.

Dorigo, W. A., D. Chung, A. Gruber, S. Hahn, T. Mistelbauer, R. M. Parinussa, C. Paulik, C. Reimer, R. van der Schalie, R. A. M. de Jeu, W. Wagner. 2016. Soil Moisture [in State of the Climate in 2015]. *Bulletin of the American Meteorological Society*, 97(8):S31–32.

ECMWF. European Centre for Medium-Range Weather Forecasts. 2015. ERA-interim project. http://www.ecmwf.int/en/research/climate-reanalysis/era-interim. Last accessed: July 12, 2015.

EEA. European Environment Agency. 2006. Corine land cover map.

Entekhabi, D., E. G. Njoku, P. E. O'Neill, Y. H. Kellogg, W. T. Crow, W. N. Edelstein, J. K. Entin et al. 2010. The Soil moisture active passive (SMAP) mission. *Proceedings of the IEEE* 98:704–716.

ESA. European Space Agency. 2016b. Climate change initiative. http://cci.esa.int. Last accessed: July 10, 2016.

ESA. 2016a. Land Cover CCI. Product user guide, version 2. Available: http://maps.elie.ucl.ac.be/CCI/viewer/download/ESACCI-LC-PUG-v2.5.pdf. Last accessed: October 19, 2016.

ESA. European Space Agency. 2016b. Fire climate change initiative. http://www.esa-fire-cci.org/. Last accessed: July 12, 2016.

European Comission. 2010. Forest fires in Europe 2009, EUR 24502 EN, Luxembourg, German: Office for Official Publications of the European Communities, p. 81.

Fischer, E. M., S. I. Seneviratne, P. L. Vidale, D. Lüthi, and C. Schär. 2007. Soil moisture-atmosphere interactions during the 2003 European summer heat wave. *Journal of Climate* 20(20):5081–5099. doi:10.1175/JCLI4288.1.

Forkel, M., K. Thonicke, C. Beer, W. Cramer, S. Bartalev, and C. Schmullius. 2012. Extreme fire events are related to previous-year surface moisture conditions in permafrost-underlain larch forests of Siberia. *Environmental Research Letters* 7(4):044021. doi:10.1088/1748-9326/7/4/044021.

Government of Canada. 2016. Canadian wildland fire information system. http://cwfis.cfs.nrcan.gc.ca/background/summary/fwi. Last accessed: July 13, 2016.

Hardy, C. C. and R. E. Burgan. 1999. Evaluation of NDVI for monitoring live moisture in three vegetation types of the Western U.S. *Photogrammetric Engineering and Remote Sensing* 65:603–610.

Joint Research Center. 2015. EFFIS. under development. http://forest.jrc.ec.europa.eu/effis/about-effis/technical-background/under-development/. Last accessed: July 12, 2016.

Jolly, W. M., M. A. Cochrane, P. H. Freeborn, Z. A. Holden, T. J. Brown, G. J. Williamson, and D. M. J. S. Bowman. 2015. Climate-induced variations in global wildfire danger from 1979 to 2013. *Nature Communications* 6. doi:10.1038/ncomms8537.

Kerr, Y., P. Waldteufel, P. Richaume, J. P. Wigneron, P. Ferrazzoli, A. Mahmoodi, A. Al Bitar et al. 2012. The SMOS soil moisture retrieval algorithm. *IEEE Transactions on Geoscience and Remote Sensing* 50(5):1384–1403.

Knorr, W., I. Pytharoulis, G. P. Petropoulos, and N. Gobron. 2011. Combined use of weather forecasting and satellite remote sensing information for fire risk, fire and fire impact monitoring. *Computational Ecology and Software* 1:112–20.

Konings, A. G., M. Piles, K. Rötzer, K. A. McColl, S. K. Chan, and D. Entekhabi. 2016. Vegetation optical depth and scattering albedo retrieval using time series of dual-polarized L-band radiometer observations. *Remote Sensing of Environment* 172:178–189. doi:10.1016/j.rse.2015.11.009.

Konings, A.G., M. Piles, N. Das, and D. Entekhabi. 2017. L-band vegetation optical depth and effective scattering albedo estimation from SMAP. Remote Sensing of the Environment 198:460–470.

Li, X., W. Song, A. Lanorte, and R. Lasaponara. 2016. Remote sensing fire danger prediction models applied to Northern China. *International Conference on Computational Science and Its Applications*. Springer International Publishing, pp. 624–633.

Martínez-Fernández, J., A. González-Zamora, N. Sánchez, A. Gumuzzio. 2015. A soil water based index as a suitable agricultural drought indicator. *Journal of Hydrology* 522:265–273. doi:10.1016/j. jhydrol.2014.12.051.

Martínez-Fernández, J., A. González-Zamora, N. Sánchez, A. Gumuzzio, C. M. Herrero-Jiménez. 2016. Satellite soil moisture for agricultural drought monitoring: Assessment of the SMOS derived Soil Water Deficit Index. *Remote Sensing of Environment* 177:277–286. doi:10.1016/j.rse.2016.02.064.

McArthur, A. G. 1967. Fire behavior in eucalypt forests. Commonwealth of Australia Forestry and Timber Bureau, Leaflet 107. Canberra, Australia.

McKee, T. B., N. J. Doesken, J. Kleist. 1993. The relationship of drought frequency and duration to time scales. *Eighth Conference on Applied Climatology*. Boston, MA: American Meteorological Society.

Merlin, O, A. Chehbouni, Y. Kerr, E. Njoku, D. Entekhabi. 2005. A combined modelling and multispectral/ multiresolution remote sensing approach for disaggregation of surface soil moisture: Application to SMOS configuration. *IEEE Transactions on Geoscience and Remote Sensing* 43(9):2036–2050.

Merlin, O., J. Walker, A. Chehbouni, Y. Kerr, 2008. Towards deterministic downscaling of SMOS soil moisture using MODIS derived soil evaporative efficiency. *Remote Sensing of Environment* 112(10):3935–3946.

Mitsopoulos, I., G. Mallinis, and M. Arianoutsou. 2014. Wildfire risk assessment in a typical Mediterranean wildland-urban interface of Greece. *Environmental Management* 55(4):900–915. doi:10.1007/s00267-014-0432-6.

Moran, M. S., T. R. Clarke, Y. Inoue, and A. Vidal. 1994. Estimating crop water deficit using the relation between surface-air temperature and spectral vegetation index. *Remote Sensing of Environment* 49:246–263.

Moreno, J. M., E. Chuvieco, A. Cruz, E. García, E. de Luis, B. Pérez, F. Rodríguez et al. 2005. Impactos sobre los riesgos naturales de origen climático. Riesgo de incendios forestales. In *Evaluación Preliminar de los Impactos en España por Efecto del Cambio Climático*. Proyecto ECCE – Informe final [Coord. J. M. Moreno; ed. Ministerio de Medio Ambiente].

Naeimi, V., K. Scipal, Z. Bartalis, S. Hasenauer, and W. Wagner. 2009. An improved soil moisture retrieval algorithm for ERS and METOP scatterometer observations. *IEEE Transactions on Geoscience and Remote Sensing* 47(7):1999–2013. doi:10.1109/TGRS.2008.2011617.

Padilla, M. and C. Vega-García. 2011. On the comparative importance of fire danger rating indices and their integration with spatial and temporal variables for predicting daily human-caused fire occurrences in Spain. *International Journal of Wildland Fire* 20(1):46–58. doi:10.1071/WF09139.

Paes do Amaral, J. M. 2000. Zonas fitogeográficas predominantes. Ministerio do Ambiente e do Ordenamento do Território, Direcção-Geral do Ambiente. Lisboa.

Piles, M., A. Camps, M. Vall-llossera, I. Corbella, R. Panciera, C. Rüdiger, Y. H. Kerr, J. Walker. 2011a. Downscaling SMOS-derived soil moisture using MODIS visible/infrared data. *IEEE Transactions on Geoscience and Remote Sensing* 49:3156–3166.

Piles, M., A. Camps, M. Vall-llossera, A. Marín, and J. Martínez. 2011b. SMOS derived soil moisture at 1 km spatial resolution and first results of its application in identifying fire outbreaks. *Oral contribution to the 1st SMOS Science Conference*, Arles, France. September 27–29, 2011b. http://earth.eo.esa.int/workshops/smos_science_workshop/SESSION_6_OTHER_APPLICATIONS/M.Piles_SMOS_Fire%20Outbreaks_.pdf

Piles, M., D. Entekhabi, A. G. Konings, K. A. McColl, N. N. Das, and T. Jagdhuber. 2016. Multi-temporal microwave retrievals of soil moisture and vegetation parameters from SMAP. *Geoscience and Remote Sensing Symposium (IGARSS)*. 2016 IEEE International, July 10–15, 2016.

Piles, M., N. Sánchez, M. Vall-llossera, A. Camps, J. Martínez-Fernández, J. Martínez, and V. González. 2014. A downscaling approach for SMOS land observations: Evaluation of high-resolution soil moisture maps over the Iberian Peninsula. *IEEE Journal of Selected Topics in Applied Earth Observations and Remote Sensing* 7(9):3845–3857. doi: 10.1109/JSTARS.2014.2325398.

Roberto, C., B. Lorenzo, M. Michele, R. Micol, and P. Cinzia. 2012. Optical remote sensing of vegetation water content. In P. S. Thenkabail and J. G. Lyon (Eds.) *Hyperspectral Remote Sensing of Vegetation*. Boca Raton, FL: CRC Press (Taylor & Francis Group).

Ross, M. A., G. E. Ponce-Campos, M. L. Barnes, J. Hottenstein, and S. Moran. 2014. Response of grassland ecosystems to prolonged soil moisture deficit. *Geophysical Research Abstracts* 16, EGU General Assembly 2014.

San-Miguel-Ayanz, J., E. Schulte, G. Schmuck, A. Camia, P. Strobl, G. Liberta, C. Giovanda et al. 2012. Comprehensive monitoring of wildfires in Europe: The European Forest Fire Information System (EFFIS). In J. Tiefenbacher (Ed.) *Approaches to Managing Disaster–Assessing Hazards, Emergencies and Disaster Impacts.* Rijeka, Croatia: InTech.

San-Miguel-Ayanz, J., E. Schulte, G. Schmuck, A. Camia. 2013. The European forest fire information system in the context of environmental policies of the European Union. *Forest Policy and Economics* 29:19–25.

Sánchez, N., A. González-Zamora, M. Piles, J. Martínez-Fernández. 2016. A new Soil moisture agricultural drought index (SMADI) integrating MODIS and SMOS products: A case of study over the Iberian Peninsula. *Remote Sensing* 8(4):287.

Sandholt, I., K. Rasmussen, and J. Andersen. 2002. A simple interpretation of the surface temperature/vegetation index space for assessment of surface moisture status. *Remote Sensing of Environment* 79:213–224.

Scaini, A., N. Sánchez, S. M. Vicente-Serrano, J. Martínez-Fernández. 2014. SMOS-derived soil moisture anomalies and drought indices: A comparative analysis using *in situ* measurements. *Hydrological Processes* 29(3):373–383. doi:10.1002/hyp.10150.

Shvetsov, E. 2013. Fire danger estimation in Siberia using SMOS data. Geophysical research abstracts 15. EGU General Assembly 2013.

Skibba, R. 2015. Assessing U.S. fire risks using soil moisture satellite data. *Earth & Space Science News* 96. doi:10.1029/2015EO042071. Published on December 17, 2015.

Syphard, A. D., V. C. Radeloff, N. S. Keuler, R. S. Taylor, T. J. Hawbaker, S. I. Stewart, and M. K. Clayton. 2008. Predicting spatial patterns of fire on a southern California landscape. *International Journal of Wildland Fire* 17:602–613.

Tomlinson, C. J., L. Chapman, J. E. Thornes, and C. Baker. 2011. Remote sensing land surface temperature for meteorology and climatology: A review. *Meteorological Applications* 18(3):296–306. doi:10.1002/met.287.

Trigo, R. M., J. M. C. Pereira, M. G. Pereira, B. Mota, T. J. Calado, C. C. Dacamara, F. E. Santo. 2006. Atmospheric conditions associated with the exceptional fire season of 2003 in Portugal. *International Journal of Climatology* 26:1741–1757.

Ulaby, F. T., R. K. Moore, A. K. Fung. 1981. Microwave Remote Sensing: Microwave remote sensing fundamentals and radiometry. Vol. 1, Norwood, MA: Artech House.

Ullah, S., A. K. Skidmore, A. Ramoelo, T. A. Groen, M. Naeem, A. Ali A. 2014. Retrieval of leaf water content spanning the visible to thermal infrared spectra. *ISPRS Journal of Photogrammetry and Remote Sensing* 93:56–64.

Van Wagner, C. E. 1987. Development and structure of the Canadian forest fire weather index system. Government of Canada, Canadian Forestry Service, Forest technical report 35. Ottawa.

Verdú, F., J. Salas, and C. Vega-García. 2012. A multivariate analysis of biophysical factors and forest fires in Spain, 1991–2005. *International Journal of Wildland Fire* 21(5):498–509. doi:10.1071/WF11100.

Vicente-Serrano, S. M., S. Beguería, and J. I. López-Moreno. 2010. A multiscalar drought index sensitive to global warming: The standardized precipitation evapotranspiration index. *Journal of Climate* 23:1696–1718.

Vilar, L., A. Camia, J. San-Miguel-Ayanz. 2015. A comparison of remote sensing products and forest fire statistics for improving fire information in Mediterranean Europe. *European Journal of Remote Sensing* 48:345–364. doi:10.5721/EuJRS20154820.

Wang, L., Y. Zhou, W. Zhou, and S. Wang. 2013. Fire danger assessment with remote sensing: A case study in Northern China. *Natural Hazards* 65:819–834. doi:10.1007/s11069-012-0391-2.

Whelan. R. J. 1995. *The Ecology of Fire.* Cambridge University Press, Cambridge.

12 Remote Sensing of Fire Effects
A Review for Recent Advances in Burned Area and Burn Severity Mapping

Ran Meng and Feng Zhao

CONTENTS

12.1 INTRODUCTION

As a primary disturbance agent, fire significantly alters ecological processes and ecosystem services around the world, driving the changes in terrestrial carbon stocks; shaping the distribution and structure of vegetation; and influencing the temporal variability in carbon, water, and energy fluxes (Bowman et al. 2009, Scott et al. 2013, Franklin et al. 2016). For example, fire-related deforestation is a net CO_2 source with a flux estimated to be 2.1 Pg C per year (Van der Werf et al. 2010), whereas postfire forest recovery is a CO_2 sink and might be enhanced by proper management (Bowman et al. 2009); the water yield of river catchments was also found to be significantly influenced by fire effects and the postfire vegetation recovery process (Benda et al. 2003, Mayor et al. 2007). Due to the importance of fire on these fundamental ecosystem processes, accurately monitoring the effects of fire events (i.e., time, location, and severity) is thus one of the central questions in ecology and natural resource management. In addition, projection of fire behavior under potential future climate also relies on the proper characterization of fire effects at local, regional, and global levels.

Burned area and burn severity are the two most widely used metrics for assessing fire effects (Turner et al. 1997 and 1999, Lentile et al. 2006, Meng et al. 2015) for calculating smoke generation and carbon consumption (Miller and Yool 2002, Randerson et al. 2012), for characterizing fire regimes (Morgan et al. 2001, Keane et al. 2003, Kasischke and Turetsky 2006), and for modeling the feedback between climate change and fire activity (Randerson et al. 2006, Westerling et al. 2006, Loehman et al. 2011, Smithwick et al. 2011, McKenzie and Littell 2016). Burned areas are usually composed of complex landscape mosaics of low, moderate, and high burn severity (Figure 12.1) because of variations in wind, topography, fuel conditions, and so on (Turner et al. 1994). The variable burn severity results in a heterogeneous pattern of fire effects including vegetation loss and soil alteration (Sugihara 2006, Lentile et al. 2006, Keeley 2009, Veraverbeke et al. 2011, Quintano et al. 2013). Burn severity refers to the degree in which an ecosystem has changed (e.g., vegetation removal, soil exposure, and soil color alteration), caused by fire disturbance. Although often used interchangeably nowadays (Keeley 2009), Lentile et al. (2006) discussed and clarified the distinctions between the term of burn severity and fire severity: fire severity refers to short-term (i.e., about within 1 year following the fire) effects on the local environment, whereas burn severity refers to both short-term and long-term (i.e., up to ten years) effects, including ecosystem response processes (e.g., vegetation recovery). Recently, Composite Burn Index (CBI, a generalized rating of postfire conditions in the field) and its variant GeoCBI have gradually become the standard protocol to measure field burn severity at landscape scale (Key and Benson 2006; De Santis et al.

FIGURE 12.1 Plot examples of unburned (first row), low (second row), moderate (third row) and high (fourth row) burn severity across *in situ* (first column) photos at May, 2016, 0.10 m aerial ortho-photos (second column) at May, 2012, 2 m WorldView-2 imagery at September, 2012, and 30 m Landsat-7 imagery spatial scales at September, 2012 in a Pine Barrens ecosystem in the Eastern United States.

2009). Specifically, as an integrated metric, (Geo)CBI averages the magnitude of change by fire across five strata from soil to vegetation canopies, then within each strata four or five variables are visually assessed and assigned a value from zero (unburned) to three (highest severity).

Remote sensing has provided a convenient and consistent way to monitor fire events and quantity fire effects across spatial scales. Remote sensing sensors measure reflected energy within specified regions of the electromagnetic spectrum, which is known as a band or bandwidth. Each band responds differently to surficial characteristics such as water, soil, and vegetation. A common practice to enhance information from target features is to combine brightness values of multiple bands, such as the red, near-infrared (NIR), and shortwave near-infrared bands. Unique spectral signatures of vegetation and burn residuals become the foundation for detecting vegetation change by fires (Figure 12.2). Since the late-1970s and 1980s, remote sensing technique has been widely used to assess how *severe* is the fire. Different variables have been measured as ground reference readily to assess burn severity from remotely sensed measurements (Morgan et al. 2014). Fire effects lead to the changes in spectral response and make the remote sensing of burn severity possible. After a fire, a dramatic reduction in visible to NIR surface reflectance (i.e., 0.4–1.3 µm) associated with the charring and removal of vegetation is the dominant signals detected by pre- and postfire sensors at moderate–coarse spatial resolution; at fine spatial scales (< 5 m), an increase in surface reflectance is likely detected, due to the deposition of white ash, as an indicator of combustion completeness (higher burn severity). With the increase in wildfire's size, severity, and frequency over recent decades, there are increasing interests in remote sensing of fire effects and the potential impact of climate change on wildfire activities. We did a series of searches in Web of Science to examine the current research on remote sensing of fire effects, with keywords such as *burned area remote sensing* and *burn severity remote sensing*. Results from such investigation show that the number of publications on fire effects has been increasing over the past decade, especially after the year 2002, in consistent with the increase in large wildfire events around the world (Figure 12.3).

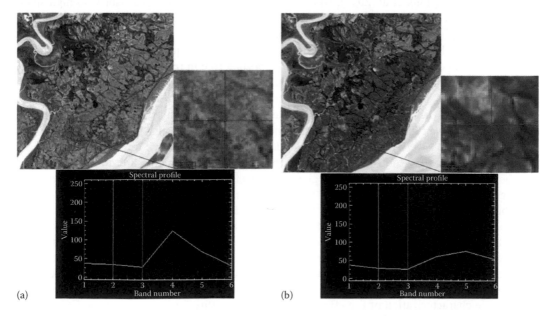

FIGURE 12.2 Vegetation spectral reflectance of Landsat bands 1–6 for (a) prefire and (b) postfire conditions. The spectral profile shows the reflectance value for the center pixel at the crosshair mark.

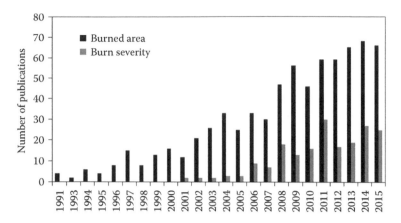

FIGURE 12.3 Number of publications related to remote sensing of burned area and burn severity in the Web of Science database from 1991 to 2015.

As fire effects can vary at different scales, one spatial or temporal scale may not be appropriate to address all objectives for assessing burn severity (Morgan et al. 2014). Over the past three decades, various satellite remote sensing-based approaches have been developed to monitor fire events at coarse, moderate, and fine resolutions. A few studies have already reviewed the application of moderate–coarse resolution remotely sensed measurement in mapping large-scale fire characteristics (Lentile et al. 2006, Chu and Guo 2013, Roy et al. 2013, Morgan et al. 2014), so here we will focus on remote sensing of fire effects at moderate and high spatial resolution in this review. A number of studies with different types of satellite imagery and approaches have been conducted for burn severity assessment. Moderate Resolution Imaging Spectroradiometer (MODIS) fire product is one of the most popular datasets for wildfire studies across the globe. The Landsat sensors provide one of the longest and widely used imagery collections for wildfire monitoring, especially for burn severity applications (Eidenshink et al. 2007); while images from newer launched sensor with high spatial resolution, such as WorldView-2 and QuickBird, also incur interest in very high spatial resolution (VHR) fire mapping (Holden et al., 2010; Meng et al., 2017).

In addition to the type and resolution of imagery used, image acquisition date, in relation to field data collection and time since fire, also plays an important role in remote sensing of burn severity: interannual phonological change of vegetation, the interaction of long-term climate patterns (i.e., drought), and regeneration trends might confuse varying fire effects. What is more, challenges still exist in the repeatable and transferable assessment of burn severity across spatial scales or fire regimes, given the limited mechanistic and predictive power of widely used but subjective descriptors of burn severity (unburned, low, moderate and high severity): thresholds on the widely used Normalized Burn Ratio (NBR)-based burn severity measurements are arbitrary and often vary between fires within the same ecoregion (Kolden et al. 2015). A new paradigm in burn severity assessment, based on a consistent and transferable quantification of burn severity (e.g., changes in carbon, water and energy fluxes), has been discussed and explored in the community recently (Morgan et al. 2014; Smith et al. 2016; Sparks et al. 2016; Meng et al. 2017). With the development of remote sensing techniques (i.e., light detection and ranging [LiDAR], hyperspectral, and VHR imagery) these years, fire measurements with high temporal, spatial, and spectral resolution become increasingly available and provide new opportunities in remote sensing of fire effects studies (Montealegre et al. 2014, Schepers et al. 2014, McCarley et al. 2017, Meng et al. 2017).

Accurate characterization of fire effects is critical for postfire forest management. Effective fire management is reliant on reliable information on which to base appropriate decisions and actions. With projected increasing occurrences of wildfires under the current climate change scenarios, there are urgent needs to better characterize the impact of fires on ecosystem dynamics and processes.

In this chapter, we discuss the recent advances of remote sensing applications in monitoring burned areas and burn severity at local to regional and global scales. In the following sections, we summarize key mapping techniques for both burned areas and burn severity, respectively, and also discuss the potential future directions in characterizing burned areas and burn severity.

12.2 REMOTE SENSING OF BURNED AREA

One of the key remote sensing measurements of fire effects is the burned area. Fires produce a significant change in the structure and the reflectance of vegetation and the soil properties within the burned area that are noticeable in the microwave, visible, and especially the infrared part of the electromagnetic spectrum (Leblon et al. 2012). In this section, we will discuss the remote sensing of burned areas, by techniques.

A variety of techniques have been employed for burned area mapping. These techniques can be grouped into five types of approaches:

1. Burned area estimation using active fire counts
2. Image classification using the spectral properties of burned residues
3. Burned areas estimation using changes in canopy cover
4. Burned areas classification using active remote sensing sensors
5. Burned area mapping using hybrid classification approach (Table 12.1)

12.2.1 BURNED AREA ESTIMATION USING ACTIVE FIRE COUNTS

Active fire count products capture the location and timing of fire burning at the time of the satellite overpass, usually as swath-based fire masks or lists of fire pixel locations and dates (Giglio et al. 2006). Globally, long-term observations of active fires made with coarse- and medium-resolution spaceborne sensors are readily available. Selected examples of these observations

TABLE 12.1

Summary of Major Burned Area Mapping Methods and Selected References

Burned Area Mapping Methods	Types	Selected References
2.1 Active fire counts	Aggregate active fire detections	Giglio et al. (2006)
		Oliva and Schroeder (2015)
2.2 Spectral change detection approach	Multitemporal composites	Chuvieco et al. (2008)
	Spectral indices (SIs)	Key and Benson (2006)
	Spectral mixture analysis (SMA)	Quintano et al. (2006)
	Machine learning classification	Petropoulos et al. (2010, 2011)
		Hudak and Brockett (2004)
	Time series change detection	Goodwin and Collett (2014)
2.3 Canopy cover change detection	Changes in Leaf Area Index	Boer et al. (2008)
	Forest cover loss	Potapov et al. (2008)
2.4 Active remote sensing	Synthetic aperture radar (SAR)	Siegert and Hoffman (2000)
		Gimeno et al. (2004)
		Kasischke et al. (2008)
	PALSAR	Polychronaki et al. (2013)
2.5 Hybrid approach	SIs + thermal	Roy et al. (1999)
	Time series change detection + machine learning classification	Zhao et al. (2015)
		Kennedy et al. (2015)
		Schroeder et al. (2015)

TABLE 12.2

Name, Equation, and References for Major Vegetation Spectral Indices (SIs) in Burned Area Mapping

Index	Equation	Reference
Normalized Difference Vegetation Index (NDVI)	$(\rho4 - \rho3)/(\rho4 + \rho3)$	Tucker et al. (1986)
Enhanced Vegetation Index (EVI)	$(\rho5 - \rho4)/(\rho5 + 6 * \rho4 - 7.5 * \rho2 + 1)$	Gao et al. (2000)
Soil Adjusted Vegetation Index (SAVI)	$((\rho5 - \rho4)/(\rho5 + \rho4 + 0.5)) * 1.5$	Huete (1988)
Normalized Burn Ratio (NBR)	$(\rho4 - \rho7)/(\rho4 + \rho7)$	Key and Benson (2006)

Note: $\rho3$, $\rho4$, and $\rho7$ represent the surface spectral reflectances as measured in Bands 3 (red band, 0.3–0.69 µm), 4 (near-infrared band, 0.76–0.90 µm), and 7 (Shortwave infrared band, 2.08–2.35 µm) of the Landsat Thematic Mapper and Enhanced Thematic Mapper Sensors

include the advanced very high resolution radiometer (AVHRR) active fire product (Li et al. 2001), along track scanning radiometer (ATSR) nighttime fire product (Schultz 2002), the MODIS global active fire product (Giglio 2010), and the visible infrared imaging radiometer suite (VIIRS) global active fire product (Schroeder et al. 2014).

Although these fire count products capture many aspects of the spatial and temporal distribution of burning, it is difficult to relate them to actual area burned due to inadequate spatial and temporal resolutions, variability in cloud cover and fuel conditions, and differences in fire behavior (Giglio et al. 2009, Oliva and Schroeder 2015). In addition, the probability of active fire detection is dependent on the fire temperature and size: small- and/or low-intensity fires may not be detected at the time of satellite overpass (Boschetti et al. 2015). Cumulative active fire detection algorithms usually underestimate the area burned in grassland and savanna ecosystems where the fires front progresses rapidly across the landscape (Roy et al. 2008, Oliva and Schroeder 2015). Conversely, active fire detection methods may overestimate the area burned for isolated fire points that are detected but very hot and smaller than the pixel dimension, for example, in certain forest ecosystems where the fuel conditions can sustain high energy fires and where the fire spread is slow relative to the satellite overpass frequency (Boschetti et al. 2015).

Several studies reported burned areas mapped from aggregated active fire detections for large fires. Many studies underestimated the burned areas due to cloud contamination and difficulty in detecting small fires with coarse satellite images. For example, Sukhinin et al. (2004) used aggregated active fire detection data from the AVHRR to estimate burned areas in Russia in 2000, underestimating the total area measured by 27% (Sukhinin et al. 2004). Oliva and Schroeder (2015) assessed the performance of the VIIRS 375 m active fire detection product for direct burned area mapping. Fire detection rates were lower for small fires (<500 ha), and mapping accuracies highly depend on ecosystem and vegetation conditions, where forested areas have higher accuracy than grassland, savannas, and agricultural areas due to differences in the duration of fires. Although active fire detection-based burned area product generally has higher commission and omission errors compared with other mapping approach, it was one of the first methods that were developed to consistently map burned areas at global scale (Table 12.2).

12.2.2 Image Classification Using Burned Areas' Spectral Properties

Fires change the spectral signatures of the land surface by reducing the cover, greenness, density, and water content of the vegetation, by partially or completely consuming surface litter fuel, and by exposing and altering the color and brightness of the soil (Lentile et al. 2006). Such changes in surface properties can often be detected as a decrease in spectral reflectance in the visible NIR and an increase in the mid-infrared (MIR) wavelengths. The vast majority of satellite-based burned area mapping studies use information on differences in spectral properties of a land surface before and after a fire.

Vegetation spectral indices (SIs) are the most widely used approach to detect fire-induced spectral reflectance changes because of the simple concept and easy computation. Table 12.2 summarizes the commonly used vegetation indices for mapping burned areas. Normalized Difference Vegetation Index (NDVI) is strongly related to above-ground biomass and as a result, the index has shown to discriminate reasonably well between burned and unburned areas (Tucker et al. 1986). Several modifications of NDVI, including the Enhanced Vegetation Index (EVI) (Gao et al. 2000), the Soil Adjusted Vegetation Index (SAVI) (Huete 1988), and Modified SAVI (MSAVI) (Qi et al. 1994), have been successfully applied in burned area mapping. NBR is commonly used for burn severity assessment, whereas its application in burned areas mapping is also widespread and has been found to outperform the other SIs in some ecoregions (Schepers et al. 2014). Modifications for NBR include differenced NBR (dNBR) and relative dNBR (RdNBR). A consistent burned area product for United States—Monitoring Trends of Burn Severity—was developed by delineating the calculated dNBR images from Landsat (Eidenshink et al. 2007). A more recently developed spectral index, Integrated Forest Index (IFI), representing the probability that a pixel is forest based on the image statistics (Huang et al. 2009, 2010, Chen et al. 2011) has also been applied to track burned areas in the Greater Yellowstone Ecosystems (Zhao et al. 2015), as well as the postfire forest spectral recovery (Zhao et al. 2016). Despite the popularity of SIs in mapping burned areas, it is still limited in its ability to accurately characterize burned area across different ecosystems due to spectral confusion with shaded surface such as cloud shadow or topography variations (Schepers et al. 2014).

Spectral mixture analysis (SMA) is a classification technique based on modeling image spectra as the linear combination of endmembers (e.g., soil, vegetation, nonvegetation, etc.) has been used to derive the fractional contribution of endmember materials to image spectra in a wide variety of applications including burned area assessment (Riaño et al. 2002, Dennison and Roberts 2003). Several studies examined the usefulness of SMA for mapping burned areas from sensors such as AVHRR, MODIS, and Landsat, and the results show that SMA could accurately identify the burned surface area, and the spatial resolution of the satellite images do not affect burned area mapping (Vafeidis and Drake 2005, Quintano et al. 2006).

Machine learning algorithms are a group of statistical algorithm that can learn and make predictions on data (Kohavi and Provost 1998). A number of studies have also investigated the utility of machine learning algorithms in classifying burned areas, such as principal components analysis (PCA) (Hudak and Brockett 2004, Maingi and Henry 2007), support vector machines (SVM) (Petropoulos et al. 2011, Zhao et al. 2015), neural networks (NN) (Shabanov et al. 2005, Petropoulos et al. 2010), spectral angle mapper (SAM) (Petropoulos et al. 2010), decision trees (Silva et al. 2005, Giglio et al. 2009, Loboda et al. 2011, Hall et al. 2016), random forests (RF), and so on. These methods were proven to be effective in mapping burned areas in many ecosystems, but spectral confusion with cloud shadow and shades from high topography variations could be difficult to separate (Petropoulos et al. 2011).

In the past decade, high and very high spatial resolution burned area mapping using spectral characteristics began to grow rapidly. Data availability from sensors such as Landsat and Worldview-2 made it possible to characterize burned area at the meter level. The processing of these imageries provides a great level of spatial details that are needed for the accurate analysis of fire damages and for the sound planning of postfire restoration measures. Mitri and Gitas (2008) used 1 m Ikonos images to map object-based burn severity in open Mediterranean forests. Holden et al. (2010) assessed burn severity using 3 m QuickBird differenced spectral index from prefire to postfire. Meng et al. (2017) endeavored to produce ecological meaningful and scalable burn severity from WorldView-2 images in an imperiled fire-dependent Pine Barren Ecosystem in Northeastern United States.

12.2.3 BURNED AREAS ESTIMATION USING CHANGES IN CANOPY COVER

Change in green vegetation cover manifests one of the most visible and ecologically significant impacts of fire. Partly limited by the data availability of vegetation cover, however, burned area mapping based on canopy cover change is not as common as the spectral-based methods. Boer et al. (2008) proposed to quantify burned area and burn severity as the change of the Leaf Area Index (LAI) of a 27,700 ha fire in Australia, by linking spectral information from satellite images with field-derived LAI estimates. LAI was found to be strongly related to NBR, and changes in NBR can largely be explained by the variations in ground measured LAI ($R^2 = 0.76$). Potapov et al. (2008) examined forest cover loss induced by wildfires by combining MODIS and Landsat imagery, and the MODIS-derived burned forest area fraction resulted in an estimated root-mean-square error (RMSE) of 2.24% and R^2 of 0.75. Hansen et al. (2013) developed global 30 m forest cover maps of the twenty-first century, and wildfire was found to be the dominant disturbance agent in regions such as forests in the boreal and the intermountain West of North America. Classification accuracies were found to be over 80% for each individual climate domain and the globe as a whole.

12.2.4 BURNED AREA MAPPING USING ACTIVE REMOTE SENSING SENSORS

Remote sensing instruments can be grouped into two types: passive and active. Although passive instruments collect information on energy that is reflected or emitted from the observed object, active instruments provide their own energy source (electromagnetic radiation) and send a pulse of energy from the sensor to the object. Active instruments then receive the radiation that is reflected or backscattered from the object.

Active sensors such as the synthetic aperture radar (SAR) and LiDAR were widely used in mapping burned areas. Many SAR-based burned area studies were carried out in the boreal forest (Kasischke et al. 1994, French et al. 1999, Siegert and Hoffmann 2000, Menges et al. 2004), but some examples for the Mediterranean area exist (Gimeno et al. 2004, Tanase et al. 2010a,2010b, Stroppiana et al. 2015). Instead of the changes in vegetation condition and structure, the detection of burned area from SAR is based on the changes in water content in the burned surface compared with the unburned areas. Burned areas tend to have high water content than unburned areas, which reduces the backscatter. Thus, burned areas appear as relatively darker objects compared with the surrounding nonaffected areas. Images from passive remote sensing instruments, such as Landsat thematic mapper (TM)/enhanced thematic mapper (ETM), are well suited for capturing horizontally distributed forest conditions, structure, and change, whereas LiDAR data are more appropriate for capturing vertically distributed elements of forest structure and change. Therefore, the integration of passive optical and active LiDAR remote sensing can often provide improved accuracies in characterizing postfire effects, especially forest recovery following fires. Ballhorn et al. (2009) used LiDAR to derive burn scar depth and carbon emissions from peatland fires in Indonesia. Goetz et al. (2010) synergized spaceborne LiDAR data and MODIS data to assess the vegetation response following fires in Alaska. Wulder et al. (2009) characterized boreal forest wildfires using multitemporal Landsat and LiDAR data. In addition, the synergies between active remote sensing sensors, as well as active and passive sensors, show promising results in burned area mapping in recent decades (Kane et al. 2014; McCarley et al. 2017).

Radar and LiDAR applications in burned area mapping are relatively narrow due to difficulty in data interpretation and limited data availability, respectively. But with their advantages compared with traditional remote sensing instruments, such as penetration through cloud and not getting saturated at high biomass levels, radar and LiDAR instruments have potentials for assessing crown bulk density (Lentile et al. 2006) and loss from underground burnings (Reddy et al. 2015), when optical remote sensing is not capable of doing so.

12.2.5 BURNED AREA MAPPING USING HYBRID ALGORITHMS

With the development of burned area mapping techniques, many hybrid algorithms have been developed to further improve mapping accuracies. Methods such as SIs, machine learning classification, and so on have been integrated in mapping burned areas in many regions (Roy et al. 2013). Information from thermal remote sensing and land cover/land use change has also been proven to be useful in many studies (Chu and Guo 2013).

Although most of the studies on burned area mapping were based on the use of optical imagery, temperature information derived from thermal remote sensing is sometimes used in combination with spectral changes to better characterize burned areas (Wooster et al. 2013). Roy et al. (1999) used AVHRR thermal channel, in combination with the NIR channel, to detect burn scar in a savanna fire in Southern Africa. Fraser et al. (2000) further integrated AVHRR thermal band with multitemporal NDVI to examine a large burned area in boreal forests (Fraser et al. 2000). Alonso-Canas and Chuveico integrated hot spot information from MODIS data and temporal trends of mEdium resolution imaging spectrometer (MERIS) reflectance bands to develop a hybrid burn area algorithm (Alonso-Canas and Chuvieco 2015). Roy and Kumar (2017) combined MODIS active fire product and random forest algorithm to classify burned areas in the Amazon basin into deforestation, maintenance, and forest fire types.

Many land cover change detection algorithms have been developed to process Landsat pixel-level time series over large areas, focusing on mapping forest land cover change and forest disturbance using methods that identify significant changes by examination of the temporal trajectory of surface reflectance or vegetation indices (Huang et al. 2010, Kennedy et al. 2010, Verbesselt et al. 2010, Hansen et al. 2013). A number of hybrid land cover change (including burned area) detection approaches have been developed that integrate those forest change detection algorithms with machine learning algorithms to attribute the causes of forest change such as wildfire or timber harvesting. Zhao et al. (2015) used time series forest change maps produced by vegetation change tracker (Huang et al. 2010) and SVM to separate fires from harvests in the Greater Yellowstone Ecosystem from 1985 to 2011. The overall classification accuracy was about 85%, and the integration of vegetation change tracker (VCT) and SVM algorithm was proven effective in mapping fires and harvests in ecosystem such as the Greater Yellowstone. Kennedy et al. (2015) integrated temporal segmentation and RF to attribute forest disturbance change agent in support of habitat monitoring in the Puget Sound region. Overall accuracy was 80%, and mapping accuracy of burned area was lower than that of the forest management activities such as harvesting. The use of hybrid approach to map burned areas often achieves high mapping accuracies, but the drawback is that this

---

TABLE 12.3
Relative Advantages and Disadvantages of Main Burned Area Mapping Methods

Burned Area Mapping Method	Pros	Cons
2.1 Active fire counts	Consistent algorithm/product available globally; Relatively high temporal resolution	Relatively high omission and commission errors; Inadequate for accurate burned area mapping (lack the ability to detect small fires)
2.2 Spectral change detection approach	Most common, relatively easy to calculate	Relatively high commission error in some ecosystems
2.3 Canopy cover change detection	Direct measurement of fire effects on canopy change, ecological meaningful	Limited data availability
2.4 Active remote sensing	Provide more information than optical remote sensing (such as for forest structural or water content)	Limited data availability; Difficulty in interpretation; Computing load too high for large area assessment
2.5 Hybrid approach	Relatively high mapping accuracy	Complex procedures and high computation loads

approach usually involves many processing steps with high computation loads (Table 12.3). With the advances of super computing resources, however, computing load might not be a limiting factor in the near future.

12.3 REMOTE SENSING OF BURN SEVERITY

The assessment of short- and long-term fire effects on local, regional, and global vegetation has been conducted using a range of remote sensing methods. Burn severity is one of the most commonly used assessments to monitor and assess the impacts of fires on local and regional environments by remotely sensed imagery. In a fire event, the burned area is usually a complex mosaic of low, moderate, and high fire intensity because of variations in wind, topography, and fuel conditions (Turner et al. 1994). Variable fire intensity results in a heterogeneous pattern of fire effects promoting landscape heterogeneity and ecosystem biodiversity; both are considered to be important for ecosystem resilience (Peterson 2002, Sugihara 2006, Lentile et al. 2006, Keeley 2009, Veraverbeke et al. 2011, Quintano et al. 2013, Wilson et al. 2015, Spasojevic et al. 2016).

In general, four different techniques (Table 12.2) were used to assess burn severity across large areas, including remotely sensed SIs (e.g., Miller et al. 2009; Norton et al. 2009; Lu et al. 2015), radiative transfer models (RTMs) (e.g., Chuvieco et al. 2006; De Santis et al. 2009), SMA (e.g., Riaño et al. 2002; Veraverbeke et al. 2012; Quintano et al. 2013, 2017), and supervised and unsupervised classification (Mitri and Gitas 2013; Chen et al. 2015). In addition to the previously mentioned techniques, as the development of new remote sensing techniques, new remotely sensed datasets including imaging spectrometer (e.g., Scheper et al. 2014), LiDAR (e.g., Wang and Glenn 2009), Radar (e.g., Kasischke et al. 2008; Tanase et al. 2010a,2010b), and VHR imagery (e.g., Holden et al. 2010; Meng et al. 2017) have also become available and applied for burn severity assessment. Recently, the hybrid approach for combining the SI and SMA or supervised classification techniques (e.g., Quintano et al. 2017; Meng et al. 2017) has also been explored for burn severity assessment (Table 12.4).

TABLE 12.4
Summary of Major Burned Severity Mapping Methods and Selected References

Burned Severity Mapping Methods	Types	Selected References
3.1 Spectral indices (SIs)	NDVI	Escuin et al. (2008)
	NBR family	Miller et al. (2007)
		Miller et al. (2009)
	RBR	Parks et al. (2014)
	SAVI family	Schepers et al. (2014)
		Arnett et al. (2015)
3.2 Spectral Mixture Analysis (SMA)	Simple linear SMA	Rogan and Franklin (2001)
	Multiple endmember SMA	Quintano et al. (2013)
3.3 Radiative Transfer Model (RTM)	Turbid RTMs	Chuvieco et al. (2006)
		De Santis et al. (2007)
	Geometric RTMs	De Santis et al. (2009)
3.4 Classification	Supervised classification	Hultquist et al. (2014)
	Unsupervised classification	Roldán-Zamarrón et al. (2006)
3.5 Hybrid approach	SMA and SIs	Meng et al. (2017)
	LiDAR-based measurements and SIs or LiDAR-based measurements	Wang and Glenn (2009) McCarley et al. (2017)
	Thermal-infrared and SIs or Thermal-infrared bands	Zheng et al. (2016) Quintano et al. (2017)

12.3.1 Representative Studies of Spectral Indices-Based Burn Severity

Among these techniques, SIs have been most widely applied to images with varying spectral and spatial resolutions (Van Wagtendonk et al. 2004, Epting et al. 2005, Miller and Thode 2007, Schepers et al. 2014). Density slicing is one of the most commonly used methods for classifying the continuous SI values into several subjective categories (e.g., unburned, low, moderate, and high). The NDVI, the NBR, or other similar SIs are frequently developed and applied in this context in various ecosystems. The capability of these SIs for assessing burn severity comes from the dramatic changes in red, NIR, MIR, shortwave infrared (SWIR) regions of the spectrum, as a result of fire (White et al. 1996; Lentile et al. 2006). A recent study indicated that SIs (e.g., dNDVI) could accurately quantify changed plant physiology caused by fire at the leaf level and thus could potentially be used for physics-based burn severity assessment in future (Smith et al. 2016; Sparks et al. 2016).

The performances of different SIs for assessing burn severity have been compared and discussed in numerous literature for various ecosystems including boreal forest (Barrett et al. 2010; Chu et al. 2016), Mediterranean forests and shrublands (Escuin et al. 2008, Harris et al. 2011), and temperate forests (Chen et al. 2011; Meng et al. 2017). The use of a single postfire image without prefire reference leads to confusion in assessing fire severity due to the spectral similarities with other sparsely vegetated areas; however, bitemporal image-based approach could also introduce additional problems, such as differences in geometric correction, atmospheric effects, illumination effects, plant phenology, and so on. The bitemporal NBR-based approaches have been predominantly used in the community and accepted as the standard spectral index to assess the burn severity across the United States going back to 1984 with the 30 m Landsat mission imagery, by the Monitoring Trends in Burn Severity Project (MTBS, http://www.mtbs.gov/; Eidenshink et al. 2007). The NBR relates to vegetation moisture content by combining NIR with SWIR reflectance. Therefore pre- and postfire NBR data are generally bitemporally differenced, resulting in the dNBR, which permits a clearer distinction between low severity and unburned regions. In addition, Miller and Thode (2007) proposed a relative version of the dNBR (RdNBR), accounting for heterogeneous landscapes with low prefire vegetative cover (Figure 12.4). RdNBR relates the fire-induced change to the prefire amount of biomass, and therefore, rather than being a measure of absolute change, it reflects the change caused by fire relative to the prefire condition. Some new SIs, such as SWIR–MIR index (SMI) (Veraverbeke et al. 2012), relativized burn ratio (RBR, a relativized version of dNBR; Parks et al. 2014), and land surface temperature (LST)-based index (Zheng et al. 2016, Quintano et al. 2017), have been developed to assess burn severity.

However, NBR and other similar SI-based methods still have limitations, potentially preventing their applicability to infer burn severity across various scales and ecosystems (see Lentile et al. 2009 for the limitations of an NBR and other similar SI-based methods in detail). In short, SI-based methods are mainly based on the statistical correlations between field measurements of burn severity (e.g., (Geo)CBI) (Key and Benson 2006; De Santis et al. 2009) and selected SIs (e.g., NBR, NDVI), but the relationships are usually nonlinear asymptotic (Lentile et al. 2009), varying in both spatial scales (Van Wagtendonk et al. 2004) and ecosystem types (Epting et al. 2005). Several authors have also stated that the spectral bands used for NBR calculation are not optimal to evaluate the degree of burning; thus, an NBR-based approach cannot be optimal for inferring both burned areas and varied burned effects (i.e., severity) (Roy et al. 2006, Smith et al. 2005).

12.3.2 Representative Studies of Spectral Mixture Analysis-Based Burn Severity

SMA is a well-known remote sensing technique used for addressing the mixed pixel issue (e.g., the mixture of vegetation, substrate, and ash in the short-term postfire environment), by quantifying the subpixel proportions of different features or classes (endmembers), which are assumed to represent the spectral variability of dominant land covers (e.g., green vegetation, nonphotosynthetic

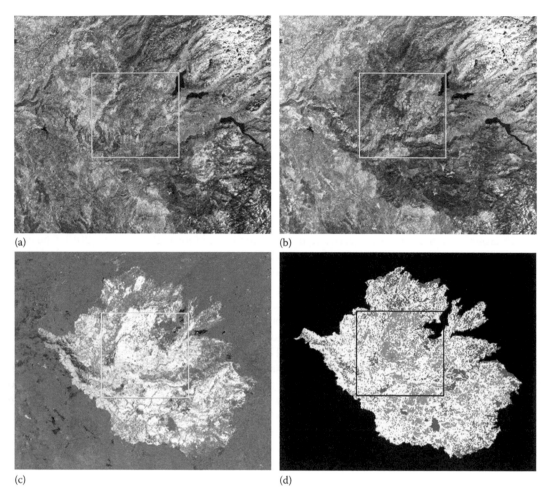

(a)   (b)

(c)   (d)

FIGURE 12.4   Prefire (a) and (b) postfire Landsat false color composite images, (c) differenced Normalized Burn Ratio (dNBR), and (d) MTBS burn severity for the 2013 Rim Fire near Yosemite National Park, California.

vegetation, and soil). Different SMA techniques exist for solving mixing problem, but linear spectral mixture analysis (LSMA) has predominantly been used for inferring burn severity (Rogan and Franklin 2001, Smith et al. 2007, Quintano et al. 2013) and for studying postfire recovery (Lentile et al. 2009, Quintano et al. 2017). The theory and limitations of LSMA are well documented in the literature (Drake et al. Mackin, and Settle 1999, Asner and Heidebrecht 2002). SMA applied to postfire images has resulted in fractional ground cover measures closely related to burning efficiency, usually implementing at least the green vegetation and char endmembers. Comparing to SI-based approaches, SMA provides a directly transferable measure between ground truth of burn severity and remotely sensed measures, and thus is inherently scalable, and other major advantages of SMA include its ability to detect the charcoal signal even in lightly burned areas that kept a strong vegetation signal, reliance on the single image data without constraints on bitemporal imagery or specific channel availability (i.e., SWIR bands are necessary for NBR calculations), and making use of the full spectra, rather than just two or three bands for SI calculations (Lentile et al. 2006, 2009, Veraverbeke and Hook 2013). Recently, an improved version of typical LSMA (i.e., Multiple Endmember SMA, MESMA) (Roberts et al. 1998) has been highlighted for burn severity studies (Veraverbeke and Hook 2013, Quintano et al. 2013, 2017, Meng et al. 2017). Different from the typical LSMA technique, MESMA uses a variable number of endmembers for unmixing pixels and takes within class spectral variability of endmembers into account (see Quitano et al. 2013 for applying MESMA for mapping burn severity in detail).

12.3.3 REPRESENTATIVE STUDIES OF RADIATIVE TRANSFER MODEL-BASED BURN SEVERITY

RTMs are physics based and can model a set of independent variables for assessing burn severity (e.g., leaf types and conditions, LAI, vegetation fraction cover [FCOV]). The main advantage of these simulation models is that they are generally applicable, which potentially enhances the applicability and intercomparability of assessments made over a wide range of ecosystems, by considering the whole spectral profile (Table 12.5). Researchers conducted turbid RTMs in forward model for simulating burn severity scenarios (spectral signatures) from a set of input parameters at both leaf and canopy level, and then they estimated (Geo)CBI values for burn severity mapping in reverse model using postfire remote sensing imagery (Chuvieco et al. 2006, Chuvieco and Kasischke 2007, De Santis and Chuvieco 2007). All of these studies found improvements in burn severity estimation, comparing to empirical fitting methods (e.g., NBR, NDVI, Tasseled Cap (TC) transformation). However, turbid RTMs assumed that each vegetation stratum was composed of a turbid medium as a homogeneous layer, but this assumption is not often true in the burned areas (De Santis and Chuvieco 2009). In order to overcome this shortcoming, geometric RTMs accounting for the canopy structure and the illumination-shadow effects were applied for improving the burn severity estimation from postfire satellite images (De Santis and Chuvieco 2009, De Santis et al. 2009). A 3D RTM has also been developed to modeling the pre- and postfire reflectance of a two-layer savanna system, after detailed field measurements of overstory (tree) and understory (grass) structural and radiometric properties from burned plots (Disney et al. 2011).

12.3.4 SUPERVISED AND UNSUPERVISED CLASSIFICATION

Image classification is one of the most common practices to extract information from remotely sensed imagery, and supervised and unsupervised classification are the two most commonly used techniques for image classification. With known pixel samples from final classification classes, supervised classification uses the spectral signature of these training samples to classify an image, whereas unsupervised classification finds spectral clusters in an image without prior knowledge. Both supervised and unsupervised classification methods have been conducted for assessing burn severity (Benson and Briggs 1978; Chuvieco and Congalton 1988). In terms of predictor variables,

TABLE 12.5
Relative Advantages and Disadvantages of Main Burn Severity Mapping Methods

Burn Severity Mapping Method	Pros	Cons
3.1 Spectral indices (SIs)	Easy to compute, especially for large area assessment	Lacks ecological meanings
3.2 Spectral mixture analysis (SMA)	Ecological meaningful assessment (direct measurement of vegetation composition change)	High computing loads, time consuming for large areas
3.3 Radiative transfer model (RTM)	Physical based, generally applicable; relatively high mapping accuracy	Complex procedures and high computing loads
3.4 Classification	Relatively easy and straightforward in computation	Lacks ecological meanings
3.5 Hybrid approach	High in mapping accuracy; integrates information from multiple sensors	Complex procedures and high computing loads

multispectral channels, SIs, image transformations (e.g., PCA and Karhunen-Loeve (KT)), SMA fractions, biophysical variables (e.g., LAI), LiDAR-based measurements, and so on have all been proved to provide critical information for classifying burn severity (Quintano et al. 2017; McCarley et al. 2017). In terms of classification algorithms, decision trees, SAM, object-oriented classification, artificial neural network, and other machining learning algorithms were applied both on single postfire imagery and on bitemporal pre- and postfire imagery for burn severity assessment (Roldán-Zamarrón et al. 2006, Hultquist et al. 2014).

12.3.5 New Remote Sensing Techniques for Burn Severity Studies

Other types of remotely sensed measurements have also been used to quantify burn severity, including thermal, hyperspectral imagery, VHR imagery, LiDAR, and SAR. With higher spatial and spectral resolution compared to commonly used multispectral imagery (e.g., Landsat mission and MODIS), hyperspectral imagery and VHR imagery datasets can distinguish finer surface features caused by fire (e.g., increased soil and ash cover, decreased vegetation cover). For example, using hyperspectral imagery, Robichaud et al. (2007) accurately map postfire soil and ash cover fractions for assessing soil erosion caused by fire. In addition, active sensor systems, such as LiDAR and SAR, also provide new power for inferring burn effects. For example, Wang and Glenn (2009) calculated the vegetation height change for burn severity estimation with an overall accuracy of 84%, from pre- and postfire LiDAR data in a rangeland ecosystem (Wang and Glenn 2009); In Mediterranean and boreal environments, Tanase et al. (2010a, 2010b and 2014) investigated the properties of SAR data for assessing burn severity and postfire recoveries, and they concluded that SAR data was useful for fire-related studies, although they have limitations. McCarley et al. (2017) recently investigated the correlations between changes in forest structure caused by wildfire derived from multitemporal LiDAR acquisitions and multitemporal spectral changes captured by Landsat imagery. Their findings suggested the limitations of SIs on detecting fire-induced changes in the topmost surface and LiDAR provide a reliable physical measure of vegetation structure and change caused by fire, commensurating with Landsat spectral measures.

However, one limitation of the operational use of SAR images as well as other advanced remote sensing techniques (i.e., LiDAR, hyperspectral, and VHR imagery) in burn severity assessment is data availability that is limited by the long revisit periods or the absence of satellite platforms, the commercial operating mode of some new satellites, and the relatively high cost of aerial-based measurements. However, the availability of new space missions, such as the planned Radarsat-3 mission (Girard et al. 2002), Global Ecosystem Dynamics Investigation LiDAR (GEDI) (Dubayah et al. 2014), Sentinel missions (Berger and Aschbacher 2012), and Hyperspectral Infrared Imager (HyspIRI) (Lee et al. 2015), will solve the problem to some extent. Further studies are still in need to assess the integration of hyperspectral, LiDAR, SAR, thermal infrared data for burn severity assessment.

12.4 FUTURE DIRECTIONS

12.4.1 New Satellite Instruments for Remote Sensing of Fire Effects

Burned area and burn severity maps are the very basic information required for modeling the impact of fires on ecosystem dynamics. Major efforts have been made to map burned area and burn severity at regional to continental scales. New satellites that are launched recently open up new avenues for characterizing fire effects across multiple spatial, temporal, and spectral scales (Wooster et al. 2012; Schroeder et al. 2016; Meng et al. 2017; McCarley et al. 2017). Synergies between data from new sensors and existing instruments also provide opportunities to characterize fire events, one of the key ecosystem processes, and its impact on ecosystem mass and energy exchanges in more detail over large areas.

There is a wealth of untapped information for use in remote sensing of fire effects using existing data, and there is a huge opportunity for burned area and burn severity mapping with new datasets. As more data become available, they will contribute to the improvement of fire effects characterization and reduction of mapping errors. However, a systematic set of defined fire variables is not present to satisfy the needs for fire mapping comparison and fire effects modeling at regional and global scales. Information on fire types (such as agricultural maintenance fires, land cover conversion fires or wildland forest fires) is important for modeling fire effects and postfire vegetation recovery, but was rarely provided, except in a few recent studies (Hall et al. 2016; Roy and Kumar 2017). Remotely sense-based burn severity is often not well linked to its ecological meanings that can be applied across different ecosystems. These gaps in fire effects characterization prevent us from advancing and unifying our understanding of fire effects at multiple spatial and temporal scales, and require more collaborative work from the fire ecology and remote sensing communities in the near future.

12.4.2 Scalable Burn Severity Maps for Improved Wildfire Monitor across Spatial Scales

Landscape distribution of burned areas and burn severity for a fire event is not homogenous, varying with biophysical conditions. This gives rise to challenges to remote sensing of burn effects. Traditional coarse and moderate resolution satellite images provide consistent and frequent burned area and burn severity maps, but they are limited in their ability to capture the heterogeneous burn conditions within the fire boundary. Medium-to-high resolution images from the Landsat sensors provide the longest available observation and widely used source for assessing burned area and burn severity. Although 30 m Landsat TM imagery provided decent resolution for regional fire assessment, higher spatial resolution data have obvious potential for quantifying fine-scale postfire effects. Images with meter-level spatial resolutions provide additional information on crown and canopy change dynamics and are well suitable for fire effects characterization at individual tree, plot, and landscape levels.

The spatial resolution of the remote sensing imagery should match the ecological scale of the fire effect of interest. Direct fire effects (e.g., Greenhouse gas emissions) have been parameterized as critical components in ESMs, but their *legacy effects*, as a result of vegetation successional trajectories, are still poorly represented and validated in the current ESMs. In the development of cutting-edge ecosystem models (ESMs), improved characterization of vegetation response to disturbance events is critical for reducing model uncertainty, under the projected climate change. Remote sensing technologies such as VHR open the window of opportunity to characterize fire effects at finer spatial scale and toward scalable burn severity mapping for improved wildfire monitoring across varying spatial scales (Figure 12.3). Meng et al. (2017) use 2 m WorldView-2 images to map burn severity and to assess fire effects at subcrown, crown, and intercrown scales, which highlight the importance of heterogeneous fire patterns captured by VHR images for understanding fine-scale fire effects on ecosystem processes (Figure 12.5).

12.4.3 Toward Ecological Meaningful Characterization of Fire Effects

Quantifying fire effects is critical for understanding the ecological impacts of fires, including assessing ecosystem resilience, mitigating postfire soil and water loss, examining postfire forest recovery, and calculating carbon dynamics (McCarley et al. 2017). Although (Geo) CBI is gradually used as a standard for field measurement of burn severity across strata in burned sites, burn severity is still a subjective measurement changing with the context. For practical purpose, burn severity is often broadly defined and partitioned into discrete classes ranging from low, moderate to high to link with space measurements, though having limited mechanistic and predictive power

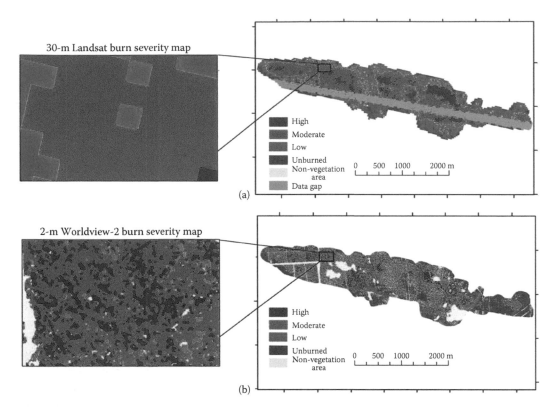

FIGURE 12.5   Example of burn severity mapping at (a) 30-m and (b) 2-m resolutions in an imperiled fire-dependent Pine Barrens Ecosystem in Northeastern United States.

(see Smith et al., 2016 for limitations of discrete severity classifications in detail). As a result, in recent years, researchers suggest to abandon the categorical descriptions of burn severity (i.e., low, moderate, and high severity) and use more ecological- or physical-based severity classifications (e.g., tree mortality percentage, live basal area, diameter of the smallest remaining branches) instead (Jain et al., 2004; Morgan et al., 2014; Smith et al., 2016). Without building the physical linkages between spectral data and quantitative measures of postfire forest change, it will be difficult to accurately calculate carbon dynamics following fires. Therefore, further studies on remote sensing of fire effects should work toward more ecological meaningful characterization of burn effects and should address the gaps in understanding the relationship between biophysical fire effects and spectral remote sensing.

ACKNOWLEDGMENT

Ran Meng was supported during the writing of this chapter by the U.S. Department of Energy, contract No. DE-SC0012704 to Brookhaven National Laboratory.

REFERENCES

Alonso-Canas, I. and E. Chuvieco. 2015. Global burned area mapping from ENVISAT-MERIS and MODIS active fire data. *Remote Sensing of Environment* 163:140–152.

Arnett, J., N. C. Coops, L. D. Daniels, and R. W. Falls, 2015. Detecting forest damage after a low-severity fire using remote sensing at multiple scales. *International Journal of Applied Earth Observation and Geoinformation* 35:239–246.

Asner, G. P. and K. B. Heidebrecht. 2002. Spectral unmixing of vegetation, soil and dry carbon cover in arid regions: Comparing multispectral and hyperspectral observations. *International Journal of Remote Sensing* 23 (19):3939–3958.

Benda, L, D. Miller, P. Bigelow, and K. Andras. 2003. Effects of post-wildfire erosion on channel environments, Boise River, Idaho. *Forest Ecology and Management* 178 (1):105–119.

Benson, M. and I. Briggs. 1978. Mapping the extent and intensity of major forest fires in Australia using digital analysis of Landsat imagery. *Proceedings of International Symposium on Remote Sensing Observation of Earth Resources*. Freiburg, Germany.

Berger, M. and J. Aschbacher. 2012. Preface: The sentinel missions—New opportunities for science. *Remote Sensing of Environment* 120:1–2.

Boer, M. M., C. Macfarlane, J. Norris, R. J. Sadler, J. Wallace, and P.F. Grierson. 2008. Mapping burned areas and burn severity patterns in SW Australian eucalypt forest using remotely-sensed changes in leaf area index. *Remote Sensing of Environment* 112 (12):4358–4369.

Boschetti, L., D. P. Roy, C. O. Justice, and M. L. Humber. 2015. MODIS–Landsat fusion for large area 30m burned area mapping. *Remote Sensing of Environment* 161:27–42.

Bowman, D. M. J. S., J. K. Balch, P. Artaxo, W. J. Bond, J. M. Carlson, M. A. Cochrane, C. M. D'Antonio, R. S. DeFries, J. C. Doyle, and S. P. Harrison. 2009. Fire in the Earth system. *Science* 324 (5926):481–484.

Chen, G., M. R. Metz, D. M. Rizzo, W. W. Dillon, and R. K. Meentemeyer. 2015. Object-based assessment of burn severity in diseased forests using high-spatial and high-spectral resolution MASTER airborne imagery. *ISPRS Journal of Photogrammetry and Remote Sensing*, 102:38–47.

Chen, X., J. E. Vogelmann, M. Rollins, D. Ohlen, C. H. Key, L. Yang, C. Huang, and H. Shi. 2011. Detecting post-fire burn severity and vegetation recovery using multitemporal remote sensing spectral indices and field-collected composite burn index data in a ponderosa pine forest. *International Journal of Remote Sensing* 32 (23):7905–7927.

Chuvieco, E., L. Giglio, and C. Justice. 2008. Global characterization of fire activity: Toward defining fire regimes from Earth observation data. *Global Change Biology* 14 (7):1488–1502.

Chu, T. and X. Guo. 2013. Remote sensing techniques in monitoring post-fire effects and patterns of forest recovery in boreal forest regions: A review. *Remote Sensing* 6 (1):470–520.

Chuvieco, E., D. Riaño, F. M. Danson, and P. Martin. 2006. Use of a radiative transfer model to simulate the postfire spectral response to burn severity. *Journal of Geophysical Research: Biogeosciences* 111 (G4) 1–15.

Chuvieco, E. and E. S. Kasischke. 2007. Remote sensing information for fire management and fire effects assessment. *Journal of Geophysical Research G: Biogeosciences* 112 (1): 1–8.

Chuvieco, E. and R. G. Congalton. 1988. Mapping and inventory of forest fires from digital processing of TM data. *Geocarto International*, 3:41–53.

De Santis, A. and E. Chuvieco. 2007. Burn severity estimation from remotely sensed data: Performance of simulation versus empirical models. *Remote Sensing of Environment* 108 (4):422–435. doi:10.1016/j.rse.2006.11.022.

De Santis, A. and E. Chuvieco. 2009. GeoCBI: A modified version of the Composite Burn Index for the initial assessment of the short-term burn severity from remotely sensed data. *Remote Sensing of Environment* 113 (3):554–562. doi:10.1016/j.rse.2008.10.011.

De Santis, A., E. Chuvieco, and P. J. Vaughan. 2009. Short-term assessment of burn severity using the inversion of PROSPECT and GeoSail models. *Remote Sensing of Environment* 113 (1):126–136. doi:10.1016/j.rse.2008.08.008.

Dennison, P. E. and D. A. Roberts. 2003. Endmember selection for multiple endmember spectral mixture analysis using endmember average RMSE. *Remote Sensing of Environment* 87 (2):123–135.

Disney, M. I., P. Lewis, J. Gomez-Dans, D. Roy, M. J. Wooster, and D. Lajas. 2011. 3D radiative transfer modelling of fire impacts on a two-layer savanna system. *Remote Sensing of Environment* 115 (8):1866–1881.

Drake, N. A., S. Mackin, and J. J. Settle. 1999. Mapping vegetation, soils, and geology in semiarid shrublands using spectral matching and mixture modeling of SWIR AVIRIS imagery. *Remote Sensing of Environment* 68 (1):12–25.

Dubayah, R., S. J. Goetz, J. B. Blair, T. E. Fatoyinbo, M. Hansen, S. P. Healey, M. A. Hofton, G. C. Hurtt, J. Kellner, and S. B. Luthcke. 2014. The global ecosystem dynamics investigation. Paper read at AGU Fall Meeting Abstracts.

Eidenshink, J. C., B. Schwind, K. Brewer, Z. L. Zhu, B. Quayle, and S. M. Howard. 2007. A project for monitoring trends in burn severity. *Fire Ecology* 3 (1):3–21.

Epting, J., D. Verbyla, and B. Sorbel. 2005. Evaluation of remotely sensed indices for assessing burn severity in interior Alaska using Landsat TM and ETM+. *Remote Sensing of Environment* 96 (3):328–339.

Escuin, S., R. Navarro, and P. Fernandez. 2008. Fire severity assessment by using NBR (Normalized Burn Ratio) and NDVI (Normalized Difference Vegetation Index) derived from LANDSAT TM/ETM images. *International Journal of Remote Sensing* 29 (4):1053–1073.

Franklin, J., J. M. Serra-Diaz, A. D. Syphard, and H. M. Regan. 2016. Global change and terrestrial plant community dynamics. *Proceedings of the National Academy of Sciences* 113 (14):3725–3734. doi:10.1073/pnas.1519911113.

Fraser, R. H., Z. Li, and J. Cihlar. 2000. Hotspot and NDVI differencing synergy (HANDS): A new technique for burned area mapping over boreal forest. *Remote Sensing of Environment* 74 (3):362–376.

French, N. H. F., L. L. Bourgeau-Chavez, Y. Wang, and E. S. Kasischke. 1999. Initial observations of Radarsat imagery at fire-disturbed sites in interior Alaska. *Remote sensing of Environment* 68 (1):89–94.

Gao, X., A. R. Huete, W. Ni, and T. Miura. 2000. Optical–biophysical relationships of vegetation spectra without background contamination. *Remote Sensing of Environment* 74 (3):609–620.

Giglio, L. 2010. MODIS collection 5 active fire product user's guide version 2.4. Science Systems and Applications, Inc, Lanham, MD.

Giglio, L., T. Loboda, D. P. Roy, B. Quayle, and C. O Justice. 2009. An active-fire based burned area mapping algorithm for the MODIS sensor. *Remote Sensing of Environment* 113 (2):408–420.

Giglio, L., G. R. Van der Werf, J. T. Randerson, G. J. Collatz, and P. Kasibhatla. 2006. Global estimation of burned area using MODIS active fire observations. *Atmospheric Chemistry and Physics* 6 (4):957–974.

Gimeno, M., J. San-Miguel-Ayanz, and G. Schmuck. 2004. Identification of burnt areas in Mediterranean forest environments from ERS-2 SAR time series. *International Journal of Remote Sensing* 25 (22):4873–4888.

Girard, R., P. F. Lee, and K. James. 2002. The RADARSAT-2&3 topographic mission: An overview. Paper Read at *Geoscience and Remote Sensing Symposium*, 2002. IGARSS'02. 2002 IEEE International.

Goetz, S. J., M. Sunt, A. Baccini, and P. S. A. Beck. 2010. Synergistic use of spaceborne lidar and optical imagery for assessing forest disturbance: An Alaska case study. *Journal of Geophysical Research G: Biogeosciences*, 115.

Goodwin, N. R. and L. J. Collett. 2014. Development of an automated method for mapping fire history captured in Landsat TM and ETM + time series across Queensland, Australia. *Remote Sensing of Environment* 148:206–221.

Hall, J. V., T. V. Loboda, L. Giglio, and G. W. McCarty. 2016. A MODIS-based burned area assessment for Russian croplands: Mapping requirements and challenges. *Remote Sensing of Environment* 184:506–521. doi:10.1016/j.rse.2016.07.022.

Hansen, M. C, P. V. Potapov, R. Moore, M. Hancher, S. A. Turubanova, A. Tyukavina, D. Thau, S. V. Stehman, S. J. Goetz, and T. R. Loveland. 2013. High-resolution global maps of 21st-century forest cover change. *Science* 342 (6160):850–853.

Harris, S., S. Veraverbeke, and S. Hook. 2011. Evaluating spectral indices for assessing fire severity in chaparral ecosystems (Southern California) using MODIS/ASTER (MASTER) airborne simulator data. *Remote Sensing* 3 (11):2403–2419.

Holden, Z.A., P. Morgan, A. M. Smith, and L. Vierling. 2010. Beyond Landsat: A comparison of four satellite sensors for detecting burn severity in ponderosa pine forests of the Gila Wilderness, NM, USA. *International Journal of Wildland Fire*, 19:449–458.

Huang, C., S. N. Goward, J. G. Masek, N. Thomas, Z. Zhu, and J. E. Vogelmann. 2010. An automated approach for reconstructing recent forest disturbance history using dense Landsat time series stacks. *Remote Sensing of Environment* 114 (1):183–198.

Huang, C., S. N. Goward, K. Schleeweis, N. Thomas, J. G. Masek, and Z. Zhu. 2009. Dynamics of national forests assessed using the Landsat record: Case studies in eastern United States. *Remote Sensing of Environment* 113 (7):1430–1442.

Hudak, A. T. and B. H. Brockett. 2004. Mapping fire scars in a southern African savannah using Landsat imagery. *International Journal of Remote Sensing* 25 (16):3231–3243.

Huete, A. R. 1988. A soil-adjusted vegetation index (SAVI). *Remote Sensing of Environment* 25 (3):295–309.

Hultquist, C., G. Chen, and K. Zhao. 2014. A comparison of Gaussian process regression, random forests and support vector regression for burn severity assessment in diseased forests. *Remote Sensing Letters* 5 (8):723–732.

Kane, V. R., M. P. North, J. A. Lutz, D. J. Churchill, S. L. Roberts, D. F. Smith, R. J. McGaughey, J. T. Kane, and M. L. Brooks. 2014. Assessing fire effects on forest spatial structure using a fusion of Landsat and airborne LiDAR data in Yosemite National Park. *Remote Sensing of Environment* 151:89–101.

Kasischke, E. S., L. L. Bourgeau-Chavez, and N. H. F French. 1994. Observations of variations in ERS-1 SAR image intensity associated with forest fires in Alaska. *IEEE Transactions on Geoscience and Remote Sensing* 32 (1):206–210.

Kasischke, E. S., L. L. Bourgeau-Chavez, and J. F. Johnstone. 2007. Assessing spatial and temporal variations in surface soil moisture in fire-disturbed black spruce forests in Interior Alaska using spaceborne synthetic aperture radar imagery—Implications for post-fire tree recruitment. *Remote Sensing of Environment* 108 (1):42–58.

Kasischke, E. S. and M. R. Turetsky. 2006. Recent changes in the fire regime across the North American boreal region—Spatial and temporal patterns of burning across Canada and Alaska. *Geophysical Research Letters* 33 (9):L09703. doi:10.1029/2006gl025677.

Kasischke, E.S., M. R. Turetsky, R. D. Ottmar, N. H. F. French, E. E. Hoy, and E. S. Kane. 2008. Evaluation of the composite burn index for assessing fire severity in Alaskan black spruce forests. *International Journal of Wildland Fire*, 17, 515–526.

Keane, R. E., G. J. Cary, and R. Parsons. 2003. Using simulation to map fire regimes: An evaluation of approaches, strategies, and limitations. *International Journal of Wildland Fire* 12 (4):309–322. doi:10.1071/WF03017.

Keeley, J. E. 2009. Fire intensity, fire severity and burn severity: A brief review and suggested usage. *International Journal of Wildland Fire* 18 (1):116–126.

Kennedy, R. E., Z. Yang, and W. B. Cohen. 2010. Detecting trends in forest disturbance and recovery using yearly Landsat time series: 1. LandTrendr—Temporal segmentation algorithms. *Remote Sensing of Environment* 114 (12):2897–2910. doi:10.1016/j.rse.2010.07.008.

Kennedy, R. E., Z. Yang, Z. Yang, J. Braaten, C. Copass, N. Antonova, C. Jordan, and P. Nelson. 2015. Attribution of disturbance change agent from Landsat time-series in support of habitat monitoring in the Puget Sound region, USA. *Remote Sensing of Environment* 166:271–285.

Key, C. H. and C. N. Benson. 2006. Landscape Assessment (LA) Sampling and Analysis Methods. In: USDA Forest Service General Technical Report RMRS-GTR-164-CD.

Kohavi, R. and F. Provost. 1998. Glossary of terms. *Machine Learning* 30 (2–3):271–274.

Kolden, C. A., A. M. S. Smith, and J. T. Abatzoglou. 2015. Limitations and utilisation of Monitoring Trends in Burn Severity products for assessing wildfire severity in the USA. *International Journal of Wildland Fire* 24 (7):1023–1028.

Leblon, B., J. San, and L. Bourgeau-Chavez. 2012. *Use of Remote Sensing in Wildfire Management*. Croatia: INTECH Open Access Publisher.

Lee, C. M., M. L. Cable, S. J. Hook, R. O. Green, S. L. Ustin, D. J. Mandl, and E. M. Middleton. 2015. An introduction to the NASA Hyperspectral InfraRed Imager (HyspIRI) mission and preparatory activities. *Remote Sensing of Environment* 167:6–19.

Lentile, L. B., Z. A. Holden, A. M. S. Smith, M. J. Falkowski, A. T. Hudak, P. Morgan, S. A. Lewis, P. E. Gessler, and N. C. Benson. 2006. Remote sensing techniques to assess active fire characteristics and post-fire effects. *International Journal of Wildland Fire* 15 (3):319–345.

Lentile, L. B., A. M. S. Smith, A. T. Hudak, P. Morgan, M. J. Bobbitt, S. A. Lewis, and P. R. Robichaud. 2009. Remote sensing for prediction of 1-year post-fire ecosystem condition. *International Journal of Wildland Fire* 18 (5):594–608.

Li, Z., Y. J. Kaufman, C. Ichoku, R. Fraser, A. Trishchenko, L. Giglio, J. Jin, and X. Yu. 2001. A review of AVHRR-based active fire detection algorithms: Principles, limitations, and recommendations. *Global and Regional Vegetation Fire Monitoring from Space, Planning and Coordinated International Effort*. 199–225.

Loboda, T. V., E. E. Hoy, L. Giglio, and E. S. Kasischke. 2011. Mapping burned area in Alaska using MODIS data: A data limitations-driven modification to the regional burned area algorithm. *International Journal of Wildland Fire* 20 (4):487–496.

Loehman, R. A., J. A. Clark, and R. E. Keane. 2011. Modeling effects of climate change and fire management on western white pine (*Pinus monticola*) in the Northern Rocky Mountains, USA. *Forests* 2 (4):832–860.

Lu, B., Y. He, and A. Tong. 2016. Evaluation of spectral indices for estimating burn severity in semiarid grasslands. *International Journal of Wildland Fire* 25 (2):147–157.

Maingi, J. K. and M. C. Henry. 2007. Factors influencing wildfire occurrence and distribution in eastern Kentucky, USA. *International Journal of Wildland Fire* 16 (1):23–33.

Mayor, A. G., S. Bautista, J. Llovet, and J. Bellot. 2007. Post-fire hydrological and erosional responses of a Mediterranean landscape: Seven years of catchment-scale dynamics. *CATENA* 71 (1):68–75.

McCarley, T. R., C. A. Kolden, N. M. Vaillant, A. T. Hudak, A. M. S. Smith, B. M. Wing, B. S. Kellogg, and J. Kreitler. 2017. Multi-temporal LiDAR and Landsat quantification of fire-induced changes to forest structure. *Remote Sensing of Environment* 191:419–432. doi:10.1016/j.rse.2016.12.022.

McKenzie, D. and J. S. Littell. 2016. Climate change and the eco-hydrology of fire: Will area burned increase in a warming western USA? *Ecological Applications*:n/a-n/a. doi:10.1002/eap.1420 27 (1): 23-36.

Meng, R., P. E. Dennison, C. Huang, M. A. Moritz, and C. D'Antonio. 2015. Effects of fire severity and post-fire climate on short-term vegetation recovery of mixed-conifer and red fir forests in the Sierra Nevada Mountains of California. *Remote Sensing of Environment* 171:311–325.

Meng, R., J. Wu, K. L. Schwager, F. Zhao, P. E. Dennison, B. D. Cook, K. Brewster, T. M. Green, and S. P. Serbin. 2017. Using high spatial resolution satellite imagery to map forest burn severity across spatial scales in a Pine Barrens ecosystem. *Remote Sensing of Environment* 191:95–109. doi:10.1016/j.rse.2017.01.016.

Menges, C. H. , R. E. Bartolo, D. Bell, and G. J. E. Hill. 2004. The effect of savanna fires on SAR backscatter in northern Australia. *International Journal of Remote Sensing* 25 (22):4857–4871.

Miller, J. D., E. E. Knapp, C. H. Key, C. N. Skinner, C. J. Isbell, R. M. Creasy, and J. W. Sherlock. 2009. Calibration and validation of the relative differenced Normalized Burn Ratio (RdNBR) to three measures of fire severity in the Sierra Nevada and Klamath Mountains, California, USA. *Remote Sensing of Environment*, 113:645–656.

Miller, J. D. and A. E. Thode. 2007. Quantifying burn severity in a heterogeneous landscape with a relative version of the delta Normalized Burn Ratio (dNBR). *Remote Sensing of Environment* 109 (1):66–80.

Miller, J. D. and S. R. Yool. 2002. Mapping forest post-fire canopy consumption in several overstory types using multi-temporal Landsat TM and ETM data. *Remote Sensing of Environment* 82 (2):481–496.

Mitri, G. H. and I. Z. Gitas. 2008. Mapping the severity of fire using object-based classification of IKONOS imagery. *International Journal of Wildland Fire* 17:431–442.

Mitri, G.H. and I. Z. Gitas. 2013. Mapping post-fire forest regeneration and vegetation recovery using a combination of very high spatial resolution and hyperspectral satellite imagery. *International Journal of Applied Earth Observation and Geoinformation* 20:60–66.

Montealegre, A. L., M. T. Lamelas, M. A. Tanase, and J. de la Riva. 2014. Forest fire severity assessment using ALS data in a Mediterranean environment. *Remote Sensing* 6 (5):4240–4265.

Morgan, P., C. C. Hardy, T. W. Swetnam, M. G. Rollins, and D. G. Long. 2001. Mapping fire regimes across time and space: Understanding coarse and fine-scale fire patterns. *International Journal of Wildland Fire* 10 (4):329–342.

Morgan, P., R. E. Keane, G. K. Dillon, T. B. Jain, A. T. Hudak, E. C. Karau, P. G. Sikkink, Z. A. Holden, and E. K. Strand. 2014. Challenges of assessing fire and burn severity using field measures, remote sensing and modelling. *International Journal of Wildland Fire* 23 (8):1045–1060.

Norton, J., N. Glenn, M. Germino, K. Weber, and S. Seefeldt. 2009. Relative suitability of indices derived from Landsat ETM+ and SPOT 5 for detecting fire severity in sagebrush steppe. *International Journal of Applied Earth Observation and Geoinformation* 11:360–367.

Oliva, P. and W. Schroeder. 2015. Assessment of VIIRS 375m active fire detection product for direct burned area mapping. *Remote Sensing of Environment* 160:144–155.

Parks, S. A., G. K. Dillon, and C. Miller. 2014. A new metric for quantifying burn severity: The relativized burn ratio. *Remote Sensing* 6:1827–1844.

Peterson, G. D. 2002. Estimating resilience across landscapes. *Conservation Ecology* 6 (1):17.

Petropoulos, G. P., C. Kontoes, and I. Keramitsoglou. 2011. Burnt area delineation from a uni-temporal perspective based on Landsat TM imagery classification using Support Vector Machines. *International Journal of Applied Earth Observation and Geoinformation* 13 (1):70–80. doi:10.1016/j.jag.2010.06.008.

Petropoulos, G. P., K. P. Vadrevu, G. Xanthopoulos, G. Karantounias, and M. Scholze. 2010. A comparison of spectral angle mapper and artificial neural network classifiers combined with Landsat TM imagery analysis for obtaining burnt area mapping. *Sensors* 10 (3):1967–1985.

Polychronaki, A., I. Z. Gitas, and A. Minchella. 2013. Monitoring post-fire vegetation recovery in the Mediterranean using SPOT and ERS imagery. *International Journal of Wildland Fire* 23 (5):631–642.

Potapov, P., M. C. Hansen, S. V. Stehman, T. R. Loveland, and K. Pittman. 2008. Combining MODIS and Landsat imagery to estimate and map boreal forest cover loss. *Remote Sensing of Environment* 112 (9):3708–3719.

Qi, J., A. Chehbouni, A. R. Huete, Y. H. Kerr, and S. Sorooshian. 1994. A modified soil adjusted vegetation index. *Remote Sensing of Environment* 48 (2):119–126.

Quintano, C., A. Fernández-Manso, O. Fernández-Manso, and Y. E. Shimabukuro. 2006. Mapping burned areas in Mediterranean countries using spectral mixture analysis from a uni-temporal perspective. *International Journal of Remote Sensing* 27 (4):645–662.

Quintano, C., A. Fernández-Manso, and D. A. Roberts. 2013. Multiple Endmember Spectral Mixture Analysis (MESMA) to map burn severity levels from Landsat images in Mediterranean countries. *Remote Sensing of Environment* 136:76–88.

Quintano, C., A. Fernandez-Manso, and D. A. Roberts. 2017. Burn severity mapping from Landsat MESMA fraction images and land surface temperature. *Remote Sensing of Environment* 190:83–95.

Randerson, J. T., Y. Chen, G. R. Werf, B. M. Rogers, and D. C. Morton. 2012. Global burned area and biomass burning emissions from small fires. *Journal of Geophysical Research: Biogeosciences* 117 (G4) 1–23.

Randerson, J. T., H. Liu, M. G. Flanner, S. D. Chambers, Y. Jin, P. G Hess, G. Pfister, M. C. Mack, K. K. Treseder, and L. R. Welp. 2006. The impact of boreal forest fire on climate warming. *Science* 314 (5802):1130–1132.

Reddy, A. D., T. J. Hawbaker, F. Wurster, Z. Zhu, S. Ward, D. Newcomb, and R. Murray. 2015. Quantifying soil carbon loss and uncertainty from a peatland wildfire using multi-temporal LiDAR. *Remote Sensing of Environment* 170:306–316. doi:10.1016/j.rse.2015.09.017.

Riaño, D., E. Chuvieco, S. Ustin, R. Zomer, P. Dennison, D. Roberts, and J. Salas. 2002. Assessment of vegetation regeneration after fire through multitemporal analysis of AVIRIS images in the Santa Monica Mountains. *Remote Sensing of Environment* 79 (1):60–71.

Roberts, D. A., M. Gardner, R. Church, S. Ustin, G. Scheer, and R. O. Green. 1998. Mapping chaparral in the Santa Monica Mountains using multiple endmember spectral mixture models. *Remote Sensing of Environment* 65 (3):267–279.

Robichaud, P.R., S. A. Lewis, D. Y. M. Laes, A. T. Hudak, R. F. Kokaly, and J. A. Zamudio. 2007. Postfire soil burn severity mapping with hyperspectral image unmixing. *Remote Sensing of Environment* 108:467–480.

Rogan, J. and J. Franklin. 2001. Mapping wildfire burn severity in southern California forests and shrublands using Enhanced Thematic Mapper imagery. *Geocarto International* 16 (4):91–106.

Roldán-Zamarrón, A., S. Merino-de-Miguel, F. González-Alonso, S. García-Gigorro, and J. M. Cuevas. 2006. Minas de Riotinto (south Spain) forest fire: Burned area assessment and fire severity mapping using Landsat 5-TM, Envisat-MERIS, and Terra-MODIS postfire images. *Journal of Geophysical Research: Biogeosciences* 111 (G4). 1–9

Roy, D. P., L. Boschetti, and S. N. Trigg. 2006. Remote sensing of fire severity: Assessing the performance of the normalized burn ratio. *Geoscience and Remote Sensing Letters, IEEE* 3 (1):112–116.

Roy, D. P., L. Boschetti, C. O Justice, and J. Ju. 2008. The collection 5 MODIS burned area product—Global evaluation by comparison with the MODIS active fire product. *Remote Sensing of Environment* 112 (9):3690–3707.

Roy, D. P., L. Boschetti, and A. M. Smith. 2013. *Satellite Remote Sensing of Fires.* Chichester, UK: John Wiley & Sons.

Roy, D. P. and S. S. Kumar. 2017. Multi-year MODIS active fire type classification over the Brazilian tropical moist forest biome. *International Journal of Digital Earth* 10 (1):54–84.

Schepers, L., B. Haest, S. Veraverbeke, T. Spanhove, J. Vanden Borre, and R. Goossens. 2014. Burned area detection and burn severity assessment of a heathland fire in Belgium using airborne imaging spectroscopy (APEX). *Remote Sensing* 6 (3):1803–1826.

Schroeder, T. A., G. G. Moisen, K. Schleeweis, C. Toney, W. B. Cohen, Z. Yang, and E. A. Freeman. 2015. Using an empirical and rule-based modeling approach to map cause of disturbance in U.S. Forests: Results and insights from the North American forest dynamics (NAFD) project. In: Stanton, Sharon M. Christensen, Glenn A. (comps.) Pushing boundaries: new directions in inventory techniques and applications: Forest Inventory and Analysis (FIA) symposium 2015. December 8–10, 2015; Portland, OR. General Technical Report PNW-GTR-931, U.S. Department of Agriculture, Forest Service, Pacific Northwest Research Station, Portland, OR, p. 239.

Schroeder, T. A., G. G. Moisen, K. Schleeweis, C. Toney, W. B. Cohen, Z. Yang, and E. A. Freeman. 2015. Using an empirical and rule-based modeling approach to map cause of disturbance in U.S. Forests: Results and insights from the North American forest dynamics (NAFD) project. In: Stanton, Sharon M. Christensen, Glenn A. (comps.) Pushing boundaries: new directions in inventory techniques and applications: Forest Inventory and Analysis (FIA) symposium 2015. December 8–10, 2015; Portland, OR. General Technical Report PNW-GTR-931, U.S. Department of Agriculture, Forest Service, Pacific Northwest Research Station, Portland, OR, p. 239.

Schroeder, W., P. Oliva, L. Giglio, and I. A. Csiszar. 2014. The New VIIRS 375m active fire detection data product: Algorithm description and initial assessment. *Remote Sensing of Environment* 143:85–96.

Schroeder, W., P. Oliva, L. Giglio, B. Quayle, E. Lorenz, and F. Morelli. 2016. Active fire detection using Landsat-8/OLI data. *Remote Sensing of Environment* 185:210–220.

Schultz, M. G. 2002. On the use of ATSR fire count data to estimate the seasonal and interannual variability of vegetation fire emissions. *Atmospheric Chemistry and Physics* 2 (5):387–395.

Scott, A. C., D. M. Bowman, W. J. Bond, S. J. Pyne, and M. E. Alexander. 2013. *Fire on Earth: An Introduction.* Chichester, UK: John Wiley & Sons.

Shabanov, N. V., K. Lo, S. Gopal, and R. B. Myneni. 2005. Subpixel burn detection in moderate resolution imaging spectroradiometer 500-m data with ARTMAP neural networks. *Journal of Geophysical Research: Atmospheres* 110 (D3) 1–17.

Siegert, F. and A. A. Hoffmann. 2000. The 1998 forest fires in East Kalimantan (Indonesia): A quantitative evaluation using high resolution, multitemporal ERS-2 SAR images and NOAA-AVHRR hotspot data. *Remote Sensing of Environment* 72 (1):64–77.

Silva, J. M. N., A. C. L. Sá, and J. M. C. Pereira. 2005. Comparison of burned area estimates derived from SPOT-VEGETATION and Landsat ETM+ data in Africa: Influence of spatial pattern and vegetation type. *Remote Sensing of Environment* 96 (2):188–201.

Smith, A. M. S., N. A. Drake, M. J. Wooster, A. T. Hudak, Z. A. Holden, and C. J. Gibbons. 2007. Production of Landsat ETM+ reference imagery of burned areas within Southern African savannahs: Comparison of methods and application to MODIS. *International Journal of Remote Sensing* 28 (12):2753–2775.

Smith, A.M., A. M. Sparks, C. A. Kolden, J. T. Abatzoglou, A. F. Talhelm, D. M. Johnson, L. Boschetti, J. A. Lutz, K. G. Apostol, and K. M. Yedinak. 2016. Towards a new paradigm in fire severity research using dose–response experiments. *International Journal of Wildland Fire* 25 (2):158–166.

Smith, A. M. S., M. J. Wooster, N. A. Drake, F. M. Dipotso, M. J. Falkowski, and A. T. Hudak. 2005. Testing the potential of multi-spectral remote sensing for retrospectively estimating fire severity in African Savannahs. *Remote Sensing of Environment* 97 (1):92–115.

Smith, A. M. S., N. A. Drake, M. J. Wooster, A. T. Hudak, Z. A. Holden, and C. J. Gibbons. 2007. Production of Landsat ETM+ reference imagery of burned areas within Southern African savannahs: Comparison of methods and application to MODIS. *International Journal of Remote Sensing* 28 (12):2753–2775.

Smithwick, E. A. H., A. L. Westerling, M. G. Turner, W. H. Romme, and M. G. Ryan. 2011. Vulnerability of landscape carbon fluxes to future climate and fire in the Greater Yellowstone Ecosystem *In Proceedings of the 10th Biennial Scientific Conference on the Greater Yellowstone Ecosystem* 11-13.

Sparks, A.M., C. A., Kolden, A. F. Talhelm, A. M. S. Smith, K. G. Apostol, D. M. Johnson, and L. Boschetti. 2016. Spectral indices accurately quantify changes in seedling physiology following fire: Towards mechanistic assessments of post-fire carbon cycling. *Remote Sensing* 8:572.

Spasojevic, M. J., C. A. Bahlai, B. A. Bradley, B. J. Butterfield, M. N. Tuanmu, S. Sistla, R. Wiederholt, and K. N Suding. 2016. Scaling up the diversity–resilience relationship with trait databases and remote sensing data: The recovery of productivity after wildfire. *Global Change Biology* 22 (4):1421–1432.

Stroppiana, D., R. Azar, F. Calò, A. Pepe, P. Imperatore, M. Boschetti, J. Silva, P. A. Brivio, and R. Lanari. 2015. Integration of optical and SAR data for burned area mapping in Mediterranean Regions. *Remote Sensing* 7 (2):1320–1345.

Sugihara, N. G. 2006. *Fire in California's Ecosystems.* Berkeley, CA: University of California Press.

Sukhinin, A. I., N. H. F. French, E. S. Kasischke, J. H. Hewson, A. J. Soja, I. A. Csiszar, E. J. Hyer, T. Loboda, S. G. Conrad, and V. I. Romasko. 2004. AVHRR-based mapping of fires in Russia: New products for fire management and carbon cycle studies. *Remote Sensing of Environment* 93 (4):546–564.

Tanase, M.A., M. Santoro, C. Aponte, and J. de la Riva. 2014. Polarimetric properties of burned forest areas at C- and L-band. *IEEE Journal of Selected Topics in Applied Earth Observations and Remote Sensing,* 7:267–276.

Tanase, M., M. Santoro, J. de La Riva, F. Pérez-Cabello, and T. Le Toan. 2010a. Sensitivity of X-, C-, and L-band SAR backscatter to burn severity in Mediterranean pine forests. *IEEE Transactions on Geoscience and Remote Sensing* 48:3663–3675.

Tanase, M. A., M. Santoro, U. Wegmüller, J. de la Riva, and F. Pérez-Cabello. 2010b. Properties of X-, C-and L-band repeat-pass interferometric SAR coherence in Mediterranean pine forests affected by fires. *Remote Sensing of Environment* 114 (10):2182–2194.

Tucker, C. J., I. Y. Fung, C. D. Keeling, and R. H. Gammon. 1986. Relationship between atmospheric CO_2 variations and a satellite-derived vegetation index. *Nature* 319:195–199.

Turner, M. G, W. H. Romme, and R. H. Gardner. 1999. Prefire heterogeneity, fire severity, and early post-fire plant reestablishment in subalpine forests of Yellowstone National Park, Wyoming. *International Journal of Wildland Fire* 9 (1):21–36.

Turner, M. G., W. H. Romme, R. H. Gardner, and W. W. Hargrove. 1997. Effects of fire size and pattern on early succession in Yellowstone National Park. *Ecological Monographs* 67 (4):411–433.

Turner, M.G., W. W. Hargrove, R. H. Gardner, and W. H. Romme. 1994. Effects of fire on landscape heterogeneity in Yellowstone National Park, Wyoming. *Journal of Vegetation Science*, 5:731–742.

Vafeidis, A. T. and N. A. Drake. 2005. A two-step method for estimating the extent of burnt areas with the use of coarse-resolution data. *International Journal of Remote Sensing* 26 (11):2441–2459.

Van der Werf, G. R., J. T. Randerson, L. Giglio, G. J. Collatz, M. Mu, P. S. Kasibhatla, D. C. Morton, R. S. DeFries, Y. Jin, and T. T. van Leeuwen. 2010. Global fire emissions and the contribution of deforestation, savanna, forest, agricultural, and peat fires (1997–2009). *Atmospheric Chemistry and Physics* 10 (23):11707–11735.

Van Wagtendonk, J. W., R. R. Root, and C. H. Key. 2004. Comparison of AVIRIS and Landsat ETM+ detection capabilities for burn severity. *Remote Sensing of Environment* 92 (3):397–408.

Veraverbeke, S. and S. J. Hook. 2013. Evaluating spectral indices and spectral mixture analysis for assessing fire severity, combustion completeness and carbon emissions. *International Journal of Wildland Fire* 22 (5):707–720.

Veraverbeke, S., S. Hook, and G. Hulley. 2012. An alternative spectral index for rapid fire severity assessments. *Remote Sensing of Environment* 123:72–80.

Veraverbeke, S., S. Lhermitte, W. W. Verstraeten, and R. Goossens. 2011. Evaluation of pre/post-fire differenced spectral indices for assessing burn severity in a Mediterranean environment with Landsat Thematic Mapper. *International Journal of Remote Sensing* 32 (12):3521–3537.

Verbesselt, J., R. Hyndman, G. Newnham, and D. Culvenor. 2010. Detecting trend and seasonal changes in satellite image time series. *Remote Sensing of Environment* 114 (1):106–115. doi:10.1016/j.rse.2009.08.014.

Wang, C. and N. F. Glenn. 2009. Estimation of fire severity using pre-and post-fire LiDAR data in sagebrush steppe rangelands. *International Journal of Wildland Fire* 18 (7):848–856.

Westerling, A. L., H. G. Hidalgo, D. R. Cayan, and T. W. Swetnam. 2006. Warming and Earlier Spring Increase Western U.S. Forest Wildfire Activity. *Science* 313 (5789):940–943.

Wilson, A. M., A. M. Latimer, and J. A. Silander. 2015. Climatic controls on ecosystem resilience: Postfire regeneration in the Cape Floristic Region of South Africa. *Proceedings of the National Academy of Sciences* 112 (29):9058–9063.

Wooster, M. J., G. Roberts, A. M. S. Smith, J. Johnston, P. Freeborn, S. Amici, and A. T. Hudak. 2013. Thermal remote sensing of active vegetation fires and biomass burning events. In *Thermal Infrared Remote Sensing*, 347–390. Springer.

Wooster, M. J., W. Xu, and T. Nightingale. 2012. Sentinel-3 SLSTR active fire detection and FRP product: Pre-launch algorithm development and performance evaluation using MODIS and ASTER datasets. *Remote Sensing of Environment* 120:236–254.

Wulder, M.A., J. C. White, F. Alvarez, T. Han, J. Rogan, and B. Hawkes. 2009. Characterizing boreal forest wildfire with multi-temporal Landsat and LIDAR data. *Remote Sensing of Environment*, 113:1540–1555.

Zhao, F., C. Huang, and Z. Zhu. 2015. Use of vegetation change tracker and support vector machine to map disturbance types in greater yellowstone ecosystems in a 1984–2010 Landsat time series. *Geoscience and Remote Sensing Letters, IEEE* 12 (8):1650–1654. doi:10.1109/LGRS.2015.2418159.

Zhao, F., R. Meng, C. Huang, M. Zhao, F. Zhao, P. Gong, L. Yu, and Z. Zhu. 2016. Long-term post-disturbance forest recovery in the greater yellowstone ecosystem analyzed using Landsat time series stack. *Remote Sensing* 8 (11):898.

Zheng, Z., Y. N. Zeng, S. N. Li, and W. Huang. 2016. A new burn severity index based on land surface temperature and enhanced vegetation index. *International Journal of Applied Earth Observation and Geoinformation*, 45:84–94.

13 Exploring the Relationships between Topographical Elements and Forest Fire Occurrences in Alberta, Canada

Masoud Abdollahi, Quazi K. Hassan,
Ehsan H. Chowdhury, and Anil Gupta

CONTENTS

13.1 INTRODUCTION

Forest fire is one of the critical natural disturbances or disasters taking place over the various ecoregions across the world. In Canada, about 8,300 forest fires occur that burn, on an average, 2.3 million hectares every year during the last 25 years (NRC 2016). In order to suppress the fire events, the Government of Canada spends between $500 million and $1 billion per year (NRC 2016). Despite some benefits of forest fires, such as regenerating healthy forest, killing diseases and insects, and balancing soil nutrient regimes (Ruokolainen and Salo 2009; Chu and Guo 2014), they exert great adverse impacts on socio-economic conditions. Thus, it is critical to study the factors behind such fire occurrences in order to develop an efficient forest fire management system to optimize the consequences of fires.

In general, there are five major influential factors that play significant role in the fire occurrences. These factors include (1) meteorological variables, such as temperature, precipitation, and relative humidity (Kral et al. 2015; Argañaraz et al. 2015; Chowdhury and Hassan 2015); (2) land cover variables comprising forest fuels' conditions (both live and dead fuel loading and moisture conditions), land cover type, biomass density, and so forth (Akther and Hassan 2011; Ager et al. 2014); (3) ignition source, which can be arson or natural, such as lightning strike (Podur et al. 2003; Martínez et al. 2009; Wang and Anderson 2010); (4) anthropogenic (human-related) factors, such as infrastructures, socioeconomic condition, population density, settlements, weekends and public holidays, and so on (Yang et al. 2007; Prasad et al. 2008; Cipriani et al. 2011); and (5) topographical elements, including elevation, slope, and aspect (Vasconcelos et al. 2001; Kalabokidis et al. 2007; Adab et al. 2013). The first four factors are usually dynamic over time and space. In contrast, topographical elements are usually temporally static and spatially variable.

In the recent time, we conducted an extensive literature review and found that a significant amount of research had been carried out in comprehending fire occurrences. In most of the studies, the five factors described in the previous paragraph were employed. As we opted to comprehend the influence of topographical elements on forest fire regimes, we further investigated studies that considered these elements. Some of the example cases are described in Table 13.1.

In general, it was evident from Table 13.1 that the topographical elements might influence the fire occurrences differently from one geographic region to another. In the Canadian context, forest fire

---

TABLE 13.1
Some Example Fire Occurrence Studies That Employed Topographical Elements as One of the Major Factors

References	Study Area	Factors	Variables	Remarks
Guo et al. 2016	Fujian, China	Topographical	Elevation	Among all the ten variables, elevation was found to be the 2nd highest influential variable.
		Meteorological	Daily precipitation, sunshine hours, daily minimum and maximum ground surface temperature, and daily mean relative humidity	
		Anthropogenic	Distance to road and settlement, density of population, and per capita gross domestic product	
Adab et al. 2013	Golestan, Iran	Topographical	Elevation, slope, and aspect	Slope, aspect, and elevation were ranked as 2nd, 3rd, and 6th influential variables, respectively, among the 6 variables of interest.
		Land cover	Vegetation moisture	
		Anthropogenic	Distance from roads and settlements	
Dlamini 2010	Swaziland	Topographical	Elevation, slope, and aspect	Elevation, slope, and aspect were found to be 2nd, 11th, and 12th influential variables, respectively, among the 12 variables of interest.
		Meteorological	Mean annual temperature and rainfall, and winter relative humidity	
		Land cover	Land cover/use and soil type	
		Anthropogenic	Human population, road and livestock density, and distance to settlements	
Catry et al. 2009	Portugal	Topographical	Elevation	Among the four variables, elevation was ranked as the 3rd influential variable.
		Land cover	Land cover type	
		Anthropogenic	Population density and distance to roads	
Kalabokidis et al. 2007	Greece	Topographical	Elevation, slope, and aspect	Slope, elevation, and aspect were the 2nd, 3rd, and 8th influential variables, respectively, among the 10 variables of consideration.
		Meteorological	Air temperature, relative humidity, and annual precipitation	
		Land cover	Vegetation cover and geology	
		Anthropogenic	Distance to roads and density of livestock	

occurrence prediction system (known as Canadian Forest Fire Danger Rating System [CFFDRS]) uses meteorological variables (i.e., air temperature, precipitation, relative humidity, and wind) and ignition sources (that include lightning and human-caused ones) (Van Wagner 1987; Wotton 2009). However, topographical elements are used in the Canadian fire behavior prediction systems to study fire situation and expansion after its ignition (Van Wagner 1987). In addition, some recent studies developed remote sensing–based system (known as Forest Fire Danger Forecasting System [FFDFS]) on incorporating meteorological variables (e.g., surface temperature and total precipitable water) and land cover–related variables (e.g., vegetation greenness, vegetation wetness, and surface wetness) (Akther and Hassan 2011; Chowdhury and Hassan 2013, 2015) and implemented it over the Canadian boreal forested regions. These studies recommended to investigate the influence of topographical elements on the fire occurrences.

Therefore, our overall objective was to explore relationship between topographical elements (i.e., elevation, slope, and aspect) and fire occurrences in the fire-prone nine forested natural subregions in the Canadian province of Alberta (see Figure 13.1 for details). It would be worthwhile

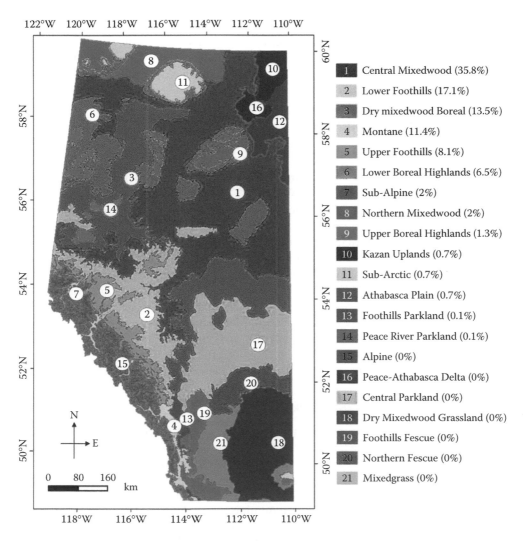

FIGURE 13.1 Map illustrating the 21 natural subregions of Alberta. The subregion-specific relative frequency of the forest fire occurrences over the period 2001–2014 is provided in the parenthesis. Note that Alberta had experienced a total of 20,588 fires over the time period of interest.

to note that a similar study (i.e., [PU et al. 2007]) was conducted to evaluate the impact of the topography on the fire occurrences through analyzing remote sensing (i.e., National Oceanic and Atmospheric Administration [NOAA's] advanced very-high-resolution radiometer [AVHRR])–derived burned area maps across North America, including Canadian forests, during the period 1989–2000. In this study, Pu et al. (2007) derived the topographical elements by using a digital elevation model (DEM) at 1×1 Km^2 spatial resolutions. These elements were then categorized into coarser classes for establishing relations with fire occurrences. Those classes included (1) elevations into four classes (i.e., ≤ 500 m, 500–1000 m, 1000–2000 m, and >2000 m), (2) slopes into three classes (i.e., $\leq 10°$, $10°–20°$, and $>20°$), and (3) aspects into nine classes (i.e., flat, north, south, east, west, northwest, northeast, southwest, and southeast). Thus, our specific objectives were as follows:

1. To evaluate whether a finer spatial resolution (i.e., 30×30 m^2) DEM–derived topographical elements with more detailed/narrower intervals would still hold similar patterns as depicted in Pu et al. (2007) in comparison with the ground-based historical dataset of the fire occurrences during the period 2001–2014.
2. To implement probability density functions (PDFs) to develop probabilistic models to predict fire occurrences by using the data during the period 2001–2008 and their validation by using an independent dataset available over the period 2009–2014.
3. Finally, to implement the best model to be found in the scope of second objective over the nine fire-dominant natural subregions in the study area in order to comprehend the spatial variability of the probabilities of fire occurrences.

13.2 MATERIALS AND METHODS

13.2.1 GENERAL DESCRIPTION OF THE STUDY AREA

We considered Alberta (i.e., one of the western Canadian provinces) as our study area. It is situated between 49°N and 60°N latitudes and 110°W and 120°W longitudes (see Figure 13.1). The province has a variable topography, where the elevations are to be found in the range of 150–3650 m above mean sea level. Climatically, this province has short summers and long and cold winters (Hassan and Rahman 2013). The mean annual temperature varies from −7.1°C to 6°C, and mean annual precipitation is in the range of 260–1710 mm (Downing and Pettapiece 2006). In addition, based on landscape properties (that include topography, climate, vegetation, soil, and geology), Alberta is divided into 21 natural subregions (see details in Figure 13.1 [Downing and Pettapiece 2006]). Among these, we only selected nine natural subregions as our focus in this study. These were (1) Central Mixedwood, (2) Lower Foothills, (3) Dry Mixedwood Boreal, (4) Montane, (5) Upper Foothills, (6) Lower Boreal Highlands, (7) Sub-Alpine, (8) Northern Mixedwood, and (9) Upper Boreal Highlands. The rationale of choosing these subregions was related to the fact that they had experienced equal to or greater than 1% of forest fires during the period 2001–2014 (see Figure 13.1 for more detailed information).

13.2.2 DATA REQUIREMENTS AND ITS PRE-PROCESSING

In this study, we employed two types of dataset. First one was historical fire occurrence spots available from Alberta Forest Service for the period 2001–2014. This dataset contained information about geolocation of the fire spots, fire start date, and the area burned among others for 20,588 fire events. Note that we considered all fire sizes (i.e., ≥0.01 hectares). The second dataset was a Shuttle Radar Topography Mission (SRTM)–derived DEM (i.e., SRTMGL1N3) at 30 m spatial resolution available from the National Aeronautics and Space Administration (NASA). The absolute vertical height accuracy for these data was 16 m at 90% confidence (Rodríguez et al. 2005). We used this dataset in order to generate elevation, slope, and aspect maps over our study area of interest. Using these two datasets, we extracted three topographical elements (that included elevation, slope, and aspect) for each fire spot (i.e., 20,122 fires) that occurred in the nine fire-dominant natural subregions during the period 2001–2014. Then, we divided these data into two components: (1) calibration dataset for model development, comprising 11,375 fire events during the period 2001–2008 and (2) validation dataset for model validation, consisting of 8,747 fire events during the period 2009–2014.

13.2.3 MODEL DEVELOPMENT AND ITS VALIDATION

In the model development phase, we employed the calibration dataset mentioned in the last sub-section and generated relative frequency distributions by diving the number of fire events in each bin/class over the total number of fire events (also known as observed probabilities) as a function of elevation, slope, and aspect (see Figure 13.2 in the Results section). On having a closer look into these frequency distributions, we found that aspect did not exhibit any distinct pattern. For elevation and slope, we assumed that they might follow Log-Logistic (LL) and Generalized Extreme Value (GEV) PDFs, respectively; this assumption was based on their visual appearances (see Figure 13.2). The generic forms of these two PDFs are as follows (Kantam et al. 2006; Rulfová et al. 2016):

LL PDF:

$$f(x) = \frac{\alpha}{\beta} \left(\frac{x - \gamma}{\beta} \right)^{\alpha - 1} \left(1 + \left(\frac{x - \gamma}{\beta} \right)^{\alpha} \right)^{-2} \tag{13.1}$$

where x is elevation at the location of the fire spot, and α, β, and γ are shape, scale, and location parameters, respectively.

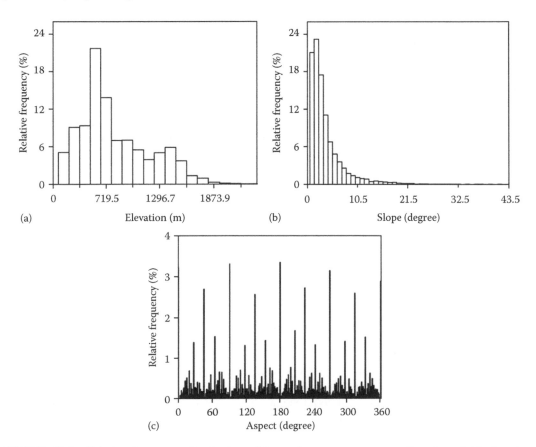

FIGURE 13.2 Relative frequency distribution of forest fire occurrences as a function of (a) elevation; (b) slope; and (c) aspect, during the period 2001–2008.

GEV PDF:

$$\left[f(x) = \begin{cases} \dfrac{1}{\sigma}\exp\!\left(-(1+\kappa z)^{\frac{-1}{\kappa}}\right)(1+\kappa z)^{-1-\frac{1}{\kappa}} & \kappa \neq 0 \\[2ex] \dfrac{1}{\sigma}\exp\!\left(-z-\exp(-z)\right) & \kappa = 0 \\[2ex] z = \dfrac{x-\mu}{\sigma} \end{cases} \right] \tag{13.2}$$

where x is the slope at the location of the fire spot, and κ, σ, and μ are shape, scale, and location parameters, respectively.

We then applied both PDFs in order to fit the observed probabilities related to both elevation and slope elements. For evaluating the goodness of fit, we used Kolmogorov–Smirnov (hereafter K–S) test. According to this test, the K–S statistic would be equal to the greatest vertical difference between the observed and modeled cumulative probabilities (see Equation 13.3, [Massey 1951; Steinskog et al. 2007]). If this test statistic would be lower than a critical value (see Equation 13.4) at a given confidence level (i.e., 95% in this study), then the null hypothesis (i.e., the observed probabilities depict the employed PDF of interest) would be acceptable.

$$K\text{–}S \text{ test statistic} = \text{maximum}\,|F_m(x) - F_o(x)| \tag{13.3}$$

$$\text{Critical value} = \frac{1.36}{\sqrt{n}} \tag{13.4}$$

where $F_m(x)$ and $F_o(x)$ are the modeled and observed cumulative probabilities, respectively, and n is total number of data.

On evaluation of the goodness of fit, we determined the suitability of a particular PDF. Thus, the approved PDF was applied over an independent validation dataset to evaluate its applicability. In such a case, we again employed the K–S test, as described in the previous paragraph.

13.2.4 Generating Probability Map for Forest Fire Occurrences

Employing the statistically approved PDF, we then generated a slope-derived probability of forest fire occurrences map at 30 m spatial resolution over the nine fire-dominant natural subregions (see Figure 13.5 for details).

13.3 RESULTS

13.3.1 Model Development and Validation

During the model development phase, we observed that the relative frequencies of elevation and slope elements corresponding to fire spots (i.e., 2001–2008) demonstrated a distinct pattern; however, the aspect did not follow any distinctive pattern (i.e., almost equally distributed in every aspect angle between 0° and 360°) (see Figure 13.2). So, we did not consider the aspect element any further and proceeded with the elevation and slope elements. In case of elevation, we found that about 99.7% of fire events fell in the altitude range of 200–1932 m, where the largest frequency was in between 546 m and 662 m (i.e., about 21.7%). In case of slope, we observed that 99.1% of the fires happened in the slope range of 0°–20°, and the maximum amount of fires occurred between 1° and 2° (i.e., 23.2%).

On deciding the relative frequencies of the fire occurrences for both elevation and slope, we fitted the elevation- and slope-derived probabilities with LL and GEV PDFs, respectively, and calculated their K–S test statistic for the fitted distribution. The K–S test statistic using the LL PDF was found to be 0.04234, which was greater than the critical value. The critical values of the K–S test statistics for both elevation and slope elements were calculated to be 0.01275 by using Equation 13.4, as there were 11,375 numbers of fire events in the calibration dataset (i.e., 2001–2008). Hence, the null hypothesis (having the same distribution for both observed and modeled elevation-derived probabilities) was rejected at the 95% confidence level. This showed that the LL PDF might not represent the best fit for elevations corresponding to fire events (see Figure 13.3), and thus, elevation values could not be modeled. In addition, the slope values were fitted using the GEV PDF, and corresponding K–S test statistic was calculated (i.e., 0.01127). It was observed that the test statistic was less than the critical value, so we did not have enough evidence to reject the null hypothesis. This result showed that the observed and modeled slope-derived probabilities followed the same distribution. The values of the PDFs parameters for both elevation and slope elements were shown in Figure 13.3.

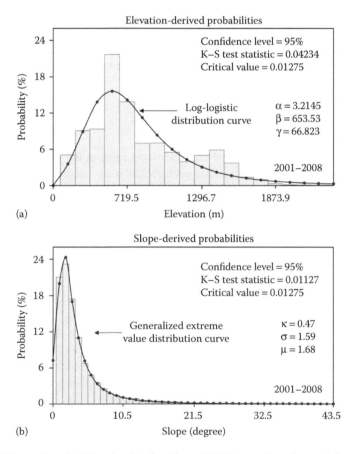

FIGURE 13.3 Fitting of probability density functions (PDFs) over the observed forest fire occurrence probabilities (i.e., relative frequencies) derived from (a) elevation and (b) slope elements, during the period 2001–2008. Kolmogorov–Smirnov test statistics and critical values for selected topographical elements at confidence level of 95%, along with the PDF parameters, are shown in the graph; α and κ are shape parameters, β and σ are scale parameters, and γ and μ are location parameters.

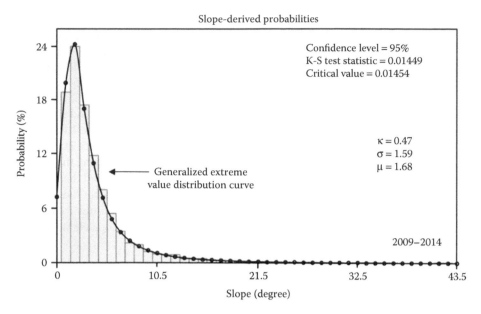

FIGURE 13.4 Comparison between observed and modeled slope-derived probabilities of the forest fire occurrences during the period 2009–2014. Kolmogorov–Smirnov test statistic, critical value at confidence level of 95%, and the probability density function parameters (i.e., κ, σ, and μ) are shown in the graph.

During the model validation phase, we used the slope value at the location of fire spots during the period 2009–2014, and probabilities of fire occurrences were calculated using the GEV distribution. The K–S test statistic for the slope-derived probabilities was found to be 0.01449, which was less than the critical value (i.e., 0.01454), so we did not have enough evidence to reject the null hypothesis. Therefore, we could model the slope values by using GEV PDF at 95% confidence level (see Figure 13.4).

13.3.2 PROBABILITY MAP FOR FOREST FIRE OCCURRENCES

Between the three topographical elements, we found that the slope values could be the best predictor for forest fire occurrences in our study area of interest. Thus, we implemented the GEV PDF over the slope map in order to generate the slope-derived probabilities of forest fire occurrences over the nine fire-dominant natural subregions (see Figure 13.5a). The result of our model showed that the probability of fire occurrence was equal to or greater than 20% in 26.2% and less than 5% in 17.5% of the study area. In addition, we observed 41.3% of the study area with a probability of fire occurrence in the range of 15–20% (see Figure 13.5b). The maximum probability was found to be 24.265%, which was at the slope 1.5°, which covered about one-fourth of our study area. In addition, it was evident that by increasing the slope values, the probability of fire occurrence would decrease (see Figure 13.5b). This result was clearly observed in steeper slopes such as mountain areas, which were mostly located in the southwestern part of Alberta, known as the Rocky Mountains (see Figure 13.5a). The only exception was observed between the slopes 0.5° and 1.5°, where the probability of fire occurrence increased by increasing the slope value (see Figure 13.5b).

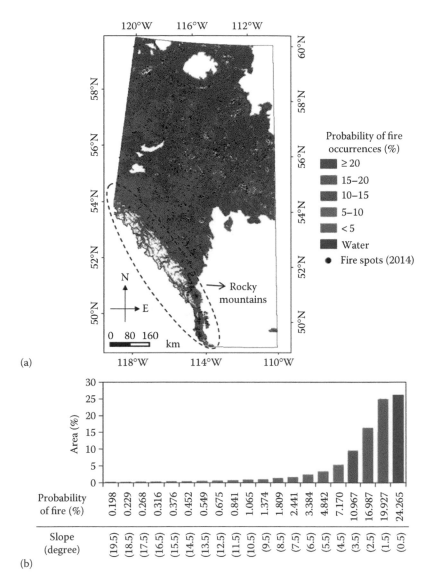

FIGURE 13.5 (a) Slope-derived probability of forest fire occurrences map over the nine fire-dominant natural subregions (see their boundaries in Figure 13.1), where the black dots represented the observed fire spots during the year 2014. (b) Percentage of area for each probability of forest fire occurrence along with its corresponding slope value. The probability of fire occurrence in each slope after 19.5° was almost close to 0% (total percentage of area was 2.6%).

13.4 DISCUSSION

Our findings (i.e., relation between fire occurrences and topographical elements) were similar with those of Pu et al. (2007), where they examined relationships between the percentage of burned area and topographical elements. The details of such similarities are illustrated in Table 13.2. However, as we employed finer intervals for the topographical elements (i.e., 115 m for elevation and 1° for both slope and aspect) in comparison with the coarser intervals used in Pu et al. (2007) (see the last paragraph of the "Introduction" section for details), we found that the forested areas with slope

TABLE 13.2

The Comparison of the Results Reported by Pu et al. (2007) and Obtained in the Current Study for All Topographical Elements

Topographical Elements (Range)	Percentage of Burned Area over North America per Pu et al. (2007)	Percentage of Fire Occurrences over Alberta per This Study
Elevation (≤1000 m)	≈85	72.5
Slope (≤10°)	≈95	93.6
Aspect (all directions)	No clear relationship	No pattern

values in the range of 0°–5° had experienced majority (i.e., 80%) of the fires. In general, flat terrain with relatively lower elevation would support most of the fire events in North America (PU et al. 2007) and elsewhere (Adab et al. 2015), which we also observed in this study (see Figure 13.6 for details). According to Figure 13.6, a significant amount of fires (i.e., 84.3%) took place where slopes were in the range of 0°–6° and elevations were lower than 1000 m. In fact, such fire occurrences might take place due to one or more of the following reasons:

- The flat regions and gentle slope areas might be more accessible for people for camping (Yang et al. 2007).
- Usually, high elevations with northern slope experience relatively cooler temperature (Tabony 1985; Akther and Hassan 2011), and such temperature regimes may disfavor the growth of the vegetation and thus reduce fuel loading. As a result, due to lack of fuel, the probability of fire occurrences may decrease (Preisler et al. 2004; Meyer et al. 2015).
- High-elevation areas usually receive relatively higher amount of rainfall and snowfall, which might potentially decrease the probability of fire occurrences, in particular to North America (PU et al. 2007).

It would be interesting to note that PDFs have been widely employed in modeling/predicting forest fire occurrences. In general, the type of PDFs was found to vary from one case to another, which

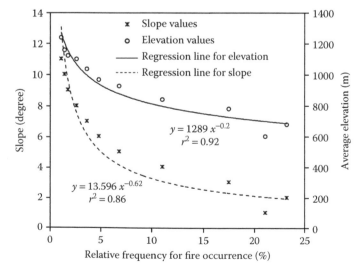

FIGURE 13.6 Relative frequency of forest fire occurrences (i.e., 2001–2008) in relation to elevation and slope elements in the nine fire-dominant natural subregions, along with power regression line for elevation and slope elements.

was also evident in this study. For example, we found that the GEV and LL PDFs were the best ones to model fire occurrences as functions of slope and elevation, respectively. On the contrary, other researchers modeled/predicted fire occurrences in (1) Iran as a function of elevation and slope, using the Pearson type III PDF (Adab et al. 2015); (2) Spain, using elevation on applying truncated-exponential PDF (González et al. 2006); and (3) the United States as a function of elevation and slope, using homogeneous Poisson PDF (Yang et al. 2007), among others.

In general, we observed the following distinct features in the slope-derived forest fire occurrence probability map. For example:

- The areas located in southwestern part of the province had lower probability (i.e., <5%) of forest fire occurrence. As shown in Figure 13.1, these areas fell within the two natural subregions (i.e., Montane and subalpine), where the average elevations were found to be 1400 m and 1750 m and slopes were found to be 9.8° and 17.4°, respectively (see Table 13.3). Note that areas with relatively higher elevations with steeper slopes would disfavor the fire occurrences, and the rationales were described in the first paragraph of "Discussion" section.
- The majority of lands with higher probability (i.e., ≥10 %) of forest fire occurrence were located in the northern part of the province, and they fell within the following four natural subregions: Central Mixedwood, Dry Mixedwood Boreal, Lower Boreal Highlands, and Northern Mixedwood. The average of elevation and slope in these natural subregions were found to be in the range of 350–675 m and 1.4°–2.5°, respectively (see Table 13.3). The areas with such lower elevations with gentle slopes might be attractive places for people to go for camping and might enhance the chances of fire occurrences.
- In general, we observed that the increase in slope would potentially decrease the probability of fire occurrences, with an exception for the slope going from 0.5° to 1.5°. The reason of such increase in probability would be related to the fact that the areas having a slope of 1.5° in comparison with 0.5° were a bit higher (i.e., 26% for 1.5° vs. 25% for 0.5°). Thus, people might have more access into these areas, which would potentially enhance the probability of fire occurrences.

Despite the effectiveness of slope-derived probability of forest fire occurrences, other known factors influencing fire occurrences (i.e., meteorological variables, land cover variables, ignition source, and anthropogenic factors) might be incorporated, as mentioned in the "Introduction" section. In addition, it would be an interesting idea to incorporate our finding within the current operational Canadian forest fire danger systems, such as CFFDRS (Van Wagner 1987), to

TABLE 13.3

The Average Elevation and Slope of Nine Fire-Dominant Natural Subregions over the Period 2001–2014 in Alberta

Number on Figure 13.1	Natural Subregions	Average Elevation (m)	Average Slope (°)
1	Central Mixedwood	525	2.1
2	Lower Foothills	950	3.8
3	Dry Mixedwood Boreal	600	2.4
4	Montane	1400	9.8
5	Upper Foothills	1300	6.9
6	Lower Boreal Highlands	675	2.5
7	Sub-Alpine	1750	17.4
8	Northern Mixedwood	350	1.4
9	Upper Boreal Highlands	825	27.6

evaluate its performance in predicting fire occurrences. In addition, we plan to integrate the generated probability map in the framework of our earlier-developed fully remote sensing–based fire occurrence prediction system (Akther and Hassan 2011; Chowdhury and Hassan 2013, 2015).

13.5 CONCLUDING REMARKS

In this paper, we explored relationships between topographical elements and fire occurrences over the nine fire-dominant natural subregions in Alberta and found that slope was the best predictor in modeling fire occurrences. The outcomes of this study would be useful for forest fire managers in defining/locating the best places to stock fire extinguishing and suppression materials in the region with high odds of fire occurrence, in order to reduce the firefighting costs (Wang and Anderson 2010; Plucinski 2011). Finally, we strongly suggest that the proposed model should be thoroughly evaluated before its implementation in other geographic locations around the world.

ACKNOWLEDGMENTS

This study was partially funded by the (1) Ministry of Science, Research and Technology of Iran via offering the PhD scholarship to M.A., and (2) Natural Sciences and Engineering Research Council of Canada Discovery Grant to Q.K.H. We would like to thank NASA and Alberta Forest Service for providing DEM and fire spots data free of cost.

AUTHOR CONTRIBUTIONS

M.A. and Q.K.H. conceived and designed the study; M.A., Q.K.H, and E.H.C. performed the experiments and analyzed the data; M.A., Q.K.H, E.H.C., and A.G. wrote the paper; and Q.K.H. supervised the entire work.

CONFLICTS OF INTEREST

The authors declare no conflict of interest.

REFERENCES

Adab, H., K.D. Kanniah, and K. Solaimani. 2013. Modeling forest fire risk in the northeast of Iran using remote sensing and GIS techniques. *Natural Hazards* 65: 1723–1743. doi:10.1007/s11069-012-0450-8.

Adab, H., K.D. Kanniah, K. Solaimani, and R. Sallehuddin. 2015. Modelling static fire hazard in a semi-arid region using frequency analysis. *International Journal of Wildland Fire* 24(6): 763–777. doi:10.1071/WF13113.

Ager, A.A., H.K. Preisler, B. Arca, D. Spano, and M. Salis. 2014. Wildfire risk estimation in the Mediterranean area. *Environmetrics* 25: 384–396. doi:10.1002/env.2269.

Akther, M.S. and Q.K. Hassan. 2011. Remote sensing-based assessment of fire danger conditions over boreal forest. *IEEE Journal of Selected Topics in Applied Earth Observations and Remote Sensing* 4(4): 992–999. doi:10.1109/JSTARS.2011.2165940.

Argañaraz, J.P., G.G. Pizarro, M. Zak, and L.M. Bellis. 2015. Fire regime, climate, and vegetation in the Sierras de Córdoba, Argentina. *Fire Ecology* 11(1): 55–73. doi:10.4996/fireecology.1101055.

Chowdhury, E.H. and Q.K. Hassan. 2013. Use of remote sensing-derived variables in developing a forest fire danger forecasting system. *Natural Hazards* 67: 321–334. doi:10.1007/s11069-013-0564-7.

Chowdhury, E.H. and Q.K. Hassan. 2015. Development of a new daily-scale forest fire danger forecasting system using remote sensing data. *Remote Sensing* 7: 2431–2448. doi:10.3390/rs70302431.

Chu, T. and X. Guo. 2014. Remote sensing techniques in monitoring post-fire effects and patterns of forest recovery in boreal forest regions: A review. *Remote Sensing* 6: 470–520. doi:10.3390/rs6010470.

Cipriani, H.N., J.A.A. Pereira, R.A. Silva, S.G. De Freitas, and L.T. De Oliveira. 2011. Fire risk map for the Serra de São Domingos municipal park, Poços de Caldas, MG. *Cerne Lavras* 17(1): 77–83. http://www.scielo.br/pdf/cerne/v17n1/v17n1a09.pdf.

Downing, D.J. and W.W. Pettapiece. 2006. Natural regions and subregions of Alberta. *Natural Regions Committee. Government of Alberta*. Publication Number T/852. https://www.albertaparks.ca/media/2942026/nrsrcomplete_may_06.pdf.

González, J.R., M. Palahí, A. Trasobares, and T. Pukkala. 2006. A fire probability model for forest stands in Catalonia (north-east Spain). *Annals of Forest Science* 63: 169–176. doi:10.1051/forest:2005109.

Hassan, Q.K. and K.M. Rahman. 2013. Applicability of remote sensing-based surface temperature regimes in determining deciduous phenology over boreal forest. *Journal of Plant Ecology* 6: 84–91. doi:10.1093/jpe/rts010.

Kalabokidis, K.D., N. Koutsias, P. Konstantinidis, and C. Vasilakos. 2007. Multivariate analysis of landscape wildfire dynamics in a Mediterranean ecosystem of Greece. *Area* 39(3): 392–402. doi:10.1111/j.1475-4762.2007.00756.x.

Kantam, R.R.L., G. Srinivasa Rao, and B. Sriram. 2006. An economic reliability test plan: Log-logistic distribution. *Journal of Applied Statistics* 33: 291–296. doi:10.1080/02664760500445681.

Kral, K.C., R.F. Limb, T.J. Hovick, D.A. McGranahan, A.L. Field, and P.L. O'Brien. 2015. Simulating grassland prescribed fires using experimental approaches. *Fire Ecology* 11(3): 34–44. doi:10.4996/fireecology.1103034.

Martínez, J., C. Vega-Garcia, and E. Chuvieco. 2009. Human-caused wildfire risk rating for prevention planning in Spain. *Journal of Environmental Management* 90: 1241–1252. doi:10.1016/j.jenvman.2008.07.005.

Massey, F.J. 1951. The Kolmogorov-Smirnov Test for Goodness of Fit. *Journal of the American Statistical Association* 46: 68–78. https://r-forge.r-project.org/scm/viewvc.php/*checkout*/pkg/literature/1951-jamsta-massey-kolmsmirntest.pdf?root=glogis.

Meyer, M.D., S.L. Roberts, R. Wills, M. Brooks, and E.M. Winford. 2015. Principles of effective USA federal fire management plans. *Fire Ecology* 11(2): 59–83. doi:10.4996/fireecology.1102059.

NRC (Natural Resources Canada). 2016. Facts about wildland fires in Canada. http://www.nrcan.gc.ca/forests/fire-insects-disturbances/fire/13143.

Plucinski, M.P. 2011. A review of wildfire occurrence research. *Bushfire Cooperative Research Centre, Australia*. http://www.bushfirecrc.com/sites/default/files/managed/resource/attachment_g_fire_occurrence_literature_review_0.pdf.

Podur, J., D.L. Martell, and F. Csillag. 2003. Spatial patterns of lightning-caused forest fires in Ontario, 1976–1998. *Ecological Modelling* 164: 1–20. doi:10.1016/S0304-3800(02)00386-1.

Prasad, V.K., K.V.S. Badarinath, and A. Eaturu. 2008. Biophysical and anthropogenic controls of forest fires in the Deccan Plateau, India. *Journal of Environmental Management* 86: 1–13. doi:10.1016/j.jenvman.2006.11.017.

Preisler, H.K., D.R. Brillinger, R.E. Burgan, and J.W. Benoit. 2004. Probability based models for estimation of wildfire risk. *International Journal of Wildland Fire* 13: 133–142. doi:10.1071/WF02061.

PU, R., Z. LI, P. Gong, I. Csiszar, R. Fraser, W. Hao, S. Kondragunta, and F. Weng. 2007. Development and analysis of a 12-year daily 1-km forest fire dataset across North America from NOAA/AVHRR data. *Remote Sensing of Environment* 108: 198–208. doi:10.1016/j.rse.2006.02.027.

Rodríguez, E., Morris, C. S., Belz, J. E., Chapin, E. C., Martin, J. M., Daffer, W., and Hensley, S. 2005. An assessment of the SRTM topographic products. Technical Report JPL D-31639, Jet Propulsion Laboratory, Pasadena, CA, pp. 143.

Rulfová, Z., A. Buishand, M. Roth, and J. Kyselý. 2016. A two-component generalized extreme value distribution for precipitation frequency analysis. *Journal of Hydrology* 534: 659–668. doi:10.1016/j.jhydrol.2016.01.032.

Ruokolainen, L. and K. Salo. 2009. The effect of fire intensity on vegetation succession on a sub-xeric heath during ten years after wildfire. *Annales Botanici Fennici* 46: 30–42. http://www.sekj.org/PDF/anbf46/anbf46-030.pdf.

Steinskog, D.J., D.B. Tjøstheim, and N.G. Kvamstø. 2007. A cautionary note on the use of the Kolmogorov-Smirnov test for normality. *Monthly Weather Review* 135: 1151–1157. doi:10.1175/MWR3326.1.

Tabony, R.C. 1985. Relations between minimum temperature and topography in Great Britain. *Journal of Climatology* 5: 503–520. doi:0196-1748/85/050503-18$01.80.

Van Wagner, C.E. 1987. Development and Structure of the Canadian Forest Fire Weather Index System. Canadian Forestry Service, Headquarters, Ottawa. Forestry Technical Report 35. http://cfs.nrcan.gc.ca/pubwarehouse/pdfs/19927.pdf.

Vasconcelos, M.J.P. de, S. Sllva, M. Tome, M. Alvim, and J.M.C. Perelra. 2001. Spatial prediction of fire ignition probabilities: Comparing logistic regression and neural networks. *Photogrammetric Engineering & Remote Sensing* 67(1): 73–81. doi:0099-1112/01/6701-73.

Wang, Y., and K.R. Anderson. 2010. An evaluation of spatial and temporal patterns of lightning- and human-caused forest fires in Alberta, Canada, 1980–2007. *International Journal of Wildland Fire* 19: 1059–1072. doi:10.1071/WF09085.

Wotton, B.M. 2009. Interpreting and using outputs from the Canadian forest fire danger rating system in research applications. *Environmental and Ecological Statistics* 16: 107–131. doi:10.1007/s10651-007-0084-2.

Yang, J., H.S. He, S.R. Shifley, and E.J. Gustafson. 2007. Spatial patterns of modern period human-caused fire occurrence in the Missouri Ozark Highlands. *Forest Science* 53(1): 1–15. http://www.nrs.fs.fed.us/pubs/jrnl/2007/nc_2007_yang_001.pdf?

14 Quantifying the Interannual Variability of Wildfire Events across Portugal for the 2014–2015 Wildfires Using the Data from the European Forest Fire Information System

Aaron Mills and Daniel Colson

CONTENTS

14.1 INTRODUCTION

Fire is a naturally occurring phenomena that has shaped the Earth's vegetation throughout its natural history (Chuvieco, 2009), and fire can be described as an integral part of the Mediterranean ecosystems. Wildland fires play an important role in the evolution, organization, and distribution of ecosystems in the Mediterranean region and globally (Arianoutsou et al. 2005; Koutsias et al. 2012). Wildfires are complex multiscale events that incorporate several key *variables. These are vegetation, climate (weather), human impact*, and *topography* (Wisner et al. 2012), all of which are assessed within the case study section of this chapter, based on two Portuguese fire seasons. Wildfires have negative effects, such as being a threat to the natural environment, wildlife, the economy and putting human life at risk (Tanase et al. 2015; Vhengani et al. 2015). Fires can lead to loss of life, infrastructure damage, and suppression costs (Keeley et al. 2008; Tanase et al. 2015). In the Mediterranean region, wildfires are regarded as one of the most threatening natural disasters to affect property and infrastructure, with them having a long and important presence in the region, intertwined with the area's history (Mayor et al. 2007; Petropoulos et al. 2011). The changes brought upon landscapes after a wildfire event can dramatically affect land cover dynamics at various spatial scales and have impacts on degradation processes, for example, soil erosion (Ireland and Petropoulos, 2015).

At a global scale, figures state that about 350 million hectares of land are annually affected by fire events (van der Werf et al. 2006). It has been observed that in the last few decades, forest fires in the Mediterranean region have increased in frequency, due to numerous climatic and anthropogenic factors (Maselli et al. 2003; Petropoulos et al. 2011). This increase in frequency and the factors outlined previously have led to significant attention being placed on wildfires in recent decades (Lentile et al. 2006). On a regional scale, nearly 90% of all wildland forest fires within the boundaries of the European Union take place in the Mediterranean (Petropoulos et al. 2011; Rosa et al. 2008). This translates to approximately 65,000 fires every year, which in turn burn, on an average, half a million hectares of forested areas (European Commission, 2010; San-Miguel-Ayanz et al. 2013). These fires within the Mediterranean region are the main source of recorded environmental damages in Southern Europe (San-Miguel-Ayanz et al. 2013). Owing to the damages caused by fire in this ecosystem, there is a need to understand the phenomena further. This has led to research of the interannual variability of wildfires, outlined in this chapter, to understand how and why wildfire occurrence and severity vary between certain years.

In the near future, it is observed by many climate models that anthropogenically enhanced climate change will lead to an increase in frequency of wildfire events (Intergovernmental Panel on Climate Change [IPCC], 2014; Ireland and Petropoulos, 2015). Climate models state that fire frequency is expected to increase with anthropogenically enhanced climate change, in particular where precipitation remains the same or is reduced (IPCC, 2014; Stocks et al. 1998). This suggests that there is potential for an increase in the risk, severity, and frequency of forest fires globally and in Europe (IPCC, 2014). The literature also suggests that an ever-changing global climate will result in more extreme weather conditions such as heat waves and drought, which will exacerbate the frequency of wildfire occurrence and increase fire severities across the Mediterranean region. In work conducted in the late 1990s, four different global climate models predicted an earlier start to the fire season, with significant increase in areas affected due to an increase in flammable fuel (Cramer and Steffen, 1997; Stocks et al. 1998). All of this indicates that wildfires and the damages that these wildfires cause are on the increase; therefore, there is a need to understand the dynamics of wildfires and the impacts that they can have on a region.

The damages caused by wildfire events and the potential for more frequent events have led to policy changes to reflect changing attitudes globally, such as in the United States (Dellasala et al. 2004). Policies toward wildfires in the European region have been driven by the factors outlined previously and also by the occurrence of large fires in the 1980s (San-Miguel-Ayanz et al. 2013). Policies toward wildfires in the European region are determined by the European Union, in particular by the establishment of the European Forest Fire Information System (EFFIS). This is a collaboration that has been ongoing since 1998. This partnership has allowed European Union member states to have uniform information on forest fires in the pan-European region (European Commission, 2015); exchanges of information on fire prevention and restoration practices, among other activities, are enabled by this collaboration. Annual reports on the state of forest fires in Europe have been produced since the year 2000. The EFFIS is the organization that provides harmonized information on forest fires and the assessment of their effects in the pan-European region (European Commission, 2015). One of the stated aims of the EFFIS is to maintain and protect European landscapes and natural heritage, while avoiding loss of human lives and minimizing the damage caused by forest fires (European Commission, 2015). This is achieved through field data collections and observations from satellite data. Therefore, this study hopes to benefit future policies enabled by EFFIS.

One of the issues with an increase in fires globally and regionally is that there is an accompanying rise in costs to monitor and suppress these events. Therefore, there is a need to understand patterns of fires and cost-efficient techniques of mapping burned areas (Lentile et al. 2006; Flannigan et al. 2009). There are a number of approaches used to evaluate the extent and damage of wildfires, with this traditionally taking place in the field. However, due to the rise in accessible Earth observation (EO) products, wildfire risk and areas affected can be mapped with relatively low labor-intensive costs over large areas (Preisler et al. 2004; Vhengani et al. 2015). The use of EO datasets has been advocated by many, as when gathering ground fire severity estimates, there is considerable effort and labor involved; EO is essential for landscape-level assessments of wildland fires (Boer et al. 2008; Miller and Yool, 2002; Tanase et al. 2015). The advantage of using EO data when exploring wildland fires is that large areas can be assessed with relative ease and cost (Cohen and Goward, 2004); in addition, inaccessible regions can be assessed (Chambers et al. 2007; Tanase et al. 2015).

Currently, remote sensing is being used to explore various wildfire studies. Remote sensing of wildfire is split into three main categories: prefire, during, and postfire. Mapping burned areas is relatively simple when using imagery of a high spatial resolution, and the extent can be observed through visual imagery (Boschetti et al. 2004). Remote sensing has reached a level of maturity today, which has allowed the development and distribution of burnt area estimates from remote sensing systems at operational scales (Kalivas et al. 2013). Burnt area products are a record of where a fire has occurred, leaving a scar on the landscape (Lentile et al. 2006). One major advantage of using burnt are products is that they allow for a direct estimate of how much area has been burned (Kaufman et al. 1997). They also have useful operational capabilities, such as burnt area maps, which allow for overall emission rates of biomass-burning events to be calculated (Wooster et al. 2005). Such operational products have been proved to be generally of high demand from research groups and communities interested in modeling the carbon cycle; understanding the relationships between fire regime and climate, as well as between atmospheric emissions and pollution resulting from fires; and understanding the impact of vegetation burning on land cover change (Patra et al. 2005; Jupp et al. 2006). The availability of operational products related to burnt area in particular can additionally provide important information on land cover change related to ecology and biodiversity and contribute significantly in better understanding postfire recovery of an affected area (Rong et al. 2004). Although EO products are constantly being developed and improved, they still pose scientific challenges based mainly on accuracy and extensive validation procedures that are undertaken. These challenges are highlighted within the case study which allows for greater context, when they are explained with reference to operational products that have been used within the study.

Wildfire occurrence describes the frequency and presence of fires, either within a certain time or across a certain space (Pinhol et al. 1998). It is a measure of how many ignitions have occurred and does not focus on the size of events (Diaz-Delgado et al. 2004). The interannual variability of wildfire occurrence looks to monitor fire events from different years and compare the data for similarities and differences (Liu et al. 2013). The first key observation that can be made is to analyze how many events occurred for each individual year, then compare them to see how the numbers vary, and then try to assess why this has occurred (Gedalof et al. 2005). Many people have analyzed interannual variability of wildfire, with varying success (Metsaranta, 2010; Mulqueeny et al. 2011; Bedia et al. 2014). Many factors influence interannual variability interchangeably, such as topography, land cover, and climate (Pausas, 2004; Costa et al. 2011). If it is a dry year, then wildfire occurrence rates are higher; this is based on vegetation (fuels) being low in moisture content, which results in a higher chance of ignition, and thus, the overall ignition rates increase (Gouveia et al. 2012). Wildfires have also been found to be more common in elevated areas, as suppression of fire in these regions tends to be lower, with more focus being placed on protecting lower elevations with higher population (Bhandary and Muller, 2009; Gray et al. 2014).

This chapter aims at exploring and demonstrating the use of EO-based burnt area products from the EFFIS in analyzing the interannual variability of wildfire events in Portugal during the years 2014–2015. As a result, the chapter is structured as follows: following the introduction, the datasets, the study site, and the method for this case study implementation are described. Then, the results and main findings of this study are systematically presented, and the patterns in the burnt area estimates from year to year, including the effect of topography and land use/cover, are discussed. The chapter closes with a summary of the main conclusions of this study, which also provides some suggestions for future continuation of the work conducted herein.

14.2 CASE STUDY: INTERANNUAL VARIABILITY OF THE PORTUGAL WILDFIRES FOR 2014–2015

14.2.1 STUDY AREA

The study area is Portugal, which is located in southwestern Europe (see Figure 14.1). The North is mainly mountainous, with the South being relatively flat in comparison. As it is a Mediterranean country, it is subject to hot, dry summers and very wet winters (Costa et al. 2011). It is also one of the hottest countries in Europe, which has resulted in increase in the wildfire occurrences over the years (Ibid.). This brings justification to the study of wildfires in Portugal, as climate change has been stated to cause increased wildfire activity in the future (Giannakopoulos et al. 2009). This has meant that an increased need for burnt area mapping and a better understanding of wildfires relationship with variables such as topography and land cover have occurred (Lentile et al. 2006).

14.2.2 DATASETS

With respect to burnt area mapping, the EFFIS provides the Rapid Damage Assessment (RDA) product. In this product, burnt area estimates are derived at 250 m spatial resolution from the daily processing of the Moderate Resolution Imaging Spectroradiometer (MODIS) Terra and Aqua MODIS visible-near infrared (VNIR) and shortwave (SWIR) data. Burnt area detection is assisted by the MODIS 1 km active fire product (Giglio et al. 2003). Burnt areas occurring in agricultural land, as defined by the Coordination of Information on the Environment (CORINE 2000) land cover

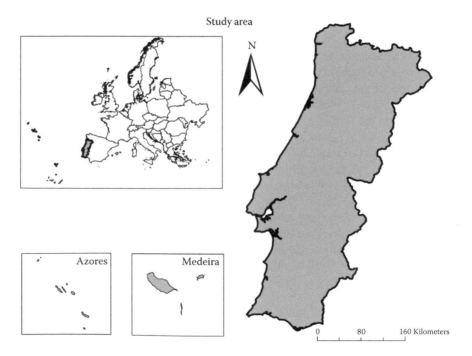

FIGURE 14.1 The study area of Portugal on which the investigation has been based. The two smaller boxes within the figure display Portuguese islands. There was no burnt area data for these areas, so they were excluded from the rest of the study.

map (JRC-EEA, 2005), are masked out during the algorithm implementation (Boschetti et al. 2008). The whole process is also assisted by visual image interpretation and by the systematic collection of fire news from various European media sources (Barbosa et al. 2006). The EFFIS RDA is being implemented since 2003. It provides the daily update of the perimeters of burnt areas in Europe for fires of about 40 hectares or larger; however, the product may also include the perimeters of burned areas of smaller dimension. The product is provided by the EFFIS via a web interface (http://effis-viewer.jrc.ec.europa.eu/wmi/viewer.html), but potential users can also request those products directly from the EFFIS. In our study, the annual burnt area estimates from the EFFIS RDA product for years 2005–2007 were acquired directly from the EFFIS team. In total, three files were provided, each corresponding to the yearly burnt area map included in our analysis. Each file was delivered in vector format (shapefile) and at LAEA-ETRS89 projection.

Land use/cover information from CORINE 2012 was utilized in the present study. We used specifically the 250 m spatial resolution raster dataset, provided at no cost from the European Environmental Agency website (http://www.eea.europa.eu/data-and-maps/data/corine-land-cover-2000-raster-1) as a georeferenced dataset at ETRS1989 datum and projection. CORINE (JRC-EEA, 2005) is a project that was created in 1985 by the European Union, with the aim to create a European land cover/use map, derived primarily from the interpretation of satellite images and ancillary data. CORINE 2000 consists of an updated version of the initial CORINE product, providing land use/cover maps of 29 European countries derived from the processing of orthorectified Landsat images.

Information on elevation was obtained from the Global Digital Elevation Model (GDEM) of the Advanced Spaceborne Thermal Emission and Reflection Radiometer (ASTER) sensor. The ASTER GDEM product was released in 2009 and was updated (version 2) at the end of 2011. It provides elevation information between 83°N and 83°S, with geographic latitude–longitude coordinates at 1 arc sec (30 m) grid. Estimated accuracies of the product are for 20 m at 95% confidence for vertical data and 30 m at 95% confidence for horizontal data (ASTER G DEM, 2009). The dataset is provided in geoTIFF format, in geographic latitude/longitude projection and WGS84/EGM96 datum. It is available at no cost to users via electronic download from the Earth Remote Sensing Data Analysis Center (ERSDAC) of Japan or the National Aeronautics and Space Administration (NASA's) Land Warehouse Inventory Search Tool (WIST, https://wist.echo.nasa.gov/~wist/api/imswelcome/). The ASTER GDEM is distributed as separate tiles of elevation covering the Earth. In our study, the tiles covering Greece were acquired from WIST.

14.2.3 Methods

Some preprocessing was necessary to standardize those before intercomparisons were performed. Most of the acquired datasets were in vector format and had been provided at LAEA-ETRS89 projection, which is also commonly used for European products distribution according to the INSPIRE (INfrastructure for SPatial InfoRmation in Europe) Directive. Based on this, a decision was made to adopt the same specifications in terms of data format and projection for all our collected datasets. All preprocessing and geospatial analysis of the spatial datasets were carried out in ENVI (v. 4.7, ITT Visual Solutions) and ArcMap (v 9.3, ESRI) software platforms. CORINE 2000 land use/cover map was resampled to a spatial resolution of 500 m. The ASTER GDEM product tiles of Greece were merged into a single file by using image mosaic. This latter dataset was subsequently reprojected to LAEA-ETRS89 projection and was resampled by the nearest neighbor to a spatial resolution of 500 m to match the EFFIS spatial resolution. CORINE 2000 land use/cover map was resampled to a spatial resolution of 500 m. The ASTER GDEM product tiles of Greece were merged into a single file by using image mosaic. This latter dataset was subsequently reprojected to the LAEA-ETRS89 projection and was resampled by the nearest neighbor to a spatial resolution of 500 m to match the MCD45A1 spatial resolution. The remainder of the paper is focused on presenting the results obtained from the intercomparisons performed between the two products, also discussing the differences observed.

14.3 RESULTS

This section will highlight the main results that were gathered for this study. Key observations to be made from Figure 14.2 (as seen previously) are that the majority of burnt areas for both years are in the North. There are also some burnt areas in central Portugal, and very few burnt areas can be observed in the South.

Figure 14.3 displays the monthly wildfire data for both years across Portugal. In 2014, 35 burnt areas were recorded, and in 2015, 177 burnt areas were recorded. A large majority of the events occurred within the prescribed fire season of the Mediterranean region (April 1–September 30). However, some wildfire events did occur outside of the fire season, with a large number of events occurring in March, especially in 2015. It is clearly apparent that more wildfire events occurred in 2015 than in 2014. Spatiotemporal observations were found for certain months of the fire season, where many fires would occur in one area (province). This is particularly noticeable where large

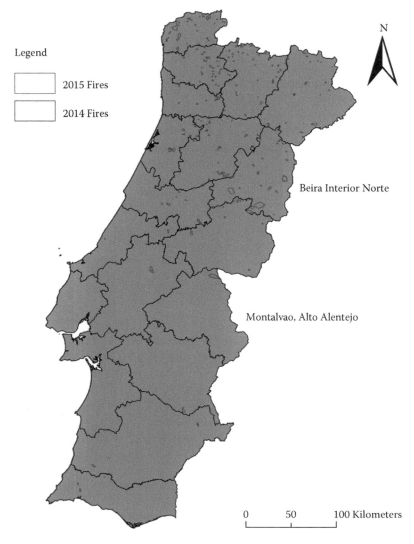

FIGURE 14.2 A map of Portugal, split into provinces. Burnt area data have been overlaid for 2014 and 2015. Fires of 2014 are outlined in orange and those of 2015 are outlined in yellow. The largest fire event for 2014 and 2015 was located in Montalvao, Alto Alentejo, and it had a burnt area of 2869 hectares. For 2015, the largest event was located at Beira Interior Norte, and it had a burnt area of 5539 hectares.

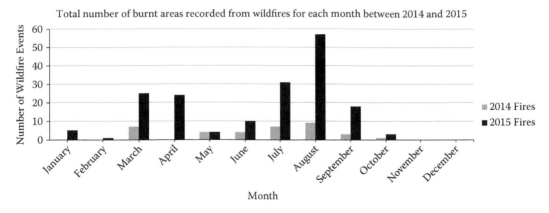

FIGURE 14.3   Graph displaying the total burnt areas for each month of the year across 2014 and 2015.

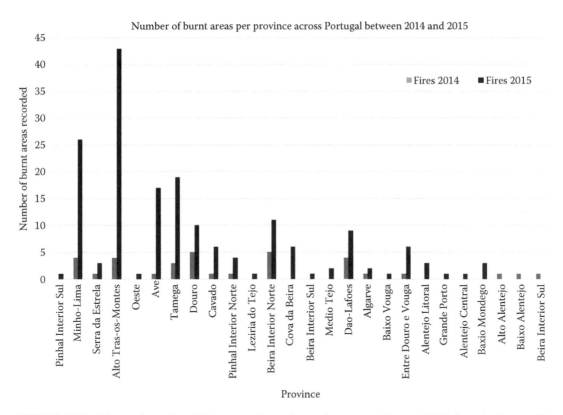

FIGURE 14.4   The total number of fire events for each province across Portugal. This has also been split between the two years, so that variability can be observed.

numbers of events occurred during the 2015 fire season in comparison with the number of the events in 2014. Figure 14.4 displays the fire data for both 2014 and 2015 across all provinces in Portugal. Some provinces clearly show high levels of wildfire activity, compared with others such as Minho-Lima and Alto Tras-os-Montes. These data could be useful if combined with land cover data to see whether the cover types in these provinces are the cause of more wildfire events. Key points to note from Figure 14.5 are that the areas where fires have occurred are heavily dominated by certain land

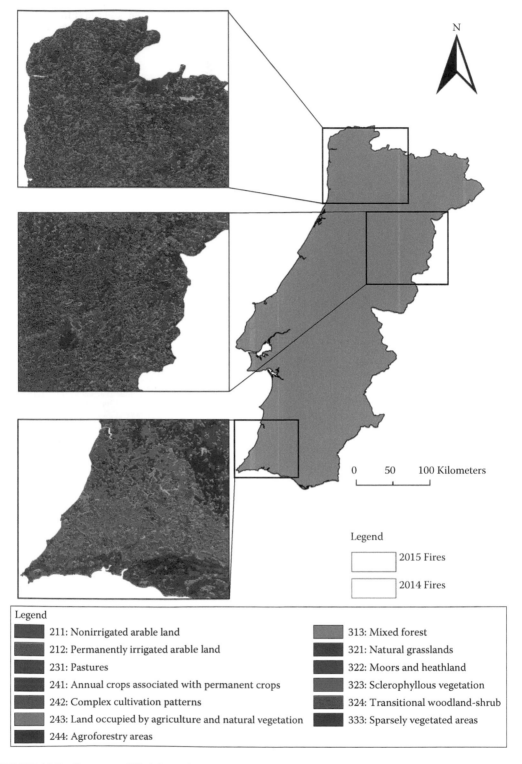

FIGURE 14.5 Image modified from the CORINE land cover (CLC) model. It highlights three key areas where wildfires have been observed and identifies the land cover types within which the burnt areas lie, for 2014–2015. (From Copernicus, 2012. CLC. Available at: http://land.copernicus.eu/pan-european/corine-land-cover/clc-2012. Accessed December 15, 2016.)

cover types. This has clearly demonstrated that there is a strong relationship between land cover type and wildfire events. It can be observed that more fires occurred where moors, heathland, and transitional woodland shrubs were dominant. The larger burn areas were in areas of moors and heathland land cover. The key observation to be made from Figure 14.6 is that most wildfire events occurred in areas where slope gradient exceeded 15%, which would imply that there is a relationship between wildfire events and topography. This also applies to elevation, as the majority of events are found in the mountainous north, where elevations are much higher.

14.4 DISCUSSION

14.4.1 INTERANNUAL AND SPATIOTEMPORAL VARIABILITY OF WILDFIRE OCCURRENCE

The interannual variability of wildfire occurrence was clearly observed within this study. Large variance was observed between 2014 and 2015 in terms of overall burnt area counts. In 2014, 35 burnt areas were identified, and 2015 saw staggering 177 burnt areas across Portugal. Key observations from statistical analysis of the two years showed that 2014 had a larger mean burnt area for wildfires at 330 hectares and 2015 had an average burnt area of only 268 hectares. This clearly indicates variability between the two years in terms of wildfire occurrence.

Spatiotemporal variability was also observed within the study. This was most noticeably observed during the fire season, where a clear increase in the frequency of wildfire events occurred. This was generally true for both years but was particularly highlighted in the 2015 fire season (see Figures 14.2 and 14.3). The fire season is from April 1 to September 30, and this is when most fire events occurred (Huesca et al. 2009). July and August particularly saw an increase in the frequency of wildfire occurrence compared with other months. These months experience the height of summer, where temperatures are at their highest. Therefore, a potential reason for this variability could be extreme climatic conditions, causing increased wildfire occurrence (Costa et al. 2011).

After analyzing climatic reports and literatures from 2015, it was found that this year played host to several large heatwaves across the Mediterranean region (World Meteorological Organization, 2016). This would explain the interannual variability between years and the spatiotemporal variability between months, particularly during the fire season (Salis et al. 2014). These heatwaves also link with other factors surrounding wildfire events, such as effects on land cover types and high temperatures in areas of increased elevation (Trigo et al. 2006). This has happened previously in Portugal, where heatwaves caused an exceptional fire season in 2003 (Ibid.).

Portugal saw a large precipitation deficit because of these heatwaves, which resulted in drought (World Meteorological Organization, 2016). This caused a lack of moisture in fuels, which increased ignition rate probabilities (Chuvieco et al. 2004). The drought peaked toward the end of July and beginning of August, which explains why July and August of 2015 saw such an increase in wildfire frequency (World Meteorological Organization, 2016). Temperatures during this period exceeded 43°C, and monthly averages for both months were 4°C higher than normal (Ibid.). This explains and supports the spatiotemporal variability that was observed between months of the fire season (Salis et al. 2014).

Smaller heatwaves were also observed in March and April of 2015, with large increases in temperature occurring compared with a normal year (World Meteorological Organization, 2016). This would explain the variability between the wildfire occurrence of the two years, as fire frequencies were much higher in 2015 across these months than in 2014 (see Figure 14.3). Extreme weather events were also frequent during 2015; this led to large lightning storms occurring frequently, which would explain why so many wildfire events occurred in 2015 compared with 2014 (Correia et al. 2016).

14.4.2 LAND COVERS RELATIONSHIP WITH WILDFIRE EVENTS

There is a clear relationship between wildfire events/occurrence and land cover within this study. The first observations that demonstrate this can be seen in Figure 14.5, where wildfire occurrence was higher in certain areas that were dominated by highly flammable land cover types. As you can see in the figure, most fires occurred in areas where moors and heathland or transitional woodland shrubs were dominant. Pereira et al. (2014) have identified how certain land cover types are more prone to fire across Europe. The relationship between land cover type and wildfire events is also supported when you observe Figure 14.4. This graph identifies the number of burnt areas per province. The highest number of events was observed in the Minho-Lima and Alto Tras-os-Montes provinces, which is unsurprising, as they are heavily dominated by moors, heathland, and transitional woodland shrubs (see Figure 14.5). This explains the spatial variability of wildfire occurrence across Portugal. Similar studies have also been found to confirm the relationship between land cover types and wildfire events/occurrence (Mermoz et al. 2005; Bajocco and Ricotta, 2008; Pereira et al. 2014). The two largest burnt areas for 2014 and 2015 identified in Figure 14.2 occurred where moors and heathland were the dominant cover type, showing that there is also a relationship between cover type and size of the wildfire events. This has been documented in similar studies across Europe (Pereira et al. 2014; Araya et al. 2016). Wildfire events are heavily influenced by the type, amount, and quality of fuel available (Smith, 2013). Wildfire size is influenced by what fuel types are present in the area, as some vegetation burn faster than others (Mermoz et al. 2005). The best way to assess this effect is by analyzing land cover maps, which can help distinguish between fuel types (Gallardo et al. 2016). Grassland and shrub areas are prone to large fire events, and forested areas generally burn in proportion to their presence (Bajocco and Ricotta, 2008). The smaller a fuel load, the slower a wildfire can spread (Mermoz et al. 2005). More fuel allows for higher burning intensities, which result in the fast spread of fire (Ibid.). Less dense fuels dry out faster, meaning that they ignite and burn faster, whereas dense fuels hold moisture and can slow a fire's progress (Yebra et al. 2013). Fuels with a high moisture content tend to burn slowly, and in times of low humidity, wildfire events are more likely to occur (Vasilakos et al. 2009).

14.4.3 RELATIONSHIP OF TOPOGRAPHY WITH WILDFIRE EVENTS

The relationship between topography and wildfire events is clearly demonstrated in Figures 14.6 and 14.7. All fire events across Portugal for both years are found in areas of high elevation. In addition, many of the events are in areas where slope gradient is high. Several studies have identified slope as being a large contributing factor to wildfire occurrence (Ryan, 2002; Marschall et al. 2016). This is due to wind mechanisms surrounding slopes that causes draft upslope, which result in the warmer air that hugs the surface, preheating fuels (Sharples, 2008). Slope also contributes to the spread of fire, which may be the reason why the larger burnt areas can be observed in areas of relatively high slope gradient (Marschall et al. 2016). As you can see from Figures 14.6 and 14.7, not only do the burnt areas lie within the areas of high slope, but these areas also coincide with high elevation.

The relationship between wildfire events and topography, particularly elevation, was also highlighted. A potential explanation is that lightning strikes are more common in high-elevated areas, which would explain the high occurrence of wildfires across Portugal's elevated regions. This idea is supported in several studies, where lightning has been found to cause the most ignitions in high-elevated areas (Narayanaraj and Wimberly, 2012; Correia et al. 2016). Another possible reason is linked to climatic conditions, where temperatures are now becoming higher in elevated regions due to climate change (Giannakopoulos et al. 2009). This could be a potential cause of numerous wildfire events occurring in these elevated regions and has been found to be the case

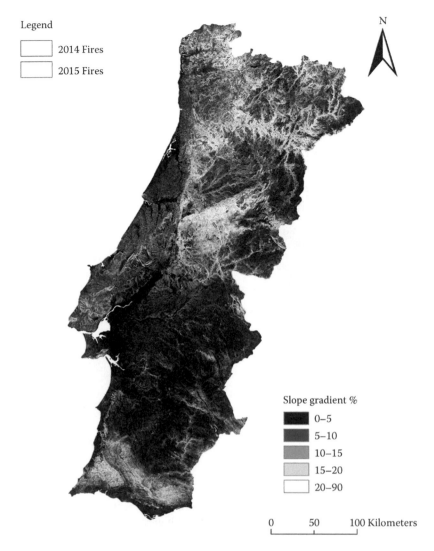

FIGURE 14.6   Figure derived from a DEM of Portugal. Slope gradient has been displayed as a percentage. Areas of slope within the figure are focused on the high-elevated areas of the country, and the burnt areas for 2014 and 2015 have been marked in orange and yellow, respectively.

in similar studies across the planet and may very well be applicable to Portugal (Westerling et al. 2006; Blouin et al. 2016).

Slope affects fire in two ways: preheating (convection and radiation) and draft (Ryan, 2002). On a slope, less dense air sits at the surface due to warmth of the earth, which allows lighter air to rise along slopes, which creates draft (Ibid.). This consequently causes wildfires to burn upslope, as wind drives them (Sharples, 2008). Another topographic influence of wildfires is aspect (Marschall et al. 2016). South-facing slopes receive high levels of solar radiation, which results in high temperatures and loss of moisture in fuels on the slope (Shakesby, 2011). These dry fuels are then at higher risk of being ignited (Chuvieco et al. 2004). Elevation is also linked with wildfire occurrence (Brosofske et al. 2007). Multiple studies give reference to this and express how a warming climate has resulted in increased wildfire occurrence (Dlamini, 2010; Mori and Johnson, 2013).

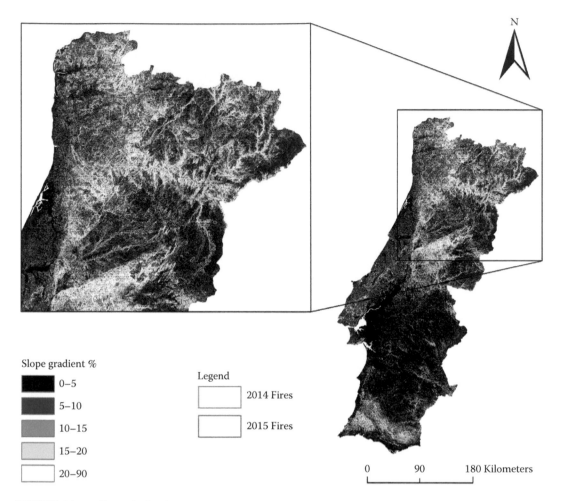

FIGURE 14.7 Figure indicating the Northern sector of Portugal, where the majority of wildfires occurred. This area is high in slope gradient, indicating an area that is of higher elevation.

14.5 CONCLUSIONS AND FUTURE RESEARCH

It was concluded that the interannual variability of wildfire occurrence between 2014 and 2015 was due to the climatic influence of heatwaves, which caused increase fire occurrence in 2015, as conditions for ignition were more favorable. Climatic variables affect spatiotemporal variability of wildfire occurrence. This explains the variability between months in the fire season for both years, but especially for 2015. With regard to land cover types, it was found that certain cover types, especially moors and heathland, had a strong relationship with wildfire events and can dictate not only the occurrence of wildfires but also the extent of these biomass-burning events. Strong relationships were also observed between topography and wildfire events, with the main conclusions being that slope is linked with increased fire size and increased fire occurrence. Elevation was also found to have a strong relationship, as areas of high elevation saw larger numbers of wildfire events. This was linked with extreme weather events, where lightning strikes were presumed to have caused them.

Future directions for this research could be to gather individual elevation data for each fire event and analyze them to support the relationship between changes in elevation and how this impacts wildfire occurrence. This would also be useful if elevation data for each event were analyzed against the elevations most susceptible to lightning strikes, to identify how strong the relationship is between elevation and the ignition of wildfire events. A temperature dataset would also be useful across the different provinces of Portugal, to further assess the spatiotemporal variability of wildfire occurrence regarding climatic conditions. Finally, having a larger dataset spanning over a larger period would allow for a more in-depth interannual variability assessment of wildfire occurrence to be conducted.

At present, there are many scientific challenges associated with operational products. These are heavily based on the accuracy of such products. For example, burnt area products are generally low in spatial, spectral, and temporal resolutions, which can cause accuracy issues. To minimize error and monitor accuracy, extensive validation of products must be carried out, which has proven to be very time-consuming. However, newly proposed sensors, which have already been and are set to be launched, are thought to have far greater operational capabilities, which will address the scientific challenges of low accuracy and allow for an easier validation of data products.

REFERENCES

Araya, Y. H., Remmel, T. K., and Perera, A. H. (2016). What governs the presence of residual vegetation in boreal wildfires? *Journal of Geographical Systems*, 18(2): 159–181.

Arianoutso, M., Gimeno, T., Kazanis D., Pausas, J., and Vallego, R. (2005). Characterization of fire vulnerable *Pinus halepensis* ecosystems in Spain and Greece. In: V. Leone and Lovreglio. (Eds.) *Proceedings of the International Workshop MEDPINE 3: Conservation, Regeneration and Restoration of Mediterranean pies and their Ecosystems*. Bari, Italy: Cheam, 2005, pp. 131–142.

ASTER GDEM (2009). ASTER GDEM Version, 1. Available at: http://www.gdem.aster.ersdac.or.jp (accessed: July 24, 2011).

Bajocco, S. and Ricotta, C. (2008). Evidence of selective burning in Sardinia (Italy): Which land-cover classes do wildfires prefer? *Landscape Ecology*, 23(2): 241–248.

Barbosa, P., Kusera, J., Strobl, P., Vogt, P., Camia, A., and SanMiguel-Ayanz, J. (2006). European Forest Fire Information System (EFFIS)—Rapid damage assessment: Appraisal of burnt area maps in southern Europe using MODIS data (2003–2005), *Forest Ecological Management*, 232: S218.

Bedia, J., Herrera, S., and Gutierrez, J. M. (2014). Determinants of inter-annual variation in the area burnt in a semiarid African savanna. *Natural Hazards and Earth Science Systems*, 14(1): 53–66.

Bhandary, U. and Muller, B. (2009). Land use planning and wildfire risk mitigation: An analysis of wildfire-burned subdivisions using high-resolution remote sensing imagery and GIS data. *Journal of Environmental Planning and Management*, 52(7): 939–955.

Blouin, K. D., Flannigan, M. D., Wang, X. L., and Kochtubajda, B. (2016). Ensemble lightning prediction models for the province of Alberta, Canada. *International Journal of Wildland Fire*, 25(4): 421–432.

Boer, M., Macfarlane, C., Norris, J., Sadler, R., Wallace, J. and Grierson, P. (2008). Mapping burned areas and burn severity patterns in SW Australian eucalypt forest using remotely-sensed changes in leaf area index. *Remote Sensing of Environment*, 112(12): 4358–4369.

Boschetti, L., Eva, H. D., Brivio, P. A., and Gregoire, J. M. (2004). Lessons to be learned from the comparison of three satellite derived biomass burning products. *Geophysical Research Letters*, L21501, 31(21).

Boschetti, L., Roy, D., Barbosa, P., Boca, R., and Justice, C. (2008). A MODIS assessment of the summer 2007 extent burned in Greece. *International Journal of Remote Sensing*, 29: 2433–2436.

Brosofske, K. D., Cleland, D. T., and Saunders, S. C. (2007). Factors influencing modern wildfire occurrence in the Mark Twain National Forest, Missouri. *Southern Journal of Applied Forestry*, 31(2): 73–84.

Chambers, J., Asner, G., Morton, D., Anderson, L., Saatchi, S., Espirito-Santo, F., Palace, M. and Souza, C. (2007). Regional ecosystem structure and function: ecological insights from remote sensing of tropical forests. *Trends in Ecology and Evolution*, 22(8): 414–423.

Chuvieco, E. (2009). Global Impacts of Fire. In: E. Chuvieco. (Ed.) *Earth Observation of Wildland Fires in Mediterranean Ecosystems*. 1st ed. London: Springer, 2009, pp. 1–11.

Chuvieco, E., Aguado, I., and Dimitrakopoulos, A. P. (2004). Conversion of fuel moisture content values to ignition potential for integrated fire danger assessment. *Canadian Journal of Forest Research*, 34(11): 2284–2293.

Cohen, W. and Goward, S. (2004). Landsat's role in ecological applications of remote sensing. *BioScience*, 54(6):. 535–545.

Copernicus, 2012. CLC (2012). European Space Agency. Available at: http://land.copernicus.eu/pan-european/corine-land-cover/clc-2012 (Accessed December 15, 2016).

Correia, S., Lourenco, A., Rio, J., Prior, V. and Moreira, N. (2016). Portuguese lightning detection network, applications and developed products. *Proceeding of the 33rd International Conference on Lightning Protection* (*ICLP*) Lisbon: IEEE, 2016.

Costa, L., Thonicke, K., Poulter, B. and Badeck, F. W. (2011). Sensitivity of Portuguese forest fires to climatic, human, and landscape variables: Subnational differences between fire drivers in extreme fire years and decadal averages. *Regional Environmental Change*, 11(3): 543–551.

Cramer, W. and Steffen, W. (1997) Forecast changes in the global environment: What they mean in terms of ecosystem responses on different time-scales. In: Huntley, B., Cramer, W., Morgan, A.V., Prentice, H.C. and Allen. J.R.M. (Eds.) *Past and Future Rapid Environmental Changes: The Spatial and Evolutionary Responses of Terrestrial Biota*. Vol. I 47. Berlin, Germany: Springer-Verlag, 1997, pp. 415–426.

Dellasala, D., Williams, J., Williams, C. and Franklin, J. (2004). Beyond smoke and mirrors: A sythesis of fire policy and science. *Conservation Biology*, 18(4): 976–986.

Diaz-Delgado, R., Lloret, F., and Pons, X. (2004). Spatial patterns of fire occurrence in Catalonia, NE Spain. *Landscape Ecology*, 19(7): 731–745.

Dlamini, W. M. (2010). A Bayesian belief network analysis of factors influencing wildfire occurrence in Swaziland. *Environmental Modelling and Software*, 25(2): 199–208.

European Commission (2010). Forest Fires in Europe 2009—EUR 24502 EN. Official Publication of the European Communities. Luxembourg: Publications Office of the European Union, 2010, pp. 1–88. http://forest.jrc.ec.europa.eu/media/cms_page_media/9/forest-fires-in-europe-2009.pdf (accessed on March 11th, 2017)

European Commission (2015). Forest Fires in Europe, Middle East and North Africa 2014. Joint Report of JRC and Directorate-General Environment. Brussels: Joint Research Centre, 2015, pp. 1–118. http://forest.jrc.ec.europa.eu/media/cms_page_media/9/Forest%20fires%20in%20Europe,%20Middle%20east%20and%20North%20Africa%202014_final_pdf.pdf (Accessed on March 11th, 2017)

Flannigan, M., Stocks, B., Turetsky, M. and Wotton, M. (2009). Impacts of climate change of fire activity and fire management in the circumboreal forest. *Global Change Biology*, 15(3): 549–560.

Gallardo, M., Gomez, I., Vilar, L., Martinez-Vega, J., and Martin, M P. (2016). Impacts of future land use/ land cover on wildfire occurrence in the Madrid region (Spain). *Regional Environmental Change*, 16(4): 1047–1061.

Gedalof, Z., Peterson, D. L., and Mantua, N. J. (2005). Atmospheric, climatic, and ecological controls on extreme wildfire years in the northwestern United States. *Ecological Applications*, 15(1): 154–174.

Giannakopoulos, C., Le Sager, P., Bindi, M., Moriondo, M., Kostopoulou, E., and Goodess, C M. (2009). Climatic changes and associated impacts in the Mediterranean resulting from a 2 degrees C global warming. *Global and Planetary Change*, 68(3): 209–224.

Giglio, L., Descloitres, J., Justice, C. O., and Kaufman, Y.(2003). An enhanced contextual fire detection algorithm for MODIS. *Remote Sensing of Environment*, 87(2): 273–282.

Gouveia, C. M., Bastos, A., Trigo, R. M., and DaCamara, C. (2012). Drought impacts on vegetation in the pre- and post-fire events over Iberian Peninsula. *Natural Hazards and Earth Systems Sciences*, 12(10): 3123–3137.

Gray, M. E., Dickson, B. G., and Zachmann, L. J. (2014). Modelling and mapping dynamic variability in large fire probability in the lower Sonoran Desert of south-western Arizona. *International Journal of Wildland Fire*, 23(8): 1108–1118.

Huesca, M., Litago, J., Palacios-Orueta, A., Montes, F., Sebastian-Lopez, A., and Escribano, P. (2009). Assessment of forest fire seasonality using MODIS fire potential: A time series approach. *Agricultural and Forest Meteorology*, 149(11): 1946–1955.

IPCC (2014). Climate Change 2014: Impacts, adaptation and vulnerability. Part B: Regional aspects. Contribution of workig group II to the fifth assessment report of the intergovernmental panel on climate change. Barros, V.R., C.B. Field, D.J. Dokken, M.D. Mastrandrea, K.J. Mach, T.E. Bilir, M. Chatterjee et al. (Eds.). Cambridge, UK and New York: Cambridge University Press, 2014, pp. 688.

Ireland, G. and Petropoulos, G. P. (2015). Exploring the relationships between post-fire vegetation regeneration dynamics, topography and burn severity: A case study from the Montane Cordillera Ecozones of Western Canada. *Applied Geography*, 56(1): 232–248.

JRC-EEA (2005): CORINE land cover updating for the year 2000: image 2000 and CLC2000. In: Lima, V. (Ed.) Products and Methods, Report EUR 21757 EN, JRC-Ispra, Available at: http: //reports.eea.europa. eu/COR0-landcover/en (accessed: July 28, 2011).

Jupp, T. E., Taylor, C. M., Balzter, H., and George, C. T. (2006) A statistical model linking Siberian forest fire scars with early summer rainfall anomalies. *Geophysical Research Letters*, 33(14): 1–5.

Kalivas, D. P., Petropoulos, G. P., Athanasiou, I. M., and Kollias, V. J. (2013). An intercomparison of burnt area estimates derived from key operational products: the Greek wildland fires of 2005–2007. *Nonlinear Processes in Geophysics*, 20(1): 1–13.

Kaufman, Y J., Tanre, D., Remer, L A., Vermote, E F., Chu, A., and Holben, B N. (1997). Operational remote sensing of tropospheric aerosol over land from EOS moderate resolution imaging spectroradiometer (MODIS). *Journal of Geophysical Research-Atmospheres*, 102(D14): 17051–17067.

Keeley, J., Brennan, T., and Pfaff, A. (2008). Fire severity and ecosystem responses following crown fires in California shrublands. *Ecological Applications*, 18(6): 1530–1546.

Koutsias, N., Arianoutsou, M., Kallimanis, A., Mallinis, G., Halley, J., and Dimopoulos, P. (2012). Where did the fires burn in Peloponnisos, Greece the summer of 2007? Evidence for a synergy of fuel and weather. *Agricultural and Forest Meteorology*, 156: 41–53.

Lentile, L B., Holden, Z A., Smith, A M S., Falkowski, M J., Hudak, A T., Morgan, P., Lewis, S A., Gessler, P E., and Benson, N C. (2006). Remote sensing techniques to assess active fire characteristics and post-fire effects. *International Journal of Wildland Fire*, 15(3): 319–345.

Liu, Y. Q., Goodrick, S. L., and Stanturf, J. A. (2013). Future US wildfire potential trends projected using a dynamically downscaled climate change scenario. *Forest Ecology and Management*, 294(1): 120–135.

Maselli, F., Romanelli, S., Bottai, L., and Zipoli, G. (2003). Use of NOAA-AVHRR NDVI images for the estimation of dynamic fire risk in Mediterranean areas. *Remote Sensing of Environment*, 86(2): 187–197.

Marschall, J M., Joseph, M., Stambaugh, M C., Jones, B C., Guyette, R P., Brose, P H. and Dey, D C. (2016). Fire regimes of remnant pitch pine communities in the ridge and valley region of Central Pennsylvania, USA. *Forests*, 7(10): 224.

Mayor, A., Bautista, S., Llovet, J., and Bellot, J. (2007). Post-fire hydrological and erosional responses of a Mediterranean landscape: Seven years of catchment-scale dynamics. *Catena*, 71(1): 68–75.

Mermoz, M., Kitzberger, T., and Veblen, T. T. (2005). Landscape influences on occurrence and spread of wildfires in Patagonian forests and shrublands. *Ecology*, 86(10): 2705–2715.

Metsaranta, J. M. (2010). Potentially limited detectability of short-term changes in boreal fire regimes: A simulation study. *International Journal of Wildland Fire*, 19(8): 1140–1146.

Miller, J. D. and Yool, S. R. (2002). Mapping forest post-fire canopy consumption in several over story types using multi-temporal Landsat TM and ETM data. *Remote Sensing of Environment*, 82(23): 481–496.

Mori, A. S. and ohnson, E. A. (2013). Assessing possible shifts in wildfire regimes under a changing climate in mountainous landscapes. *Forest Ecology and Management*, 310(1): 875–886.

Mulqueeny, C. M., Goodman, P. S., and O'Connor, T. G. (2011). Determinants of inter-annual variation in the area burnt in a semiarid African savanna. *International Journal of Wildland Fire*, 20(4): 532–539.

Narayanaraj, G. and Wimberly, M. C. (2012). Influences of forest roads on the spatial patterns of human- and lightning-caused wildfire ignitions. *Applied Geography*, 32(2): 878–888.

Patra, P. K., Ishizawa, M., Maksyutov, S., Nakazawa, T., and Inoue, G. (2005). Role of biomass burning and climate anomalies for land-atmosphere carbon fluxes based on inverse modeling of atmospheric CO_2. *Global Biogeochemical Cycles*, 19(GB3005): 11–15.

Pausas, J. G. (2004). Changes in fire and climate in the eastern Iberian Peninsula (Mediterranean basin). *Climate Change*, 63(3): 337–350.

Pereira, M. G., Aranha, J., and Amraoui, M. (2014). Land cover fire proneness in Europe. *Forest Systems*, 23(3): 598–610.

Petropoulos, G., Kontoes, C., and Keramitsoglou, I. (2011), Burnt area delineation from a uni-temporal perspective based on Landsat TM imagery classification using support vector machines. *International Journal of Applied Earth Observation and Geoinformation*, 13(1): 70–80.

Pinhol, J., Terradas, J., and Lloret, F. (1998). Climate warming, wildfire hazard, and wildfire occurrence in coastal eastern Spain. *Climate Change*, 38(3): 345–357.

Preisler, H. K., Brillinger, D. R., Burgan, R. E., and Benoit, J. (2004). Probability based models for estimation of wildfire risk. *International Journal of Wildland Fire*, 13(2): 133–142.

Rong, R. L., Kaufman, J., Hao, W. M., Salmon, J. M., and Gao, B.-C. (2004). A technique for detecting burn scars using MODIS data. *IEEE Transactions on Geoscience and Remote Sensing*. 42(6): 1300–1308.

Rosa De La, J. M., Gonzalez-Perez, J. A., Gonzalez-Vazquez, R., Knicer, H., Lopez-Capel, E., Manning, D. A. C., and Gonzalez-Vila, F. J. (2008). Use of pyrolysis/GC-MS combined with thermal analysis to monitor C and N changes in soil organic matter from a Mediterranean fire affected forest. *Catena*, 74(3): 296–303.

Ryan, K. C. (2002). Dynamic interactions between forest structure and fire behaviour in boreal ecosystems. *Silva Fennica*, 36(1): 13–39.

Salis, M., Ager, A. A., Finney, M. A., Arca, B. and Spano, D. (2014). Analysing spatiotemporal changes in wildfire regime and exposure across a Mediterranean fire-prone area. *Natural Hazards*, 71(3): 1389–1418.

San-Miguel-Ayanz, J., Schulte, E., Schmuck, G., and Camia, A. (2013). The European Forest Fire Information System in the context of environmental policies of the European Union. *Forest Policy and Economics*, 29: 19–25.

Shakesby, R. A. (2011). Post-wildfire soil erosion in the Mediterranean: Review and future research directions. *Earth-Science Reviews*, 105(3–4): 71–100.

Sharples, J. J. (2008). Review of formal methodologies for wind-slope correction of wildfire rate of spread. *International Journal of Wildland Fire*, 17(2): 179–193.

Smith, K. (2013). Environmental Hazards: Assessing Risk and Reducing Disaster. London: Routledge, 2013.

Stocks, B. J., Fosberg, M. A., Lynham, T. J., Mearns, L., Wotton, B. M., Yang, Q., Jin, J-Z. et al. (1998). Climate change and forest fire potential in Russian and Canadian boreal forests. *Climate Change*. 38(1): 1–13.

Tanase, M., Kennedy, R., and Aponte, C. (2015). Fire severity estimation from space: A comparison of active and passive sensors and their synergy for different forest types. *International Journal of Wildland Fire*, 24(8): 1062–1075.

Trigo, R M., Pereira, J M C., Pereira, M G., Mota, B., Calado, T J., Dacamara, C C. and Santo, F E. (2006). Atmospheric conditions associated with the exceptional fire season of 2003 in Portugal. *International Journal of Climatology*, 26(13): 1741–1757.

Van der Werf, G., Randerson, J., Giglio, L., Collatz, G., Kasibhatla, P., and Arellano, A. (2006). Interannual variability of global biomass burning emissions from 1997 to 2004. *Atmospheric Chemistry and Physics Discussions*, 6(2): 3175–3226.

Vasilakos, C., Kalabokidis, K., Hatzopoulos, J. and Matsinos, I. (2009). Identifying wildland fire ignition factors through sensitivity analysis of a neural network. *Natural Hazards*, 50(1): 125–143.

Vhengani, L., Frost, P., Lai, C., Booi, N., van den Dool, R. and Raath, W. (2015). Multitemporal burnt area mapping using Landsat 8: Merging multiple burnt area indices to highlight burnt areas. In: *2015 IEEE International Geoscience and Remote Sensing Symposium* (IGARSS), pp. 4153–4156.

Westerling, A. L., Hidalgo, H. G., Cayan, D. R., and Swetnam, T. W. (2006). Warming and earlier spring increase western US forest wildfire activity. *Science*, 313(5789): 940–943.

Wisner, B., Gaillard, J. C., and Kelman, I. (2012). *The Routledge Handbook of Hazards and Disaster Risk Reduction*. London: Routledge, pp. 1–858.

Wooster, M. J., Roberts, G., Perry, G. L. W., and Kaufman, Y. J. (2005). Retrieval of biomass combustion rates and totals from fire radiative power observations: FRP derivation and calibration relationships between biomass consumption and fire radiative energy release. *Journal of Geophysical Research-Atmospheres*, D24311, 110(D24).

World Meteorological Organization. (2016). Annual Bulletin on the Climate in WMO Region VI - Europe and Middle East 2015, Geneva, Switzerland: WMO.

Yebra, M., Dennison, P E., Chuvieco, E., Riano, D., Zylstra, P., Hunt, E R., Danson, F M., Qi, Y., and Jurdao, S. (2013). A global review of remote sensing of live fuel moisture content for fire danger assessment: Moving towards operational products. *Remote Sensing of Environment*, 136(1): 455–468.

Section IV

Remote Sensing of Floods

15 Satellite Remote Sensing of Floods for Disaster Response Assistance

Guy J.-P. Schumann

CONTENTS

15.1 INTRODUCTION

Over the past decade, floods around the world have set new records and caused unprecedented damage[*] in many countries, including China, India, Malawi, England, Thailand, the Philippines, and the United States. These events cover spatial scales well beyond what we observed in the past, and they frequently surpass traditional regional flood management and disaster response capabilities (Schumann, 2016).

Mostly from a scientific standpoint, for about 40 years, with a proliferation over the last two decades, remote sensing data, primarily in the form of satellite imagery and altimetry, have been used

---

[*] Munich RE: https://www.munichre.com/site/corporate/get/documents_E-111857635/mr/assetpool.shared/Documents/5_ Touch/_NatCatService/Significant-Natural-Catastrophes/2015/1980_2015_Ueberschwemmungen_eco_e.pdf

to study floods, floodplain inundation, and river hydrodynamics. Instruments onboard spaceborne platforms and algorithms for turning image data into useful geospatial information about floods are numerous. Instruments that record flood events may operate in the visible to infrared and microwave ranges of the electromagnetic spectrum.

Owing to the limitations posed by adverse weather conditions during flood events, radar (microwave range) sensors are invaluable for monitoring floods; however, images of flooding in the visible and infrared ranges are very widely used, given the long history of satellite missions operating in that spectral range, and, since these images are more apt to the human eye, retrieving useful information from these types of images is often more straightforward.

During recent years, scientific contributions in the field of remote sensing of floods have increased considerably, and science has presented innovative research and methods for retrieving information content from multiscale coverages of disastrous flood events all over the world. Progress has been transformative, and the information obtained from remote sensing of floods is becoming mature enough not only to be integrated with computer simulations of flooding to allow better prediction but also to assist flood response agencies in their operations.

Furthermore, this advancement has led to a number of recent and upcoming satellite missions that are already transforming current procedures and operations in flood modeling and monitoring as well as our understanding of river and floodplain hydrodynamics globally. Global initiatives that utilize remote-sensing data to strengthen support in managing and responding to flood disasters (e.g., the International Charter and the Dartmouth Flood Observatory [DFO]), primarily in the developing nations, are becoming established and recognized by many nations that are in need of assistance because traditional ground-based monitoring systems are sparse and declining.

In the past decade, many articles in the scientific literature and numerous book chapters have reviewed remote sensing for flood management, focusing on research and applications, including data processing and integration, with the help of computer simulations and model predictions of floods (Carbonneau and Piégay, 2012; Schumann, 2015; Klemas, 2015; and Rahman and Di, 2016). This rapidly growing attention is reflecting the growing value that remote sensing can offer in flood monitoring and management.

The challenge now lies in ensuring sustainable and interoperable use and optimized distribution of remote-sensing products and services for science, as well as in ensuring operational assistance. In the context of the ever-increasing trend of remote-sensing applications in flood science and flood risk management, this chapter will review and critically examine the current status of remote sensing of floods for disaster response assistance.

15.2 MAPPING AND MONITORING FLOODS WITH SATELLITES

Deriving information about the area or extent of permanent water bodies, flood inundation area, and shoreline extent from remote sensing is generally much simpler than deriving information of other variables in hydrology, such as soil moisture and discharge. Information about the surface area of permanent water bodies and associated change can be used in a variety of applications, ranging from simple mapping and monitoring of water bodies to more complex water quality assessments of lakes and reservoirs. Data on inundation area and extent are commonly used to assess the magnitude and extent of a flood, with the aim to support relief services and to calibrate and validate flood inundation models.

Clearly, the mapping of permanent water bodies can be done with most satellite imaging platforms (Figure 15.1) at almost any time, obtaining the area and extent of a flood; however, it is still rather opportunistic, and certain conditions on both the Earth's surface and in the atmosphere during an event (such as emergent flooded vegetation and persistent cloud cover) may restrict suitable data-acquisition technology only to a few remote-sensing instruments, such as synthetic aperture radar (SAR). However, generally speaking, the area and extent of surface water may be measured with a variety of visible band sensors (e.g., Landsat, Moderate Resolution Imaging Spectroradiometer [MODIS], and Sentinel-2)

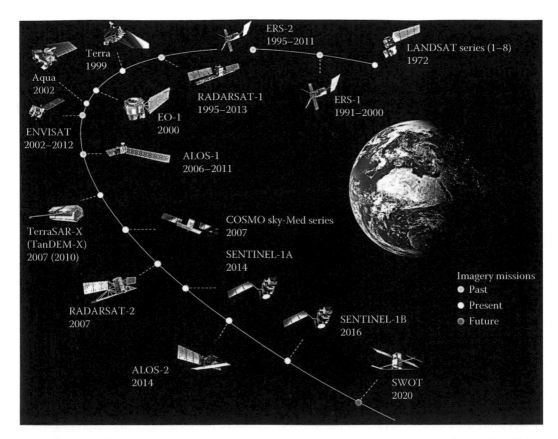

FIGURE 15.1 Past, current, and future satellite missions that carry instruments for imaging floods. Note that only missions that are commonly used to capture floods are shown. (After Schumann, G. J.-P. and Domeneghetti, A., *Hydrol. Process.*, 30, 2891–2896, 2016.)

with different repeat frequencies (Marcus and Fonstad, 2008) and by SAR (e.g., COSMO-SkyMed, TerraSAR-X, Radarsat-2, and Sentinel-1A and -1B), with varying degrees of success (Schumann and Moller, 2015) and very limited operational applications (Schumann et al. 2016), with the only notable exception being the near real-time (NRT) MODIS flood mapping effort by the National Aeronautics and Space Administration (NASA) and the DFO (http://floodobservatory.colorado.edu).

Given the strong inverse relationship between spatial resolution and revisit time for satellites, monitoring floods from space in NRT or operationally is currently only possible through either low-resolution imagery or satellite constellations. For instance, revisit times for imagery of ~100 m or lower spatial resolution are in the order of just up to a few days and, as with most publicly funded satellite missions, images can be obtained usually at no cost and with a latency of only several hours. Hence, this type of spaceborne data can be used for monitoring major floods on medium to large rivers. Schumann et al. (2012) noted that for basin areas down to around 10,000 km², flood waves usually take several days to transit through the catchment river network, and thus, there is a reasonable chance of floodplain inundation coinciding with a satellite overpass. In smaller basins with shorter flood wave travel times, the probability of imaging a flood decreases proportionately and acquisitions become increasingly opportunistic, such that even wide swath systems could not be relied on for operational monitoring. For finer-resolution systems, the same issue occurs, but here, revisit times can be up to 35 days, and so, one can only be guaranteed to capture flooding imagery in the very largest river basins such as the Amazon, which have a mono-modal annual flood pulse that lasts for several months. Therefore, in the majority of river basins, the chances of imaging a flood with a high-resolution system becomes vanishingly small, except for missions in constellation or when activating the International Charter (https://www.disasterscharter.org).

Problems with optical imagery include cloud cover (and fire smoke), restriction to daylight operations, and low spatial resolution for sensors with high temporal coverage intervals (e.g., MODIS); however, such sensors are applied to global monitoring of flood inundation area with success, as demonstrated by the DFO, using MODIS images as much as twice daily. This database currently represents the only global observed record of flood events. In addition, at the global scale, multisatellite historic observations of open water surfaces over a large spectrum (Prigent et al. 2012), ranging from visible (advanced very-high-resolution radiometer [AVHRR]) to microwave wavelengths, both passive (special sensor microwave/imager [SSMI]) and active (ERS scatterometer), have been used to map wetland dynamics over time, albeit at spatial resolution too coarse to be of significance to local decision making, floodplain management, or disaster assistance, unless downscaled to the appropriate spatial resolution.

The SAR is usually preferred for flood mapping, given its relatively high (1–3 m is now possible) spatial resolution (compared with most civil-sector satellite missions operating in the optical range) and its near-all-weather as well as day and night operating capabilities. However, there are a number of challenges related to SAR image geometry and processing that make its use in civil applications still somewhat less widespread than optical sensors, despite attractive advantages, particularly during flood events.

The magnitude of the deteriorating effects in a SAR (flood) image is a function of wavelength, incidence angle, and polarization. Incidence angle refers to the angular deviation of the incident signal from nadir, while polarization describes the direction at which materials reflect signals and SAR sensor receive these signals (Ulaby et al. 1982). Both these properties impact the ability to discriminate features or conditions of the Earth's surface.

Many image-processing algorithms exist to map flooding on a SAR image (see, e.g., Matgen et al. [2011] for a concise review on the different methods), but difficulties in interpreting a SAR image may arise from a variety of sources: complex signal backscatter (e.g., diffuse and volume backscatter), inadequate wavelength and/or polarizations, remaining geometric image distortions, and multiplicative noise. During flooding, wind roughening of the water surface and protruding vegetation can complicate the imaging process. Moreover, in built environments, the structure of rectangular surfaces, for example, buildings, is such that the wave is returned to the SAR antenna and thus may cause complete sensor saturation, resulting in white image pixels (corner reflectors), or dihedral corner reflectors (i.e., a corner reflector of two sides, creating signal bounce) in conjunction with often-inadequate spatial resolution make it very challenging to extract flooding from urban areas. Of course, for obvious reasons, this would be desirable when using remote sensing for flood management, and despite these considerable challenges, some studies have used dihedral corner reflection to assist flood detection in urban areas (Mason et al. 2014).

15.3 SELECTED APPLIED RESEARCH EXAMPLES

Satellites series and sensors such as MODIS, Landsat, EO-1, and Sentinel-1 and -2, can document regional floodplain inundation, and high-resolution satellites, such as TerraSAR-X, Radarsat-2, and COSMO-SkyMed, or commercially operated very-high-resolution sensors from the air and space can provide city block-level data, including damage assessment capability; note that the latter has also been demonstrated with high-resolution satellite SAR sensors[*]

The following sections will illustrate a state-of-the-art application example for optical as well as SAR imagery and radiometry data. Finally, integration with computer models of flood inundation is also outlined. These sections focus on applied research examples, whereas Section 15.4 describes the existing tools for operational use and flood disaster response assistance.

---

[*] http://www.jpl.nasa.gov/spaceimages/details.php?id5PIA17687

15.3.1 Optical Imagery

Probably, the most notable applications of optical imagery for flood detection on a daily or bi-daily basis are the NASA-funded NRT MODIS flood mapping effort and, similarly to this, the Dartmouth Flood Observatory (DFO), which will be described in detail in Section 15.4.

In terms of applied research and big data analytics, a state-of-the-art example is the Water Observations from Space (WOfS, Mueller et al. 2016), a web mapping service (WMS) displaying historical surface water observations derived from satellite imagery for all of Australia for the period 1987 up to the present. The WOfS displays the detected surface water from the Australia-wide Landsat 5 and Landsat 7 satellite imagery archives. The aim of the WOfS is to better understand where water is usually present, where it is rarely observed, and where inundation of the surface has been occasionally observed by the satellites (Figure 15.2).

Surface water is detected using a water detection algorithm based on a decision tree classifier and a comparison methodology using a logistic regression, which provided an understanding of the confidence in the classification (Mueller et al. 2016). The water detected for each location is summed through time and then compared with the number of clear observations of that location (i.e., observations not affected by cloud, shadow, or other quality issues). The result is a percentage value of the number of times water was observed at a location.

As with all optical satellite sensors, common limitations or errors in classification mostly include cloud shadow, snow, and flat rooftops of large building being labeled as water and small water bodies remaining undetected. In addition, important to note is that this is a historic observation dataset that spans a certain time period, so differences in river and landscape geomorphology as well as in the built environment before 1987 in this case, and likely changes in those settings in the future, would, of course, alter the location of water.

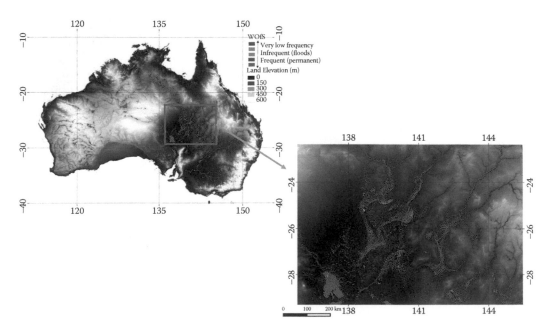

FIGURE 15.2　The WOfS open-source dataset showing the number of times water was detected between 1987 and 2014 by Landsat 5 and 7. Areas frequently observed as containing water (such as permanent lakes and reservoirs) are shown in purple and blue, down through green to areas not often observed to contain water (such as floods) in yellow, and finally to areas rarely observed as containing water in red. (Courtesy of Geoscience Australia, Canberra, Australia.)

Nevertheless, this dataset represents a unique example of big data analytics and data interoperability. More recently, a number of efforts using online processing platforms, such as Google's Earth Engine, have started processing large amounts of satellite imagery to derive meaningful geophysical information, such as global deforestation rates (Hansen et al. 2013) or a mask of global surface water (Pekel et al. 2014), which is very difficult to produce on regular desktop computers.

15.3.2 Synthetic Aperture Radar Imagery

Probably, the best-known SAR flood image research studies that propose an automated classification algorithm are based on the TerraSAR-X image of the 2007 summer flood in and around the town of Tewkesbury, England. The first study using this case study to illustrate a fully automated flood mapping procedure was performed by Martinis et al. (2009), followed by Mason et al. (2012), and then by Giustarini et al. (2013). Figure 15.3 illustrates the results of all three studies. The reason for choosing this particular site and event is that many auxiliary datasets for algorithm development and validation are available, such as a Light Detection and Ranging (LiDAR) digital elevation model (DEM), flooded aerial photography, and an accurate, high-resolution two-dimensional flood inundation model simulation of the event (Schumann et al. 2011).

In their application, Martinis et al. proposed an automatic NRT flood detection approach (resulting map shown in Figure 15.3a), which combines histogram thresholding and segmentation-based classification, specifically oriented to the analysis of single-polarized very-high-resolution SAR satellite images. In most cases, thresholding is applied to an image to obtain a binary classified image of *wet* and *dry* pixels, with respect to detecting the flooded areas. Since local gray-level changes may not be distinguished by global thresholding techniques in detailed, large satellite scenes, Martinis et al. integrated thresholding into a split-based approach, where the derived global

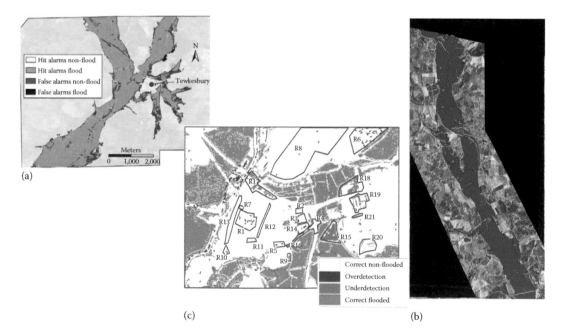

FIGURE 15.3 Flood maps produced by three different automated SAR flood mapping algorithms: (a) DLR's algorithm by Martinis et al. (2009); (b) Mason et al.'s (2012) algorithm; and (c) the algorithm by Matgen et al. (2011), as applied by Giustarini et al. (2013). Note that all three algorithms were applied to the same TerraSAR-X image of the Tewkesbury (England, the United Kingdom) summer floods of 2007.

threshold is integrated into a multiscale segmentation, thereby combining the advantages of small-, medium- and large-scale per-parcel segmentations (Martinis et al. 2009).

Classification of flooded area can be augmented and improved by using a high-resolution DEM, such as the one proposed by Mason et al. (2010, 2012) for the same study site and event, using a LiDAR DEM. Mason et al. (2010) suggested the use of a LiDAR digital surface model (DSM) within the urban area, including buildings, to account for areas of misclassification due to layover and shadow effects in the SAR scene. These areas were generated from the DSM before the flood mapping process by using the German Aerospace Center's (DLR's) in-house SAR simulator and were then ingested in the classification algorithm. Flooding was first delineated with a region-growing active contour model based on local metrics (also termed *snake* after Horritt [1999]). The regional image statistics from these areas were then applied locally in the urban area and fine-tuned by using only seeding regions in the low-lying areas of the urban floodplain, as derived from the DSM (Schumann and Moller, 2015). Region growing within the urban area was achieved by image dilation and labeling, followed by erosion. Mason et al. (2012) used the same image but suggested an object-oriented classification, such as the distance of trees and hedgerows from flooded areas that should be classified as flooded (based on compactness rule) but are not and the distance of shadow/layover areas from flooded pixels if these fall within a certain water surface height threshold determined from the digital terrain model (DTM) (resulting map shown in Figure 15.3b).

Giustarini et al. (2013) suggested a fully automated approach, adopted from Matgen et al. (2011) operational procedure (for details, see Section 15.4.3.1), based on comparing the actual flood image backscatter distribution function with a theoretical gamma distribution, thereby objectively selecting the threshold backscatter value that distinguishes flooded from non-flooded surfaces. This classifier has the advantage of working in an operational setting where no user intervention is desirable. Although this is not a requirement, classification can be improved by ingesting a non-flood image with the exact same orbit and sensor parameters as the flood image. Limited to the urban area only, their final flood map showed a classified correct score of 82%, but the main issue of shadow/layover in dense urban areas still remained (resulting map shown in Figure 15.3c).

In addition, for operational flood mapping from very-high-resolution COSMO-SkyMed (CSK) SAR images that observed the evolution of a flood event in Albania in January 2010, Pulvirenti et al. (2011), though using a non-urban test case, also used a DEM in addition to land-cover data. These datasets were used to determine degrees of membership to the flood class based on similar rules to Mason et al. (2012). The class membership rules were then used to guide a fuzzy logic classifier applied to the multitemporal COSMO-SkyMed flood images. A similar fuzzy logic classifier was also employed by Martinis et al. (2013) in a modified version of their previously developed split-based processor for flood mapping from high-resolution SAR imagery (Martinis et al. 2009; outlined earlier). In the operational procedure at DLR for Satellite-based Crisis Information (ZKI), this modified SAR classifier is automatically triggered by a systematic detection of potential flood events, using daily-acquired medium spatial resolution optical data from the MODIS satellite sensor. More details of this operational mapping tool are given in Section 15.4.3.2.

15.3.3 RADIOMETRY

To some extent, passive microwave radiometry can be used to map inundated area (Schroeder et al. 2015), but the low resolution (25 km), resulting from the large angular beam, limits the applicability of these sensors. De Groeve (2010) showed that passive microwave-based flood extent corresponds well with gauged flood hydrographs when river overtopping occurs; however, the signal-to-noise ratio is highly affected by variable local conditions, such as specific river bank geometry configuration that may prevent variations in width with rising water levels.

The experimental Global Flood Detection System (GFDS[*]) hosted by the Global Disaster Alert and Coordination System (GDACS) monitors floods worldwide by using NRT satellite data. Surface water extent is observed using passive microwave remote sensing (advanced microwave scanning radiometer-EOS AMSR-E) and tropical rainfall measuring mission (TRMM) sensors). When surface water increases significantly (anomalies with probability of <99.5%), the system flags it as a flood. Time series are calculated in more than 10,000 monitoring areas, along with small-scale flood maps and animations.

15.4 EXAMPLES OF OPERATIONAL SYSTEMS AND TOOLS

15.4.1 THE DARTMOUTH FLOOD OBSERVATORY AND NASA's NEAR-REAL-TIME FLOOD MAPPING

The DFO (http://floodobservatory.colorado.edu) conducts global remote-sensing-based flood mapping and measurements in NRT and archives this information. The primary satellite sensor for this is the MODIS instrument onboard NASA's Aqua and Terra satellites. The DFO is most known for its rapid flood mapping with MODIS (Brakenridge and Anderson, 2006), but during high-impact flood disasters, it also maps flooding from other satellites, such as EO-1, the Landsat series, and SAR satellite missions, and aggregates these maps to a number of map formats that assist flood response teams through situational awareness across large-scale coverages (Figure 15.4a). The observatory offers two map series accessible from the global index: *Current Flood Conditions*, providing daily, satellite-based updates of surface water extent, and the *Global Atlas of Floodplains*, a remote-sensing record of floods, from 1993 to 2015.

The DFO also performs global hydrological modeling, which it integrates with its global surface water mapping, primarily for calibrating its global *River Watch*[†] virtual stations (Brakenridge et al. 2012) that gauge the state of potential large river flooding daily based on changes in the brightness temperature of the passive microwave signal onboard the AMSR-E, TRMM, AMSR-2 and Global Precipitation Measurement (GPM) sensors (Figure 15.5). Collaborating and partnering with a number of humanitarian and flood disaster emergency management agencies, such as the United Nations World Food Program (UN WFP), ensures maximum utility of the information. These systems have been sustained by grants and contracts among others from NASA, the European Commission, the World Bank, and the Latin American Development Bank.

The NASA's NRT flood mapping (http://oas.gsfc.nasa.gov/floodmap) is similar to the DFO, and since 2012, it feeds its automated product to the DFO. The LANCE processing system at NASA Goddard provides such products typically within a few hours of satellite overpass. As with the DFO, open water is detected by using a ratio of MODIS bands in the visible and near-infrared range at 250 m spatial resolution. The impact of clouds is minimized by compositing images typically over 2 or more days. Flooding is classified as anomaly to a reference water layer denoting *normal* water extent.

15.4.2 THE GLOBAL FLOOD MONITORING SYSTEM

Real-time quasi-global hydrological calculations at 1/8th degree and 1 km resolution inundation simulations are performed with the University of Maryland's Global Flood Monitoring System (GFMS; Wu et al. [2012], http://flood.umd.edu), which is a NASA-funded experimental system that uses real-time TRMM Multisatellite Precipitation Analysis (TMPA) data and now the iMERG product from the Global Precipitation Measurement (GPM) mission (Figure 15.4b). This system also issues flood forecasts with 4- to 5-day lead time, based on numerical weather prediction (NWP)

---

[*] http://www.gdacs.org/flooddetection
[†] http://floodobservatory.colorado.edu/DischargeAccess.html

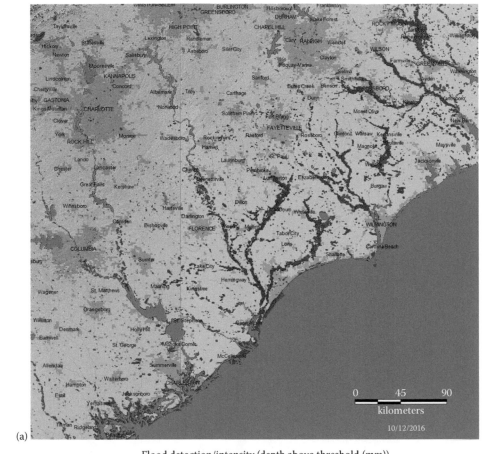

(a)

Flood detection/intensity (depth above threshold (mm))
12Z Oct 12, 2016

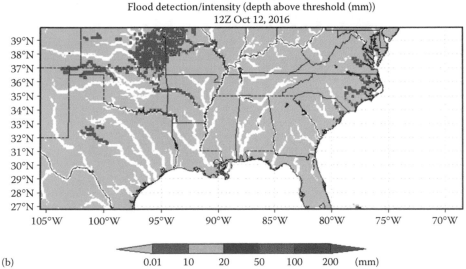

(b)   0.01   10   20   50   100   200   (mm)

FIGURE 15.4 (a) Flood map (Southeastern U.S. Coast and vicinity) displayed during the Hurricane Matthew flood disaster in October 2016. Red is flood water during past 14 days from MODIS 250 m data. Dark red is flooding on date shown from Landsat 8. Green is previous flooding, year 2000 to present. Dark blue is permanent surface water. (b) A snapshot of the NRT computation of the Global Flood Monitoring System (GFMS) showing inundation for the same event on October 12, 2016, at 1 km resolution.

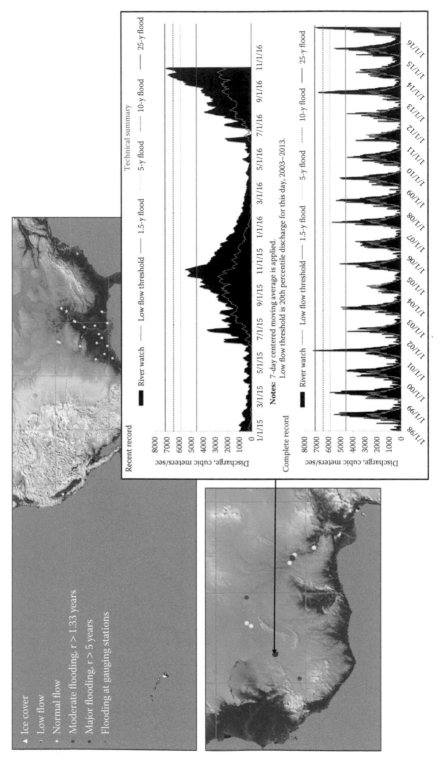

FIGURE 15.5 Extract from the DFO's global *River Watch* sites, showing an example of the radiometry-derived discharge time series and associated return periods for the Niger River in West Africa.

precipitation, and mapping of inundation at a 3-hour time step. Although at a relatively low resolution, the timeliness of the system and its forecast capability make it very attractive to flood relief services and flood disaster response organizations worldwide, such as the UN WFP.[*]

The extent of a flood, as estimated by the GFMS during large events, is regularly validated against satellite flood maps from the DFO. Accuracy is, of course, not great all the time and in all locations, but the system does typically very well in estimating the extent and identifying which rivers are affected. In general, over certain types of complex terrain, satellites can underestimate precipitation, and ground information for the model, including topography, is also limited in accuracy. In countries such as the United States and Europe, where geospatial data and services for event observation are abundant and other local datasets and monitoring stations provide high-accuracy information, the GFMS can add value by forecasting situational awareness over very large scales. However, for areas and countries without extensive weather and monitoring networks on the ground, the satellite view of floods, as delivered by the GFMS and the DFO, is often the only source of actionable information.

15.4.3 OTHER RAPID MAPPING SYSTEMS

15.4.3.1 Automated SAR Flood Mapping with ESA's G-POD

The European Space Agency (ESA) hosts a SAR-based mapping tool on its Grid Processing on Demand (G-POD) system (http://gpod.eo.esa.int), which is currently operated in testing mode but will soon be freely available to end users, who can query the ESA SAR database for a flood image and retrieve an automatically generated flood map (Figure 15.6).

The mapping algorithm calibrates a statistical distribution of *open water* backscatter values of SAR images of floods. Then, a radiometric thresholding provides the seed region for a

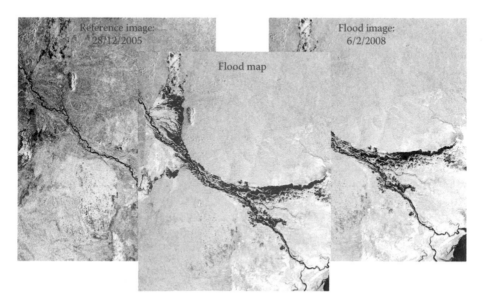

FIGURE 15.6 Flood map extracted from a satellite SAR image of the early February flood event on the Lower Zambezi River by using Matgen et al.'s (2011) algorithm, fully automated on ESA's G-POD service.

---

[*] https://pmm.nasa.gov/articles/improving-flood-predictions-gpm

region growing process. Change detection is included as an additional step, which minimizes overdetection of flooded area (Matgen et al. 2011). Using two case studies, evaluation showed the same performance as the optimized manual approaches. Their automated flood mapping algorithm works on different SAR image modes and resolutions.

15.4.3.2 DLR's ZKI

The DLR's ZKI monitors flood disasters by tasking its TerraSAR-X satellite, and, as part of the International Charter, it has access to other tasked satellites that may provide relevant crisis data. In addition, the data archive is searched for matching predisaster satellite scenes. According to the DLR[*], during the first 6 hours after the activation of ZKI, reference maps based on archive satellite data providing a first overview of the affected area can be made available to relief organizations. The newly acquired postdisaster satellite data is used to assess and monitor the ongoing crisis situation, that is, delineate the affected areas (see Section 15.3.2 and Figure 15.7a) and estimate the damages caused by the disaster.

15.4.3.3 NASA JPL's ARIA

The ARIA Center (https://aria.jpl.nasa.gov), a joint venture cosponsored by the California Institute of Technology (Caltech) and by NASA through the Jet Propulsion Laboratory (JPL), plans to provide the infrastructure to generate imaging products in NRT that can improve situational awareness for disaster response. The ARIA Center also plans to provide automated imaging and analysis capabilities necessary to keep up with the imminent increase in raw data from geodetic imaging missions planned for launch by NASA, as well as international space agencies. Analyses of these data sets are currently handcrafted following each event, such as shown in Figure 15.7b, and may not be generated rapidly enough for an operational response assistance during natural disasters.

15.4.3.4 Rapid Flood Mapping from NOAA's VIIRS Sensor

The Visible Infrared Imaging Radiometer Suite (VIIRS) instrument is one of the five major Earth observation (EO) instruments onboard the National Oceanic and Atmospheric Administration's (NOAA's) S-NPP and JPSS satellites, essentially a continuation of NOAA's AVHRR legacy sensors. With a very large swath width of 3060 km, it provides full daily coverage, both in the day and night sides of the Earth. The VIIRS has 22 spectral bands, including 16 moderate spatial resolution bands at 750 m pixel spacing at nadir, 5 imaging resolution bands at 375 m at nadir, and 1 panchromatic band with 750 m spatial resolution.

Using the VIIRS and the coastal flooding caused by Hurricane Sandy as a test case, Sun et al. (2016) present an approach to estimate the extent of large-scale floods in an operational context (Figure 15.7c). The approach estimates the water fraction from VIIRS 375-m imager data through mixed-pixel linear decomposition and a dynamic nearest-neighbor search method. By using the reflectance characteristics of the VIIRS visible channel, near-infrared channel, and shortwave infrared channel, the method dynamically searches the nearby land and water end members.

As an optional postprocessing step, based on simple physical characterization of water spreading, the low-resolution flood map from the VIIRS can be extrapolated to a much higher spatial resolution by using topographic information from a digital elevation model, in that case, for instance, to 30 m pixel spacing.

---

[*] https://www.zki.dlr.de/mission

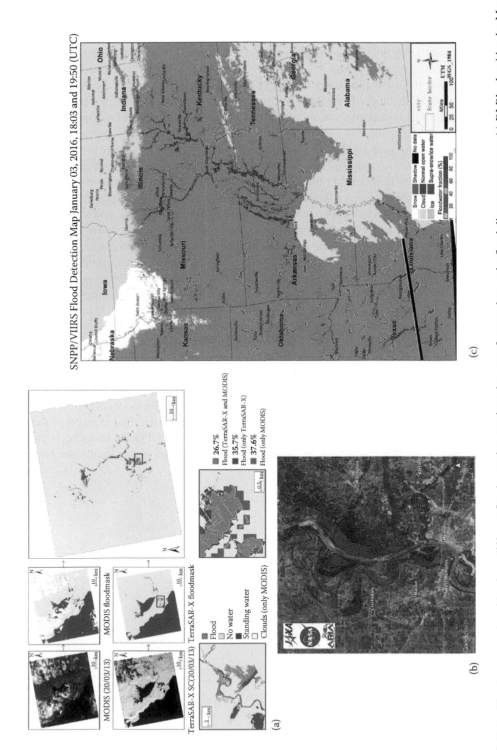

FIGURE 15.7 Flood maps produced by three different flood mapping services that are often used to assist flood disaster response: (a) DLR's algorithm by Martinis et al. (2013), as implemented at ZKI. (b) In response to the U.S. Midwest floods along the Mississippi River in December 2015/January 2016, the ARIA project produced this flood extent map from an ALOS-2 SAR image. (c) The same U.S. Midwest flood event was mapped with the VIIRS sensor on January 3, 2016, using the algorithm by Sun et al. (2016).

15.5 IMPROVING DISASTER RESPONSE ASSISTANCE

15.5.1 REQUIREMENTS

Despite the notable applied research efforts reported in the previous section, it is apparent in many case study reports and during flood disasters that important needs, from the decision-making stand-point, are not (yet) directly met by existing satellite technology or, at least, not by any one single data product. This was the main conclusion of a paper published on this topic by Schumann et al. (2016), including several coauthors from flood response organizations and satellite operators and space agency managers. They reported outcomes of the Texas flood disaster of May–June 2015 and as an immediate action suggested a scoping workshop to be held that attempts to lay out a plan for better coordinating the U.S.-wide and global flood response assistance with EO data and products.

A first such NASA-funded flood response workshop was held in June 2016, with the objective to enable a unique dialogue between EO mission technology and science, capacity-building, and the flood response community, in order to foster better coordination in flood response worldwide. Many organizations already have ongoing missions, programs, initiatives, and research funding to provide flood-monitoring and response services (e.g., the DFO), as well as image and computer simulation products (e.g., the GFMS), during an event. However, it is obvious that with the prolifera-tion of information (see, for instance, the commentary by Schumann and Domeneghetti [2016] for a discussion on this topic), it is difficult to coordinate systems, organizations, and people during a single event, let alone multiple simultaneous events or successive large events (Schumann, 2016). In addition, each of those systems is oftentimes developed with a specific purpose, at least at the out-set of its funding and thus may provide a unique capability, which requires effective coordination. Therefore, a new *community of practice* (Figure 15.8) is needed that will set out to improve EO data and products to better assist flood disaster response (Schumann, 2016).

Having said that, the situation in assisting flood response with EO data, in the United States and elsewhere in the world, has recently taken a more positive turn, where, for instance, NASA's Applied Sciences Disasters program at headquarter level is trying to aim for better coordinating its many different image releases and product deliveries during disasters (for detailed example of these products and services, see previous section). In this context, not only the DFO, for instance, is post-ing the MODIS flood maps that it produces, but also its main website page (http://floodobservatory.

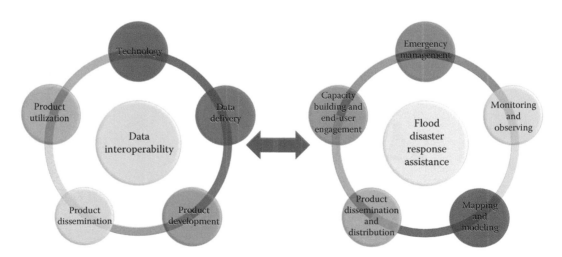

FIGURE 15.8 Graphical illustration of the importance of data and product interoperability, showing the dif-ferent components that are typically needed to get seamlessly from a data acquisition technology to product utilization. Such an interoperability mechanism needs to be adopted by the many actors involved in flood disaster response, in particular when using EO-derived information to assist.

colorado.edu) now displays other related services and products, such as the model outputs from the NASA/University of Maryland's GFMS (see previous section) and NRT satellite precipitation maps, on an event-by-event basis.

In addition, Europe, at the European Commission level, has developed a satellite- and model forecast-aided disaster response capability. The European Commission's (EC's) Copernicus program, previously known as the Global Monitoring for Environment and Security (GMES) service, is establishing a European capacity for EO. Its Emergency Management Service (EMS, http://emergency.copernicus.eu) is providing a number of operational services during disasters, including flood mapping from satellite images, in particular SAR, acquired, for instance, during activation of the International Charter (https://www.disasterscharter.org). Most recently, Copernicus EMS was activated during the devastating Louisiana summer 2016 floods* and for the Hurricane Matthew flood disaster in October 2016,† for flood mapping from SAR images in both instances.

In fact, both the above-mentioned flood disasters (Louisiana and Hurricane Matthew in 2016) are prime examples of the notable progress in a now-much-improved flood disaster response coordination that in these two cases included not only flood response teams and traditional products and services in the United States, such as river streamflow forecasts, aerial photography, and satellite images from the International Charter, but also images and product services from all major space agencies around the world as well as the EC Copernicus EMS. Not only this coordinated effort helped in getting more useful products in the right hands, but it also considerably shortened the latency between data acquisition and product delivery. This, in turn, ensured more efficient and more effective field operations and a higher degree of interoperability (Figure 15.4), which are ever more important in an era of geospatial data proliferation (Schumann and Domeneghetti, 2016).

15.5.2 Challenges

The two events described previously could serve as benchmark cases on how this *new community of practice* could and probably should operate in the future. Of course, this requires resources, both in time and money, and can be extremely challenging, particularly in the event of simultaneous disasters or large successive events that have manifested themselves several times in the last decade, particularly when looking at floods.

For instance, it is very common that decision makers would desire at least daily status updates on the affected regions and at high-enough spatial resolution that can resolve the necessary scale to assess local infrastructure assets at risk. In addition, in terms of EO images, clouds, fog, rain, and vegetation are frequently posing problems for accurate flood mapping at optical wavelengths and also to some extent at radar wavelengths. Furthermore, for flood relief operations and many other applications, river reaches tend to be monitored and studied at much smaller scale than that typically acquired with wide-swath EO imagery, and therefore, monitoring those reaches for situational awareness does actually require much finer spatial resolutions. Very-fine-resolution (<5 m) images also become a prerequisite when monitoring and modeling urban areas, where most assets at the risk of flooding are located and where city-block scale often dictates the ability to model or monitor flood inundation patterns accurately (Schumann et al. 2012). Here, airborne repeat overpasses are commonly the preferred choice, but from space, constellations of multiple fine-resolution SAR systems may present a possible solution or even an alternative. For instance, COSMO-SkyMed can get a 3 m image sequence with a time from request to acquisition of the first image of 26–50 hours and then subsequent images at 12-hour intervals. This capability is now further enhanced with the recent constellation of Sentinel-1A and -1B and coupled with a program such as Copernicus or an

---

* http://emergency.copernicus.eu/mapping/list-of-components/EMSR176
† http://emergency.copernicus.eu/mapping/list-of-components/EMSR185

operational system such as ESA's G-POD Fast Access to Imagery for Rapid mapping Exploitation (FAIRE) (http://gpod.eo.esa.int), SAR-derived flood maps can be made available to the users several hours after acquisition (see Section 15.4.3.1). This form of rapid delivery of fine resolution, actionable information is now technically feasible and might be a common form of dissemination in the near future.

As argued by Schumann et al. (2016), although systematic, high-resolution, and lower-resolution but wide-area observations are now possible during flood disaster, the *full potential* of EO data can only be unlocked when combining all these data in an intelligent way. This major challenges are still clearly posed during every large flood disaster and, as further argued by Schumann et al. (2016) in their account of the Texas 2015 flood disaster, are, in part, to assemble data from a variety of sources and to embed any local higher-resolution flood images within the larger and time-extended regional context of an expanding large-scale flood event.

With all this wealth of EO products and services produced and disseminated, there is still a noticeable underutilization of this information. Most of the reasons for this situation relate to the relative novelty of these types of data. Generally, very limited time and personnel are available during an emergency situation to understand, process, and handle new types of geospatial datasets. In addition, limited NRT data accessibility, bandwidth, and sharing capacity make product and data distribution cumbersome. Further, oftentimes, incompatibility between user platforms and geospatial data formats hinders more widespread use of new EO products and services, or data availability may be simply unknown and data latency may be inadequate. However, the largest gap that needs to be bridged is the limited understanding by scientists and engineers about the end user's product and timing needs, which are also discussed by Hossain et al. (2016) as a widespread concern in the applied Earth sciences arena.

Although all these challenges represent a clear limitation in fulfilling the needs of the decision maker, they create an opportunity for innovation to develop products that deliver better actionable information.

15.5.3 PERSPECTIVES

In order to address the many challenges that exist in making EO data more readily useable and actionable for assisting flood disaster response between data availability, the scientific community should seek closer collaboration with end users. This is also strongly advocated by Hossain et al. (2016). According to Schumann and Domeneghetti (2016), a step in the right direction would be to build a *one-stop shop* (i.e., data portal) dedicated to the remote sensing of floods. The idea of a data clearinghouse or one-stop shop is not new (see, e.g., the USGS Hazards Data Distribution System [HDDS] and related services, all under the U.S. Geological Survey [USGS] Emergency Operations portal, https://hdds.usgs.gov) and is also a top priority action item that came out of the NASA Flood Response Workshop mentioned earlier (Schumann, 2016; Table 15.1).

This platform could collect EO imagery and products and synthesize knowledge as well as data from past events and experiences. Decision makers need to be able to pull data and products from this portal at low bandwidth and latency and request tailored information layers, as needed for their operations. Other relevant information could be made available alongside remote-sensing data of floods, such as output layers from models such as those produced by flood forecast or NRT event models. Those model computations can then be complemented and verified by social media streams and help pinpoint target regions for satellite image acquisition and delivery of flood products. Of course, this functionality requires the highest level of interoperability (Figure 15.8), since it needs to integrate seamlessly with end-user operation systems and platforms and ideally also be accessible on any device. In this context, strengthening public–private partnerships would allow access to high-end capabilities and leveraging advanced data interoperability standards and service protocols developed by the geospatial industry sector.

TABLE 15.1
Top Priorities for the Flood Response Community

Emergency Management	Monitoring and Observing	Mapping and Modeling	Product Dissemination and Distribution	Capacity Building and End-user Engagement
Push data and products out in 12- to 24-hour intervals, within capabilities (ask for assistance with resources, as needed, and manage expectations)	One-stop shop (should also include future acquisitions), where communities can pull rather than push data and products	Automated polygon generation of flood disaster location to target EO data and products, especially at the international level	Single access point (one-stop shop) that allows automated product delivery system	Build trust in the products and report value to community: more than one-stop shop need to have products that are tailored to user needs and allowed to report feedback

Source: Schumann, G.J., *EOS*, 97, 2016.

Regardless of where this one-stop shop resides and who manages it, it would not only better organize and structure data availability, thereby clarifying existing confusion over data and products but also help meet expectations and add further value to the various EO products and services. At the same time, the end-user community should have the opportunity to provide feedback on data and products, which, in turn, should be used to improve the different types of information disseminated.

15.6 CONCLUSIONS

In recent years, there has been a significant increase in the number of satellite instruments that can be used to map floods and produce actionable information. The ability to monitor floods with sensors onboard satellites has been known for quite some time now (at least since the early 1970s), and over those years, there has been much stimulating research in this area, and significant progress has also been achieved in fostering our understanding of the ways in which remote sensing can support and advance flood modeling, even flood forecasting (Revilla-Romero et al. 2015), and assist in flood disaster response operations.

This article reviewed the utility of remote sensing from satellites to map and monitor floods, with the aim to assist disaster response activities. Examples of applications in different landscape settings and at various spatial and temporal scales have been illustrated and critically discussed. Many satellite missions are collecting data that can inform directly or indirectly about water bodies and flood inundation processes. This data proliferation has shifted the research and application fields in the area of remote sensing of floods from a data-poor (prior 2000) to a data-rich (post 2000) environment (Bates, 2012). Consequently, innovative methods and products from these data have been developed, which led not only to better understanding of flood processes at various spatial and temporal scales but also to global initiatives and applications that utilize and promote remote sensing for improved decision-making activities, particularly in developing nations and during emergencies.

Global-scale initiatives and end-user oriented applications are now becoming established and are also recognized by many nations that are in need of assistance because traditional ground-based monitoring systems are sparse and in decline. The value that remote sensing can offer is growing rapidly, and many challenges lie ahead. New sensor technologies, for instance, light-weight small satellites and drones, will soon add many terabytes of new data every day, and as a result, innovative and powerful online data analytics platforms are being offered to retrieve actionable information from these data.

It is clear that we have entered an era of big data and the Internet of Things, in which everyone and everything is connected across networks, transmitting and receiving an overload of information. For remote sensing of floods, the grand challenge now lies in ensuring sustainable and interoperable use as well as optimized distribution of remote-sensing products and services for science and end-user applications and for operational flood disaster assistance (Schumann, 2017). In addition, a top priority is the need for end-user-driven validation cases to make satellite-based products and services more credible to the decision maker. At the same time, it is paramount to manage expectations, and if satellite-based applications are to achieve the required readiness level for decision making, scientists and engineers need to be clear about what exactly science and technology can offer and what the capabilities of the many products and services being offered mean to the end users.

REFERENCES

Bates, P. D. (2012). Integrating remote sensing data with flood inundation models: How far have we got? *Hydrological Processes*, 26: 2515–2521. doi:10.1002/hyp.9374.

Brakenridge, G. R., S. Cohen, A. J. Kettner, T. De Groeve, S. V. Nghiem, J. P. M. Syvitski, and B. M. Fekete (2012). Calibration of satellite measurements of river discharge using a global hydrology model. *Journal of hydrology*, 475: 123–136, doi:10.1016/j.jhydrol.2012.09.035.

Brakenridge, R. and E. Anderson (2006). MODIS-based flood detection, mapping and measurement: The potential for operational hydrological applications. In J. Marsalek, G. Stancalie, and G. Balint (Eds.) *Transboundary Floods: Reducing Risks through Flood Management, NATO Science Series IV Earth and Environmental Sciences, NATO Advanced Research Workshop on Transboundary Floods— Reducing Risks through Flood Management*, vol. 72, Baile Felix, Romania: NATO, pp. 1–12.

Carbonneau, P. E. and Piégay, H. (2012). *Fluvial Remote Sensing for Science and Management*. Chichester, UK: Wiley-Blackwell, p. 458.

De Groeve, T. (2010). Flood monitoring and mapping using passive microwave remote sensing in Namibia. *Geomatics, Natural Hazards and Risk*, 1(1): 19–35.

Giustarini, L., R. Hostache, P. Matgen, G. J. P. Schumann, P. D. Bates and D. C. Mason (2013) A change detection approach to flood mapping in urban areas using TerraSAR-X. In *IEEE Transactions on Geoscience and Remote Sensing*, 51(4): 2417–2430. doi:10.1109/TGRS.2012.2210901.

Hansen, M. C., P. V. Potapov, R. Moore, M. Hancher, S. A. Turubanova, A. Tyukavina, D. Thau et al. (2013). High-resolution global maps of 21st-century forest cover change. *Science*, 342: 850–853.

Horritt, M. S. (1999). A statistical active contour model for SAR image segmentation. *Image and Vision Computing*, 17 (3–4): 213–224.

Hossain, F., Serrat-Capdevila, A., Granger, S., Thomas, A., Saah, D., Ganz, D., Mugo, R. et al., (2016). A global capacity building vision for societal applications of earth observing systems and data: Key questions and recommendations. *Bulletin of the American Meteorological Society*, 97: 1295–1299. doi:10.1175/BAMS-D-15-00198.1.

Klemas, V. (2015). Remote sensing of floods and flood-prone areas: An overview. *Journal of Coastal Research*, 31(4): 1005–1013.

Marcus, W. A. and Fonstad, M. A. (2008). Optical remote mapping of rivers at sub-meter resolutions and watershed extents. *Earth Surface Processes and Landforms*, 33: 4–24.

Martinis, S., Twele, A., Strobl, C., Kersten, J., and Stein, E. (2013). A multi-scale flood monitoring system based on fully automatic MODIS and TerraSAR-X processing chains. *Remote Sensing*, 5: 5598–5619.

Martinis, S., Twele, A., and Voigt, S. (2009). Towards operational near real-time flood detection using a split-based automatic thresholding procedure on high resolution TerraSAR-X data. *Natural Hazards and Earth System Sciences*, 9: 303–314, doi:10.5194/nhess-9-303-2009.

Mason, D. C., Davenport, I. J., Neal, J. C., Schumann, G. J. P., and Bates P. D. (2012). Near real-time flood detection in urban and rural areas using high-resolution synthetic aperture radar images. *IEEE Transactions on Geoscience and Remote Sensing*, 50(8): 3041–3052. doi: 10.1109/TGRS.2011.2178030.

Mason, D. C., Giustarini, L., Garcia-Pintado, J., Cloke, H. L. (2014). Detection of flooded urban areas in high resolution Synthetic Aperture Radar images using double scattering. *International Journal of Applied Earth Observation and Geoinformation*, 28: 150–159. doi:10.1016/j.jag.2013.12.002.

Mason, D. C., Speck, R., Devereux, B., Schumann, G. J.-P., Neal, J. C., and Bates, P. D. (2010). Flood detection in urban areas using TerraSAR-X. *IEEE Transactions on Geoscience and Remote Sensing*, 48: 882–894.

Matgen, P., Hostache, R., Schumann, G., Pfister, L., Hoffmann, L., and Savenije, H. (2011). Towards an automated SAR-based flood monitoring system: Lessons learned from two case studies. *Physics and Chemistry of the Earth*, 36: 241–252.

Mueller, N., Lewis, A., Roberts, D., Ring, S., Melrose, R., Sixsmith, J., Lym- Burner, L., McIntyre, A., Tan, P., Curnow, S., and Ip, A. (2016). Water observations from space: Mapping surface water from 25 years of Landsat imagery across Australia. *Remote Sensing of Environment*, 174: 341–352.

Pekel, J. F., Cottam, A., Clerici, M., Belward, A., Dubois, G., Bartholome, E. and Gorelick, N. (2014). A Global Scale 30m Water Surface Detection Optimized and Validated for Landsat 8. American Geophysical Union, Fall Meeting 2014, abstract #H33P-01.

Prigent, C., Papa, F., Aires, F., Jiménez, C., Rossow, W. B., and Matthews E. (2012). Changes in land surface water dynamics since the 1990s and relation to population pressure. *Geophysical Research Letters*, 39: L08403, doi:10.1029/2012GL051276.

Pulvirenti, L., Pierdicca, N., Chini, M., and Guerriero, L. (2011). An algorithm for operational flood mapping from Synthetic Aperture Radar (SAR) data using fuzzy logic. *Natural Hazards and Earth System Sciences*, 11: 529–540, doi:10.5194/nhess-11-529-2011.

Rahman, S. and Di, L. (2016). The state of the art of spaceborne remote sensing in flood Management. *Natural Hazards*, 85: 1223–1248. doi 10.1007/s11069-016-2601-9.

Revilla-Romero, B., Hirpa, F. A., Thielen-del Pozo, J., Salamon, P., Brakenridge, R., Pappenberger, F., De Groeve, T. (2015). On the use of global flood forecasts and satellite-derived inundation maps for flood monitoring in data-sparse regions. *Remote Sensing*, 7(11): 15702–15728.

Schroeder, R., McDonald, K. C., Chapman, B. D., Jensen, K., Podest, E., Tessler, Z. D., Bohn, T. J., Zimmermann, R. (2015). Development and evaluation of a multi-year fractional surface water data set derived from active/passive microwave remote sensing data. *Remote Sensing*, 7(12): 16688–16732.

Schumann, G. J.-P. (2015). Preface: Remote sensing in flood monitoring and management. *Remote Sensing*, 7(12): 17013–17015.

Schumann, G. J-P. (2016). Flood response using Earth observation data and products. *Eos*, 97, doi:10.1029/2016EO060741.

Schumann, G. J-P. (2017). Remote sensing of floods. *Natural Hazard Science Oxford Research Encyclopedias*. Oxford University Press. doi:10.1093/acrefore/9780199389407.013.265. In review.

Schumann, G. J.-P., Bates, P.D., Di Baldassarre, G. and Mason, D.C. (2012), The use of radar imagery in riverine flood inundation studies, pp. 115–140. In: P. E. Carbonneau and H. Piégay (Eds.) *Fluvial Remote Sensing for Science and Management* ('*Advancing River Restoration and Management*' Series), Chichester, UK: Wiley-Blackwell, p. 458.

Schumann, G. J.-P. and Domeneghetti, A. (2016). Exploiting the proliferation of current and future satellite observations of rivers. *Hydrological Processes*, 30: 2891–2896. doi:10.1002/hyp.10825.

Schumann, G. J.-P., Frye, S., Wells, G., Adler, R., Brakenridge, R., Bolten, J., Murray, J. et al. (2016). Unlocking the full potential of Earth Observation during the 2015 Texas flood disaster. *Water Resources Research*, 52(5): 3288–3293.

Schumann, G. J.-P. and Moller, D. K. (2015). Microwave remote sensing of flood inundation. *Physics and Chemistry of the Earth, Parts A/B/C*, 8384: 84–95.

Schumann, G. J.-P., Neal, J. C., Mason, D. C., and Bates, P. D. (2011). The accuracy of sequential aerial photography and SAR data for observing urban flood dynamics, a case study of the UK summer 2007 floods. *Remote Sensing of Environment*, 115: 2536–2546.

Sun, D., Li, S., Zheng, W., Croitoru, A., Stefanidis, A. and Goldberg, M. (2016). Mapping floods due to Hurricane Sandy using NPP VIIRS and ATMS data and geotagged Flickr imagery. *International Journal of Digital Earth*, 9(5): 427–441. doi: 10.1080/17538947.2015.1040474.

Ulaby, F. T., Moore, R.K. and Fung, A.K. (1982). *Microwave Remote Sensing: Active and Passive, Vol. II— Radar Remote Sensing and Surface Scattering and Emission Theory*, Reading, MA: Addison-Wesley, Advanced Book Program, 1982, 609 pages.

Wu, H., Adler, R.F., Hong, Y., Tian, Y., and Policelli F. (2012). Evaluation of global flood detection using satellite-based rainfall and a hydrologic model. *Journal of Hydrometeorology*, 13, 1268–1284.

16 Usefulness of Remotely Sensed Data for Extreme Flood Event Modeling
A Study Case from an Amazonian Floodplain

Sebastien Pinel, Joecila Santos Da Silva, C. R. Fragoso Jr.,
J. Rafael Cavalcanti, Jeremie Garnier, Frederique Seyler,
Stephane Calmant, David Motta Marques,
and Marie-Paule Bonnet

CONTENTS

16.1 INTRODUCTION

The Amazon basin, with an area of 6.2 million km², is the largest watershed in the world (Figure 16.1a). The Amazon River has a mean discharge of 170,000 m³s⁻¹, with a minimum and maximum of 60,000 m³s⁻¹ and 270,000 m³s⁻¹, respectively (Gallo and Vinzon, 2005). The Amazon River runs through more than 6,700 km. It contributes 17% of the freshwater inputs to the global ocean (Callede et al. 2010; Richey et al. 1986). Located between latitudes 5°N and 20°S and between longitudes 50°W and 80°W, the Amazon basin covers 4/10 of South America and 5% of the worldwide continents. It is spread over several countries: 63% of the total area is in Brazil, 16% in Peru, 12% in Bolivia, 5.6% in Colombia, 2.3% in Equateur, 0.6% in Venezuela, and 0.2% in French Guyana (Goulding et al. 2003).

Lowland basin flatness, with slopes ranging from 3.7 cm/km (2900–4000 km upstream) to 1.6 cm/km (800–1020 km upstream), joined to high tidal (>10 m in Manaus), favors the emergence of large alluvial floodplains (*várzeas*) and wetlands (Birkett et al. 2002). The latter are commonly defined as inland areas that are periodically inundated or permanently waterlogged, including lakes, rivers, estuaries, and freshwater marshes. In the Amazon basin lowlands (altitude <500 m), wetlands and floodplains cover a huge area, recently estimated to 800,000 km² (Melack and Hess, 2011), from which 12% are located along the Solimões Amazon mainstream.

Amazonian floodplains play an important role in the spread of flows—mitigation of the flood wave (Paiva et al. 2011, 2012; Richey et al. 1989)—and also in sediment transfer (Bourgoin et al. 2007; Mangiarotti et al. 2013). They influence cycles of a large number of chemical elements (Bonnet et al. 2016; Melack et al. 2004; Moreira-turcq et al. 2013). The moving littoral in the aquatic terrestrial transition zone (ATTZ) compounds a complex mosaic of habitats, changing in space and time, conducive to biodiversity (Junk and Wantzen, 2004; Parolin et al. 2004; Tockner and Stanford, 2002). It also prevents from stagnation and enables a rapid recycling of organic matter and nutrients, thereby explaining the large productivity of these systems (Junk, 1997). Fish-bearing waters and soil fertility higher than that in upland regions form attractive conditions for human settlement. For these reasons and because of their relative accessibility, Amazon floodplains have long been colonized (Dufour, 1990; Grennand and Bahri, 1990). The nutrient-rich water inflow sustains human activities such as agriculture, animal husbandry, and forestry. Hydroclimatic seasonality rhythms agricultural and fishing activities and more generally local social life (Bommel et al. 2016).

In the past five decades, considerable public investment has encouraged economic growth in the Amazon region, leading to not only record expansion of the agricultural sector but also important development within the mining sector and hydroelectric dams (Ferreira et al. 2014) and a rapid population increase, which passed from 6 million to 20 million in Brazilian Amazonia from 1960 to 2010 (IBGE, 2016). This economic development has modified landscapes in multiple ways, resulting in many regions being characterized as a mosaic of disturbed forests, agriculture fields, and pastures.

Human impacts have also resulted in changes in freshwater ecosystems, with impacts on terrestrial and aquatic systems often being felt across large distances (Castello et al. 2013). The Brazilian government has engaged in finding a compromise between ecosystems preservation and development in the Amazon region. Consequently, policy makers have acted against deforestation and also pursued poverty alleviation strategies, with the perspective that this would contribute to limit environmental damages and mitigate the vulnerability of local Amazonian populations. A strong action plan has drastically reduced the agriculture expansion into forests (including protected areas, law enforcement, land regulation, markets and credit restriction (Arima et al. 2014; Nepstad et al. 2014)). The National Plan on Climate Change, signed in 2008, planned to reduce the deforestation average rate of the period 1996–2005 by 80% by 2020. In 2012, this goal was almost achieved through coercive measures to stop deforestation, with convincing results since 2005 (annual deforestation fell from 27.772 km² in 2004 to 4.656 km² in 2012). However, existing management policies, including the protected area network, still fail to protect freshwater

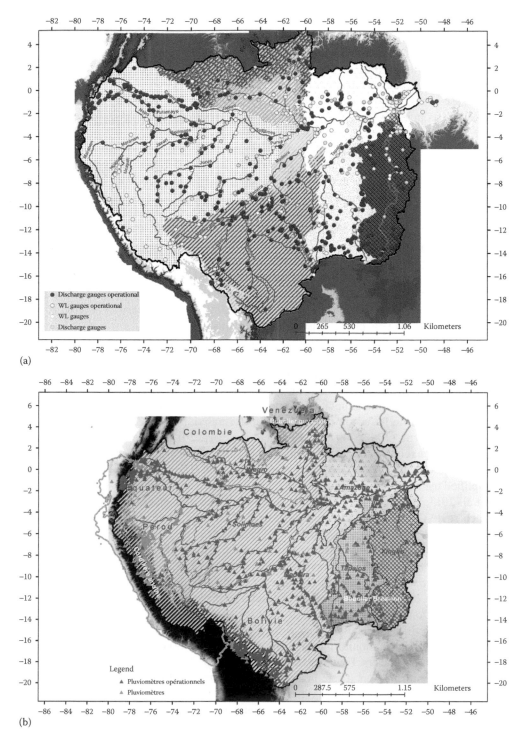

FIGURE 16.1 (a) Amazonian water discharge and level gauges. (b) Amazonian rain gauges.

ecosystems. These policies are insufficient to guarantee the conservation of hydrological connectivity (Castello et al. 2013), and they are increasingly threatened by dam construction (Ferreira et al. 2014) waterways projects (Soito and Freitas, 2011) and increasing investments in oil exploitation (Zurita-Arthos and Mulligan, 2013).

Combined with global warming and a spatiotemporal redistribution of precipitation (Nobre et al. 2013), land-use changes across Amazonia have resulted in widespread loss of soils and biodiversity and marked shifts in biogeochemical cycles, seriously threatening the functioning and biodiversity of terrestrial and aquatic Amazonians ecosystems. Although the interactions and impact of man-made land transformation on the local to regional climate and their feedbacks are still in debate, changes in precipitation and discharge are already observed by the most deforested regions of the Amazon. It provides evidence of the potential shift in vegetation and further feedback on climate and river corridor dynamics (Davidson et al. 2012; Funatsu et al. 2012). This could partly explain the rapid succession of extreme hydrological events (droughts in 2005 and 2010, and floods in 2009, 2012, and 2014) (Marengo et al. 2011; Zeng et al. 2008) in the last decade. Gloor et al. (2013) recently showed that the Amazon basin has presented wetter climatic conditions since 1990.

Finally, anthropogenic pressures and climatic changes alter the hydrological regime of Amazonian rivers, thus threatening the whole river-floodplain ecosystem (Junk et al. 2010). To better understand the interplay between water circulation and ecological characteristics and to predict their evolution in terms of probable changes in river regime, it is necessary to develop appropriate models to reproduce the flow characteristics in floodplains.

Studies of Amazonian floodplain hydrology have included water balances of individual lakes (Lesack and Melack, 1995; Bonnet et al. 2008, 2011, 2016). These studies have shown that floodplain water balance was influenced by different water sources, whose contribution to floodplain water mixture varies seasonally and in function of floodplain morphologies. Several modelling approaches have been used to simulate flood propagation in the Amazon. Richey et al. (1989) proposed a routing scheme based on the Muskingum formulation for a 2000 km reach. Decharme et al. (2008), Coe et al. (2002; 2007), and Paiva et al. (2011, 2013) succeeded in regional flooding modeling by using a relative simple formulation to describe the lateral exchanges between the mainstream and its floodplain. Models, coupling one-dimensional (1D) simulation with two-dimensional (2D) simplified hydrodynamic formulation in floodplain, such as LISFLOOD-FP (Bates et al. 2010; Baugh et al. 2013; Hunter et al. 2005), have been developed and successfully applied at medium-sized reach scale (hundreds of kilometers) (Baugh et al. 2013; Rudorff et al. 2014a; Trigg et al. 2009; Wilson et al. 2007) and at regional scale (Yamazaki et al. 2013; 2011). Wilson et al. (2007) and Trigg et al. (2012) highlighted the importance of the drainage network floodplain. Baugh et al. (2013) and Yamazaki et al. (2012) proposed digital elevation model (DEM) adjustment algorithms in order to simulate realistic water exchanges between river channels and floodplains. Rudorff et al. (2014a) provided a detailed analysis of the hydraulic controls of flooding of the Curuaí floodplain on the lower Amazon.

However, due to difficult access and huge extent, data scarcity and low accuracy are often major problems to be faced, when monitoring Amazonian wetlands. Most of the meteorological stations in Brazil are located along rivers. Few data are available on floodplains. This distribution leaves large areas without any information. There is also a lack of water level (WL) gauges. On the Brazilian side, the National Water Agency (ANA) has made efforts to increase the number of flow and rain gauges, but the density remains the lowest in the country. Calmant et al. (2009) counted an average of one WL gauge in the Amazon basin for 7,200 km² (Figure 16.1b). It is also noteworthy that *in situ* gauges' maintenance is complicated and costly. Their number is decreasing as in others countries in the world. In addition, political relations inside the country and between Amazon countries have influence on data collection and exchange. If Brazilian hydrological data are easily available, this is not the case with other countries. Moreover, getting detailed topographic data or ecological data is even more challenging in this region.

Hence, analyses of hydrological processes and full forecast are limited by the availability of data with appropriate spatial and temporal distribution. In this context, monitoring systems based on remotely sensed observations are an efficient alternative.

Indeed, beside the traditional use of satellite-based imagery for soil, land-cover and land-use mapping (Lu et al. 2012; Yengoh et al. 2014), or wetlands types and flood extent (Chapman et al. 2015; Martinez and Le Toan, 2007; Sippel et al. 1998), the use of remotely sensed products for hydrology has been significantly intensified in the last decades.

Even though it has been originally conceived for ocean studies, satellite radar altimetry has been proven to give reliable WL over lakes and rivers (Birkett et al. 2002; Calmant and Seyler, 2006; Cretaux et al. 2011; Frappart et al. 2005; Roux et al. 2008, 2010; Santos da Silva et al. 2010, among others). It has given a great impulse for monitoring large and relatively poor gauged basins (Santos et al. [2014]). A valuable information for hydrological modeling validation (Coe and Costa, 2002; Getirana et al. 2010; Paiva et al. 2013) and for improving model previsions (Paiva et al. 2012) has been shown. Combined with validated modeled discharge, it enabled to generate rating curves all over the Amazon basin, thereby providing the basis for a fully spatially distributed monitoring system in nearly real time (Paris et al. 2016). Rainfall is also estimated with a reasonable accuracy, for instance, from the Tropical Rainfall Measuring Mission (TRMM) (Huffman and Bolvin, 2014) or Global Satellite Mapping of Precipitation (GSMaP) (GSMaP, 2013) (Nerini et al. 2015; Satge et al. 2015). Several works relate the possibility to use data from the Moderate Resolution Imaging Spectroradiometer (MODIS) to obtain evapotranspiration estimation (Mu et al. 2007, Velpuri et al. 2013). Other information on the dynamics of water stock has been obtained with gravimetric mission, the Gravity Recovery and Climate Experiment (GRACE) (Schmidt et al. 2008; Ramillien et al. 2008). Among the most significant advances for hydrology, it is noteworthy that the production of global DEMs is essential to numerically capture flow direction and watershed limits (Jenson and Domingue, 1988). In the last decade, three nearly global DEMs of high resolution have been freely released. The National Aeronautics and Space elevation data Administration (NASA) and the National Geospatial-Intelligence Agency (NGA) of the United States released the Shuttle Radar Topography Mission (SRTM) (Farr et al. 2007, Rodriguez et al. 2006). The NASA and the Ministry of Economy, Trade, and Industry (METI), Japan, released the first GDEM version in 2009 and the second version in 2011 (Tachikawa et al. 2011). Ultimately, the AW3D30 was released in 2015 by the Japan Aerospace Exploration Agency (JAXA) (Tadono et al. 2016).

These different products have improved the knowledge of floodplains dynamics. Combining radar altimetry with flood extent deduced from imagery enabled retrieving of floodplain storage dynamics (Frappart et al. 2008, 2005); insights on water circulation within the floodplain (Alsdorf, 2003; Alsdorf et al. 2000) were obtained from synthetic aperture radar (SAR) and interferometric data analysis, and exchanged fluxes between the mainstream were estimated from a combination of satellite products (Alsdorf et al. 2010).

In this chapter, we present a methodological framework to set a hydrodynamic model at local scale, integrating a broad range of remotely sensed data. The model is applied to simulate the 2009 flooding hazard, one of the major flood events ever recorded. According to the methodological framework, Earth observation (EO) products are used in each step, namely input data production, calibration, and validation. Input data production consists of obtaining reliable and distributed WLs in the floodplain and mainstream, topographic data, and flood extent maps. Boundary conditions include conditions of WL issued from altimetry and conditions of flow generated from a hydrological model (Bonnet et al. [in revision]). Calibration is performed on Manning's roughness coefficients during the first year of simulation. The latter are determined from derived SAR images. Finally, we validated the model in terms of vertical accuracy, using WLs, as well as in terms of horizontal accuracy, comparing simulated flood extent with maps deduced from available remote-sensed product imagery.

The wide range of useful remotely sensed data for our purpose and the hydrodynamic model used for the study are presented in the first section. Sections 16.5 and 16.6 give details of the different methods used to produce relevant information for modelling. Calibration and validation steps are described in Sections 16.7 and 16.8.

16.2 THE STUDY SITE

The Janauacá Floodplain is located in the low Amazon basin between 3.200°S and 3.250°S and between 60.230°W and 60.130°W, along the right margin of the Solimões River, approximately 40 km upstream from its junction with the Rio Negro (Amazon state, Brazil) (Figure 16.1).

It is composed of one lake connected with the Solimões River, the major affluent of the Amazon River. Rich in nutrients and suspended solids, this river is commonly considered as a whitewater river (Moquet et al. 2015; Sioli, 1984). In contrast, small streams (*igarapés*), draining the south of the watershed, present properties closer to black waters, which are rich in dissolved organic matter. According to the rain gauge data in Manacapuru (3.317°S, 60.583°W), at about 40 km from the study, the mean annual rainfall is 1976 mm/y. The river WL mean annual fluctuation reaches 12.2 m, at the Manacapuru WL gauge, when considering the period 2006–2011. The Solimões River has a mono-modal flood phase, with WL usually starting to rise mid-November until mid-June, when the recession phase starts. According to altimetric data, the water surface slope between Manacapuru and VS VSR is, on the average, about 2.2 cm/km.

The Janauacá watershed is divided into two municipalities (Manaquiri and Carreiro). Between 2010 and 2015, population increased on average of 20% (IBGE, 2016), attesting to the local dynamism of the region. Population growth is accompanied by landscape modifications, such as augmentation of non-forested landscapes in the watershed (Drapeau et al. 2011). As mentioned in the introduction, population has mainly rural activity, which is rhythmed by hydrological seasonality. Even if the local population is used to important changes from low water (LW) to high water (HW), the last extreme flood events (especially 2009 and 2012) have provoked material damages and have had direct consequences on local life.

16.3 THE FLOOD 2009–2010

At the regional scale, Filizola et al. (2014) detailed the flood event of 2009. A combination of regional-scale climatic events (Marengo et al. 2012) and unusual flood mechanism can explain this hazard. Large positive anomalies of sea surface temperatures (SSTs) in the tropical South Atlantic Ocean anomalously maintained the intertropical convergence zone (ITCZ) in the South, producing greater rainfall. Rainfall led to advanced flood in the western part of the river. The latter was this year almost in phase with the flood peak of the southern tributaries. The backwater effect was amplified, and it resulted in higher stages and water quantity in the Amazon mainstream. Stage recorded at Óbidos, the lowermost gauge station in the Amazon, was the highest (10.83 m) registered since the beginning of the measurement (1928). With a recorded stage of 10.67 m, the 2012 flood event did not reach the higher stage at Óbidos gauge. Nevertheless, measured water stage at Manaus (central Amazon) was paradoxically higher than the one measured in 2009, such as measured flow at Óbidos (260,000 m^3s^{-1}). Such a flood event impacts the socioeconomic life of the Amazon. In 2009, 38 Brazilian municipalities states declared to be flooded along the Amazon River and its tributaries (Filizola et al. 2014).

At a local scale, the RL1 stages recorded inside the Janauacá floodplain at RL1 are 24.0 m and 24.3 m for the years 2009 and 2012, respectively. Average peak stage over the period 2006–2012 is 22.3 m. In 2009, rising water lasted 241 days, against an averaged value of 181 days over the period 2006–2012. Whole hydrological year lasted 398 days. Tidal amplitude was 11.6 m for the year 2009, whereas the average over 2006–2012 tidal amplitude was 9.9 m. Paradoxically, local rain during the hydrological year 2008–2009 is not elevated. The *agência nacional das águas* (ANA) provided daily rainfall through the stations labeled 00359005 (3.100°S, 59.994°W) and 00359007, both located at Manaus, and 00360001 (3.317°S, 60.583°W), located at Manacapuru. We computed daily rainfall following the Thiessen polygon method applied on Manacapuru and Manaus data. This method led to an estimation of 1469 mm during the hydrological year 2008–2009. Average value over the period 2006–2011 is 1980 mm/y. The EO TRMM data (Huffman and Bolvin, 2014) provide an estimation of 1544 mm during the hydrological year 2008–2009.

16.4 THE MODEL

The IPH-TRIM3D-PCLake model (Fragoso et al. 2009), also known as IPH-ECO, freely available at www.ipheco.org, is a three-dimensional (3D) hydrodynamic module coupled with an ecosystem module. It describes the main physical (water temperature and density, velocity fields, and free-water elevation), chemical, and biological (e.g., nutrients and trophic structure) processes existing in the aquatic ecosystem. Since its first release, some improvements have been added to the model, such as a resuspension flux that is a function of wind fetch and the day length as a function of latitude (Fragoso et al. 2011).

The IPH-ECO hydrodynamic module solves the Reynolds-averaged Navier–Stokes equations by using a semi-implicit discretization on a structured staggered grid (Casulli and Cheng, 1992; Cheng et al. 1993). The non-linear convective terms existing in the Tidal, Residual, Intertidal Mudflat (TRIM) solution (Cheng et al. 1993) are solved by using an explicit Eulerian-Lagrangian finite-difference scheme. To increase the stability and accuracy of the numerical discretization, the θ-method is used (Casulli and Cattani, 1994). The horizontal eddy viscosity is calibrated manually, and the vertical eddy viscosity is parameterized by using an empirical relationship (Pacanowski and Philander, 1981).

The model was successfully applied to simulate phytoplankton dynamics in a large shallow Brazilian lake (Fragoso et al. 2008) and then extended to evaluate trophic dynamics and aquatic metabolism in a large shallow aquatic environment (Cavalcanti et al. 2016; Fragoso et al. 2011). In addition, the model was used to simulate a complex river–lake interface (Pereira et al. 2013; Cavalcanti et al. 2016) and as a tool to understand the complex dynamics of a biomanipulated Danish lake (Pereira et al. 2013). Currently, the model is being applied to simulate estuarine regions' hydrodynamics, deep reservoirs' carbon dynamics, and floodplain lakes' dynamics.

In this study, we used a 2D horizontal (vertically averaged) representation of the floodplain. It differs from the common approach, as the floodplain is fully simulated, and not seen as a flooded zone, mainly depending on the river simulation, where flows are only guided by bed slopes.

16.5 PRECURSORY WORK: GATHERING DATA

16.5.1 Generating and Correcting *In Situ* (Real or Virtual) Gauges

Accuracy of hydraulic model especially relies on boundary conditions of high confidence. In this study, part of the boundary conditions and reference data used for calibration and validation are in part composed of WL.

16.5.1.1 Ground Gauges

The WLs of Solimões River at the Manacapuru gauge (3.317°S, 60.583°W), labeled 14100000, located 50 km upstream of the study zone (Figure 16.2), were obtained from the ANA (http://hidroweb.ana. gov.br/). This gauge has been leveled against EGM08 by differential Global Positioning System (GPS).

In addition, two WLs gauges were installed in the floodplain at "RL1" (3.424°S, 60.264°W) and "RL2" (3.368°S, 60.193°W) in the thalweg connecting the lake and the Solimões (Figure 16.2). These gauges were leveled with the help of satellite radar altimetry level time series data, according to the method described in Santos da Silva et al. (2010), and validated by high-precision bi-frequency

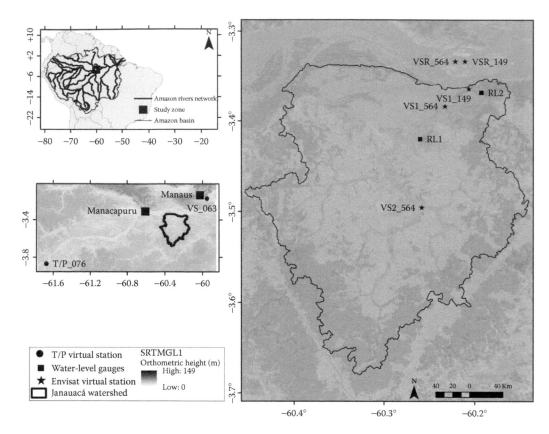

FIGURE 16.2 Study site with the locations of available river stage data and the virtual stations with background SRTM Global 1 arc-second (SRTMGL1).

GPS stations. Finally, level time series were reported to the EGM08 geoid. A series of daily values is available beginning from September 1, 2006 to December 31, 2012.

16.5.1.2 Virtual Stations

In this study, we used data from two different radar altimeters, namely T/P and ENVISAT/RA-2. They were used to produce WL time series in several locations of the floodplain and in the mainstream, following (Roux et al. 2010; Santos da Silva, 2010).

The NASA and the *Centre National d'Etudes Spatiales* (CNES) launched the T/P altimeter. Data were collected from 1992 to 2004. The T/P altimeter has a 10-day repeat cycle. The intertrack distance is 315 km at the equator. Birkett et al. (2002) reported accuracies ranging from 10 cm to several meters (root-mean-square error [RMSE] 1.1 m). Two VSs were created from the T/P passes, crossing the Solimões River in the vicinity of the Manacapuru gauge: VS_076 (3.863°S, 61.685°W), 147 km upstream of the gauge, and VS_063 (3.172°S, 59.947°W), 84 km downstream of the gauge.

Within the Earth Observation Program (EOP), the ENVISAT satellite, launched by the European Spatial Agency (ESA), collected data from 2002 to 2012. The latter was the follow-on mission for the ERS-1 and ERS-2 missions (Zhang, 2009). ENVISAT embarked 10 instruments (Wehr and Attema, 2001), including a nadir radar altimeter (RA-2). RA-2 is a high-precision nadir radar that operates at two frequencies (ESA, 2007): Ku-band (13.575 GHz) and S-band (3.2 GHz). ENVISAT flied on a sun-synchronous circular orbit with an inclination of 98.5° and a 35-day repeat period. It completed a global cover of the Earth between latitudes of ±81.5° and the orbit ground track has an intertrack distance of approximately 80 km at the equator (Louet, 2000).

With 10 years of services, ENVISAT allowed numerous hydrological modeling over the Amazon (e.g., Getirana et al. 2010, 2013; Paiva et al. 2013; and Paris et al. 2016), monitoring studies (Getirana et al. 2009; Santos da Silva et al. 2014; Seyler et al. 2008; and Da Silva et al. 2012), and other derived studies (Bonnet et al. 2016 and Pinel et al. 2016). Regarding altimetry accuracy in the Amazon, comparison at crossovers with *in situ* gauges showed that the quality of the series can be highly variable, with results ranging from 12 cm to several meters (Santos da Silva, 2010; Santos da Silva et al. 2010).

In this study, satellite radar altimetry ENVISAT mission provided five additional VSs. VS1_564 (3.392°S, 60.239°W), VS2_564 (3.493°S, 60.253°W), and VS1_149 (-3,375 °S, 60.202°W) were located in the floodplain, whereas VSR_564 (3.325°S, 60.220°W) and VSR_149 (3,358°S, 60.214°W) were located in the Solimões (Figure 16.2).

16.5.1.3   Water Level Data Consolidation

As a whole, we dispose of the records from the official ANA network at Manacapuru gauge and from two *in situ* stations (RL1 and RL2) installed in the floodplain. Altimetry provides five VSs (VS1_564, VS2_564, VS1_149, VSR_149, and VSR_564) (Figure 16.2). However, gauge elements' displacements or errors in reading or reporting data are always possible. The comparison between altimetry data and *in situ* records enabled to correct or fulfill *in situ* measurements, when necessary, giving more confidence in the *in situ* WL series, as mentioned in (Santos da Silva et al. 2014). In this paper, these authors highlighted how altimetry helps in managing the Brazilian ANA network. Common technic to improve the quality of an *in situ* gauge located in the river is to compare it with two VSs (upstream and downstream).

Altimetry helped in cleaning the ANA records. It was indeed necessary to have confident data at Manacapuru to properly estimate the slope along the reach from Manacapuru to the Janauacá connecting channel mouth. We also needed to estimate WL in Janauacá at the acquisition dates of the Japanese Earth Resources Satellite-1 (JERS-1) images used in this study (see Section 16.5.2). As shown in Figure 16.3a, the series at the Manacapuru gauge presents noisy variations during years 1994 and 1995. From the T/P series SV_076 and SV_063, we estimated the WL on October 10, 1995, to be 12.20 m against 10.42 m, as given by the Manacapuru gauge series. The same reasoning for the HW level gave a difference of 0.2 m between altimetry estimation and Manacapuru gauge level. As the T/P altimetry data have a vertical accuracy of about 0.5 m (Santos da Silva, 2010), we did not alter the level provided by the ANA gauge for the HW level. Then, by building linear regression between the Manacapuru gauge and RL1 over the LW and HW periods, we estimated the WL in the floodplain at RL1 gauge to be 13.5 m on October 10, 1995, and 22.1 m on May 27, 1996.

Altimetry enabled filling and cleaning of RL1 and RL2 gauges. The latter were specially constructed for research needs. Compared with ANA gauges, quality robustness were inferior. Maintenance was not as sharp as the ANA gauges. In addition to human report error, unstable background (muddy moving soil) caused displacements. Hence, the initial leveling did not stand for the whole study, and several leveling corrections were necessary along the study period. RL1 is located between VS1_564 and VS2_564. We selected the dates for which both stations have data and retained only those for which the WL difference was less than 5 cm. The leveling value at RL1 was obtained as the average of the differences between WL measured at the gauge and the average value measured by altimetry.

We also generated four daily VSs. In the Solimões, in front of the Janauacá floodplain, VSR_149 and VSR_564 were distant of 1.16 km. We merged these data into a single, one VSR. We linearly correlated these records with those situated 40 km upstream at Manacapuru place to obtain daily VSR (Figure 16.3b and 3c), following (Roux et al, 2010) methodology. In the floodplain, we linearly correlated RL1 to VS1_564, RL1 to VS2_564, and RL2 to VS1_149 to generate daily WL VS1_564, VS2_564, and VS1_149, respectively (Figure 16.3f).

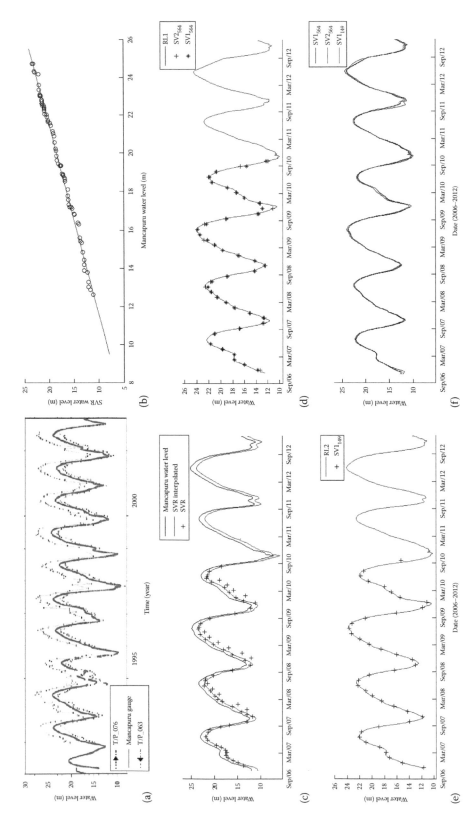

FIGURE 16.3 Combination of *in situ* water level and altimetry: (a) Temporal series of Manacapuru gauge, SV_076 and SV_063. (b) Linear regression between Manacapuru and VSR gauges. (c) Temporal series of Manacapuru gauge and VSR. (d) Temporal series of RL1, VS1_564, and VS2_564. (e) Temporal series of RL2 VS_149 and SVR149. (f) Daily temporal series of VS1_564, VS2_564, and VS1_149.

16.5.2 Generating Flood Extent Map

In perspective of generating the DEM and validating the simulation, we built map of inundation extent for expected range of WLs in the floodplain.

16.5.2.1 Available Remotely Sensed Products

16.5.2.1.1 Synthetic Aperture Radar Imagery

Hess et al. (2015) produced a dual-season mapping of wetlands inundation and vegetation for the central Amazon basin. It is available at the NASA's Earth Observing System Data and Information System (EOSDIS) website (http://reverb.echo.nasa.gov/). In this product, wetland areas have been defined as land areas that are seasonally or permanently waterlogged, including lakes, rivers, estuaries, and freshwater marshes. The Hess et al. (2015) product is based on mosaicked L-band SAR imagery, acquired by JERS-1 during two periods: August–September 1995 and May–August 1996 for LW and HW, respectively. The dual-season approach provided 15 classes of vegetation. Information about the acquisition date and the corresponding level of both images are summarized in Table 16.1.

The maps were validated using high-resolution, geocoded digital videography collected during aerial surveys at HW and LW periods. Thematic accuracy of the wetlands mask was estimated to be 95%. Inundation maps issued from the wetland map are commonly used for assessment of accuracy in modeling studies of inundation in the central Amazon basin (Alsdorf et al. 2007; Baugh et al. 2013; Coe et al. 2007; Rudorff et al. 2014b; Trigg et al. 2009; Yamazaki et al. 2012). Some studies reported an overestimation of the inundation at LW stages (Rudorff et al. 2014a; Pinel et al. 2015).

The product named ALOS SCANSAR classification and imagery was released by NASA and JAXA through the program Inundated Wetlands—Earth Science Data Record (IW-ESDR). Data are available at the global monitoring of wetlands extent and dynamics website (http://wetlands.jpl.nasa.gov/). Extensive multitemporal L-band ALOS ScanSAR data acquired bi-monthly by the

---

TABLE 16.1
List of the Inundation Maps

Images Date	Hydrologic Period	Corresponding Water Level (m)	Source
October 10, 1995	Low waters	13.5	Hess et al. 2015
May 27, 1996	High waters	22.1	Hess et al. 2015
January 25, 2007	Rising waters	17.3	ALOS SCANSAR imagery
March 12, 2007	Rising waters	17.8	ALOS SCANSAR imagery
March 17, 2007	Rising waters	17.8	ALOS SCANSAR imagery
April 27, 2007	Rising waters	20.2	ALOS SCANSAR imagery
July 28, 2007	High waters	21.5	ALOS SCANSAR imagery
August 2, 2007	High waters	21.3	ALOS SCANSAR imagery
December 18, 2007	Rising waters	14.9	ALOS SCANSAR imagery
September 14, 2008	Flushing waters	16.6	ALOS SCANSAR imagery
August 4, 2008	High waters	21.00	ALOS SCANSAR imagery
November 29, 2011	Low waters	11.5	Landsat 5 TM
October 20, 2006	Low waters	13.3	Landsat 5 TM
September 2, 2006	Flushing waters	18.4	Landsat 5 TM
July 24, 2009	High waters	23.5	Landsat 5 TM
August 25, 2009	High waters	21.5	Landsat 5 TM
October 09, 2009	Flushing waters	19.8	Landsat 5 TM

PALSAR instrument onboard ALOS were used to classify the inundation state for South America (Chapman et al. 2015). Horizontal resolution is 100 m at the equator. For the study zone that we were interested in, nine classified images were useful (Table 16.1).

16.5.2.1.2 Optical Products

The United States Geological Survey (USGS) and NASA have managed the Landsat missions since 1972. Recently, the Landsat archive has been made freely available (Woodcock et al. 2008). Landsat 5 Thematic Mapper (1984–2013) imagery has moderate spatial resolution (30 m) and provides multispectral images (seven or eight bands), with a short revisit interval (16 days). We selected six Landsat images relatively free of clouds. Two images were representative of the LW period, with water elevations at the RL1. The third image was representative of the flushing period. Other images were acquired during the year 2009.

Water areas were distinguished from non-water areas computationally by using the normalized difference ratio between the mid-infrared band (1.55–1.75 μm) and the visible band (0.52–0.60 μm). The latter has already been used by Toivonen et al. (2007) to map open water areas over a 2.2 million km² portion of the western Amazon. Each image has been associated with a specific threshold. The remaining errors, such as individual clouds and cloud shadows, were corrected by using high thresholds. However, this led to underestimation of the water bodies. To attenuate this underestimation, we ranked the classification regarding the associated WL and decided that a pixel that is *water* for one classification has to stay *water* in the upper-ranked classifications. Inversely, on pixel that is non-water for one classification has to stay *non-water* in the lower-ranked classifications.

16.5.3 GENERATING THE TOPOGRAPHIC DATA

16.5.3.1 Available Data for Digital Elevation Model Construction

Obtaining reliable topography is another condition for realistic flood modeling. It also has to be seamless from dry land to the beds of water bodies. However, detailed collection is challenging over large wetland environments. Currently, the construction of such large floodplain DEMs is achieved through integration of land data, generally issued from remote-sensing methods, and bathymetric data acquired during several field camps at different water stages. In addition, dense vegetation prevents classical airborne DEM in fully reaching the bare earth elevation. Indeed, in the Amazon lowland basin, (Carabajal and Harding, 2006) estimated the SRTM elevation to be 40% of the distance from the canopy top to the ground.

16.5.3.1.1 Earth Observation Data

The global 1 arc-second SRTM V3.0 dataset (SRTMGL1) is a joint product of the NGA and NASA. Data were collected during 11 days in February 2000 by using dual Spaceborne Imaging Radar and dual X-band SAR. From these data, a near-global DEM was generated, downloadable at NASA's EOSDIS website. Since 2015, the SRTMGL1 had been freely released for South America. The SRTM data products have been validated on continental scales: the absolute and relative vertical accuracies over South America are 6.2 m and 5.5 m, respectively (Rodriguez et al. 2006). Rudorff et al. (2014a) noted a local negative bias of 4.4 m in an Amazonian floodplain. Satge et al. (2015) reported a negative bias of 7.2 m for the Andean Plateau region.

Besides the bias introduced by interferometric errors, the SRTM data present an elevation ranging above the bare earth and below the maximum canopy height (Brown et al. 2010; Carabajal and Harding, 2006), because of the incapacity of C-band radar in reaching the bare earth. (Carabajal and Harding, 2006) estimated the vertical height accuracy to 22.4 m in the lowland Amazon basin. Original data are referenced to the World Geodetic System 84 (WGS84) ellipsoid and the Earth Gravitational Model 1996 (EGM96) geoid. The EGM96 geoidal undulations were replaced by the EGM08 ones (Pavlis et al. 2013).

Water surfaces have low radar backscatter. Hence, water bodies and coastlines are not well defined. As a by-product of the water body editing, a water mask was generated: the SRTM Water Body Data (SWBD). It presents the same coverage as SRTMGL1 and is available in 1° by 1° tile from NASA's EOSDIS website. Lehner et al. (2006) reported that this dataset presents some inconsistencies. These are partly explained by the fact that water body depiction required ancillary data sources, such as Landsat 5 data, that were collected much earlier than the shuttle mission.

The NASA produced ICESAT/GLAS data, collected from 2003 to 2009 by the Geoscience Laser Altimeter System (GLAS) on board of the Ice Cloud and land Elevation Satellite laser altimeter (ICESat). The GLAS footprint size was approximately 65 m in diameter, spaced by 170 m along track and several tens of kilometers across tracks. The ICESat/GLAS products are available at NASA's EOSDIS website. The ICESat/GLAS mission was initially launched for monitoring icecaps, and it does not penetrate water surfaces. Nonetheless, literature reported its capability in monitoring other land covers. Baghdadi et al. (2011) showed the ability to monitor a lake with a vertical accuracy of 5 cm. Many studies also used ICESat/GLAS as ancillary data to validate DEM (Satgé et al. 2015, 2016; Schutz et al. 2005; Zwally et al. 2002).

In this study, we used the last released v34 of GLA14 Global Land-Surface, specific for land-surface elevation. Data were referenced to the TOPEX/Poseidon ellipsoid and EGM96 geoid. We selected data acquired at LW, between October and January, outside of SWDB boundaries and inside herb classes of the wetlands map. The original geoid was replaced by the EGM08.

The vegetation offset correction required vegetation characteristics. Simard et al. (2011) released a global map of forest canopy height at 1 km spatial resolution. This map is based on data from the MODIS on NASA's Terra and Aqua satellites and from the ICESat/GLAS. The Simard et al. (2011) map was created by regressing ICESat RH100 (relative height) canopy height measurements with global grids of annual mean precipitation, precipitation seasonality, annual mean temperature, temperature seasonality, elevation, and percentage tree cover. It is distributed as an 890 MB GeoTIFF file on the website http://lidarradar.jpl.nasa.gov/. This product presents an accuracy of 6.1 m compared with measurements at 66 Fluxnet sites.

16.5.3.1.2 Bathymetric In Situ Data

Bathymetry field campaigns were performed. Most of the data were acquired during a field trip organized during June 2012, when water stage in the floodplain raised exceptionally to a level of 24.3 m. *In situ* depth data were acquired with an Acoustic Doppler Profiler Current (ADCP, 12 Hz, Teledyne RD Instruments) linked to a GPS or an echo sounder linked to a GPS station. The ADCP and echo sounder data were checked against manual measurements by using a ruler and showed an overall good agreement between them. Other *in situ* bathymetric data were acquired in May 2008 and August 2006. Errors in the bathymetric grid are expected in both depth and position accuracy. According to the constructor, an ADCP has a vertical precision of 1 cm in 99% of the measurements, far less below the errors induced by field conditions (instruments onboard). In this study, we estimated that waves can generate errors in the order of 0.2 m. According to the constructor, the position error is less than 15 m for 95% of the measurements for both instruments. This distance is less than the usual grid mesh (30 m for the SRTMGL1). Examining survey data, we excluded all the data that had no consistency with the others.

16.5.3.2 Correction of the Interferometric Bias and Vegetation Offset

Pinel et al. (2015) proposed a systematic approach over Amazonian floodplains to generate topographic data (Figure 16.4). The method removes the vegetation signal, addressing its heterogeneity by combining estimates of vegetation height and a land cover map. They improved this approach by interpolating the first results with drainage network, field, and altimetry data to obtain a hydrologically conditioned DEM. Data needed for these methods are initial DEM (SRTMGL1), land cover map (Hess et al. (2015) wetlands map and SWBD, height vegetation map (Simard et al. 2011), *in situ* elevation of land, and underwater split-independent Ground Control Point (GCP) datasets: GCP_I and GCP_V for interpolation and validation phases, respectively.

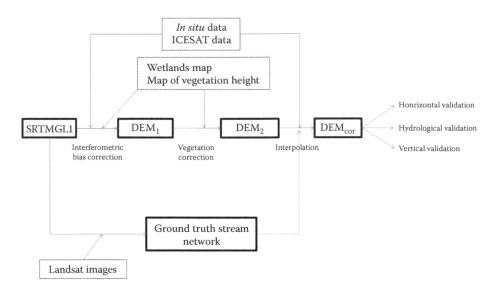

FIGURE 16.4 Overview of the method to generate the topographic input.

First, focusing on a window, including a large area around the floodplain and the river, the process begins with the correction of the interferometric bias (Rodriguez et al. 2006). Bare soil areas were deduced from the wetland map. The bias was computed as the mean difference between *in situ* and GCP_I data over bare soil and the SRTMGL1 elevations. Bias was uniformly reported over the entire study site domain. Finally, an intermediary DEM was obtained, denoted as DEM_1.

To correct the positive vegetation bias, the wetland map and SWBD classes were used to split the study area into several regions, where the correction offset was computed considering the type of vegetation and constrain on elevation related to the flooded status. As a whole, 16 regions were identified. In the study area, WLs at the time of images acquisition were estimated to be 13.5 m and 22.1 m for the LW and HW seasons, respectively. The DEM_ADJ was obtained by subtracting an elevation offset to the DEM1 pixel within each region and by turning all pixels that do not respect the elevation constraints into NODATA. Forested classes (ID41, ID42, ID43, and ID51) needed special attention to determine vegetation subtract offset by using the Simard et al. (2011) product. New DEM generated was named DEM_2.

The third step consisted of merging DEM_2 with the GCP_I dataset. The resulting dataset was interpolated by using the ANUDEM v5.3 algorithm (Hutchinson et al. 2011), constrained by a drainage network, to produce the corrected DEM (DEM_{COR}). The input *ground truth* stream network was formed by applying the commonly used D8 algorithm (Jenson and Domingue, 1988) to SRTMGL1, subsequently improved by the inspection of Landsat images.

The DEMs improvements were assessed vertically (against the GCP_V), horizontally (against maps of flood extent gathered during the previous step), and hydrologically (through watershed and river network assessments).

16.5.3.3 Application to the Janauacá Floodplain

The first step led in the estimation of the interferometric bias to be 2.0 m (standard deviation [SD] = 4.1 m). An offset of this magnitude was applied to increase the SRTMGL1 elevations. This bias is half the value encountered by Rudorff et al. (2014a) in another floodplain, located approximately 700 km downstream of the Janauacá floodplain. However, as reported by Rodriguez et al. (2006), the interferometric bias is expected to vary from place to place.

The vegetation bias presented a mean value of 5.9 m (SD = 6.9 m) and 7.4 m (SD = 7.3 m) over the whole study area and the highland zone, respectively. Carabajal and Harding (2006) reported that

TABLE 16.2
DEM Vertical Accuracy Assessment against 10 GCP Dataset and River Network Assessments

DEM	Mean (m)	SD (m)	RMSE (m)	Roughness (m)	Outlet (Boolean)	Connectivity (Boolean)	GTSN Matching Index (%)
SRTMGL1	−0.4	4.7	4.8	1.5	0	1	58
DEMCOR	0.1	1.7	1.7	0.9	1	1	83

SRTM elevation is located at approximately 40% of the distance from the canopy top to the ground. Hence, we estimated that the mean canopy height was 12.3 m over the highland zone. This value is consistent with the canopy heights found in studies of tree species in central Amazonian floodplain forests (Schöngart et al. 2010; Wittmann et al. 2002). In another floodplain, Rudorff et al. (2014a) applied a lower vegetation offset of 1.4 m, but this area includes a larger proportion of savanna and secondary vegetation than the Janauacá floodplain. It is also much smaller than the value of 22.4 m estimated by Carabajal and Harding (2006) or suggested in regional models (23 m and 17 m, proposed by Coe et al. [2007] and Paiva et al. [2011], respectively).

The vertical accuracy assessments, against the GCP_V dataset for the generated DEM and the original SRTMGL1, as well as the roughness criteria, are gathered in Table 16.2. Vertical validation against GCP shows an RMSE reduction of 64% (from 4.8 m to 1.7 m). Focusing on hydrological agreement, DEM$_{COR}$ presents the best matching with the ground truth stream network (GTSN) (83%), against percentages lower than 58% for the SRTMGL1. It also presents the right outlet position. Assuming a hypothesis of horizontality in the floodplain, flood extent accuracy, controlled against generated flood maps, stresses improvements in the LW and HW periods (+10% and +27%, respectively) (see Pinel et al. [2015] for more details).

16.6 MODELING PROJECT SETTING

Setting the model consists of defining the modeled zone, the resolution, and the boundary conditions. Area of simulation is chosen as the region, subject to intense variations of open water extent, mainly the downstream part of the floodplain and the main *igarapé* (Figure 16.5). Owing to high computational costs, spatial resolution needs to be stepped down. Bilinear resampling has effects on roughness and slopes, and curvatures has effects on the inside channels. To overcome resampling effects, we redraw and burnt the channels.

We first restrained the computing domain to the flooded area dropped down by 71% of the cell number of model. Hence, we resampled the modeled zone from 31 m to 278 m. The resampling has side effects on the roughness (+166%), on bankfull levels, and on floodplains channels (curvature and bed slope). The first simulation attempts showed that the model failed in draining the floodplain during flushing waters and LWs. Several studies mention the importance of channels systems that organize the flows inside floodplains (Rudorff et al. 2014a; Trigg et al. 2012). Taking into account these studies, we first artificially burned the DEM, creating a channel network (*channelized DEM*). Based on the bathymetric field dataset, we burned channels in view of generating realistic slopes (*channelized with slopes*). The field trip highlighted that depth in the Solimões can reach up to −35 m (against −16 m in the main thalweg). Hence, the last modification was to artificially burn the ending cells of the main thalweg (*channelized with slopes and Solimões depth*). We controlled that each of the above-mentioned step allowed improving the accuracy during the first hydrological cycle (Figure 16.6). For instance, controlled against RL2, RMSE between simulated and observed waters varies from 1.32 m to 0.33 m for simulated WL based on the original DEM and based on a *Channelized with slopes and Solimões depth* DEM, respectively.

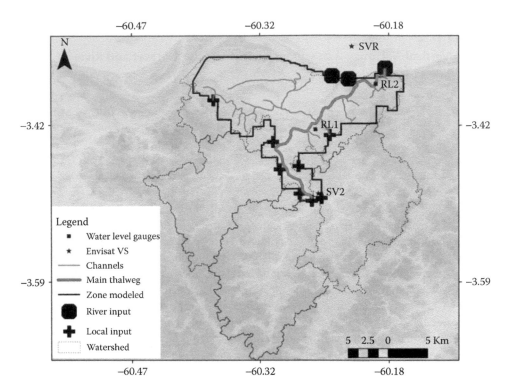

FIGURE 16.5   Model domain and location of the boundary conditions.

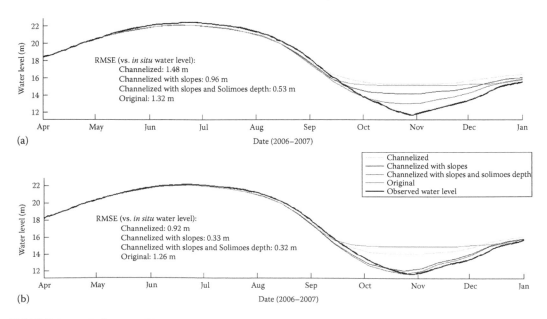

FIGURE 16.6   Influence of burning processes over the simulation. (a) Water level simulated and observed at RL1, and (b) water level simulated and observed at RL2.

Two types of boundary conditions are needed: ones based on flow over the catchment and the others based on WL of the Solimões River. Hess's wetlands map provided the details of the main water sources from the upstream catchment. It allowed the partitioning of the upper catchment in eight subzones. Bonnet et al. (in revision), through a hydrological model, estimated the flow over upland part of the catchment from ANA rainfall gauges. For each *igarapé* determined through the wetland map, we associated a subcatchment. Flow generated over the latter was computed as a ratio of the global upland flow.

Visual inspection of the Landsat images allowed to determine the beginning and ending dates of overflood and the bankfull cells. Two places were identified: WLBC1 (WL Boundary Condition 1) underwater for a WL of 20 m at 49 km from Manacapuru and WLBC2 underwater for a level of 20.5 m at 47 km from Manacapuru. At these cells (WLBC1/2), bankfull elevations were modified in order to match with the flooding overbank. Level water along this zone was estimated by using the recorded WL at Manacapuru gauge and a mean slope (2.2 cm/km) of this Solimões reach.

Visual inspection of the Landsat images allows delimitating the zones of overflow from the river into the floodplain. As described in the presentation of the study zone, a thalweg continuously makes the connection between the floodplain and the river (Figure 16.3). This channel acts alternatively as an inlet or as an outlet in function of the period in the hydrological cycle (Bonnet et al. [in revision]). Thus, a WL condition is required here (WLBC3). Luckily, the junction of this thalweg and the river is located under an ENVISAT track. Hence, the VS generated at this place (VSR) stands for boundary condition.

16.7 CALIBRATION

Simulation is developed in view of modeling extreme hydrological year of 2008–2009. Simulated WLs at RL1 and RL2 gauges from June 2007 (HWs) to January 2008 (LWs) are analyzed in order to calibrate the model through RMSE between simulated and *in situ* records. Calibration is realized against Manning's roughness coefficient. A map of roughness is generated from EO product. Values of the Hess's wetlands map are grouped by vegetation main type. Each type of vegetation leads to a specific Manning's coefficient. As much as we have vegetation types, we have different Manning's values. We separately disturbed the Manning's values in order to check the influence of each vegetation type on the model.

For the study site, we grouped the EO Hess's wetlands classes into four distinct ones: water/herb, shrubs, flooded forest, and forest in spatial proportions of 53%, 12%, 22%, and 21%, respectively. Guided by published values for wooded floodplains (Arcement Jr and Schneider, 1989; Chow, 1959), we assigned a range of Manning's roughness values (Table 16.3). We first assigned the mean of the range to each class: 0.03 for a range 0.02–0.04, 0.045 for a range 0.03–0.06, 0.1 for a range 0.05–0.15, and 0.15 for a range 0.1–0.2 for water/herb, shrubs, flooded forest, and forest, respectively. We separately disturbed each Manning's class by changing the manning by minimum and maximum values. For each dataset, we analyzed the RMSE of first year of simulation at RL1 and RL2 places (Table 16.3). Depending on the assigned Manning's coefficient, we obtained RMSE value ranges from 0.5 m to 0.59 m at RL1 and RMSE value ranges from 0.28 to 0.32 m at RL2 place. Finally, the calibration led to the following assignment for the Manning's values: 0.032, 0.042, 0.14, and 0.18 for water/herb, shrubs, flooded forest, and forest, respectively (Figure 16.7). For these coefficients, RMSE between observed and simulated elevations at RL1 and RL2 places reaches the lower values encountered during the calibration phase. These are near the values widely used in hydraulics studies in the Amazon. For instance, Rudorff et al. (2014a) assigned Manning's values of 0.14 and 0.10 for forest and shrubs classes, respectively. Their calibration yielded a value of 0.031 in the Solimões. Calibration by Trigg et al. (2009) resulted in a value of 0.034 in the Purus River.

TABLE 16.3

Analysis Sensibility over Manning's Roughness Coefficients

Vegetation Type	Area Percentage (%)	Manning's Roughness Dataset								Manning's Chosen
Water/herb	53	0.02	0.04	0.03	0.03	0.03	0.03	0.03	0.03	0.032
Shrubs	12	0.04	0.04	0.02	0.06	0.04	0.04	0.04	0.04	0.042
Flooded forest	22	0.1	0.1	0.1	0.1	0.05	0.15	0.15	0.15	0.14
Forest	21	0.15	0.15	0.15	0.15	0.15	0.15	0.1	0.2	0.18
RL1 RMSE		0.51	0.54	0.55	0.5	0.59	0.5	0.53	0.52	0.5
RL2 RMSE		0.32	0.28	0.28	0.32	0.27	0.28	0.32	0.28	0.28
Mean RMSE		0.415	0.41	0.415	0.41	0.43	0.39	0.425	0.4	0.39

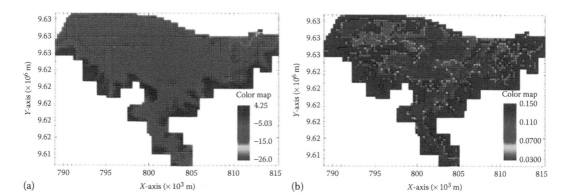

FIGURE 16.7 Input grid data: (a) Channelized with slopes and Solimões depth DEM and (b) Manning roughness map.

16.8 VALIDATION

The vertical WL accuracy assessment was estimated by comparison with the four gauges located inside the floodplain (RL1, RL2, VS1_564, and VS2_564) and against the VSR (limit between the Solimões and floodplain). We computed the following statistics: the correlation coefficient, the RMSE, the mean, and the SD of the difference between the observed and simulated data.

The simulated inundation extents are compared to flood maps derived from the 2009 Landsat images. The agreement between the flooding extent deduced from the DEM and imagery is calculated by using the following classical skills scores (Paiva et al. 2011; Wilks, 2006): The threat score (TS) measures the model's accuracy with a perfect score of 100; the bias index (BIAS) indicates the type of error (overestimation or underestimation); the false alarm ratio (FAR) measures the overestimation of the flooded areas deduced from the DEM with a perfect score of 0, and the missed flooded areas ratio (MFR) measures the underestimation of the flooded areas deduced from the DEM with a perfect score of 0. Those scores are determined by using the following relations:

$$TS = 100\left(\frac{a}{a+b+c}\right)$$

$$BIAS = 100\left(1 - \frac{a+b}{a+c}\right)$$

$$FAR = 100\left(\frac{b}{a+b}\right)$$

$$MFR = 100\left(\frac{c}{a+c}\right)$$

Where, a represents the total area that is both mapped and predicted by the model as inundated, b is the predicted but not mapped inundated area, and c is the mapped but not predicted inundated area.

Simulated water elevations were compared with data from the *in situ* and VSs in the floodplain (Figure 16.8 and Table 16.4).

Global RMSE and correlation coefficient between the simulated and observed data at the fifth stations over the entire period were 0.22 m and 0.99, respectively. It showed that calibration over the first hydrological year was successful. Compared with the accuracy of the used DEM and the

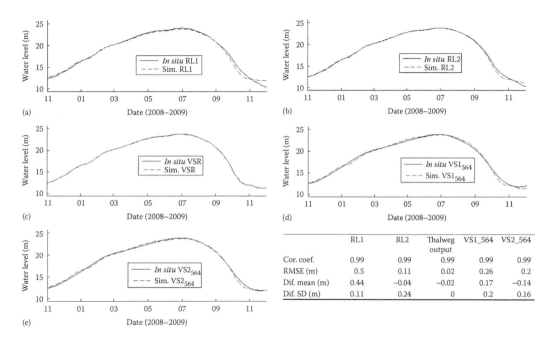

	RL1	RL2	Thalweg output	VS1_564	VS2_564
Cor. coef.	0.99	0.99	0.99	0.99	0.99
RMSE (m)	0.5	0.11	0.02	0.26	0.2
Dif. mean (m)	0.44	−0.04	−0.02	0.17	−0.14
Dif. SD (m)	0.11	0.24	0	0.2	0.16

FIGURE 16.8 Water level comparison graphics at the different places and global statistics at each place. (a) RL1: simulated and observed water level. (b) RL2: simulated and observed water level. (c) Thalweg output (VSR): simulated and observed water level (d) South zone (VS1564): simulated and observed water level and (e) Mid-Thalweg (VS2564): simulated and observed water level.

TABLE 16.4

WL Vertical Accuracy Assessment

	All Station	*In situ* Station	Virtual Station	Low Water	Rising Water	High Water	Flushing Water
Correlation coefficient	0.99	0.99	0.99	0.95	0.99	0.99	0.99
RMSE (m)	0.22	0.31	0.23	0.38	0.13	0.15	0.33
Differential mean (m)	0.08	0.20	0.015	−0.11	−0.0	0.07	0.06
Differential SD (m)	0.14	0.18	0.18	0.35	0.13	0.10	0.28

amplitude of the flood wave (11.6 m), RMSE value is low. This suggests that IPH-ECO was able to reproduce well the variations of water-surface elevations in the floodplain. Rudorff et al. (2014a) yielded an RMSE of 0.24 m when simulating the WL in another Amazonian floodplain. However, statistics indexes can vary depending on the station's nature: an RMSE value of 0.31 m with a correlation coefficient of 0.99 when comparing with both *in situ* stations, against an RMSE value of 0.99 with a correlation coefficient of 0.23 when comparing with VS. As SVR is a boundary condition, statistics for this gauge improve the statistics when comparing VS with *in situ* station. Figure 16.8 highlights that we can expect a statistics variation, depending on the water period. Setting global RMSE values as references, Table 16.4 presents that RMSE values can vary from +71 % at LWs to −43% at rising waters. This highlights that the ability of the model to reproduce the floodplain flows depends on the period of the hydrological year.

The other important source of errors can also come from boundary conditions. Those of water flow type are derived from rainfall through a hydrological model developed in a work currently in revision (Bonnet et al. [in revision]). VS1_564 located in the south of the basin is mostly dependent on the water issued from the upstream subcatchments. Good statistics (RMSE = 0.26 m) obtained at VS1 place tends to positively validate these premodeled data. However, the model seems to prove that at LW, these water inputs are overestimated. Lower accuracy (RMSE = 0.38 m and correlation coefficient = 0.95) obtained at LW can be linked to a draining problem related to an unadapted resolution. Such a coarse resolution cannot take into account all the channels that drive the flows (Trigg et al. 2012). Rudorff et al. (2014a) mentioned the importance of the channels that organize the flooding and flushing periods.

Regarding the evaluation of the flood extent, the values of the different skill scores (TS, BIAS, MFR, and FAR) obtained by comparing simulation results with flood maps are reported in Table 16.5. We consider the areas predicted by the model but not mapped as inundated to be overpredictions and the areas mapped but not predicted by model as inundated to be underpredictions. Two validations of flood extent correspond to HWs (July 24, 2009 and August 25, 2009) and one to flushing waters (September 10, 2009).

From a general point of view, simulation overpredicts the inundation in the floodplain and the skill scores are consistent from a date to another. The TS remains the highest in the case of the highest WL (TS = 71 for the date July 24, 2009). Variations in the accuracy (TS index) are little. Indeed, the lowest TS value is 63, obtained for the date August 25, 2009. The BIAS values are similar and range from −35 to −53 for the dates July 24, 2009, and August 25, 2009. The average negative BIAS value suggests that the model overpredicted the flooding extent. This fact is corroborated by a high FAR and a low MFR. Regarding the spatial distribution, Figure 16.9a through 9c clearly spotlights that simulated waters are in overprediction. Comparing with similar hydraulics studies in the Amazon basin, Wilson et al. (2007a) yielded to an accuracy of 73 at the HW level. The matching result of Yamazaki et al. (2012) reached 60 at the HW level. On a global scale, Paiva et al. (2011) found a model performance of 70 at the HW level.

TABLE 16.5

Agreement between Simulated and ALOS-1/PLASAR Mapping of Inundation Extent

Date Acquisition	TS	BIAS	FAR	MFR
July 24, 2009	71	−35	28	2
August 25, 2009	63	−53	36	2
October 09, 2009	65	−41	33	5

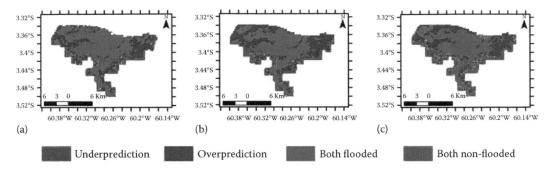

| | Underprediction | | Overprediction | | Both flooded | | Both non-flooded |

FIGURE 16.9 Comparisons of inundation extent obtained by IPH-ECO simulation against inundation maps at different dates: (a) July 24, 2009, (b) August 25, 2009, and (c) September 10, 2009.

Overestimation of the simulated flood extents can be attributed to an underestimation to a pixel misclassification during the process of mapping the Landsat images. Detecting water under vegetation with the Landsat images is challenging (Toivonen et al. 2007; Yamazaki et al. 2015). Indeed, most of the overestimated pixels lie around the Janauacá Lake (Figure 16.9a through 9c), where woody areas are encountered according to the wetlands map. Overestimation of the simulated flood extents can also be linked to an underrepresentation of the channels network. Such a coarse resolution cannot take into account the entire channels that drive the flows. Trigg et al. (2012) and Rudorff et al. (2014a) mentioned the importance of the channels that organize the flooding and flushing periods.

16.9 CONCLUSIONS

Floodplains are among the most diverse and productive ecosystems in the world. They ensure several hydrological and ecological functions for the basin. Today, these sensitive zones are facing increasing human anthropization and augmentation in frequency and intensity of extreme flood events. How they will evolve with these progressive changes is still difficult to predict. In this context, hydrodynamic models are attractive tools for studying water circulation patterns in the floodplain and exchange with mainstream. However, they require relatively high-quality data of topography, land cover, WLs, and water flows to produce realistic results. In the large unmonitored regions, such as the low Amazonian basin, remotely sensed data appear as a solution to gather input data in view of hydrodynamics modeling.

Parallelly, the recent and ongoing proliferation of free EO data brings the challenge of integrating the many heterogeneous geospatial datasets to monitor and model in view of effective information management.

In this study, we aimed at reproducing one of the largest flood events (2009–2010) ever recorded for a floodplain of the low Amazon basin. Stage recorded at Óbidos, the lowermost gauge station in the Amazon, was the highest registered (10.83 m). We detailed a systematic procedure step by step to generate relevant and reliable datasets for model setting, calibration, and validation. All the steps involve various EO data products: altimetry, airborne DEM, land cover derived from SAR imagery, vegetation height map derived from altimetry (Light Detection and Ranging [LIDAR]) and optical imagery, inundation maps derived from SAR, and optical images.

Dataset generation was divided into three steps. The first step was to map the zone in terms of WL. Altimetry furnished five VSs in the zone. We use the latter to correct local national official data, to generate *in situ* daily virtual WL in the mainstream. Two ground-stage stations were installed in the area for research needs in the floodplain. Altimetry allowed in filling records gap and correcting gauge elements displacements. By interpolation of VSs, two additional daily VSs were generated. The second step led to building and gathering 14 flood extent maps. The third step was to generate an accurate DEM. Here, we exhibited and applied a method presented in a recently published work (Pinel et al. 2015). For this, using ICESat and *in situ* data, we began to detect and rectify a negative bias of 2.0 m noising the SRTMGL1. Then, we removed the vegetation signal by a coupled analysis of EO products: two maps of the vegetation and of the vegetation height map. Over the flooded forested zones, analysis led to subtract an average percentage of 53% from the vegetation height. We built drainage network, generated from the SRTM, and then rectified with the help of satellite imagery. Inputting the adjusted DEM, the drainage network, *in situ* data, and altimetry data in ANUDEM v5.3, we generated an adjusted and conditioned DEM. Improvements between original SRTMGL1 and DEM were validated vertically (+64%), horizontally (+24% and +18% for LW and HW periods, respectively), and hydrologically (+43%).

Setting the model includes defining the computing domain and boundary conditions. In view of reducing computation time, we restricted the watershed area to a modeled zone (−71% in terms of cells), resampled the computation grid from 1 arc-second to 3 arc-second. To overcome resampling effects, we redrew and burned the channel network inside the floodplain. These topographic modifications dropped down the RMSE from 1.32 m to 0.33 m between observed and simulated elevations. Visual inspection of EO images allowed the identification of boundary conditions. Those of WL type, located in the Solimões, were derived from interpolated data from *in situ* WL given by ANA and by altimetry (SVR). Those of water flow type, located in the floodplain, were derived from rainfall through a hydrological model developed in a previous work (Bonnet et al. 2017).

We performed the calibration over roughness coefficient against WL at RL1 and RL2 during the 2007–2008 hydrological year. A wetlands map, provided by Hess et al. (2015), derived from JERS-1 images, was used to establish a land cover map of four zones. On each zone, a commonly used Manning's value was proposed. Each value was separately altered to investigate the influence of each vegetation type. Finally, calibration phase led to Manning's values of 0.032, 0.042, 0.14, and 0.18 for the water/herb, shrubs, flooded forest, and *terra firme* zones, respectively.

Extreme flood event (2009–2010) was validated in terms of vertical accuracy against altimetrics and *in situ* WLs and in terms of horizontal accuracy, comparing simulated flood extent against inundation map. Simulation WLs presented a high correlation level (r = 0.99) and an RMSE of 0.22 m (compared with a WL fluctuation of 11.6 m). However, depending on the location of the reference, the RMSE varied: 0.5 m for RL2 located near the river and 0.11 m for RL1 located in the center of the lake. Accuracy also depended on the hydrological periods, and RMSE can range from 0.13 m at rising waters to 0.38 m during LWs. Regarding horizontal accuracy, simulation tended to overpredict the inundation in the floodplain. The skill scores, over the three studied dates, were similar. Horizontal accuracy remained the highest in the case of comparison, with the images corresponding with the highest WL.

The presented case demonstrates how remote-sensing data can be integrated with flood models. Since 1998, the Copernicus program, the world largest EO program, aims at providing continuous and accurate EO data. The ongoing Sentinel series, the future mission of surface water and ocean topography (SWOT), and the numerous studies on integrating remotely sensed data with flood modeling spotlight that there is now a real will of space agencies to fortify the support that satellite missions can offer over water monitoring. Although progresses have already been achieved in recent years, research in comprehension of the ways in which remote sensing can support flood monitoring, modeling, and management is still an active open field.

ACKNOWLEDGMENTS

Sebastien Pinel would like to thank Dr. Tanvir Islam and Pr. George Petropoulos for giving him the opportunity for collaboration. Above all, Sebastien Pinel is grateful to all the authors for contributing their work. This research was conducted in the framework of the CNPq (*Conselho Nacional de Desenvolvimento Científico e Tecnológico*, Brazil), IRD (*Institut de Recherche pour le Developpement*) n° 490634/2013-3, and LMI- OCE (*Laboratoire Mixte International Observatoire des Changements Environementaux*). It was supported by four research programs: HIDRAS (CNPq), Dinâmica Fluvial do

Sistema Solimões/Amazonas (CPRM), ANZIM (CNES/TOSCA), BIODIVA (GUYAMAZON, IRD, CIRAD, French embassy and FAPEAM). Sebastien Pinel is also grateful to CAPES (*Coordenação de Aperfeiçoamento de Pessoal de Nível Superior, Brazil*) for financial support.

AUTHOR CONTRIBUTIONS

Marie-Paule Bonnet, Joecila Santos Da Silva, Frédérique Seyler, David Motta Marques, and Stephane Calmant conceived and designed the experiments; Marie-Paule Bonnet, Joecila Santos Da Silva, Stephane Calmant, and Sebastien Pinel performed the part relative to altimetry; Sebastien Pinel and Marie-Paule Bonnet analyzed the data; Ruberto Fragoso Jr. and Rafael J. Cavalcanti helped in building and running the hydrodynamic model. Marie-Paule Bonnet, Jeremie Garnier, and Sebastien Pinel wrote the chapter. All authors contributed to resolving problems during the data processing.

CONFLICTS OF INTEREST

The authors declare no conflict of interest.

BIBLIOGRAPHY

Alsdorf, D. (2003). Water storage of the central Amazon floodplain measured with GIS and remote sensing imagery. *Annals of the Association of American Geographers, 93*(1), 55–66.

Alsdorf, D., Bates, P. D., Melack, J. M., Wilson, M. D., and Dunne, T. (2007). Spatial and temporal complexity of the Amazon flood measured from space. *Geophysical Research Letters, 34*(8), L08402. doi:10.1029/2007GL029447.

Alsdorf, D., Han, S.-C., Bates, P. D., and Melack, J. M. (2010). Seasonal water storage on the Amazon floodplain measured from satellites. *Remote Sensing of Environment, 114*(11), 2448–2456. doi:10.1016/j.rse.2010.05.020.

Alsdorf, D., Melack, J. M., Dunne, T., Mertes, L., Hess, L., and Smith, L. (2000). Interferometric radar measurements of water level changes on the Amazon flood plain. *Nature, 404*(March), 174–177. doi:10.1038/35004560.

Arcement Jr, G. J. and Schneider, V. R. (1989). Guide for selecting Manning's roughness coefficients for natural channels and flood plains. *Technical Report, Geological Survey Water-Supply, United States Government Printing Office, Washington*, 38. doi:Report No. FHWA-TS-84-204.

Arima, E. Y., Barreto, P., Araújo, E., and Soares-Filho, B. (2014). Public policies can reduce tropical deforestation: Lessons and challenges from Brazil. *Land Use Policy, 41*, 465–473.

Baghdadi, N., Lemarquand, N., Abdallah, H., and Bailly, J. S. (2011). The relevance of GLAS/ICESat elevation data for the monitoring of river networks. *Remote Sensing, 3*, 708–720. doi:10.3390/rs3040708.

Bates, P. D., Horritt, M. S., and Fewtrell, T. J. (2010). A simple inertial formulation of the shallow water equations for efficient two-dimensional flood inundation modelling. *Journal of Hydrology, 387*(1–2), 33–45. doi:10.1016/j.jhydrol.2010.03.027.

Baugh, C., Bates, P. D., Schumann, G., and Trigg, M. (2013). SRTM vegetation removal and hydrodynamic modeling accuracy. *Water Resources Research, 49*(9), 5276–5289. doi:10.1002/wrcr.20412.

Birkett, C. M., Mertes, L., Dunne, T., Costa, M. H., and Jasinski, M. J. (2002). Surface water dynamics in the Amazon Basin: Application of satellite radar altimetry. *Journal of Geophysical Research: Atmospheres, 107*(October 2001), LBA 26-1–LBA 26-21. doi:10.1029/2001JD000609.

Bommel, P., Coudel, E., and Haentjens, E. (n.d.). Livelihoods of local communities in an Amazonian floodplain coping with global changes: From role-playing games to hybrid simulations to involve local stakeholders in participatory foresight study at territorial level.

Bonnet, M. P., Barros, W.., Martinez, J. M., Seyler, F., Moreira-Turcq, P., Cochonneau, G., … Seyler, P. (2008). Floodplain hydrology in an Amazon floodplain lake (Lago Grande de Curuai). *Journal of Hydrology, 349*(1), 18–30.

Bonnet, M. P., Garnier, J., Barroux, G., Boaventura, G. R., and Seyler, P. (2016). Biogeochemical functioning of Amazonian floodplains: The case of Lago Grande de Curuai. In *Riparian Zones: Characteristics, Management Practices and Ecological Impacts*, Environmental Research Advances, Vol. 14 (Spring). Hauppauge, New York: Nova Science Publishers, pp. 1–14.

Bonnet, M. P., Lamback, B., Boaventura, G. R., Oilveira, E., Seyler, F., Calmant, S. et al. (2011). Impact of the 2009 exceptional flood on the flood plain of the Solimões River. *Conceptual and Modelling Studies of Integrated Groundwater, Surface Water, and Ecological Systems Proceedings of Symposium H01, 1*(July), pp. 1–7.

Bonnet, M. P., Pinel, S., Garnier, J., Bois, J., Bonaventura, G., and Seyler, P. (2017). Amazonian floodplain water balance based on hydrologic and electrical conductivity data analyses and modelling, *Hydrological Processes*, (November), 1–38. Available at: http://doi.wiley.com/10.1002/hyp.11138.

Bourgoin, L. M., Bonnet, M. P., Martinez, J. M., Kosuth, P., Cochonneau, G., Moreira-turcq, P., Guyot, J. L. et al. (2007). Temporal dynamics of water and sediment exchanges between the Curuaí floodplain and the Amazon River, Brazil. *Journal of Hydrology, 335*(1–2), 140–156. doi:10.1016/j.jhydrol.2006.11.023.

Brown, G., Sarabandi, K., and Pierce, L. (2010). Model-based estimation of forest canopy height in red and austrian pine stands using shuttle radar topography mission and ancillary data: A proof-of-concept study. *IEEE Transactions on Geoscience and Remote Sensing, 48*(3 PART 1), 1105–1118.

Callede, J., Cochonneau, G., Alves, F. V., Gummadi, V., Guimarães, V., and De Oliveira, E. (2010). Les apports en eau de l'Amazone à l'Océan Atlantique. *Revue Des Sciences de L'eau, 23*(3), 247. Retrieved from http://id.erudit.org/iderudit/044688ar.

Calmant, S. and Seyler, F. (2006). Continental surface waters from satellite altimetry. *Comptes Rendus - Geoscience, 338*(14–15), 1113–1122. doi:10.1016/j.crte.2006.05.012.

Calmant, S., Seyler, F., and Cretaux, J. F. (2009). Monitoring continental surface waters by satellite altimetry. *Surveys in Geophysics, 29*(4–5), 247–269. doi:10.1007/s10712-008-9051-1.

Carabajal, C. C. and Harding, D. (2006). SRTM C-band and ICESat laser altimetry elevation comparisons as a function of tree cover and relief. *Photogrammetric Engineering & Remote Sensing, 72*(3), 287–298.

Castello, L., Mcgrath, D. G., Hess, L., Coe, M. T., Lefebvre, P., Petry, P., Macedo, M.N. et al. (2013). The vulnerability of Amazon freshwater ecosystems. *Conservation Letters, 6*, 217–229.

Casulli, V. and Cattani, E. (1994). Stability, accuracy and efficiency of a semi-implicit method for three-dimensional shallow water flow. *Computers & Mathematics with Applications, 27*(4): 99–112.

Casulli, V. and Cheng, R. T. (1992). Semi-implicit finite difference methods for three-dimensional shallow water flow. *International Journal for Numerical Methods in Fluids, 15*(6), 629–648. Retrieved from http://www.scopus.com/inward/record.url?eid=2-s2.0-0027009126&partnerID=40&md5 = 378003d9bf 7ccc3a33a4f7ff71512219.

Cavalcanti, J. R., da Motta-Marques, D., and Fragoso, C. R. (2016). Process-based modeling of shallow lake metabolism: Spatio-temporal variability and relative importance of individual processes. *Ecological Modelling, 323*, 28–40.

Chapman, B., McDonald, K., Shimada, M., Rosen, P. a., Schroeder, R., and Hess, L. (2015). Mapping regional inundation with spaceborne L-band SAR. *Remote Sensing, 7*(5), 5440–5470. doi:10.3390/rs70505440.

Cheng, R. T., Casulli, V., and Gartner, J. W. (1993). Tidal, residual, intertidal mudflat (TRIM) model and its applications to San Francisco Bay, California. *Estuarine, Coastal and Shelf Science, 36*(3), 235–280. Retrieved from http://linkinghub.elsevier.com/retrieve/pii/S0272771483710164.

Chow, V. Te. (1959). *Open-Channel Hydraulics*. New York: McGraw-Hill Book Company, p. 728.

Coe, M. T. and Costa, M. H. (2002). Long-term simulations of discharge and floods in the Amazon Basin, *107*(August), 1–17.

Coe, M. T., Costa, M. H. and Howard, E. A. (2007). Simulating the surface waters of the Amazon River basin: Impacts of new river geomorphic and flow parameterizations. *Hydrological Processes, 2553*(October 2007), 2542–2553. doi:10.1002/hyp.

Coomes, O. T., Lapointe, M., Templeton, M., and List, G. (2016). Amazon river flow regime and flood recessional agriculture: Flood stage reversals and risk of annual crop loss. *Journal of Hydrology, 539*, 214–222. doi:10.1016/j.jhydrol.2016.05.027.

Cretaux, J. F., Calmant, S., Del Rio, R. A., Kouraev, A., Berge, M., and Maisongrande, P. (2011). Lakes studies from satellite altimetry. In *Coastal Altimetry* (pp. 509–533). Berlin Heidelberg: Springer.

Da Silva, J. S., Seyler, F., Calmant, S., Rotunno Filho, O. C., Roux, E., Araújo, A. A. M., and Guyot, J. L. (2012). Water level dynamics of Amazon wetlands at the watershed scale by satellite altimetry. *International Journal of Remote Sensing*.

Davidson, E. A., de Araújo, A. C., Artaxo, P., Balch, J. K., Brown, I. F., C. Bustamante, M. M., Coe, M. T. et al. (2012). The Amazon basin in transition. *Nature, 481*(7381), 321–328.

Decharme, B., Douville, H., Prigent, C., Papa, F., and Aires, F. (2008). A new river flooding scheme for global climate applications: Off-line evaluation over South America. *Journal of Geophysical Research, 113*(D11). doi:10.1029/2007JD009376.

Drapeau, G., Mering, C., and Ronchail, J. (2011). Variabilite hydrologique et vulnerabilite des populations du Lago Janauaca (Amazonas, Brasil). *Confins, 11*, 2–17. Retrieved from http://confins.revues.org/6904.

Dufour, D. L. (1990). Use of tropical rainforests by native Amazonians. *BioScience, 40*(9), 652–659. doi:10.2307/1311432.

ESA. (2007). *ENVISAT RA2/MWR Product Handbook*. Paris, France: *European Spatial Agency*.

Ferreira, J., Aragão, L. E. O. C., Barlow, J., Barreto, P., Berenguer, E., Bustamante, M., … Soares-Filho, B. (2014). Environment and Development. Brazil's environmental leadership at risk. *Science, 346*(6210), 706–7. doi:10.1126/science.1260194.

Filizola, N., Latrubesse, E. M., Fraizy, P., Souza, R., Guimarães, V., and Guyot, J. L. (2014). Was the 2009 flood the most hazardous or the largest ever recorded in the Amazon? *Geomorphology, 215*, 99–105.

Fragoso, C. R., Marques, D. M. L. M., Collischonn, W., Tucci, C. E. M., and van Nes, E. H. (2008). Modelling spatial heterogeneity of phytoplankton in Lake Mangueira, a large shallow subtropical lake in South Brazil. *Ecological Modelling, 219*(1–2), 125–137.

Fragoso, C. R., Motta Marques, D. M. L., Ferreira, T. F., Janse, J. H., and Van Nes, E. H. (2011). Potential effects of climate change and eutrophication on a large subtropical shallow lake. *Environmental Modelling and Software, 26*(11), 1337–1348. doi:10.1016/j.envsoft.2011.05.004.

Fragoso, C. R., Van Nes, E. H., Janse, J. H., and da Motta Marques, D. (2009). IPH-TRIM3D-PCLake: A three-dimensional complex dynamic model for subtropical aquatic ecosystems. *Environmental Modelling and Software, 24*(11), 1347–1348. doi:10.1016/j.envsoft.2009.05.006.

Frappart, F., Papa, F., Famiglietti, J., Prigent, C., Rossow, W. B., and Seyler, F. (2008). Interannual variations of river water storage from a multiple satellite approach: A case study for the Rio Negro River basin. *Journal of Geophysical Research, 113*(D21), D21104. doi:10.1029/2007JD009438.

Frappart, F., Seyler, F., Martinez, J. M., León, J. G., and Cazenave, a. (2005). Floodplain water storage in the Negro River basin estimated from microwave remote sensing of inundation area and water levels. *Remote Sensing of Environment, 99*, 387–399. doi:10.1016/j.rse.2005.08.016.

Funatsu, B. M., Dubreuil, V., Claud, C., Arvor, D., and Gan, M. A. (2012). Convective activity in Mato Grosso state (Brazil) from microwave satellite observations: Comparisons between AMSU and TRMM data sets. *Journal of Geophysical Research Atmospheres, 117*(16), D16109.

Gallo, M. N. and Vinzon, S. B. (2005). Generation of overtides and compound tides in Amazon estuary. *Ocean Dynamics 55*, 441–448).

Getirana, A. and Peters-Lidard, C. (2013). Estimating water discharge from large radar altimetry datasets. *Hydrology and Earth System Sciences, 17*(3), 923–933. doi:10.5194/hess-17-923-2013.

Getirana, A., Bonnet, M. P., Calmant, S., Roux, E., Rotunno Filho, O. C., and Mansur, W. J. (2009). Hydrological monitoring of poorly gauged basins based on rainfall-runoff modeling and spatial altimetry. *Journal of Hydrology, 379*, 205–219. doi:10.1016/j.jhydrol.2009.09.049.

Getirana, A., Bonnet, M. P., Rotunno Filho, O. C., Collischonn, W., Guyot, J. L., Seyler, F., and Mansur, W. J. (2010). Hydrological modelling and water balance of the Negro River basin: evaluation based on in situ and spatial altimetry data. *Hydrological Processes, 24*(22), 3219–3236. doi:10.1002/hyp.7747.

Gloor, M., Brienen, R. J. W., Galbraith, D., Feldpausch, T. R., Schöngart, J., Guyot, J. L., Espinoza, J. C. et al. (2013). Intensification of the Amazon hydrological cycle over the last two decades. *Geophysical Research Letters, 40*(9), 1729–1733. doi:10.1002/grl.50377.

Grennand, P. and Bahri, S. (1990). L'Agriculture de várzea et le paysannat d'Amazonie. *Actes du colloque international de Toulouse des 13 et 14 décembre 1990*, 181–185.

GSMaP. (2013). User's guide for global rainfall map in near-real-time by JAXA global rainfall watch (GSMaP _ NRT), (July).

Hess, L., Melack, J. M., Affonso, A. G., Barbosa, C. C. F., Gastil-Buhl, M., and Novo, E. (2015). Wetlands of the lowland Amazon basin: Extent, vegetative cover, and dual-season inundated area as mapped with JERS-1 Synthetic Aperture Radar. *Wetlands, 35*, 745–756. doi:10.1007/s13157-015-0666-y.

Huffman, G. J. and Bolvin, D. T. (2014). TRMM and other data precipitation data set documentation, (May), 1–42.

Hunter, N. M., Horritt, M. S., Bates, P. D., Wilson, M. D., and Werner, M. G. F. (2005). An adaptive time step solution for raster-based storage cell modelling of floodplain inundation. *Advances in Water Resources, 28*(9), 975–991. doi:10.1016/j.advwatres.2005.03.007.

Hutchinson, M. F., Xu, T., and Stein, J. A. (2011). Recent progress in the ANUDEM elevation gridding procedure. In T. Hengel, I. S. Evans, J. P. Wilson, and M. Gould (Eds.), *Geomorphometry* (pp. 19–22).

IBGE. (2016). http://www.cidades.ibge.gov.br/xtras/uf.php?lang=&coduf=13&search=amazonas.

Jenson, S. K. and Domingue, J. O. (1988). Extracting topographic structure from digital elevation data for geographic information system analysis. *Engineering, 54*(11), 1593–1600. doi:0099-1112/88/ 5411-1593$02.25/0.

Junk, W. (1997). *General aspects of floodplain ecology with special reference to Amazonian floodplains.* In W. J. Junk (Ed.)*The Central-Amazonian Floodplain: Ecology of a Pulsing System,* Vol. 126. Berlin, Germany: Springer Verlag. Retrieved from http://books.google.com/books?hl=pt-BR&lr=&id=1eMWT wL4rbEC&pgis=1.

Junk, W. and Wantzen, K. M. (2004). The flood pulse concept: new aspects, approaches and applications—An update. In *Proceedings of the Second International Symposium on the Management of Large Rivers for Fisheries* (Vol. 2, pp. 117–149).

Junk, W., Piedade, M. T. F., and Wittmann, F. (2010). *Amazonian Floodplain Forest Ecophysiology, Biodiversity and Sustainable Management.* Dordrecht, the Netherlands: Springer.

Lehner, B., Verdin, K. L., and Jarvis, A. (2006). HydroSHEDS Technical Documentation v1.0, 1–27. Retrieved from http://hydrosheds.cr.usgs.gov/webappcontent/HydroSHEDS_TechDoc_v10.pdf.

Lesack, L. F. W. and Melack, J. M. (1995). Flooding hydrology and mixture dynamics of lake water derived from multiple sources in an Amazon f.pdf. *Water Resources Research, 31,* 329–345.

Louet, J. (2000). Envisat system & mission. In *European Space Agency, (Special Publication) ESA SP* (pp. 2296–2298).

Lu, D., Batistella, M., Li, G., Moran, E., Hetrick, S., Freitas, C. D.,Dutra, L. V. et al. (2012). Land use/cover classification in the Brazilian Amazon using satellite images. *Pesquisa Agropecuária Brasileira, 47*(9), 1185–1208. doi:10.1590/S0100-204 × 2012000900004.

Mangiarotti, S., Martinez, J. M., Bonnet, M. P., Buarque, D. C., Filizola, N., and Mazzega, P. (2013). Discharge and suspended sediment flux estimated along the mainstream of the Amazon and the Madeira Rivers (from in situ and MODIS Satellite Data). *International Journal of Applied Earth Observation and Geoinformation, 21,* 341–355. doi:10.1016/j.jag.2012.07.015.

Marengo, J. A, Tomasella, J., Alves, L. M., Soares, W. R., and Rodriguez, D. A. (2011). The drought of 2010 in the context of historical droughts in the Amazon region. *Geophysical Research Letters, 38*(12). doi:10.1029/2011GL047436.

Martinez, J. M. and Le Toan, T. (2007). Mapping of flood dynamics and spatial distribution of vegetation in the Amazon floodplain using multitemporal SAR data. *Remote Sensing of Environment, 108*(3), 209–223. doi:10.1016/j.rse.2006.11.012.

Melack, J. M. and Hess, L. (2011). Remote sensing of the distribution and extent of wetlands in the Amazon basin. In W. J. Junk and M. Piedade (Eds.), *Amazonian Floodplain Forests: Ecophysiology, Ecology, Biodiversity and Sustainable Management* (pp. 1–28). Berlin: Ecological StudiesSpringer. doi:10.1007/978-90-481-8725-6_3.

Melack, J. M., Hess, L., Gastil, M., Forsberg, B. R., Hamilton, S. K., Lima, I. B. T., and Novo, E. (2004). Regionalization of methane emissions in the Amazon Basin with microwave remote sensing. *Global Change Biology, 10*(5), 530–544.

Moquet, J., Guyot, J., Crave, A., and Viers, J. (2015). Amazon river dissolved load: Temporal dynamics and annual budget from the Andes to the ocean. *Environmental Science and Pollution Research,* 23(12), 11405–11429. doi:10.1007/s11356-015-5503-6.

Moreira-turcq, P., Bonnet, M. P., Amouroux, D., Bernardes, M. C., Lagane, C., Maurice-Bourgoin, L. Oliveira, T. C. et al. (2013). Seasonal variability in concentration, composition, age, and fluxes of particulate organic carbon exchanged between the floodplain and Amazon River. *Global Biogeochemical Cycles,* 27(1), 119–130. doi:10.1002/gbc.20022.

Mu, Q., Heinsch, F. A., Zhao, M., and Running, S. W. (2007). Development of a global evapotranspiration algorithm based on MODIS and global meteorology data. *Remote Sensing of Environment, 111,* 519–536. doi:10.1016/j.rse.2007.04.015.

Nepstad, D., McGrath, D., Stickler, C., Alencar, A., Azevedo, A., Swette, B., Bezerra, T et al. (2014). Slowing Amazon deforestation through public policy and interventions in beef and soy supply chains. *Science, 344*(6188), 1118–1123.

Nerini, D., Zulkafli, Z., Wang, L.-P., Onof, C., Buytaert, W., Lavado, W., and Guyot, J.-L. (2015). A comparative analysis of TRMM-rain gauge data merging techniques at the daily time scale for distributed rainfall-runoff modelling applications. *Journal of Hydrometeorology, 16*(5), 2153–2168. Retrieved from http://journals.ametsoc.org/doi/full/10.1175/JHM-D-14-0197.1#.VhZoNW9_e4A.mendeley

Nobre, C. A., Obregón, G. O., Marengo, J. A, Fu, R., and Poveda, G. (2013). Characteristics of Amazonian climate: Main features. In *Amazonia and Global Change* (pp. 149–162). Washington, DC: Wiley Blackwell.

Pacanowski, R. C. and Philander, S. G. H. (1981). Parameterization of vertical mixing in numerical models of tropical oceans. *Journal of Physical Oceanography, 11*(11), 1443–1451.

Paiva, R. C. D., Buarque, D. C., Collischonn, W., Bonnet, M. P., Frappart, F., Calmant, S., and Bulhões Mendes, C. A. (2013). Large-scale hydrologic and hydrodynamic modeling of the Amazon River basin. *Water Resources Research, 49*(3), 1226–1243. doi:10.1002/wrcr.20067.

Paiva, R. C. D., Collischonn, W., and Bonnet, M. P. (2012). On the sources of hydrological prediction uncertainty in the Amazon. *Hydrology and Earth System Sciences, 16*, 1–20.

Paiva, R. C. D., Collischonn, W., and Buarque, D. C. (2011). Validation of a full hydrodynamic model for large-scale hydrologic modelling in the Amazon. *Hydrological Processes, 27*(3), 333–346. doi:10.1002/hyp.

Paiva, R. C. D., Collischonn, W., and Tucci, C. E. M. (2011). Large scale hydrologic and hydrodynamic modeling using limited data and a GIS based approach. *Journal of Hydrology, 406*(3–4), 170–181. doi:10.1016/j.jhydrol.2011.06.007.

Paiva, R. C. D., Collischonn, W., Bonnet, M. P., de Gonçalves, L. G. G., Calmant, S., Getirana, A., and Santos Da Silva, J. (2013). Assimilating in situ and radar altimetry data into a large-scale hydrologic-hydrodynamic model for streamflow forecast in the Amazon. *Hydrology and Earth System Sciences, 17*(3), 2929–2946. doi:10.5194/hess-17-2929-2013.

Paris, A., Dias, R., Paiva, R. C. D., Santos Da Silva, J., Medeiros, D., Ufrgs, I. P. H., Calmant, S. et al.(2016). Stage-discharge rating curves based on satellite altimetry and modeled discharge in the amazon basin legos, Université de Toulouse, CNRS, CNES, IRD, UPS, Toulouse, France LMI OCE IRD/UNB Campus Darcy Ribeiro, Brasilia, Brazil CESTU / UEA, Mana. *Water Resources Research, 52*, 3787–3814.

Parolin, P., Ferreira, L., Albernaz, A., and Almeida, S. (2004). Tree species distribution in Várzea forests of Brazilian Amazonia. *Folia Geobotanica, 39*(4), 371–383.

Pavlis, N. K., Holmes, S. A., Kenyon, S. C., and Factor, J. K. (2013). Erratum: Correction to the development and evaluation of the earth gravitational model 2008 (EGM2008). *Journal of Geophysical Research: Solid Earth, 118*.

Pereira, F. F., Fragoso, C. R., Uvo, C. B., Collischonn, W., and Motta Marques, D. M. L. (2013). Assessment of numerical schemes for solving the advection-diffusion equation on unstructured grids: Case study of the Guaíba River, Brazil. *Nonlinear Processes in Geophysics, 20*(6), 1113–1125. doi:10.5194/npg-20-1113-2013.

Ramillien, G., Bouhours, S., Lombard, A., Cazenave, a, Flechtner, F., and Schmidt, R. (2008). Land water storage contribution to sea level from GRACE geoid data over 2003–2006. *Global and Planetary Change, 60*(3–4), 381–392.

Richey, J. E., Meade, R. H., Salati, E., Devol, A. H., Nordin, C. F., and Santos, U. M. (1986). Water discharge and suspended sediment concentrations in the Amazon river: 1982–1984. *Water Resources Research, 22*(5), 756–764. doi:10.1029/WR022i005p00756.

Richey, J. E., Mertes, L., Dunne, T., Victoria, R. L., Forsberg, B. R., Tancredi, A. C. F. N. S., and Oliveira, E. (1989). Sources and routing of the Amazon river flood wave. *Global Biogeochemical Cycles, 3*(3), 191. doi:10.1029/GB003i003p00191.

Rodriguez, E., Morris, C. C., and Belz, J. J. (2006). A global assessment of the SRTM performance. *Photogrammetric Engineering and Remote Sensing, 72*(3), 249–260. Retrieved from http://www.asprs.org/a/publications/pers/2006journal/march/2006_mar_249-260.pdf.

Roux, E., Cauhope, M., Bonnet, M. P., Calmant, S., Vauchel, P., and Seyler, F. (2008). Daily water stage estimated from satellite altimetric data for large river basin monitoring. *Hydrological Sciences Journal, 53*, 81–99.

Roux, E., Santos da Silva, J., Cesar Vieira Getirana, A., Bonnet, M. P., Calmant, S., Martinez, J. M.,Getirana, A. (2010). Producing time series of river water height by means of satellite radar altimetry—A comparative study. *Hydrological Sciences Journal, 55*(1), 104–120. doi:10.1080/02626660903529023.

Rudorff, C. D. M., Melack, J. M., and Bates, P. D. (2014a). Flooding dynamics on the lower Amazon floodplain: 1. Hydraulic controls on water elevation, inundation extent, and river-floodplain discharge. *Water Resources Research, 50*(1), 619–634. doi:10.1002/2013WR014091.

Rudorff, C. D. M., Melack, J. M., and Bates, P. D. (2014b). Flooding dynamics on the lower Amazon floodplain: 2. Seasonal and interannual hydrological variability. *Water Resources Research, 50*(January), 635–649. doi:10.1002/2013WR014714.

Santos da Silva, J. (2010). *Application de l Altimetrie Spatiale a l etude des Processus Hydrologique dans les zones Humides du Bassin Amazonnien.*

Santos da Silva, J., Calmant, S., Seyler, F., Moreira, D., Oliveira, D., and Monteiro, A. (2014). Radar altimetry aids managing gauge networks. *Water Resources Management, 28*(3), 587–603. doi:10.1007/s11269-013-0484-z.

Santos da Silva, J., Calmant, S., Seyler, F., Rotunno Filho, O. C., Cochonneau, G., and Mansur, W. J. (2010). Water levels in the Amazon basin derived from the ERS 2 and ENVISAT radar altimetry missions. *Remote Sensing of Environment, 114*(10), 2160–2181. doi:10.1016/j.rse.2010.04.020.

Satge, F., Bonnet, M. P., Calmant, S., and Cretaux, J. F. (2015). Accuracy assessment of SRTM v4 and ASTER GDEM v2 over the 2 Altiplano's watershed using ICESat/GLAS data. *International Journal of Remote Sensing, 36,* 465–488. doi:10.1163/2352-0248_edn_a0251000.

Satge, F., Bonnet, M. P., Gosset, M., Molina, J., Lima, W., Zolá, P., Timouk, F. et al. (2015). Assessment of satellite rainfall products over the Andean plateau. *Atmospheric Research, 167,* 1–14. doi:10.1016/j.atmosres.2015.07.012.

Schmidt, R., Petrovic, S., Güntner, A., Barthelmes, F., Wünsch, J., and Kusche, J. (2008). Periodic components of water storage changes from GRACE and global hydrology models. *Journal of Geophysical Research: Solid Earth, 113*(8).

Schöngart, J., Wittmann, F., and Worbes, M. (2010). Biomass and net primary production of central Amazonian floodplain forests. In *Amazonian Floodplain Forests Ecophysiology Biodiversity and Sustainable Management* (Vol. 210, pp. 347–388). doi:10.1007/978-90-481-8725-6_18.

Schutz, B., Zwally, H. J., Shuman, C. A., Hancock, D., and DiMarzio, J. P. (2005). Overview of the ICESat mission. *Geophysical Research Letters* 32 (21), 1–4.

Seyler, F., Calmant, S., da Silva, J., Filizola, N., Roux, E., Cochonneau, G., Vouchel, P. et al. (2008). Monitoring water level in large transboundary ungauged basins with amtimetry: The example of Envisat over the Amazon basin. *Remote Sensing of Inland, Coastal, and Oceanic Waters,* p. 715017–715017. doi:10.1117/12.813258.

Seyler, F., Oliveira, J. De, Pfeffer, J., Santos, J., Leon, J. G., Frappart, F., … Bonnet, M. P. (2017). In S. Vignudelli and Kostianoy, A. G. (Eds.) *Inland Water Altimetry.* Berlin, Germany: Springer-Verlag.

Sioli, H. (1984). The Amazon and its main afluents: Hydrography, morphology of the river courses and river types. In *The Amazon Liminology and Landscape Ecology of a Mighty Tropical River and Its Basin,* pp. 127–165. Dordrecht, the Netherlands: Springer. doi:10.1007/978-94-009-6542-3_5.

Sippel, S. J., Hamilton, S. K., Melack, J. M., and Novo, E. (1998). Passive microwave observations of inundation area and the area/stage relation in the Amazon River floodplain. *International Journal of Remote Sensing, 19,* 3055–3074. doi:10.1080/014311698214181.

Soito, J. L. D. S., and Freitas, M. A. V. (2011). Amazon and the expansion of hydropower in Brazil: Vulnerability, impacts and possibilities for adaptation to global climate change. *Renewable and Sustainable Energy Reviews, 15*(6), 3165–3177. doi:10.1016/j.rser.2011.04.006.

Tachikawa, T., Kaku, M., Iwasaki, A., Gesch, D., Oimoen, M., Zhang, Z., Danielson, J J. et al. (2011). ASTER global digital elevation model version 2–Summary of validation results. *NASA Land Processes Distributed Active Archive Center,* 27. Retrieved from https://lpdaacaster.cr.usgs.gov/GDEM/Summary_GDEM2_validation_report_final.pdf.

Tadono, T., Nagai, H., Ishida, H., Oda, F., Naito, S., Minakawa, K., and Iwamoto, H. (2016). Generation of the 30 M-mesh global digital surface model by ALOS PRISM. *ISPRS - International Archives of the Photogrammetry, Remote Sensing and Spatial Information Sciences, XLI-B4,* 157–162. doi:10.5194/isprsarchives-XLI-B4-157-2016.

Tockner, K., and Stanford, J. A. (2002). Riverine flood plains: Present state and future trends. *Environmental Conservation, 29*(3), 308–330.

Toivonen, T., Mäki, S., and Kalliola, R. (2007). The riverscape of Western Amazonia–A quantitative approach to the fluvial biogeography of the region. *Journal of Biogeography, 34,* 1374–1387). doi:10.1111/j.1365-2699.2007.01741.x.

Trigg, M., Bates, P. D., Wilson, M., Schumann, G., and Baugh, C. (2012). Floodplain channel morphology and networks of the middle Amazon River. *Water Resources Research, 48*(October), 1–17. doi:10.1029/2012WR011888.

Trigg, M., Wilson, M. D., Bates, P. D., Horritt, M. S., Alsdorf, D., Forsberg, B. R., and Vega, M. C. (2009). Amazon flood wave hydraulics. *Journal of Hydrology, 374*(1–2), 92–105. doi:10.1016/j.jhydrol.2009.06.004.

Wehr, T. and Attema, E. (2001). Geophysical validation of Envisat data products. *Advances in Space Research, 28*(1), 83–91.

Wilks, D. S. (2006). *Statistical Methods in the Atmospheric Sciences.Meteorological Applications,* (Vol. 14). Amsterdam, the Netherlands: Elsevier Academic Press.

Wilson, M. D., Bates, P. D., Alsdorf, D., Forsberg, B. R., Horritt, M. S., Melack, J. M., Frappart, F. et al. (2007). Modeling large-scale inundation of Amazonian seasonally flooded wetlands. *Geophysical Research Letters, 34*(15), L15404. doi:10.1029/2007GL030156.

Wittmann, F., Anhuf, D., and Funk, W. J. (2002). Tree species distribution and community structure of central Amazonian várzea forests by remote-sensing techniques. *Journal of Tropical Ecology, 18*(2002), 805–820. doi:10.1017/S0266467402002523.

Woodcock, C., Allen, R., Anderson, M., Belward, A., Bindschadler, R., Cohen, W., and Gao, F. (2008). Free access to landsat imagery. *Science, 320, 1011.*

Yamazaki, D., Baugh, C., Bates, P. D., Kanae, S., Alsdorf, D., and Oki, T. (2012). Adjustment of a spaceborne DEM for use in floodplain hydrodynamic modeling. *Journal of Hydrology, 436–437*, 81–91. doi:10.1016/j.jhydrol.2012.02.045.

Yamazaki, D., De Almeida, G. A. M., and Bates, P. D. (2013). Improving computational efficiency in global river models by implementing the local inertial flow equation and a vector-based river network map. *Water Resources Research, 49*(11), 7221–7235. doi:10.1002/wrcr.20552.

Yamazaki, D., Kanae, S., Kim, H., and Oki, T. (2011). A physically based description of floodplain inundation dynamics in a global river routing model. *Water Resources Research, 47*(4). doi:10.1029/2010WR009726.

Yamazaki, D., Trigg, M. A., and Ikeshima, D. (2015). Development of a global ~90m water body map using multi-temporal Landsat images. *Remote Sensing of Environment, 171*, 337–351.

Yengoh, G. T., Dent, D., Olsson, L., Tengberg, A. E., and Tucker, C. J. (2014). The use of the Normalized Difference Vegetation Index (NDVI) to assess land degradation at multiple scales: A review of the current status, future trends, and practical considerations. *Lund University Center for Sustainability Studies (LUCSUS), and the Scientific and Technical Advisory Panel of the Global Environment Facility (STAP/GEF)*, 47. doi:10.1007/978-3-319-24112-8.

Zeng, N., Yoon, J.-H., Marengo, J. A., Subramaniam, A., Nobre, C. A., Mariotti, A., and Neelin, J. D. (2008). Causes and impacts of the 2005 Amazon drought. *Environmental Research Letters, 3*(1), 14002. doi:10.1088/1748-9326/3/1/014002.

Zhang, M. (2009). Satellite radar altimetry for inland hydrologic studies. Ohio State University. Division of Geodetic Science.

Zurita-Arthos, L. and Mulligan, M. (2013). Multi-criteria GIS analysis and geo-visualisation of the overlap of oil impacts and ecosystem services in the Western Amazon. *International Journal of Geoinformatics, 9*(2), 45–52.

Zwally, H. J., Schutz, B., Abdalati, W., Abshire, J., Bentley, C., Brenner, A., Bufton, J. et al. (2002). ICESat's laser measurements of polar ice, atmosphere, ocean, and land. *Journal of Geodynamics, 34*, 405–445. doi:10.1016/S0264-3707(02)00042-X.

17 Large-Scale Flood Monitoring in Monsoon Asia for Global Disaster Risk Reduction Using MODIS/EOS Data

Youngjoo Kwak

CONTENTS

17.1 INTRODUCTION

Natural disaster risk under climate change is an inevitable threat to sustainable development, as major flood disasters have been frequent in both developing and advanced countries. Causing widespread devastation, with massive economic damage and loss of human lives, flood disasters hamper economic growth and accelerate poverty, particularly in most developing countries. Globally, this trend will likely continue owing to an increase in flood magnitude and lack of preparedness for extreme events (World Bank 2013). Since the early twenty-first century, risk reduction of natural disasters has been globally recognized as a common goal and has been included in the eight Millennium Development Goals (MDGs) and the Sustainable Development Goals (SDGs) to adapt to climate change. The United Nations Office for Disaster Risk Reduction (UNISDR) (2015) reported SDGs that pertain to disaster risk reduction and action to strength at local, national, regional, and global levels in priority areas. In 2015, the Sendai Framework for Disaster Risk Reduction was adopted by 187 national governments and international organizations as the first of the post-2015 international agreements and as the basis for a risk-informed and resilient future. The Sendai Framework particularly emphasized the importance of "geospatial and space-based technologies and related services and maintaining and strengthening in-situ and remotely-sensed earth and climate observations" to support national measures for understanding disaster risk and successful disaster risk communication (UNISDR 2015). In line with these efforts, the monitors and governors of global river floods pay attention to international scientific and policy communities for their support to facilitate evidence-based policy making for disaster risk reduction.

Around 40% of the world's poor countries live on transboundary river-basin systems in South Asia. At the same time, many floodplains in this region have been experiencing a rising number of flood disasters. In particular, the Economic Social Commission for Asia and Pacific (ESCAP 2015) reported, based on data from the World Resources Institute (2015), that 10 riparian countries, where transboundary river-basin floods occur frequently, are disproportionally influenced by the impacts of large-scale flood disasters in the Asia-Pacific region (World Water Assessment Programme [WWAP] 2009). For instance, India has the largest number of flood-exposed population (approximately 4.84 million), followed by Bangladesh (approximately 3.48 million) and China (approximately 3.28 million) (Hanson et al. 2011, UNISDR 2015, ESCAP 2015). In terms of annual gross domestic product (GDP) affected by floods, Bangladesh has the highest percentage (4.75%) of the country's total GDP, followed by Cambodia (3.42%) and Afghanistan (2.58%) (Dilley et al. 2005, ESCAP 2015). Kwak (2012b) demonstrated that the Asia-Pacific region will be exposed to higher flood risks throughout the twenty-first century than ever before, because more extreme rainfall will lead to greater flood inundation depths in many areas. Figure 17.1 shows the maximum extent of inundation risk areas (blue-colored areas) in transboundary river basins and nationwide large flood areas affected by the maximum daily discharge of 50-year return period based on the Gumbel distribution from the hydrological block-wise use of TOPMODEL (BTOP) under the present climate conditions (1980–2004). Figure 17.1 also shows the representative transboundary river basins in

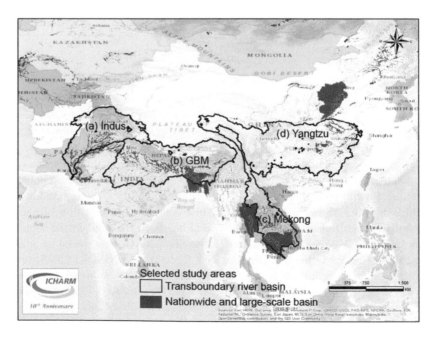

FIGURE 17.1 Maximum extent of potential flood-inundated area (blue color) in transboundary river basins (black line) and nationwide large flood areas (orange color) in Monsoon Asia.

the area enclosed by the black line. Therefore, the management of transboundary river basins is a very important driver for sustainable development and international water resource.

Flood detection and mapping are one of the traditional themes of satellite-based remote sensing, using optical images, such as Landsat (Flood Mapping, https://oas.gsfc.nasa.gov/Landsat) and the advanced very-high-resolution radiometer (AVHRR, the National Oceanic and Atmospheric Administration [NOAA]), and synthetic aperture radar (SAR) images, such as X-band TerraSAR-X, X-band Cosmo-SkyMed, C-band RADARSAT-2, and L-band ALOS/PALSAR-1 and -2 (Mason et al. 2010, Giustarini et al. 2013, Refice et al. 2014, Evans et al. 2010). Considering transboundary river-basin floods and nationwide flood monitoring, floodwater detection is a more specific area of challenge in relation to the selection of suitable sensors, and many methods have been developed (Alsdorf et al. 2003, Brakenridge et al. 2012, Verbesselt et al. 2012).

Satellite remote sensing has opened a new era to pursue global flood estimation, which is particularly important for hard-to-access, remote and transboundary areas. Since its launch in December 1999, the Moderate Resolution Imaging Spectroradiometer (MODIS) has been one of the main contributors to progress in near real-time (NRT) monitoring and global mapping. One of the greatest benefits of the MODIS instrument is its capability to broadcast raw data directly all over the world. In addition, it stores raw data for later download and has a wide swath range (approximately 2330 km), with good temporal and moderate spatial resolutions (250, 500, and 1000 m) (Pinheiro et al. 2007). Flood mapping systems utilizing MODIS multispectral sensors are now able to generate NRT flood maps, with a global coverage on a daily basis. The National Aeronautics and Space Administration (NASA) is also in charge of the Land, Atmosphere Near-Real-Time Capability for EOS (Earth Observing System) (LANCE) and supports application users interested in monitoring a wide variety of Earth's land surface. In 2011, the MODIS NRT Global Flood Mapping System was released as a 3-day product by NASA and the Dartmouth Flood Observatory (DFO). It uses a water detection algorithm based on a reflectance ratio of MODIS bands 1 and 2 and a threshold on band 7 to provisionally identify pixels as water (Brakenridge and Anderson 2006, DFO 2008).

A reliable flood monitoring system at a global scale is in great demand by a variety of national and international agencies for disaster risk reduction and management. Disaster risk reduction strategies and policies should also be developed to systematically incorporate a framework for the use of NRT satellite data, which offer significant advantages for quick response to emergency situations soon after a disaster. In January 2015, Japan's space policy was established as the new *Basic Plan for Space Policy* emphasizing the development and utilization of space, because Japan has been faced with increasing demands for safety and security, including recovery from the Great East Japan Earthquake (Cabinet Office [CAO] 2015). Remote-sensing technology should play a central role in this effort for more effective and accurate disaster risk reduction and restoration. Online flood mapping systems have been created by NASA, which not only provide fundamental observational information but also produce such maps with a rapid mapping technique (NASA 2007, 2015). These online data are downloadable for monitoring nationwide flood disasters all over the world, with high spatial and temporal resolution. Therefore, disaster managers and other end users will be able to monitor floods and evaluate larger-scale flood risk by accessing these flood maps and related products.

Remote-sensing-based index algorithms have been designed to detect surface water in a conceptually simple way, relying mainly on spectral indices, such as the Normalized Difference Water Index (NDWI) (McFeeters 1996, Gao 1996, Ji et al. 2009) and the Land Surface Water Index (LSWI) (Chandrasekar et al. 2010). With worldwide applicability, the MODIS sensor has three spectral bands that are sensitive to water and soil moisture: near-infrared (NIR, band 2: 841–876 nm) and shortwave infrared (SWIR, band 6: 1628–1652 nm, and SWIR, band 7: 2105–2155 nm). In order to acquire a better detection capability, spectral indices are necessary to detect surface water in a spectrally normalized way; for example, the Modified Land Surface Water Index (MLSWI), which was a new index developed from NDWI and LSWI, specifically for floodwater detection (Kwak et al. 2012a, 2014a).

This study focuses on large-scale flood monitoring during Asia's monsoon season by using EOS/MODIS data. Although it is obviously hard to identify distributed floodwaters, the ultimate goal of this study is to instantly produce a nationwide flood map, while maximizing the utility of MODIS time series data for the spatial and temporal dynamics of inundation areas, covering the whole of the Asia as well as all global surface for disaster risk reduction. Therefore, we developed an improved algorithm of the MODIS-derived water index based on a reflectance ratio of MODIS bands 2 and 7 for floodwater detection immediately after a disaster, and then, we confirmed the performance of the water detection algorithm even in different conditions such as flood time, flood duration, and topography.

17.2 DATA USED

17.2.1 MODIS

The MODIS instruments, onboard NASA's EOS/Terra and Aqua satellites, provide twice-daily near-global coverage, with differences in Terra's and Aqua's orbits resulting in different viewing and cloud-cover conditions. The Terra equatorial overpass is at 10:30 a.m. local solar time and Aqua is at 1:30 p.m. The frequent acquisition of the MODIS onboard the Aqua and Terra satellite platforms enables efficient monitoring of the seasonal change of land cover such as water bodies and vegetation.

For nationwide and large-scale NRT flood mapping, MODIS is the most suitable for detecting floodwater because of high frequent acquisition, high observation coverage, low view angle, the absence of clouds or cloud shadow, and aerosol loading at 500 m spatial resolution (the U.S. Geological Survey [USGS] LPDAAC 2012, NASA 2015). Recently, Kwak et al. (2015) conducted a feasibility study on the development of a rapid flood mapping system in case of Bangladesh by using MODIS products and level-three 8-day composite surface reflectance products in the sinusoidal projections, MOD09A1 (Terra) and MYD09A1 (Aqua), respectively. Both studies used the same products as the best observations during an 8-day period. The quality of the MOD09A1 products was checked for cloud cover in reference to the quality layer of the Quality Assessment (QA) Science Datasets (SDS) as quality indicator. The images were acquired from the five case studies of representational catastrophic floods in Asia from 2010 to 2016 (Figure 17.1): the Indus River flood of Pakistan in 2010, the Chao River flood of Thailand in 2011, the Mekong River flood of Cambodia in 2011, the Ganges-Brahmaputra-Meghna (GBM) River flood of Bangladesh in 2015, and the Hai River flood of China in 2016. Table 17.1 shows the list of MODIS data from data pool of USGS-NASA Land Processes Distributed Active Archive Center (USGS LPDAAC 2012) (https://lpdaac.usgs.gov/data_access/data_pool).

TABLE 17.1
MODIS Data (MOD09A1: Terra and MYD09A1: Aqua)

River Basins	Countries	Scale	Observed Date	Tile
Indus	Pakistan	Transboundary	August 5–12, 2010 and August 21–28, 2010	h23-24v05-06
GBM	Bangladesh	Transboundary	August 5–12, 2007, September 6–13, 2015 and September 14–21, 2015	h26v06
Chao Phraya	Thailand	Nationwide	October 24–31, 2011	h27v07
Mekong	Cambodia	Transboundary	September 22–30, 2011	h28v07
Hai	China	Nationwide	July 19–26, 2016	h26-27v05

17.2.2 ALOS OBSERVATION

The Advanced Land Observing Satellite (ALOS) was launched by the Japan Aerospace Exploration Agency (JAXA) in 2006, and it carries three remote-sensing instruments: an L-band Polarimetric Synthetic Aperture Radar (PALSAR), an along-track 2.5 m Panchromatic Resolution Stereo Mapper (PRISM), and an Advanced Visible and Near-Infrared Radiometer, type 2 (AVNIR-2). The ALOS AVNIR-2 is a visible and NIR radiometer with four bands for observing land and coastal zones. The data types of AVNIR-2 and PALSAR are 8- and 16-bit unsigned integers and have spatial resolutions of 10 m and 12.5 m, with absolute accuracy of 19.8 m and 0.76 dB, respectively (Tadono et al. 2009, Shimada et al. 2009, JAXA 2011). The AVNIR-2 and PALSAR images from ALOS in this study were employed in order to verify the detection of surface-water products, including floodwater from MODIS-derived MLSWI. The PALSAR level 1.5 images, which were acquired on August 5, 2010, were captured around the conjunction area between the Indus and Kabul Rivers of Pakistan and used for validation in a pixel-based comparison after image classification with the flood areas extracted by a MODIS-derived hybrid detection algorithm. The AVNIR-2 images, captured on December 23, 2006 (preflood time), and August 10, 2007 (flood time), were also used for validation in the Brahmaputra River flood areas of the Sirajganj district (approximately 2480 km^2), Bangladesh.

17.2.3 ELEVATION DATA

The Shuttle Radar Topography Mission (SRTM) digital elevation model (DEM) data (DEM 15 s, with a 450 m spatial resolution, 1 m vertical resolution, and 6.2 m absolute height error in Eurasia) is an international project spearheaded by the National Geospatial-Intelligence Agency (NGA) and NASA (Rodriguez et al. 2005, Farr et al. 2007). The void-filled DEM in this research was acquired from Hydrological data based on the SHuttle Elevation Derivatives at multiple Scales (HydroSHEDS), originating from a combination of the SRTM-3 and Digital Terrain Elevation Data (DTED®)-1 for regional and global-scale applications (Lehner et al. 2006). As a reference data, we also used the river network data from HydroSHEDS for confirming the detected major river. In this study, DEM data were used to compensate for the weakness of detecting spatial flood areas at 15 arc-second spatial resolution (approximately 500 m at the equator) and 1 m vertical resolution.

17.3 METHODOLOGY

17.3.1 NATIONWIDE EMERGENCY FLOOD MONITORING FRAMEWORK

For effective nationwide flood mapping in a transboundary river basin, a framework of emergency activities is essential. We applied a framework of national environment monitoring, focusing on disaster risk response, which was recently developed by the Ministry of Land, Infrastructure, Transport, and Tourism of Japan. In this study, we developed a technical approach to assist in selecting and providing risk information as part of the new framework (Figure 17.2). For this technical approach, we invented hybrid floodwater detection algorithm for flood inundation mapping, which eliminates complexity and ambiguity of spectral characteristics, so that hazard maps will be easy to understand for decision makers of national and local governments, for example, in differentiating flooded from non-flooded areas. We selected MODIS data to ensure a nationwide coverage of emergency flood detection in this framework.

17.3.2 SPECTRAL CHARACTERISTICS OF FLOODWATER

The reflectance characteristics of land covers become complicated during flooding due to mixture of land types. In particular, turbid water albedo increases significantly during flooding, with a maximum reflectance peak moving toward band 1 (red: 620–670 nm) because silt and debris are

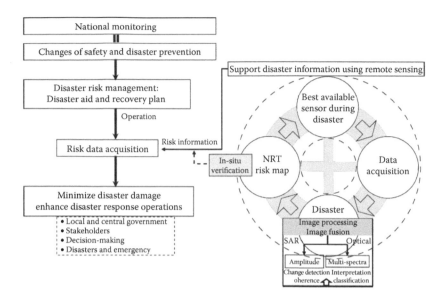

FIGURE 17.2 Conceptual framework of national flood monitoring for supporting emergency activities.

concentrated in the water; in contrast, an increase in soil moisture causes a decrease in soil albedo. To discriminate water from other types of land cover, that is, soil and vegetation, the reflectance rate of band 2 (NIR) increases when clean water becomes turbid during a flood. Consequently, the reflectance rates of bands 6 and 7 (SWIR) are lower than those of the other bands in the case of some types of surface water, such as clean water, muddy water, and turbid water (Kwak et al. 2014a, 2015). Based on the theory of experimental reflectance characteristics, NIR and SWIR can be used to devise new floodwater indices. Therefore, the design of a spectral floodwater index is based on the fact that water absorbs energy at NIR (841–876 nm) and SWIR (1628–1652 nm and 2105–2155 nm) wavelengths.

17.3.3 Hybrid Floodwater Detection Algorithm

In this study, we developed a hybrid floodwater detection method with a *tracking* algorithm, sequentially using MLSWI and SRTM DEM. This algorithm is designed to track the trajectory of floodwater flowing in eight directions to adjacent pixels with the same or lower DEM. The MLSWI from the MODIS bands was applied to floodwater detection after analyzing inland water characteristics, with a focus on the sensitivity of water indices. The MLSWI combining bands 2 and 7 was validated and compared with two water indices of $NDWI_{1,6}$ $(\rho_{Red} - \rho_{SWIR})/(\rho_{Red} + \rho_{SWIR})$ and $LSWI_{2,6}$ $(\rho_{NIR} - \rho_{SWIR})/(\rho_{NIR} + \rho_{SWIR})$. An equation of "1–NIR" to emphasize the effect of changes in NIR used to derive MLSWI from the MODIS bands is as follows:

$$MLSWI_{2,7} = \frac{(1 - \rho_{NIR}) - \rho_{SWIR}}{(1 - \rho_{NIR}) + \rho_{SWIR}} \qquad (17.1)$$

Where, ρ_{NIR} and ρ_{SWIR} are atmospherically corrected surface reflectance for their respective MODIS bands: band 1 (red: 620–670 nm), band 2 (NIR: 841–876 nm), and bands 6 and 7 (SWIR: 1628–1652 nm and 2105–2155 nm, respectively).

After detecting floodwater by means of selected MLSWI, a hybrid floodwater detection algorithm (Kwak et al. 2014b) was applied to each pixel to calculate the floodwater boundary, based

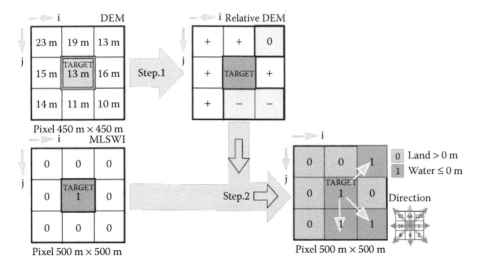

FIGURE 17.3   Schematic tracking algorithm of floodwater from the the MLSWI and DEM.

on the difference in height between a flooded pixel and a non-flooded pixel. Figure 17.3 illustrates how to detect flooded pixels around the target pixel. Floodwater was assumed to flow in eight directions to any adjacent pixel with the same or lower DEM; that is, if floodwater_DEM $\leq$ 0 m, the adjacent pixel is given the value of 1 as a flooded pixel. The MLSWI alone is not complete in detecting floodwater from a mixture of water and vegetation within a moderate spatial resolution of 500 m. In addition, it is not sufficiently capable of accurately detecting floodwater over a large area in its solo application. In order to compensate for these limitations of the MLSWI, the hybrid floodwater detection algorithm was applied to nationwide flood mapping as a supplementing approach by using the eight-direction flow scheme, solely based on remotely sensed data sources via open internet, that is, DEM and MODIS. This hybrid floodwater detection algorithm made it possible to identify flood areas more accurately than the single use of the MLSWI or any indicator.

17.3.4 Verification of Floodwater from ALOS Observation and Hydroinformation

For validating the hybrid floodwater detection algorithm coupled with MLSWI and DEM, ambiguities of flooded pixels were examined by using high-spatial-resolution ALOS images, AVNIR-2 (10 m) and PALSAR (15 m), and observed data from the representative gauge stations (point data) along the Brahmaputra River of Bangladesh and the Indus River of Pakistan, respectively. Cross-validation using high-spatial-resolution ALOS images and ground-based water levels was performed in order to estimate the optimal threshold values of the MLSWI and to identify flooded areas more accurately.

First, the AVNIR-2 (10 m) and PALSAR (15 m) were used to verify floodwater and non-flooded areas in the Sirajganj district along the Brahmaputra River of Bangladesh and near the Chashma Barrage on the Indus River in Mianwali District of Punjab Province of Pakistan. The floodwater pixels of the MLSWI were confirmed by a comparison with the water pixels from Normalized Difference Vegetation Index (NDVI_2,1 = (ro_Red − ro_NIR)/(ro_Red + ro_NIR)) threshold-based classification, focusing on NIR (band 4: 760–890 nm) of AVNIR-2, and by a comparison with the water pixels from backscattering threshold-based classification (−13 to −15 dB), focusing on water scattering of single L-band PALSAR image. For evaluating floodwater map products, Kappa coefficient (0.0 < K $\leq$ 1.0, Landis and Koch 1977, Smeeton 1985) was used for comparison of inundated areas (Kwak et al. 2015). It is an index to estimate the degree of agreement; for example, when the value is less than 1.0, it implies that the agreement is not perfect between two rasters:

$$K = \frac{P_0 - P_e}{1 - P_e} \tag{17.2}$$

$$P_e = \frac{(a_1 \times b_1 + a_0 \times b_0)}{n \times n} \tag{17.3}$$

Where, P_0 is the overall accuracy, which is the ratio of matched pixels; P_e stands for the probability of random agreement, including both inundated (a_1 and b_1) and non-inundated (a_0 and b_0) pixels in MODIS-derived MLSWI and ALOS data, respectively; and n is the total number of compared pixels.

Next, flooded and non-flooded areas were cross-validated with hydrological data from river water gauging stations. Such data are crucial in cross-validation, because remarkable changes are usually observed in the extent of detected floodwater from the time when the water level exceeded the water danger level or the flood reached the danger level. In this respect, hydrological gauging stations play an important role in recording current maximum water levels and flood stages in monitoring river flood situations. Water level data were collected at the gauge stations of the Bangladesh Water Development Board (BWDB) and Pakistan Meteorological Department (PMD) during the flood season from June to September in 2007 and 2010, respectively.

17.4 RESULTS AND VALIDATION

17.4.1 VALIDATED FLOOD AREAS

Figure 17.4 presents inundation extent maps, with spatial distributions for two cases of flooding. The pixel-based classification of the MLSWI, generated by MODIS images from MLSWI 2 and 7

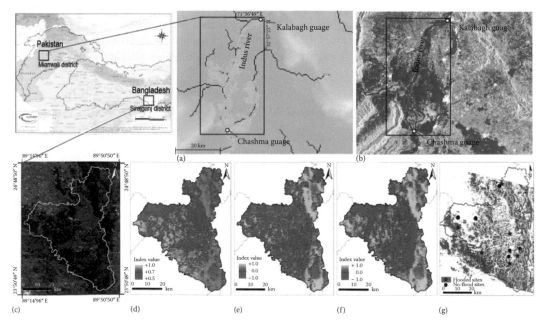

FIGURE 17.4 Maximum extent inundation map: (a) from the MLSWI (threshold = 0.85) (August 5–12, 2010) (floodwater in blue pixels), and (b) a comparison of the MLSWI with ALOS PALSAR (cyan pixels) in the Indus River, Pakistan, and in-time flooding situation from (c) composite RGB images of MODIS on August 5–12, 2007, (d) MODIS-derived MLSWI, (e) LSWI, (f) NDWI, and (g) NDVI threshold classification of ALOS AVNIR-2 in the Sirajganj district, respectively.

TABLE 17.2

Hydrological Data during the Floods at Representative Gauge Stations

		Flood Peak Events			
Basin	Station	Water Level (MSL, meter)	Danger Level (MSL, meter)	Flood Travel Time	Date
Indus, Pakistan	Kalabagh	211.53	198.12	N/A	July 30, 2010
Indus, Pakistan	Chashma	197.51	198.12	48 h (2 days)/60 km	August 01, 2010
Brahmaputra, Bangladesh	Sirajganj	14.95	13.35	99 h (4 days)/140 km	August 02, 2007

Source: Data from PMD (2011) and Bangladesh Water Development Board (BWDB), Flood Forecasting and Warning Centre (FFWC), Annual flood report 2015. Available: http://www.ffwc.gov.bd (accessed February 20, 2016), 2015.

(Figure 17.4d), was confirmed by comparisons to be superior to the other water indices of the LSWI (Figure 17.4e) and the NDWI (Figure 17.4f) during 5–12 August 2007.

The flooded areas were also verified by comparing MODIS-derived results with ALOS AVNIR-2 captured on August 10, 2007 (Figure 17.4g), and PALSAR data captured on August 5, 2010 (Figure 17.4b). Although the flooded areas smaller than a resolution of 500 m were not detected accurately in the case of MODIS, the MLSWI products are in moderate agreement with the high-spatial-resolution ALOS AVNIR-2 (10 m) and PALSAR images (10 and 15 m spatial resolution, respectively) for estimating the most vulnerable areas near the major river. The overall accuracy of 78% was achieved with a Kappa coefficient (K) of 0.57, confirming a moderate agreement due to a different water fraction caused by the different spatial resolution, especially in a mixed area. In addition, the accuracy of floodwater contained some errors originating from the accuracy of DEM, such as overestimated height of the terrain due to gaps of forests, buildings, and other artificial structures.

In addition, Table 17.2 shows a clear relationship between the peak water level (exceeding the danger level) and the detected flooded areas during the floods at the representative gauge stations in both countries. We clearly confirmed that flood propagation was in good agreement with the timing of the water level exceeding the water danger level, for example, which is 13.35 m above mean sea level (MSL, meter) in the case of the 2007 flood event at the Sirajganj station. The hybrid floodwater detection algorithm confirmed that the total inundated area also increased at the peak of water level on August 1, 2007, as the water level exceeded the danger level on July 19, 2007. These validations appear consistent with the maximum flood areas (August 5–12) in Figure 17.4d, from the hybrid floodwater detection algorithm coupled with MLSWI and DEM after the peak water level on August 1, 2007, and show the universal superiority of floodwater detection.

17.4.2 Nationwide Flood Mapping from the MODIS/EOS Observation

This study focused on transboundary river-basin floods and used a nationwide comprehensive approach to detect floodwater for flood mapping, based on characterization of flood detection indices. We estimated flooded areas by using an optimal threshold of the MLSWI (red-colored pixels in Figure 17.5, ranging between 0.75 and 0.85 in the five cases) and then improved the accuracy of detecting flooded areas by using a hybrid floodwater detection algorithm (blue-colored pixels in Figure 17.5a and 5c). The variation of optimal threshold was sensitive to the water fraction of mixture pixels covered with water, vegetation, and soil. It was difficult to standardize the optimal threshold for all cases in a single formula, and we found that the dependence on the spectral characteristics of land surface was the main reason why the value of the MLSWI fluctuated with mixture pixels during a flood. The selected case studies of flood mapping represented the MLSWI's variation divided into two threshold groups: an optimal threshold of 0.75 resulted for Chao Phraya River

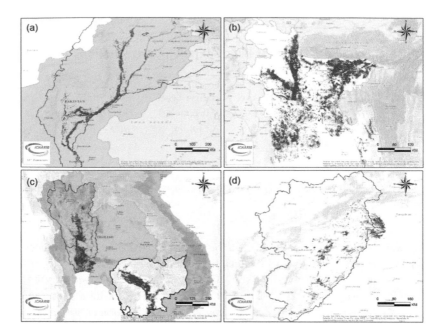

FIGURE 17.5 Nationwide flood map from the hybrid floodwater detection algorithm, coupled with the MLSWI and DEM in the five selected flood cases. (a) An optimal threshold of 0.85 for the Indus River of Pakistan flood during August 22–28, 2010; (b) GBM of Bangladesh flood during September 6–21, 2015; (c) the Chao Phraya River flood during October 24–31, 2011, and the Lower Mekong River flood during September 22–30, 2011; and (d) the Hai River flood during July 19–26, 2016; flooded areas from the MLSWI (red-colored pixels) and improved flooded areas from the hybrid algorithm (blue-colored pixels).

of Thailand, Mekong River of Cambodia, and Hai River of China, and 0.85 resulted for the Indus River of Pakistan and GBM River of Bangladesh.

The two groups can be characterized from spectral and topographical aspects of the MLSWI. First, in terms of spectral characteristics, the more soil and vegetation a pixel contains, the further the optimal threshold of the MLSWI decreases during a flood. Next, in terms of topographical characteristics, the value of the optimal threshold of the MLSWI was larger in lowland floodplains than in alluvial fan floodplains near mountainous areas, because of land surface reflectance from soil and forests. A typical example is Cambodian lowland as part of the Mekong Delta, most of which is used for agriculture. Excluding microtopographic patterns, the area can be considered as nearly uniform land in the floodplain. Therefore, floodwater can be detected better during a flood at 0.75 in the floodplain rather than at 0.85, which is the optimal threshold for alluvial fan floodplains near mountainous areas.

As a result, the maximum inundation areas covered at least 18,000 km² (12% of the total area of Pakistan, transboundary river flood) during the Indus River flood in 2010, at least 29,900 km² (32% of the total area of Bangladesh, transboundary river flood) during the Bangladesh flood in the GBM River in 2015, at least 4,000 km² (nationwide flood) during the Chao Phraya River flood in 2011, at least 23,000 km² (12% of the total area of Cambodia, transboundary river flood) during the Lower Mekong River flood in 2011, and at least 1,700 km² (nationwide flood) during the Hai River flood in 2016.

Improved up-to-date nationwide flood maps are in great demand from relevant government ministries and agencies right after flood disasters in developing countries. Although they have an obligation to do their best to collect risk information from all sectors, it is obviously hard to identify the distribution and location of flood-related damage in temporal and spatial distribution after flooding at local, national, and transboundary levels. In order to improve satellite-detected flood mapping at national or larger scales, data limitation is the main difficulty, and it needs to be resolved by applying different approaches, including some using a single sensor, while others using multisensors via advanced international disasters charts such as Sentinel Asia Constellation. At the same time, a new algorithm of water detection should be developed for image fusion-based flood change detection from SAR and optical multispectrometers, because open surface water and turbid water, including muddy variations, are hard to discriminate clearly due to weather conditions, especially clouds over mixed areas during a flood. It is also important to verify satellite-detected flood areas with ground-based truth and in situ data from intergovernmental collaboration, since floods can cause not only domestic issues but also international issues involving transboundry river basins. Despite these limitations, the flood mapping introduced in this study can be a globally applicable approach for any scale of countries exposed to damage and risk from extreme river floods to improve their emergency response framework and data availability.

17.5   CONCLUSIONS

The study found that the hybrid floodwater detection algorithm coupled with the MLSWI and DEM can be a comprehensive and universal approach in the framework for nationwide emergency flood monitoring and that it will be a valuable tool to risk managers and decision makers, despite the limitation of data availability. We confirmed that this hybrid detection algorithm can identify floodwater, and thus flood hazard areas, in the application of the monsoon Asia. The main conclusions of the study can be summarized as follows.

First, an improved floodwater index, the MLSWI, should be applicable to large flood detection in any floodplain over the monsoon Asia and reduce ambiguity in detecting floodwaters. In particular, SWIR (2105–2155 nm) reflectance is more stable to discriminate inland surface water from land cover as an optimal water indicator. Although the hybrid floodwater detection algorithm is a straightforward methodology, we found that the MLSWI can directly detect floodwaters from the reflectance of multitemporal MODIS during flood events. The nationwide flood mapping was evaluated and validated through pixel-to-pixel comparison with observed high-spatial-resolution satellite images (10~15 m) and recorded hydrological data. We confirmed that these results were reasonable and acceptable as a flood hazard map after comparing them with the results of the observed data.

Second, we confirmed that nationwide flood maps can be quickly provided as a low-cost solution to stakeholders, governments at all levels, policy makers, and administrative agencies in various relevant sectors of their countries. This type of map will play an important role in providing disaster information for effective emergency response in the early stage of a flood disaster before obtaining validated data at the regional and local levels. The balance between accuracy and timeliness should be taken into account in detecting and estimating floodwater and its extent, because emergency task forces need as much disaster information as possible as soon as possible to better understand actual flood situation.

The next challenge is to develop broadly scalable products and contribute a practical tool for an NRT flood monitoring system with time-series analysis, while overcoming the lack of data and limitations, such as weather problems, spatial resolution considering a mixed-water pixel, floodwater depth, and evidence-based comparable data. In this respect, it will be one of the appropriate solutions to adopt multitemporal change detection and water fraction calculation based on a multispectral linear mixture approach to improve the accuracy of floodwater detection in a pixel of MODIS data. Future work will continue to contribute to global flood monitoring as well as to NRT flood risk mapping, for building better emergency response and economic development as an innovative solution.

ACKNOWLEDGMENTS

We acknowledge the use of data products or imagery from the Land, Atmosphere Near real-time Capability for EOS (LANCE) system operated by the NASA/GSFC/Earth Science Data and Information System (ESDIS) with funding provided by NASA/HQ. This work was supported by the Japan Society for the Promotion of Science (JSPS) KAKENHI Grant-in-Aid for Scientific Research B: 15H05136.

REFERENCES

Alsdorf, D., Lettenmaier, D., Vorosmarty, C. 2003. The need for global, satellite based observations of terrestrial surface waters. *Eos Trans. Am. Geophys. Union.* 84:275–276.

Bangladesh Water Development Board (BWDB), Flood Forecasting and Warning Centre (FFWC), Annual flood report 2015. Available: http://www.ffwc.gov.bd (accessed February 20, 2016).

Brakenridge, G.R., Anderson, E. 2006. MODIS-based flood detection, mapping, and measurement: The potential for operational hydrological applications. In: Marsalek, J. et al. (Eds.) *Transboundary Floods: Reducing the Risks Through Flood Management*, Dordrecht, the Netherlands: Springer-Verlag, 16 p.

Brakenridge, G.R., Syvitski, J.P.M., Overeem, I., Stewart-Moore, J.A., Kettner, A.J., Westerhoff, R. 2012. Global mapping of storm surges, 2002-present and the assessment of coastal vulnerability. *Nat. Hazards* 66:1295–1312.

Cabinet Office (CAO). 2015 Established by Strategic Headquarters for Space Policy, Government of Japan, Basic Plan on Space Policy. http://www8.cao.go.jp/space/plan/plan-eng.pdf (accessed November 5, 2015).

Chandrasekar, K., Sesha Sai, M., Roy, P., Dwevedi, R. 2010. Land Surface Water Index (LSWI) response to rainfall and NDVI using the MODIS Vegetation Index product. *Int. J. Remote Sens.* 31:3987–4005.

Dartmouth Flood Observatory (DFO). 2012. *Dartmouth Atlas of Global Flood Hazard*; 2008. Available: http://www.dartmouth.edu/~floods/index.html (accessed April 15, 2012).

Dilley M. et al., 2005. *Natural Disaster Hotspots: A Global Risk Analysis*. Washington, DC: World Bank.

Evans T. L. et al., 2010. Using ALOS/PALSAR and RADARSAT-2 to Map Land Cover and Seasonal Inundation in the Brazilian Pantanal, *IEEE J. Sel. Top. Appl. Earth Obs. Remote Sens.* 3(4):560–575.

Farr, T. G. et al. 2007. The shuttle radar topography mission. *Rev. Geophys.* 45, RG2004, doi:10.1029/2005 RG000183.

Gao, B. 1996. NDWI—A normalized difference water index for remote sensing of vegetation liquid water from space. *Remote Sens. Environ.* 58:257–266.

Giustarini L. et al., 2013. A change detection approach to flood mapping in urban areas using TerraSAR-X. *IEEE Trans. Geosci. Remote Sens.* 51(4):2417–2430.

Hanson S. et al., 2011. A global ranking of port cities with high exposure to climate extremes. *Clim. Change* 104:89–111.

Japan Aerospace Exploration Agency (JAXA). The Advanced Land Observing Satellite "DAICHI" (ALOS), Tokyo, Japan, 2011. Available: http://global.jaxa.jp/projects/sat/alos/ (accessed on December 20, 2014).

Ji, L., Zhang, L., Wylie, B. 2009. Analysis of dynamic thresholds for the normalized difference water index. *Photogramm. Eng. Remote Sens.* 75:1307–1317.

Kwak, Y., Arifuzzanman, B., Iwami, Y. 2015. Prompt proxy mapping of flood damaged rice fields using MODIS-derived indices. *Rem. Sens.* 7(12):15969–15988.

Kwak, Y., Park, J., Yorozuya, A., Fukami, K. 2012a. Estimation of flood volume in Chao Phraya river basin: Thailand from MODIS images coupled with flood Inundation level. *Intern. Geosci. Remote Sens. Sympo. (IGARSS)*, 2012 IEEE International, 887–890.

Kwak, Y., Park, K., Fukami, J. 2014a. Near real-time flood volume estimating from MODIS time-series imagery in the Indus River basin. *IEEE J. Sel. Top. Appl. Earth Obs. Remote Sens.* 7:578–586.

Kwak, Y., Park, K., Fukami, J. 2014b. Estimating a floodwater from MODIS time series and SRTM DEM data. *Artificial Life Robotics* 19(1):95–102.

Kwak, Y., Takeuchi, K., Fukami, J., Magome, J. 2012b. A new approach to flood risk assessment in Asia-Pacific region based on MRI-AGCM outputs. *Hydrol. Res. Lett.* 6:55–60.

Landis, J., Koch, G. 1977. The measurement of observer agreement for categorical data. *Biometrics* 33:159–174.

Lehner, B., Verdin, K., Jarvis, A. HydroSHEDS, Technical Documentation V1.0, World Wildlife Fund US, 2006. Available: http://hydrosheds.cr.usgs.gov/ (accessed December 15, 2012).

Mason D. C. et al., 2010. Flood detection in urban areas using TerraSAR-X, *IEEE Trans. Geosci. Remote Sens.* 48(2):882–894.

McFeeters, S. K. 1996. The use of the Normalized Difference Water Index (NDWI) in the delineation of open water features. *Int. J. Remote Sens.* 17:1425–1432.

National Aeronautics and Space Administration (NASA). Flooding in Bangladesh from the MODIS Rapid Response System, 2007. Available: http://earthobservatory.nasa.gov/NaturalHazards/view.php?id=18492& eocn=image&eoci=related_image (accessed January 10, 2014).

National Aeronautics and Space Administration (NASA). Near Real-Time (NRT) Global MODIS Flood Mapping, January 2015. Available: http://oas.gsfc.nasa.gov/floodmap/ (accessed January 3, 2015).

The Pakistan Meteorological Department (PMD). Flood Forecasting Division, 2011. Available: http://ffd.pmd. gov.pk/cp/floodpage.htm (accessed January 31, 2011).

Pinheiro, A. C. T., Descloitres, J., Privette, J. L., Susskind, J., Iredell, L.; Schmaltz, J. 2007. Near-real time retrievals of land surface temperature within the MODIS rapid response system. *Remote Sens. Environ.* 106:326–336.

Refice A. et al., 2014. SAR and InSAR for flood monitoring: Examples with COSMO-SkyMed data. *IEEE J. Sel. Top. Appl. Earth Obs. Remote Sens.* 7(7):2711–2722.

Rodriguez, E., C.S. Morris, J.E. Belz, E.C. Chapin, J.M. Martin, W. Daffer, S. Hensley, 2005. An assessment of the SRTM topographic products, Technical Report JPL D-31639, Jet Propulsion Laboratory, Pasadena, California, 143 pp.

Shimada, M., Isoguchi, O., Tadono, T., and Isono K. 2009. PALSAR radiometric and geometric calibration. *IEEE Trans. Geosci. Remote Sens.* 47(12):3915–3932.

Smeeton N. C. 1985. Early history of the kappa statistic. *Biometrics* 41:795.

Tadono,T., Shimada, M., Murakami, H., and Takaku J. 2009. Calibration of PRISM and AVNIR-2 Onboard ALOS Daichi. *IEEE Trans. Geosci. Remote Sens.* 47(12):4042–4050.

UN Economic and Social Commission for Asia and the Pacific (ESCAP). 2015. Disasters in Asia and the Pacific: 2015 Year in Review, http://www.unescap.org/our-work/ict-disaster-risk-reduction (accessed June 21, 2016).

UN Office for Disaster Risk Reduction (UNISDR). 2015. Sendai framework for disaster risk reduction 2015–2030. Geneva, Switzerland: World Conference on Disaster Risk Reduction in Sendai of Japan, Scientific and Technical Advisory Group. http://www.unisdr.org/we/inform/publications/43291 (accessed May 3, 2015).

The UN Office for Disaster Risk Reduction (UNISDR) and CRED, 2015. The human cost of weather related disasters 1995–2015, p.30 The Centre for Research on the Epidemiology of Disasters (CRED)'s Emergency Events Database (EM-DAT).

United Nations Office for Disaster Risk Reduction (UNISDR). Living with risk: A global review of disaster reduction. New York: UNISDR 2004.

US Geological Survey (USGS). Land Processes Distributed Active Archive Center. Available: https://lpdaac. usgs.gov/ (accessed July 30, 2012).

Verbesselt, J., Zeileis, A., Herold, M. 2012. Near real-time disturbance detection using satellite image time series. *Remote Sens. Environ.* 123:98–108.

World Bank. World development report 2014: Risk and opportunity, managing risk for development, World Bank: Washington, DC 2013.

World Water Assessment Programme (WWAP). 2009. The united nations world water development report 3: Water in a changing world. Paris, France: UNESCO Publishing, 2009. Available: http://www.unesco. org/water/wwap/wwdr/wwdr3/ (accessed October 20, 2014).

18 Introducing Flood Susceptibility Index Using Remote-Sensing Data and Geographic Information Systems

Empirical Analysis in Sperchios River Basin, Greece

Nikolaos Stathopoulos, Kleomenis Kalogeropoulos,
Christos Polykretis, Panagiotis Skrimizeas, Panagiota
Louka, Efthimios Karymbalis, and Christos Chalkias

CONTENTS

18.1   INTRODUCTION

Natural disasters have important impacts in many countries worldwide, with a large number of deaths, destructions in technical works and infrastructure, and dislocations of population. Moreover, due to the significant effects of climate change, these impacts are expected to rise in the upcoming years for a lot of countries. Although, technology and science have developed significantly nowadays, natural disasters keep having disastrous economic, environmental, and human results at a global scale. The study of these phenomena is constant during the last decades and presents augmentation trends (Smith and Petley, 2009). In order to manage adequately their effects, it is of high importance to synthesize risk and hazard maps for both natural and artificial environments (Maantay and Maroko, 2009; Smith, 2014). Flood phenomena are the most devastating and severe among natural disasters and affect more than 75 million people globally (Smith, 2001). In extreme flood events, it is important to manage rapidly the magnitude of the flood effects and land uses covered with water (Wang, 2004). Flood mapping is a useful tool for improving short- and long-term assistance in the affected areas directly after the event. The flood danger/risk can be defined at many scales, from universal to local ones.

In the beginning of the 1980s, geographers argued that flood maps do not influence significantly people's perception on floods and at the same time do not offer a satisfactory research design. Based on the Canadian Flood Damage Reduction Program, even though flood awareness has risen among the post-map groups, it is not due to the maps (Handmer, 1980; Handmer and Milne, 1981). Many methods were tested in an attempt to improve this deficiency. Use of remote sensing (RS) is not a new idea. In 1980, a research discussed the application of RS for disaster warning and procedures such as flood mapping; in addition, flood assessment was included (Deekshatulu et al. 1980). Twelve years later, Hubert-Moy et al. (1992) suggested that Landsat Thematic Mapper (TM) spatial resolution is compatible with small-sized catchment areas. This study also supported the fact that image processing and flow analysis can simulate the flood conditions several days after the peak of the event, even though the repeatability of satellites is low.

The usefulness of synthetic aperture radar (SAR) data was addressed within a research in 1996. This study managed to compare satellite data on river flooding, with photographic records (obtained from a light aircraft) aiming to demonstrate the accuracy of the SAR technique (Biggin and Blyth, 1996). In the United Arab Emirates (U.A.E.), in 2002, in an attempt to deal with erosion caused by flash floods, coupling of RS data and GIS techniques was used (Torab, 2002). Light Detection and Ranging (LIDAR) data were processed in 2004 to identify coastal impacts and sea-level rise due to climate change, on Prince Edward Island (PEI). In this study, Webster et al. (2004) demonstrated that validated digital elevation models (DEMs) derived from airborne LIDAR data can be used efficiently for mapping flood risk and hazard zones in coastal areas.

Pelletier et al. (2005) attempted an integrated approach, combining four different but complementary methods: (a) two-dimensional raster-based hydraulic modeling, (b) change detection from satellite images, (c) flood inundation field-based mapping, and (d) surficial geologic mapping. Each method had different spatial detail but provided important information for a complete and sufficient assessment, thus leading to the creation of a probabilistic flood hazard map by modeling several events. In the same year, the North Atlantic Treaty Organization (NATO) developed a project for monitoring extreme flood events in Romania and Hungary. For improving the existing local operational flood hazard assessment and monitoring, Earth observation (EO) data were used in the project. Optical and microwave data acquired from U.S. DMSP/Quikscat, RADARSAT, LANDSAT-7/TM, EOS-AM TERRA/MODIS and ASTER satellite sensors were used to extract information such as accurate updated digital maps of the hydrographical network, land cover/land use, extent of the flooded areas, multitemporal maps concerning the flood dynamics, hazard maps for the flooded areas and the affected zones, and so on (Stancalie and Craciunescu, 2005).

The following year, Dewan et al. (2006) used multidate RADARSAT SAR and geographic information system (GIS) data in order to delineate flood hazard. It was accomplished by using SAR data for estimating flood frequency and water depth, which were the basis for synthesizing flood hazard maps. The creation of the maps was based on a ranking matrix, with land-cover, geomorphic, and elevation data as GIS layers.

The RS data were also used in many studies in 2007. For example, satellite data such as Landsat 7 enhanced thematic mapper (ETM)+ images and Shuttle Radar Topography Mission (SRTM) DEM were combined with a distributed rainfall–runoff model in an attempt to map flood hazard on urban canyons. For this to be accomplished, Landsat 7 ETM+ images were classified (supervised classification), thus creating the land-cover map, whereas flow characteristics and topography were extracted by analyzing the SRTM DEM and the geographic extent of the drainage lines by the topographic map (Tapia-Silva et al. 2007). Two years later, Pradhan (2009) used logistic regression on RS data via GIS-based analysis, to delineate risk areas and produce flood susceptibility maps. The study concerned Malaysian possible flood areas, and its goal was to present them in a susceptibility map produced by statistical GIS modeling. The methodology was to create a geodatabase populated by all the necessary spatial data (topographical, geological, land cover, hydrological, DEM, Global Positioning System [GPS], and precipitation data), which were later processed to calculate the rating of each factor by using logistic regression. The final flood susceptibility map was synthesized by overlaying the produced layers (of the factors).

Pulvirenti et al., in 2011, proposed an algorithm for flooded area mapping, applied on SAR data. The main idea was to integrate radar data, describing the flooded areas along with simple hydraulic information via fuzzy logic. Radar data, and specifically L, C, and X frequency bands, were the algorithm's inputs, aided by the land-cover and DEM data. In order to validate the mapping process COSMO-SkyMed, very-high-resolution X-band SAR data were used (Pulvirenti et al. 2011). Tehrany et al. (2013) also published an assessment, about prediction efficiency for two other methods on flood susceptibility mapping, which are (a) rule-based decision tree (DT) and (b) combination of frequency ratio (FR) and logistic regression (LR) statistical methods. In parallel, Gioti et al. (2013) worked on a GIS-based runoff model for flash floods, using high-resolution DEM and meteorological data.

A time series analysis of satellite imagery for flooded areas took place in 2014. The areas covered by water were extracted by satellite, images, and as a result, an image for each flood event came up, presenting the maximum extent for each case. Furthermore, the produced maps were processed in order to calculate the relative frequency of inundation (RFI), along with a Landsat-5/7 time series (from 1989 to 2012) imagery analysis (Skakun et al. 2014). Flood mapping is a crucial element of flood risk management. In Europe, the European Commission (EC) requested from the member states to prepare two types of maps, flood hazard maps and flood risk maps. Flood hazard maps depict the extent and expected water depths/levels of a flooded area in three different scenarios: a low-probability scenario or extreme events, a medium-probability scenario (at least with a return period of 100 years), and, if appropriate, a high-probability scenario. Flood risk maps must also be prepared for the flooded areas under these three scenarios, presenting the potential population, economic and environmental risks due to flood events, and other possible information that member states may find useful to include, for instance, other sources of pollution (Directive2007/60/EC). To support the transition from traditional flood defense strategies to a flood risk management approach at a basin scale in Europe, the European Union (EU) has adopted this directive. Based on the EU directive, De Moel et al. (2009) attempted to record the flood mapping practices followed in 29 European countries and study the available maps and the ways that were used. The conclusion was that roughly half of the countries had adequate maps covering their complete extent, whereas no more than another third had maps that cover only significant areas. Five countries were found to possess a very small number or even no flood maps.

This study introduces a methodology for flood hazard assessment by using RS data (SAR data from Sentinel-1, Aster global digital elevation map (GDEM), and so on), free RS and GIS software (Quantum GIS (QGIS) and European space agency (ESA) sentinel application platform (SNAP), advanced geo-computations, and statistical techniques. The main objective of this project is to present a new method, which is based on Flood Susceptibility Index (FSI) mapping, for flood hazard susceptibility assessment.

18.2   DATA AND METHODOLOGY

18.2.1   RESEARCH FRAMEWORK

The implementation of the proposed method (Figure 18.1) requires two main categories of spatial data: geographical background data related to flooding and floods inventory. The general research strategy is to calculate the density of floods within classes of each background factor and then to

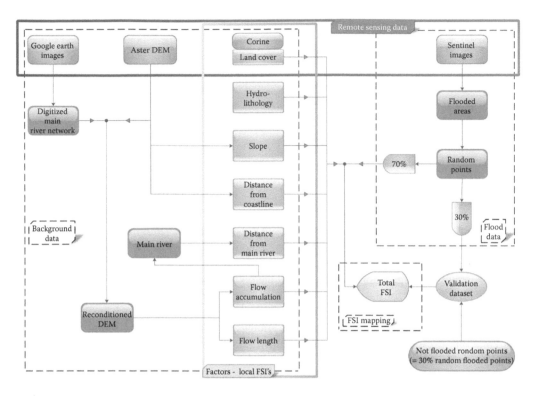

FIGURE 18.1   Flowchart of the methodology.

use these values for the estimation of integrated FSI. Accordingly, the mapping of this index can be used for the flood susceptibility zonation. Thus, the research framework includes (a) spatial database creation of the flood-related factors; (b) categorization of these factors; (c) extraction of the flooded areas by analyzing RS imagery; (d) calculation of the flooded areas within each factor category, by using GIS techniques (the output of this step is the calculation of the FSI for each category); (e) integration of FSI values in order to calculate total FSI for each mapping unit; and finally (f) mapping/zonation of the total FSI in the study area.

18.2.2   STUDY AREA

Sperchios River basin (Figure 18.2) is in Central Greece and is bounded by Mount Othrys (1,727 m) on the North, by Vardousia (2,437 m), Oeta (2,152 m), and Kallidromo (1,372 m) mountains on the South, and by Tymfristos mountain (2,316 m) on the West. Maliakos Gulf is the eastern limit of the basin. The basin covers an area of about 2116 km2, with an average altitude of 810 m (Kakavas, 1984).

Sperchios River, with a main river channel of about 82.5 km, originates from the eastern sides of Tymfristos mountain and follows a west–east flow direction, discharging into Maliakos Gulf. The main channel of Sperchios River is recharged by both perennial and ephemeral flow streams. The river's valley, in the two-thirds of its length, has steep slopes, which give to the river a rather mountainous-torrential character, with crucial flooding peaks and very intense sediment yield. On the contrary, at the last third of its course, Sperchios gradually becomes a lowland river, crossing low-altitude areas, often causing severe flooding (Koutsogiannis, 2007).

The delta of the river covers about 200 km^2 and is continuously prograding with a unique rate compared with other Greek river deltas. This progradation rate, according to Maroukian et al. (1995), was estimated approximately in 130 acres annually, with augmentation trends in the last 150–200 years.

The type of the hydrographic network is characterized in general as elongated dendritic, and the main channel of Sperchios is a seventh-order stream per Strahler's ordering system.

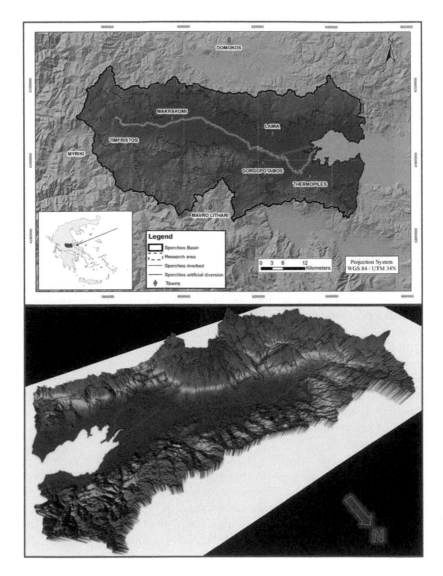

FIGURE 18.2   Study area (Sperchios River basin).

The climate ranges from dry to semi-humid.

The main land uses refer to agricultural areas with significant natural vegetation and complex cultivation systems (Stathopoulos et al. 2013).

Regarding the geology of Sperchios River basin, it is mainly composed of igneous and sedimentary rocks (Kallergis et al. 1970; Marinos et al. 1957, 1962, 1963, 1967; Papastamatiou, 1960, 1962). The main geological formations in the west part of the basin are limestones, flysch, and clastic sediments. In the north, northeast, and southeast parts of the basin, and specifically in Kallidromo mountain, the main rocks are limestone and schist-crests with ophiolites. In the south part (Oiti mountain), the prevailing geological formations are limestones and, in a limited extent, flysch formation.

The 450 km² of the lowland area of the river's basin consists of the most recent Quaternary deposits, which constitute the 20% of the basin's area. These sediments consist of conglomerates, lacustrine deposits, Quaternary alluvial deposits, scree, talus cones, and Holocene alluvial deposits.

The total thickness of these deposits, in the area of Sperchios delta, exceeds 1000 m (Stathopoulos et al. 2011).

Sperchios River basin and Maliakos Gulf constitute an asymmetric tectonic rift.

The geological formations are divided into four relevant categories according to their permeability (Kakavas, 1984):

1. *Permeable* (carbonate rocks, conglomerates, coarse-grained sediments, recent deposits of Sperchios riverbed, coarse-grained formations of alluvial, etc.)
2. *Medium-permeable* (deep deposits of Sperchios river, several petrologic types of the ophiolitic complex, etc.)
3. *Semi-permeable* (weathered territorial mantle of flysch and of schist-chert, and so on)
4. *Impermeable* (formations of flysch, old deposits of Sperchios, the deposits of mineral-thermal springs, etc.)

18.2.3 FLOOD EVENT

The upper-atmosphere synoptic circulation was driven by an extensive low geopotential heights centered in Northern Europe. The 500-hPa analysis of 1800 UTC on January 31, 2015 (Figure 18.3a), shows a strong northern airflow over West Europe, bringing cold air mass over West Mediterranean. This air mass, after a long sea track over the relatively warmer Mediterranean Sea, is then forced to move over Greece due to the strong west–southwest flow dominating over central and east Mediterranean.

This upper-atmosphere synoptic circulation, corresponding to an extensive low-pressure area at the surface (Figure 18.3b, mean sea level (MSL)), triggered and deepened mesoscale lows over Northern Ionian Sea–Northwest Greece, which then, associated (or not) with cold fronts, gave a lot of precipitation passing over West Greece, with the greatest intensity recorded on the evening of January 31.

Mediterranean depressions typically produce high rainfall intensities over West Greece due to the intense relief. The air mass in this kind of weather disturbances, after crossing the Mediterranean Sea, except of being supplied with abundant moisture, also becomes unstable. Furthermore, as this moist and unstable air is continuously forced to rise over the mountains of Pindus, it gives rise to showers and thunderstorms, mainly in the windward side. In the east part of the mainland, the rainfall amounts, although relatively limited, may also result in extensive flooding (Stathis et al. 2005).

In Table 18.1, we can see the 24-hour precipitation height for the 2-day period (January 31 to February 1, 2015) in four meteorological stations in the area of interest (Figure 18.2).

The two maxima of this episode (in the afternoon on January 31 and in the early hours on February 1) are clearly represented in the diagram of the 3-hour accumulated precipitation, recorded at the Makrakomi meteorological station (Figure 18.4).

The studied rainfall event is extreme and rare, thus leading to overestimation of floods in Sperchios River basin. It is assumed that estimating FSI for this event covers floods caused by less severe rainfall events and consequently floods of smaller spatial extent.

18.2.4 DATA DESCRIPTION

The proposed FSI methodology focuses on using open-source data as primary inputs of this procedure. Loyal to this idea, the main data reservoirs were RS projects, such as Aster, Sentinel-1, Google Earth, and Coordination of Information on the Environment (CORINE), as well as open government databases. Table 18.2 presents all the primary data that were used, the purpose for which the data were used, and the sources from which the data were acquired.

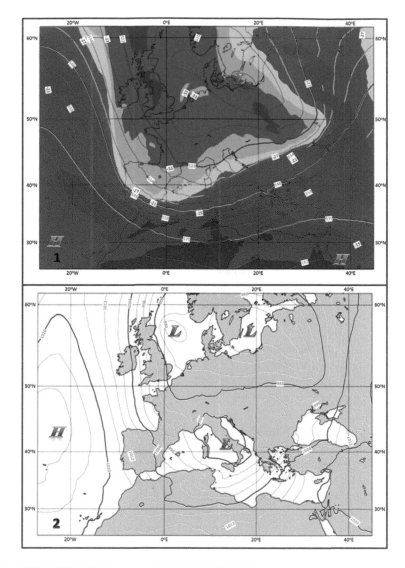

FIGURE 18.3 (1) ECMWF 500hPa analysis 31-01-2015/1800UTC, and (2) ECMWF MSL (Mean Sea Level) analysis 31-01-2015/1800UTC.

TABLE 18.1

Twenty-Four-Hour Accumulated Precipitation at Meteorological Stations in the Area of Interest

Meteorological Station (NOA)	Altitude (m)	24-Hour Accumulated Precipitation Height (mm) January 31, 2015	24-Hour Accumulated Precipitation Height (mm) February 1, 2015
Makrakomi	125.00	24.4	18.0
Domokos	570.00	12.2	6.4
Myriki	1045.00	69.0	48.6
Mavro Lithari	1250.00	80.4	43.2

Source: http://www.meteo.gr/meteoplus/Gmap.cfm

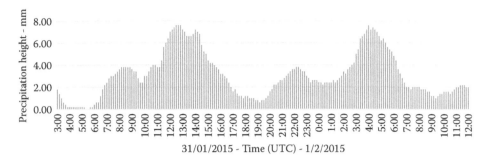

FIGURE 18.4 Meteorological Station Makrakomi (NOA)—3-hour accumulated precipitation.

TABLE 18.2
Primary Data Information

Data—Projects	Purpose	Source
Aster GDEM	Digital elevation model	http://reverb.echo.nasa.gov/reverb/
SENTINEL–1 images	Extraction of flooded areas	https://scihub.copernicus.eu/dhus/#/home
CORINE	Land cover—use	http://land.copernicus.eu/pan-european/ corine-land-cover/clc-2012
Google Earth images	Digitization of main river network	https://www.google.com/earth/
IGME geological maps	Digitization of geological formations	http://www.igme.gr/index.php
Greece basins—shapefiles	Sperchios River basin limits	http://geodata.gov.gr/dataset

From Aster[*] project, by METI[†] and NASA[‡] (http://asterweb.jpl.nasa.gov/GDEM.ASP), GDEM V2[§] data were used. In addition, from SENTINEL-1 mission, which is the European Radar Observatory for the Copernicus (2012) joint initiative of the EC and the European Space Agency (ESA), SENTINEL-1A Level 1 GRD IW HR Dual Polarization data images were acquired, where (https://sentinel.esa.int/web/sentinel/missions/sentinel-1):

- Level 1 is the data-processing level for common (most) users
- GRD is ground-range-detected products
- IW is the interferometric wide swath
- HR is high resolution
- Dual polarization indicates VV + VH or HH + HV (where V = Vertical and H = Horizontal)

Land-cover data were acquired from CORINE project, which also belongs to Copernicus initiative (the pan-European component is coordinated by the European Environment Agency). The most recent land-cover data (2012) that were used for this study have the following characteristics (http://land.copernicus.eu/pan-european/corine-land-cover/view):

- Satellite data: IRS P6 LISS III and RapidEye dual date
- Time consistency: 2011–2012
- Geometric accuracy, satellite data: 25 m or less
- Minimum mapping unit/width: 25 ha/100 m

[*] Advanced Spaceborne Thermal Emission and Reflection Radiometer
[†] Ministry of Economy, Trade, and Industry, Japan
[‡] National Aeronautics and Space Administration, United States
[§] Global Digital Elevation Model Version 2, October 2011

- Geometric accuracy, CLC: better than 100 m
- Thematic accuracy, CLC: 85% or more

The main river network, in order to depict the contemporary regime of the study area, was digitized by using Google Earth images. Furthermore, the geological formations of the study area were digitized from the Institute of Geology and Mineral Exploration (IGME) 1:50,000 Geology Maps. Finally, the extent (limits) of the studied basin was acquired from the open database Geodata. gov.gr (the delimitation of all the national hydrological basins is a result of national projects).

Aster GDEM and Sentinel 1 images are raster datasets, whereas CORINE land cover (polygon), main river network (polyline), geological formations (polygon), and basin's extent (polygon) are vector shapefiles.

18.2.5 DATA HOMOGENIZATION

The homogenization of the primary data was a crucial stage for the implementation of the proposed method. This stage is constituted by (a) reprojection of all datasets in WGS 1984 UTM 34N projection in order to match DEM's projection system, (b) rectification of all raster layers in order to correspond to DEM's pixel size (~28.5 m), and (c) masking of all layers according to the extent of the basin. With these preprocessing actions, we ensured the integration of all datasets in the next stages of the proposed method.

18.2.6 FLOOD DATA—REMOTE-SENSING ANALYSIS

To create a layer depicting the total extent of the basin that was either permanently or periodically flooded throughout the extreme rainfall event until the basin drained out, six Sentinel-1 images were acquired and processed (via ESA SNAP software), covering days 2, 3, 8, 9, 14, and 15 of February 2015.

The preprocessing (A, B, and C), processing (D), and postprocessing (E) steps that were followed, for each image, are (www.un-spider.org, steps for flood mapping using Sentinel imagery):

A. *Spatial subset*: The region of interest is selected, as the total area covered by the images is significantly wider, thus discarding the unnecessary data and making the processing faster (Figure 18.5A1 and A2).
B. *Radiometric calibration*: The backscatter (the signal that was reflected from the surface back to the direction where it came from) coefficient values are calibrated (Figure 18.5B).
C. *Speckle filtering*: The *salt and pepper* effect (grainy texture caused by random constructive and destructive interferences from the multiple scattering returns that occurs within each resolution cell) is reduced. As speckle is a form of noise, which degrades the quality of an image and may make interpretation (visual or digital) more difficult, it is generally desirable to reduce it before interpretation and analysis. In this case study, Lee Filter with window size 7 × 7 (after trials) was applied (Figure 18.5C).
D. *Binarization*: A threshold value is selected (after trials), in order to separate water from non-water, by analyzing the logarithmic histogram of the filtered backscatter coefficient (different for each image), as presented in Figure 18.5D1 and D2. The histogram shows two peaks of different magnitude. Low values of the backscatter correspond to water class, and high values correspond to non-water class. For example, the threshold value for February 2, 2015, image is $2.22^{\wedge}E-2$. The threshold value is applied in a band math expression in order to binarize the image (The image's backscatter pixel values that are lower than the threshold are multiplied by 255 and represent water objects, while the higher ones are multiplied by 0 (non-water)). The binary image is presented in Figure 18.5D3.
E. *Range-Doppler terrain correction*: As the obtained image is in the geometry of the sensor, it is reprojected to the geographic projection (geometric correction), as shown in Figure 18.5E. For this final step, SRTM3Sec DEM is used—bilinear interpolation as resampling method and WGS 1984 UTM 34N as projection system.

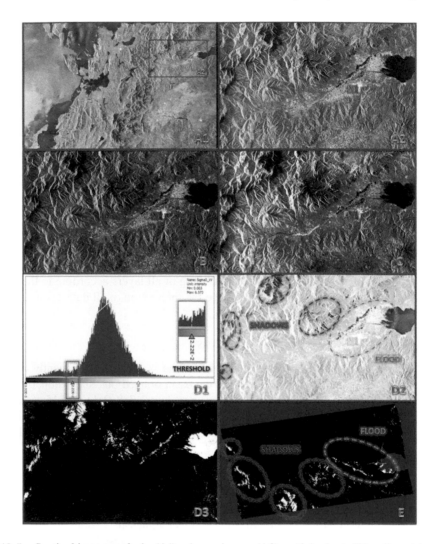

FIGURE 18.5 Sentinel image analysis: (A1) primary image, (A2) spatial subset, (B) radiometric calibration, (C) speckle filtering, (D1) histogram—threshold, (D2) separation of water/non-water, (D3) binarization, and (E) range-Doppler terrain correction.

The final binary raster images are converted to vector shapefiles and then merged, creating this way a final vector dataset, with the total flooded extent of the basin as an integration of all flooded areas for the complete duration of the flood event. It must be noted that before merging the shapefiles, a small number of polygons, from each shapefile, was manually removed, as it corresponded in *shadows* created by the satellite's acquisition angle and the morphology of the study area (the backscatter values of shadows and water are similar). This was not made arbitrarily, as both the sun's position and the satellite's angle were known for the specific acquisition date and time of each image.

18.2.7 GEOGRAPHIC INFORMATION SYSTEM PROCESS

It is of high importance for the proposed analysis to create a hydrologically valid DEM, representing the current hydrologic regime of the basin. For this, the first step was the digitization of the main river network, using the most resent images available from Google Earth. Alongside, the primary DEM was processed in order to fill the existing sinks. Finally, the digitized main river network and the filled DEM were combined and co-processed by using a GIS DEM reconditioning method.

The reconditioning method modifies a DEM by imposing linear features onto it (burning/fencing). It is an implementation of the AGREE method developed at the University of Texas at Austin in 1997; it requires setting a *stream buffer,*[*] a *smooth drop/raise value,*[†] and a *sharp drop/raise value*[‡] (Arc Hydro Tools Version 2.0 Tutorial (ESRI, 2011)). After many trials, the values for these parameters that produced a hydrologically valid DEM are *5 pixels*, 5 m, and 25 m accordingly. The main concept is to create a hydrologically valid DEM and, at the same time, to preserve the real morphological characteristics of the basin.

It must also be noticed that the studied river network has an important particularity, which posed specific obstacles during the basin's hydrological analysis. In the final part of the riverbed, an artificial channel is constructed, which diverts part of the river's flow and is mostly used for flooding management and irrigation purposes. The diversion was digitized along with the main river network and was included in the DEM reconditioning step. Nevertheless, this fact results in creating two outlet points for Sperchios River in Maliakos Gulf. When attempting to calculate the basin's flow accumulation, a false layer is created, as the algorithm is set to work only for one outlet point. To overcome this problem, the valid flow accumulation layer was created by combining five different flow accumulation layers that were created by using the two outlet points separately. The next step involved the creation of the flow length layer. This layer was reclassified to discard areas with flow length more than 10 km. These areas are mountainous with progressively increasing slopes.

Finally, the conditional-background factors, related to flooding, were created. Slope, flow accumulation, and flow length layers were produced from DEM analysis; distance layers were produced from main river network and coastline vectors (polyline shapefiles); and hydrolithology layer was produced from geology layer after classification of the geological formations according to their permeability and hydraulic conductivity (K coefficient).

18.2.8 TOTAL FIXED SUSCEPTIBILITY INDEX CALCULATION

The FSI defines the importance of a factor category on flood occurrence according to spatial distribution of the pixels of the considered factors and the flooded pixels. This method calculates the FSI for each category of all factors (e.g., land cover, lithology, and slope), which are selected for the case study. Thus, the FSI for the factor category j (FSI$_j$) is defined as follows:

$$\text{FSI}_j = \ln \left[\frac{\left(\dfrac{\text{Npix}(S_j)}{\text{Npix}(N_j)} \right)}{\left(\dfrac{\sum \text{Npix}(S_j)}{\sum \text{Npix}(N_j)} \right)} \right] \tag{18.1}$$

where:
Npix (S_j) is the number of flooded pixels in factor category j
Npix (N_j) is the total number of pixels in the same-factor category

Thus, the FSI presents the relative susceptibility to flood occurrence. If a category is highly correlated to flooding, the area associated with this category will have a high positive FSI value.

---

[*] The number of cells around the linear feature for which the smoothing will occur.

[†] The amount (in vertical units) that the linear feature will be dropped (if the number is positive) or the fence extruded (if the number is negative). This value will be used to interpolate the DEM into the buffered area (between the boundary of the buffer and the dropped /raised vector feature).

[‡] The additional amount (in vertical units) that the linear feature will be dropped (if the number is positive) or the fence extruded (if the number is negative). This results in additional burning/fencing on top of the smooth buffer interpolation and needs to be performed to preserve the linear features used for burning/fencing.

A negative FSI value for a specific category is an indicator of low flooding density in this class. Thus, high positive FSI values indicate FS density in this class much higher than the average, and high negative FSI values indicate FS density much lower than the average. Consequently, for a causal factor to be useful for flood susceptibility mapping, its categories should provide a range of FSI values. The overall susceptibility, S, for each pixel is defined as

$$S = \frac{1}{n} \sum_{i=1}^{n} \text{FSI}_i \qquad (18.2)$$

where:

FSI$_i$ is the susceptibility for the factor i

n is the total number of the factors

A similar concept has been used for the assessment of landslide susceptibility (Van Westen, 1997; Tien Bui et al. 2011).

In this case study, in order to calculate local FSI values for each layer and determine the final layers that were used for the total FSI (TFSI) computation, the following steps were followed:

1. *Categorization of all flood-related factors:* The *Natural Breaks* (*Jenks*) categorization (five categories) was implemented (by using a GIS-based function) for the factors with continuous values (slope, flow accumulation, flow length, distance from coastline, and distance from main river network). *Natural Breaks* (*Jenks*) method is a data classification method that minimizes variance within groups of data and maximizes variance between groups of data (Jenks, 1967). For land cover and hydrolithology, the thematic classes of the nominal scale were preserved (Table 18.3).
2. *Creation of flood events (points):* The total flood polygon, produced by Sentinel images, was converted in raster layer, which was then converted in point features.
3. *Random selection of 70% of flood points:* From the flood points, 70% were randomly selected, whereas the rest 30% were kept as a validation set.
4. *Export all layer class values for each flood point:* The corresponding class value of all layers or factors were exported for each one of the flood points.
5. *Calculation of local FSI values for each class of all layers:* The number of flood points corresponding to each class was calculated for every layer. This number was divided by the total number of pixels of each class, thus resulting in flood density per class for every layer. Local FSI values for each class of every layer were calculated by dividing flood density per class by the total density (all flood points divided by the total number of pixels) of each layer. The calculation of FSI for all the categories of factors (local FSI) was based on the implementation of Equation 18.1. Table 18.4 presents the calculation of local FSI values for flow length layer (as an example), and Table 18.5 presents all local FSI values for every layer.

TABLE 18.3

Classification of Thematic Layers

Thematic Class		Raster
Hydrolithology (Formations)	**Land Cover**	**Class**
Very low up to no permeability	Water bodies	1
Very low permeability	Wetlands	2
Low permeability	Artificial surfaces	3
Medium permeability	Agricultural areas	4
High up to very high permeability	Forest and semi-natural areas	5

TABLE 18.4

Local FSI Calculation for Flow Length Layer

Flow Length Class	Flood Points	Pixels	Density Per Class	Total Density	FSI
1	13,588	302,426	0.044929999	0.011396261	1.37
2	1,616	278,138	0.005810066	0.011396261	−0.67
3	551	274,459	0.002007586	0.011396261	−1.74
4	339	282,384	0.001200493	0.011396261	−2.25
5	14	276,039	5.07175E-05	0.011396261	−5.41

TABLE 18.5

Local FSI Values Per Class, for Each Factor

FSI ＼ Classes	1	2	3	4	5
Slope	0.36	−4.84	—	—	—
Flow accumulation	0.00	−2.03	—	0.90	1.73
Flow length	1.37	−0.67	−1.74	−2.25	−5.41
River distance	0.91	−0.75	−1.97	−2.40	—
Coastline distance	0.88	0.03	−2.75	−4.65	—
Land cover	2.62	3.51	−1.55	0.21	−1.61
Hydrolithology	−4.48	−4.06	−7.96	0.86	−5.71
DEM	0.83	−5.77	—	—	—
Aspect	0.05	0.03	−0.03	0.06	−0.14
TWI[a]	−1.21	−0.03	0.00	−0.09	−0.22

[a] Topographic Wetness Index (TWI), also referred as Compound Topographic Index (CTI).

6. *Correlation tests of layers:* Range and standard deviation values of FSI were calculated for all factors, in order to interpret the importance of each factor (He and Beighley, 2008). After the statistical correlation tests among these layers (for every pair of factors), some of them were excluded (DEM, aspect, and Topographic Wetness Index). The layers used for the process should not be correlated with each other; thus, when two correlated layers were found, the one that was kept was that with the wider range of local FSI values, while standard deviation values were also taken under consideration. The final layers or factors used for the process are as follows:

- Slope
- Flow accumulation
- Flow length
- Distance from main river network
- Distance from coastline
- Land cover
- Hydrolithology

7. Finally, by using GIS overlay analysis and Equations 18.1 and 18.2, the integrated flood susceptibility map was created by combining the FSI values of the final selected factors (equal overlay of layers or factors). This map was classified into five classes (*very low, low, moderate, high,* and *very high* susceptibility), based on *Natural Breaks (Jenks)* classification method (Foumelis et al. 2004; Wati et al. 2010).

18.3 RESULTS AND DISCUSSION

18.3.1 FLOOD SUSCEPTIBILITY INDEX MAPPING

Figure 18.6 presents the local FSI values for all the factors that compose the total FSI layer. Some factor or layers have less than five FSI classes, as there are classes without flooded areas, and thus the FSI value cannot be calculated. Hydrolithology is the layer with the wider range of values, whereas the layer with the shortest range is the one concerning the distance from the main river.

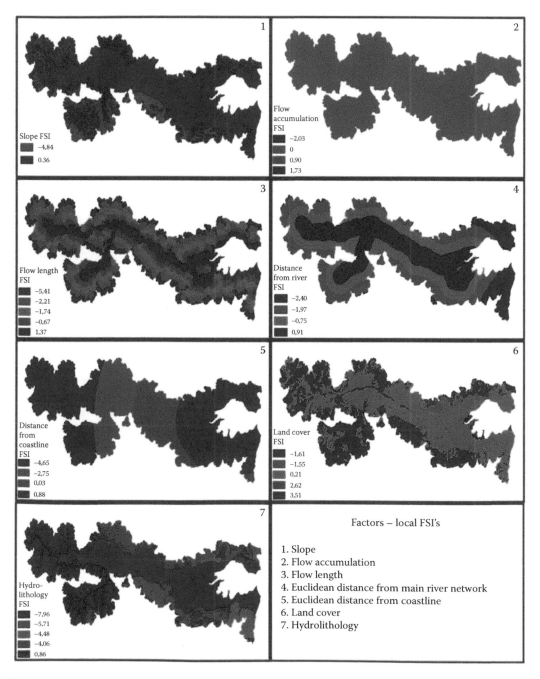

FIGURE 18.6   Factors—local FSI values.

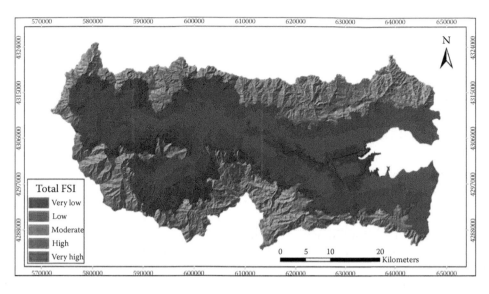

FIGURE 18.7 Total FSI.

As can be deduced from Figure 18.6, FSI values are higher in flat or low-slope areas such as the river's valley (Figure 18.6.1). Hydrologically, higher FSI classes concern high-flow accumulation and low-flow length values, thus referring to the middle and final parts of the riverbed (Figure 18.6.2), as well as to a narrow zone covering both sides of the main river axis (Figure 18.6.3). Areas close to the main river (Figure 18.6.4) and areas close to the coastline (Figure 18.6.5) belong to high-FSI classes. Based on land cover, high FSI values are found in the final part of the riverbed, whereas the highest ones are seen on the deltaic part and specifically in the wetlands created in the areas where the river's outlets meet the coastline (Figure 18.6.6). Finally, high FSI values are in the Quaternary formations of the basin (mainly alluvial deposits with high permeability rate but quick saturation), which cover the highest extent of the river's valley (Figure 18.6.7).

Total FSI layer (Figure 18.7) presents the highest-susceptibility class over the last part of Sperchios riverbed and around the outlet of the two estuaries. The next-highest-susceptibility class is found in the second part of the basin's valley, covering a small area around the river's flow path and around its two riverbeds (artificial and natural).

These high-flood-risk parts of the basin, as expected, meet most of the flood susceptibility criteria. Namely, these are flat areas that (a) are covered by alluvial deposits, (b) include the river's highest flow-accumulation downstream part, (c) are in short distance and low flow-length in both sides of the natural riverbed and its diversion, and finally (d) are next to the coastline. The areas with the highest susceptibility are found in the deltaic wetlands.

18.3.2 RECEIVER OPERATING CHARACTERISTICS ANALYSIS

The validation step is essential to know the predictive value of the model (Remondo et al. 2003). A standard validation method, known as receiver operating characteristics (ROC) analysis, was performed in order to evaluate the overall performance of the FSI model in the study area. This method is considered a powerful tool for the validation of predictive models and has been widely used to provide estimates of their performance (Frattini et al. 2010).

In ROC analysis, the sensitivity of the model is shown as a function of the specificity. The sensitivity refers to a percentage of positively predicted cases among the whole positive observations (Althuwaynee et al. 2014):

$$\text{Sensitivity} = \frac{n(TP)}{[n(TP) + n(FN)]} \tag{18.3}$$

where:
  $n(TP)$ is the number of the true positive predictions
  $n(FN)$ is the number of the false-negative predictions

On the other hand, the specificity refers to a percentage of negatively predicted cases among the whole negative observations:

$$\text{Specificity} = \frac{n(TN)}{n(TN) + n(FP)} \tag{18.4}$$

where:
  $n(TN)$ is the number of the true negative predictions
  $n(FP)$ is the number of the false-positive predictions

A true positive is a prediction of flood at a location where the flood occurred, whereas a false positive is a prediction of flood at a location where the flood did not occur.

The ROC graph consists of two axes: y-axis represents the sensitivity and x-axis represents the 1–specificity. Thus, high sensitivity indicates a high number of true positives (correct predictions), whereas high specificity (low 1–specificity difference) indicates a low number of false positives (Cervi et al. 2010). The corresponding ROC curve shows the ability of the model to correctly discriminate between positive and negative observations in the validation space (Montrasio et al. 2011). The area under the ROC curve (AUC) characterizes the quality of a model and is often used when a general measure of its predictive capability is desired. In practice, AUC value ranges from 0.5 to 1.0. The ideal model performs at a value close to 1.0 (perfect fit), whereas a value close to 0.5 indicates inaccuracy in the model (random fit) (Polykretis et al. 2015).

In the current study, the remaining 6,905 pixels (rest 30% flood points) from the flood-occurrence area and an equal number of pixels from the flood-not-occurrence area (validation dataset) were used for the validation of the model output. Subsequently, the 13,810 pixels were matched with the relative flood susceptibility categories of the final map.

Then, based on this matching, the ROC curve was drawn and the AUC value was calculated for the proposed model.

As can be deduced from the validation analysis, the selected layers or factors attribute to flood susceptibility (TFSI) in a very high level of accuracy. The AUC value of 0.965 indicates an excellent prediction ability of the model (Figure 18.8).

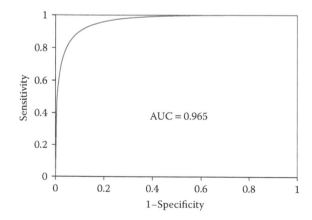

FIGURE 18.8 ROC curve validation analysis.

18.4 CONCLUSIONS

The need to efficiently cope with natural hazards has led many researchers to carry out studies on prediction and impact assessment. The contemporary needs for low-budget but efficient applications and management tools are met by the constantly developing modern technology. Thus, advances in RS technology, open spatial databases, and new modern technologies such as GIS tend to replace costly and time-consuming field work, traditional techniques, and extremely difficult and costly data-collecting methods.

The proposed FSI methodology uses free RS datasets and open spatial databases, as well as non-commercial software packages such as QGIS and SNAP. Moreover, the construction of the spatial database was supported by low-cost or freeware spatial datasets.

This method is based on the mapping of past flood events and on the construction of a spatial database with background information (conditional factors). In this case, the assumption that past flood events are located within the extent of a severe event was taken into consideration. Specifically, the studied rainfall event was extreme and very rare, thus leading to overestimation of floods in the study area. Therefore, it was assumed that estimating FSI for this event will include floods caused by less severe rainfall events and consequently floods of smaller spatial extent. Accordingly, it is substantial to analyze the appropriate RS imagery in order to map this event.

A key aspect of the proposed process is the need to create a hydrologically valid DEM and, at the same, to preserve the real morphological characteristics of the area under investigation. This fact constitutes the pre-process of the DEM and its derivatives that are essential and necessary. The validation of the method showed very good flood susceptibility prediction properties of the model. These results show that the proposed FSI mapping is a solid and easily implemented approach for flood susceptibility at regional scale.

The FSI can and will be further tested and developed by incorporating additional and different conditioning factors (e.g., triggering factors such as rainfall), by working on different scales of analysis (e.g., higher spatial resolution), and by applying it in areas with various geomorphological characteristics, climatic characteristics, and so on, the results of which will be compared and evaluated. Furthermore, another highly interesting implementation of the proposed method would be the development of FSI models, by examining different change scenarios of the conditioning factors (e.g., changes in land cover, in flow length after a technical work, in permeability after extensive transport of fine sediment materials, and so on). The next step for the present case study is a multitemporal analysis of FSI, covering as many past flood events as can be found and specified, thus creating not only a spatial but also a temporal FSI correlation.

It must be noticed that the proposed method does not replace a high-scale hydrological analysis, which is based on detailed and actual runoff data, thus providing highly accurate flood risk assessments. Conversely, it is a regional-scale application, which provides fast and reliable estimations of flood susceptibility.

The empirical analysis indicated that FSI could be useful for management at regional scale (flood susceptibility zonation) and for the indication of areas where detailed hydrological analyses should be implemented.

Finally, EO data are used as a key input for estimating FSI; thus, any further evolution concerning spatial analysis, availability, and so on, will help significantly in developing the proposed methodology and the accuracy of its results. The EO data combined with GIS techniques are widely used worldwide for setting up policies for management of natural hazards and extreme phenomena. Future development on EO projects should focus on improving spatial resolution of RS data and on acquiring data in denser time intervals.

ACKNOWLEDGMENTS

The authors of this chapter thank The European Centre for Medium-Range Weather Forecasts (ECMWF) for providing the primary data in order for the two meteorological analysis maps of Figure 18.3 to be created.

REFERENCES

AGREE—DEM surface reconditioning system. Retrieved from www.ce.utexas.edu/prof/maidment/GISHYDRO/ferdi/research/agree/agree.html

Althuwaynee, O. F., Pradhan, B., Park, H. J., and Lee, J. H. 2014. A novel ensemble decision tree-based CHi-squared Automatic Interaction Detection (CHAID) and multivariate logistic regression models in landslide susceptibility mapping. *Landslides*, 11(6), 1063–1078.

Biggin, D. S. and Blyth, K. 1996. A comparison of ERS-1 satellite radar and aerial photography for river flood mapping. *Journal of the Chartered Institution of Water and Environmental Management*, 10(1), 59–64.

Cervi, F., Berti, M., Borgatti, L., Ronchetti, F., Manenti, F., and Corsini, A. 2010. Comparing predictive capability of statistical and deterministic methods for landslide susceptibility mapping: A case study in the northern Apennines (Reggio Emilia Province, Italy). *Landslides*, 7(4), 433–444.

Copernicus. 2012. CORINE Land Cover. Retrieved from http://land.copernicus.eu/pan-european/corine-land-cover/view.

De Moel, H., Van Alphen, J., and Aerts, J. C. J. H. 2009. Flood maps in Europe–Methods, availability and use. *Natural Hazards and Earth System Science*, 9(2), 289–301.

Deekshatulu, B. L., Lohani, B. N., and Narayan, L. R. A. 1980. Disaster warning and assessment by remote sensing. *Proceedings of the Southeast Asian Conference on Soil Engineering*, pp. 819–824.

Dewan, A. M., Kumamoto, T., and Nishigaki, M. 2006. Flood hazard delineation in Greater Dhaka, Bangladesh using an integrated GIS and remote sensing approach. *Geocarto International*, 21(2), 33–38.

Directive 2007/60/EC on the assessment and management of flood risks in all available languages (OJ L288, 6.11.2007, p.27).

ESA. SENTINEL-1. Retrieved from https://sentinel.esa.int/web/sentinel/missions/sentinel-1

ESRI. 2011. Arc Hydro Tools—Tutorial. Retrieved from http://downloads.esri.com/archydro/archydro/tutorial/doc/arc%20hydro%20tools%202.0% 20-%20tutorial.pdf

Foumelis, M., Lekkas, E., and Parcharidis, I. 2004. Landslide susceptibility mapping by GIS-based qualitative weighting procedure in Corinth area. Proceedings of the 10th international congress, Thessaloniki. *Bulletin of the Geological Society of Greece*, 36, 904–912.

Frattini, P., Crosta, G., and Carrara, A. 2010. Techniques for evaluating the performance of landslide susceptibility models. *Engineering Geology*, 111(1–4), 62–72.

Gioti, E., Riga, C., Kalogeropoulos, K., and Chalkias, C. 2013. A GIS-based flash flood runoff model using high resolution DEM and meteorological data. *EARSeL eProceedings*, 12(1), 33–43.

Handmer, J. W. 1980. Flood hazard maps as public information: An assessment within the context of the Canadian flood damage reduction program. *Canadian Water Resources Journal*, 5(4), 82–110.

Handmer, J. W. and Milne, J. 1981. Flood maps as public information. *Proceedings Floodplain Management Conference*, Canberra, May 1980 (Australian Water Resources Council, Conference Series 4 Australian Government Publishing Service, Canberra), pp. 1–26.

He, Y. and Beighley, R. E. 2008. GIS-based regional landslide susceptibility mapping: A case study in southern California. *Earth Surface Processes and Landforms*, 33, 380–393.

Hubert-Moy, L., Ganzetti, I., Bariou, R., and Mounier, J. 1992. Maps of flooded areas in Ille-et-Vilaine through remote sensing [Une cartographie des zones inondables en Ille-et-Vilaine par teledetection]. Norois, 155, pp. 337–347.

Jenks, G. F. 1967. The data model concept in statistical mapping. *International Yearbook of Cartography*, 7, 186–190.

Kakavas, N. I. 1984. Hydrological water balance of Sperchios river basin. PhD thesis, IGME, Athens, Greece.

Kallergis, G., Koch, K. E., and Nicolaus, H. J. 1970. Geological maps of Greece, sheet "Sperchias," scale 1:50.000. IGME Publications, Greece.

Koutsogiannis, D., (YP.AN., E.M.P., I.G.M.E., K.E.P.E.). 2007. Project plan of water resources management of the country. Ministry of Development—Directory of Water Dynamic and Natural Resources, Athens.

Maantay, J. and Maroko, A. 2009. Mapping urban risk: Flood hazards, race, and environmental justice in New York. *Applied Geography*, 29(1), 111–124.

Marinos, G., Anastopoulos, I., Maratos, G., Melidonis, N., and Andronopoulos, V. 1957. Geological maps of Greece, sheet "Anavra," scale 1:50.000. IGME Publications.

Marinos, G., Anastopoulos, I., Maratos, G., Melidonis, N., and Andronopoulos, V. 1962. Geological maps of Greece, sheet "Leontarion," scale 1:50.000. IGME Publications.

Marinos, G., Anastopoulos, I., Maratos, G., Melidonis, N., and Andronopoulos, V. 1963. Geological maps of Greece, sheet "Stylis," scale 1:50.000. IGME Publications.

Marinos, G., Anastopoulos, I., Maratos, G., Melidonis, N., Andronopoulos, V., and Papastamatiou, I. 1967. Geological maps of Greece, sheet "Lamia," scale 1:50.000. IGME Publications.

Maroukian, H., Gaki-Papanastassiou, K., Pavlopoulos K., and Zamani, A. 1995. Comparative geomorphological observations in the Kalamas delta in western Greece and the Sperkhios delta in eastern Greece. In: Rapport Commission International Mer Méditerranée, 34, 110.

Montrasio, L., Valentino, R., and Losi, G. L. 2011. Towards a real-time susceptibility assessment of rainfall-induced shallow landslides on a regional scale. *Natural Hazards and Earth System Sciences*, 11, 1927–1947.

NASA. ASTER global digital elevation map announcement. Retrieved from http://asterweb.jpl.nasa.gov/GDEM.ASP

National Observatory of Athens. Precipitation data. Retrieved from www.meteo.gr.

Papastamatiou, I., Tataris, A., Vetoulis, D., Katsikatsos, G., Lalexos, N., and Eleutheriou, A. 1962. Geological maps of Greece, sheet "Amfikleia," scale 1:50.000. IGME Publications.

Papastamatiou, I., Tataris, A., Vetoulis, D., Mpornovas, I., and Xristodoulou, G. 1960. Geological maps of Greece, sheet "Amfissa," scale 1:50.000. IGME Publications.

Pelletier, J. D., Mayer, L., Pearthree, P. A., House, P. K., Demsey, K. A., Klawon, J. E., and Vincent, K. R. 2005. An integrated approach to flood hazard assessment on alluvial fans using numerical modeling, field mapping, and remote sensing. *Bulletin of the Geological Society of America*, 117 (9–10), 1167–1180.

Polykretis, C., Ferentinou, M., and Chalkias, C. 2015. A comparative study of landslide susceptibility mapping using landslide susceptibility index and artificial neural networks in the Krios River and Krathis River catchments (northern Peloponnesus, Greece). *Bulletin of Engineering Geology and the Environment*, 74(1), 27–45.

Pradhan, B. 2009. Flood susceptible mapping and risk area delineation using logistic regression, GIS and remote sensing. *Journal of Spatial Hydrology*, 9(2), 1–18.

Pulvirenti, L., Pierdicca, N., Chini, M., and Guerriero, L. 2011. An algorithm for operational flood mapping from synthetic aperture radar (SAR) data using fuzzy logic. *Natural Hazards and Earth System Science*, 11 (2), 529–540.

Remondo, J., González-Díez, A., Díaz de Terán, J. R., and Cendrero, A. 2003. Landslide susceptibility models utilizing spatial data analysis techniques: A case study from the Lower Deba Valley, Guipúzcoa (Spain). *Natural Hazards*, 30(3), 267–279.

Skakun, S., Kussul, N., Shelestov, A., and Kussul, O. 2014. Flood hazard and flood risk assessment using a time series of satellite images: A case study in Namibia. *Risk Analysis*, 34(8), 1521–1537.

Smith, K. 2001. *Environmental Hazards Assessing Risk and Reducing Disaster*, 3rd ed. 11 newfetter lane, London, UK: Routledge.

Smith, K. 2014. *Regions of Risk: A Geographical Introduction to Disasters*. 1st ed. Abingdon, UK: Routledge.

Smith, K. and Petley, D. N. 2009. *Environmental Hazards: Assessing Risk and Reducing Disaster*, 5th ed. Routledge: Abingdon, UK.

Stancalie, G. and Craciunescu, V. 2005. Contribution of earth observation data supplied by the new satellite sensors to flood disaster assessment and hazard reduction. *Geo-information for Disaster Management*, 1315–1332.

Stathis, D., Dafinka I., Balafoutis, C., and Makrogiannis, T. 2005. Orographic effect on heavy rainfall in Chalkidiki Penisula (Greece) induced by a Mediterranean cold front: A case cdudy on 7 to 8 of October 2000. *Croatian Meteorological Journal*, 40(40), 490–493.

Stathopoulos, N., Koumantakis, I., Vasileiou, E., and Markantonis, K. 2011. Quality regime of water resources in Anthele area, in Sperchios River Delta, Fthiotida Prefecture, Greece. *Proceedings of 9th Hydrogeological Conference*, October 5–8, Kalavrita, Greece.

Stathopoulos, N., Rozos, D., and Vasileiou, E. 2013. Water resources management in Sperchios river basin, using SWOT analysis. *Proceedings of 13th International Congress of the Geological Society of Greece* with the title: "Exploration & Exploitation of Mineral Resources," September 5–8, Chania, Greece.

Tapia-Silva, F. O., Nuñez, J. M., and López-López, D. 2007. Using SRTM DEM, Landsat ETM+ images and a distributed rainfall-runoff model to define inundation hazard maps on urban canyons. *Proceedings, 32nd International Symposium on Remote Sensing of Environment: Sustainable Development Through Global Earth Observations*, 4 p.

Tehrany, M. S., Pradhan, B., and Jebur, M. N. 2013. Spatial prediction of flood susceptible areas using rule based decision tree (DT) and a novel ensemble bivariate and multivariate statistical models in GIS. *Journal of Hydrology*, 504, 69–79.

Tien Bui, T., Lofman, O., Revhaug, I., and Dick, O. 2011. Landslide susceptibility analysis in the Hoa Binh province of Vietnam using statistical index and logistic regression. *Natural Hazards*, 59, 1413–1444.

Torab, M. M. 2002. Flood-hazard mapping of The Hafit Mountain slopes–The eastern province of United Arab Emirates (U.A.E.). *Bulletin of the Society of Cartographers*, 36(2), 39–44.

Van Westen, C. J. 1997. Statistical landslide hazard analysis. ILWIS 2.1 for Windows application guide, ITC Publication, Enschede, 73–84.

Wang, Y. 2004. Using Landsat 7 TM data acquired days after a flood event to delineate the maximum flood extent on a coastal floodplain. *International Journal of Remote Sensing*, 25, 959–974.

Wati, S. E., Hastuti, T., Widjojo, S., and Pinem, F. 2010. Landslide susceptibility mapping with heuristic approach in mountainous area: A case study in Tawangmangu sub district, Central Java, Indonesia. *International Archives of the Photogrammetry, Remote Sensing and Spatial Information Science*, 38, 248–253.

Webster, T. L., Forbes, D. L., Dickie, S., and Shreenan, R. 2004. Using topographic lidar to map flood risk from storm-surge events for Charlottetown, Prince Edward Island, Canada. *Canadian Journal of Remote Sensing*, 30(1), 64–76.

19 Satellite-Based Precipitation for Modeling Floods
Current Status and Limitations

Yiwen Mei, Efthymios, E. I. Nikolopoulos,
and Emmanouil N. Anagnostou

CONTENTS

19.1 INTRODUCTION

Accurate measurement of surface precipitation is of great importance for the monitoring, forecasting, and early warning of flood hazard. Conventional ground-based measurements for quantifying precipitation include observations from rain gauge and weather radar networks (Michaelides et al. 2009). Rain gauge networks provide accurate pointwise precipitation measurements but suffer from lack of space–time representation of precipitation. On the other hand, weather radar networks provide precipitation estimates with high spatiotemporal resolutions (i.e., 1–4 km and 5–15 min) but are subjected to uncertainties arising from variations in rainfall drop size distribution, beam blockage, beam overshooting, beam filling, hardware calibration, and random sampling error (Berne and Krajewski, 2013; Delrieu et al. 2014; Kirstetter et al. 2015). A viable solution to the accuracy issues of precipitation measurement is to combine rain gauges with weather radar, but this cannot be applied to typically mountainous areas, which are characterized by large radar beam blockages due to the orographic effect (Kirstetter et al. 2015; Prat and Nelson, 2015).

Integration of satellite-precipitation products with hydrologic models can be used to study catchment flood response in areas with inaccurate or scare ground-based measurements. Precipitation estimation from satellite remote sensors is uninhibited by topography and thus can provide coherent global-scale estimates at high space (as fine as 0.04°) and time (as fine as half-hourly) resolution. High-temporal-resolution satellite-precipitation retrievals are based on a combination of observations from the visible–infrared (VIS–IR) spectrum on geostationary (GEO) satellites and the polar-orbiting microwave sensors (PMW) on low-Earth-orbiting (LEO) satellites, or either the VIS–IR or PMW observations. The IR sensors have high sampling frequency, but precipitation estimates are based on cloud-top temperature observations. The PMW sensors observe directly the hydrometeor content present within the atmospheric column but have less temporal sampling frequency. Although the combinations of VIS–IR and PMW retrievals provide high spatiotemporal precipitation estimates, the ground-gauge information is often incorporated with the remote-sensing-based measurements to improve the accuracy of retrievals. The ground-gauge adjustments are often done on monthly basis to reduce the systematic bias in remote-sensing precipitation estimations.

Studies have demonstrated that satellite-precipitation estimation is subject to uncertainties from either the sensor observations or the assumptions used in the retrieval algorithms. The existence of precipitation uncertainties limits the hydrologic applications of the satellite products, as these uncertainties propagate to the hydrologic simulations. Therefore, assessing the accuracy of satellite-precipitation estimates and their corresponding use in simulating hydrologic variables are critical for advancing satellite-based hydrologic applications. The aim of this chapter is to provide the reader with an overview of the current status of satellite-precipitation-driven flood modeling, including evaluation of the error in precipitation estimates and analysis of error propagation in flood modeling. This chapter is structured as follows: in Section 19.2, we introduce some of the most widely used satellite-precipitation products, along with findings from their corresponding ground validation studies. Section 19.3 provides an overview of factors affecting the accuracy of satellite-precipitation-driven flood modeling. Summary and conclusion are reported in Section 19.4.

19.2 SATELLITE PRECIPITATION

19.2.1 Overview

The vast and continuous advancement of space-borne sensors and precipitation retrieval algorithms over the last two decades have made available a number of satellite-precipitation products that have been used for local, regional, and global-scale studies (Wu et al. 2014; Li et al. 2015; Derin et al. 2016). Some of the most frequently used products that are available in quasi-global scale and involve distinct differences in their corresponding retrieval algorithms are listed in Table 19.1. The Tropical Rainfall Measuring Mission (TRMM) Multisatellite Precipitation Analysis (TMPA) 3B42 is a combined VIS–IR and PMW product available from the National Aeronautics and Space Administration (NASA) Goddard Space Flight Center (GSFC) (Huffman et al. 2007; Huffman et al. 2010). Specifically, the PMW precipitation estimates are calculated and then calibrated with respect to the TRMM Combined Instrument and merged. These calibrated PMW precipitation estimates are used to calibrate the IR-based estimates, and then, the IR- and PMW-based estimates are combined. A bias-correction procedure based on monthly rain gauge is implemented as the last step. The 3B42 products are available in real time and research-grade post analysis, based on the inclusion of the gauge adjustment. A climatological calibration algorithm (CCA) has been included in the 3B42 since October 2014, owing to the degradation of TRMM precipitation radar (Huffman et al. 2010). These products are available at 0.25°/3-hourly resolution.

The Precipitation Estimation from Remotely Sensed Information using Artificial Neural Networks (PERSIANN) is an IR-based product developed by the University of California, Irvine (Hsu et al. 1997; Sorooshian et al. 2000). The PERSIANN system scans and extracts features from the IR cloud images with a moving window. The extracted features are classified by the self-organizing feature map (SOFM) algorithm into groups representing the cloud-top characteristics. The features are mapped to rain rate by a multivariate linear function to output precipitation estimates at 0.25°/3-hourly resolution. A bias-corrected version of PERSIANN is also available. This product maintains the total monthly precipitation estimates of the Global Precipitation Climatology Project (GPCP) product and the spatiotemporal patterns of the original PERSIANN (Adler et al. 2003; Huffman et al. 2009). Hong et al. (2004) proposed the PERSIANN-CCS (Clouds Classification System) based on a procedure similar to the PERSIANN product. The PERSIANN-CCS extracts

TABLE 19.1

A List of Satellite-Precipitation Products Involved in This Chapter

Product	Source	Type	Resolution	
			Space	Time
TMPA-3B42-RT	NASA GFSC	VIS-IR + PMW	0.25°	3-hour
TMPA-3B42		VIS-IR + PMW + G		
NRL-blended	U.S. NRL	VIS-IR + PMW	0.25°	3-hour
PERSIANN	University of	VIS-IR	0.25°	3-hour/6-hour
PERSIANN-adj	California, Irvine	VIS-IR + G		
PERSIANN-CCS		VIS-IR + R	0.04°	1-hour
H-E	NOAA STAR	VIS-IR + R	0.04°	Quarter/half-hour
CMORPH	NOAA CPC	PMW	0.25/0.072°	Half-hour/3-hour
CMORPH-adj		PMW + G		
CMORPH-KF		VIS-IR + PMW	0.072°	Half-hour
GSMaP-MVK	JAXA	PMW	0.1°	1-hour
GSMaP-MVK-adj		PMW + G		
IMERG	NASA GFSC	VIS-IR + PMW	0.1°	Half-hour
IMERG-adj		VIS-IR + PMW + G		

Note: G and R represent information from gauge and radar rainfall, respectively. Products with adj as suffix are the gauge-adjusted products.

the cloud features based on the segmentation of the IR cloud images and then classifies the cloud features by the SOFM algorithm. The cloud-top brightness temperature (T_b) and rainfall relationships are calibrated for the clusters by using gauge-adjusted radar rainfall datasets at hourly resolution. The PERSIANN-CCS has a higher space–time resolution available at 0.04°/hourly.

The National Oceanic and Atmospheric Administration (NOAA) Climate Prediction Center (CPC) morphing technique (CMORPH) is a PMW-only global-scale satellite-precipitation product (Joyce et al. 2004). It is based on the propagation of PMW-derived precipitation estimates by the IR-derived cloud system advection vectors (CSAVs). Specifically, the precipitation estimates are derived from the PMW observations, and CSAVs are generated based on the GEO–IR imagery at half-hourly interval. The PMW precipitation estimates are propagated and morphed spatially by a time-weighted linear interpolation. A bias-corrected version of CMORPH is also available by matching the probability distribution function (PDF) with the interpolated daily gauge data over land (Xie et al. 2011). These two versions of CMORPH are available at 0.072°/half-hourly and 0.25°/3-hourly resolutions. Another type of CMORPH available is the Kalman filter-based CMORPH (CMORPH-KF), which also includes the PMW-calibrated IR-based precipitation estimates as part of the inputs for propagation when no PMW observations are available. The PMW-calibrated IR-based precipitation estimates are used to update the propagated PMW estimates through the use of the Kalmar filter (Joyce and Xie, 2011). Besides, more PMW sensors are included in the CMORPH-KF algorithm compared with the original CMORPH. This product is available with 0.072°/half-hourly resolution.

Another PMW-based satellite-precipitation product is the Global Satellite Mapping of Precipitation (GSMaP), produced by the Japan Aerospace Exploration Agency (JAXA). The GSMaP-MVK (Motion Vector Kalman filter) product applies a similar approach to propagate the PMW estimates by the motion vector of clouds derived from the IR imagery. Compared with CMORPH, GSMaP-MVK uses a Kalman filter to refine the propagated PMW precipitation estimates based on the IR T_b data and surface precipitation. Following the same adjustment scheme as the CMORPH, the gauge-adjusted GSMaP is also produced (Mega et al. 2014). These two GSMaP products are available at 0.1°/hourly resolution.

The current state of the art in global satellite-precipitation estimates is related to the recently launched Global Precipitation Measurement (GPM) mission (Hou et al. 2014). An example of GPM-related products is the Integrated Multisatellite Retrievals for GPM (IMERG) product (Huffman et al. 2015). This product draws on strengths from three prior multisatellite algorithms, namely the TMPA (Huffman et al. 2007), the CMORPH-KF (Joyce and Xie, 2011), and the PERSIANN-CCS (Hong et al. 2004). The PMW estimates are intercalibrated and merged to create the precipitation estimates at 0.1°-by-0.1° half-hourly resolution. These PMW estimates are morphed by the PMW-calibrated GEO–IR images, following the Kalman filter framework, as developed in Joyce and Xie (2011). The calibrated GEO–IR images are created by following the procedures adopted in the PERSIANN-CCS system. For gaps of the PMW estimation longer than 90 min, precipitation rates are defined by the GEO–IR estimates. The gauge-based information is also infused to the IMERG product to create the gauge-adjusted IMERG by following the same procedures as those for the TMPA (Huffman et al. 2007).

Other example satellite-precipitation products listed in Table 19.1 are two combined VIS–IR and PMW estimations and a VIS–IR-only estimation. The U.S. Naval Research Laboratory (NRL)-blended satellite-precipitation technique is a combined VIS–IR- and PMW-based estimation available at 0.25°/3-hourly (Turk and Miller, 2005). Precipitation estimations are retrieved using the cloud-top temperature based on the lookup table, mapping the T_b with the PMW precipitation estimates. The Hydro-Estimator (H-E) is a GEO–IR-based precipitation estimation provided by the NOAA Center for Satellite Applications and Research (STAR) (Scofield and Kuligowski, 2003). The H-E algorithm uses a power law function to fit the radar rain rate with observed T_b. Data from numerical weather prediction models are utilized to correct for evaporation of raindrops, topographic influence on rainfall, and other factors.

19.2.2 Evaluation of the Satellite-Precipitation Products

The ground-based evaluation of satellite-precipitation products is an essential component in satellite-based hydrologic applications, because it provides information on the relative accuracy of products, thus allowing to identify the advantages and limitations of the various satellite-precipitation estimates with respect to different hydrologic applications. The importance of the evaluation of satellite-precipitation datasets is reflected in a large number of investigations that have been conducted over different regions of the globe (see Maggioni et al. [2016] for a review), and it has furthermore motivated the scientific community to establish scientific groups dedicated to this task, for example, the International Precipitation Working Group (Turk and Bauer, 2006). Validation of satellite-precipitation products is carried out by comparing rainfall estimates to those derived from ground-based networks. Observations from rain gauge networks are widely used for validation of satellite-precipitation products at a global scale (Stampoulis and Anagnostou, 2012; Chen et al. 2013; Cattani et al. 2016; Derin et al. 2016). National mosaics of weather radar and rain gauge rainfall products are typically available through operational National Weather Services (Lin and Mitchell, 2005; Kamiguchi et al. 2010; Boudevillain et al. 2011; Figueras i Ventura and Tabary, 2013; Zhang et al. 2015), which in several studies have been used to validate satellite-precipitation products. For example, the Stage IV radar/gauge precipitation product (Lin and Mitchell, 2005) has been used to evaluate satellite-precipitation products over the continental United States (AghaKouchak et al. 2011; Nikolopoulos et al. 2015; Dis et al. 2016; Zhang et al. 2016; to name a few).

The strong regional dependence of the satellite-precipitation accuracy, as it is nicely portrayed in the recent work of Derin et al. (2016) and depicted in Figure 19.1, shows clearly that the accuracy of satellite-precipitation products varies across regions and products. This suggests that there is no single product that can be identified as a *global optimum* (even with the adjustments from ground-gauge information), and therefore, region-specific evaluations are needed to gain a better understanding of the performance of different dataset (Derin et al. 2016).

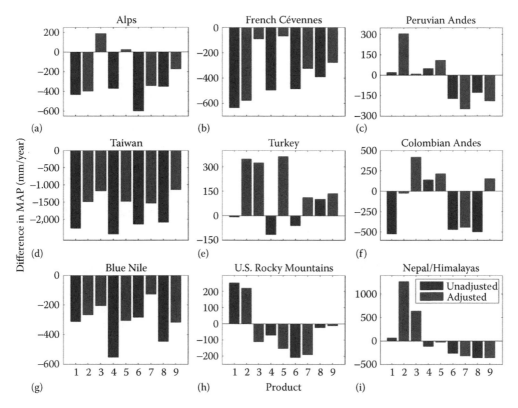

FIGURE 19.1 Differences in mean annual precipitation of (a) the Alps, (b) the French Cévennes, (c) the Peruvian Andes, (d) Taiwan, (e) Turkey, (f) the Colombian Andes, (g) the Blue Nile, (h) the U.S. Rocky Mountains, and (i) Nepal/the Himalayas for every satellite-precipitation product. The numbers on x-axis stands for the 3B42-RT, 3B42-CCA, 3B42-v7, PERSIANN, PERSIANN-adj, CMORPH, CMORPH-adj, GSMaP-MVK, and GSMaP-MVK-adj. (Based on Derin, Y. et al., *J. Hydrometeor.*, 17, 1817–1836, 2016, Figure 4.)

A number of evaluation studies of various satellite-precipitation products have been (and are still being) carried out. Some of those are described later as examples to provide the reader with an overview of past and recent work on this. Cattani et al. (2016) categorized the area of East Africa into eight clusters representing different precipitation climates and conducted assessments of six satellite products to the Global Precipitation Climatology Centre Full Data Reanalysis, version 6, product; they showed that the error characteristics of satellite precipitation are different for the eight clusters. Salio et al. (2015) evaluated six satellite-precipitation products by a 5414-station network over five subregions of South America. Their results revealed different patterns on the systematic and random components of error over different seasons for the five subregions. Bharti and Singh (2015) evaluated the gauge-adjusted 3B42, version 7, product over the Himalayan region with respect to the India Meteorological Department (IMD) rain gauge network and concluded that the magnitudes of error reduce from the low- to high-elevation bands. Mei et al. (2014) compared four satellite-precipitation products to a dense rain gauge network (1 gauge per 53 km²) over the mountainous basin in the Eastern Italian Alps; the results demonstrated lower magnitudes of systematic bias for basins with area greater than 500 km². More studies on the evaluation of satellite-precipitation products are available based on different study areas over the globe in the literature. Chen et al. (2013) evaluated the real-time and post-real-time 3B42, versions 6 and 7, products over China with respect to the China daily Precipitation Analysis Products (CPAP); they found that the satellite products over-/underestimated precipitation in the arid/humid region; the scores of probability of detection (POD) and critical success index (CSI) were low for rainfall rates greater than 50 mm/day over the arid region. Stampoulis and Anagnostou (2012) conducted the analysis of the 3B42, version 6, and CMORPH products with respect to 825 meteorology stations covering the European Continent; they concluded that increasing rainfall intensity leads to an increasing underestimation of both satellites products. AghaKouchak et al. (2011) evaluated four satellite-precipitation products with respect to the Stage IV radar rainfall measurement over the South Great Plain. The study reviewed that the near-real-time CMORPH, PERSIANN, and 3B42, version 6, product over-/underestimated precipitation during the warm/cold season, whereas the gauge-adjusted 3B42, version 6, product showed the opposite trend.

A general finding from the literature on the accuracy of satellite-precipitation products is the regional dependency of products' error characteristics and that there is no single product that can be considered the best. The accuracy of satellite-precipitation products is related to the types of precipitation measurements involved in the retrieval algorithm (Michaelides et al. 2009; Nikolopoulos et al. 2015; Duan et al. 2016) and can vary due to the physiography of the study areas (geomorphology, land use, land cover, precipitation climate, and so on) (Yong et al. 2015; Derin et al. 2016). Besides, the designated space–time evaluation scales and resolutions affect the products' performance (Vergara et al. 2013).

19.3 CONTROL FACTORS IN SATELLITE-BASED FLOOD APPLICATIONS

Precipitation estimated from satellite remote sensing constitutes a viable solution for hydrologic simulations over data-scare regions (e.g., mountainous region and tropical rain forest). However, the satellite-precipitation products are subjected to systematic and random errors that propagate to corresponding hydrologic simulations. Owing to the nonlinear transformation of rainfall to other hydrologic variables (e.g., soil moisture and runoff), the error in simulated variables can be reduced or increased. The properties of the error propagation are regional-dependent and can vary based on the designated scales of modeling and the characteristics of the targeted hydrologic variable. Satellite-rainfall error-correction procedures commonly based on reference precipitation information can potentially improve the hydrologic performance of satellite products, but the degree of effectiveness depends again on several factors (e.g., density of in situ observations and topography). Recalibration of hydrologic models to account for errors in satellite-precipitation estimates has been used as an alternative approach to improve the hydrologic performance. All these aspects are discussed in more detail in the following section.

19.3.1 REGIONAL DEPENDENCY

The hydrologic performance of satellite-precipitation products can vary from region to region, based on the local physiography, the dominant precipitation type, and other climatic factors. For example, hydrologic modeling over complex-terrain basins is more challenging in terms of the simulation accuracy because of the high spatiotemporal variability of rainfall and high heterogeneity in runoff-generation routing processes. Tobin and Bennett (2010) studied the flood response of the San Pedro River basin in Arizona (1036–2885 m a.s.l.), a typical open-book basin, with mountains riming the eastern and western boundaries. The results show reasonable performance of the reference flow simulation compared with the observation but significant overestimation by the 3B42-RT, version 6, product (Table 19.2). Another case study conducted over the Siloam basin in Arkansas, a basin that is free of major complications, such as orographic influences (285–590 m a.s.l), significant snow accumulation, and stream regulations, indicates significantly better performance by the 3B42-RT, version 6 (Behrangi et al. 2011). This simple comparison, involving similar scale of basins and length of study period (3 years for both studies), provides clear evidence on the variability and regional dependence in performance of satellite-based flood simulation studies.

Another factor that is embedded in the observed regional difference is the availability and quality of the ground-gauge networks. A comparison of the error metrics derived from the gauge-adjusted 3B42, version 6, products between the Gilgel Abay study and the San Pedro and Siloam basin demonstrates that the inclusion of ground-based information in satellite-precipitation

TABLE 19.2
Error Metrics of Flow Simulation Derived with Respect to the Observed Flow Taken from Tobin and Bennett (2010), Behrangi et al. (2011), and Bitew and Gebremichael (2011)

Basin (km²)	Elevation (m a.s.l.)	Reference MRE	Reference NSI	3B42-RT, Version 6 MRE	3B42-RT, Version 6 NSI	3B42, Version 6 MRE	3B42, Version 6 NSI
San Pedro (1971)	1036–2885	8	0.9	165	−1.4	8	0.9
Siloam (1489)	285–590	−9.9	0.84	14.5	0.62	−6.8	0.71
Gilgel Abay (1656)	1880–3530	−2	0.76	−20	0.76	−75	<0

Note: MRE and NSI stand for the mean relative error and Nash Sutcliffe Index.

retrievals plays different role for the areas (Table 19.2). Bitew and Gebremichael (2011) showed that for Gilgel Abay, flow simulations using the gauge adjusted 3B42, version 6, product yielded higher underestimation compared with its real-time counterpart. However, the use of gauge-adjusted 3B42, version 6, clearly benefited the hydrologic simulation for the other study basins in Tobin and Bennett (2010) and Behrangi et al. (2011). Previous works of Gourley et al. (2011) and Wilk et al. (2006) have revealed that the highly sparse distributed feature of rain gauges (such as that of the Gilgel Abay case) over mountainous terrain can introduce systematic bias to the gauge-adjusted satellite precipitation.

19.3.2 SPACE–TIME SCALES OF APPLICATIONS

The target space–time scales and resolution of the hydrologic modeling system are critical aspects that control the performance in flow simulations. In general, evaluation of streamflow simulations at coarser temporal resolutions (e.g., monthly) result in better performance (Behrangi et al. 2011; Meng et al. 2014; Li et al. 2015). Tong et al. (2014) and Meng et al. (2014) simulated the flow for the Upper Yellow Basin (121,972 km² basin area). Their results indicate that, at monthly scale, the values of Nash Sctuliffe Index (NSI) of flow simulation are higher than the daily NSI for both the reference- and satellite-driven scenarios (Table 19.3). Similar results are demonstrated from Behrangi et al. (2011), conducted for the Siloam basin. The study compares the results of flow simulations at 6-hourly and monthly resolutions and is able to reveal decreases in root-mean-square error (RMSE) values from 6-hourly to monthly results.

TABLE 19.3
Impact on Hydrologic Performance Indices from the Temporal Aggregation

	Upper Yellow Basin (121,972 km²) Reference	Upper Yellow Basin (121,972 km²) 3B42, Version 6	Siloam Basin (1489 km²) Reference	Siloam Basin (1489 km²) 3B42, Version 6	Siloam Basin (1489 km²) CMORPH	Siloam Basin (1489 km²) PERSIANN
6-hourly			22.03	25.97	64.24	26.88
Daily	0.524	0.632				
Monthly	0.744	0.804	6.57	6.92	18.35	10.71

Source: Statistics are taken from Tong, K. et al., Evaluation of satellite precipitation retrievals and their potential utilities in hydrologic modeling over the Tibetan Plateau, *J. Hydrol.*, 2014, 519: 423–437, and Behrangi, A. et al., 2011. Hydrologic evaluation of satellite precipitation products over a mid-size basin. *J. Hydrol.*, 2011, 397(3–4): 225–237.
Note: The respective statistics used in the Upper Yellow and Siloam study are the Nash Sutcliffe Index and root mean square error (mm/h).

The spatial scale of the basin has also been demonstrated to play an important role in the performance of satellite-based flood simulations (Nikolopoulos et al. 2010; Maggioni et al. 2013; Mei et al. 2016a). The general finding is that larger amount of error can be buffered as the size of the basin increases due to the increased space–time filtering of precipitation by the catchment surface processes. Studies on the nested sub-basins of the Tar-Pamlico river basin (five sub-basins from 529 to 5,709 km^2), low-elevation mild-slope basin, have shown that the random error in runoff simulations decreases with the increase in basin area (Maggioni et al. 2013; Vergara et al. 2013). Maggioni et al. (2013) also shows that the increase in basin area leads to decrease in the ratio of random error of flow over random error of precipitation, indicating higher degree of error-dampening effects in larger basin scales. The decrease in random error of flow simulation with basin scale is also revealed by Mei et al. (2016a) on the mountainous nested Upper Adige river basin (16 sub-basins from 255 to 6,967 km^2) in Eastern Italian Alps.

19.3.3 FLOOD CHARACTERISTICS

The performance of satellite-based flood modeling varies considerably with season, which subsequently relates to the seasonal characteristics of rainfall and corresponding flood response. This is exemplified by the study of Mei et al. (2016a), which evaluated separately the flood simulations for warm (May–August) and cold (September–November) periods in Eastern Italian Alps. Specifically for their study, it was shown that performance of satellite-based simulations in terms of the systematic bias, the random error, and correlation coefficient was superior for the warm period. Kim et al. (2016) evaluated the utilities of four satellite-precipitation products in runoff modeling for the Soyang Dam basin in South Korea. Their study basin is characterized by well-defined cold, dry winters and hot, wet summers, and they concluded that the biases in rainfall are directly dictated to the biases in flow simulation for the wet period, whereas the uncertainties in flow simulations during the dry period are attributed to the false snowfall estimations by the model.

The performance of satellite-based flood modeling is different for flood types and the flood event properties of interest. Figure 19.2 exhibits the mean relative error (MRE) of four flood event properties for two different types of flood event available from a flood event database for basins located in Eastern Italian Alps (Zoccatelli et al. 2016). The figure indicates more severe underestimations (overestimations) with larger (lower) variability in the simulations of cumulative rainfall, flow volume, and flood peak (time lag) parameter for the flash flood events compared with the long rain flood events. Besides, the gauge adjustments are able to bring up the magnitude of flow simulations, which in turn modulate the underestimations in the cumulative rainfall, flow volume, and flood peak parameter by the CMORPH- and PERSIANN-based simulations; they also lead to overestimations in these parameters by the TMPA-based simulations. For the time lag parameter, consistent pattern is shown among products (with or without gauge adjustment), referring to the fact that effects from gauge adjustments on event timing are negligible. A recent study focusing on the use of satellite precipitation in modeling different types of floods from the same flood event database is provided in Mei et al. (2016b). Their results revealed that the long rain flood events exhibited higher magnitudes of random error than the flash flood events for the event hyetographs and hydrographs. In addition, higher degrees of dampening effects on the random error are found for the long rain flood cases. The study also reveals that the error-dampening effect tends to be more linear to events with higher runoff coefficient, which suggests the impact of soil wetness condition on rainfall error propagation, as it was shown by (Nikolopoulos et al. 2013).

19.3.4 IMPROVING SATELLITE-PRECIPITATION PRODUCTS FOR FLOOD STUDIES

Clearly, the various evaluation studies suggest that further improvement in the accuracy of satellite-precipitation data in terms of their magnitude and space–time distribution patterns is needed to enhance the usability of these products for flood-related studies. Numerous efforts have focused on

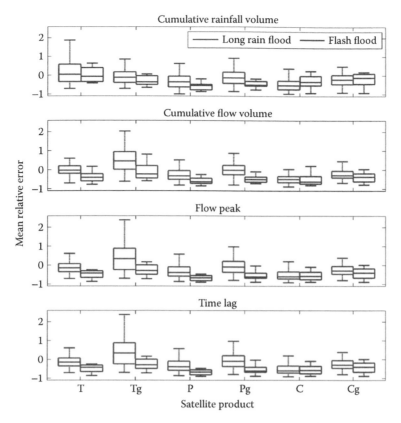

FIGURE 19.2 Mean relative error of flood event properties for different types of flood events. T, P, and C stand for the TMPA-3B42, version 7, PERSIANN, and CMORPH satellite products. Products with and without the lowercase g represent the near-real-time and gauge-adjusted products.

deriving procedures for correcting satellite-precipitation estimates. In general, these procedures involve adjustment of satellite-precipitation estimates, using additional precipitation information from ground-based sources (radar and rain gauge), atmospheric models, and other auxiliary atmospheric and land surface. Nikolopoulos et al. (2013) carried out a mean field bias (MFB) adjustment based on the available radar-rainfall field. Results from the study demonstrated slight improvement in hydrologic simulations. Similar method to remove the MFB was investigated in Habib et al. (2014) over the Gilgel Abay. The study paid attention to three different schemes for retrieving the MFB with respect to the interpolated rain gauge data used. These schemes included a space–time fixed (TSF) one, a time varied (TV) one, and a space–time varied (TSV) one (see Habib et al. [2014] for details). Results indicated improvements from all the adjustment schemes; however, the space–time varied one outperformed the other two in terms of the improved accuracy. Tobin and Bennett (2010) developed a filter based on autoregression (AR) to adjust the satellite products with respect to information from rain gauge and radar precipitation. Results indicated lower degree of MRE when the satellite products are adjusted to the radar measurements. However, despite the apparent improvement in satellite-precipitation estimates, results on the hydrologic simulations for many cases maintained significant discrepancies (MRE ranged from −90% to 40%) compared with the corresponding reference (Table 19.4). This suggests that error in flood simulation can be significantly attributed to the random error component as well as to resolution factors (Nikolopoulos et al. 2010; Maggioni et al. 2013; Vergara et al. 2013).

Application of stochastic error modeling for satellite-precipitation correction is another methodology that has been followed by several researchers (Maggioni et al. 2013; Müller and

TABLE 19.4
MREs of the Hydrologic Simulations of Satellite-Precipitation Products' Statistical Bias Adjustments

Basin (Area, km²)	Type	3B42, Version 6	CMORPH	PERSIANN-CCS
Fella (623)	Original	−98	−95	−99.7
	MFB	−93	−86	−94
	MFB$_{dyn}$	−94	−90	−90
San Pedro (1971)	Original	165	63	
	AR filter to gauge	35	41	
	AR filter to MPE	16	34	
Gilgel Abay (1656)	Original		26.6	
	TSF		23.8	
	TV		23.6	
	TSV		16.9	

Thompson, 2013; Falck et al. 2015). Falck et al. (2015), Nikolopoulos et al. (2010), and Maggioni et al. (2013) investigated the hydrologic uses of satellite precipitation ensemble produced by the two-dimensional satellite rainfall error model (SREM2D, Hossain and Anagnostou [2006]). Their results indicated that the flow simulations derived from the SREM2D-corrected precipitation ensembles are able to resolve the biases in flow simulations driven by the unperturbed satellite precipitation. In addition, Maggioni et al. (2013) revealed that the random error is larger for the ensemble precipitation than the corresponding runoff simulations in small-scale basins. However, this is reversed as the basin scale increases. Nikolopoulos et al. (2010) showed that there exists a linear relationship on systematic error derived between the ensemble mean of rainfall and the mean of ensemble-driven flow simulations.

A novel approach for correcting satellite-precipitation products by matching the PDF between satellite-rainfall estimates and high-resolution atmospheric simulations was proposed in Zhang et al. (2016), who carried out a study for six hurricane landfall precipitation events over 20 basins in the southern Appalachian. The study assessed the hydrologic potential of CMORPH adjusted by the weather research and forecasting (WRF) atmospheric model. The evaluation of corresponding hydrologic simulations showed that the WRF-adjusted CMORPH nearly eliminated biases in the simulations of peak runoff and cumulative flow volume (5% for peak runoff and nearly zero bias for cumulative flow volume) compared with the original CMORPH (33% and 29% underestimation) and gauge-adjusted CMORPH (12% and 10% underestimation). This method improves considerably the hydrologic simulations and appears more effective than the common gauge-based adjustment approach.

19.3.5 MODEL RECALIBRATION

The optimum parameters of hydrologic modeling for a basin are typically retrieved by using the ground-based precipitation measurement as reference. The optimized parameters are considered as the synthetic true parameter set, representing the hydrologic response of the basin, and are directly applied on the different satellite-precipitation products for simulations. This often leads to deterioration in the resulted streamflow simulations, because the two precipitation products are different from each other in characteristics. Model recalibration with respect to the individual satellite-precipitation products consists of another avenue for improving the performance of satellite-based hydrologic simulation. In their recent work, Skinner et al. (2015) showed the potential of improving satellite-based ensemble hydrologic simulations if model recalibration is considered for the whole ensemble. However, model recalibration for different rainfall input has to be treated with caution because many studies have reported that the recalibrated model parameters may reach values that exceed a realistic range of the parameters. Nikolopoulos et al. (2013) showed that the recalibrated saturated hydraulic conductivity parameters (K_{sat} in mm/h) are more than three times of the K_{sat} derived from the radar-driven calibration because of the severe underestimation from satellite products. Xue et al. (2013) also showed that severe underestimation of 3B42, version 6, resulted in model parameters that are significantly different than the optimum gauge-derived parameters, in order to allow the 3B42, version 6, forced model to capture the hydrologic response.

Studies on large-scale basins also report similar issues on model recalibration. Yong et al. (2012; 2010) integrates the real-time and post-real-time 3B42, version 6, with the three-layer variable infiltration capacity (VIC-3L) hydrologic model to simulate flow for the Laohahe basin. Results suggested that the recalibrated parameters for the gauge-adjusted 3B42 were closer to the rain gauge ones. However, due to the severe overestimation of rainfall from the real-time 3B42, the soil layer thickness parameter was forced to take value above the upper bound (2 m) to increase the water storage capacity of the soil and thus compensate for the overestimation in precipitation.

19.4   CONCLUSIONS

The integration of satellite-precipitation products with hydrologic models for modeling floods is a topic of worldwide concern. In this chapter, we provided an overview of the past and recent studies over the globe that highlight the current state and limitations of satellite-precipitation-driven flood predictions. Uncertainties in flow simulations driven by the satellite precipitation depend on a wide variety of factors, including regional geomorphology and climate characteristics, the scales of application, as well as the typology of the targeted floods. Results suggest that the space–time scale of application exerts an important control on the performance of satellite-based hydrologic simulations. Large basins (>1000 km²) and coarse temporal scales (i.e., monthly) seem to be a favorable setting for skillful satellite-based hydrologic predictions. Flood modeling applications over complex terrains remains the most challenging task, given that both precipitation and runoff generation are subject to high space–time variability in these areas. Moreover, different flood types are associated with different performance characteristics, suggesting that seasonality and precipitation/flood event characteristics should be considered when developing error correction procedures for satellite rainfall.

Error correction procedures of satellite-precipitation products and recalibration of hydrologic models are typical approaches aiming to improve the performance of satellite-based hydrologic simulations, but their efficiency is not consistent. Some new approaches that involve high-resolution numerical weather prediction simulations for correcting satellite estimates over complex terrains look promising, but they need to be tested at several regions and climates to conclude on their applicability at a large, global scale. Recalibration of hydrologic model with individual satellite products can improve the performance of the model, but the model parameters are often forced to take unrealistic values. Thus, this should not be treated as a typical method for improving the hydrologic simulation accuracy.

The integration of remotely sensed precipitation with hydrologic models has been the focus of recent hydro-meteorological extremes studies (e.g., flash flood, debris flow, and drought), which is a vastly growing field, owing to the more frequent occurrences of these extreme events and the increased availability of high-resolution satellite-precipitation products. As new satellite missions are launched with new and advanced sensors (Kerr et al. 2001; Entekhabi et al. 2010; Hou et al. 2014), more hydrologic variables can be targeted from the Earth observation systems. Combining satellite observations from different sensors will allow us to develop improved correction algorithms that will better characterize and correct errors in satellite-precipitation estimates. Thus, the use of satellite-precipitation products in observing and predicting catastrophic floods and associated hazards correlates well with the development of integrated products in the GPM era. An example is the IMERG product (Huffman et al. 2015), available since the launch of the GPM satellite in February 2014. This novel algorithm is expected to overcome some issues in precipitation estimation encountered in the TRMM-era precipitation products (e.g., coverage of sensors and quantification of snow) and will provide estimations with higher accuracy at finer resolution, which could potentially advance flood-modeling applications worldwide.

REFERENCES

Adler, R. F. et al., 2003. The version-2 global precipitation climatology project (GPCP) monthly precipitation analysis (1979–Present). *J. Hydrometeor.*, 4(6): 1147–1167.

AghaKouchak, A. et al., 2011. Evaluation of satellite-retrieved extreme precipitation rates across the central United States. *J. Geophys. Res.*, 116: D02115.

Behrangi, A. et al., 2011. Hydrologic evaluation of satellite precipitation products over a mid-size basin. *J. Hydrol.*, 397(3–4): 225–237.

Berne, A. and Krajewski, W. F., 2013. Radar for hydrology: Unfulfilled promise or unrecognized potential? *Adv. Water Res.*, 51: 357–366.

Bharti, V. and Singh, C., 2015. Evaluation of error in TRMM 3B42V7 precipitation estimates over the Himalayan region. *J. Geophys. Res. Atmos.*, 120(24): 12458–12473.

Bitew, M. M. and Gebremichael, M., 2011. Assessment of satellite rainfall products for streamflow simulation in medium watersheds of the Ethiopian Highlands. *Hydrol. Earth Syst. Sci.*, 15(4): 1147–1155.

Boudevillain, B. et al., 2011. The Cévennes-Vivarais Mediterranean hydrometeorological observatory database. *Water Resour. Res.*, 47(7): W07701.

Cattani, E., Merino, A. and Levizzani, V., 2016. Evaluation of monthly satellite-derived precipitation products over East Africa. *J. Hydrometeor.*, 17(10): 2555–2573.

Chen, S. et al., 2013. Similarity and difference of the two successive V6 and V7 TRMM multisatellite precipitation analysis performance over China. *J. Geophys. Res. Atmos.*, 118(23): 13060–13074.

Delrieu, G., Bonnifait, L., Kirstetter, P.-E. and Boudevillain, B., 2014. Dependence of radar quantitative precipitation estimation error on the rain intensity in the Cévennes region, France. *Hydrolog. Sci. J.*, 59(7): 1308–1319.

Derin, Y. et al., 2016. Multiregional satellite precipitation products evaluation over complex Terrain. *J. Hydrometeor.*, 17(6): 1817–1836.

Dis, M. O., Anagnostou, E. N. and Mei, Y., 2016. Using high-resolution satellite precipitation for flood frequency analysis: Case study over the Connecticut River Basin (In press). *J. Flood Risk Manag.*

Duan, Z., Liu, J., Tuo, Y., Chiogna, G. and Disse, M., 2016. Evaluation of eight high spatial resolution gridded precipitation products in Adige Basin (Italy) at multiple temporal and spatial scales. *Sci. Total Environ.*, 573: 1536–1553.

Entekhabi, D. et al., 2010. The soil moisture active passive (SMAP) mission. *Proc. IEEE*, 98(5): 704–716.

Falck, A. S. et al., 2015. Propagation of satellite precipitation uncertainties through a distributed hydrologic model: A case study in the Tocantins–Araguaia basin in Brazil. *J. Hydrol.*, 527: 943–957.

Figueras i Ventura, J. and Tabary, P., 2013. The new French operational polarimetric radar rainfall rate product. *J. Appl. Meteor. Climatol.*, 52(8): 1817–1835.

Gourley, J. J. et al., 2011. Hydrologic evaluation of rainfall estimates from radar, satellite, gauge, and combinations on Ft. Cobb Basin, Oklahoma. *J. Hydrometeor.*, 12(5): 973–988.

Habib, E. et al., 2014. Effect of bias correction of satellite-rainfall estimates on runoff simulations at the source of the Upper Blue Nile. *Remote Sens.*, 6(7): 6688–6708.

Hong, Y., Hsu, K.-L., Sorooshian, S. and Gao, X., 2004. Precipitation estimation from remotely sensed imagery using an artificial neural network cloud classification system. *J. Appl. Meteor.*, 43(12): 1834–1853.

Hossain, F. and Anagnostou, E. N., 2006. A two-dimensional satellite rainfall error model. *IEEE T. Geosci. Remote*, 44(6): 1511–1522.

Hou, A. Y. et al., 2014. The global precipitation measurement mission. *Bull. Amer. Meteor. Soc.*, 95(5): 701–722.

Hsu, K.-l., Gao, X., Sorooshian, S. and Gupta, H. V., 1997. Precipitation estimation from remotely sensed information using artificial neural networks. *J. Appl. Meteor.*, 36(9): 1176–1190.

Huffman, G. J., Adler, R. F., Bolvin, D. T. and Gu, G., 2009. Improving the global precipitation record: GPCP Version 2.1. *Geophys. Res. Lett.*, 36(17): L17808.

Huffman, G. J., Adler, R. F., Bolvin, D. T. and Nelkin, E. J., 2010. The TRMM multi-satellite precipitation analysis (TMPA). In: M. Gebremichael and F. Hossain (Eds.) *Satellite Rainfall Applications for Surface Hydrology.* New York: Springer, pp. 3–22.

Huffman, G. J. et al., 2007. The TRMM multisatellite precipitation analysis (TMPA): Quasi-global, multiyear, combined-sensor precipitation estimates at fine scales. *J. Hydrometeor.,* 8(1): 38–55.

Huffman, G. J. et al., 2015. NASA Global Precipitation Measurement (GPM) Integrated Multi-satellitE Retrievals for GPM (IMERG), Greenbelt, MD: NASA.

Joyce, R. J., Janowiak, J. E., Arkin, P. A. and Xie, P., 2004. CMORPH: A method that produces global precipitation estimates from passive microwave and infrared data at high spatial and temporal resolution. *J. Hydrometeor.,* 5(3): 487–503.

Joyce, R. J. and Xie, P., 2011. Kalman filter–based CMORPH. *J. Hydrometeor.,* 12(6): 1547–1563.

Kamiguchi, K. et al., 2010. Development of APHRO_JP, the first Japanese high-resolution daily precipitation product for more than 100 years. *Hydrolog. Res. Let.,* 4: 60–64.

Kerr, Y. H. et al., 2001. Soil moisture retrieval from space: The soil moisture and ocean salinity (SMOS) mission. *IEEE Trans. Geosci. Rem Sens.* 39(8): 1729–1735.

Kim, J. P. et al., 2016. Hydrological utility and uncertainty of multi-satellite precipitation products in the mountainous region of South Korea. *Remote Sens.,* 8(7): 608.

Kirstetter, P.-E. et al., 2015. Probabilistic precipitation rate estimates with ground-based radar networks. *Water Resour. Res.,* 51(3): 1422–1442.

Lin, Y. and Mitchell, K. E., 2005. 1.2 the NCEP stage II/IV hourly precipitation analyses: Development and applications. *19th Conference. Hydrology*, San Diego, CA: American Meteorological Society.

Li, Z. et al., 2015. Multiscale hydrologic applications of the latest satellite precipitation products in the Yangtze river basin using a distributed hydrologic model. *J. Hydrometeor.,* 16(1): 407–426.

Maggioni, V., Meyers, P. C. and Robinson, M. D., 2016. A review of merged high-resolution satellite precipitation product accuracy during the tropical rainfall measuring mission (TRMM) era. *J. Hydrometeor.,* 17(4): 1101–1117.

Maggioni, V. et al., 2013. Investigating the applicability of error correction ensembles of satellite rainfall products in river flow simulations. *J. Hydrometeor,* 14(4): 1194–1211.

Mega, T. et al., 2014. Gauge adjusted global satellite mapping of precipitation (GSMaP_Gauge). In: *XXXIth URSI*, Beijing, IEEE. pp. 1–4.

Mei, Y. et al., 2016. Error analysis of satellite precipitation-driven modeling of flood events in complex Alpine Terrain. *Remote Sens.,* 8(4): 293.

Mei, Y., Nikolopoulos, E. I., Anagnostou, E. N. and Borga, M., 2016. Evaluating satellite precipitation error propagation in runoff simulations of mountainous basins. *J. Hydrometeor.,* 17(5): 1407–1423.

Mei, Y., Nikolopoulos, E. I., Anagnostou, E. N. and Marco, B., 2014. Error analysis of satellite rainfall products in mountainous basins. *J. Hydrometeorol.,* 15(5): 1778–1793.

Meng, J. et al., 2014. Suitability of TRMM satellite rainfall in driving a distributed hydrological model in the source region of Yellow River. *J. Hydrol.,* 509: 320–332.

Michaelides, S. et al., 2009. Precipitation science: Measurement, remote sensing, climatology and modelling EGU08 precipitation European Geosciences Union General Assembly. *Atmos. Res.,* 94(4): 512–533.

Müller, M. F. and Thompson, S. E., 2013. Bias adjustment of satellite rainfall data through stochastic modeling: Methods development and application to Nepal. *Adv. in Water Resour.,* 60: 121–134.

Nikolopoulos, E. I., Anagnostou, E. N. and Borga, M., 2013. Using high-resolution satellite rainfall products to simulate a major flash flood event in Northern Italy. *J. Hydrometeor.,* 14(1): 171–185.

Nikolopoulos, E. I., Bartsotas, N. S., Anagnostou, E. N. and Kallos, G., 2015. Using high-resolution numerical weather forecasts to improve remotely sensed rainfall estimates: The case of the 2013 Colorado flash flood. *J. Hydrometeor.,* 16(4): 1742–1751.

Nikolopoulos, E. I. et al., 2010. Understanding the scale relationships of uncertainty propagation of satellite rainfall through a distributed hydrologic model. *J. Hydrometeor.,* 11(2): 520–532.

Prat, O. P. and Nelson, B. R., 2015. Evaluation of precipitation estimates over CONUS derived from satellite, radar, and rain gauge data sets at daily to annual scales (2002–2012). *Hydrol. Earth Syst. Sci.,* 19(4): 2037–2056.

Salio, P., Hobouchian, M. P., García Skabar, Y. and Vila, D., 2015. Evaluation of high-resolution satellite precipitation estimates over southern South America using a dense rain gauge network. *Atm. Res.,* 163: 146–161.

Scofield, R. A. and Kuligowski, R. J., 2003. Status and outlook of operational satellite precipitation algorithms for extreme-precipitation events. *Weather Forecast.,* 18(6): 1037–1051.

Skinner, C. J., Bellerby, T. J., Greatrex, H. and Grimes, D. I. F., 2015. Hydrological modelling using ensemble satellite rainfall estimates in a sparsely gauged river basin: The need for whole-ensemble calibration. *J. Hydrol.*, 522: 110–122.

Sorooshian, S. et al., 2000. Evaluation of PERSIANN system satellite–Based estimates of tropical rainfall. *Bull. Amer. Meteor. Soc.*, 81(9): 2035–2046.

Stampoulis, D. and Anagnostou, E. N., 2012. Evaluation of global satellite rainfall products over Continental Europe. *J. Hydrometeorol.*, 13(2): 588–603.

Tobin, K. J. and Bennett, M. E., 2010. Adjusting satellite precipitation data to facilitate hydrologic modeling. *J. Hydrometeor.*, 11(4): 966–978.

Tong, K., Su, F., Yang, D. and Hao, Z., 2014. Evaluation of satellite precipitation retrievals and their potential utilities in hydrologic modeling over the Tibetan Plateau. *J. Hydrol.*, 519: 423–437.

Turk, F. J. and Miller, S. D., 2005. Toward improved characterization of remotely sensed precipitation regimes with MODIS/AMSR-E blended data techniques. *IEEE Trans. Geosci. Remote.* 43(5): 1059–1069.

Turk, J. and Bauer, P., 2006. The international precipitation working group and its role in the improvement of quantitative precipitation measurements. *Bull. Amer. Meteorol. Soc.*, 87(5): 643–647.

Vergara, H. J. et al., 2013. Effects of resolution of satellite-based rainfall estimates on hydrologic modeling skill at different scales. *J. Hydrometeorol.*, 15(2): 593–613.

Wilk, J. et al., 2006. Estimating rainfall and water balance over the Okavango river basin for hydrological applications. *J. Hydrol.*, 331(1–2): 18–29.

Wu, H. et al., 2014. Real-time global flood estimation using satellite-based precipitation and a coupled land surface and routing model. *Water Resour. Res.*, 50(3): 2693–2717.

Xie, P., Yoo, S.-H., Joyce, R. J. and Yarosh, Y., 2011. Bias-corrected CMORPH: A 13-Year analysis of high-resolution global precipitation. Vol. 13, EGU.

Xue, X. et al., 2013. Statistical and hydrological evaluation of TRMM-based Multi-satellite Precipitation Analysis over the Wangchu Basin of Bhutan: Are the latest satellite precipitation products 3B42V7 ready for use in ungauged basins? *J. Hydrol.*, 499: 91–99.

Yong, B. et al., 2010. Hydrologic evaluation of multisatellite precipitation analysis standard precipitation products in basins beyond its inclined latitude band: A case study in Laohahe basin, China. *Water Resour. Res.*, 46(7): W07542.

Yong, B. et al., 2012. Assessment of evolving TRMM-based multisatellite real-time precipitation estimation methods and their impacts on hydrologic prediction in a high latitude basin. *J. Geophys. Res.*, 117(D9): D09108.

Yong, B. et al., 2015. Global view of real-time TRMM multisatellite precipitation analysis: Implications for its successor global precipitation measurement mission. *Bull. Amer. Meteor. Soc.*, 96(2): 283–296.

Zhang, X., Anagnostou, E. N. and Vergara, H., 2016. Hydrologic evaluation of NWP-adjusted CMORPH estimates of hurricane-induced precipitation in the Southern Appalachians. *J. Hydrometeor.*, 17(4): 1087–1099.

Zhang, J. et al., 2015. Multi-radar multi-sensor (MRMS) quantitative precipitation estimation: Initial operating capabilities. *Bull. Amer. Meteor. Soc.*, 97(4): 621–638.

Zoccatelli, D., Parajka, J., Gaál, L., Blöschl, G. and Borga, M., 2014. A process flood typology along an Alpine transect: Analysis based on observations and modelling approaches. In EGU General Assembly Conference Abstracts, 16.

Section V

Remote Sensing of Storms

20 Application of Remote-Sensing Images for Post-Wind Storm Damage Analysis

Sudha Radhika, Yukio Tamura, and Masahiro Matsui

CONTENTS

20.1 INTRODUCTION

From the days of cave dwellings to the present day, wind damage to structures and buildings has been a fact of life. The bar chart shown in Figure 20.1, provided by the Munich Re Geo Risks Research, illustrates the world's major natural disasters (1980–2013). It shows a dramatic increase in number of catastrophes from 1987 onward. Majority of the catastrophes are due to meteorological

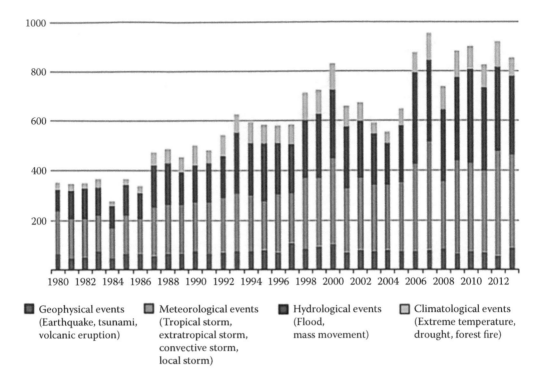

FIGURE 20.1 Bar chart showing the number of natural catastrophe worldwide (1980–2013). (Courtesy of Munich Re Geo Risks Research, Munich, Germany.)

and hydrological events. Among the meteorological events, the major share of losses goes to tropical cyclones, or Hurricanes: "Hugo" (1989), "Andrew" (1992), "Georges" (1998), and so on. These are large storms that form in the Earth's tropical regions just north or south of the equator and travel toward the west and spin in a circular manner about the eye, or the center, of the storm. They result in storm surges, causing eroding of shorelines, flooding areas far inland, and finally causing economic losses and losses of life.

Not only tropical cyclones but also tornadoes account for much of the economic and human losses. A downward rotating air column, formatting into a funnel shape, violently pulling up dust and debris as the wind force increases near the Earth's surface, is the general characteristic of a matured tornado. Such tornadoes finally result in a catastrophic damage to both life and building structures in their path.

It is necessary to obtain detailed damage information from field investigations after strong winds such as tropical cyclones and tornadoes, but this is time-consuming. For quicker responses after catastrophic damage, automated computational investigation becomes inevitable, in order to provide instant help to large damaged areas and to access remote damage areas. Exploitation of remote-sensing technology along with image-processing techniques and the latest pattern-recognition knowledge create a new route for such automated computational investigations. Fast tracking of such tornado-damaged path from wind-borne debris (Radhika et al. 2012) and identification of damaged roofs of buildings by change-detection method, using pre-storm and post-storm imageries (Womble et al. 2007; Radhika et al. 2015), could cover a wider area of investigation at a faster rate.

Past researches have been done on other types of natural disasters such as earthquakes by using aerial images. Hasegawa et al. (2000) identified damaged buildings by maximum likelihood classification method on the Kobe Earthquake in Japan, whereas Mitomi et al. (2001) used the same methods for analysis of the Gujarat Earthquake, India, from aerial television images. Sumer et al. (2004)

used a watershed segmentation method to identify damaged buildings from post-earthquake aerial images, and Ozisik (2004) used edge detection methods from pre- and post-earthquake satellite imageries and post-aerial imageries of Turkey earthquakes. A major contribution to earthquake building-damage detection using satellite imagery was also done by Matsuoka and Yamazaki (2000), using the backscattering coefficient of the European remote sensing satellite–synthetic aperture radar (ERS-SAR) satellite images pre-and post-event of the 1995 Hyogoken-Nanbu (Kobe) Earthquake, Japan. A shadow-based detection of building damage from QuickBird Satellite imagery performed by Vu et al. (2005) was another remarkable contribution to remote-sensing-based natural disaster building-damage investigation.

Many researches have also been done on other natural disasters, such as on wild fires by Ambrosia et al. (1998), on floods by Groeve et al. (2009), on landslides by Danneels et al. (2008), and so on, by computational identification, using low-resolution satellite images. However, in all these methods, both pre- and post-disaster images were necessary.

Tornado-damaged-path tracking was done from low-resolution Landsat satellite imagery by change detection from pre-and post-storm imageries by Myrint et al. (2008) and Thomas et al. (2002). Myrint et al. used a nearest-neighbor-classifier approach, whereas Thomas et al. used a principal component analysis.

The introduction of high-resolution satellite imageries has created a breakthrough in identification of disaster-affected building structures, for rescue purposes as well as for reconstruction. In Womble et al. (2007), Lakshminarasimhan (2004), and Womble (2005), computational analyses were done on building damage detection, mainly using statistical analysis on image pixel radiance value. They classified the damage to building structures into a remote-sensing scale (RS scale).

However, damage tracking by a change-detection algorithm from both pre- and post-disaster imageries faced many difficulties. Immediate availability of both pre- and post-disaster imageries of the same location, need for an error-free image registration procedure for aligning both pre- and post-disaster imageries in order to detect the changes, and so on, are some of the major difficulties that will affect the accuracy of the detection. These difficulties can be minimized if the detection is performed by using only post-storm imagery. Therefore, in the current research, tornado-damaged path is detected from post-storm images alone, which solved the difficulty of non-availability of pre-storm images. After a tornado, there will be a lot of wind-borne debris deposits. A particular pattern is identified for this debris deposit when it is taken as an image by texture-wavelet analysis, using a two-dimensional biorthogonal wavelet. Introduction of this texture-wavelet analysis instead of conventional statistical analysis to identify damaged buildings after the passage of a hurricane was proved efficient by using both pre- and post-storm imageries (Radhika et al. 2015).

This chapter also describes a novel technique for rapid and accurate damage estimation by using post-storm RS imagery alone. Once damaged buildings are identified, the extent of damage to building roofs is estimated, by calculating the percentage area of damage to roof structures, which is very difficult to obtain in a ground survey. This makes damage identification more informative. More accurate and faster damage identification will save more lives and enable more building structures to be restored faster.

20.2 METHODOLOGY

Once a natural disaster occurs, post-storm imagery is acquired and texture-wavelet analysis (Radhika et al. 2010, 2011b, c) is performed to detect wind-borne debris deposit area, leading to tracking of the disaster area. Once the disaster area is tracked, the buildings are segmented from the imagery. From the segmented-building image, wavelet features are extracted and classification of damaged and non-damaged buildings is done by using artificial neural network (ANN) and support vector machine (SVM) classification. In the segmented roof imagery, it is observed that damaged roof areas show a particular pattern when compared with undamaged smooth image portions of roof structures. These patterns can be successfully recognized by applying texture-wavelet analysis

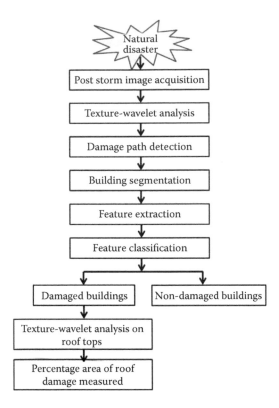

FIGURE 20.2 Flow chart showing detection of damaged buildings and estimation of degree of damage to building roofs.

on segmented-building roof image portions; that is, on the buildings images, which are classified as damaged buildings, wavelet-based edge extraction is performed, texture of the broken roof area is identified, and the percentage area of roof damage is estimated.

Validation is performed by using visual interpretation and conventional field survey information. The entire methodology is shown as a flow chart in Figure 20.2.

20.3 POST-STORM IMAGE ACQUISITION

The current work utilizes a combination of both low- and high-resolution remote sensing images.

Two different disaster events were selected for analysis, as shown below:

1. Aerial imagery after Saroma, Japan Tornado, 2006.
2. Satellite imagery after U.S. Hurricane "Charley" at Punta Gorda in 2004.

20.3.1 Aerial Imagery after Saroma, Japan Tornado, 2006

Aerial image data of the damaged area after a deadly tornado hit at Saroma town in Hokkaido, on November 7, 2006, was acquired. The three-band natural-color high-resolution (10 cm/pixel) aerial imagery, provided by the Saroma-cho Local Government, is shown in Figure 20.3a. This RGB imagery is of 1/2500 scale. The damage covered an area of roughly 200 m*1500 m and was almost in a straight line (Cao et al. 2006). Field survey data were also collected for validation.

(a)

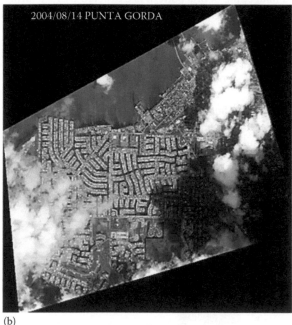

2004/08/14 PUNTA GORDA

(b)

FIGURE 20.3 (a) Post-storm aerial imagery after Saroma, Japan Tornado (2006). Courtesy of Shin Engineering Consultants Co. Ltd. (b) Post-storm satellite imagery after U.S. Hurricane "Charley" at Punta Gorda (2004).

20.3.2 Satellite Imagery after U.S. Hurricane "Charley" at Punta Gorda, 2004

Quick Bird satellite imagery of the disaster location of Punta Gorda, Florida, after Hurricane Charley provided wider information about the area where it had a severe impact. The landfall of Hurricane Charley took place on Friday, August 13, 2004, on the southwest coast of Florida at Charlotte Harbor. A multispectral post-storm imagery was taken by optical probes of Quick Bird imagery satellite on August 14, 2004, just a day after landfall. This satellite imagery used in this research was purchased from Digital Globe™ and was licensed and provided by the Remote Sensing Technology Center of Japan (RESTEC); it is shown in Figure 20.3b. This imagery is of 2.44 m/pixel resolution. The revisit time of the satellite was 1 ~ 3.5 days, depending on the latitude, which makes the damage detection faster; that is, the image will be available very soon after a disaster has occurred.

20.4   DAMAGED-PATH DETECTION BY TEXTURE-WAVELET ANALYSIS

Strong-wind-damaged location is tracked automatically from debris deposits in a post-storm image by using texture-wavelet analysis by tracing image portions that represent the wind- borne debris deposits (Radhika et al. 2012).

Immediately after the disaster, the post-storm imagery is acquired and wavelet-based edge extraction is performed on the input imagery. This process divides the imagery into one low-frequency coefficient (approximate coefficient) matrix and three high-frequency co efficient (horizontal, vertical, and diagonal coefficient) matrices. The information on wind-borne debris lies within the three high-frequency imagery (horizontal, vertical, and diagonal coefficient matrices), as the debris deposit has sudden gradient change due to the irregularity of the damaged-building deposits. Therefore, the image is reconstructed from horizontal, vertical, and diagonal coefficient matrices and the low-frequency matrix; that is, the approximate co-efficient matrix is removed. From the reconstructed image, the debris pattern is extracted by measuring the degree of distribution of the broken edges in the debris deposits, that is, by calculating the standard deviation. The debris deposit path has the maximum standard deviation due to its irregular distribution. For separating the extracted debris pattern, image segmentation is performed using Otsu's thresholding method, and noise removal is performed by filtering. Thus, the tornado-damaged path is tracked by texture-wavelet analysis. A step-by-step process of texture-wavelet analysis for path detection is shown in Figure 20.4.

In order to automate the identification of the disaster location, for faster recovery, user-friendly automatic damage area detection system is designed, as shown in Figure 20.5.

Figure 20.6 shows the results of a damaged path detected for different disasters from satellite/aerial imagery. Once the exact location of the disaster is tracked from aerial/satellite imagery, an immediate automatic survey report for emergency aid can be provided.

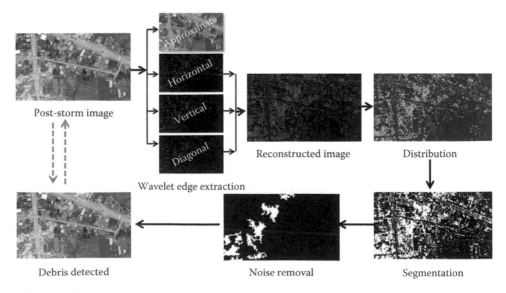

FIGURE 20.4   Step-by-step results for strong-wind-damaged path detection from low-resolution post-storm imagery (debris detected are masked in red).

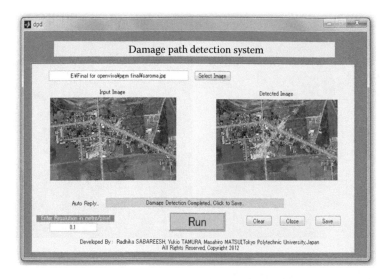

FIGURE 20.5   Automatic damaged-area detection system designed for immediate tracking of disaster area.

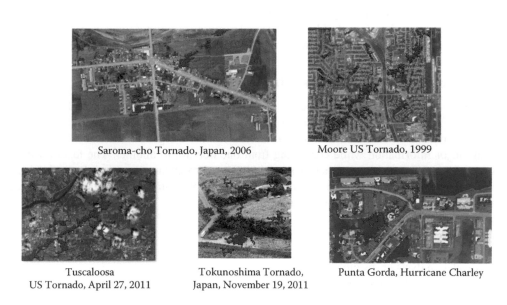

FIGURE 20.6   Wind-damaged locations identified for different imageries (detected path shown in red).

20.5   BUILDING SEGMENTATION

Once the debris path is detected from low-resolution imagery, high-resolution imagery segment is cropped from the image at the exact location where the damage has occurred (Figure 20.7a) and buildings are more concentrated. Building segmentation is performed using edge detection and color invariance property (Radhika et al. 2011a), and buildings are separated as shown (Figure 20.7b).

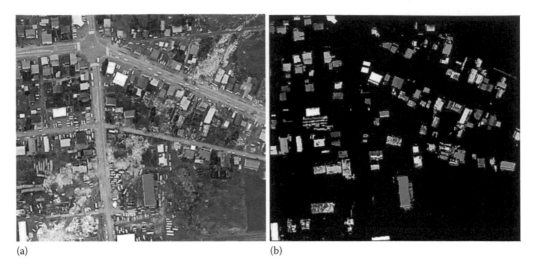

(a)   (b)

FIGURE 20.7   (a) Damaged area cropped. (b) Damaged houses segmented.

20.6 FEATURE EXTRACTION

Features are extracted from the segmented post-storm roof images. In Radhika et al. (2015), the data used for feature extraction were the change-detected portions of wind-damaged roof images. In the present work, a wavelet-based feature extraction approach is used on post-storm roof images alone.

20.6.1 WAVELET-BASED EDGE FEATURE EXTRACTION

In all the feature-extraction methods used in the past for different types of fault classification (Sabareesh et al. 2006; Douglas et al. 2005), either spatial (time) domain information or frequency domain information is available, but the two are not available together. Hence, there is a chance of losing major information while converting from one domain to another. The feature extraction using wavelet transform aids in gathering spatial (time) and frequency domain information together (Radhika et al. 2009). In the current case of remote sensing of digital images, wavelet-based feature extraction is performed by using two-dimensional discrete wavelets. The best wavelet, which is capable of identifying the broken edges, was selected as biorthogonal 3.7 wavelet (Radhika et al. 2012).

The wavelet-based feature-extraction method includes two main steps:

1. Wavelet analysis or decomposition
2. Wavelet synthesis or reconstruction by using high-frequency information (coefficients)

The wavelet has two scaling functions $(\phi(x), \tilde{\phi}(x))$ and two wavelet functions $(\varphi(x), \tilde{\varphi}(x))$. Analysis or decomposing using wavelets basically performs filtering and downsampling by using the decomposing filters $l(k)$ and $h(k)$ of the wavelets. As the post-storm image is a 2D signal, the filtering and downsampling are performed first horizontally and then vertically (Kannan et al. [2010]). There are $l(k)$ and $\tilde{l}(k)$ low-pass impulse responses (low-pass filters) and $h(k)$ and $\tilde{h}(k)$ high-pass impulse responses (high-pass filters). Of these, $l(k)$ and $h(k)$ are used for decomposition or analysis, and $\tilde{l}(k)$ and $\tilde{h}(k)$ are quadrature mirror filters and are used for reconstruction or synthesis.

The basic decomposition steps are described in Figure 20.8, where:

$$I_{LL}(x, y) = f(I(x, y), I_L(x, y), l(k)) \text{(Approximate Coefficient)} \tag{20.1}$$

$$I_{LH}(x, y) = f(I(x, y), I_L(x, y), h(k)) \text{(Horizontal Detailed Coefficient)} \tag{20.2}$$

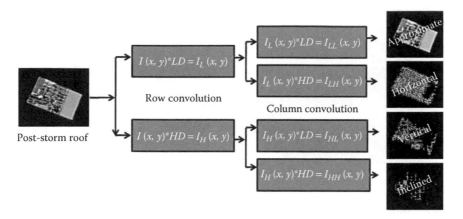

FIGURE 20.8   Wavelet decomposition on post-storm image of a building roof.

$$I_{HL}(x, y) = f(I(x, y), I_H(x, y), l(k))(\text{Vertical Detailed Coefficient}) \tag{20.3}$$

$$I_{HH}(x, y) = f(I(x, y), I_H(x, y), h(k))(\text{Inclined Detailed Coefficient}) \tag{20.4}$$

as explained by Feng et al. (2000). In the image, *LD* and *HD* correspond to the decomposition filters $l(k)$ and $h(k)$, respectively; $I_L(x, y)$ is the low-pass coefficient matrix; and $I_H(x, y)$ is the high-pass coefficient matrix given by Equations 20.5 and 20.6 of the image $I(x, y)$ of size $M \times P$ (Pajares and de la Cruz 2004).

$$I_L(x, y) = \frac{1}{N} \sum_{k=0}^{N-1} l(k) \cdot I((2x + k) \bmod M, y) \tag{20.5}$$

$$I_H(x, y) = \frac{1}{N} \sum_{k=0}^{N-1} h(k) \cdot I((2x + k) \bmod M, y) \tag{20.6}$$

For $x = 0, 1, 2 \dots M/2–1$ and $y = 0, 1, 2, \dots P–1$

The approximate coefficients contribute to low-frequency information, which gives information on smoother surfaces, that is, non-damaged roof portions, whereas the three detailed coefficients give information on high-frequency information, that is, broken edges, which show a sudden gradient change in pixel radiance information. The wavelet decomposition is performed till the third level (Akhtar et al. 2008), as the high-frequency information saturates at this level.

Finally, the low-frequency information is removed and the high-frequency information is aggregated together to extract the features. This aggregation of high-frequency information is performed by wavelet synthesis or reconstruction, as shown in Figure 20.9. In this figure, *HR* corresponds to $\tilde{h}(k)$, as the synthesis filters are $\tilde{l}(k)$ and $\tilde{h}(k)$. Equation 20.7 explains the reconstruction by using the three detailed high-frequency pieces of information, using $\tilde{h}(k)$ (Feng et al. 2000; Lixin et al. 2004).

$$E(x, y) = f(\tilde{I}_{LH}(x, y), \tilde{I}_{HL}(x, y), \tilde{I}_{HH}(x, y)) \tag{20.7}$$

From the wavelet-reconstructed image, both statistical and image histogram features are extracted. The image after reconstruction with high-frequency information alone has more prominent edge information than before, which increases the accuracy of the extracted statistical features. Statistical features such as standard deviation and entropy depend on the distribution and the randomness of the edge pixels, respectively.

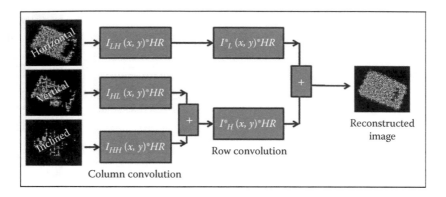

FIGURE 20.9   Wavelet reconstruction on post-storm image of the building roof.

20.6.2 STANDARD DEVIATION

The broken edges of roof surfaces show higher distribution, that is, more gradient change in the pixel radiance information, which results in higher standard deviation than the portion of the roof with intact smooth surfaces. Thus, as the area of the damage portions increases, the standard deviation also increases.

The standard deviation of red, green, and blue layers; the hue, saturation, and vision information; and the image intensity information of the wavelet-reconstructed post-storm building images are calculated. The variation of the standard deviation of all the damaged-building segmented images with respect to the degree of damage is plotted. The best features that contributed to efficient classification for each building are plotted against percentage area of roof damage obtained by texture-wavelet analysis and are shown in Figure 20.10. A comparison of the variation of the features, using the conventional change-detection method and texture-wavelet analysis, is also shown in Figure 20.10.

It is observed that as the percentage area of roof damage increases, the standard deviation also increases approximately when texture-wavelet analysis is used.

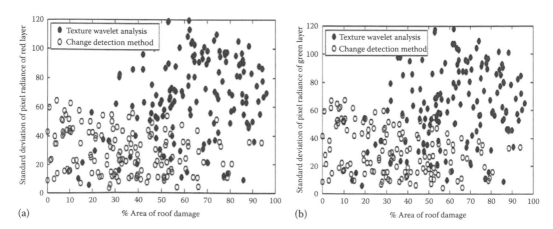

FIGURE 20.10   (a) Standard deviation of pixel radiance of red layer. (b) Standard deviation of pixel radiance of green layer.   (*Continued*)

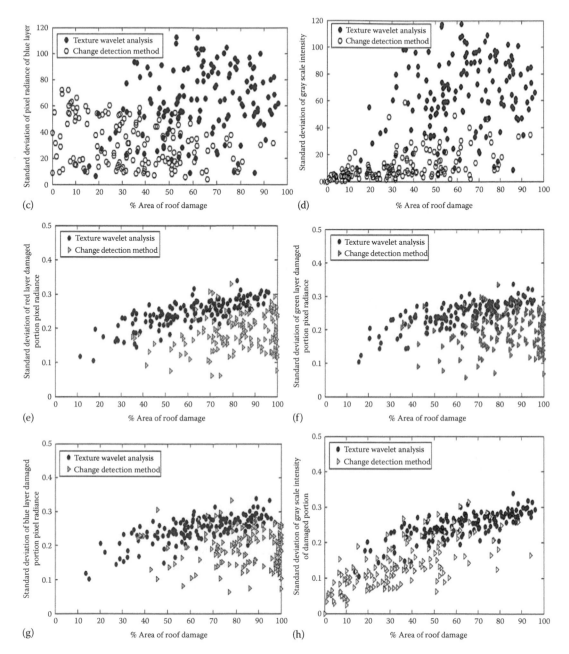

FIGURE 20.10 (Continued) (c) Standard deviation of pixel radiance of blue layer. (d) Standard deviation of gray-scale intensity. (e) Damaged-portion standard deviation of red-layer pixel radiance. (f) Damaged-portion standard deviation of green-layer pixel radiance. (g) Damaged-portion standard deviation of blue-layer pixel radiance. (h) Damaged-portion standard deviation of gray-scale intensity.

20.6.3 ENTROPY

Randomly arranged broken portions of a damaged roof image show high entropy values when compared with other non-broken roof image portions, where the pixels are regularly arranged. The variation of the entropy of all the damaged-building segmented images with respect to the degree of damage is plotted. The best features that contributed to efficient classification for each building

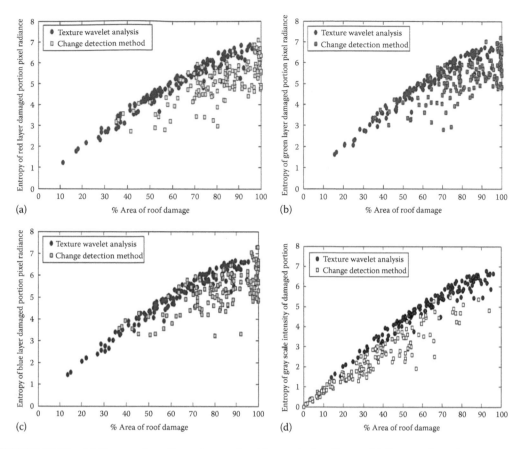

FIGURE 20.11 (a) Damaged-portion entropy of red-layer pixel radiance. (b) Damaged-portion entropy of green-layer pixel radiance. (c) Damaged-portion entropy of blue-layer pixel radiance. (d) Damaged-portion entropy of gray-scale intensity.

are plotted against percentage area of roof damage obtained by texture-wavelet analysis and are shown in Figure 20.11. A comparison of the variation of the features, using the conventional change-detection method and texture-wavelet analysis, is also shown in Figure 20.11.

Therefore, the measure of randomness, entropy, also shows an ascending nature as the damaged area increases.

20.6.4 Histogram

Once the damage pixels are identified, the histogram of the broken edges is measured, which is one of the 15 features used for classifying damaged buildings from non-damaged ones.

20.6.5 Observation

Standard deviations of the pixel radiance of the three visible layers, that is, red, green, and blue, and also the standard deviation and entropy of the gray-scale intensity obtained by the wavelet-based feature-extraction method for all the damaged buildings have shown an almost linear increase as the percentage area of roof damage obtained by automatic texture wavelet analysis increases. However, for conventional features extracted by the change-detection method, a scattered relation with the percentage area of roof damage was obtained. This linear relation between extracted features and the automatically obtained percentage area of damage by texture wavelet analysis aid in a faster and more precise identification of damaged-building structures.

20.7 FEATURE CLASSIFICATION

20.7.1 Artificial Neural Network Classification

Neural network is one of the approaches for forecasting and validating by using computer models, with some of the architecture and processing capabilities of the human brain. The technology that attempts to achieve such results is called neural computing or artificial neural networks (ANN). It mimics biological neurons by simulating some of the workings of the human brain. An ANN is made up of processing elements called neurons that are interconnected in a network. The way in which the information is processed and intelligence is stored depends on the architecture and algorithms of ANN. The ANN is characterized by the topology, weight vector, and activation functions. It has three layers, namely an input layer that receives signals from some external source, a hidden layer that does the processing of the signals, and an output layer that sends processed signals back to the external world. The main advantage of the ANN is its ability to learn patterns in very complex systems. Through learning or self-organizing processes, it translates the inputs into desired outputs by adjusting the weights given to signals between neurons.

The proposed method diagnoses a damaged building by using the ANN. A multilayered feed-forward neural network trained with error backpropagation was used. The backpropagation of an ANN assumes that there is a supervision of learning of the network. The features extracted for both damaged and non-damaged building are given to train the ANN. On the trained ANN, features are given in order to classify the unknown data into damaged or non-damaged building images. Results of ANN classification are detailed in Section 20.9.

20.7.2 Support Vector Machine Classification

Classification is also performed by using Support Vector Machine (SVM), as SVM was proved to be an efficient classifier in earlier researches (Radhika et al. 2009; Sabareesh et al. 2006). In SVM, the selected features are arranged in an input space. This input space is mapped into a high-dimensional dot product space called feature space, and in the feature space, the optimal hyperplane is determined to maximize the generalization ability of the classifier. As the classification includes only two categories, damaged roofs and non-damaged roofs, a biclassifier SVM is enough. A biclassifier optimization theory is utilized to obtain the optimal hyperplane with maximum percentage margin of separation and with minimum error, as shown in Equation 20.8.

$$\underset{w,y}{\text{Min}}\left(\frac{\sqrt{w^T w}}{2} + ve^T y\right) \qquad (20.8)$$

The class decided for the new set of building samples is based on Equation 20.9.

$$f(x) = \text{sign}(w^T x - \gamma) \qquad (20.9)$$

If $f(x)$ is positive, the new set of features belongs to a particular class; otherwise, it belongs to the other, depending on the training information feed.

Classification is performed from the wavelet-extracted features, and damaged buildings are identified. Results of SVM classification are detailed in Section 20.9. The validation procedure is done by visual inspection of the damaged buildings and from information collected from field survey.

20.8 TEXTURE-WAVELET ANALYSIS ON POST-STORM DAMAGED-BUILDING IMAGES

The following steps are involved in texture-wavelet analysis of the extracted wavelet edge and reconstructed post-storm roof images:

1. Wavelet-based edge feature extraction
2. Measure of distribution of the broken edges in the image
3. Segmentation using Otsu's thresholding method
4. Percentage area of damage

Wavelet-based edge feature extraction using biorthogonal 3.7 wavelet on post-storm images are already performed and discussed in detail in Section 20.6.1.

20.8.1 MEASURE OF DISTRIBUTION OF BROKEN EDGES IN IMAGE

After wavelet-based edge extraction, the reconstructed roof image includes only high-frequency information, where the broken edge information is more obvious. However, in rare cases, the edges $E(x, y)$ derived in Equation 20.7 extracted by wavelet-based edge extraction sometimes extract non-broken edges, for example, as shown in Figure 20.12. Edges of the intact roof portions are detected (shown within grey rectangle), as there is no common portion-deletion procedure as in change-detection methods (Canny 1986).

In order to extract the broken edges from a reconstructed image, the measure of distribution of broken edges is calculated. The broken roof edges show a higher distribution rate than long non-broken edges in a reconstructed image, as the broken edge extraction using biorthogonal 3.7 wavelet enhances the edge properties in the reconstructed image. A 2D standard deviation is performed using a 3×3 window on the edge-extracted imagery, and the central pixel radiance is replaced by the calculated standard deviation. Figure 20.13 shows the intensity of distribution of the broken edges of a reconstructed post-storm imagery.

20.8.2 SEGMENTATION USING OTSU'S THRESHOLDING METHOD

The damaged area is segmented from the rest by using Otsu's thresholding method (Otsu, 1979). In this method, the objects (damaged area) to be segmented are separated from the background (remaining part of the roof) by initializing a threshold level, formulating the objectives, and finally treating segmentation process as an optimization problem, that is, by maximizing any one of these objectives by maximizing the separability between the two classes, objects, and the background. The damaged portion segmented using Otsu's segmentation method is shown in Figure 20.14.

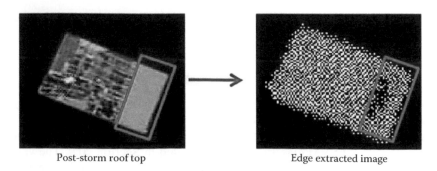

Post-storm roof top   Edge extracted image

FIGURE 20.12   Edge-extracted sample roof top with intact edges detected.

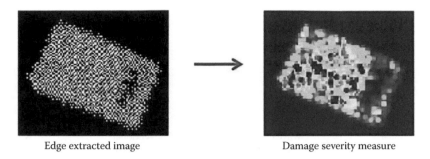

Edge extracted image   Damage severity measure

FIGURE 20.13 Measure of distribution of broken edges or damage severity measure (regions shown in grey show maximum distribution).

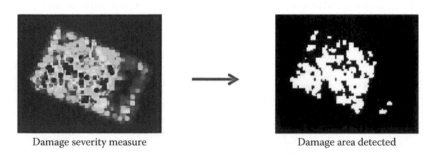

Damage severity measure   Damage area detected

FIGURE 20.14 Segmentation performed by Otsu's thresholding method.

20.8.3 Percentage Area of Damage

The percentage area of damage is calculated from the damaged portions thus segmented, by normalizing the damaged area with the total roof area, as shown in Equation 20.10.

$$\% P_D = \left(\frac{A_D}{A_T} \right) \times 100 \tag{20.10}$$

where:
P_D is the percentage area of roof damage
A_D is the damaged roof area
A_T is the total roof area

20.9 RESULTS AND DISCUSSION

To compare the efficiency of wavelet-based feature extraction, conventional feature extraction is also performed on the same raw data. Classification using SVM and ANN is also performed on the features extracted by conventional method and wavelet-based feature extraction method.

20.9.1 Classification Results

A precise identification of 80% efficiency for satellite images and 82% efficiency for aerial images, using artificial neural network (ANN), and 88% efficiency for satellite images and 90% efficiency for aerial images, using support vector machine (SVM)-based classification, is obtained when wavelet-based feature extraction is used. However, using conventional feature extraction, only very

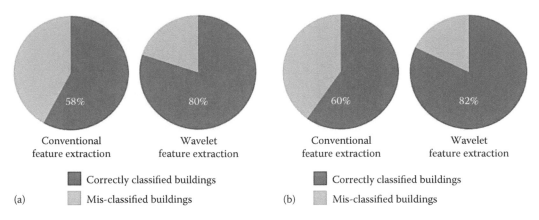

FIGURE 20.15 Percentage efficiency of classification of ANN for conventional and wavelet-extracted features: (a) for satellite imagery and (b) for aerial imagery.

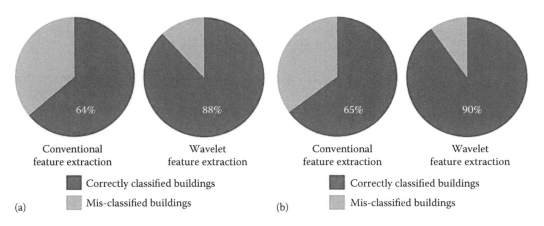

FIGURE 20.16 Percentage efficiency of classification of SVM for conventional and wavelet-extracted features: (a) for satellite imagery and (b) for aerial imagery.

low efficiency of 58% for satellite images and 60% for aerial images, using artificial neural network (ANN), and 64% for satellite images and 65% for aerial images, using support vector machine (SVM), is obtained. The results are shown in Figures 20.15 and 20.16.

The higher percentage of classification efficiency results in accurate damage identification. Wavelet-based feature-extraction methods provided a more accurate identification of damaged buildings than conventional feature-extraction methods. At the same time, a better classification was performed by SVM than by ANN.

20.9.2 PERCENTAGE AREA OF ROOF DAMAGE

Damage gradation of four different roof image samples by calculating their percentage area of roof damage from post-storm imagery alone, using texture-wavelet analysis, is shown in Figure 20.17. Similarly, texture wavelet analysis is performed on all the segmented roof images, and damage gradation is obtained.

Correlation between this automatically calculated percentage area of roof damage by texture-wavelet analysis and visually interpreted roof damage is calculated in order to measure the accuracy of texture-based wavelet analysis. Correlation factors are calculated for both satellite and aerial

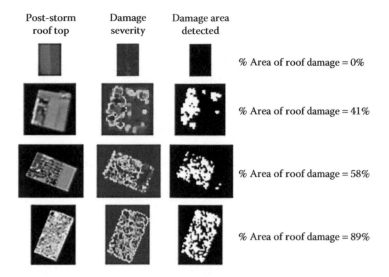

FIGURE 20.17　Damage gradation of percentage area of roof damage by using post-storm imagery alone.

TABLE 20.1
Correlation Calculated for All Building Samples

	Correlation Factor			
	Between Canny Edge Detection and Visual Interpretation		**Between Texture-Based Wavelet Analysis and Visual Interpretation**	
	Pre- and Post-Storm	**Post-Storm Alone**	**Pre- and Post-Storm**	**Post-Storm Alone**
% area of roof damage on satellite imagery	0.12	0.09	0.78	0.75
% area of roof damage on aerial imagery	0.52	0.45	0.85	0.8

imageries and tabulated in Table 20.1. A higher positive correlation value is also observed for percentage area of roof damage automatically obtained by texture-based wavelet analysis than by Canny edge detection.

The damage estimation from post-storm imagery alone has shown almost equal correlation factor as that of modified wavelet-based change-detected method, where both pre- and post-storm images are used. In addition, damage estimation from low-resolution satellite imagery showed a comparable correlation factor with the high-resolution aerial imagery, which reflects the efficiency of texture-wavelet analysis even with lower-resolution imagery.

The conventional change-detection method that used Canny edge detection showed a positive correlation but with a very low correlation factor with the visually measured data. The validation is performed by comparing the data with conventional methods such as field investigation data as well as by manually measured data through visual interpretation.

20.10　SUMMARY

Wavelet-based feature extraction was performed directly on raw data of segmented post-storm building images, and damaged buildings were identified with 88% efficiency by using SVM and with 80% efficiency by using ANN for 2.44 m/pixel resolution satellite imagery, whereas the efficiency using SVM was 90% and that using ANN was 82% for 10 cm/pixel resolution aerial imagery. This method proved to be more efficient and accurate than conventional feature-extraction methods.

From the identified damaged buildings, damage estimation is performed by calculating the percentage area of roof damage by using texture-wavelet analysis on post-storm imageries. A correlation factor of 0.75 for satellite imagery and 0.8 for aerial imagery is obtained when compared with the manually obtained visual interpretation data. This validates the accuracy of the automated detection of the percentage area of damaged roof structures. The high-resolution aerial imagery showed better correlation than satellite imagery in both cases, but with the use of texture-wavelet analysis, even with lower-resolution satellite imagery, a better correlation was obtained.

Thus, a cost-effective precise automatic detection of damaged roof tops and the percentage area of damage from post-storm images alone aid in a rapid risk-reduction attempt, soon after disastrous strong-wind damage and immediate reconstruction by prioritization.

REFERENCES

Akhtar, P., Ali, T., J., Bhatti, M. I., and Muqeet, M. A. 2008. A framework for edge detection and linking using wavelets and image fusion. *Proceedings of the 2008 Congress on Image and Signal Processing*, Vol. 1.

Ambrosia, V. G., Buechel, W. W., Brass, J. A., Peterson, J. R., Davies, R. H., Kane, R. J., and Spain, S. 1998. An integration of remote sensing, GIS, and information distribution for wildfire detection and management. *Journal of the American Society for Photogrammetry and Remote Sensing*, 64, 977–985.

Canny, J. 1986. A computational approach to edge detection. *IEEE Transactions on Pattern Analysis and Machine Intelligence*, PAMI-8, 6, 679–698.

Cao, S., Yoshida, A., Matsui, M., Tamura, Y., Kobayashi, F., Kikuchi, H., and Sassa, K. 2006. Tornado damage in Saroma. *Wind Effect Bulletin*, 7, 5.

Danneels, G., Havenith, H., Strom, A., and Pirard, E. 2008. Landslide detection methods, inventory analysis and susceptibility mapping applied to the Tien Shan, Kyrgyz Republic. *Proceedings of the First World Landslide Forum*, Tokyo, 426–429.

Douglas, H. and Pillay, P. 2005. The impact of wavelet selection on transient motor current signature analysis. *IEEE International Conference on Electric Machines and Drives*. IEEE , pp. 80–85.

Feng, L., Suen, C. Y., Tang, Y. Y., and Yang, L. H. 2000. Edge extraction of images by reconstruction using wavelet decomposition details at different resolution levels. *International Journal of Pattern Recognition and Artificial Intelligence*, 14, 6, 779–793.

Groeve, T. and Riva, P. 2009. Early flood detection and mapping for humanitarian response. *Proceedings of the 6th International ISCRAM Conference*, Gothenburg, Sweden.

Hasegawa, H., Aoki, H., Yamazaki, F., Matsuoka M., and Sekimoto, I. 2000. Automated detection of damaged buildings using aerial HDTV images. *Proceedings of the IEEE 2000 International Geosciences and Remote Sensing Symposium*. IEEE, CD-ROM, pp. 310–312.

Kannan, K., Arumuga, S., Perumal, and Arulmozhi, K. 2010. Area level fusion of multi-focused images using multi-stationary wavelet packet transform. *International Journal of Computer Applications*, 2, 1, 88–95.

Lakshminarasimhan, V. S. 2004. Image based assessment of windstorm damage. Thesis, Texas Tech University, Lubacco, TX.

Lixin, Sheny and Qiyu, Sun, 2004. Bi-orthogonal wavelet system for high-resolution image reconstruction. *IEEE Trans. Signal Process.*, 52, 7, 1997–2011.

Matsuoka, M. and Yamazaki, F. 2000. Characteristics of satellite SAR images in the areas damaged by earthquakes. *Proceedings of the IEEE 2000 International Geosciences and Remote Sensing Symposium*, 6, 2693–2696.

Mitomi, H., Saita, J., Matsuoka, M., and Yamazaki, F. 2001. Automated damage detection of buildings from aerial television images of the 2001 Gujarat, India earthquake. *Proceedings of IEEE International Geoscience and Remote Sensing Symposium*, pp. 147–149.

Myrint, S. W., Yuan, M., Cerveny, R. S., and Giri, C. P. 2008. Comparison of remote sensing image processing techniques to identify tornado damage areas from LandSat TM data. *Sensors*, 8, 1128–1156.

Otsu, N., 1979. A threshold selection method from gray-level histograms. *IEEE Transactions on Systems, Man and Cybernetics*, 9, 1, 62–66.

Ozisik, D. 2004. Post earthquake damage assessment using satellite and aerial video imagery. Thesis submitted to International Institute for Geo-information Science and Earth Observation, Enschede, the Netherlands.

Pajares, G. and de la Cruz, J. M. 2004. A wavelet-based image fusion tutorial. *Journal of the Pattern Recognition* 37, 1855–1872.

Radhika, S., Sabareesh, G. R., Jagadanand, G., and Sugumaran, V. 2009. Precise wavelet for current signature in 3 phase IM. *Journal on Expert Systems with Applications*, 37, 450–455.

Radhika, S., Tamura, Y., and Matsui, M. 2010. Using wavelets as an effective alternative tool for wind disaster detection from satellite images. *Proceedings of Computational Wind Engineering*, Chapel, Hill, NC, May 23–27.

Radhika, S., Tamura, Y., and Matsui, M. 2011a. Automated detection of tornado damage to building structures from aerial imageries using color invariant features. *Proceedings of 13th International Conference on Wind Engineering*, Amsterdam, the Netherlands, July 10–15.

Radhika, S., Tamura, Y., and Matsui, M. 2011b. Post-storm satellite images to trace Tornado damage path from the Wind Borne Debris Deposits. *Proceedings of Second Asia/Oceania Meteorological Satellite Users' Conference*, Tokyo, Japan, December 6–9.

Radhika, S., Tamura, Y., and Matsui, M. 2011c. Tracking the foot prints of tornado damages from post-storm aerial imageries by texture-wavelet analysis images. *Proceedings of 5th International Symposium on Wind Effects on Buildings and Urban Environment (ISWE5) Wind Hazard Resilient Cities: New Challenges, 2011*, Shinjuku, Tokyo, Japan, March 7–8, 2011.

Radhika, S., Tamura, Y., and Matsui, M. 2012. Use of post-storm images for automated tornado-borne debris path identification using texture-wavelet analysis. *Journal of Wind Engineering & Industrial Aerodynamics*, 107, 202–213.

Radhika, S., Tamura, Y., and Matsui, M. 2015: Cyclone damage detection on building structures from pre- and post-satellite images using wavelet based pattern recognition. *Journal of Wind Engineering & Industrial Aerodynamics*, 136, 23–33.

Sabareesh, G. R., Sugumaran, V., and Ramachandran, K. I. 2006. Fault diagnosis of a taper roller bearing through histogram features and proximal support vector machines. *Proceedings of IEEE International Conference on Signal and Image Processing*, Hubli, India.

Sheny, L. and Sun, Q. 2004. Bi-orthogonal wavelet system for high-resolution image reconstruction. *IEEE Transactions on Signal Processing*, 52(7), 1997–2011.

Sumer, E. and Turker, M. 2004. Building damage detection from post-earthquake aerial images using watershed segmentation in Golcuk, Turkey. *Proceedings of XXth International Society for Photogrammetry and Remote Sensing (ISPRS'04) Congress, Commission VII*, Istanbul, Turkey, July 12–23.

Thomas, L., Nancy, P., Jonathan, C., Robert, G., Kelley, L., and Timothy, O. 2002. Assessing tornado damage via analysis of multi temporal LandSat 7 ETM+ data. *Proceedings of Annual conference of the American society for Photogrammetry and Remote sensing (ASPRS)*, Washington, DC, April 21–27.

Vu, T.T., Matsuoka, M., and Yamazaki, F. 2005. Shadow analysis in assisting damage detection due to earthquakes from QuickBird imagery. Commission VII, WG II/5.

Womble, J. A. 2005. Remote-sensing applications to windstorm damage assessment. Thesis, Texas Tech University, Lubbock, TX.

Womble, J. A., Adams, B. J., and Mehta, K. C. 2007. Automated building damage assessment using remote sensing imagery. *Proceedings of the ASCE/SEI Structures Congress*, Long Beach, CA.

BIBLIOGRAPHY

Soman, K. P. and Ramachandran, K. I. 2004. *Insight into Wavelets from Theory to Practice*. Prentice –Hall of India Private Limited, Delhi, India. October 30, 2004, 305 Pages.

21 Analyzing Tropical Cyclones over India Using Precipitation Radar

Devajyoti Dutta, A. Routray, and Prashant K. Srivastava

CONTENTS

21.1 INTRODUCTION

A tropical cyclone (TC) is a symmetric system characterized by a low pressure at the center and numerous thunderstorms that produce strong winds and heavy rain causing flash flood (Henderson-Sellers et al. 1998). The TCs form over large, relatively warm ocean. They get energy through evaporation of water from the ocean surface. This warm, moist air (water vapor) rises and cools to saturation; ultimately, it condenses into clouds and rain. During this processes, it releases latent heat, and this heat feeds TCs. Latent heat provides more energy and transforms to a violent storm. The TCs are associated with strong wind, torrential rains, and storm surges, which create havoc along the coastal regions. The heavy downpours can cause significant flooding, and surges also result in extensive coastal flooding. The TC track and intensity forecasting is an important component for disaster warnings and mitigation efforts (Mohapatra et al. 2013).

The North Indian Ocean (NIO), including the Bay of Bengal (BoB) and the Arabian Sea (AS), experiences two TC seasons, that is, post-monsoon season and pre-monsoon season. During the post-monsoon season (October–December), maximum number of TCs occurr as compared with the pre-monsoon season (April–early June). The damage steadily increases along the coastal region due to the TCs. The track forecast errors are relatively high over the NIO compare with those over the Atlantic and Pacific Oceans (Mohapatra et al. 2013). The accurate prediction of track and intensity of the TCs well in advance is now a challenge for all research and operational communities. The NIO is one of the important basins contributing about 7% of the global annual tropical storms (WMO Technical Report 2008). The frequency and landfalling of the TCs are more over BoB, with the development of about four TCs per year, and hence cause more disasters than the TCs at the AS (IMD 2008). The previous studies (Mohanty and Gupta 1997; Gupta 2006) show that synoptic and statistical methods have limitations in TC track and intensity prediction over the NIO. Observational networks and numerical weather prediction (NWP) models have the potential to provide forecast for genesis, intensity, and movement of TC for disaster mitigation and warning. Observational and modeling studies have established a link between the TC intensification and deep convection (Schubert and Hack 1982; Hendricks et al. 2004; Kelley et al. 2004; Montgomery et al. 2006; Reasor et al. 2009; Guimond et al. 2010; Montgomery and Smith 2011; Jiang 2012).

Observations of TCs are limited, because they spend most of their lifetime over oceans. Therefore, remote sensing is an important tool to detect a cyclone and to study its characteristics. The use of microwave portion of the spectrum has the advantage that microwave radiation penetrates the clouds. The precipitation-sized drops interact strongly with microwave radiation, which allows their detection by microwave radiometers. The main disadvantage of microwave precipitation monitoring and estimation is that the radiometers have poorer spatial and temporal resolution compared with infrared (IR) and visible spectral bands. Furthermore, there are two types of microwave remote-sensing approaches to study the precipitation associated with cyclone: active and passive. In the active type, the transmitted signals are received from the target by backscattering process. Doppler Weather Radar (DWR), Precipitation Radar (PR), microwave radar altimeters, scatterometers, and so on, are active microwave sensors. In active microwave sensing, the characteristics of scattering can be derived from the radar cross-section, calculated from the received power, antenna parameters and the relationship between them, and physical characteristics of an object. The passive types receive the microwave radiation emitted from the target. The microwave radiometer is one of the passive microwave sensors. The passive microwave remote sensing can be understood with the help of radiative transfer theory (Chandrasekhar 1960), based on the law of Rayleigh Jeans. In passive microwave remote sensing, the characteristics of an object can be detected from the relationship between the received power and the physical characteristics of the object, by using the parameter known as brightness temperature (Tb). The temperature of the black body that radiates the same radiant energy as an observed object is called the *brightness temperature* of the object. The Tropical Rainfall Measuring Mission (TRMM) satellite had both passive and active sensors, viz. TRMM measurements from the TRMM Microwave Imager (TMI), PR, Lightning Imaging Sensor (LIS), and Visible and Infrared Scanner (VIRS). The detailed specification of the sensors boarded in the TRMM satellite is shown in Table 21.1. The high-resolution PR made it possible for the first time that TCs in all ocean basins can be viewed from orbit. The TRMM measurements from the TMI, PR, LIS, and VIRS have provided valuable sources of information for the study on TC's structure, intensity, and intensity change. The convective structure in different regions of TCs is fundamental, and it is important for the community of TC research. In addition, TC intensity is linked with satellite-based ice-scattering signatures and inner-core areal mean rainfall (Rao and MacArthur 1994; Cecil and Zipser 1999).

TABLE 21.1
Details of Various Sensors Boarded in the TRMM Satellite

Parameters	TMI	TRMM-PR	VIRS
Frequency	10.75 GHz (V, H), 19.35 GHz (V, H), 22.235 GHz (V), 85 GHz (V, H)	13.8 GHz (H)	0.63, 1.6, 3.75, 10.8, and 12 μm
Data products	1B11 (TMI brightness temperature), 2A12 (hydrometeor profile product)	1C21 (reflectivities), 2A25 (rainfall rate and profile), 2A23 (radar rain characteristics)	1B01 (radiance)
Swath	~880 km	~215 km	~830 km
Resolution	11 km × 8 km	5 km (H) and 0.25 km (V)	2 km (nadir)

TABLE 21.2
Detail of Tropical Cyclones over the North Indian Ocean

TC Name	Basin Name	Duration	Peak Intensity (km/h)
Ogni	Bay of Bengal	October 27–30, 2006	65
Thane		December 25–30, 2011	85
Helen		November 19–23, 2013	65

The major objective of this chapter is to study the evolution and convective rain bands associated with the TCs from the available remote-sensing observations. For this purpose, three TCs that occurred over BoB are considered (Table 21.2), and details of the synoptic conditions associated with the TCs can be found from the Regional Specialized Meteorological Centre (RSMC's) reports of the particular year (http://www.rsmcnewdelhi.imd.gov.in/).

21.2 DESCRIPTION OF OBSERVATIONAL SYSTEMS

There are major improvements in TC track forecasting in the past decade. However, TC intensity change and rainfall prediction are still challenging, and these have immense practical importance. Tropical cyclone intensification, especially rapid intensification (RI), is one of the important and necessary conditions in TC research. There is an increasing demand for more accurate and precise quantitative precipitation forecast (QPF), along with longer lead times.

Satellite-based near-real-time precipitation observations provide new opportunity to develop global hazard prediction techniques. The capability of microwave sensors to monitor and measure the precipitation from microwave brightness temperature data was initiated with the launch of Electrically Scanning Microwave Radiometer (ESMR) instruments on Nimbus-5, with a center frequency of 19.35 GHz. The significant works on identification of rain systems and retrieval of rain parameters were started with SSM/I sensors of the Defiance Meteorological Satellite Program (DMSP) satellite, that is, F8-F14.

There are different satellite rainfall products worldwide, for example, TMPA, CMORPH, PERSIANN, PERSIANN-CCS, and so on, by different agency. The National Aeronautics and Space Administration (NASA's) Goddard Earth Sciences Data and Information Services Center (GES DISC) released the Tropical Rainfall Measuring Mission (TRMM) Multisatellite Precipitation Analysis (TMPA) products (resolution 025° × 0.25°) with near-global (50°S–50°N) coverage. The algorithm combines multiple independent precipitation estimates from the TMI, Advanced Microwave Scanning Radiometer for Earth Observing Systems (AMSR-E), Special Sensor Microwave Imager (SSMI), Special Sensor Microwave Imager/Sounder (SSMIS), Advanced Microwave Sounding Unit (AMSU), Microwave Humidity Sounder (MHS), and microwave-adjusted merged geoinfrared (IR). All input microwave data are intercalibrated to TRMM Combined Instrument (TCI) precipitation estimates (TRMM product 3B31).

The Climate Prediction Center (CPC) morphing technique (CMORPH) satellite produces global precipitation analyses at very high spatial and temporal resolution. This technique uses precipitation estimates that have been derived from low orbiter satellite microwave observations exclusively and whose features are transported via spatial propagation information that is obtained entirely from geostationary satellite IR data.

The current operational Precipitation Estimation from Remotely Sensed Information using Artificial Neural Network (PERSIANN) system developed by the Center for Hydrometeorology and Remote Sensing (CHRS) at the University of California, Irvine (UCI), uses neural network function approximation procedures to compute an estimate of rainfall rate at each 0.25° × 0.25° pixel of the IR brightness temperature image provided by geostationary satellites.

The PERSIANN-Cloud Classification System (PERSIANN-CCS) is a real-time global high-resolution ($0.04° \times 0.04°$) satellite precipitation product developed by the CHRS at the UCI. The PERSIANN-CCS system enables the categorization of cloud-patch features based on cloud height, areal extent, and variability of texture estimated from satellite imagery.

The TRMM has provided important information that are relevant to the Global Precipitation Measurement (GPM) mission. A lot of research information was also learned from the TRMM, which is relevant to the operational use of GPM data.

21.2.1 THE TRMM SATELLITE

The TRMM is the first mission dedicated to measuring tropical and subtropical rainfall through microwave and visible IR sensors and includes the first space-borne rain radar. The TRMM's orbit ranges between 35°N and 35°S of the equator, allowing the TRMM to fly over each position on the Earth's surface at a different local time each day. The data from this kind of orbit can be used to calculate the rain variations over a 24-hour period. By use of a low-altitude orbit of 217 miles (350 km), the TRMM complements the state-of-the-art instruments and provides accurate measurements of rainfall.

The TRMM satellite carries five instruments (Table 21.1), with both passive and active sensors: TMI, PR, VIRS, Clouds and the Earth's Radiant Energy System (CERES), and LIS (Kummerow et al. 1998). In this study, two sensors' (TMI and PR) data are considered to analyze the TCs. The scanning geometries of these two sensors are discussed in the following subsections.

21.2.1.1 TRMM Microwave Imager

The TMI is a passive sensor designed to provide quantitative rainfall information over a wide swath. By measuring the minute amounts of microwave energy emitted by the Earth and its atmosphere, TMI quantifies the water and the rainfall intensity in the atmosphere. It consists of nine frequency channels, that is, 10 GHz (V), 10 GHz (H), 19 GHz (V), 19 GHz (H), 21 GHz (V), 37 GHz (V), 37 GHz (H), 85 GHz (V), 85 GHz (H). It is a relatively small instrument that consumes little power. The TMI measures the intensity of thermally emitted radiance by an object in units of temperature at nine frequency channels. It is because there is a correlation between the intensity of the radiation emitted and the physical temperature of the radiating body. It is defined as the temperature of a black body that emits the same intensity as measured (Ulaby et al. 1981). T_b is found by inverting the Planck's function. The Planck's function can be expressed in wavelength as

$$B_\lambda(T) = \frac{2hc^2}{\lambda^5(\exp(hc/k_B T\lambda)-1)} \quad (21.1)$$

where:
λ is the wavelength
h is the Planck's constant
k_B is the Boltzmann's constant
c is the velocity of light
T is the absolute temperature of a black body

The brightness temperature is found by inverting the Planck's function as

$$T_b = \frac{C_2}{\lambda \ln\left[1+\dfrac{C_1}{\lambda^5 I_\lambda}\right]} \quad (21.2)$$

Where, $C_1 = 1.1911 \times 10^8$ W m^{-2} sr^{-1} μm^4 and $C_2 = 1.4388 \times 10^4$ K μm.

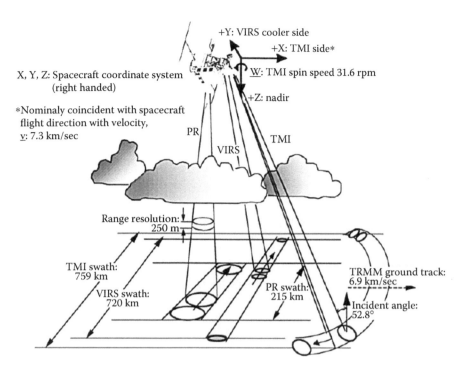

FIGURE 21.1 Scanning geometry of TRMM sensors. (Adapted from Kummerow, C.D. et al., *J. Atmos. Oceanic Technol.*, 15, 809–817, 1998. With permission.)

The TMI antenna is an offset parabola, with an aperture size of 61 cm and a focal length of 50.8 cm. The antenna beam views the Earth's surface with a *nadir* angle of 49°, which results in an incident angle of 52.8° at the Earth's surface (Figure 21.1). The standard level 1B11 TMI data are given as single—sample effective field of view (EFOV) (Kummerow et al. 1998). The Earth science data in each scan line consists of latitude and longitude values, along with brightness temperatures for the 208 EFOVs at 85 GHz and 104 EFOVs at all the remaining lower-frequency channels.

21.2.1.2 The Precipitation Radar

The PR is the first space-borne instrument designed to provide three-dimensional maps of storm structure. The measurements yield information on the intensity and distribution of the rain, rain type, storm depth, and the height at which the snow melts into rain. The PR is a 128-element active-phased array system operating at 13.8 GHz. The PR uses a frequency agility technique to obtain 64 independent samples with a fixed pulse repetitive frequency of 2776 Hz, in which a pair of 1.6 μs pulses differing in frequency by 6 MHz is transmitted.

The PR antenna scans in the cross-track direction over ±17° (~220 km swath), which gives data in 49 FOVs.

21.2.2 Utilized Data Product of the TRMM Sensors

There are different data products from level 1 to level 3. The PR product of levels 1 and 2 has a resolution of 5 × 5 km, and the TMI product has a resolution of 5 × 11 km at 85 GHz. A detailed description of these data product can be found in TRMM Science User–Interface Control Specification, volume numbers 3 and 4. The present study utilizes levels 1 and 2 data product, viz. 1B11 (TMI brightness temperature), 2A25 (PR rainfall rate and profile), and 2A23 (radar rain characteristics). The 1B11 data product gives brightness temperature at different frequency channels (10.65, 19.35, 21.3, 37.0, and 85.5 GHz, with both horizontal and vertical polarizations and only vertical polarization at 21.3 GHz). The 2A25 data product contains different parameters viz. near-surface rain rate, near-surface reflectivity, stratiform, and convective rain. The primary aspects of PR retrieval are the (1) precipitation classification, which is facilitated by the high vertical resolution (250 m) reflectivity profile measurements; (2) an inversion algorithm controlled by a surface reference technique for path-integrated attenuation (Meneghini et al. 2000); and (3) a reflectivity relationship, with parameters differentiated for convective and stratiform rain regimes (Iguchi et al. 2000). The 2A23 data product gives the bright band information, storm height, and convective and stratiform classifications.

21.2.3 DOPPLER WEATHER RADAR

Pulse Doppler techniques are increasingly applied in weather radars to characterize severe weather systems, with astounding success (Doviak and Zrnić 1993).

21.2.3.1 Doppler Shift

The principle of Doppler weather radar is based on Doppler effect. *Shift in frequency caused by moving sources of sound is directly proportional to speed of the source.* Doppler radar compares the received signal with the frequency of the transmitted signal and measures the frequency shift, giving the speed of the target. For a radar with wavelength λ observing a target at range r, if radar signal is transmitted with initial phase of $[\varphi_0]$, then the phase of returned signal φ_t will be $[\varphi_0 - 4\pi r\,(t)/\lambda]$. If the target is moving with respect to the radar with a radial velocity v_r, the phase of the signal varies, and we have:

$$\frac{d\Phi(t)}{dt} = \omega_d = 2\,\pi\,f_d = -\frac{4\pi}{\lambda}\frac{dr(t)}{dt} = -\frac{4\pi}{\lambda}v_r \tag{21.3}$$

Thus, the frequency of the echo has a shift due to the Doppler effect, $f_d = -2\,v_r/\lambda$. The Doppler frequencies of atmospheric targets do not exceed a few kilohertz. Therefore, they are too small with respect to the transmitted frequencies to measure them directly. Doppler signal is derived from the comparison of transmitted and received signals. Because the duration of one pulse is only $\tau = 0.5-2\ \mu s$, it is not possible to determine Doppler frequency directly from this one pulse, but it is possible from phases of several consecutive pulses. The fast Fourier transformation or the method called pulse-pair algorithm (calculation of autocorrelation coefficients of measured phases) is used for calculating Doppler frequency from series of phases (Doviak and Zrnić 1993).

21.2.3.2 Doppler Dilemma

The maximum Doppler frequency that can be measured unambiguously is equal to half of the pulse repetition frequency, $f_{dmax} = f_r/2$ (also called the Nyquist frequency). Consequently, the maximum unambiguous velocity is equal to

$$V_{dmax} = \frac{f_r\lambda}{4} \tag{21.4}$$

Thus, the unambiguously measured velocity interval is equal to $(-v_{dmax},\ v_{dmax})$. The higher velocity $(v > v_{dmax})$ will be interpreted as the velocity from the opposite direction $-v_{dmax} + (v-v_{dmax})$. This ambiguity is called äliasing. To detect higher velocities unambiguously, either operate the radar at a higher wavelength (but then it will not be able to detect a part of cloudiness composed of small particles) or increase the pulse repetition frequency.

Because the velocity speed of radar pulse is approximately equal to the speed of light, c, the maximum unambiguous range (defined as a range from which the backscattered signal can be received before next pulse's transmission) is given by $r_{max} = c/2\,f_r$. The echo of a strong target at range $r > r_{max}$ is interpreted as a new pulse echo target at range $r-r_{max}$. The combination of this relation with the equation for maximum unambiguous velocity gives the equation (also called equation of Doppler dilemma) that describes the relationship between maximum unambiguous velocity and maximum unambiguous range for Doppler radar measurements. Unambiguous measurement of high velocities is thus possible only at small ranges, and measurements at large range are unambiguous for only lower velocities.

$$V_{dmax}r_{max} = \frac{c\lambda}{8} \tag{21.5}$$

21.2.3.3 Radar Reflectivity Factors

Radar meteorologists need to relate the reflectivity, η, which is a general radar terminology for the backscattering cross-section per unit volume, to the factors that have meteorological significance (Doviak and Zrnic 1993). For spherical water drops that have small diameter compared with wavelength (i.e., in the Rayleigh approximation), η can be expressed as

$$\eta = \frac{\pi^5}{\lambda^4}|K_w|^2 Z \tag{21.6}$$

Where, $Z = \frac{1}{\Delta V}\sum_i D^6 = \int_0^\infty N(D,r)D^6 dD$ is the reflectivity factor (21.7), ΔV = unit volume, and D = diameter of the hydrometeor.

The unit of reflectivity is $mm^6 m^3$, and it is generally expressed in logarithm as

$$dBZ = 10\log\left(\frac{Z \ mm^6}{1 \ m^3}\right) \tag{21.8}$$

21.2.3.4 Doppler Weather Radar System Configuration

For the present study, the observations from the DWR at Shriharikota (13.66°N and 80.23°E), Chennai (13.04°N and 80.17°E), and Machilipatnam (16.09°N and 81.12°E), India, are considered. Figure 21.2 shows the schematic diagram of the operational mode of the DWR. The DWR operates in S-band (2.8 GHz frequency/10 cm wavelength) and is normally configured as a fixed station. The DWR consists of a high-power coherent transmitter, a pencil beam antenna with 1° beam width, and very low sidelobe levels. It is steerable in azimuth and elevation. A state-of-the-art digital signal

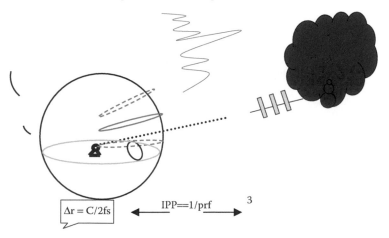

FIGURE 21.2 Schematic diagram of operational mode of the DWR.

processor extracts the three essential base products, viz. reflectivity (η), mean velocity (V), and spectral width (σ) of hydrometeors, from the log/linear channels of the receiver.

21.3   RESULTS AND DISCUSSION

In this study, remote-sensing approach is used to analyze the four TCs (Table 21.2). The IR data are used to define deep convection by calculating the number of pixels, with brightness temperatures colder than a given threshold. The cirrus cloud causes problem in finding out convective intensity. Since here, the near-surface rain is grouped by a PR 2A25, the cirrus contamination is reduced.

Radar reflectivity depends on hydrometeor phase and the sixth power of hydrometeor diameter for Rayleigh scattering. Reflectivity, therefore, responds preferentially to the largest particles in a sample volume. High reflectivities below the freezing level indicate a large liquid water content, whereas high values above the freezing level indicate super-cooled liquid raindrops or large ice particles, which can only reach those altitudes by substantial updrafts. Without strong updrafts, reflectivity decreases rapidly with height above the freezing level (Zipser and LeMone 1980; Szoke et al. 1986).

A case study of the cyclonic storm Ogni on October 25–27, 2006, was carried out by considering the near-simultaneous observations from DWR, TRMM-PR, and TMI. First of all, the plan position indicator (PPI) plots of reflectivity and radial velocity, as observed from DWR, are presented in Figure 21.3 (a and b, respectively). The spiral pattern of the cyclonic system is very much clear from these PPI plots. The total area covered by this event is around $16 \times 10^4 \text{ km}^2$. The maximum radial velocity observed was of the order of 30 m/s, both inward and outward with respect to the DWR beam. The positive radial velocity indicates the direction of the wind away from the beam, and the negative radial velocity indicates the direction of the wind toward the beam. For the same sample record, the vertical cross-section of the cyclonic event is shown in Figure 21.4. The two bands of high reflectivity are clearly visible, with heights of 5 km and 7 km, respectively. The eye is identified with the help of the absence of reflectivity regimes at around 60–70 km from the DWR. The near-simultaneous observations from the DWR (reflectivity in dBZ), TMI (brightness temperature in K), and TRMM-PR (reflectivity in dBZ) are shown in Figure 21.5 (a through c, respectively). The DWR and TMI are able to capture the full cyclonic event, whereas the TRMM-PR is able to capture the small portion, only of the same systems, due to its smaller swath. Figure 21.6a shows the frequency distribution of the total radar pixel area at various dBZ ranges, as observed from the DWR, and Figure 21.6b shows the frequency distribution of the total pixel area at various rain intensity ranges,

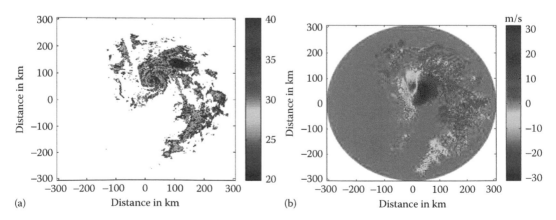

FIGURE 21.3   (a) PPI plot of radar reflectivity (dBZ) for cyclonic event on October 29, 2006. (b) Corresponding PPI plot of radial velocity, as observed from the DWR.

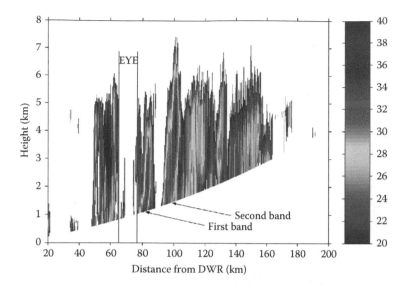

FIGURE 21.4 Vertical cross-section of cyclonic event on October 29, 2006, showing the eye and spiral bands of the system.

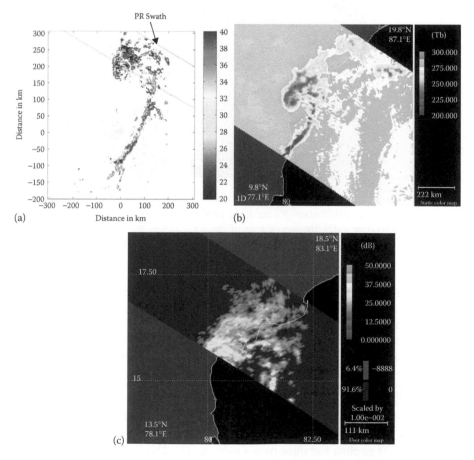

FIGURE 21.5 Near-simultaneous observations of the same cyclonic event by (a) DWR (reflectivity, dBZ), (b) TMI (brightness temperature, K), and (c) partial observation of the system by TRMM-PR (reflectivity dBZ).

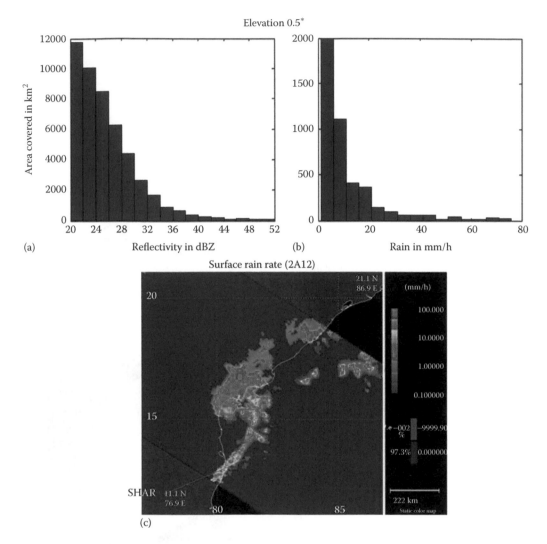

FIGURE 21.6   (a) Frequency distribution of area of the total radar pixels at various dBZ ranges, as observed from the DWR. (b) Frequency distribution of area of the total pixels at various rain intensity ranges, as estimated by the DWR. (c) Spatial plot of rain intensity from TMI for the same event.

as estimated by the DWR. Figure 21.6c shows the PPI plot of rain intensity from TMI for the same event. In the rain intensity domain, there is a reasonably good agreement between the DWR and TMI. Two distinct modes of heat released are observed for stratiform region, with upper height heating and lower height cooling (Figure 21.7a). Heat released for the stratiform region is ~5 k/h at ~5 km and cooling of value ~1.5 k/h at ~2 km. Convective region, on the other hand, shows a net heating for all levels, with maximum release of heat of the order ~20 k/h (Figure 21.7b). Although the heat released in stratiform region is less compared with convective, as noticed in Figure 21.3, stratifom region covers a very large area compared with convective, and thus, it also plays a significant role in atmospheric circulation.

The TC Thane was originated in the BoB on December 25, 2011. The TC Thane was associated with winds up to 137 km/h and hit Tamil Nadu, India, on December 30, 2011. During that duration of the TC, the TRMM satellite had three passes over the TC. The TRMM passed on December 26, 2011, at 13:05 UTC over the TC. That time, the storm's wind speed was about 40 knots. The PR was able to capture a large portion of the storm surrounding the north side (Figure 21.8a). Figure 21.8a

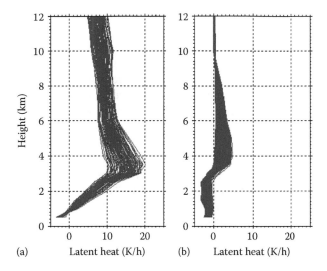

FIGURE 21.7   Vertical profiles of latent heat for storm height over (a) land (convective region), and (b) ocean (stratiform region).

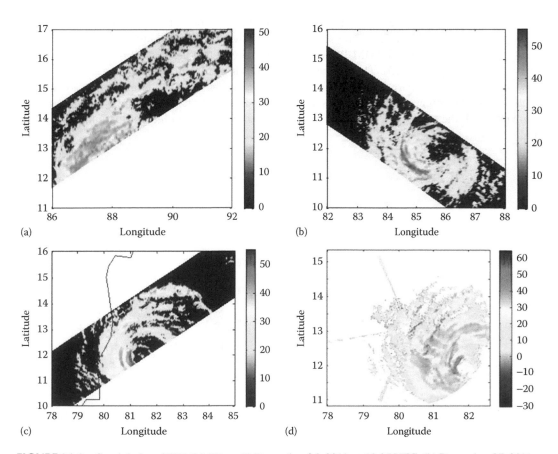

FIGURE 21.8   Spatial plot of TRMM-PR on (a) December 26, 2011, at 13:05 UTC, (b) December 27, 2011, at 21:59 UTC, and (c) December 29, 2011, at 11:53 UTC. (d) PPI plot at 0.5^0 elevation from Chennai DWR on December 29, 2011, at 11:10 UTC.

clearly depicted a developing stage of the storm. The TMI rain shows a circular eye-type structure near the storm's center. On 21:59 UTC of December 27, 2011, the TRMM-PR showed continuous bands of intense convective storms occurring around rainbands (Figure 21.8b). The PR reflectivity shows more than 50 dBZ in some areas of the storm, proving that heavy rainfall occurred in that area. The vertical cross-section showed that convective towers reached up to a height of more than 16 km (Figure 21.9a). The frequency distribution of rain area pixel is shown in Figure 21.9b. It is seen that maximum occurrence of rain intensity is 20 mm/h. From 20 to 40 mm/h, there is a sharp decrease in the occurrence of rain intensity, and after that, it decreases gradually. Highest intensity, 120 mm/h, is observed by the PR. The TC had the wind speed of about 75 knots on December 29, 2011. On that day at 11:53 UTC, the TRMM pass was observed. The PR showed multiple spiral intense convective rainbands southwest of the TC center (Figure 21.8c). The reflectivity values go to more than 55 dBZ, showing the proof of heavy rainfall (Roy et al. 2010). The TC Thane made landfall on December 30, 2011, near Pondicherry. The system was tracked by the DWR Chennai from December 28 evening till the landfall (Figure 21.8d). The DWR Chennai radar pictures show that the system had circular-open eye.

The TC Helen formed over the BoB and made landfall in a couple of days along the coast of southeastern India. The TRMM satellite shows scatter convective rainbands over the northern and eastern quadrants of the TC on November 19, 2013. The TC was located just 180 nautical miles south-southeast of Visakhapatnam, India, near 15.0°N and 84.5°E at 1500 UTC on November 19, 2013, with maximum sustained winds around 35 knots and moving to the west at 8 knots/14.8 km/h. On that day, the TRMM swath passed over that region and the TC was clearly seen. Satellite imagery showed that the low-level center is organized and there is convection (building thunderstorms) flaring around the storm's center. The scattered convective bands around the northern quadrant of the storm are detected. The TRMM-PR satellite data showed that the rainfall rates were as high as 120 mm/h (see Figure 21.10). The TC came to the DWR at Machilipatnam range on 02 UTC of November 21, 2013. Figure 21.11 shows the life cycle of the TC at different times, as observed by the DWR. The highest reflectivity is seen up to 50 dBZ, and the stratiform region is dominated compared with the convective portion, which mean that cyclone is in the dissipation stage (Houze et al. 2004).

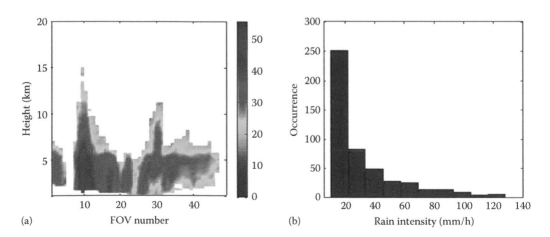

FIGURE 21.9 (a) Vertical cross-section of cyclonic event on December 27, 2011, showing the eye and spiral bands of the system. (b) Frequency distribution of area of the total pixels at various rain intensity ranges, as estimated by PR.

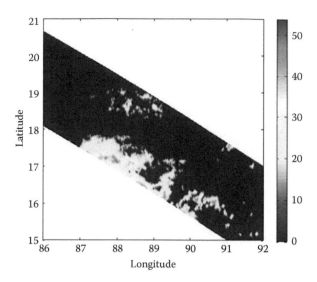

FIGURE 21.10 Spatial plot of TRMM-PR on November 19, 2013, at 13:05.

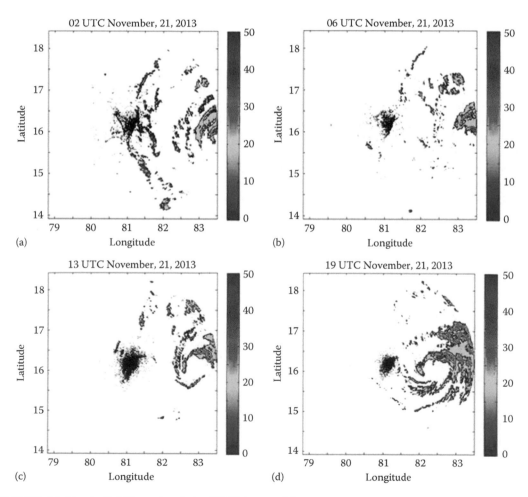

FIGURE 21.11 (a–f) Life cycle of the TC Helen at different UTCs, as seen by Machilipatnam DWR.
(*Continued*)

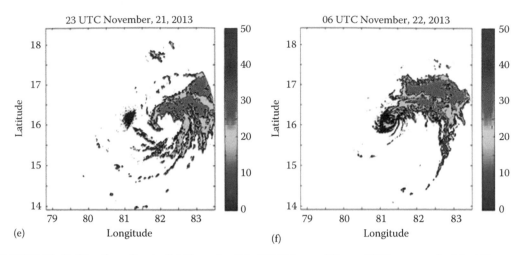

FIGURE 21.11 (Continued)   (a–f) Life cycle of the TC Helen at different UTCs, as seen by Machilipatnam DWR.

21.4   CONCLUSIONS

The evolution of TCs over the NIO is analyzed through the available remote-sensing observations. In this study, we utilized the TRMM-PR, TMI, and ground-based DWR. For the convective regime, the storm heights are characterized by high reflectivity of 40 dBZ or more. The maximum radial velocity observed was of the order of 30 m/s, both inward and outward with respect to the DWR. The convective and stratiform regions in the TCs are associated with high- and low-intensity rainfall, respectively. The convective regime is characterized by the release of significant amount of latent heat at all heights in the troposphere compared with the stratiform regime, where cooling is observed at lower height (~1–2 km). Although the heat released in the stratiform region is less compared with the convective, it is noticed that the stratifom region covers a very large area compared with convective and the stratiform region plays a significant role in atmospheric circulation. The vertical cross-sections of the convective towers reached up to a height of more than 16 km. High reflectivity up to 55 dBZ and corresponding rains of around 120 mm/h were observed. The maximum rainfall was 120 mm/h, and the maximum value of occurrence was 20 mm/h.

A long-term, strategic program of applied research will address many complex problems regarding the National Oceanic and Atmospheric Administration (NOAA's) use of space-based precipitation information to improve modeling, forecasting, and climate applications. The TRMM is one of the very successful missions of NASA. The TRMM-PR provided significant insight into the information about the structure of TCs. However, the TRMM precipitation radar can only detect moderate to high rainfall rates because of low sensitivity. In addition, the swath of the PR is less. The life cycle of TC characteristics is difficult to analyze if the PR is less repetitively.

There many international initiatives to utilize the satellite dataset for disaster management. The International Charter on Space and Major Disasters aims at providing a unified system of space data acquisition and delivery to those affected by natural or manmade disasters anywhere in the globe, through authorized users. Indian Space Research Organization (ISRO) joined the Charter in 2002 as a member and plays an important role in providing remote-sensing imageries to the global community, for the major disasters.

ISRO has joined Sentinel Asia Project initiated by the Japan Aerospace Exploration Agency (JAXA) in 2007. Sentinel Asia (SA) is a *voluntary and best-efforts-basis initiative* led by the Asia-Pacific Regional Space Agency Forum (APRSAF) to share disaster information in the Asia-Pacific region on the Digital Asia (Web-GIS) platform and to make the best use of Earth observation satellites' data for disaster management in the Asia-Pacific region.

REFERENCES

Cecil, D. J. and E. J. Zipser, 1999. Relationships between tropical cyclone intensity and satellite-based indicators of inner core convection: 85-GHz ice-scattering signature and lightning. *Mon. Weather. Rev.*, 127, 103–123.

Chandrasekhar S., 1960. *Radiative Transfer*, New York: Dover Publications.

Doviak, R. J. and D. S. Zrnić, 1993. *Doppler Radar and Weather Observation*, San Diego, CA: Academic Press.

Guimond, S. R., G. M. Heymsfield, and F. J. Turk, 2010. Multiscale observations of Hurricane Dennis (2005): The effects of hot towers on rapid intensification. *J. Atmos. Sci.*, 67, 633–654.

Gupta, A., 2006. Current status of tropical cyclone track prediction techniques and forecast errors. *Mausam*, 57, 151–158.

Henderson-Sellers, A., Zhang, H., Berz, G., Emanuel, K., Gray, W., Landsea, C., Holland, G. et al. 1998. Tropical cyclones and global climate change: A post-IPCC assessment. *Bull. Amer. Meteorol. Soc.*, 79, 19–38. doi:10.1175/1520.

Hendricks, E. A., M. T. Montgomery, and C. A. Davis, 2004. Therole of "vortical" hot towers in the formation of tropical cyclone Diana (1984). *J. Atmos. Sci.*, 61, 1209–1232.

Houze, R.A., 2004. Mesoscale convective systems. *Rev. Geophys.*, 42, RG4003.

Iguchi, T., T. Kozu, R. Meneghini, J. Awaka, and K. Okamoton, 2000. Rain-profiling algorithm for the TRMM precipitation radar. *J. Appl. Meteorol.*, 39, 2038–2052.

IMD Atlas, 2008. Tracks of storms and depressions in the Bay of Bengal and the Arabian Sea, India Meteorological Department, New Delhi, India

Jiang, H., 2012. The relationship between tropical cyclone intensity change and the strength of inner core convection. *Mon. Weather Rev.*, 140, 1164–1176.

Kelley, O. A., J. Stout, and J. B. Halverson, 2004. Tall precipitation cells in tropical cyclone eyewalls are associated with tropical cyclone intensification. *Geophys. Res. Lett.*, 31, L24112, doi:10.1029/2004GL021616.

Kummerow C. D, W. Barnes, T. Kozu, J. Shiue, and J. Simpson, 1998. The tropical rainfall measuring mission (TRMM) sensor package, *J. Atmos. Oceanic Technol.*, 15, 809–817.

Meneghini R., T. Iguchi, T. Kozu, L. Liao, K. Okamoto, J. A. Jones, and J. Kwiatkowski, 2000. Use of surface reference technique for path attenuation estimates from the TRMM precipitation radar, *J. Appl. Meteorol.*, 39, 2053–2070.

Mohanty, U. C. and A. Gupta, 1997. Deterministic methods for prediction of tropical cyclone tracks. *Mausam*, 48, 257–272.

Mohapatra, M., B. K. Bandyopadhyay, and D. P. Nayak, 2013. Evaluation of operational tropical cyclone intensity forecasts over North Indian Ocean issued by India Meteorological Department. *Nat. Hazards*, 68, 433–451.

Montgomery, M. T., M. E. Nicholls, T. A. Cram, and A. B. Saunders, 2006. A vortical hot tower route to tropical cyclogenesis. *J. Atmos. Sci.*, 63, 355–386.

Rao, G. V. and P. D. MacArthur, 1994. The SSM/I estimated rainfall amounts of tropical cyclones and their potential in predicting the cyclone intensity changes. *Mon. Weather Rev.*, 122, 1568–1574.

Reasor, P. D., M. Eastin, and J. F. Gamache, 2009. Rapidly intensifying Hurricane Guillermo (1997). Part I: Low-wavenumber structure and evolution. *Mon. Weather. Rev.*, 137, 603–631.

Schubert, W. H. and J. J. Hack, 1982. Inertial stability and tropical cyclone development. *J. Atmos. Sci.*, 39, 1687–1697.

Roy, S. S., V. Lakshmanan, S. K. Roy Bhowmik, and S. B. Thampi, 2010. Doppler weather radar based now-casting of cyclone Ogni, *J. Earth Syst. Sci.*, 119, 183–199.

Szoke, E. J., E. J. Zipser, and D. P. Jorgensen, 1986. A radar study of convective cells in mesoscale systems in GATE. Part I: Vertical profile statistics and comparison with hurricanes. *J. Atmos. Sci.*, 43, 182–198.

Ulaby F. T., R. K. Moore, and A. K. Fung, 1981. *Microwave Remote Sensing: Active and Passive*, Vol.1, Boston, MA: Artech House.

World Meteorological Organization Technical Document, 2008, *Tropical cyclone operational plan for the Bay of Bengal and the Arabian Sea*, Document No. WMO/TDNo. 84, 1–1.

Zipser, E. J. and M. A. LeMone, 1980. Cumulonimbus vertical velocity events in GATE. Part II: Synthesis and model core structure. *J. Atmos. Sci.*, 37, 2458–2469.

22 Radar Rainfall Estimates for Debris-Flow Early Warning Systems

Effect of Different Correction Procedures on the Identification of Intensity–Duration Thresholds

F. Marra, E. I. Nikolopoulos, J. D. Creutin, and M. Borga

CONTENTS

22.1 INTRODUCTION

Debris flows (DFs) are rapidly flowing, gravity-driven mixtures of roughly equal parts of sediment and water, in which a broad distribution of grain size, commonly including gravel and boulders, is mixed vertically (Iverson 2005). In fact, DFs are among the most devastating natural disasters in mountainous regions (Dowling and Santi 2014), and their occurrence has increased during the last decades (Dietrich and Krautblatter 2016). Development of DF forecasting and early warning systems is of great economical and societal importance and requires accurate knowledge on the triggering mechanisms and their corresponding characteristics (Borga et al. 2014). Early warning systems for rainfall-induced DFs are based on combining information about DF susceptibility in the region under consideration with rainfall measurements and forecasts (Hong and Adler 2007; Tiranti et al. 2014; Berenguer et al. 2015). Assessment of DF susceptibility is generally carried out by relating the occurrence of DF with a number of variables controlling DF initiation to identify the locations more prone to future events. When the information about the spatial variability of

DF susceptibility is neglected, early warning systems make use of rainfall thresholds, intended as rainfall conditions that, when reached or exceeded, are likely to result in DF (Guzzetti et al. 2008). Starting with the works by Caine (1980) and Innes (1983), rain intensity and duration were recognized as the characteristic properties of rainfall triggering landslides and DF, which lead to the development of rainfall intensity–duration (ID) thresholds that have since been used widely for identification of landslides/DF occurrence at the local, regional, and global scales (Guzzetti et al. 2008; Brunetti et al. 2010; Tiranti et al. 2014; Vennari et al. 2014; Rosi et al. 2015; Iadanza et al. 2016; Bel et al. 2016).

Concerning rainfall monitoring and forecasting, Alfieri et al. (2012) gave an extensive review of the benefits and limitations of the different rainfall inputs used in the context of early warning systems. The DFs are very small-scale phenomena that are frequently triggered by rainfall extremes (e.g., due to stationary convective thunderstorms) at scales that are generally not well resolved by numerical weather prediction models or low-resolution rain gauge networks. More specifically, rainfall estimates from rain gauges can be affected by large uncertainties, caused by the insufficient sampling of the triggering rainfall, mainly due to the combination of two factors: (1) rain gauges in the mountainous context are generally scarce and located at low elevations (e.g., in the valley floors); (2) DFs are generally initiated in the head part of the catchments by strong convective storms (Stoffel et al. 2011; Borga et al. 2014; Nikolopoulos et al. 2015b).

Owing to these reasons, it is expected that rainfall thresholds derived from rain gauge observations will be associated with considerable uncertainty. In fact, recent work by Nikolopoulos et al. (2014) showed that uncertainty in rain gauge estimation results in a systematic underestimation of the identified rainfall thresholds, leading to important degradation of the performance under operational conditions. Furthermore, Nikolopoulos et al. (2015a) showed that the problem persisted even when more complex (than reference rain gauge) interpolation techniques are considered for spatial rainfall estimation.

A potential solution to the observational limitations of rain gauges lies on remote-sensing observations. More specifically, the high spatial and temporal resolutions of weather radar (~1 km^2 × 5 min for a common C-Band) offer the unique advantage of estimating rainfall above the DF-triggering area. Nevertheless, the quantitative accuracy provided by radar is an important concern, particularly in mountainous regions (Germann et al. 2006). Correction algorithms have been proposed and tested for the main sources of error, generally focusing on flood and flash flood events (Pellarin et al. 2002; Krajewski et al. 2006; Villarini and Krajewski 2010; Gourley et al. 2011), but DF-triggering events pose different challenges, mainly related to the small size of the involved catchments, sometimes even less than 1 km^2 (D'Agostino and Marchi 2001). So far, only few studies took advantage of radar rainfall estimates for DF observation or forecasting, generally using standard products merged with rain gauge data (Wieczorek et al. 2000; David-Novak et al. 2004; Chen et al. 2007; Chiang and Chang 2009; Saito et al. 2010).

The present study aims at (1) assessing radar rainfall estimates for DF storm events, and (2) analyzing the benefits and limitations of different radar and rain gauge rainfall estimation scenarios for the derivation of rainfall thresholds.

Section 22.2 describes the study area and data. Section 22.3 assesses the radar correction procedure and introduces the radar rainfall scenarios. Section 22.4 derives ID thresholds from different sources of rainfall data. Finally, Section 22.5 summarizes the messages of the chapter.

22.2   STUDY AREA

The present research is based on the Upper Adige river basin (9700 km^2), in the Eastern Italian Alps. The significant societal risks of the area, marked by a large number of casualties and important damages caused by DF (Salvati et al. 2010), together with the availability of a long-term, accurate catalog of DF events, make this area unique for such a study. The region is characterized by complex topography, with elevation ranging from 200 m to almost 4000 m a.s.l.

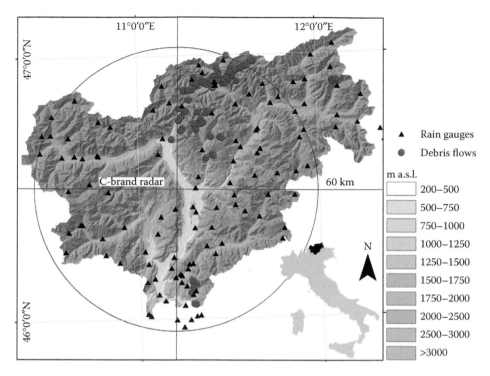

FIGURE 22.1 Map showing the study area. Orography of the Adige river basin closed at Trento is shown with colors. Rain gauge and debris flow locations are shown with black triangles and green circles, respectively. The weather radar location with a 60-km-range circle is also presented.

(Figure 22.1). Metamorphic rocks and calc-schists, prasinites, and serpentinites characterize the western part of the region, whereas dolomites and limestones prevail in the eastern part (Norbiato et al. 2009b). Mean annual precipitation varies from 400–700 mm of the internal alpine area to 1300–1800 mm of the southern and northern areas. The seasonal distribution of rainfall is influenced by western Atlantic airflows and southern circulation patterns (Frei and Schär 1998) and shows two peaks in the summer and fall seasons. The DFs typically occur (90% of the observed DFs) during the summer (Nikolopoulos et al. 2015b), when the precipitation regime is dominated by mesoscale convective systems and localized thunderstorms (Norbiato et al. 2009a; Mei et al. 2014).

A catalog listing more than 400 DFs is available for the area, starting from the year 2000 (Comiti et al. 2014). It can be considered complete, except for very small failures (<700 m³) that stopped upstream or remained hidden under the forest canopy. The catalog provides information about the initiation point of DFs, with accuracy up to 50 m.

The region is monitored by a rain gauge network, with spatial density ~1/80 km² and temporal resolution of 5 min, and a C-band, Doppler weather radar located in a central position and providing rainfall estimates with 1 km², 5-minute resolutions (Figure 22.1). Rain gauge data quality has been examined, rejecting suspicious measurements. Technical features of the radar instrument are reported in Table 22.1.

The study is based on seven storm events that triggered a total of 117 DFs, causing casualties and significant damages. The events, among the most severe that hit the region (Destro et al. 2017), occurred during the summer months and can be considered representative of the DF seasonality of the area. The characteristics of the storm events are reported in Table 22.2.

TABLE 22.1

Technical Characteristics of the Weather Radar

Range	120 km
Peak power	307.0 kW
Wavelength	5.3 cm (C-band)
Pulse	0.8 s
Antenna gain	45.8 dB
Radial resolution	250 m
Azimuthal resolution	0.8°–0.9°
Elevation geometry	1°, 2°, 3°, 4°, 5°, 6°, 8°, 10°, 13°, 16°, 19.5°, 24°
Beam width (3 dBZ)	0.8°

TABLE 22.2

Characteristics of the Examined Storms

Date	Triggered Debris Flows	Duration (h)	Rain Gauges	Maximum Gauge Rainfall Depth (mm)	Maximum Gauge Hourly Rainfall [mm h^{-1}]
August 1, 2005	5	2	7	5.6	5.6
June 20–21, 2007	13	20	14	102.1	25.8
July 16–17, 2009	7	27	8	150.1	95.5
July 30, 2009	15	9	14	54.0	53.0
September 4, 2009	6	12	17	122.5	23.8
August 14–15, 2009	7	13	25	160.1	47.4
August 4, 2012	64	8	8	86.0	30.6

22.3 RADAR RAINFALL

22.3.1 RAINFALL ESTIMATION PROCEDURE

Radar data are provided by Ripartizione Protezione Civile, Provincia Autonoma di Bolzano, already cleaned from ground echoes. Antenna pointing accuracy is verified by cross-correlating observed and simulated reflectivity fields of ground echoes (Rico-Ramirez et al. 2009). The elaboration of radar estimates then undergoes three steps:

> *C.1. Physically based corrections*: Radar raw data are corrected considering the effects of several sources of error: (1) we accounted for attenuation due to the wetting of the radome under heavy rainfall following the procedure reported in Marra (2013) and Marra et al. (2014), that is, comparing reflectivity of dry ground echoes in the presence and absence of rainfall over the radome during a study event (2.7 dBZ two-way wet radome attenuation was observed); (2) numerical simulations of radar beam propagation over a digital terrain model of the radar domain were used to compute the fraction of the pulse volume blocked by the orography (Pellarin et al. 2002); (3) signal attenuation in heavy rain was corrected by using the Mountain Reference Technique, based on the procedure reported in Bouilloud et al. (2009); (4) vertical variations of reflectivity have been accounted for by using the inverse procedure developed by Andrieu and Creutin (1995), applied to the event scale accumulation, owing to the limited spatial and temporal extension of the studied events.

C.2. Derivation of rain rate: Reflectivity (*Z*) from the lowest elevation scan with less than 30% of blocked beam was converted into rain rate (*R*) by using the classic power law Z–R relationship (Marshall et al. 1955; Battan 1973; Uijlenhoet 2001). In this study, the precise Z–R relationship used was of the form $Z = 308 \cdot R^{1.5}$, whose parameters have been empirically derived, comparing radar reflectivity with rain gauge rainfall estimates, and independently verified for isolated convective events observed over the area (Anagnostou et al. 2010).

C.3. Bias adjustment: Radar estimates are adjusted according to the mean field bias observed on event basis with respect to rain gauge measurements (Marra et al. 2014).

A detailed description of the procedure and the algorithms can be found in Marra et al. (2014). It is worth noting that the complete procedure provides accurate reanalysis data, capitalizing on the retrospective nature of the study and allowing the identification, correction, and quantification of errors that are generally untreatable in real-time conditions.

22.3.2 Assessment of the Radar Rainfall Estimation Procedure

Accuracy of radar estimates is assessed by statistical comparison with rain gauge observations on total event rainfall depths, in order to minimize the uncertainties involved in the comparison of rain gauge (sampling area of 400 cm^2) and radar (sampling area of 1 km^2) data for short time intervals (Gires et al. 2014).

Three statistical parameters, fractional standard error (*FSE*), that is, the root mean square error normalized over the average rain gauge estimate, normalized bias (*NB*), and Pearson's correlation coefficient (*CC*) are computed for each event, as follows (Marra et al. 2014):

$$\text{FSE} = \frac{\sqrt{\frac{1}{N}\sum_i^N (r_i - g_i)^2}}{\frac{1}{N}\sum_i^N g_i} \quad (22.1)$$

$$\text{NB} = \frac{\frac{1}{N}\sum_i^N (r_i - g_i)}{\frac{1}{N}\sum_i^N g_i} \quad (22.2)$$

$$\text{CC} = \frac{\text{Cov}(r, g)}{\sigma(r) \cdot \sigma(g)} \quad (22.3)$$

Where, *r* and *g* refer to the arrays of radar and gauge observations, Cov and σ correspond to covariance and standard deviation terms, respectively.

Figure 22.2 presents the statistical comparison parameters calculated after the sequential application of the correction algorithms described previously.

Raw radar data show significant underestimation with minimum, mean, and maximum values of NB equal to −81%, −62%, −53%, respectively, FSE equal to 1.1, 1.3, 1.8, respectively, and CC equal to 0.37, 0.59, 0.90, respectively. Wet radome attenuation contributes only slightly to the improvement due to the non-synchronicity between rain over the radar and over the study regions. The same holds true for the beam blockage correction, owing to the fact that raw radar estimates are obtained by using the lowest least-blocked elevation available, hence partially accounting for this problem. Significant improvement is provided by the correction for attenuation in heavy rain, which improves all the statistics, with FSE: 0.81, 1.0, 1.3, NB: −53%, −20%, +14%, and CC: 0.51, 0.69, 0.91. The Vertical Profile of Reflectivity (VPR) correction increases the accuracy for the events, characterized by the worst performances: FSE ranges in a limited interval: 0.81, 0.96, 1.0; NB: −43%, −13%,

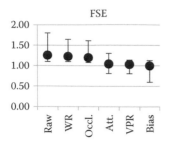
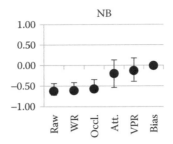
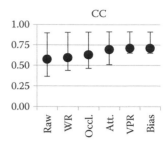

FIGURE 22.2 Radar-gauge comparison statistics calculated during the correction procedure. The dots and the vertical bars identify the mean value and the range of values over the seven rainfall events, respectively. Raw: raw estimates. Correction algorithms are applied sequentially: WR: wet radome; Occl: beam blockage; Att: signal attenuation; VPR: vertical profile of reflectivity; Bias: mean field bias adjustment. (Redrawn from Marra, F. et al., *J. Hydrol.*, 519, 1607–1619, 2014. With permission.)

+14%; and CC: 0.65, 0.71, 0.91. Mean field bias adjustment eliminates the NB, further reducing the FSE. The total improvement provided by the correction chain is clear, with an average 20% decrease in the FSE and 24% increase in the CC.

22.3.3 Radar Rainfall Scenarios

Four radar rainfall scenarios, characterized by different requirements in terms of correction procedures and rain gauge data, are prepared. The choice of the scenarios aims at covering potential situations in which not all radar correction procedures can be implemented or in which rain gauge data may not be available for the adjustment. They will form the basis for investigating the impact of radar rainfall estimation accuracy on ID threshold estimation.

> *R.I—raw radar rainfall*: Only C.2 is applied to the data. This scenario can be easily implemented in real time, with no need for rain gauge data.
>
> *R.II—raw radar rainfall adjusted for mean field bias*: C.2 and C.3 are applied to the data. This scenario is conditional to the availability of rain gauge data, but computational requirements are minimal, and the obtained rainfall accumulations are unbiased with respect to rain gauge measurements. Rainfall spatial patterns are unchanged with respect to R.I.
>
> *R.III—corrected radar rainfall*: C.1 and C.2 are applied to the data. This scenario considers the effects of a number of sources of error, leading to improved rainfall spatial patterns, and can be applied unconditionally to the availability of rain gauge data. The complexity of the computational sequence prevents its use in real time, and a bias on average rainfall amount may still arise.
>
> *R.IV—corrected radar rainfall adjusted for mean field bias*: The three correction steps are sequentially applied. This scenario offers accuracy in the identification of rainfall spatial patterns and unbiased rainfall accumulations and is expected to provide the most accurate spatial representation of the triggering rainfall. Therefore, hereinafter, this scenario will be used as a reference.

22.3.4 Accuracy of Rainfall Estimates at the Debris Flows' Locations

The total rainfall estimated by radar (R.IV) for three example events is shown in Figure 22.3. Several DF locations correspond to rainfall peaks. At the same time, some DFs occurred away from the core of the storms. Moreover, rainfall fields are characterized by very high spatial gradients, with 100% variations in less than 1 km distance. The problem of measuring the triggering rainfall by using rain gauges emerges clearly: rainfall peaks are rarely sampled by rain gauges, and few of

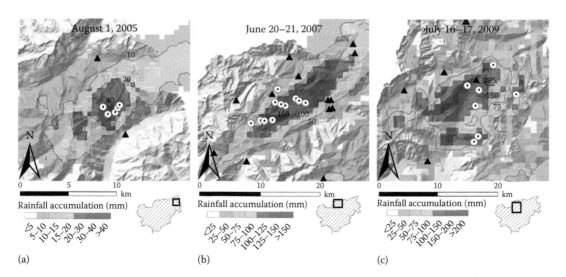

(a)   (b)   (c)

FIGURE 22.3 Rainfall depth fields estimated by radar (R.IV) for three example storm events. Dotted circles represent the location of triggered debris flows, and black triangles represent the location of rain gauges. (Redrawn from Marra, F. et al., *J. Hydrol.*, 519, 1607–1619, 2014. With permission.)

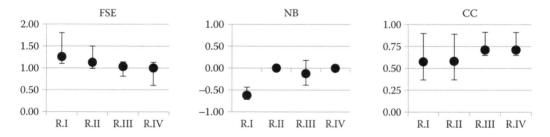

FIGURE 22.4 Comparison statistics obtained, extending their meaning to the comparison of rainfall esti-mated at the DF locations from R.I to R.III and from rain gauges (G) when R.IV is used as a reference. The dots and the vertical bars identify the mean value and the range of values over the seven rainfall events, respectively. (Redrawn from Marra, F. et al., *J. Hydrol.*, 519, 1607–1619, 2014. With permission.)

them are able to capture interesting features of the storm, even when a number of rain gauges are present within 10 km from the peak.

Figure 22.4 reports the comparison statistics, extending their meaning to the comparison of rain-fall estimated at the DF locations from R.I to R.III and from rain gauges (G) when R.IV is used as a reference. Following a common operational practice (Guzzetti et al. 2008), rain gauge estimates of DF rainfall are obtained by using measurements from the reference rain gauge. R.III consistently outperforms the other rainfall estimates for all three metrics. It is very interesting to note that R.III (which is not gauge-adjusted) is characterized by lower bias than the gauge-adjusted R.II. Rain gauge rainfall estimate performance is similar to that of R.II for FSE and NB, but the values of CC are dramatically lower.

22.4 INTENSITY–DURATION THRESHOLDS

A widely used rainfall ID threshold consists of a power law, linking average intensity, I (or, equiva-lently, depth), and duration, D, of DF-triggering rainfall events, according to

$$I = \alpha \cdot D^{\beta} \tag{22.4}$$

Where, α and β are the parameters that adapt the power law model to the empirical data.

The parameters of the model are estimated using the *frequentist* method (Brunetti et al. 2010) by using 5% exceedance probability level (Marra et al. 2014; Vennari et al. 2014; Nikolopoulos et al. 2015a). Rainfall events are defined as rainy periods (>0.1 mm h^{-1}) separated by hiatuses (<0.1 mm h^{-1}) of at least 24 hours (Guzzetti et al. 2008).

Nikolopoulos et al. (2014) calculated a regional ID threshold from the DF included in the catalog, using rainfall measurements from the nearest rain gauge, finding that rainfall intensities exceeding $I = 1.12\, D^{-0.49}$ (*I* in mm h^{-1} and *D* in hours) are likely to trigger a DF.

22.4.1 INTENSITY–DURATION THRESHOLDS FOR DIFFERENT RAINFALL ESTIMATION METHODS

In this section: (1) we assess the potential of radar rainfall estimates obtained with different requirements in terms of computation and data needs (the radar rainfall scenarios) and (2) we quantify the effect of the point sampling by rain gauges on the representation of spatially variable rainfall fields that characterize DF events (Marra et al. 2016; Destro et al. 2017).

The ID thresholds are identified using rainfall estimates above the DF locations from the four radar rainfall scenarios (R.I–R.IV) as well as from the nearest rain gauge (G) and from the rainfall estimated by radar over the nearest rain gauge location (R.G). In the case of the nearest rain gauge and of the gauge-colocated radar estimates, we followed the operational procedure adopted in the area, using constraints on rainfall duration (>2 h) and intensity (>1 mm h^{-1}), in order to avoid unreliable thresholds due to critical gauge sampling issues (Vessia et al. 2014). The obtained results are reported in Figure 22.5 and Table 22.3 and analyzed by using the ID threshold derived from R.IV as a reference.

The range of rainfall intensity provided by the reference relationship R.IV ($I = 8.57\, D^{-0.48}$) is relatively low with respect to thresholds reported for the Alpine range by Guzzetti et al. (2007). However,

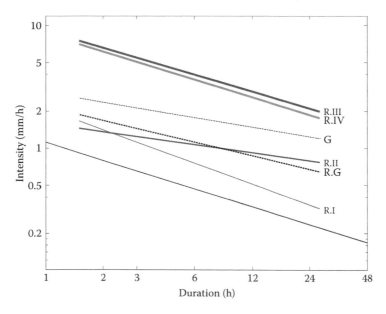

FIGURE 22.5 Intensity–duration thresholds derived from different rainfall estimates. Radar rainfall scenarios R.I–R.III are shown with solid lines, whereas the nearest rain gauges and the colocated radar estimates are shown with dashed lines. The reference threshold (R.IV) is shown in gray color, and the regional threshold identified from rain gauge data for the whole debris flow catalog by Nikolopoulos et al. (2014) is shown as a light solid-black line.

TABLE 22.3

***ID* Parameters Derived Using Different Rainfall Estimation Methods and Relative Error Calculated with Respect to the Reference *ID* Threshold Line**

Rainfall Estimation Method	α Parameter	Δα/α	β Parameter	Δβ/β
R.IV (reference)	8.57	–	0.48	–
R.I	2.10	−75%	0.57	+19%
R.II	1.59	−81%	0.22	−54%
R.III	9.07	+6%	0.46	−4%
S.G	2.84	−67%	0.26	−46%
R.IV.G	2.18	−75%	0.37	−23%
Nikolopoulos et al. 2014	1.12	−87%	0.49	+2%

the severity of a threshold should be assessed with respect to the local climatology of rainfall extremes. This aspect has been investigated for the here-presented dataset by Destro et al. (2017).

Underestimation affecting radar rainfall in R.I affects mainly the parameter α (−75% relative error). Underestimation of parameter α for R.II is large (−81%), but in this case, the error in the exponent β is also important (54%), meaning that the bias adjustment of raw radar data is not able to provide useful point information corresponding to DF locations. The ID thresholds for R.III are similar to the reference up to 6% relative error. This provides an interesting indication on the feasibility of using rain gauge–unadjusted corrected radar estimates for *ID* identification and has important practical implications, because rain gauges may be unavailable for the radar analysis of DF-triggering rain events, due to the small spatial and temporal scale, which characterizes these events.

Both multiplicative and scale parameters derived from G show important underestimation (−67% and −46%, respectively), and similar results (−75% and −23%, respectively) are observed for R.G. Such a result is related to the observed decrease of rainfall with distance from the DF. This result also suggests that the large differences observed between radar and gauge thresholds are likely associated with the spatial variability of rainfall around DF, which may systematically be leading to underestimation of rainfall when the measurement is operated away from the DF locations. Marra et al. (2016) analyzed this aspect in detail, finding that the event-cumulated rainfall fields systematically exhibit a peak corresponding or close to the DF initiation points, with rain depth decreasing with the distance, and showing that the use of log transformations on such fields explains the underestimation of the thresholds derived from rain gauge data.

22.5 CONCLUSIONS AND FUTURE WORK DIRECTIONS

This chapter analyzed the potential of radar rainfall estimates, opposed to the commonly used rain gauge estimates, for inclusion into early warning systems of DFs. More specifically, we analyzed the use of radar rainfall estimates, either corrected and not corrected for a number of errors sources, for the identification of rain ID thresholds. Two objectives are considered: (1) assessment of radar rainfall estimates for DF-triggering storm events, and (2) evaluation of ID rainfall thresholds obtained from rain gauge data with respect to thresholds obtained from radar rainfall estimates subject to different correction procedures.

We elaborated radar data, combining physical and empirical adjustments. We assessed the procedure step by step and obtained a set of radar and rain gauge rainfall estimation scenarios. The scenarios were characterized by increasing accuracy and requirements in terms of data and elaborations, and we assessed their ability to reproduce the DF-triggering thresholds.

Three important messages are provided by this work:

1. Raw radar estimates significantly underestimate convective rainfall (−62% bias with respect to rain gauge measurements), mainly due to the problem of attenuation in heavy rain. An adequate correction procedure, including physically based and empirical correction algorithms, strongly improves the estimates, with average CC increasing by 24% and FSE decreasing by 20%.
2. Raw radar estimates significantly underestimate the *ID* threshold, and bias adjustment of radar data does not improve the thresholds. On the contrary, implementation of physically based correction algorithms allows us to obtain the threshold parameters with less than 6% error. This suggests that, for the purpose of identifying DFs' thresholds, radar estimates can be used without need of rain gauge data.
3. The *ID* thresholds derived from rain gauge data or rain gauge-colocated radar observations are strongly underestimated, underlining the importance of the spatial distribution of triggering rainfall.

Results of this work highlight the unique ability of weather radar to estimate rainfall above the triggering locations, further stressing the added value of weather radar rainfall estimates for the threshold-based forecasting of DFs and, in general, for DF early warning.

Future efforts on the use of remote-sensed rainfall for DF early warning systems should be focused on three main directions: (1) implementation of real-time correction procedures able to provide accurate rainfall estimates at the relevant scales, (2) development of regional DF-triggering rainfall catalogs from corrected radar archives, and (3) investigation of the potential of satellite-based rainfall estimates for DF-warning applications. Particular attention should be paid on the newly launched Global Precipitation Measurement (GPM) mission that has made available high-resolution (0.1°/30 min) precipitation estimates at the global scale.

ACKNOWLEDGMENTS

This work is supported from EU FP7 Marie Curie Actions IEF project PIEF-GA-2011-302720) (HYLAND, http://intra.tesaf.unipd.it/cms/hyland/). We acknowledge Ripartizione Opere Idrauliche, Autonomous Province of Bolzano (Italy), for providing access to the DFs catalog, and Ufficio Idrografico, Autonomous Province of Bolzano (Italy), for providing hydrological and meteorological data and access to the weather radar information. This work is also a contribution to the HyMeX program (HYdrological cycle in the Mediterranean EXperiment), and the authors acknowledge the HyMeX database teams (the North Eastern Italy Hydrometeorological Observatory). We would like to thank the editors George Petropoulos and Tanvir Islam and two anonymous reviewers for helping us in improving the chapter. F. Marra sincerely thanks Prof. Efrat Morin for her support.

REFERENCES

Aleotti, P., 2004. A warning system for rainfall-induced shallow failures. *Engineering Geology* 73(3–4): 247–265.

Alfieri, L., P. Salamon, F. Pappenberger, F. Wetterhall, and J. Thielen. 2012. Operational early warning systems for water-related hazards in Europe. *Environmental Science and Policy* 21: 35–49. doi:10.1016/j.envsci.2012.01.008.

Anagnostou, M. N., J. Kalogiros, E. N. Anagnostou, M. Tarolli, A. Papadopoulos and M. Borga. 2010. Performance evaluation of high-resolution rainfall estimation by X-band dual polarization radar for flash flood applications in mountainous basins. *Journal of Hydrology* 394 (1–2): 4–16. doi:10.1016/j.jHydrology2010.06.026.

Andrieu, H. and J. D. Creutin. 1995. Identification of vertical profiles of radar reflectivity for hydrological applications using an inverse method. Part I: Formulation. *Journal of Applied Meteorology* 34: 225–239.

Battan, L. J. 1973. *Radar Observation of the Atmosphere.* Chicago, IL: The University of Chicago Press, 324 pp.

Bel, C., F. Liébault, O. Navratil, N. Eckert, H. Bellot, F. Fontaine, and D. Laigle. 2016. Rainfall control of debris-flow triggering in the Réal Torrent, Southern French Prealps. *Geomorphology.*

Berenguer, M., D. Sempere-Torres, M. Hürlimann. 2015. Debris-flow forecasting at regional scale by combining susceptibility mapping and radar rainfall. *Natural Hazards and Earth System Sciences* 15(3): 587–602. doi:10.5194/nhess-15-587-2015.

Borga, M., M. Stoffel, L. Marchi, F. Marra, M. Jakob. 2014. Hydrogeomorphic response to extreme rainfall in headwater systems: Flash floods and debris flows. *Journal of Hydrology* 518: 194–205. doi:10.1016/j.jhydrol.2014.05.022.

Bouilloud, L., G. Delrieu, B. Boudevillain, M. Borga and F. Zanon. 2009. Radar rainfall estimation for the post-event analysis of a Slovenian flash-flood case: Application of the mountain reference technique at C-band frequency. *Hydrology and Earth System Sciences* 13: 1349–1360. doi:10.5194/hess-13-1349-2009.

Brunetti, M. T., S. Peruccacci, M. Rossi, S. Luciani, D. Valigi and F. Guzzetti. 2010. Rainfall thresholds for the possible occurrence of landslides in Italy. *Natural Hazards and Earth System Science* 10(3): 447–458. doi:10.1007/978-94-007-4336-6_22.

Caine, N. 1980. The rainfall intensity-duration control of shallow landslides and debris flows. *Geografiska Annaler. Series A. Physical Geography* 62A: 23–27. doi:10.2307/520449

Comiti, F., L. Marchi, P. Macconi, M. Arattano, G. Bertoldi, M. Borga, F. Brardinoni, M. Et al. 2014. A new monitoring station for debris flows in the European Alps: first observations in the Gadria basin. *Natural Hazards* 73(3): 1175–1198. doi:10.1007/s11069-014-1088-5

Chen, C.Y., L. Y. Lin, F. C. Yu, C. S. Lee, C. C. Tseng, A. H. Wang and K. W. Cheung. 2007. Improving debris flow monitoring in Taiwan by using high-resolution rainfall products from QPESUMS. *Natural Hazards* 40(2): 447–461. doi:10.1007/s11069-006-9004-2.

Chiang, S. H. and K. T. Chang. 2009. Application of radar data to modeling rainfall-induced landslides. *Geomorphology* 103(3): 299–309. doi:10.1016/j.geomorph.2008.06.012

D'Agostino, V. and L. Marchi. 2001. Debris flow magnitude in the Eastern Italian Alps: Data collection and analysis. *Physics and Chemistry of the Earth* 26(9): 657–663. doi:10.1016/S1464-1917(01)00064-2.

David-Novak, H. B., E. Morin and Y. Enzel. 2004. Modern extreme storms and the rainfall thresholds for initiating debris flows on the hyperarid western escarpment of the Dead Sea. *Israel Bulletin of the Geological Society of America* 116(5–6): 718–728. doi:10.1130/B25403.2

Destro, E., F. Marra, E.I. Nikolopoulos, D. Zoccatelli, J.D. Creutin and M. Borga. 2017. Spatial estimation of debris flows-triggering rainfall and its dependence on rainfall severity. *Geomorphology,* 278: 269–279. doi:10.1016/j.geomorph.2016.11.019

Dietrich, A and M. Krautblatter. 2016. Evidence for enhanced debris-flows activity in the Northern Calcareous Alps since 1980s (Plansee, Austria). *Geomorphology.* Published online. doi:10.1016/j.geomorph.2016.01.013.

Dowling, C.A. and P. M. Santi. 2014. Debris flows and their toll on human life: A global analysis of debris-flow fatalities from 1950 to 2011. *Natural Hazards* 71: 203–227. doi:10-1007/s11069-013-0907-4.

Frei, C. and C. Schär. 1998. A precipitation climatology of the Alps from high-resolution rain-gauge observations. *International Journal of Climatology* 18(8): 873–900.

Germann, U., G. Galli, M. Boscacci, and M. Bolliger. 2006. Radar precipitation measurement in a mountainous region. *Quarterly Journal of the Royal Meteorological Society* 132 (618 A): 1669–1692. doi:10.1256/qj.05.190.

Gires, A., I. Tchiguirinskaia, D. Schertzer, A. Schellart, A. Berne and S. Lovejoy. 2014. Influence of small scale rainfall variability on standard comparison tools between radar and rain gauge data. *Atmospheric Research* 138: 125–138. doi:10.1016/j.atmosres.2013.11.008.

Gourley, J. J., Y. Hong, Z. L. Flamig, J. Wang, H. Vergara and E. N. Anagnostou. 2011. "Hydrologic evaluation of rainfall estimates from radar, satellite, gauge, and combinations on Ft. Cobb basin, Oklahoma. *Journal of Hydrometeorology* 12(5): 973–988. doi:10.1175/2011JHM1287.1.

Guzzetti, F., S. Peruccacci, M. Rossi and C. P. Stark. 2007. Rainfall thresholds for the initiation of landslides in central and southern Europe. *Meteorology and Atmospheric Physics* 98, 239–267. doi:10.1007/s00703-007-0262-7.

Guzzetti, F., S. Peruccacci, M. Rossi, and C. P. Stark. 2008. The rainfall intensity–duration control of shallow landslides and debris flows: An update. *Landslides* 5: 3–17. doi:10.1007/s10346-625 007-0112-1.

Hong, Y. and R. F. Adler. 2007. Towards an early-warning system for global landslides triggered by rainfall and earthquake. *International Journal of Remote Sensing* 28: 3713–3719. doi:10.1080/01431160701311242.

Innes, J. L. 1983. Debris flows. *Progress in Physical Geography* 7: 469–501.

Iadanza, C., A. Trigila and F. Napolitano. 2016. Identification and characterization of rainfall events respon-sible for triggering of debris flows and shallow landslides. *Journal of Hydrology* 541: 230–245. doi:10.1016/j.jhydrol.2016.01.018.

Iverson, R. 2005. Debris flow mechanics. In M. Jakob and O. Hungr (Eds.) *Debris Flow Hazards and Related Phenomena*, pp. 105–134. Berlin, Germany: Springer.

Krajewski, W. F., A. A. Ntelekos and R. Goska. 2006. A GIS-based methodology for the assessment of weather radar beam blockage in mountainous regions: Two examples from the US NEXRAD network. *Computer and Geosciences* 32: 283–302. doi:10.1016/j.cageo.2005.06.024

Marra, F. 2013. Procedura integrata di analisi e correzione delle osservazioni radar per la stima di precipi-tazioni intense in ambiente alpino. PhD diss., University of Padova (in Italian).

Marra, F., E.I. Nikolopoulos, J.D. Creutin, M. Borga. 2014. Radar rainfall estimation for the identifica-tion of debris-flow occurrence thresholds. *Journal of Hydrology* 519: 1607–1619. doi:10.1016/j.jhydrol.2014.09.039.

Marra, F., E. I. Nikolopoulos, J. D. Creutin, M. Borga. 2016. Space–time organization of debris flows-triggering rainfall and its effect on the identification of the rainfall threshold relationship. *Journal of Hydrology* 541: 246–255. doi:10.1016/j.jhydrol.2015.10.010.

Marshall, J. S., W. Hitschfeld, and K. L. S. Gunn. 1955. Advances in radar weather. *Advances in Geophysics* 2: 1–56.

Mei, Y., E. N. Anagnostou, E. I. Nikolopoulos and M. Borga. 2014. Error analysis of satellite rainfall prod-ucts in mountainous basins. *Journal of Hydrometeorology* 15(5): 1778–1793. doi:dx.doi.org/10.1175/JHM-D-13-0194.1.

Nikolopoulos, E. I., M. Borga, J. D. Creutin and F. Marra. 2015a. Estimation of debris flow triggering rainfall: Influence of rain gauge density and interpolation methods. *Geomorphology* 243: 40–50. doi:10.1016/j.geomorph.2015.04.028.

Nikolopoulos, E. I., M. Borga, F. Marra, S. Crema and L. Marchi. 2015b. Debris flows in the Eastern Italian Alps: Seasonality and atmospheric circulation patterns. *Natural Hazards and Earth System Sciences* 15: 647–656. doi:10.5194/nhess-15-647-2015.

Nikolopoulos, E. I., S. Crema, L. Marchi, F. Marra, F. Guzzetti, M. Borga. 2014. Impact of uncertainty in rain-fall estimation on the identification of rainfall thresholds for debris flow occurrence. *Geomorphology* 221: 286–297. doi:10.1016/j.geomorph.2014.06.015.

Norbiato, D., M. Borga, R. Dinale. 2009a. Flash flood warning in ungauged basins by use of the flash flood guidance and model-based runoff thresholds. *Meteorological Applications* 16: 65–75. doi:10.1002/met.126.

Norbiato, D., M. Borga, R. Merz, G. Blöschl and A. Carton. 2009b. Controls on event runoff coefficients in the eastern Italian Alps. *Journal of Hydrology* 375(3–4): 312–325. doi:10.1016/j.jhydrol.2009.06.044.

Pellarin, T., G. Delrieu, G. M. Saulnier, H. Andrieu, B. Vignal, and J. D. Creutin. 2002. Hydrologic visibility of weather radar systems operating in mountainous regions: Case study for the Ardèche catchment (France). *Journal of Hydrometeorology*, 3: 539–555.

Rico-Ramirez, M. A., E. Gonzalez-Ramirez, I. Cluckie and D. W. Han. 2009. Real-time monitoring of weather radar antenna pointing using digital terrain elevation and a Bayes clutter classifier. *Meteorological Applications* 16: 227–236. doi:10.1002/met.112.

Rosi, A., D. Lagomarsino, G. Rossi, S. Segoni, A. Battistini and N. Casagli 2015. Updating EWS rainfall thresholds for the triggering of landslides. *Natural Hazards* 78(1): 297–308. doi://doi.org/10.1007/s11069-015-1717-7.

Saito, H., D. Nakayama and H. Matsuyama. 2010. Relationship between the initiation of a shallow landslide and rainfall intensity—Duration thresholds in Japan. *Geomorphology* 118(1–2): 167–175. doi:10.1016/j.geomorph.2009.12.016.

Salvati, P., C. Bianchi, M. Rossi and F. Guzzetti. 2010. Societal landslide and flood risk in Italy. *Natural Hazards and Earth System Science* 10(3): 465–483. doi:10.5194/nhess-10-465-2010.

Stoffel, M., M. Bollschweiler and M. Beniston. 2011. Rainfall characteristics for periglacial debris flows in the Swiss Alps: Past incidences—Potential future evolutions. *Climatic Change* 105: 263–280. doi:10.1007/s10584-011-0036-6.

Tiranti, D., R. Cremonini, F. Marco, A. R. Gaeta and S. Barbero. 2014. The DEFENSE (debris Flows trig-gEred by storms - nowcasting system): An early warning system for torrential processes by radar storm tracking using a geographic information system (GIS). *Computers and Geosciences* 70: 96–109. doi:10.1016/j.cageo.2014.05.004.

Uijlenhoet, R. 2001. Raindrop size distributions and radar reflectivity–Rain rate relationships for radar hydrol-ogy. *Hydrology and Earth System Sciences* 5(4): 615–628. doi:10.5194/hess-5-615-2001.

Vennari, C., S. L. Gariano, L. Antronico, M. T. Brunetti, G. Iovine, S. Peruccacci, O. Terranova, F. Guzzetti. 2014. Rainfall thresholds for shallow landslide occurrence in Calabria, southern Italy. *Natural Hazards and Earth System Sciences* 14: 317–330. doi:10.5194/nhess-14-317-2014.

Vessia, G., Parise, M., Brunetti, M. T., Peruccacci, S., Rossi, M., Vennari, C., Guzzetti. 2014. F. Automated reconstruction of rainfall events responsible for shallow landslides. *Natural Hazards and Earth System Sciences*, 14(9): 2399–2408.

Villarini, G. and W. F. Krajewski. 2010. Review of the different sources of uncertainty in single polarization radar-based estimates of rainfall. *Surveys in Geophysics* 31(1): 107–129. doi:10.1007/s10712-009-9079-x.

Wieczorek, G. F., B. A. Morgan and R. H. Campbell. 2000. Debris-flow hazards in the blue ridge of central Virginia. *Environmental and Engineering Geoscience* 6(1): 3–23. doi:10.2113/gseegeosci.6.1.3.

Section VI

Remote Sensing of Landslides

23 A Review of Unmanned Aerial Vehicles, Citizen Science, and Interferometry Remote Sensing in Landslide Hazards
Applications in Transportation Routes and Mining Environments

Panagiotis Partsinevelos, Zacharias Agioutantis, Achilleas Tripolitsiotis, and Nathaniel Schaefer

CONTENTS

23.1 INTRODUCTION

The term *landslide* can be defined as the *movement of a mass of rock, earth, or debris down a slope* (Cruden, 1991, p. 28). Cruden and Varnes (1996) proposed a landslide classification system, whereas almost 20 years later, the advancement in understanding the triggering mechanisms involved led Hungr et al. (2014) to propose a new classification system. This broad definition of landslides, in conjunction with the fact that most landslides are triggered by other natural hazards, causes an underestimation of the economic losses induced from landslides (Kjekstad and Highland, 2009). For example, in the NatCatService[*] report of the 10 costliest events worldwide between 1980 and 2015, the overall losses of the 2011 Thailand floods and landslides are evaluated together

---

[*] https://www.munichre.com/en/reinsurance/business/non-life/natcatservice/significant-natural-catastrophes/index.html [accessed: 10-Oct-2016]

TABLE 23.1

Total Deaths and Affected People from Landslides and Mass Movement Events, per Continent, for the Period 2000–2016

Continent	Events Count	Total Deaths	Total Affected
Africa	27	901	46,658
America	62	2312	260,010
Asia	216	11,281	4,028,293
Europe	11	204	2,852
Oceania	8	153	11,095

Source: EM-DAT database.

(amounting to about US$ 43,000M), whereas EM-DAT: The International Disaster Database (Guha et al. 2016) classifies the same event as a flood. Therefore, it is often difficult to determine the direct environmental, societal, and economical impacts of landslides.

Table 23.1 presents the results obtained from the EM-DAT database from 2000 to 2016 with respect to the total deaths and total number of people affected due to landslides and mass movement events per continent.

For this period, it is evident that Asia is most affected by landslide hazards, followed by America, whereas Europe comes in fourth place. This does not necessarily mean that Europe is less vulnerable to landslide risks: Europe faced the second highest number of fatalities in the twentieth century (Haque et al. 2016).

Copernicus, the European system for Earth monitoring, consists of Earth observation (EO) satellites whose data are complemented with *in situ* observations to provide users with environmental and security related services on land, marine, climate change, atmosphere, emergency management, and security. According to a recent Copernicus market report (PwC 2016), a 40% added value in the European economy is expected through the investment in the Copernicus program. Similar large-scale EO programs are supported by other countries (i.e., the United States and China), whereas international collaboration for the exploitation of EO capabilities takes place via international initiatives and organizations (such as the Group of Earth Observation, the International Program on Landslides, the International Consortium on Landslides, etc.).

Since 2000, the European Commission Framework Programmes for Research and Technology (FP5, FP6, FP7, and now H2020) supported the execution of 134 projects related to landslides, with a budget exceeding 185M Euro. These are not the only research projects funded in Europe for landslide disaster mitigation. There exists a plethora of funding organizations and entities (i.e., the Directorate-General for European Civil Protection and Humanitarian Aid Operations [DG ECHO]), not to mention funds from the European Structural Funds and/or national support that sustain landslide-related projects.

Landslides constitute a major disaster on a global scale, and even a simple search on the Scopus database on scientific publications revealed that more than 27,000 articles contained the term *landslide* in their title, abstract, or keywords. About 81% of these scientific works were published during the period 2000–2016 (Figure 23.1).

Generally, a review paper on remote-sensing techniques for landslide studies is published about once every decade (Mantovani et al. 1996; Metternicht et al. 2005; Scaioni et al. 2014). Hence, we have explored the Scopus database for the period 2014–2017 by using *landslide* and *remote sensing* as search terms. A literature search revealed that more than 600 scientific documents have been published since the last review paper of 2014.

Figure 23.2 illustrates the distribution of these documents into three main groups identified in Scaioni et al. (2014): (1) landslide recognition, (2) landslide monitoring, and (3) landslide hazard assessment and prediction.

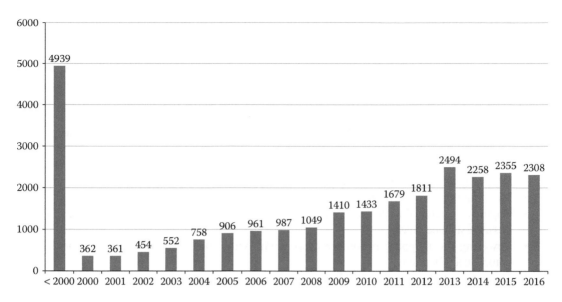

FIGURE 23.1   The publications registered in the Scopus database with the term *landslide* in title, abstract or keywords.

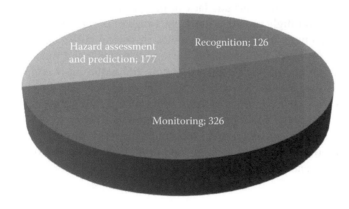

FIGURE 23.2   Number of documents published after 2014, included in the Scopus database, that contain the terms *landslide* and *remote sensing* in either their title, abstract, or keywords, following the categorization of Scaioni et al. (2014).

This chapter's contribution and differentiation from previous studies are justified by our focus on two emerging scientific areas: the unmanned aerial vehicles (UAVs) (Section 23.2) and the citizen's science (Section 23.3). In addition, review on two critical application areas, that is, transportation networks and the mining industry, where remote sensing has a proven and emerging track record in connecting land deformation with hydrometeorological hazards, is provided in Sections 23.4 and 23.5, respectively.

23.2   UNMANNED AERIAL VEHICLES AND LANDSLIDES

Unmanned aerial vehicles or systems (UAVs or UAS) have been used for an abundance of applications extending throughout various geoscience-related sectors. Even though the most prominent techniques and operations are quite old, the last decade's simple UAV hardware implementations became quite popular, giving an opportunity for widespread use. In the landslide application area, UAVs offer many advantages over other forms of data collection and processing, such as

- Scanning an area of interest on demand almost anytime, providing high temporal resolution: Since natural hazards often occur unexpectedly, it is quite important to access the areas under distress in near real time. Furthermore, this multi-epoch data collection allows trend and flow dynamics monitoring toward the understanding of the landslide life.
- Mapping the affected areas in very high-spatial resolution that may reach under centimeter magnitude.
- Representing the corresponding geomorphology in the form of a digital surface model (DSM) as a continuous space and not sparsely over selected fixed points.
- Mapping inaccessible or dangerous areas (Figure 23.3).
- Capturing occluded areas from various angles, making sure that coverage will minimize possible surface holes.
- Lowering the mapping procedure costs along with equipment for common landslide-mapping applications.
- Offering real-time data visualization, collection, and near-real-time processing and assessment.
- Carrying multiple sensors for customization.
- Facilitating other types of analyses, including fissure recognition and assessment.

The main disadvantages generally include lack of penetration capabilities through vegetation and other obstacles and a short flight time that limits the range of coverage.

In the literature (Scopus), most landslide studies involving UAVs share a series of commonalities. The majority of these studies involve custom or commercial micro- or mini multicopter UAVs that use common off-the-shelf cameras, mainly for acquiring overlapping imagery toward photogrammetric-vision-based DSM construction. Image processing involves proprietary open or developed software based on structure from motion, matching algorithms, photogrammetric bundle adjustment, and camera calibration implementations (Lowe 1999; Westoby et al. 2012). In terms of navigation, UAVs cover the areas of interest in an automated or human-operated mode. In Table 23.2, a series of related papers demonstrate similarities and variations in terms of landslide geometries, equipment, and corresponding results.

(a)

(b)

FIGURE 23.3 A geotechnical study for an inaccessible rockfall-prone area along a transportation corridor (a) has been supported by a UAV survey to derive the area's DSM (b).

TABLE 23.2

Selected Publications on UAV-Assisted DSM Construction for Landslide Monitoring

Landslide Information	UAV Equipment	Mission Aspects	Results	Citation
• Slovenia, Potoška planina landslide • Geological, tectonic, and hydrological conditions • 30°–70° slope	• Quadrocopter Microdrone MD4-1000 • Survey Drone 01hexacopter	• 22.5 months' monitoring • Flight altitude of 40 m	• Change assessments for an area of 1417 m² • 0.9 m and 17.9 m displacement for 22.5 months	Peternel et al. 2016
• Spain, La Guardia de Jaén • Heavy rainfall • Average slope of 15%	• Eight-rotor AscTec Falcon 8 with Sony NEX 5N, Asctec Trinity • ATyges FV-8, with Canon Powershot G12	• 4 years' monitoring, Flight altitude of 100 m and 120 m • GSD lower than 5 cm	• Displacement decimeters to meters by year in x and y, and from 1 m to 10 m/year in z • Measurement accuracy: cm level	Fernandez et al. 2016
• United Kingdom, Hollin Hill, North Yorkshire • Average slope of 12°	• Mini fixed-wing UAV (Quest UAV 300) with Panasonic Lumix DMC-LX5	• 2 years • Flight altitude 80–120 m	• Displacement sensitivity in z: 9 cm • Measurement accuracy: several cm level	Peppa et al. 2016
• Spain, Jaén	• Falcon 8 with Sony Nex-5N camera	• Two epochs (with four months of difference)	• About 0.5–2.5 m displacement • Measurement accuracy: cm level	Mozas-Calvache et al. 2016
• Jordan, Salhoub, Al-Juaidieh • Heavy rainfall • Roadway • Slope 10°–70°	• DJI Phantom 2 with a GoPro Hero 3+	• One epoch • Automated scarp detection • 266 m × 185 m area	• Measurement accuracy: several cm level	Al-Rawabdeh et al. 2016b
• Taiwan, Nantou County, Meiyuan Shan • Rainfall brought by Typhoon Sinlaku	• Fixed-wing drone with a Canon 500D camera	• 8-year large-scale landslide with an area of about 0.9 km² • Flying height of 2500–2750 m	• Measurement accuracy: 5 cm in x and y and 35 cm in z	Hsieh et al. 2016
• Tasmania, Huon valley • 35 slope • 125 m long and 60 m wide	• Oktokopter micro-UAV • Flight duration of 5–10 min with a Canon 550D	• 4 years • 40 m above ground level • 5675 m²	• Displacement • 12 m movement in x and y and 3 m in z • Measurement accuracy: 3–5 cm in x, y, and z	Turner et al. 2015

(Continued)

TABLE 23.2 (*Continued*)
Selected Publications on UAV-Assisted DSM Construction for Landslide Monitoring

Landslide Information	UAV Equipment	Mission Aspects	Results	Citation
• Italy • Planar landslide Narzole (CN) • Rockfall in Germanasca Valley (TO)	• V-shaped, eight motors configuration, with Panasonic DMC-LX3	• UAV flights' altitudes 30–70 m • Mapping areas in the order of <1 km^2	• Absolute positioning accuracies: 10 cm	Torrero et al. 2015
• Czech Republic, Zlínský, • Slope 10°–15° • Area: 5,000 m^2	• Multirotor system with six propellers	• UAV flights' altitudes 120 and 80 m	• Measurement accuracy: 3 cm in x, y, and z	Marek et al. 2015
• France, Super-Sauze, Barcelonnette Basin (Southern French Alps), • Average slope 25° • Total volume 750,000 m^3 • Displacement velocities range 0.01–0.4 m/day	• Custom quad-rotor with Praktica Luxmedia 8213	• May 2007 and October 2008 • Flight altitude of 200 m to provide a ground resolution of approximately 0.06 m per pixel	• Horizontal displacements of 7–55 m between a high-resolution airborne ortho-photo of May 2007 and a UAV-based orthomosaic of October 2008 • Average precisions for all the GCPs were 0.079, 0.079, and 0.185 m in x, y, and z, respectively	Niethammer et al. 2011
• China, Qinglingou slope, Three Gorges Reservoir • 51,000 m^2 • Average slope 34°	• Multirotor UAV with 18 megapixel digital camera	• Altitude 100 ~ 150 m	• Average GSD: 4.5 cm • Mean projection error of bundle block adjustment: 0.32 pixel • RMS error of x, y, and z directions of four GCPs: 0.12, 0.12, 0.43 m, respectively	Lin et al. 2016
• Austria, Pechgraben village • Heavy rainfalls • Area of 70 ha, including a volume of several million cubic meters of material, • Moving rates between 0.5 m and 1 m per week	• BARF MikroKopter OktoXL with Canon EOS 650D DSLR	• A period of >1 year • Ground resolution of 10 cm • 100 m height • GSD 1.7–2.2 cm	• DSM resolution of up to 8 cm • GCP accuracy: 11.6 ± 4.9 cm • Movement <0.5 cm/day shows no specific signature	Lindner et al. 2015

As can be easily established, most studies provide multitemporal DSMs, with subpixel accuracy corresponding to a precision of a few centimeters in the horizontal and vertical coordinate space, by using ground control point (GCP) measurements through real-time kinematics (RTK) equipment. Landslides are mainly triggered by heavy rainfall, typhoons (Hsieh et al. 2016), or earthquakes.

Other landslide-related studies include methodologies for displacement modeling, by relating the resulting DSMs from different epochs through correspondence of single common points, multiple point sets (Fernandez et al. 2015), or linear features (Mozas-Colvache et al. 2016). Point cloud registration for multi-epoch DSMs is shown in Al-Rawabdeh et al. (2016a). Furthermore, fissures can be mapped and assessed through image processing toward landslide flow prediction (Niethammer et al. 2012; Stumpf et al. 2013). Another small subset of studies involves optimized distribution of GCPs, along with experiments on camera line of sight direction, either perpendicular to the slope or under classic vertical geometry (Giordan et al. 2015; Carvajal-Ramírez et al. 2016).

Apart from multi-epoch landslide mapping, a few studies demonstrate UAV usability on postdisaster assessment or simple optical visualizations to support real-time decisions (Liu et al. 2015). A gasoline helicopter-type UAV capable of carrying either a laser scanner and optical camera bundle or a hyperspectral pushbroom scanner is the most versatile yet costly system that we have encountered (Gallay et al. 2016).

It is evident that throughout the related studies, the main methodologies are quite common. It is surprising that there is no widespread involvement of multispectral cameras that may reveal several attributes (e.g., soil moisture and land use) affecting the landslide dynamics and assist prediction (Gallay et al. 2016). In addition, UAVs are mainly used in their standardized commercial or simple implementation form, and no real-time and on-the-fly processes are apparent (Tripolitsiotis et al. 2017).

There is no extensive experimentation or theoretical approach to optimize the mapping parameters that support many replications in the literature. Most studies use the error estimations given by the software, without real evidence on overall quality, since GCPs and validation are quite scarce and known photogrammetric restrictions are not tackled or quantified (Peppa et al. 2016). We would expect that a standardized methodology should be adopted, for the quite straightforward implementations presented.

23.3   CITIZEN SCIENCE AND LANDSLIDES

Remote sensing is the science of obtaining information about objects or areas from a distance, typically from aircrafts.[*] Given the limitations on the spatial and temporal coverage of satellite and (manned) aerial platforms, utilization of unmanned aerial platforms was the subject of our review in the previous section. Although UAVs significantly enhance the operational capabilities and performance of landslide investigations, there is still an important missing element to fill the puzzle of landslide management (Figure 23.4)—a piece that will provide ground truth verification of satellite/aerial measurements and thus improve the respective classification algorithms and also provide real-time monitoring capabilities of an evolving disaster.

Citizen science is often used to depict individuals, communities, or networks of citizens who participate in data collection, analysis, and dissemination in a specific domain of science (Goodchild 2007). This is mainly supported by modern devices (i.e., smartphones and tablets) and the second generation of the World Wide Web (web 2.0), which emphasizes the ability of people to collaborate and share information online via social media, blogging, and web-based communities (Techopedia). The emergency data management is one of the applications identified (Kotovirta et al. 2015) for the utilization of citizen science for EO.

---

[*] http://oceanservice.noaa.gov/facts/remotesensing.html [accessed: 17-Jan-2017]

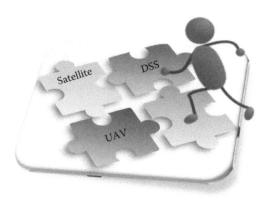

FIGURE 23.4   Although satellite and aerial systems are important constituents in landslide decision support systems (DSS), it is the active participation of citizens that will significantly enhance the performance and efficiency of landslide management systems.

Whereas in traditional EO, images are captured by satellites or airplanes, in citizen science, this task is performed by the engaged citizens' smart devices. Geolocation is performed through the device's internal global navigation satellite system (GNSS). However, it is not merely about taking pictures; sensing of atmospheric conditions via a smartphone has become a reality, since new types of sensors are continuously integrated into smart devices.

It is evident that remote sensing has much to gain from citizen science (Foody 2015). This is the reason why specific projects have been funded to identify synergies between citizen science and crowdsourcing for observations from satellites (Mazumdar et al. 2016), addressing the need of space mission stakeholders to validate satellite measurements on the ground (Mazumdar et al. 2017). This validation exercise was also the subject for several remote-sensing applications, such as forest biomass monitoring (Elmore et al. 2016, Molinier et al. 2016) and land cover/land use (Bayas et al. 2016). But, what about the connection between landslides, citizen science, and remote sensing? How can these three domains help each other?

The first way is via a geographical information system interface, using satellite imagery as background, to report on the occurrence of a landslide. Typical examples of scientists asking citizens to track landslides include: (1) Did You See It?[*] program initiated by the U.S. Geological Survey Landslide Hazards Program and (2) the Global Landslide Catalog[†] where citizens are allowed to make edits and report new landslides (Kirschbaum 2015). Similarly, the National Landslide Database of Great Britain gathers data, among others, through social media and other online resources (Pennington et al. 2015).

Another interesting example of the synergetic use of EO data with smart devices for emergency risk management and landslides includes some of the mobile apps developed under the framework of *MyGEOSS* project: *DisasterHub* is a mobile app where EO data are used to detect a geohazard and inform the users that are close to that geohazard, while the users may provide *in situ* information (text messages, videos, etc.) to the app (Tsironis et al. 2016). Moreover, the *RescueNET* mobile app (Figure 23.5) permits users that are players in the emergency management chain (victims, doctors, volunteers, etc.) to locate and interact with all other users and geotag information in a 3D geographic information system (GIS) mobile platform with base maps from EO products (http://digitalearthlab.jrc.ec.europa.eu/apps).

---

[*] https://ccsinventory.wilsoncenter.org/#projectId/124 [accessed: 18-Jan-2017]
[†] http://ojo-streamer.herokuapp.com/ [accessed: 21-Jan-2017]

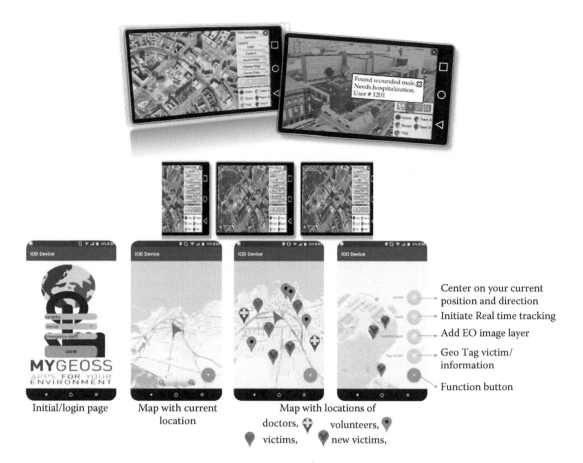

FIGURE 23.5 The general concept of the RescueNET app (up), and screen shots of the app (down).

Landslide research and citizen science are relatively new concepts and Scopus database returns just 12 documents, whereas the number of related documents rise to 50 and 78 under the *floods* and *earthquake* terms, respectively. For example, in Lee (2016), the volunteered geospatial information (VGI) GeoPortal concept is presented, and the author supports that through such a collaborative system, where any sign of a landslide or rockfall event may be reported by the citizens, enabling public authorities to take immediate actions.

Accordingly, Galizia et al. (2014) presented the Distributed Research Infrastructure for Hydro-Meteorology (DRIHM), where researchers, environmental agencies, and citizens access and combine hydrometeorological data and models.

Environmental sensing using smartphones (Aram et al. 2012) is another concept that, in conjunction with citizen science, can significantly contribute to a dense network of hydrometeorological parameters that may trigger a landslide. However, even if every citizen in a landslide-prone area is equipped with a smartphone that has weather-sensing capabilities, calibration of these measurements should be performed to attain a certain degree of accuracy and reliability (Zamora et al. 2015).

23.4 APPLICATION AREA: TRANSPORTATION NETWORKS

Transportation corridors (roads, railways, etc.) along landslide-prone areas constitute a significant challenge to the engineering experts involved in their design. This is because these dangerous areas cannot always be avoided, while at the same time, it is practically impossible to apply mitigation measures along the entire corridor length (Ferrero and Migliazza 2013). Road and railway stakeholders from all over the world employ different approaches to enhance public and infrastructure safety and also to reduce road closure and interruption of smooth transportation (Zumbrunnen et al. 2017).

The impact of landslides on transportation networks has been a subject of several studies (Figure 23.6). For example, Reichenbach et al. (2002) performed a qualitative analysis of the impact of landslides on the Umbria, Italy, regional transponder network. Their effort was to correlate the vulnerability of the road network to different types of landslides (i.e., rockfall, shallow landslide, or deep-seated landslide). A more quantitative effort to estimate the economic impact of four landslide events that occurred in Scotland in 2004 and 2007 is presented in Winter et al. (2016). These impacts are categorized as direct, direct consequential, and indirect consequential economic impacts. A similar categorization on the 2010 Colorado, United States landslides is followed by Highland (2012): direct costs are associated with the costs for repair, replacement, or maintenance resulting from property damages, whereas indirect impacts include, among others, loss of productivity in all economic sectors, reduction in real-estate values, losses of human or animal productivity, and so on.

The regional-economic impact of the 75 m × 100 m rockslide that took place on May 13, 2013, at the Felbertauem Road in Austria was presented in Pfurtscheller and Genovese (2016). A nationwide (for the UK) assessment of landslide disruptions on road transportation network is given in Postance et al. (2017), where it is estimated that a single event (the 2007 A83 landslide) costs about 1.4M Euro over a 15-day closure, whereas indirect economic losses of the order of 40,000 Euro per day will burden the national economy for each of the 152 road segments that are susceptible to landslides. Given the high impact that landslides have on transportation networks, the remaining section will review the way in which remote sensing contributes to landslide-related studies performed on such networks.

At this point, we should define the differences between landslide inventory, susceptibility, hazard maps, and risk maps (Figure 23.7). *Inventory* stands for the location, dimensions, and geographical extent of past events. *Susceptibility* refers to the relative spatial likelihood for the occurrence of landslides of a specific type and volume (van Westen 2008). In addition, *hazard maps* indicate the possibility of landslides occurring throughout a given area (USGS 2016). Finally, landslide *risk* assesses the potential damage to persons and property, accounting for temporal and spatial probability and vulnerability (Fell et al. 2008). An effort to harmonize the use of these terms has been made through the European FP7 SAFELAND project (Corominas et al. 2011).

(a)   (b)   (c)

FIGURE 23.6 Different types of earth displacement affect transportation networks: (a) embankment failure in Cyprus; (b) rockfall events interrupting traffic at Topolia Gorge, Crete, Greece; and (c) small-scale landslide may be as destructive as large ones.

Risk
Temporal/Spatial Probability

○ Hazard and
  ○ Measure of probability
  ○ For life loss
  ○ For property loss

Hazard
Potential for landslide event

○ Susceptibility and
  ○ Possibility of occurrence
  ○ Within a given period of time

Susceptibility
Learning from the past

○ Inventory and
  ○ Rock/soil strength
  ○ Steepness
  ○ Geologic conditions
  ○ ...

Inventory
Traceability of past events

○ Location, dimensions,
  geographical extent
○ Volume, activity, classification
○ Date

FIGURE 23.7 Illustration of the different landslide maps and their relation.

Remote sensing is vital for the development of detailed landslide *inventories*. Their tremendous potential has already been identified (Nichol and Wong 2005). A review on multispectral, multimission remote-sensing techniques for the creation of landslide inventory maps is given in Guzzetti et al. (2012). This landslide mapping can be performed using Google Earth-derived satellite images (Mihir and Malamud 2014), high-resolution stereoscopic pair images from GeoEye-1 (Murillo-García et al. 2014; Youssef et al. 2016), or more advanced techniques such as the persistent scatterer (PS) interferometry (Righini et al. 2012).

Regarding the generation of landslide susceptibility, Chen et al. (2017) have provided a literature review on diverse statistical methods to correlate landslide-triggering factors and the landslide occurrence. Comparison between these diverse approaches is, indicatively, provided in several studies (Eker et al. 2014; Trigila et al. 2015; Vakhshoori and Zare 2016; Wang et al. 2016, etc.).

In most of the studies focused on landslides and transportation networks, satellite images are used to derive land use parameters related to landslide susceptibility mapping. Chiu et al. (2016) applied the modified multiphase segmentation method to segment the Normalized Difference Vegetation Index (NDVI), as obtained from SPOT satellite images. This work focused on the Li-Shing Estate Road in Nantou County, Taiwan. The NDVI, as calculated by SPOT imagery, is also used in Shou and Lin (2016) to identify landslides along the Nantou County Road #89, Taiwan. In their analysis, they argue that the best landslide indentification results are obtained when the greenness index, the NDVI and the slope angles receive values: 0.3%, <0.2% and >20% rescpectively, provides the best landslide identification results for the particular area. The SPOT images have also been employed (Opiso et al. 2016) to estimate land-type parameters in their landslide susceptibility map of the Cagayan de Oro-Bukidnon-Davao city route corridor in Philippines.

Alexakis et al. (2014) analyzed Landsat 5 TM, Landsat 7 ETM+, and QuickBird images to derive land use, faults, and road networks to be used in their landslide susceptibility analysis for the transportation network of a study area in Cyprus. Satellite image analysis was also performed to develop a land use/land cover map for the year 2020 to be used in landslide susceptibility map projection for that year. Kanwal et al. (2016) considers roads as a potential landslide-triggering factor, owing to the change that they induct to the stability of the slope. To this extent, they have used Landsat-8 images not only for land cover mapping but also for road network extraction. The Indian Remote Sensing (IRS) satellite images have been used for route planning when designing new roads in mountainous areas (Saha et al. 2005) and land use mapping (Ramesh and Anbazhagan 2015).

The impact that the revisit time of radar missions has on efficient landslide monitoring affecting transportation is examined in Singhroy et al. (2015). The authors investigate how the revisit time impacts different types of landslides and conclude that the 4-day acquisition plan for the Radarsat Constellation Mission will limit the loss of coherence between the radar images and permit monitoring of smaller geological subunits of the same landslide.

An interesting application of remotely sensed images is presented in Poreh et al. (2016). The authors use the railways as PSs, thus enabling a direct estimation of their deformation rate in landslide-prone areas. The high-resolution Cosmo-SkyMed X-band images were found to be superior compared with the low-resolution images such as European remote-sensing satellite (ERS) and environmental satellite-advanced synthetic aperture radar (Envisat-ASAR) for monitoring the stability of the Campania, Italy, railways.

Castagnetti et al. (2014) studied the Collagna landslide, North Apennines, Italy, that disrupted National Road 63 in December 2008. The airborne laser scanning technique was implemented to support a rockslide inventory upgrade, whereas ground-based synthetic aperture radar (GB-InSAR) located more than 1 km away provided real-time rockslide-monitoring capabilities. The latter instrument has also been employed in monitoring of the rockslide that caused the interruption of Regional Road 65 in Monte Beni, Tuscany, Italy.

Landslide hazard detection from airborne Light Detection and Ranging (LIDAR) data at the Muskingum State Route, Zanesville, Ohio, the United States, has been the subject of several studies (Toth et al. 2013; Mora et al. 2014a, 2014b). These aerial surveys focused mainly on deriving very accurate digital elevation models of the study area and subsequently extracting slope stability parameters. Miller et al. (2012) also applied airborne laser scanning and multispectral aerial imagery on a railway corridor, focusing mainly on assessing earthwork slope stability hazard in the study area.

23.5 APPLICATION AREA: MINING INDUSTRY

Mining and quarrying are recognized to be among the anthropogenic causes of landslides.[*] A growing public interest in mining-induced landslides has been reported (Zhao 2016). Major landslides as the Bingham Canyon, Utah, the United States, certainly contribute to this rising public attention. On April 10, 2013, Bingham Canyon Mine, experienced the largest, non-volcanic rock avalanche in the history of North America (the National Aeronautics and Space Administration [NASA] Observatory). Fortunately, no victims were reported, as the operating company issued a warning some hours before the collapse. This warning was partially based on the monitoring results of a ground-based interferometric radar system that had been installed several months before the incident.

Underground mining operations are well known to induce subsidence on the ground surface while rainfall and/or mine aquifer recharge saturate the geological strata over these mines when they are abandoned (Iannacchione and Vallejo 1995). To this end, this section will review how the radar interferometry technique has been employed for monitoring ground subsidence caused by (in)active underground but also open pit mines. Radar interferometry or interferometric synthetic aperture radar (InSAR) is a powerful technique to derive surface deformation as well as elevation mapping in submillimeter scale. Its concept has been briefly presented in Milliano (2016), whereas a more in-depth analysis is given in Hanssen (2001).

A satellite (or airplane) payload emits electromagnetic waves, usually in the C-, L-, and X-band, and information such as phase and intensity is recorded on the reflection from the Earth's surface (Raucoles et al. 2007). Interferometric Synthetic Aperture Radar utilizes the phase information acquired by the synthetic aperture radar (SAR) instrument and compares it to the image of the same area at different times (McCandless and Jackson 2004). Although complications are possible in

---

[*] http://www.ga.gov.au/scientific-topics/hazards/landslide/basics/causes [accessed: 11-Nov-2016]

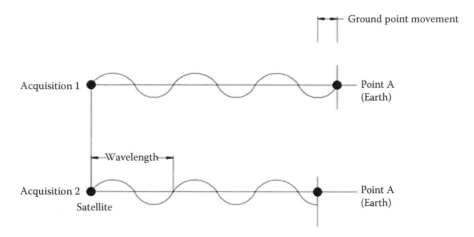

FIGURE 23.8 Illustration of phase shift due to ground movement.

practice, this method is rooted in a simple understanding of how wave transmission works. When an electromagnetic wave is transmitted to the ground, the reflected signal will have an associated phase that is dependent on the distance of travel for the signal. At some time later, if the ground point moves, the transmitted signal will return with an altered phase value (Figure 23.8). The change in the distance will be observable as a phase difference or phase shift when the emitted signal and the returned signal are compared. This change is usually presented as an interferogram (Figure 23.9).

Besides the phase difference due to land deformation, there exist some other sources that may result in a phase difference, such as the atmosphere, the topography, instrument noise, and so on. After correcting for these additional contributions, the deformation-related phase difference $\Delta\varphi_{int}$ is only analogous along the line of sight of the satellite Δr and also related to the radar wavelength (λ) (McCandless and Jackson 2004).

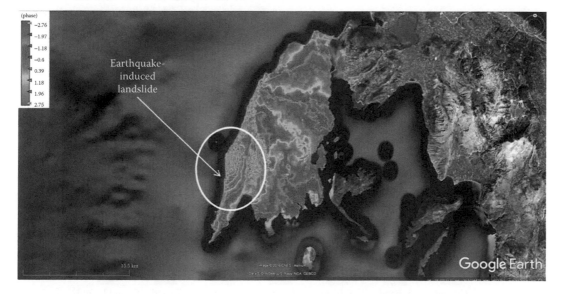

FIGURE 23.9 A 6.1 M earthquake-induced landslide took place on , November 17, 2015, at Lefkada Island, Greece. The displacement (white circle) has been captured by two Sentinel-1A images captured on November 5, 2015, and November 17, 2015, before and just after the landslide.

$$\Delta\varphi_{int} = \frac{4}{\lambda}\Delta r$$

Obviously, the wavelength determines the maximum displacement that can be observed from a satellite (or airborne) radar system. Currently, the majority of these payloads operate in the C-band, whereas the L- (i.e., ALOS-2) and X-bands (i.e., TerraSAR-X) have also demonstrated their applicability and usefulness. The evolution of satellite missions that carry payloads necessary for the implementation of InSAR processing is illustrated in Figure 23.8, whereas Figure 23.9 presents the distribution of mining-related InSAR studies according to the band (C-, L-, or X-band) employed. Not surprisingly, given that C-band satellites have been in operation 15 years longer than L-band, it seems that L-band is most suitable for InSAR studies related to mining operations. This can be justified by the radar wavelength employed, as it is to small changes of the coherence between the two acquisitions (Figures 23.10 and 23.11).

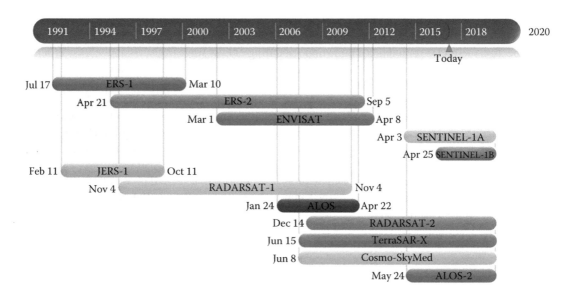

FIGURE 23.10 The history of InSAR satellite missions.

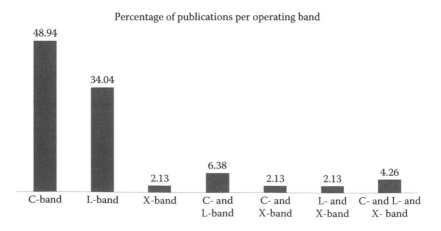

FIGURE 23.11 Percentage of studies published for InSAR analysis.

After almost 20 years of InSAR data processing and analysis and given the technological achievements in the same period, a plethora of algorithms to process these radar observations have been developed. Osmanoglu et al. (2016) provide a comprehensive review of methods and algorithms for time series analysis of InSAR data. From their analysis, it seems that the most popular, in scientific publications, are the Interferometric Point Target Analysis, followed by Permanent Scatterer InSAR and Stanford Method for Persistent Scatterers (StaMPS).

23.5.1 Mining as Landslide-Triggering Source

Liu et al. (2014) stated the importance of monitoring subsidence caused by mining activities that take place in mountainous areas, as these activities may trigger landslides. This mountainous environment poses some problems in conventional InSAR analysis, and the authors performed a thorough evaluation of different techniques (differential interefemetric synthetic aperture radar (DInSAR), persistent scatterer interferometry (PSI), small baseline subset (SBAS), tomography synthetic (Tomo-SAR)) as well as their combination by using radar images derived by Advanced Land Observing Satellite (ALOS) (L-band), Envisat (C-band), and TerraSAR-X (X-band). They tested their methodology on the Xishan coal mine area in the People's Republic of China, which is actually composed of five active coal mines.

The Wieliczka Salt Mine, Poland, is located at the edge of the Carpathian Mountains. Thus, the potential relation between the underground mining activities and landslide incidents that occurred on the surface should be investigated. Perski performed DInSAR and PSI analyses, using ERS-1 and ERS-2 imagery, to discriminate observed terrain deformations to three different types: (1) salt mining, (2) water inflows to the mine, and (3) landslides (Perski et al. 2009). The authors concluded that rapid land deformation caused by water inflow could not be determined via InSAR analysis. On the contrary, InSAR clearly identified slow subsidence due to convergence of mine caverns. Results on landslide investigations were incomplete, mainly due to the direction of the landslide and the lack of appropriate number of ERS imagery for applying the PSI technique.

A rockslide avalanche, which occurred on the Turtle Mountain, Alberta, Canada, in 1903, was responsible for more than 70 casualties. For almost 100 years, rumors pointed out that the underground mining activities were responsible for triggering this rockslide. Mei et al. (2008) employed the PSI technique by using Radarsat-1 images to provide scientific evidence and quantify the subsidence due to the abandoned Bellevue mine that lies underground the Frank Slide area. Their results indicated a 3.1 mm/y subsidence induced by the underground coal mines from April 2004 to October 2006, and they concluded that the coal mine subsidence could be considered as one of the triggering factors for the 1903 Frank Slide rock avalanche.

23.5.2 Dewatering Mining Activities and Remote Sensing

Woldai et al. (2009) used InSAR analysis to examine surface deformation signals associated with the extensive dewatering activities performed in the Pipeline open-pit mine, Nevada, the United States. The objective of these dewatering activities is to keep the groundwater level below the mining operations level. The authors employed the conventional differential (or two-pass) InSAR methodology, using ERS-1/2 imagery along with *in situ* pumping and geological data. They argue that InSAR is a suitable technique to support mining operators in the hydrogeological design studies and, in particular, in answering the question where fissures or faulting might develop.

Tripolitsiotis et al. (2014) investigated whether precursory signs of the ground fissures observed in the vicinity of the active Mavropigi, Greece, open-pit coal mine could be captured by InSAR analysis. Employing the DInSAR technique on ALOS PALSAR images, the authors demonstrated the applicability of this remote-sensing technique to provide early signs of land deformation well before tension cracks became visible.

The Gippsland Basin in Australia has been the subject of study of Ng et al. (2015). The basin hosts not only mining but also oil and natural gas extraction as well as farming. Besides irrigation and industrial use, the local aquifer is also extracted for dewatering the open-pit and as part of oil and gas extraction. An advanced time series analysis based on the SqueeSAR algorithm has been applied by using ALOS PALSAR images. Through their analysis, the authors managed to identify land deformations attributed to mining activities and groundwater extraction for mine dewatering. In addition, they identified five more areas that need further investigation to enhance the understanding of the processes that take place in the basin.

Bozzano et al. (2015) correlated the spatial and temporal evolution of a subsidence process monitored at the Acque Albule Basin, Rome, Italy, an area of approximately 30 km^2. Open-pit travertine mines constitute the main economic industry. The quarry areas increased about seven times in the period between 1954 and 2005, while the quarry floors were lowered from 14 m in 1993 to 18 m in 2005 below the initial ground level. Pumping of the water table to keep it below the quarry floors is a possible anthropogenic source of the subsidence in the area. The authors performed historical advanced DInSAR analyses of images captured as back as 1992 by ERS and Envisat satellite missions. They adopted the PS but also quasi-PS InSAR techniques, and they combined their results with groundwater numerical modeling assisted by periodic piezometric measurements. Using the historical dataset of radar images, they managed to conclude that the groundwater level variations trigger subsidence over the area, whereas the geological setting defines the subsidence magnitude.

23.5.3 Aquifer Recharge at Abandoned Mines Triggers Land Uplift

When operating, underground mines may induce land subsidence, either because of the collapse of high extraction panels or because of dewatering activities. In Limburg, the Netherlands, the land subsidence monitored during the operation of the coal mines was estimated in the order of several meters (Caro-Cuenca et al. 2013). After the closure and the end of pumping, it was normal for the aquifer to recharge with water and for the formation to swell and to start to rise. In order to protect nearby and connected mines from flooding, some pumping activities continued for several more years. Caro-Cuenca et al. (2013) employed the PSI technique and ERS-1, ERS-2, and Envisat images to investigate surface deformations and correlate them with dewatering activities. Not surprisingly, they found that the observed land uplift was due to the recharge of the aquifer.

Vervoort (2016) investigated the surface movement above the abandoned Houthalen, Belgium, underground coal mine by using the C-band ERS 1/2 and Envisat-ASAR radar images. The permanent scatterer technique for analysis of these images was employed. Both subsidence and uplift rates were observed, and an effort was made to correlate them with the caving above the mined-out areas.

Raucoules et al. (2008) applied the conventional differential (or two-pass) InSAR methodology to monitor postmining subsidence in the Nord-Pas-de-Calais Coal Basin in France. The end of the mine exploitation and subsequently the water pumping resulted in a rise in the aquifer. Using ERS-1 and ERS-2, the interferometric analysis revealed a deformation range from 2 cm to 4.5 cm for the 1992–2000 period.

Graniczny et al. (2015) performed a detailed InSAR analysis of the Upper Silecian Coal Basin in Poland. They applied both DInSAR and PS-InSAR techniques for different datasets, namely ERS1/2, Envisat, ALOS, and TerraSAR-X. Among other significant results, the authors concluded that an uplift of the abandoned mining area is attributed to groundwater recharge, leading to an increase of hydrostatic pressure in the mine aquifer and stress in the overburden.

Qin and Perissin (2015) compared interferograms created by C-band (ERS) and L-band (ALOS) radar missions over the city of Evansville and Boonville, Indiana, the United States, by employing the PS technique. The objective was to investigate whether subsidence was present due to the area's abandoned coal mines. It was determined that for the study period ranging from 2009 to 2011, the areas under examination were stable.

Herrera et al. (2010) employed the coherent pixel technique (CPT) to map and monitor ground movements in the vicinity of an open-pit metal mining area in La Union, Murcia, Spain. Because of the triggering action of rain, slope stability studies of the abandoned mine areas have to be performed. Images from ERS and Envisat missions were utilized in CPT analysis, and a qualitative correlation of the rainfall intensity with an increase of deformation rates monitored by CPT was possible.

23.5.4 RADAR INTERFEROMETRY TO SUPPORT SAFETY OF MINING OPERATIONS

The contribution of the freeze-and-thaw cycle of permafrost on mining stability was the subject of investigation reported by Rheault et al. (2015). The authors applied the DInSAR technique, using Radarsat-2 and TerraSAR-X imagery for the mining area of Nanisivik, Canada. Their analysis demonstrated that displacement maps (with centimeter accuracy) can be created, thus ensuring the safety of mining activities.

Wempen and McCarter (2017) analyzed L-band and X-band SAR data from ALOS and TerraSAR-X missions, respectively, using the differential InSAR methodology to identify longwall mine subsidence in the Wasatch Plateau, Utah, the United States. The authors argue that subsidence magnitudes are better measured in the L-band, whereas X-band results are more affected by signal saturation and temporal decorrelation.

Chatterjee et al. (2016) investigated land subsidence due to underground coal activities in the Jharia Coalfield, Jharkhand, India, using Radarsat-2 C-band SAR images and employing the DInSAR technique. The authors propose a multistep semiautomated approach to confront temporal-terrain decorrelation noise.

Liu et al. (2015) proposed a phase unwrapping method that copes with the noisy data in three different operating coal mines in China. They have used TerraSAR-X images and applied the standard D-InSAR technique. The authors argue that the modified Cubature Kalman filtering (CKF) phase unwrapping methods are well suited for areas with high noise and large phase gradients, as is the case with operating coal mines.

23.6 CONCLUSIONS

Landslides affect infrastructure, transportation routes, and even human lives. Their dynamics are quite versatile in terms of triggering mechanisms and scale, and thus, several practices have been implemented to monitor and assess them based on geodetic, geotechnical, remote sensing, and hybrid instrumentation. Remote sensing offers several approaches to model, monitor, and predict landslides from space, air, and land. Along the various remote-sensing applications concerning landslide hazards, in the last few years, two relatively new practices have emerged: (a) the wide use of UAVs, mainly for the DSM construction of the landslide area and (b) citizen science contribution for data collection and analysis.

The UAVs offer many advantages for landslide and rockfall monitoring in terms of mapping in high spatial resolution, repeatability and on-demand monitoring, inaccessible area mapping, low cost, customization, fissure and morphology recognition, and so on. Although typical DSM processing is outdated, there is still no widely accepted methodology for distributing the GCPs or measuring the DSM-related errors. There is a need to employ the full UAV potential by using varying sensors and multispectral analysis and provide real-time and on-board processing for more autonomous and standardized applications.

Concerning citizen science, there is a need for robustness and citizen's awareness to promote its wide use, which are certainly out of the scientific scope of the core remote-sensing fields. Thus, both of these two technological tools offer great potential and need further development.

Furthermore, two application studies on remote sensing of landslides are demonstrated: those of transportation routes and mining. Given the high impact that landslides have on transportation networks, remote-sensing techniques have been used for landslide inventory, susceptibility, hazard maps, and risk maps. Satellite images are used to derive land use parameters (e.g., NDVI and slope) and fault formations. Radar missions are affected by the revisit time of satellites, and thus, airborne and ground-based sensors have been used.

Underground mining operations assisted by rainfall and/or mine aquifer recharge induce subsidence on the ground surface. The radar interferometry technique employed for monitoring ground subsidence has proven to be a powerful technique to derive surface deformation as well as elevation mapping in submillimeter scale. Satellite InSAR analysis cannot always monitor and discriminate fast landslide phenomena, since image temporal acquisition is not continuous and registration is not always successful.

InSAR is a suitable technique to monitor subsidence in the mining hydrogeological circle, since dewatering activities are compulsory to keep the water under the mining level. It also provides early signs of land deformation well before tension cracks become visible. It is generally shown that there is an abundance of landslide incidents that are mapped, monitored, and analyzed through varying radar interferometry techniques. Under this scheme, centimeter-level subsidence can be monitored, enhancing the understanding of the the life cycle of landslides and their triggering mechanisms and safety support.

BIBLIOGRAPHY

Al-Rawabdeh, A., Al-Gurrani, H., Al-Durgham, K., Detchev, I., He, F., El-Sheimy, N., and Habib, A. (2016a). A robust registration algorithm for point clouds from UAV images for change detection. Paper presented at the *International Archives of the Photogrammetry, Remote Sensing and Spatial Information Sciences—ISPRS Archives*, January, 2016. pp. 765–772. doi:10.5194/isprsarchives-XLI-B1-765-2016.

Al-Rawabdeh, A., He, F., Moussa, A., El-Sheimy, N., and Habib, A. (2016b). Using an unmanned aerial vehicle-based digital imaging system to derive a 3D point cloud for landslide scarp recognition. *Remote Sensing*, 8(2), 95. doi:10.3390/rs8020095

Alexakis, D. D., Agapiou, A., Tzouvaras, M., Themistocleous, K., Neocleous, K., Michaelides, S., and Hadjkimitsis, D. G. (2014). Integrated use of GIS and remote sensing for monitoring landslides in transportation pavements: The case study of Paphos area in Cyprus. *Natural Hazards*, 72(1), 119–141. doi:10.1007/s11069-013-0770-3.

Aram, S., Troiano, A., and Pasero, E. (2012). Environment sensing using smartphone. *Sensors Applications Symposium (SAS), 2012 IEEE*. Brescia, Italy: IEEE, pp. 110–113. doi:10.1109/SAS.2012.6166275.

Balaji, P. (2011). *Estimation and Correction of Tropospheric and Ionospheric Effects on Differential SAR Interferograms*. Enschede, the Netherlands: University of Twente.

Bayas, J. L., See, L., Fritz, S., Sturn, T., Perger, C., Dürauer, M., Karner, M. et al. (2016). Crowdsourcing in-situ data on land cover and land use using gamification and mobile technology. *Remote Sensing*, 8, 905. doi:10.3390/rs8110905.

Bozzano, F., Esposito, C., Franchi, S., Mazzanti, P., Perissin, D., Rocca, A., and Romano, E. (2015). Understanding the subsidence process of a quaternary plain by combining geological and hydrogeological modelling with satellite InSAR data: The Acque Albule Plain case study. *Remote Sensing of Environment*, 168, 219–238.

Caro Cuenca, R., Hooper, M., and Hanssen, A. (2013). Surface deformation induced by water influx in the abandoned coal mines in Limburg, The Netherlands observed by satellite radar interferometry. *Journal of Applied Geophysics*, 88, 1–11.

Carvajal-Ramírez, F., Agüera-Vega, F., and Martínez-Carricondo, P. J. (2016). Effects of image orientation and ground control points distribution on unmanned aerial vehicle photogrammetry projects on a road cut slope. *Journal of Applied Remote Sensing*, 10(3), 034004–034004. doi:10.1117/1.JRS.10.034004.

Castagnetti, C., Bertacchini, E., Corsini, A., and Rivola, R. (2014). A reliable methodology for monitoring unstable slopes: The multi-platform and multi-sensor approach. In M. Ulrich, and K. Schulz (Ed.), *Proceedings SPIE 9245, Earth Resources and Environmental Remote Sensing/GIS Applications V*, (p. 92450J). Amsterdam, the Netherlands. doi:10.1117/12.2067407

Chatterjee, R., Singh, K., Thapa, S., and Kumar, D. (2016). The present status of subsiding land vulnerable to roof collapse in the Jharia Coalfield, India, as obtained from shorter temporal baseline C-band DInSAR by smaller spatial subset unwrapped phase profiling. *International Journal of Remote Sensing*, 37, 176–190.

Chen, W., Xie, X., Peng, J., Wang, J., and Hong, H. (2017). GIS-based landslide susceptibility modelling: A comparative assessment of kernel logistic regression, Naïve-Bayes tree, and alternating decision tree models. *Geomatics, Natural Hazards and Risk*, 1–24. doi:10.1080/19475705.2017.1289250.

Chiu, C., Peng, C.-P., Huang, Y., and Wang, C.-P. (2016). Monitoring landslide phenomena along Li-Shing estate road in Nantou County of Central Taiwan by applying an object-oriented segmentation approach. In S.-J. Chao, K.-L. Pun, and X. Cui (Eds.), *Fourth Geo-China International Conference. GSP 261*, pp. 57–64. Shandong, China: American Society of Civil Engineers.

Corominas, J., Mavrouli, O.-C., and Modaressi, H. (2011). *D2.4: Guidelines for landslide susceptibility, hazard and risk assessment and zoning*. SAFELAND project deliverable.

Cruden, D. M. (1991). A simple definition of a landslide. *Bulletin of the International Association of Engineering Geology*, 43, 27–29.

Cruden, D. M., and Varnes, D. J. (1996). Landslides: Investigation and mitigation. Chapter 3: Landslide types and processes. Transportation Research Board Special Report, Issue 247.

Eker, A. M., Dikmen, M., Cambazoğlu, S., Düzgün, Ş. H., and Akgün, H. (2014). Evaluation and comparison of landslide susceptibility mapping methods: A case study for the Ulus district, Bartin, northern Turkey. *International Journal of Geographical Information Science*, 29, 132–158. doi:10.1080/13658816.2014.953164.

Elmore, A. J., Stylinski, C. D., and Pradhan, K. (2016). Synergistic use of citizen science and remote sensing for continental-scale measurements of forest tree phenology. *Remote Sensing*, 8, 502. doi:10.3390/rs8060502.

Fakhri, F. (2013). Long and short term monitoring of ground deformation in Thessaly basin using space-based SAR Interferometry. PhD dissertation, Department of Geography, Harokopio University of Athens.

Fell, R., Corominas, J., Bonnard, C., Cascini, L., Leroi, E., and Savage, W. Z. (2008). Guidelines for landslide susceptibility, hazard and risk zoning for land use planning. *Engineering Geology*, 102, 85–98. doi:10.1016/j.enggeo.2008.03.022.

Fernández, T., Pérez, J. L., Cardenal, F. J., López, A., Gómez, J. M., Colomo, C., Delgado, J. and Sánchez, M. (2015). Use of a light UAV and photogrammetric techniques to study the evolution of a landslide in jaén (southern spain). *Paper presented at the International Archives of the Photogrammetry, Remote Sensing and Spatial Information Sciences–ISPRS Archives*, 40(3W3), 241–248. doi:10.5194/isprsarchives-XL-3-W3-241-2015.

Ferrero, A. and Migliazza, M. (2013). Landslide transportation network and lifelines: Rockfall and debris flow. In A. M. Ferrero, M. Migliazza, C. Margottini, P. Canuti, and K. Sassa (Eds.) *Landslide Science and Practice. Volume 6: Risk Assessment, Management and Mitigation*. Berlin, Germany: Springer, pp. 161–170. doi:10.1007/978-3-642-31319-6_23.

Foody, G. M. (2015). Citizen science in support of remote sensing research. *2015 IEEE International Geoscience and Remote Sensing Symposium*. Milan, Italy: IEEE, pp. 5387–5390. doi:10.1109/IGARSS.2015.7327053.

Gallay, M., Eck, C., Zgraggen, C., Kanuk, J., and Dvorný, E. (2016). High resolution airborne laser scanning and hyperspectral imaging with a small UAV platform. *Paper Presented at the International Archives of the Photogrammetry, Remote Sensing and Spatial Information Sciences–ISPRS Archives*, 823–827. doi:10.5194/isprsarchives-XLI-B1-823-2016.

Galizia, A., D'Agostino, D., Quarati, A., Zereik, G., Roverelli, L., Danovaro, E., Clematuis, A. et al. (2014). Towards an interoperable and distributed e-Infrastructure for Hydro-Meteorology: The DRIHM project. In D. P. Ames, N. W. Quinn, and A. E. Rizzoli (Ed.), *Proceedings of the 7th International Congress on Environmental Modelling and Software*, 1, pp. 407–414. San Diego CA: International Environmental Modelling and Software Society.

Giordan, D., Manconi, A., Tannant, D. D., and Allasia, P. (2015). UAV: Low-cost remote sensing for high-resolution investigation of landslides. *Paper presented at the International Geoscience and Remote Sensing Symposium (IGARSS)*, 5344–5347. doi:10.1109/IGARSS.2015.7327042.

Goodchild, M. F. (2007). Citizens as sensors: The world of volunteered geography. *GeoJournal*, 69(4), 211–221. doi:10.1007/s10708-007-9111-y.

Graniczny, M., Colombo, D., Kowalski, Z., Przyiucka, M., and Zdanowski, A. (2015). New results on ground deformation in the Upper Silesian Coal Basin (southern Poland) obtained during the DORIS Project (EU-FP 7). *Pure and Applied Geophysics*, 172, 3029–3042.

Guha, D., Below, R., and Hoyois, P. (2016, December 19). *Disasters List*. (Université Catholique de Louvain, Brussules, Belgium) Retrieved January 25, 2017, from EM-DAT: The CRED/OFDA International Disaster Database: www.emdat.be.

Guzzetti, F., Mondini, A. C., Cardinali, M., Fiorucci, F., Santangelo, M., and Chang, K.-T. (2012). Landslide inventory maps: New tools for an old problem. *Earth-Science Reviews*, 112, 42–66 doi:10.1016/j. earscirev.2012.02.001.

Hanssen, R. F. (2001). *Radar Interferometry*. New York: Kluwer Academic Publishers.

Haque, U., Blum, P., da Silva, P. F., Andersen, P., Pilz, J., Chalov, S. R., Malet, J. P. et al. (2016). Fatal landslides in Europe. *Landslides*, 13(6), 1545–1554. doi:10.1007/s10346-016-0689-3.

Herrera, G., Tomas, R., Vicente, F., Lopez-Sanchez, J., Mallorqui, J., and Mulas, J. (2010). Mapping ground movements in open pit mining areas using differential SAR interferometry *International Journal of Rock Mechanics and Mining Sciences*, 47, 1114–1125.

Highland, L. M. (2012). Landslides in Colorado, USA: Impacts and Loss Estimation for the Year 2010. Reston, VA: U.S. Geological Survey, Open-File Report 2012–1204.

Hsieh, Y., Chan, Y., and Hu, J. (2016). Digital elevation model differencing and error estimation from multiple sources: A case study from the meiyuan shan landslide in Taiwan. *Remote Sensing*, 8(3), 199. doi:10.3390/rs8030199.

Hungr, O., Leroueil, S., and Picarelli, L. (2014). The Varnes classification of landslide types, an update. *Landslides*, 11(2), 167–194. doi:10.1007/s10346-013-0436-y.

Iannacchione, A., and Vallejo, L. (1995). Factors affecting the slope stability of Kentucky's abandoned mine lands. In D. A. Schultz (Ed.) *Rcok Mechanics*. Rotterdam, the Netherlands: Balkema, pp. 837–842.

Lin, H., Huang, H., Lv, Y., Du, X., and Yi, W. (2016). Micro-UAV based remote sensing method for monitoring landslides in three gorges reservoir, China. *Paper presented at the International Geoscience and Remote Sensing Symposium (IGARSS)*, 4944–4947. doi:10.1109/IGARSS.2016.7730290.

Lindner, G., Schraml, K., Mansberger, R., and Hübl, J. (2016). UAV monitoring and documentation of a large landslide. *Applied Geomatics*, 8(1), 1–11. doi:10.1007/s12518-015-0165-0.

Lowe, D. G. (1999). Object recognition from local scale-invariant features. *Paper Presented at the Proceedings of the IEEE International Conference on Computer Vision*, 2, 1150–1157.

Kanwal, S., Atif, S., and Shafiq, M. (2016). GIS based landslide susceptibility mapping of northern areas of Pakistan, a case study of Shigar and Shyok Basins. *Geomatics, Natural Hazards and Risk*, 1–19. doi:10. 1080/19475705.2016.1220023.

Kirschbaum, D. (2015). Using citizen science to grow a global landslide catalog. *2015 GSA Annual Meeting*. Baltimore, MA: Geological Society of America. Retrieved from https://gsa.confex.com/gsa/2015AM/webprogram/Paper263427.html.

Kjekstad, O. and Highland, L. (2009). Economic and social impacts of landslides. In O. Kjekstad, L. Highland, K. Sassa, and P. Canuti (Eds.), *Landslides- Disaster Risk Reduction*. Berlin, Germany: Springer-Verlag, pp. 573–587. doi:10.1007/978-3-540-69970-5_30.

Kotovirta, V., Toivanen, T., Tergujeff, R., Häme, T., and Molinier, M. (2015). Citizen science for Earth observation: Applications in environmental monitoring and disaster response. *36th International Symposium on Remote Sensing of Environment.XL-7/W3*, pp. 1221–1226. Berlin, Germany: The International Archives of the Photogrammetry, Remote Sensing and Spatial Information Sciences. doi:10.5194/isprsarchives-XL-7-W3-1221-2015.

Lee, S. (2016). Implementation of VGI-based geoportal for empowering citizen's geospatial observatories related to urban disaster management. *XXIII ISPRS Congress.Volume XLI-B2*. Prague, Czech Republic: The International Archives of the Photogrammetry, Remote Sensing and Spatial Information Sciences, pp. 621–623. doi:10.5194/isprsarchives-XLI-B2-621-2016.

Liu, D., Shao, Y., Liu, Z., Riedel, B., Sowter, A., Niemeier, W., and Bian, Z. (2014). Evaluation of InSAR and TomoSAR for monitoring deformations caused by mining in a mountainous area with high resolution satellite-based SAR. *Remote Sensing*, 6, 1476–1495.

Liu, W., Bian, Z., Liu, Z., and Zhang, Q. (2015). Evaluation of a cubature Kalman filtering-based phase unwrapping method for differential interferograms with high noise in coal mining areas. *Sensors*, 15, 16336–16357.

LZWG/AGS. (2007, March). Guideline for landslide susceptibility, hazard and risk zoning. *Journal and News of the Australian Geomechanics Society*, 42(1), 13–36.

Mantovani, F., Soeters, R., and Van Westen, C. (1996). Remote sensing techniques for landslide studies and hazard zonation in Europe. *Geomorphology*, 15(3–4), 213–225. doi:10.1016/0169-555X(95)00071-C.

Marek, L., Miřijovský, J., and Tuček, P. (2015). Monitoring of the shallow landslide using UAV photogrammetry and geodetic measurements. G. Lollino, D. Giordan, G. B. Crosta, J. Corominas, R. Azzam, J. Wasowski, and N. Sciarra (Eds.) *Engineering Geology for Society and Territory—Volume 2: Landslide Processes*, pp. 113–116. Cham, Germany: Springer.

Mazumdar, S., Wrigley, S., Ciravegna, F., Pelloquin, C., Chapman, S., de Vendictis, L., Grandoni, D., Ferri, M., and Bolognini, L. (2016). Crowd4Sat: Executive summary, *European Space Agency: AO/18068/14/F/ MOS.*

Mazumdar, S., Wrigley, S., Ciravegna, F., Pelloquin, C., Chapman, S., de Vendictis, L.,,... Bolognini, L. (2016). Crowd4Sat: Executive Summary.

McCandless, S. and Jackson, C. (2004). Principles of synthetic aperture radar. *InSAR Marine Users Manual,* 1–23.

McClusky, S. and Tregoning, P. (2013). Background paper on subsidence monitoring and measurement with a focus on coal seam gas (CSG) activities. Canberra, Australia: Research School of Earth Sciences, The Australian National University, 1–43.

Mei, S., Poncos, V., and Froese, C. (2008). Mapping millimetre-scale ground deformation over the underground coal mines in the Frank Slide area, Alberta, Canada using spaceborne InSAR technology. *Canadian Journal of Remote Sensing,* 34, 113–134.

Metternicht, G., Hurni, L., and Gogu, R. (2005). Remote sensing of landslides: An analysis of the potential contribution to geospatial systems for hazard assessment in mountainous environments. *Remote Sensing of Environment,* 98, 284–303. doi:10.1016/j.rse.2005.08.004.

Mihir, M. and Malamud, B. (2014). Identifying landslides using Google Earth. FP7 LAMPRE Project. Retrieved from https://goo.gl/jG5WrG

Miller, P. E., Mills, J. P., Barr, S. L., Birkinshaw, S. J., Hardy, A. J., Parkin, G., and Hall, S. J. (2012). A remote sensing approach for landslide hazard assessment on rngineered slopes. *IEEE Transactions on Geoscience and Remote Sensing,* 50(4), 1048–1056. doi:10.1109/TGRS.2011.2165547.

Milliano, S. D. (2016). Satellite radar interferometry. GIM International.

Molinier, M., López-Sánchez, C. A., Toivanen, T., Korpela, I., Corral-Rivas, J. J., Tergujeff, R., and Häme, T. (2016). Relasphone—Mobile and participative in situ forest biomass measurements supporting satellite image mapping. *Remote Sensing,* 8(10), 869. doi:10.3390/rs8100869.

Mora, O. E., Lenzano, M. G., Toth, C. K., and Grejner-Brzezinska, D. A. (2014a). Analyzing the effects of spatial resolution for small landslide susceptibility and hazard mapping. *ISPRS Technical Commission I Symposium.Volume XL-1,* pp. 293–300. Denver, CO: The International Archives of the Photogrammetry, Remote Sensing and Spatial Information Sciences. doi:10.5194/isprsarchives-XL-1-293-2014.

Mora, O. E., Toth, C. K., Grejner-Brzezinska, D. A., and Gabriela Lenzano, M. (2014b). A probabilistic approach to landslide susceptibility mapping using multi-temporal airborne LIDAR data. *ASPRS 2014 Annual Conference.* Louisville, KY: ASPRS. Retrieved from http://www.asprs.org/a/publications/proceedings/Louisville2014/Mora.pdf.

Mozas-Calvache, A. T., Pérez-García, J. L., Fernández-del Castillo, T., Gómez-López, J. M., and Colomo-Jiménez, C. (2016). Analysis of landslides based on displacements of lines. *XXIII ISPRS Congress.Volume XLI-B7.* Praque, Czech Republic: The International Archives of the Photogrammetry, Remote Sensing and Spatial Information Sciences, pp. 549–555. doi:10.5194/isprsarchives-XLI-B7-549-2016.

Murillo-García, F. G., Fiorucci, F., and Alcántara-Ayala, I. (2014). Development of a landslide inventory for a region in Mexico using very high resolution satellite stereo-images. In K. Sassa, P. Canuti, and Y. Yin (Eds.), *Landslide Science for a Safer Geoenvironment.* Cham, Germany: Springer International Publishing, pp. 821–828. doi:10.1007/978-3-319-05050-8_127.

Ng, A., Ge, L., and Li, X. (2015). Assessments of land subsidence in the Gippsland Basin of Australia using ALOS PALSAR data. *Remote Sensing of Environment,* 159, 86–101.

Nichol, J. and Wong, M. S. (2005). Satellite remote sensing for detailed landslide inventories using change detection and image fusion. *International Journal of Remote Sensing,* 26(9), 1913–1926. doi:10.1080/0 1431160512331314047.

Niethammer, U., James, M. R., Rothmund, S., Travelletti, J., and Joswig, M. (2012). UAV-based remote sensing of the super-sauze landslide: Evaluation and results. *Engineering Geology,* 128, 2–11. doi:10.1016/j.enggeo.2011.03.012.

Opiso, E. M., Puno, G. R., and Detalla, A. L. (2016). Landslide susceptibility mapping using GIS and FR method along the Cagayan de Oro-Bukidnon-Davao City route corridor, Philippines. *KSCE Journal of Civil Engineering,* 20(6), 2506–2512. doi:10.1007/s12205-015-0182-x.

Osmanoglu, B., Sunar, F., Wdowinski, S., and Cabral-Cano, E. (2016). Time series analysis of InSAR data: Methods and trends. *ISPRS Journal of Photogrammetry and Remote Sensing,* 115, 90–101.

Pennington, C., Freeborough, K., Dashwood, C., Dijkstra, T., and Lawrie, K. (2015). The national landslide database of Great Britain: Acquisition, communication and the role of social media. *Geomorphology,* 249, 44–51. doi:10.1016/j.geomorph.2015.03.013.

Peppa, M. V., Mills, J. P., Moore, P., Miller, P. E., and Chambers, J. E. (2016). Accuracy assessment of a uav-based landslide monitoring system. *Paper Presented at the International Archives of the Photogrammetry, Remote Sensing and Spatial Information Sciences–ISPRS Archives*, 41, 895–902. doi:10.5194/isprsarchives-XLI-B5-895-2016.

Perski, Z., Hanssen, R., Wojcik, A., and Wojciechowski, T. (2009). InSAR analyses of terrain deformation near the Wieliczka Salt Mine, Poland. *Engineering Geology*, 106, 58–67.

Peternel, T., Kumelj, Š., Oštir, K., and Komac, M. (2017). Monitoring the potoška planina landslide (NW slovenia) using UAV photogrammetry and tachymetric measurements. *Landslides*, 14(1), 395–406. doi:10.1007/s10346-016-0759-6.

Pfurtscheller, C. and Genovese, E. (2016). The Felbertauern landslide of 2013: Impact on transport network, regional economy and policy decisions. *Geophysical Research Abstracts*, 18. Retrieved from http://meetingorganizer.copernicus.org/EGU2016/EGU2016-1752-1.pdf

Poreh, D., Iodice, A., Riccio, D., and Ruello, G. (2016). Railways' stability observed in Campania (Italy) by InSAR data. *European Journal of Remote Sensing*, 49(1), 417–431. doi:10.5721/EuJRS20164923.

Postance, B., Hillier, J., Dijkstra, T., and Dixon, N. (2017). Extending natural hazard impacts: an assessment of landslide disruptions on a national road transportation network. *Environmental Research Letters*, 12(1), 014010. doi:10.1088/1748-9326/aa5555.

PwC. (2016). Copernicus: Market report. Europe direct. Luxembourg, Europe: Publications Office of the European Union. doi:10.2873/827100.

Qin, Y. and Perissin, D. (2015). Monitoring underground mining subsidence in South Indiana with C- and L-Band InSAR Technique. *IEEE International Geoscience and Remote Sensing Symposium*. Milan, Italy, pp. 294–297.

Ramesh, V. and Anbazhagan, S. (2015). Landslide susceptibility mapping along Kolli hills Ghat road section (India) using frequency ratio, relative effect and fuzzy logic models. *Environmental Earth Sciences*, 73, 8009–8021. doi:10.1007/s12665-014-3954-6.

Raucoles, D., Colesanti, C., and Carnec, C. (2007). Use of SAR Interferometry for detecting and assessing ground subsidence. *Comptes Rendes Geoscience*, 289–302.

Raucoles, D., Le Mouelic, S., Carnec, C., and Guise, Y. (2008). Monitoing post-mining subsidence in the Nord-Pas-de-Calais coal basin (France): Comparison between interferometric SAR results and levelling. *Geocarto International*, 339, 287–295.

Reichenbach, P., Ardizzone, F., Cardinali, M., Galli, M., Guzzetti, F., and Salvati, P. (2002). Landslide events and their impact on the transportation network in the Umbria Region, Central Italy. *Proceedings of the 4th EGS Plinius Conference held at Mallorca, Spain, October 2002*. Mallorca, Spain: Universitat de les Illes Balears. Retrieved from http://www.uib.cat/depart/dfs/meteorologia/ROMU/informal/proceedings_4th_plinius_02/PDFs/Reichenbach_et_al.pdf.

Rheualt, M., Bourobi, Y., Sarago, V., Nguyen-Xuan, P., Bugnet, P., Gosselin, C., and Benoit, M. (2015). Integrated SAR technologies for monitoring the stability of mine sites: Application using TerraSAR-X and RadarSAT-2 images. *36th International Symposium on Remote Sensing of Environment*, 40(7), 1057.

Righini, G., Pancioli, V., and Casagli, N. (2012). Updating landslide inventory maps using Persistent Scatterer Interferometry (PSI). *International Journal of Remote Sensing*, 33(7), 2068–2096.

Saha, A. K., Arora, M. K., Gupta, R. P., Virdi, M. L., and Csaplovics, E. (2005). GIS-based route planning in landslide-prone areas. *International Journal of Geographical Information Science*, 19(10), 1149–1175. doi:10.1080/13658810500105887

Sandwell. (n.d.). *Limitations and Noise Sources of Future InSAR Missions*. Retrieved from Presentation: http://files.scec.org/s3fs-public/may31_1025_Sandwell.pdf.

Scaioni, M., Longoni, L., Melillo, V., and Papini, M. (2014). Remote sensing for landslide investigations: An overview of recent achievements and perspectives. *Remote Sensing*, 6(10), 9600–9652. doi:10.3390/rs6109600.

Shou, K.-J. and Lin, J.-F. (2016). Multi-scale landslide susceptibility analysis along a mountain highway in Central Taiwan. *Engineering Geology*, 212, 120–135. doi:10.1016/j.enggeo.2016.08.009.

Singhroy, V., Li, J., and Charbonneau, F. (2015). High resolution rapid revisit InSAR monitoring of surface deformation. *Canadian Journal of Remote Sensing*, 41(5), 458–472. doi:10.1080/07038992.2015.1104638.

Stumpf, A., Malet, J. P., Kerle, N., Niethammer, U., and Rothmund, S. (2013). Image-based mapping of surface fissures for the investigation of landslide dynamics. *Geomorphology*, 186, 12–27. doi:10.1016/j.geomorph.2012.12.010.

Techopedia. (n.d.). Web 2.0. Retrieved February 10, 2017, from Techopedia: https://www.techopedia.com/definition/4922/web-20.

Torrero, L., Seoli, L., Molino, A., Giordan, D., Manconi, A., Allasia, P., and Baldo, M. (2015). The use of micro-uav to monitor active landslide scenarios. In G. Lollino, A. Manconi, F. Guzzetti, M. Culshaw, P. Bobrowsky, and F. Luino (Eds.) *Engineering Geology for Society and Territory—Volume 5: Urban Geology, Sustainable Planning and Landscape Exploitation*, pp. 701–704. Cham, Germany: Springer.

Toth, C. K., Mora, O. E., Lenzano, G., and Grejner-Brzezinska, D. A. (2013). Landslide hazard detection from LIDAR data. *ASPRS 2013 Annual Conference*. Baltimore, MA: ASPRS. Retrieved from https://goo.gl/uLuFJk.

Trigila, A., Iadanza, C., Esposito, C., and Scarascia-Mugnozza, G. (2015). Comparison of logistic regression and random forests techniques for shallow landslide susceptibility assessment in Giampilieri (NE Sicily, Italy). *Geomorphology*, 249, 119–136. doi:10.1016/j.geomorph.2015.06.001

Tripolitsiotis, A., Steiakakis, C., Papadaki, E., Agioutantis, Z., Mertikas, S., and Partsinevelos, P. (2014). Complementing geotechnical slope stability and land movement analysis using satellite DInSAR. *Central European Journal of Geosciences*, 6, 56–66.

Tripolitsiotis, A., Prokas, N., Kyritsis, S., Dollas, A., Papaefstathiou, I., and Partsinevelos, P. (2017). Dronesourcing: A modular, expandable multi-sensor UAV platform for combined, real-time environmental monitoring. *International Journal of Remote Sensing*, 38(8–10), 2757–2770. doi:10.1080/01431161.2017.1287975.

Tsironis, V., Herekakis, T., Tsouni, A., and Kontoes, C. (2016). DisasterHub: A mobile application for enabling crowd generated data fusion in Earth Observation disaster management services. Geophysical Research Abstracts, p. 18. Retrieved from http://meetingorganizer.copernicus.org/EGU2016/EGU2016-17311.pdf

Turner, D., Lucieer, A., and de Jong, S. M. (2015). Time series analysis of landslide dynamics using an unmanned aerial vehicle (UAV). *Remote Sensing*, 7(2), 1736–1757. doi:10.3390/rs70201736.

USGS. (2016). USGS FAQs. Retrieved March 1, 2017, from USGS: https://www2.usgs.gov/faq/node/2613.

Vakhshoori, V. and Zare, M. (2016). Landslide susceptibility mapping by comparing weight of evidence, fuzzy logic, and frequency ratio methods. *Geomatics, Natural Hazards and Risk*, 7(5), 1731–1752. doi:10.1080/19475705.2016.1144655.

van Westen, C. (2008). Landslide Susceptibility Assessment. *Mountain Risks Intensive Course*. Barcelona, Spain: Université Caen Normandie. Retrieved from https://goo.gl/YCDbxE.

Vervoort, A. (2016). Surface Movement above an underground coal longwall mine after closure. *Natural Hazards and Earth System Sciences*, 16, 2107–2121.

Wang, L.-J., Guo, M., Sawada, K., Lin, J., and Zhang, J. (2016). A comparative study of landslide susceptibility maps using logistic regression, frequency ratio, decision tree, weights of evidence and artificial neural network. *Geosciences Journal*, 20(1), 117–136. doi:10.1007/s12303-015-0026-1.

Wempen, J. and McCarter, M. (2017). Comparison of L-band and X-band differential interferometric synthetic aperture radar for mine subsidence monitoring in central Utah. *International Journal of Mining Science and Technology*, 27, 159–163.

Westoby, M. J., Brasington, J., Glasser, N. F., Hambrey, M. J., and Reynolds, J. M. (2012). 'Structure-from-motion' photogrammetry: A low-cost, effective tool for geoscience applications. *Geomorphology*, 179, 300–314. doi:10.1016/j.geomorph.2012.08.021.

Winter, M. G., Shearer, B., Palmer, D., Peeling, D., Harmer, C., and Sharpe, J. (2016). The economic impact of landslides and floods on the road network. *Procedia Engineering*, 143, 1425–1434. doi:10.1016/j.proeng.2016.06.168.

Woldai, T., Oppliger, G., and Taranik, J. (2009). Monitoring dewatering induced subsidence and fault reactivation using interferometric synthetic aperture radar. *International Journal of Remote Sensing*, 30, 1503–1519.

Youssef, M., Al-kathery, M., Pradhan, B., and El-sahly, T. (2016). Debris flow impact assessment along the Al-Raith Road, Kingdom of Saudi Arabia, using remote sensing data and field investigations. *Geomatics, Natural Hazards and Risk*, 7(2), 620–638. doi:10.1080/19475705.2014.933130.

Zamora, J. F., Kashihara, S., and Yamaguchi, S. (2015). Calibration of smartphone-based weather measurements using pairwise gossip. *The Scientific World Journal, 2015* (Article ID 494687), 9. doi:10.1155/2015/494687.

Zhao J., Xiao, J., Lee, M. L., Ma, Y. (2016). Discrete element modeling of a mining-induced landslide. *SpringerPlus*, 5, 1633.

Zumbrunnen, T., Thuro, K., and König, S. (2017). Dealing with natural hazards along federal and state roads in Bavaria. *Geomechanics and Tunnelling*, 10(1), 34–46. doi:10.1002/geot.201600072.

24 Landslide Susceptibility Assessment Mapping
A Case Study in Central Greece

George D. Bathrellos, Dionissios P. Kalivas, and Hariklia D. Skilodimou

CONTENTS

24.1 INTRODUCTION

Landscape of the Earth has a complex evolution and is the result of the interactions involving surface progress, climate, tectonic activity, and human activity. Natural hazards are physical phenomena that occur worldwide and contribute to the evolution of Earth's landscape (Skilodimou et al. 2014). Their associated consequences can lead to the damage of both the natural and man-made environment. When these consequences have a major impact on human life, natural hazards are called natural disasters. Therefore, proper planning and management of natural disasters are essential to minimize the loss of human life and reduce the economic consequences. In this context, maps that provide information on the spatial distribution of natural hazards are important tools for planners and environmental managers when selecting favorable locations for land use development (e.g., Peng et al. 2012; Papadopoulou-Vrynioti et al. 2014; Youssef et al. 2015).

Natural hazard phenomena vary in magnitude, frequency, speed, and duration. Thus, detailed knowledge, about the evaluation of these events in an area, is crucial to management of natural disasters. In the last decades, Earth observation (EO) data such as aerial photography, satellite imagery, and Global Positioning System (GPS) data have become integral means for the evaluation of natural hazard events. Moreover, geographic information system (GIS) is an excellent tool in the spatial analysis, assessment, and mitigation of various natural hazard phenomena (Lu et al. 2011; Youssef et al. 2011; Papadopoulou-Vrynioti et al. 2013; Chousianitis et al. 2016). These current geo-spatial technologies are very useful to estimate future hazard occurrences, identify vulnerability of communities to hazards, and in disaster preparation and response. Therefore, nowadays, EO and GIS have become necessary tools in addressing natural disaster (Van Westen 2013).

Landslide susceptibility assessment is an important process for prediction and management of natural disasters; it is also a necessary step for natural and urban planning government policies worldwide. During the current decades, the use of landslide susceptibility maps for land use planning has drastically increased. The aim of these maps is to rank different parts of an area according to the degree of actual or potential landslide hazard. Thus, planners are capable of selecting favorable sites for urban and rural development for preventing from landslide hazards. The reliability of those maps depends mostly on the available data used, as well as applied methodology for the hazard estimation (Parise 2001; Carrara et al. 2003).

Several methods have been developed to assess a landslide susceptibility map. The choice of appropriate method depends on the nature of the problem, the observation scale, and data availability. They can be separated in qualitative and quantitative approaches. Qualitative landslide susceptibility assessments may include either detailed landslide inventory maps or a heuristic analysis. The inventory maps can be prepared by spatial and temporal record of landslide events, interpretation of aerial photographs or satellite images, and field observations. These maps are the basis for the creation of landslide density maps or landslide isopleth maps, which represent a landslide susceptibility assessment map. Although this approach provides a quantitative measure on landslide spatial distribution, it cannot give estimations on future landslides unless they have already occurred. By the heuristic or knowledge-based approach, landslide data are not required, and expert opinions are used to rank the importance of the factors influencing the landslide events. The advantage of this method is that a rough assessment can be made without landslide inventory, whereas the main weakness is the high subjectivity and uncertainty of weightings and ratings of the variables. Generally, qualitative methods are more suitable to produce landslide susceptibility assessment maps in large areas such as at national level, where the quality and quantity of the available data are not enough for quantitative analysis (Dai et al. 2002; Castellanos Abella and vanWesten 2008; Galli et al. 2008; Rozos et al. 2011).

Quantitative landslide susceptibility methods are divided into deterministic and statistical approaches. Deterministic or physically based approaches express physical processes leading to the landslides and are based on slope instability analysis as well as on simple physical laws. These models do not require long-term landslide data, are based on sound physical models, and provide a single landslide hazard value at a given space and moment of time. The cons of deterministic approaches are the high degree of simplification, which is typically necessary for their use, and the difficult evaluation of predictive models. Another problem of these models is that they require high accuracy of input parameters, and often, this is impossible. They are more suitable to assess landslide susceptibility in small areas (Mercogliano et al. 2013; Terlien et al. 2013). Statistical approaches determine the numerical correlation of causative factors and landslide events. They can be categorized into bivariate statistical analysis, which compares each factor with the existing landslide distribution, and multivariate statistical analysis, which considers relative contribution of each factor to the total landslide susceptibility. The main pros of bivariate models are that they render quantitative and objective measures on landslide susceptibility, whereas the main problems are that they assume independence of input parameters and require landslide inventory maps. The multivariate models allow the evaluation of comparative contribution of each factor in landslide occurrences and predict the spatial landslide occurrence at high reliability. When they are used in a black-box manner, then it is possible to provide misleading results. Statistical models are suitable for landslide susceptibility mapping at medium scales of 1:20,000–1:100,000 (Yalcin 2008; Cervi et al. 2010; Mancini et al. 2010; Akgun 2012; Pardeshi et al. 2013).

Landslide events are based on various physical factors. Therefore, several methods of landslide susceptibility mapping focus on (a) the compilation of a landslide inventory, (b) the determination of the physical factors that are directly or indirectly correlated with slope instability factors, (c) the selection of the rate-weighting system of all instability factors and of the individual classes of values of each factor, (d) the overall estimation of the relative role of factors in producing landslides, (e) the final susceptibility zoning by classifying the land surface according to different hazard degrees, and (f) the verification of the produced map by comparing the landslide susceptibility zones with recorded landslide occurrences. To this direction, EO data and GIS provide a significant assistance. Aerial photographs and satellite images are effective tools in recording, mapping, and monitoring of landslide occurrences. Geographic information system capabilities and facilities allow the spatial analysis and processing of instability factors and landslide events, assignment of weights, and production of landslide susceptibility zonation map (Bathrellos et al. 2009; Yilmaz 2009; Rozos et al. 2013).

Multicriteria analysis methods are decision-support tools for the solution of complex decision problems. The analytical hierarchical process (AHP) is a quantitative and multicriteria methodology designed for hierarchical representation of a decision-making problem (Saaty 1977, 2006). The AHP has gained wide application in land use suitability (Thapa and Murayama 2008; Bathrellos et al. 2012; Panagopoulos et al. 2012) and particularly in landslide susceptibility analysis (e.g., Ayalew et al. 2004; Yoshimatsu and Abe 2006; Rozos et al. 2011; Pourghasemi et al. 2012).

Moreover, the integration of the AHP in a GIS is able to improve decision-making methodology with powerful visualization and mapping capabilities and to facilitate the production of hazard maps. However, the AHP method does not have the ability to identify the uncertainty associated with spatial outputs (Bathrellos et al. 2013).

This chapter will deliberate on the landslide susceptibility assessment mapping. In central Greece, landslide events are widespread, inflicting significant damages on settlements and road networks. Therefore, a mountainous part of this area has been selected as the case study.

The main scope of this study is to present a method in order to produce a landslide susceptibility map. The major factors affecting landslide occurrences of the study area were estimated. The AHP was implemented to support the evaluation of these factors, whereas the data processing and the compilation of landslide susceptibility map were performed in a GIS environment. A sensitivity analysis of the factors was made in order to examine the effect of their variations on the spatial and quantitative distribution of landslide susceptibility assessment areas. Finally, a landslide inventory map was used for the verification of the landslide susceptibility map.

24.2 STUDY AREA

The case study area is located in the mountainous part of Karditsa Prefecture, western Thessaly, central Greece. This particular area was selected because landslide events have occurred many times in this area and have repeatedly caused serious damage at sections of the settlements and at the road network. It covers 144 km², with altitudes varying from 220 to 2,000 m above sea level. The southern mountain range of Pindus forms the higher region of the area. The upper stream of the Pamisos River, which is a tributary of the Pinios River, flows through the study area; the drainage network is well developed, with a significant surface runoff (Figure 24.1).

The geological structure of the area comprises Alpine and post-Alpine formations. The Alpine formations belong to two main stratigraphic zones, which are the Pindos zone and the Koziakas zone. The post-Alpine formations are molassic formations of the Mesohellenic trench and Quaternary deposits, which cover the beds of rivers and streams. The study area is composed of (a) coarse grained loose to semicoherent Quaternary deposits consisting of mainly pebbles and gravels of varying-sized gravels, cones, and scree; (b) Miocene molassic clastic formations of Fanari series;

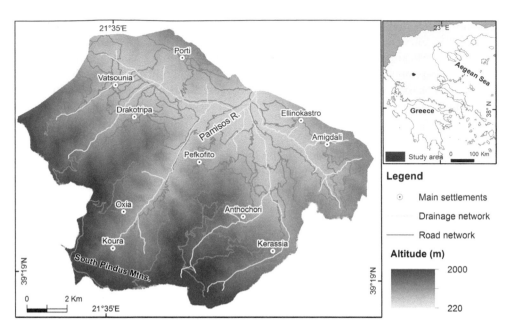

FIGURE 24.1   The location map of the study area, with altitudes, drainage network and road network.

(c) flysch formations, whose main members are sandstones, siltstones, and more rarely conglomerates; (d) carbonate rocks consisting of Middle Triassic—Cretaceous limestones; and (e) Middle—Upper Jurassic clay—chert formations (IGME 1993).

The climate is Mediterranean, with a rainy period that begins in October and ends in May.

24.3   METHODOLOGY

24.3.1   Description of Datasets

This study was carried out using the following:

- Topographic map (1:50,000 scale) from the Hellenic Military Geographical Service (HAGS)
- Geological map of the study area (1:50,000 scale) from the Institute of Geology and Mineral Exploration (IGME 1993)
- Mean annual precipitation data covering a time period of 25 years, from the Public Power Corporation (PPC) stations
- Satellite image via Google Earth and Landsat 7/Copernicus, with an acquisition date of July 2015
- Field work data involving observations on landslide sites

A spatial database was created, and ArcGIS 10.0 software was used to process the collected data.

24.3.2   Description of Method

24.3.2.1   Instability Factors

The instability factors that were taken into account for the creation of the landslide susceptibility map have been based on extended field observations and literature (i.e., Ayalew et al. 2004; Rozos et al. 2011; Bathrellos 2014; and Youssef et al. 2015). The selected factors are (1) lithology,

(2) distance from tectonic elements, (3) slope angle, (4) slope aspect, (5) rainfall, (6) altitude, (7) land use, (8) distance from roads, and (9) distance from streams.

24.3.2.1.1 Lithology

The bedrock geology is an important factor for landslide occurrences. A lithological map of the study area was generated based on the existing geological map (scale 1:50,000) (IGME 1993). The lithological formations were digitized and joined according to their engineering geological behavior, with respect to landslide manifestation. Hence, they are classified into five categories: (a) Quaternary formations, including coarse grained loose to semicoherent deposits, (b) molassic formations consisting of clastic materials, (c) flysch, (d) carbonate rocks, and (e) clay—chert formations (Figure 24.2).

24.3.2.1.2 Distance from Tectonic Elements

Tectonic activity increases landslide events by creating steep slopes and sheared, weakened rocks. The tectonic elements of the study area involve thrusts, overthrusts, and faults. These features were collected from the literature (IGME 1993) and inserted in the GIS database as line layer. Buffer zones were drawn around them at distances of 50, 100, 150, and 200 m (Figure 24.3).

24.3.2.1.3 Slope Angle

Slope angle has an effect on the slope stability, increasing the landslide hazard. Contours with 20 m intervals and height points were digitized from topographic map (scale 1:50,000) and saved as line and point layers correspondingly. A digital elevation model (DEM) was derived from the digitized elevation data by using 3D Analyst extension of ArcGIS, and the slope layer was extracted from the DEM. The slopes were classified into five classes: (1) less than 5°, (2) 5°–15°, (3) 15°–30°, (4) 30°–45°, and (5) more than 45° (Figure 24.4).

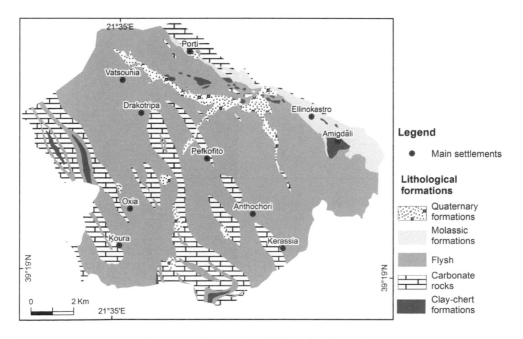

FIGURE 24.2 Map showing the spatial distribution of lithological formations.

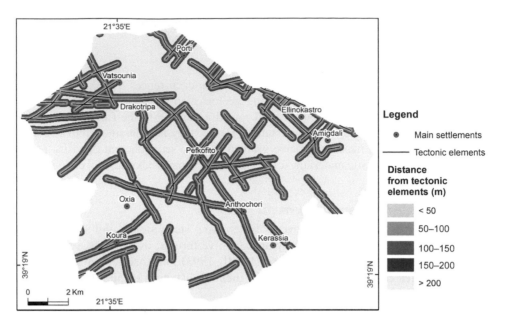

FIGURE 24.3 Map showing the distance from tectonic elements.

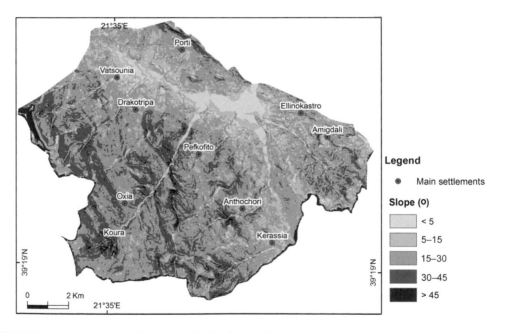

FIGURE 24.4 Map showing the spatial distribution of slopes.

24.3.2.1.4 Slope Aspect

The aspect map was derived from the DEM. It was classified into eight categories: (1) 0°–45°, (2) 46°–90°, (3) 91°–135°, (4) 136°–180°, (5) 181°–225°, (6) 226°–270°, (7) 271°–315°, and (8) 316°–360° (Figure 24.5).

24.3.2.1.5 Rainfall

Precipitation is among the most common triggering factors for landslide events. The precipitation data, which were collected from the PPC and covered a time period of 25 years, were evaluated

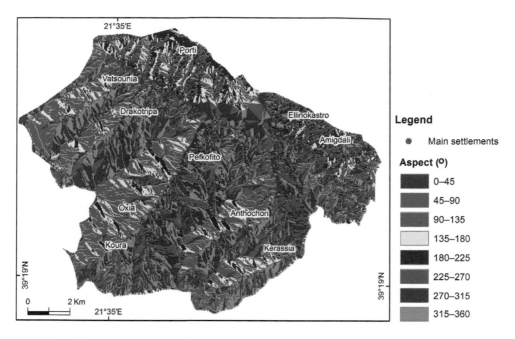

FIGURE 24.5   Map showing the spatial distribution of slope aspect.

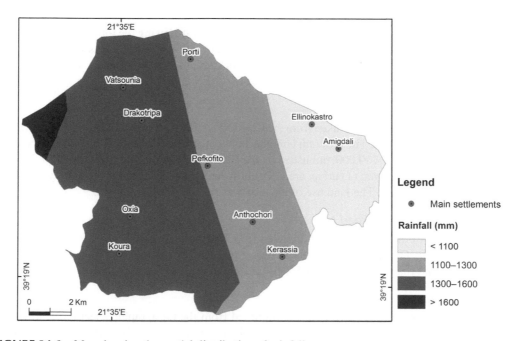

FIGURE 24.6   Map showing the spatial distribution of rainfall.

and used for the synthesis of the mean annual precipitation distribution map of Greece. The mean annual rainfall of the study area incorporated to the spatial database constitutes a digitized part of this map. According to this map, the annual precipitation of the study area ranges from 920 to 1700 mm. This map was separated into four classes: (1) less than 1100 mm, (2) 1100–1300 mm, (3) 1300–1600 mm, and (4) more than 1600 mm (Figure 24.6).

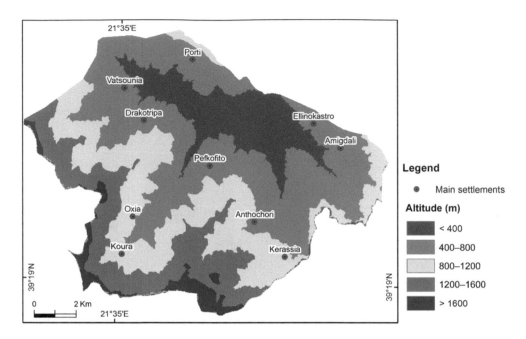

FIGURE 24.7   Map showing the spatial distribution of altitude.

24.3.2.1.6   Altitude

The altitude map was derived from the DEM, which was generated using the topographic map (scale 1:50,000) as the elevation data source. The categorization of altitude into five classes, that is (1) less than 400 m a.s.l., (2) 400–800 m a.s.l., (3) 800–1200 m a.s.l., (4) 1200–1600 m a.s.l., and (v) more than 1600 m a.s.l., was based on the morphology of the study area (Figure 24.7).

24.3.2.1.7   Land Use

The land use of the study area was taken from the CORINE 2012 Land Cover (CLC) map, Copernicus Program (Copernicus 2016). The program contains land cover data for Europe, including land cover class description at scale 1:100,000, published by the European Commission. The CORINE land use map was classified as follows: (1) urban area, (2) cultivated area, (3) forests, (4) shrubby areas, and (5) bare area (Figure 24.8). The land use of the area was saved as polygon layer.

24.3.2.1.8   Distance from Road Network

The road network of the study area was digitized from topographic map (scale 1:50,000) and saved as a line layer in the GIS database. Similar to the case of tectonic elements, four buffer zones were constructed around roads at distances of 50, 100, 150, and 200 m (Figure 24.9).

24.3.2.1.9   Distance from Streams

The drainage network of the study area was digitized from the topographic sheet (scale 1:50,000) and saved as line layer. The streams were classified by using the Strahler's method. The streams of the third- and fourth-order streams continuously modify the slopes of the rivers, and thus, they can be considered a factor influence in landslide occurrence. For the examination of this factor, buffer zones were created around the bed of the rivers and the streams of the area, at distances of 50, 100, 150, and 200 m (Figure 24.10).

24.3.2.2   Rating of the Classes of the Instability Factors

A primary step in the process of landslide susceptibility assessment is the classification of all the factors. Therefore, the classes of the involved factors have to be standardized to a uniform

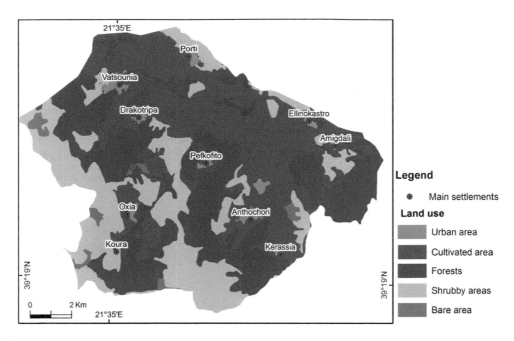

FIGURE 24.8   Map showing the land use of the study area.

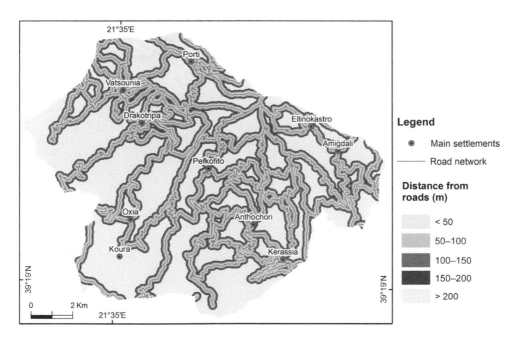

FIGURE 24.9   Map showing the distance from road network.

suitability-rating scale. The standardization method used in the analysis was consistently based on a five-grade scale. Integer numbers, ranging from 0 to 4 (Bathrellos et al. 2016), were assigned to every class corresponding to different hazard levels. Therefore, the class that was rated as 0 represented the most stable conditions (very low landslide hazard), and the one rated as 4 represented the most favorable conditions for slope failure (very high landslide hazard). Table 24.1 shows the rating of the classes of each factor.

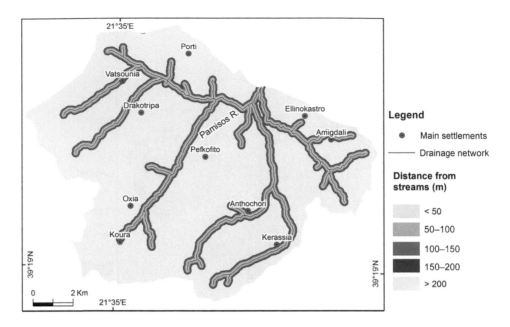

FIGURE 24.10 Map showing the distance from streams.

TABLE 24.1
The Selected Factors Involved in the Landslide
Susceptibility Assessment, Their Classes, and Their Ratings

Factors	Classes	Rating
Lithology	Quaternary formations	3
	Molassic formations	2
	Flysch	4
	Carbonate rocks	0
	Clay—chert formations	1
Distance from tectonic elements (m)	<50	4
	51–100	3
	101–150	2
	151–200	1
	>200	0
Slope angle (°)	0–5	0
	5–15	1
	15–30	2
	30–45	3
	>45	4
Slope aspect (°)	180–225	0
	135–180	1
	90–135, 225–270	2
	45–90, 270–315	3
	0–45, 315–360	4
Rainfall (mm)	<1100	2
	1100–1300	3
	1300–1600	4
	>1600	4

(Continued)

TABLE 24.1 (*Continued*)
The Selected Factors Involved in the Landslide
Susceptibility Assessment, Their Classes, and Their Ratings

Factors	Classes	Rating
Altitude (m)	<400	2
	400–800	3
	800–1200	4
	1200–1600	4
	>1600	1
Land use	Urban area	0
	Cultivated area	3
	Forest area	1
	Shrubby area	2
	Bare area	4
Distance from roads (m)	<50	4
	51–100	3
	101–150	2
	151–200	1
	>200	0
Distance from streams (m)	<50	4
	51–100	3
	101–150	2
	151–200	1
	>200	0

24.3.2.3 Weighting of the Instability Factors—Analytical Hierarchical Process Method

The rules of the AHP method were applied to get the final weights for each factor. The first step in the AHP is the computation of the pairwise comparison matrix, where each entry represents the relative significance of a factor to the others. The relative importance between two factors is measured according to a numerical scale from 1 to 9. The correlation between the numerical values and the intensity of importance are as follows: 1 = equal importance, 2 = weak or slight, 3 = moderate importance, 4 = moderate plus, 5 = strong importance, 6 = strong plus, 7 = very strong, 8 = extremely strong, and 9 = of extreme importance. Inversely, less important variables were rated between 1 and 1/9 (Saaty 1977, 2006).

The method requires normalization of all factor weights by the following equation:

$$\sum_{i=1}^{n} W_i = 1 \tag{24.1}$$

It is important to verify the consistency of each table matrix after the calculation of the weight values. Therefore, the implication of each one was checked with the consistency ratio (CR):

$$CR = CI/RI \tag{24.2}$$

Where, RI is the random index, which was developed by Saaty (1977), and it is a constant that depends on the order of the matrix, and the CI is calculated by the equation:

$$CI = \lambda_{\max} - n / n - 1 \tag{24.3}$$

TABLE 24.2

The Weighting Coefficient of Every Factor and the Consistency Ratio for Landslide Susceptibility Assessment

	F1	F2	F3	F4	F5	F6	F7	F8	F9	Weights, W_i
F1	1	2	1/3	3	1/2	7	3	2	3	0.145
F2		1	1/3	2	1/2	6	2	1	1	0.085
F3			1	5	2	9	5	3	4	0.275
F4				1	1/5	6	1/2	1/4	1/5	0.040
F5					1	7	4	3	2	0.186
F6						1	1/5	1/8	1/8	0.016
F7							1	1/3	1/3	0.049
F8								1	1/2	0.093
F9									1	0.111

CR=0.05

Note: F1 = lithology, F2 = distance from tectonic elements, F3 = slope angle, F4 = slope aspect, F5 = rainfall, F6 = altitude, F7 = land use, F8 = distance from roads, and F9 = distance from streams.

Where, λ_{max} is the largest eigenvalue of the matrix, and n is the order of the matrix. This ratio is used in order to avoid the creation of any incidental judgment in the matrix, and when CR value is less than 0.1, an acceptable level of consistency has been achieved. The CR values are less than 0.1, which means that the corresponding matrixes have an acceptable level of consistency.

The classes of every adopted factor, their ratings, the calculations of the weighting coefficient, and the CR are given in Table 24.2. All the pairwise comparisons, the eigenvectors, the weights, and the CR were calculated by using the Expert Choice 2000 software.

24.3.2.4 The Overall Landslide Susceptibility Index

The overall score of the basic landslide susceptibility assessment for the study area was calculated with the correlation of the estimated factors. This correlation was performed by using the weighted linear combination method, according to the following mathematical operator:

$$LS = \sum_{i=1}^{n} W_i X_i \tag{24.4}$$

Where:

LS is landslide susceptibility index
n is the number of the factors
W_i is the weight of the factor
i and X_i is the rating of the factor i

After the application of the above-mentioned equation, the landslide susceptibility assessment map was produced.

24.3.2.5 Uncertainty Analysis

Uncertainty plays an important role in natural hazard assessment and land use suitability estimation (Bathrellos et al. 2013; Van Westen et al. 2014). The AHP method has limitations to determinate the

TABLE 24.3
The Changes of Weighting Coefficient
Values (ΔW_i) of Every Factor

Factor	ΔW_i
Lithology	0.029
Distance from tectonic elements	0.017
Slope angle	0.055
Slope aspect	0.008
Rainfall	0.037
Altitude	0.003
Land use	0.010
Distance from roads	0.019
Distance from streams	0.022

uncertainty. Bathrelos et al. (2016, 2017) applied the AHP method to evaluate geoenvironmental factors to assess the urban flood hazard and natural hazard maps to estimate sites for urban development; they determined the uncertainty involved in the technique by introducing an uncertainty in weighting coefficient of the adopted factors.

In this concept, the influence of uncertainty of the adopted factor weights on the landslide susceptibility assessment was examined, and two more scenarios were developed. The error ΔS produced by independent errors ΔW_i in the weighting coefficient values is given by:

$$\Delta S = \sqrt{\sum_{i=1}^{n}(\Delta W_i X_i)^2} \qquad (24.5)$$

Each weighting coefficient value was altered 20%, without time step, from the original factor weight that was used for the basic landslide susceptibility assessment. The changes of weight values (ΔW_i) of every factor are given in Table 24.3.

Equation 24.5 was applied to calculate the error (ΔS). Then, it was multiplied by 1.96 to compute 95% confidence level of the level of the LS values. This process led to the creation of a map that was added and subtracted from the basic landslide susceptibility map to estimate the upper and lower LS values at 95% confidence level, respectively. Thus, two maps representing two extreme scenarios of maximum and minimum LS values were produced for landslide susceptibility assessment.

24.3.2.6 Landslide Inventory Map

The landslide inventory map compilation involves the following steps: (a) landslides were recorded from previous works (Bathrellos 2014), (b) landslide locations were recognized on satellite image, and (c) the manifested landslides were verified and mapped by field work. The landslide occurrences were digitized as point layer. The landslides were used for the compilation of the landslide inventory map (Figure 24.11), which was further used for the verification of the landslide susceptibility assessment map.

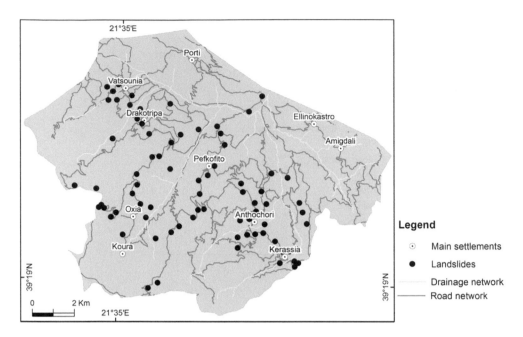

FIGURE 24.11   The landslide inventory map of the study area.

24.4   RESULTS

24.4.1   LANDSLIDE SUSCEPTIBILITY ASSESSMENT MAP

The implementation of landslide susceptibility assessment methods leads to the production of a landslide susceptibility map. The land surface of this map is usually categorized into classes that represent different landslide susceptibility levels (Nefeslioglu et al. 2013; Park et al. 2013).

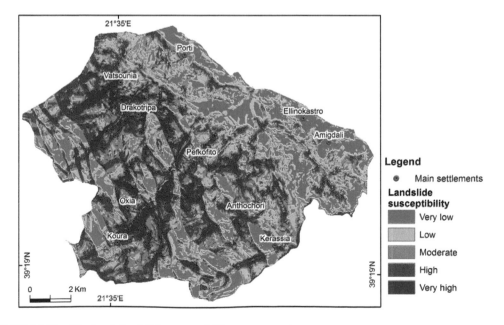

FIGURE 24.12   The basic landslide susceptibility assessment map.

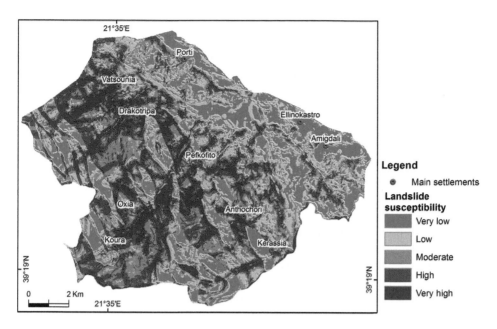

FIGURE 24.13 The maximum value of landslide susceptibility assessment map.

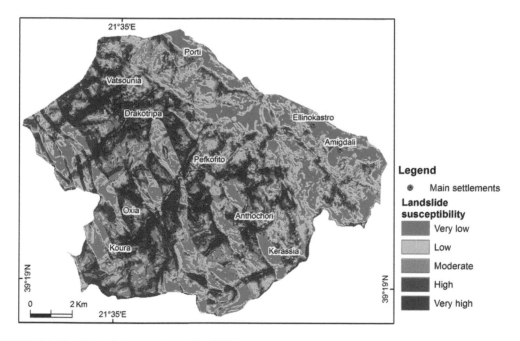

FIGURE 24.14 The minimum value of landslide susceptibility assessment map.

The results of the landslide susceptibility assessment are given in the maps of the Figures 24.12 through 24.14. Three alternative maps were derived for different scenarios. The calculated landslide susceptibility index values of the three landslide susceptibility assessment maps were categorized by using the quantile method of classification. The study area of each map was classified into five sections corresponding to very high, high, moderate, low and very low landslide susceptibility zone.

The first applied scenario was the application of the AHP method, which led to the basic landslide susceptibility (LSb) assessment map (Figure 24.12). Regarding the spatial distribution of the five

TABLE 24.4

Percentages of Each Landslide Susceptibility Zone in the Cases of Basic, Maximum, and Minimum Scenarios for the Study Area

Landslide Susceptibility Zones/Percentage of Total Study Area	LSb	LSmax	LSmin
Very low	19.4	19.3	19.6
Low	20.2	19.8	20.4
Moderate	20.2	20.4	20.6
High	20.4	20.6	19.7
Very high	19.8	19.9	19.8

landslide susceptibility zones, the areas of very high and high landslide susceptibility are located mostly at the northwestern, western, and southwestern parts of the study area.

The second map (LSmax) represents the maximum value of the landslide susceptibility assessment of each pixel (Figure 24.13). Similar to the LSb assessment map, the very high and high landslide hazard zones are largest in the northwestern, western, and southwestern parts of the study area.

The minimum-value scenario is expressed in the third map (LSmin) of Figure 24.14. The very high and high landslide susceptibility zones of this map have the largest spatial distribution in the same parts of the previous two maps.

The percentages of the five susceptibility zones, in relation to the entire extent of the study area, for the three alternative maps are given in Table 24.4.

24.4.2 VERIFICATION OF THE LANDSLIDE SUSCEPTIBILITY ASSESSMENT MAP

Testing the produced landslide susceptibility map is of great importance for the reliability and usefulness of the method followed. A landslide susceptibility assessment map is usually verified by using the spatial distribution of landslide events that affected the study area (Yilmaz 2009; Pradhan and Lee 2010).

In this context, 82 sites of landslide occurrences were examined throughout the study area. Field work revealed that the majority of the landslides manifested in the loose formations of the study area.

The applied verification process was checked to the superposition of the landslide occurrence map over the basic landslide susceptibility assessment map. The applied procedure proved that the majority of the landslides are located within the limits of the areas presenting high to very high landslide susceptibility values.

Table 24.5 presents the numerical distribution of the landslides into the landslide susceptibility zones of the map. According to this table, 78% of the landslides are located within the limits of the high to very high susceptibility classes, and 15% are located into the moderate landslide susceptibility area.

TABLE 24.5

Number and Percentage of Landslide Events into Each Landslide Susceptibility Zone

Landslide Susceptibility Zone	Number of Landslides	Landslides (%)
Very low	3	3.7
Low	3	3.7
Moderate	12	14.6
High	30	36.6
Very high	34	41.4
Total	82	100.0

24.5 DISCUSSION

In the present study, a landslide susceptibility assessment map was generated by using the AHP method, GIS tools, and EO data. In this map, we identified zones of five hazard levels: very low, low, moderate, high, and very high.

The EO data can be effective in support of detailed landslide inventories mapping. Moreover, radar and optical EO data such as interferometric synthetic aperture radar (InSAR) and object-based image analysis (OBIA) methods can contribute toward improving our ability for landslide monitoring and during all phases of emergency management: mitigation, preparedness, crisis, and recovery (Casagli et al. 2016). To this direction, satellite image was exploited for recording, rapid mapping, and monitoring of landslide events, and thus, a landslide inventory map was produced.

Concerning the produced landslide susceptibility map, the very high and high landslide susceptibility zones are mainly observed in the northwestern, western, and southwestern parts of the study area. The percentage of these two susceptibility zones, in relation to the total area, is about 41% (Table 24.4). This area suffers from many landslides due to the prevailing geological and morphological conditions. According to Bathrellos et al. (2009), this is a mountainous area with rather steep slopes, and it consists of flysch sediments, prone to landslide occurrences. On the contrary, in the eastern part of the study area, very few regions of limited area are classified into areas of high and very high landslide susceptibility.

Although the AHP presents limitations to determinate the uncertainty, it appears to be a powerful and practical tool in the case of multiobjective decision-making problems (Chen et al. 2010; Bathrellos et al. 2013; Van Westen et al. 2014). For this reason, two more scenarios were developed to examine the influence of the uncertainty on the landslide susceptibility assessment results. The uncertainty analysis of the proposed methodology proved that the weighting coefficient of slope angle presents the greatest variation, whereas the altitude has the lowest variation among all the factors involved in the landslide susceptibility assessment. The fluctuations of the weighting coefficients a little involves the spatial distribution of the hazard zones. The results of the uncertainty analysis (Table 24.4) showed no significant variations in the percentages of the landslide susceptibility zones among the LSb, LSmax, and LSmin maps. The percentages of the landslide susceptibility zones for the LSmin map have a slightly changed in comparison with the LSb map. The highest variation of the percentages is observed in the moderate and high landslide susceptibility zones. Consequently, the analysis showed no significant differences in the spatial and quantitative distribution of the landslide susceptibility zones. This fact indicates robust behavior for the predictions of the applied method.

The combination of the different maps may produce a map that does not include the actual hazard in the area (Kappes et al. 2012). For this reason, the basic landslide susceptibility map was verified by using 82 sites of landslide occurrences. The results established that the vast majority of the landslide events, 78% (Table 24.5), occurred within the limits of the high to very high susceptibility zones. Consequently, the produced landslide susceptibility assessment map presents a satisfactory agreement between the landslide susceptibility zones and the spatial distribution of landslide phenomena.

At local scale, the proposed method identifies the areas that are susceptible or not susceptible to landslides. Subsequently, the adopted approach determines favorable places for urban development. Engineers, planners, decision makers, environmental managers, and local authorities may utilize the proposed procedure in spatial planning and landslide hazard mitigation policy.

In addition, the sites of the existing urban areas of a region that are located in non-safe areas regarding landslides may be identified. Therefore, proper remedial measures able to prevent and reduce the consequences of landslides occurrences for an area may be considered.

The selection of the proper remedial measures constructions is important to prevent the consequences of landslide occurrences (Rozos et al. 2013). Thus, future works, including high-resolution radar and EO data, are essential to evaluate and record landslide events in detail. The increasing availability of EO data offered by new sensors will enhance the possibility to cover wider areas and to obtain more updated data for landslide events, with improved spatial and temporal resolutions. Information derived from up-to-date satellite images of areas prone to landslide occurrences, combined with the vulnerability of the exposed elements at landslide hazard (i.e., buildings and roads), will be very useful to propose the proper protection measures. Moreover, this process will offer a significant assistance in recovery of the areas affected by landslides.

24.6 CONCLUSIONS

In the present study, landslide susceptibility assessment maps were produced via the AHP method, GIS, and EO data. Satellite image was exploited for the rapid mapping of the landslide events of the study area. The adopted procedure was implemented in a mountainous part of western Thessaly, central Greece.

The results of the applied method highlighted that the areas of very high and high landslide susceptibility are located mostly in the northwestern, western, and southwestern parts of the study area. This is a mountainous area with rather steep slopes, and it consists of flysch sediments, prone to landslide occurrences.

The uncertainty analysis of the proposed method proved that the fluctuations of weighting coefficients a little involves the spatial distribution of the landslide susceptibility zones. Slight differences on the spatial and quantitative distribution of the landslide susceptibility zones were observed, indicating robust behavior for the model predictions.

The landslide susceptibility map showed that sufficient correlation with landslide occurrences affected the study area. This indicates that the results of the proposed method are reliable.

In regional studies, the applied procedure can be definitely used for the localization of sites susceptible to landslides. Therefore, the proposed methodology and the landslide susceptibility assessment map should be taken into account by the local authorities, engineers, and planners to adopt policies and strategies aiming toward landslide hazard prevention and mitigation.

In the present work, EO data were very effective in support of detailed and rapid landslide mapping and monitoring. In future works, the increasing availability of EO data offered by new sensors will improve our ability for landslide mapping and monitoring. Additional, up-to-date satellite images of affected and prone areas will provide reliable information for the selection of the proper remedial measures to prevent the consequences of landslides and the recovery of areas affected by landslides.

REFERENCES

Akgun, A. 2012. A comparison of landslide susceptibility maps produced by logistic regression, multi-criteria decision, and likelihood ratio methods: A case study at İzmir, Turkey. *Landslides* 9(1):93–106.

Ayalew, L., Yamagishi, H., and N. Ugawa 2004. Landslide susceptibility mapping using GIS-based weighted linear combination, the case in Tsugawa area of Agano River, Niigata Prefecture, Japan. *Landslides* 1(1):73–81.

Bathrellos, G.D. 2014. Mechanisms of landslide occurrences in the flysch of the Pindos geotectonic zone and prevention measures: Areas of Trikala prefecture Ropoto - Kotroni–Vatsounia (east slopes of the massif Karavoula, Pindos Range, Central Greece). Mcs thesis, Agricultural University of Athens.

Bathrellos, G.D., Kalivas, D.P., and H.D. Skilodimou 2009. Landslide susceptibility mapping models, applied to natural and urban planning, using G.I.S. *Journal Estudios Geológicos* 65(1):49–65.

Bathrellos, G.D., Gaki-Papanastassiou, K., Skilodimou, H.D., Papanastassiou, D., and K.G. Chousianitis 2012. Potential suitability for urban planning and industry development by using natural hazard maps and geological – geomorphological parameters. *Environmental Earth Sciences* 66(2):537–548.

Bathrellos, G.D., Gaki-Papanastassiou, K., Skilodimou, H.D., Skianis, G.A., and K.G. Chousianitis 2013. Assessment of rural community and agricultural development using geomorphological–geological factors and GIS in the Trikala prefecture (Central Greece). *Stochastic Environmental Research and Risk Assessment* 27(2):573–588.

Bathrellos, G.D., Karymbalis, E., Skilodimou, H.D., Gaki-Papanastassiou, K., and E.A. Baltas, 2016. Urban flood hazard assessment in the basin of Athens Metropolitan city, Greece. *Environmental Earth Sciences* 75(4):319.

Bathrellos, G.D., Skilodimou, H.D., Chousianitis, K., Youssef, A.M. and B. Pradhan 2017. Suitability estimation for urban development using multi-hazard assessment map. *Science of the Total Environment* 575:119–134.

Carrara, A., Giovanni, C., and P. Frattini 2003. Geomorphological and historical data in assessing landslide hazard. *Earth Surface Processes and Landforms* 28:1125–1142.

Casagli, N., Cigna, F., Bianchini, S. et al. 2016. Landslide mapping and monitoring by using radar and optical remote sensing: Examples from the EC-FP7 project SAFER. *Remote Sensing Applications: Society and Environment* 4:92–108.

Castellanos Abella, E.A. and C.J. van Westen 2008. Qualitative landslide susceptibility assessment by multicriteria analysis: A case study from San Antonio del Sur, Guantánamo, Cuba. *Geomorphology* 94 (3):453–466.

Cervi, F., Berti, M., Borgatti, L. et al. 2010. Comparing predictive capability of statistical and deterministic methods for landslide susceptibility mapping: A case study in the northern Apennines (Reggio Emilia Province, Italy). *Landslides* 7(4):433–444.

Chen, Y., Yu, J., and S. Khan 2010. Spatial sensitivity analysis of multi-criteria weights in GIS-based land suitability evaluation. *Environmental Modelling & Software* 25:1582–1591.

Chousianitis, K., Del Gaudio, V., Sabatakakis, N. et al. 2016. Assessment of earthquake-induced landslide hazard in Greece: From Arias intensity to spatial distribution of slope resistance demand. *Bulletin of the Seismological Society of America* 106(1):174–188.

Copernicus, 2016. Copernicus land monitoring service. Available: http://land.copernicus.eu.

Dai, F.C., Lee, C.F., and Y.Y. Ngai 2002. Landslide risk assessment and management: An overview. *Engineering Geology* 64: 65–87.

IGME (Institute of Geology and Mineral Exploration) 1993. Geological maps of Greece, scale 1:50,000.

Galli, M., Ardizzone, F., Cardinali, M. et al. 2008. Comparing landslide inventory maps. *Geophysical Journal of the Royal Astronomical Society* 94:268–289.

Kappes, M.S., Keiler, M., von Elverfeldt, K., and T. Glade 2012. Challenges of analyzing multi-hazard risk: A review. *Natural Hazards* 64(2):1925–1958.

Lu, P., Stumpf, A., and N. Kerle 2011. Object–oriented change detection for landslide rapid mapping. *IEEE Geoscience and Remote Sensing Letters* 8 (4):701–705.

Mancini, F., Ceppi, C., and G. Ritrovato 2010. GIS and statistical analysis for landslide susceptibility mapping in the Daunia area, Italy. *Natural Hazards and Earth System Sciences* 10:1851–1864.

Mercogliano, P., Segoni, S., Rossi, G. et al. 2013. Brief communication a prototype forecasting chain for rainfall induced shallow landslides. *Natural Hazards and Earth System Sciences* 13:771–777.

Nefeslioglu, H.A., Sezer, E.A., Gokceoglu, C., and Z. Ayas 2013. A modified analytical hierarchy process (M-AHP) approach for decision support systems in natural hazard assessments. *Computers & Geosciences* 59:1–8.

Panagopoulos G.P., Bathrellos G.D., Skilodimou H.D., and F.A. Martsouka 2012. Mapping urban water demands using multi-criteria analysis and GIS. *Water Resources Management* 26(5):1347–1363.

Papadopoulou-Vrynioti, K., Bathrellos, G.D., Skilodimou, H.D., Kaviris, G., and K. Makropoulos 2013. Karst collapse susceptibility mapping considering peak ground acceleration in a rapidly growing urban area. *Engineering Geology* 158:77–88.

Papadopoulou-Vrynioti, K., Alexakis, D., Bathrellos, G.D., Skilodimou, H.D., Vryniotis, D., and E. Vassiliades 2014. Environmental research and evaluation of agricultural soil of the Arta plain, western Hellas. *Journal of Geochemical Exploration* 136:84–92.

Parise, M. 2001. Landslide mapping techniques and their use in the assessment of the landslide hazard. *Physics and Chemistry of the Earth* (C) 26(9):697–703.

Park, S., Choi, C., Kim, B. and J. Kim 2013. Landslide susceptibility mapping using frequency ratio, analytic hierarchy process, logistic regression, and artificial neural network methods at the Inje area, Korea. *Environmental Earth Sciences* 68(5):1443–1464.

Peng, S.H., Shieh, M.J., and S.Y. Fan 2012. Potential Hazard Map for Disaster Prevention Using GIS-Based Linear Combination Approach and Analytic Hierarchy Method. *Journal of Geographic Information System* 4:403–411.

Pourghasemi, H.R., Pradhan, B. and C. Gokceoglu, 2012. Application of fuzzy logic and analytical hierarchy process (AHP) to landslide susceptibility mapping at Haraz watershed, Iran. *Natural Hazards* 63(2):965–996.

Pradhan, B. and S. Lee 2010. Delineation of landslide hazard areas on Penang Island, Malaysia, by using frequency ratio, logistic regression, and artificial neural network models. *Environmental Earth Sciences* 60(5):1037–1054.

Pardeshi, S.D., Autade, S.E., and S.S. Pardeshi 2013. Landslide hazard assessment: Recent trends and techniques. *SpringerPlus* 2:523. http://www.springerplus.com/content/2/1/523.

Rozos, D., Bathrellos G.D., and H.D. Skilodimou 2011. Comparison of the implementation of Rock Engineering System (RES) and Analytic Hierarchy Process (AHP) methods, based on landslide susceptibility maps, compiled in GIS environment. A case study from the Eastern Achaia County of Peloponnesus, Greece. *Environmental Earth Sciences* 63(1):49–63.

Rozos, D., Skilodimou, H.D., Loupasakis, C., and G.D. Bathrellos 2013. Application of the revised universal soil loss equation model on landslide prevention. An example from N. Euboea (Evia) Island, Greece. *Environmental Earth Sciences* 70(7):3255–3266.

Saaty, T.L. 1977. A scaling method for priorities in hierarchical structures. *Journal of Mathematical Psychology* 15(3):234–281.

Saaty, T.L. 2006. Rank from comparisons and from ratings in the analytic hierarchy/network processes. *European of Operational Research* 168:557–570.

Skilodimou, H.D., Bathrellos, G.D., Maroukian, H., and K. Gaki-Papanastassiou 2014. Late Quaternary evolution of the lower reaches of Ziliana stream in south Mt. Olympus (Greece) *Geografia Fisica e Dinamica Quaternaria* 37(1):43–50.

Terlien, M. T., Van Westen, C. J., and T. W. van Asch 2013. Deterministic modelling in GIS-based landslide hazard assessment. In A. Carrara and F. Guzzetti (Eds.) *Geographical Information Systems in Assessing Natural Hazards*, Vol. 5. London: Springer.

Thapa, R.B. and Y. Murayama 2008. Land evaluation for peri-urban agriculture using analytical hierarchical process and geographic information system techniques: A case study of Hanoi. *Land Use Policy* 25:225–239.

Van Westen, C. J. 2013. Remote sensing and GIS for natural hazards assessment and disaster risk management. In J.F. Schroder and M.P. Bishop (Eds.) *Treatise on Geomorphology*, pp. 259–298. San Diego CA: Academic Press.

Van Westen, C., Kappes, M.S., Luna, B.Q., Frigerio, S., Glade, T., and J.P Malet 2014. Medium-scale multi-hazard risk assessment of gravitational processes. In T. Van Asch, J. Corominas, S. Greiving, J.P. Malet, and S. Sterlacchini (Eds.) *Mountain Risks: From Prediction to Management and Governance*. Dordrecht, the Netherlands: Springer, pp. 201–231.

Yalcin, A. 2008. GIS-based landslide susceptibility mapping using analytical hierarchy process and bivariate statistics in Ardesen (Turkey): Comparisons of results and confirmations. *Catena* 72(1):1–12.

Yilmaz, I. 2009. Landslide susceptibility mapping using frequency ratio, logistic regression, artificial neural networks and their comparison: A case study from Kat landslides (Tokat-Turkey). *Computers and Geosciences* 35(6):1125–1138.

Yoshimatsu, H. and S. Abe 2006. A review of landslide hazards in Japan and assessment of their susceptibility using Analytical Hierarchy Process (AHP) method. *Landslides* 3:149–158.

Youssef, A.M., Pradhan, B., and E. Tarabees 2011. Integrated evaluation of urban development suitability based on remote sensing and GIS techniques: Contribution from analytic hierarchy process. *Arabian Journal of Geosciences* 4(3):463–473.

Youssef, A.M., Pradhan, B., Al-Kathery, M., Bathrellos, G.D., and H.D. Skilodimou 2015. Assessment of rockfall hazard at Al-Noor Mountain, Makkah city (Saudi Arabia) using spatio-temporal remote sensing data and field investigation. *Journal of African Earth Sciences* 101:309–321.

Index

Note: Page numbers followed by f and t refer to figures and tables respectively.

T - #0159 - 111024 - C550 - 254/178/26 - PB - 9780367572730 - Gloss Lamination